Plant Lipid Metabolism

Plant Lipid Metabolism

edited by

Jean-Claude Kader

and

Paul Mazliak

KLUWER ACADEMIC PUBLISHERS

DORDRECHT / BOSTON / LONDON

A C.I.P. Catalogue record for this book is available from the Library of Congress.

ISBN 0-7923-3250-4

Published by Kluwer Academic Publishers,
P.O. Box 17, 3300 AA Dordrecht, The Netherlands.

Kluwer Academic Publishers incorporates
the publishing programmes of
D. Reidel, Martinus Nijhoff, Dr W. Junk and MTP Press.

Sold and distributed in the U.S.A. and Canada
by Kluwer Academic Publishers,
101 Philip Drive, Norwell, MA 02061, U.S.A.

In all other countries, sold and distributed
by Kluwer Academic Publishers Group,
P.O. Box 322, 3300 AH Dordrecht, The Netherlands.

Printed on acid-free paper

Printed in the Netherlands

TABLE OF CONTENTS

SECTION 2 : MEMBRANES

SECTION 3 : LIPID OXIDATION AND DEGRADATION

SECTION 4 : ISOPRENOIDS

SECTION 5 : LIPIDS AND ENVIRONMENT

SECTION 6 : OIL SEED AND GENE TECHNOLOGY

PREFACE

Research on the lipid metabolism of plants, yeasts and algae has attracted much attention in recent years. This increasing interest stems from the fact that lipids play not only major roles in membrane structure and function but are also involved in various intracellular and extracellular signalling processes often linked to environmental conditions. In addition, the recent possibilities to modify the composition of triacylglycerols opened fascinating perspectives for adapting the quality of storage lipids to industrial needs. All these developments were made possible thanks to the use of the tools of molecular biology and molecular genetics, mainly by taking *Arabidopsis thaliana* and cyanobacteria as models.

This explains why the 11th International Meeting on Plant Lipids, which was held in the Carré des Sciences in Paris, France, from 26 June to 1st July 1994, attracted 285 scientists from 26 countries, giving 13 plenary lectures, 41 conferences and presenting 165 posters. From these communications, 161 papers were selected and collected into the present book. These papers are arranged in six sections, reflecting the present-day state-of-the-arts of research in the field of plant lipid metabolism.

We are very grateful to the generous support by research organizations, University Paris 6 and private corporations (see the following list). We also acknowledge the tremendous efforts of our collaborators, particularly C. Cantrel, A. Chammai, D. Chervin, A. Chicha, J. Davy de Virville, C. Demandre, D. Fontaine, A.M. Galle, M. Gawer, M. Grosbois, F. Guerbette, N. Guern, T. Guillot-Salomon, A. Jolliot-Croquin, M. Joseph, A-M. Justin, O. Kader, A. Oursel, and M. Wicker.

We are especially indebted to Mrs M. Poittevin for her pleasant and efficient secretarial work all over the meeting.

Paris, August 1994 Jean-Claude Kader and Paul Mazliak

This book is dedicated to our beloved colleague, Terry Galliard, aged 53, founder of this series of meetings on plant lipids, who sadly passed away on 31st March 1993, despite his brave fight with an illness, which was first diagnosed two years ago.

Terry's academic career began with the award of a state scholarship from his scholar school to study chemistry at Birmingham University. After securing his B.Sc in 1960, he progressed to the Medical Biochemistry Department, where he then graduated with a Ph.D. in 1963. A year later, he won a Fulbright Scholarship to work in the University of California with professor Paul Stumpf where he studied ricinoleic biosynthesis. Terry returned to the U.K. in 1966 to work at the Food Research Institute at Norwich. He made major contributions, still cited, on plant lipases and acyl hydrolases. It was in 1977 that Terry joined the Lord Rank Research Centre. At the head of the Biochemistry Department, he adapted his research to cereal science and became editor of the Journal of Cereal Science, developing the high reputation of this journal. Terry was for several years an Honorary lecturer, and then honorary reader in the School of Biological Sciences at the University of East Anglia. In 1990, the title of Visiting Professor was conferred on him by King's College, London University.

To remember him, it was decided that a Terry Galliard lecture will be given in each Plant Lipid Meeting. Professor Norio Murata had the honor to give, in Paris, the first one.

Terry was a gentle and quiet colleague, generous and thoughtful of others. We miss him very much.

Paris, August 1994

Jean-Claude Kader Paul Mazliak

The following institutions, research organizations, local authorities and private companies are gratefully acknowleged for their generous support

Ministère de l' Enseignement Supérieur et de la Recherche

Université Pierre et Marie Curie (Paris 6)

C.N.R.S. (Centre National de la Recherche Scientifique)

Comité National de la Recherche Scientifique

I.N.R.A. (Institut National de la Recherche Agronomique)

Rhône-Poulenc Agrochimie

Pioneer-France

C.E.T.I.O.M. (Centre d'Etude Technique Interprofessionnel des Oléagineux Métropolitains)

L.V.M.H. (Louis Vuitton Moët Hennessy)

O.N.I.D.O.L. (Office National Interprofessionnel des Oléagineux)

Section 1

Fatty Acid Synthesis

THE CYANOBACTERIAL DESATURASES: ASPECTS OF THEIR STRUCTURE AND REGULATION.

Norio Murata, Shoichi Higashi, Hajime Wada, Toshio Sakamoto, Marie-Helene Macherel, David Macherel, Yasushi Tasaka and Dmitry Los
National Institute for Basic Biology,
Okazaki 444, Japan

Introduction

Desaturases introduce double bonds into fatty acids. They are important in the regulation of the degree of unsaturation of membrane glycerolipids and, thus, in the ability of certain organisms to tolerate low temperatures [1,2,3,4,5]. There are three types of desaturase, as follows. (1) Acyl-CoA desaturases introduce double bonds into fatty acids bound to coenzyme A, and these enzymes are bound to the endoplasmic reticulum in animal, yeast and fungal cells. (2) Acyl-ACP desaturases introduce double bonds into fatty acids that are bound to acyl-carrier protein and they are present in the stroma in plant chloroplasts. (3) Acyl-lipid desaturases introduce double bonds into fatty acids that have been esterified to glycerolipids, and they are bound to the membranes of plant and cyanobacterial cells. This last type of desaturase is the most efficient regulator of the extent of the unsaturation of membrane lipids in response to changes in temperature.

The acyl-lipid desaturases can be further classified into subgroups according to their electron donors. One subgroup, present in the endoplasmic reticulum of plant cells, uses cytochrome b_5 as the electron donor [6,7]. The other, present in chloroplastic and cyanobacterial thylakoid membranes, uses ferredoxin as the electron donor [8,9]. A unique characteristic of the desaturases is that they recognize exact positions within long various carbon chains at which double bonds are specifically introduced.

Abbreviation; X:Y (Z_1,Z_2,Z_3,Z_4), fatty acid containing X carbon atoms with Y double bonds in the *cis* configuration at positions Z_1,Z_2,Z_3 and Z_4 counted from the carboxyl terminus.

J.-C. Kader and P. Mazliak (eds.), Plant Lipid Metabolism, 3–8.

In the present communication, we summarize our recent attempts aimed at the physiological, biochemical and molecular-biological characterization of acyl-lipid desaturases in cyanobacteria.

Substrate specificity of desaturases.

Cyanobacterial strains can be classified into four groups [10]. Strains in Group 1. [e.g., *Synechococcus* sp. PCC 7942 (*Anacystis nidulans* R2) and *Mastigocladus laminosus*] only introduce a double bond at the Δ9 position of a fatty acid. Strains in Group 2 (e.g., *Synechococcus* sp. PCC 7002, *Anabaena variabilis*, *Plectonema boryanum* and *Nostoc muscorum*) can introduce double bonds at the Δ9, Δ12 and ω3 positions. Strains in Group 3 (e.g., *Synechocystis* sp. PCC 6714 and *Spirulina platensis*) can also introduce three double bonds but these are at the Δ6, Δ9 and Δ12 positions. Strains in Group 4 (e.g., *Synechocystis* sp. PCC 6803 and *Tolypothrix tenuis*) can introduce double bonds at the Δ6, Δ9, Δ12 and ω3 positions.

To study desaturation reactions in terms of the positions within the carbon chain at which the double bonds are introduced by the desaturases, we induced *Synechocystis* sp. PCC 6803 to synthesize odd-numbered fatty acids by feeding cells with heptanoic acid (C_7). The most prevalent unsaturated fatty acids of membrane lipids synthesized in these cells were 17:4(6,9,12,14), 18:4(6,9,12,15) and 19:4 (6,9,12,16). These results demonstrate that double bonds appeared at positions 6, 9, and 12, counted from the carboxyl terminus, regardless of the chain length of the fatty acid. However, the fourth double bond was located at position 3, counted from the methyl terminus. From these physiological results, we designated the four desaturases Δ6, Δ9, Δ12 and ω3 acyl-lipid desaturases, respectively [11].

Molecular cloning of desaturases.

Among the four acyl-lipid desaturases of *Synechocystis* sp. PCC 6803, the gene for the Δ12 desaturase was the first cloned and was designated *desA* [12]. We recently cloned the *desB* gene for the ω3 desaturase [13] and the *desC* gene for the Δ9 desaturase [14], and Reddy et al. [15] have cloned the *desD* gene for the Δ6 desaturase. In each case, the gene was cloned from *Synechocystis* sp. PCC 6803. The number of amino acids in the putative products of translation of the *desA*, *desB* and *desD* genes ranges from 350 to 360, whereas the putative product of the *desC* gene contains 318 amino acids. It is unlikely that a transit peptide is included in the product of translation of the *desA* gene.

The *desC* gene has also been isolated from *Anabaena variabilis*. The deduced amino acid sequences of the Δ9 acyl-lipid desaturases of cyanobacteria are similar to those of the Δ9 acyl-CoA desaturases from the rat, the mouse and yeast, with similarity scores of about 25%. Moreover, histidine residues, which are potential ligands of Fe atoms, are well conserved among the Δ9 acyl-lipid and the Δ9 acyl-CoA desaturases. The amino acid sequence of the ω3 desaturase, deduced from the *desB* gene [13], resembles those of the ω3 desaturases from higher plants, with similarity scores of 45-50%.

We have compared the deduced amino acid sequences of the four desaturases from *Synechocystis* sp. PCC 6803. The similarity score is highest (28%) between the Δ12 and the ω3 desaturases. The similarity scores between other combinations are no more than 15%. However, there are several well-conserved regions of amino acid sequence, in particular, the histidine-cluster motifs, HxxxH and two sets of HxxHH.

Overexpression of the *desA* gene in *E. coli* and site-directed mutagenesis.

The coding region of the *desA* gene was introduced into the pET-3a vector and the vector was used to transform *E. coli* strain BL21(DE3). Upon induction with isopropyl-1-thio-β-D-galactoside (IPTG), a large amount of Δ12 desaturase with an apparent molecular mass of 36 kDa, was produced, as indicated by SDS-polyacylamide gel electrophoresis. When the *E. coli* cells were supplied with 18:1(9) in the culture medium, a homogenate of the cells in which the *desA* gene had been overexpressed was active in desaturating 18:1(9) to 18:2(9,12) [16]. When the transformed cells were not supplied with exogenous 18:1(9), there was no detectable desaturation activity in the homogenate of transformed cells that had been treated with IPTG [16].

To understand the importance of the conserved histidine residues in the Δ12 desaturase [17], we changed His90, His109, His129, His287 and His290 to arginine residues by site-directed mutagenesis. Each mutation eliminated the desaturase activity [Macherel M-H, Macherel D, Wada H, Murata N, unpublished].

Overexpression of the *desC* gene in *E. coli*.

The *desC* genes from *Synechocystis* sp. PCC 6803 and *A. variabilis* were introduced into pET-3a and overexpressed in *E. coli* under the control of bacteriophage T7 RNA polymerase. The cells were supplied with 18:0, which is present at very low levels in *E. coli* cells under normal growth conditions. The externally supplied 18:0 was esterified at both the *sn*-1 and *sn*-2 positions of the glycerol moiety of phosphatidylglycerol and phosphatidylethanolamine. The overexpressed Δ9 acyl-lipid desaturase of *Synechocystis* sp. PCC 6803 and of *A. variabilis* was active in converting 18:0 to 18:1(9) at the *sn*-1 position but not at the *sn*-2 position, and it did not desaturate 16:0 at the *sn*-1 or the *sn*-2 position. These observations lead us to conclude that the Δ9 acyl-lipid desaturases from *Synechocystis* sp. PCC 6803 and *A. variabilis* are specific to 18:0 and the *sn*-1 position but are non-specific with respect to the head group.

Temperature-dependent regulation of gene expression of desaturases.

A decrease in the ambient temperature induces desaturation of fatty acids of membrane lipids [18]. We have demonstrated that the desaturation results from the low-temperature-induced regulation of expression of genes for desaturases [19]. We have also obtained evidence that a decrease in the fluidity of plasma membranes of *Synechocystis* sp. PCC 6803 by catalytic hydrogenation can mimic a decrease in temperature in regulating the level of *desA* mRNA [20].

We recently compared changes in the levels of mRNAs for all four desaturases from *Synechocystis* sp. PCC 6803 upon a decrease of ambient temperature, and we found that the level of mRNA transcribed from the *desC* gene stayed constant regardless of the temperature. The levels of mRNAs transcribed from the *desB* and the *desD* genes, which encode the ω3 and Δ6 desaturases, respectively, responded to a decrease in temperature most rapidly. The *desA* gene for the Δ12 desaturase responded rather slowly to a change in temperature.

We compared the lifetimes of mRNAs for the four desaturases at 22°C and 34°C. The half-lives of mRNAs transcribed from the *desA*, *desB* and *desD* genes at 22°C were 20, 10, 20 minutes, respectively, whereas those at 34°C were less than one-tenth of these values [Los DA, Murata N, unpublished]. These observations suggest that the levels of the mRNAs are regulated in response to temperature by changes in the rates of degradation of the mRNAs. The contribution of transcriptional regulation is still in question. By contrast, the half-life of *desC* mRNA was about 10 min at 22°C and at 34°C.

Acknowledgements

The work summarized here was supported, in part, by Grants-in-Aid for Scientific Research on Priority Area (nos. 042273102 and 04273103) from the Ministry of Education, Science and Culture, Japan, to NM. MHM was supported by the Postdoctoral Fellowship Programme of the European Community.

References

1. Murata N. Low-temperature effects on cyanobacterial membranes. J Bioenergetics Biomembranes 1989;21:61-75.
2. Gombos Z, Wada H, Murata N. Unsaturation of fatty acids in membrane lipids enhances tolerance of the cyanobacterium *Synechocystis* PCC6803 to low-temperature photoinhibition. Proc Natl Acad Sci USA 1992;89:9959-9963.
3. Gombos Z, Wada H, Murata N. The recovery of photosynthesis from low-temperature photoinhibition is accelerated by the unsaturation of membrane lipids: a mechanism of chilling tolerance. Proc Natl Acad Sci USA 1994;in press.
4. Murata N, Ishizaki-Nishizawa O, Higashi S, Hayashi H, Tasaka Y, Nishida I. Genetically engineered alteration in the chilling sensitivity of plants. Nature 1992;356:710-713.
5. Wada H, Gombos Z, Murata N. Contribution of membrane lipids to the ability of the photosynthetic machinery to tolerate temperature stress. Proc Natl Acad Sci USA 1994;91:4273-4277.
6. Kearns EV, Hugly S, Somerville CR. The role of cytochrome b_5 in Δ12 desaturation of oleic acid by microsomes of safflower (*Carthamus tinctorius* L.). Arch Biochem Biophys 1991;284:431-436.
7. Smith MA, Cross AR, Jones TG, Griffiths WT, Stymne S, Stobart K. Electron-transport components of the 1-acyl-2-oleoyl-*sn*-glycero-3-phosphocholine Δ^{12}-desaturase (Δ^{12}-desaturase) in microsomal preparations from developing safflower (*Carthamus tinctorius* L.) cotyledons. Biochem J 1990;272:23-29.
8. Schmidt H, Heinz E. Involvement of ferredoxin in desaturation of lipid-bound oleate in chloroplasts. Plant Physiol 1990;94:214-220.

9. Wada H, Schmidt H, Heinz E, Murata N. *In vitro* ferredoxin-dependent desaturation of fatty acids in cyanobacterial thylakoid membranes. J Bacteriol 1993;175:544-547.
10. Murata N, Wada H, Gombos Z. Modes of fatty-acid desaturation in cyanobacteria. Plant Cell Physiol 1992;33:933-941.
11. Higashi S, Murata N. An *in vivo* study of substrate specificities of acyl-lipid desaturases and acyltransferases in lipid synthesis in *Synechocystis* PCC6803. Plant Physiol 1993;102:1275-1278.
12. Wada H, Gombos Z, Murata N. Enhancement of chilling tolerance of a cyanobacterium by genetic manipulation of fatty acid desaturation. Nature 1990;347:200-203.
13. Sakamoto T, Los DA, Higashi S, Wada H, Nishida I, Ohmori M, Murata N. Cloning of ω3 desaturase from cyanobacteria and its use in altering the degree of membrane-lipid unsaturation. Plant Mol Biol 1994;in press.
14. Sakamoto T, Wada H, Nishida I, Ohmori M, Murata N. Δ9 acyl-lipid desaturases of cyanobacteria: molecular cloning and substrate specificities to fatty acids, *sn*-positions and polar head groups. J Biol Chem 1994;in press.
15. Reddy AS, Nyccio ML, Gross LM, Thomas TL. Isolation of a Δ^6-desaturase gene from the cyanobacterium *Synechocystis* sp. PCC6803 by gain-of-function expression in *Anabaena* sp. strain PCC7120. Plant Mol Biol 1993;27:283-300.
16. Wada H, Macherel MH, Murata N. The *desA* gene of the cyanobacterium *Synechocystis* sp. strain PCC6803 is the structural gene for Δ12 desaturase. J Bacteriol 1993;175:6056-6058.
17. Sakamoto T, Wada H, Nishida I, Ohmori M, Murata N. Identification of conserved domains in the Δ12 desaturase of cyanobacteria. Plant Mol Biol 1994;24:643-650.
18. Wada H, Murata N. Temperature-induced changes in the fatty acid composition of the cyanobacterium, *Synechocystis* PCC6803. Plant Physiol 1990;92:1062-1069.
19. Los D, Horváth I, Vigh L, Murata N. The temperature-dependent expression of the desaturase gene *desA* in *Synechocystis* PCC6803. FEBS Lett 1993;318:57-60.
20. Vigh L, Los DA, Horváth I, Murata N. The primary signal in the biological perception of temperature: Pd-catalyzed hydrogenation of membrane lipids stimulated the expression of the *desA* gene in *Synechocystis* PCC6803. Proc Natl Acad Sci USA 1993;90:9090-9094.

MEMBRANE LIPID STRUCTURE AND PLANT FUNCTION: WHAT ARE THE RELATIONSHIPS?

John Browse, Jonathan Lightner and Michele McConn, *Institute of Biological Chemistry, Washington State University, Pullman, WA 99164-6340, USA*

Membranes are of central importance to all biology. Membranes not only define the limits of cells but also allow for the subcellular compartmentation of various biochemical functions. The ability to establish chemical potential gradients across membranes is essential to photosynthesis, the primary means of energy capture in the biosphere. Explaining membrane lipid diversity is a central problem of membrane biology. In plants, as well as in animals, membrane lipids are extremely heterogeneous in the molecular species present. However, our capacity to explain how the structure lipid molecules affects membrane function is very limited. In this paper we shall summarize recent information about polyunsaturated lipid synthesis and discuss two mutant lines of *Arabidopsis* that have expanded our understanding of the roles of lipids in plant biology.

In higher plants there are two distinct pathways by which the acyl-ACP products of plastid fatty acid synthesis are utilized for the biosynthesis of glycerolipids (Fig. 1). The final products of fatty acid synthesis and the soluble 18:0-ACP desaturase are 16:0-ACP and 18:1-ACP. These can be used either in the chloroplast envelope to produce chloroplastic lipids by way of the prokaryotic pathway, or they can be exported to the ER as CoA esters and used in the eukaryotic pathway to produce phospholipids characteristic of the various extrachloroplast membranes of the cell. In *Arabidopsis*, a portion of the PC produced by the eukaryotic pathway is returned to the chloroplast and used in the production of chloroplast lipids. Because of the difficulties involved in studying the desaturases by traditional biochemical techniques, much of our current understanding of the mechanisms and regulation of desaturation in plants has come through genetic means. Mutants of *Arabidopsis* with altered fatty acid composition have been isolated [1]. Of these six are known to be deficient in fatty acid desaturation reactions and are illustrated at the putative sites of each enzymatic lesion in Fig. 1. The mutations that define four of these loci were originally called *fadA*, *fadB*, *fadC*, and *fadD*, but these have now been renamed *fad4*, *fad5*, *fad6*, and *fad7*, respectively. There are two mutants that primarily affect desaturation of the extrachloroplast lipids: *fad2* and *fad3* [2,3]. There are five mutants which affect chloroplast lipid desaturation: *fad4*, *fad5*, *fad6*, *fad7* and *fad8* [4-8]. Two of the chloroplast desaturases are highly substrate specific. The *FAD4* gene product controls a Δ3 desaturase that inserts a *trans* double bond into 16:0 of PG while the *FAD5* gene product is responsible for the synthesis of Δ7 16:1 on MGD and possibly DGD. In contrast, the other two chloroplast desaturases act on acyl chains with no apparent specificity for the length of the fatty acid chain (16- or 18-carbon), its point of attachment to the glycerol backbone (*sn*-1 or *sn*-2) or for the nature of the lipid head group. The 16:1/18:1 desaturase is encoded by the *FAD6* gene while two 16:2/18:2

J.-C. Kader and P. Mazliak (eds.), Plant Lipid Metabolism, 9–14.

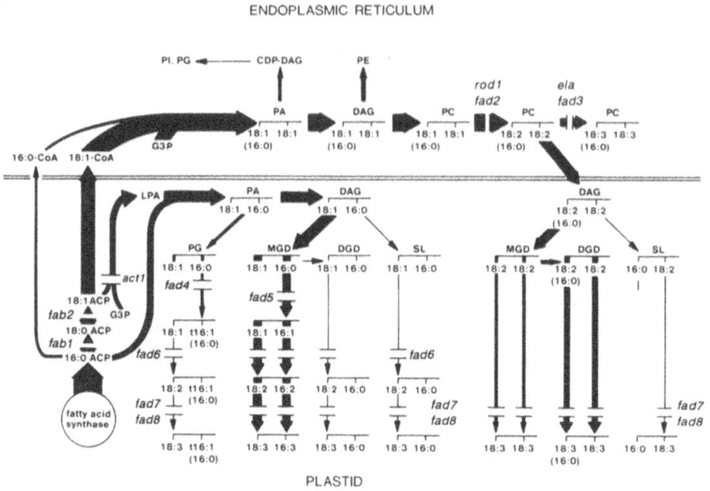

Fig. 1. The two pathways of lipid synthesis in *Arabidopsis*. Mutations are shown.

isozymes are encoded by *FAD7* and *FAD8*. The *fab1* and *fab2* mutants of *Arabidopsis* are characterized by increased levels of 16:0 or 18:0, respectively, in seed and leaf tissues [9,10,11]. In the *fab1* mutant, the biochemical defect appears to be a reduction in the activity of the condensing enzyme responsible for the elongation of 16:0 to 18:0. In *fab2*, it is assumed that 18:0-ACP desaturase activity is reduced (Table 1). In both mutants the saturated fatty acids are incorporated into all the major membrane glycerolipids. Both the mutants appear to be leaky so that the changes in overall membrane fatty acid composition are moderately small. Nevertheless, the changes do have profound effects on the biology of these plants (see below).

Isolation of most of the genes encoding membrane bound desaturases has been possible due to the recent advances in molecular biology and genetics, especially with the creation of T-DNA tagged mutants of *Arabidopsis*. Utilizing this T-DNA tagging method, *FAD2* was cloned [12]. *FAD3* was cloned by chromosome walking [13] as well as by T-DNA tagging. The following desaturase genes, *FAD6* [14], *FAD7* [15], and *FAD8* [16], were cloned by homology to the above genes. The ability of the cloned genes to complement the fatty acid mutations demonstrates that the *Arabidopsis* mutants are defective in the structural genes [17].

Physiology and Cell Biology of Membrane Lipid Composition

The central issue with respect to the role of glycerolipids in membranes is framed by the observations that each membrane of the cell has a characteristic and distinct complement of glycerolipid types and that within a single membrane, each class of lipids has a distinct fatty acid composition. This diversity implies that differences in lipid structure are important for membrane function. However, despite considerable effort, the details of this structure/function relationship have remained elusive in large part because there have not been instructive biological examples available in which changes in lipid composition have been clearly shown to have specific effects on membrane processes. The *Arabidopsis* lipid mutants have provided such examples and an understanding of the mutant phenotypes at a mechanistic level will undoubtedly advance the field considerably. Most of the lipid mutants are indistinguishable from wild type *Arabidopsis* when grown at 22°C and phenotypes only become apparent at higher and lower temperatures. These phenotypes are summarized in Table 2. In the rest of this paper we will consider the unusual biology of two *Arabidopsis* lines - the *fab2* mutant and the *fad3 fad7 fad8* triple mutant. The overall leaf fatty acid compositions of these lines are given in Table 1.

Table 1. Leaf fatty acid compositions of wild type (WT) *Arabidopsis*, the *fab2* line and the *fad3 fad7 fad8* triple mutant.

	WT	*fab2*	triple
16:0	15	14	15
16:1	3	2	3
16:2	-	-	**9**
16:3	14	6	-
18:0	1	**14**	1
18:1	3	3	6
18:2	14	18	**65**
18:3	48	42	-

Table 2. Responses of *Arabidopsis* lipid mutants to high and low temperature relative to the performance of wild type plants.

Mutant line	Observed phenotypes
fad4	None detected.
fad5	Enhanced growth and thermotolerance of PS electron transport at high T. Leaf chlorosis, reduced growth, impaired chloroplast development at low T.
fad6	Enhanced thermotolerance of PS electron transport at high T. Low T as *fad5*.
fad7	Reduced chloroplast size and altered chloroplast ultrastructure at temperatures (>25°C) at which fatty acid composition is affected.
act1	Enhanced growth and slightly enhanced thermostability of PS at high T.
fad2	Greatly reduced stem elongation at 12°C. Death of plants at 6°C.
fad3	None detected.
fab1	Death of plants after prolonged exposure to 2°C.
fab2	Dwarf phenotype at 22°C is ameliorated at temperatures above 35°C.

Increased Stearate Causes the Dramatic Dwarf Phenotype of *fab2* Plants

The most striking feature of the *fab2* mutant is not its lipid composition but an extreme dwarf growth habit. As shown in Fig. 2, *fab2* plants are less than one-tenth the size of wild type plants. Although the dwarf phenotype is extreme, the mutant retains normal body pattern and chronology of development. The number of rosette leaves at the onset of flowering was the same in wild type and mutant plants. The dwarf size of *fab2* plants is a consequence of reduced expansion of many, but not all, cell types. The failure of mesophyll and epidermal cells to enlarge produces a dramatic "brick-wall" appearance in the mutant that contrasts with the characteristic leaf anatomy evident in the wild type. On the other hand, stomatal guard cells and trichomes of the *fab2* leaves are comparable in size to the same cell types of wild type plants [11].

This dramatic phenotype initially seemed hard to explain on the basis of the relatively modest increase in leaf 18:0 levels and we originally thought it must be a result of a

Fig. 2. Wild type and *fab2* plants.

second, unrelated mutation. As a first test of the possibility that the dwarf phenotype was unrelated to the altered lipid composition, we analyzed 223 segregating F_2 progeny from a cross of *fab2* x wild type and found perfect cosegregation of the two characters. To further investigate the relationship between the high 18:0 levels of *fab2* plants and the dwarf phenotype, we screened a remutagenized population derived from the *fab2* line for plants in which the dwarf phenotype was suppressed. From a screen of 20,000 M_1 plants, one individual was identified as having a putative wild type growing sector. The M_2 progeny derived from self-pollination of this sector were scored for growth habit and leaf fatty acid composition. Three classes of progeny were found (wild type growth, <3% 18:0; intermediate size, 4-8% 18:0; dwarf growth, >10% 18:0) and the frequency distribution for these classes was a good fit to the 1:2:1 hypothesis. Also, of 102 F_2 progeny from a cross between a normal growing M_2 plant (homozygous suppressor) and wild type, 6 displayed the dwarf phenotype - a proportion consistent with the Mendelian segregation of two unlinked loci. These data indicate that the second mutation at a locus we designated *shs* is a semidominant intergenic suppressor that is unlinked to the *fab2* locus. The observation that the suppressor mutation simultaneously restores normal lipid composition and normal growth provides definitive evidence that the two traits are causally related.

The mechanisms through which physical parameters determine the function of biological membranes are poorly understood. However, with respect to the fatty acid components of membrane glycerolipids, it is generally accepted that three interrelated features are of primary importance: the phase behavior of the lipid, the fluidity characteristic contributed to the lipid bilayer, and the shape of the molecule. These characteristics are influenced by the identity of the lipid head group but, in all cases, the thermodynamic behavior is such that the physical effects of increased stearate in membrane lipids would be predicted to be offset by increased temperature. Growth temperatures in the range 15°-30°C did not alter the appearance of *fab2* plants but at 36°C the dwarf phenotype was substantially corrected. Indeed, the mutant appeared to grow better than wild type *Arabidopsis* at this temperature. The internal structure of mutant leaves was also restored to normal at 36°C. In contrast, the 18:0 content of leaves of *fab2* plants grown at 36°C was the same as that of mutant plants grown at 22°C [11]. These results are consistent with the dwarf phenotype being caused by an effect of high 18:0 on membrane structure and function.

A Triple Mutant Lacking 16:3 and 18:3

Chloroplast membranes have a characteristic and unusual fatty acid composition. Typically, 18:3, or a combination of 18:3 and 16:3, fatty acids account for approximately two-thirds of all the thylakoid membrane fatty acids and over 90% of the fatty acids of MGD, the most abundant chloroplast lipid. The atypical fatty acid Δ3,*trans*-hexadecenoate (*trans*-16:1) is present as a component of the major thylakoid phospholipid, PG. The fact that these and other characteristics of chloroplast lipids are common to most or all higher plant species suggests that the lipid fatty acid composition is important for maintaining photosynthetic function. Furthermore, the thylakoid membrane is the site of light absorption and oxygen production. The free radicals that are by-products of these reactions will stimulate oxidation of the polyunsaturated fatty acids. Since this might be expected to mediate against a high degree of unsaturation, it has been inferred that there is a strong selective advantage to having such high levels of trienoic fatty acids in thylakoid membranes. Many different approaches have been used in attempts to understand the significance of membrane fatty acid composition to photosynthetic function. These include the correlation of events during chloroplast development, physical studies of model membrane systems, reconstitution of photosynthetic components with lipid mixtures, alteration of lipids *in situ* by heat stress, lipase treatment, or hydrogenation of unsaturated fatty acids, and alteration of lipids *in vivo* by chemical inhibitors or molecular genetic techniques. However, in general, these approaches have not been successful in establishing unequivocal relationships between membrane form and function. Indeed, some of the relationships that have been proposed are now untenable in view of the results we have obtained from *Arabidopsis* mutants.

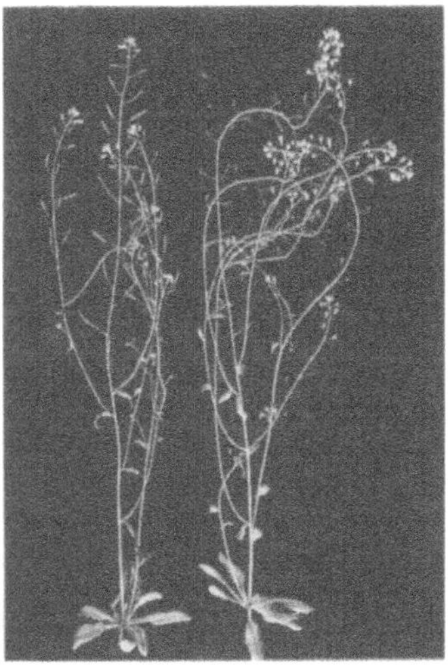

Fig. 3. Wild type (left) and the *fad3 fad7 fad8* mutant.

There are three gene products in *Arabidopsis* that mediate the synthesis of trienoic fatty acids. The *FAD7* and *FAD8* genes encode chloroplast isozymes while the *FAD3* gene product is the endoplasmic reticulum desaturase. Mutations at one or two of these loci still produce plants with considerable amounts of 18:3. We have recently generated triple mutants, *fad3 fad7 fad8*, that contain no detectable 18:3 in their leaf lipids. These plants are capable of robust autotrophic growth under at 22°C (Fig. 3). This finding very graphically demonstrates that trienoic fatty acids do not have any crucial role in maintaining photosynthesis in higher plant chloroplasts. This is not to say that trienoic fatty acids are irrelevant to photosynthetic function. Conservation of the high trienoic content of chloroplasts through the evolutionary time scale attests to their importance, but clearly their role is more subtle than expected. A totally unanticipated consequence of the lack of 18:3 and 16:3 lipids is the fact that *fad3 fad7 fad8* plants are male sterile. Pollen grains produced on triple mutant plants are not viable, based on their inability to take up vital stains, even though they are mophologically similar to wild type pollen grains.

Viable pollen with a *fad3 fad7 fad8* genotype must have been produced on the *(-/-fad3:(-/-fad7:(+/-)fad8* plants from which the triple mutant progeny were obtained in a Mendelian proportion. This observation indicates that it is the genotype of the maternal tissue that mediates the male-sterile phenotype. Indeed, we have been able to induce seed production on triple mutant plants (960 seeds from 3 plants) by spraying them during flowering with the sodium soap of Δ9,12,15-18:3 (α-linolenate). Control treatments including water and the sodium soaps of 18:2 and Δ6,9,12-18:3 (γ-linolenate) did not result in any seed set.

References

1. Somerville C, Browse J. Plant lipids: Metabolism mutants and membranes. Science 1991;252:80-87.
2. Miquel M, Browse J. *Arabidopsis* mutants deficient in polyunsaturated fatty acid synthesis. Biochemical and genetic characterization of a plant oleoyl-phosphatidylcholine desaturase. J Biol Chem 1992;267:1502-1509.
3. Browse J, McConn M, James D, Miquel M. Mutants of *Arabidopsis* deficient in the synthesis of α-linolenate. Biochemical and genetic characterization of the endoplasmic reticulum linoleoyl desaturase. J Biol Chem 1993;268:16345-16351.
4. Browse JA, McCourt PJ, Somerville CR. A mutant of *Arabidopsis* lacking a chloroplast-specific lipid. Science 1985;227:763-765.
5. Kunst L, Browse J, Somerville C. A Mutant of *Arabidopsis* deficient in desaturation of palmitic acid in leaf lipids. Plant Physiol 1989;90:943-947.
6. Browse J, Kunst L, Anderson S, Hugly S, Somerville CR. A mutant of *Arabidopsis* deficient in the chloroplast 16:1/18:1 desaturase. Plant Physiol 1989;90:522-529.
7. Browse JA, McCourt PJ, Somerville CR. A mutant of *Arabidopsis* deficient in $C_{18:3}$ and $C_{16:3}$ leaf lipids. Plant Physiol 1986;81:859-864.
8. McConn M, Hugly S, Somerville C, Browse J. A mutation at the *fad8* locus of *Arabidopsis* identifies a second chloroplast ω-3 desaturase. Plant Physiol 1994;(submitted).
9. James Jr DW, Dooner HK. Isolation of EMS-induced mutants in *Arabidopsis* altered in seed fatty acid composition. Theor Appl Genet 1990;80:241-245.
10. Wu J, James Jr DW, Dooner HK, Browse J. A mutant of *Arabidopsis* deficient in elongation of palmitic acid. Plant Physiol 1994;(in press).
11. Lightner J, James Jr DW, Dooner HK, Browse J. Altered body morphology is caused by increased stearate levels in a mutant of *Arabidopsis*. Plant J 1994;(in press).
12. Okuley J, Lightner J, Feldmann K, Yadav N, Lark E, Browse J. The *Arabidopsis FAD2* gene encodes the enzyme that is essential for polyunsaturated lipid synthesis. Plant Cell 1994;6:147-158.
13. Arondel V, Lemieux B, Hwang I, Gibson S, Goodman HM, Somerville C. Map-based cloning of a gene controlling omega-3 fatty acid desaturase in *Arbidopsis*. Science 1992;258:1353-1355.
14. Hitz WD, Carlson TJ, Booth R, Kinney AJ, Stecca KL, Yadav NS. Cloning of a higher plant plastid ω-6 fatty acid desaturase cDNA and its expression in a cyanobacterium. Plant Physiol 1994;105:635-641.
15. Iba K, Gibson S, Nishiuchi T, Fuse T, Nishimura M, Arondel V, Hugly S, Somerville C. A gene encoding a chloroplast ω-3 fatty acid desaturase complements alterations in fatty acid desaturation and chloroplast copy number of the *fad7* mutant of *Arabidopsis thaliana*. J Biol Chem 1993;268:24099-24105.
16. Gibson S, Arondel V, Iba K, Somerville CR. Temperature regulated expression of a gene encoding a chloroplast omega-3 desaturase from *Arabidopsis*. Plant Physiol (submitted).
17. Browse J, Somerville C. Glycerolipid Metabolism: Biochemistry and Regulation. Ann Rev Plant Physiol Plant Mol Biol 1991;42:467-506.

ROLE OF POLYUNSATURATED FATTY ACIDS IN GROWTH AND DEVELOPMENT OF *ARABIDOPSIS*

Martine Miquel and John Browse, *Institute of Biological Chemistry, Washington State University, Pullman, WA 99164, USA.*

Mutants of *Arabidopsis* with altered fatty acid composition provide the means for a better understanding of the biochemistry and regulation of plant lipid synthesis. Moreover, the mutants deficient in polyunsaturated fatty acids offer a model system to investigate the biological roles of polyunsaturated membrane lipids. Here we summarize work on mutants which show a substantial decrease in the amount of 18:2 and 18:3 and a concomitant increase in the amount of 18:1 compared to wild-type plants. The deficiency in polyunsaturated lipids was found in all tissues of the plant. Lipid analyses and *in vitro* assays indicate that the mutant plants lack the endoplasmic reticulum 18:1 phosphatidylcholine desaturase responsible for 18:1 desaturation on the eukaryotic pathway. Despite the pronounced effect on the fatty acid composition of the non-chloroplast membranes of the cell, the mutants show growth characteristics at 22°C that were very similar to wild-type. By contrast, studies on growth and development at temperatures toward the low end of the physiological range demonstrate a requirement of substantial levels of membrane polyunsaturation for low-temperature survival of a higher eukaryote. Our observations are not consistent with a gross disruption of membrane structure but instead suggest limited defects in the function of plant membranes.

Biochemical And Genetic Characterisation Of The *fad2* Mutants

The leaf lipids of the mutants we describe here contain reduced levels of polyunsaturated 18-carbon fatty acids and an increased proportion of 18:1 suggesting that the mutant is deficient in the desaturation of 18:1 into 18:2 [1]. Since the amounts of 16-carbon fatty acids are very similar between the mutant and the wild-type, we hypothesize that the mutation affects the endoplasmic reticulum 18:1 desaturase on the eukaryotic pathway (Figure 1) [2]. This is consistent with the findings that the roots and seeds of those mutants also have decreased levels of 18:2 and 18:3 and a concomitant increase in the level of 18:1 as compared with wild-type [1]. Quantitative analysis of the fatty acid composition of individual lipids demonstrated that all the major phospholipids of the extrachloroplast membranes are affected by the mutation whereas the chloroplast lipids show fatty acid compositions only slightly different from those of wild-type plants (Figure 2). Direct enzyme assays confirmed that the mutation involves a specific loss of the 18:1 phosphatidylcholine desaturase that is the major enzyme responsible for the synthesis of polyunsaturated phospholipids [3,4]. Despite this deficiency in 18:1-

J.-C. Kader and P. Mazliak (eds.), Plant Lipid Metabolism, 15–17.

desaturase activity, mutant plants contained relatively high levels of 18:3 in their leaf phospholipids. This finding is interpreted as additional evidence that considerable two-way exchange of lipid occurs between the chloroplast and the endoplasmic reticulum and that this exchange allows the chloroplast desaturases to provide lipids containing 18:3 to the extrachloroplast compartment, thus alleviating the deficiency in 18:1-desaturase activity (Figure 1).

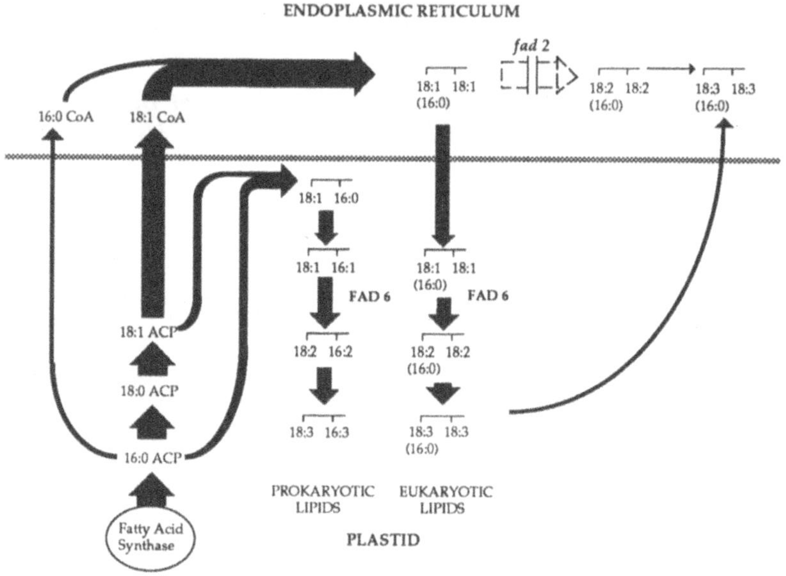

Figure 1: Abbreviated scheme of polyunsaturated lipid synthesis in *fad2* mutant plant cells. The dashed lines show the wild-type activity of the endoplasmic reticulum 18:1 desaturase.

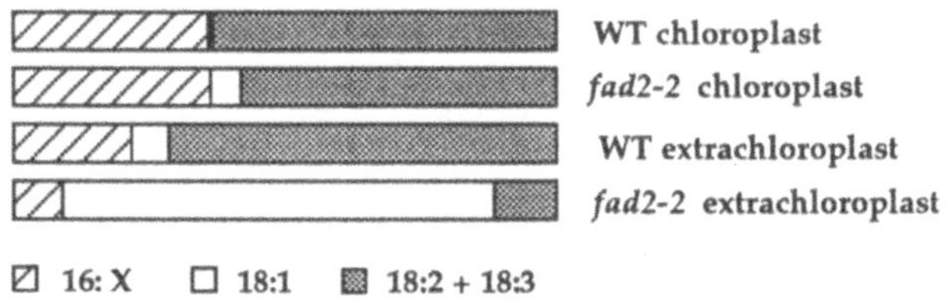

Figure 2: Fatty acid composition of membrane lipids for wild-type and *fad2-2* plants. 16:X indicates the 16-carbon fatty acids containing 0 to 3 double bonds.

Several mutant lines have been isolated from independently mutagenized populations. The lack of genetic complementation following reciprocal crosses between five of those lines indicates that they contain mutations at the same locus. Crosses between the mutants and the wild-type demonstrate that the wild-type allele is incompletely dominant, and that the altered fatty acid composition is due to a single nuclear mutation at a locus designated *fad2* [3].

Polyunsaturated Lipids Are Required For Low Temperature Survival in *Arabidopsis*

When grown at 22°C, 140 μmole of quanta.m^2.s constant illumination, the *fad2* mutants show growth characteristics that are very similar to wild-type. By contrast, at 12°C, the *fad2-2* mutants fail to undergo stem elongation during reproductive growth although they produce normal flowers and fertile seeds [5]. Because the *fad2-4* allele exhibits the same phenotype when grown at 12°C, this suggests a role for polyunsaturated lipids in insuring correct cellular responses to tissue-specific growth signals. After transfer to 6°C, the rosette leaves of the mutants gradually die, and the plants are inviable [5]. These different responses of the mutant plants at 12°C and 6°C suggest that distinct functions may be affected at these two temperatures. The gradual development of symptoms at 6°C and other lines of evidence [5] argue against a general collapse of membrane integrity as the cause of the lethal phenotype. Rather, they indicate that the decrease in polyunsaturated membrane lipids may initially have relatively limited effects in disrupting cellular function.

Lipid Composition And Germination In *Arabidopsis*

The endoplasmic reticulum 18:1 desaturase is the main enzyme responsible for polyunsaturated lipid synthesis in developing seeds of oil crops [4]. A comparison of wild-type and *fad2* seeds developing on heterozygous (*FAD2*/-) plants is used as a model for genetically engineered high oleate oilseeds of species such as soybean and canola [6]. When *fad2* seeds develop at normal temperatures (22°C), they show high viability comparable to wild-type seeds. When a portion of seed development takes place at 6°C, germination of the wild-type siblings remains high but germination of *fad2* segregants declines considerably. This is true even when exposure to low temperature is limited to the final stages of seed filling and maturation. Compared to wild-type, fully viable *fad2* seeds produced at 22°C have reduced lipid contents and are slower to germinate at 10°C and 6°C. Taken together, these results indicate that for some oilseed species at least, molecular genetic manipulation of oleate levels in the oil may result in plant lines with unacceptable performance in the field.

1. Miquel M, Browse J. Mutants of *Arabidopsis* deficient in 18:1-PC desaturation. In: Quinn PJ, Harwood JL, editors. Plant lipid biochemistry, structure and utilization. London: Portland Press Limited, 1990: 456-458.
2. Browse J, Somerville C. Glycerolipid synthesis: biochemistry and regulation. Annu Rev Plant Physiol Plant Mol Biol 1991; 42: 467-506.
3. Miquel M, Browse J. *Arabidopsis* mutants deficient in polyunsaturated fatty acid synthesis. J Biol Chem 1992; 267: 1502-1509.
4. Okuley J, Lightner J, Feldmann K, Yadav N, Lark E, Browse J. *Arabidopsis FAD2* gene encodes the enzyme that is essential for polyunsaturated lipid synthesis.Plant Cell 1994; 6: 147-158.
5. Miquel M, James Jr D, Dooner H, Browse J. *Arabidopsis* requires polyunsaturated lipids for low-temperature survival. Proc Natl Acad Sci USA 1993; 90: 6208-6212.
6. Miquel M, Browse J. High oleate seeds fail to develop at low temperature. Plant Physiol 1994; in press.

MEMBRANE BOUND DESATURASES AND HYDROXYLASES: STRUCTURE FUNCTION STUDIES

John Shanklin[*], Edward J. Whittle[*], and Brian G. Fox[+] [*]Dept of Biology, Brookhaven National Laboratory, Upton, NY 11973 USA and [+]Enzyme Institute, University of Wisconsin-Madison, Madison, WI 53705

The motif: $\mathbf{H\ X_{(3\ or\ 4)}H\ X_{(25\text{-}31)}\ H\ X_{(2\ or\ 3)}\ HH\ X_{(137\text{-}179)}\ H\ X_{(2\ or\ 3)}\ HH}$ is found in all membrane bound desaturases and in alkane hydroxylase, but is absent from almost all other proteins in GenBank. The eight conserved Histidines are all catalytically essential in the rat Δ^9 desaturase because individual conversion of any of these to Alanines prevents it from complementation of the phenotype of a yeast Δ^9 desaturase mutant. Topological analysis suggests both desaturases and hydroxylases have a common organization in which the Histidines are within soluble domains on the cytoplasmic face of the membrane. One possible function for these Histidines would be to coordinate the iron atoms in the active site. Our Mossbauer spectroscopy of the alkane hydroxylase from *Pseudomonas oleovorans* suggests that the iron-containing active site is a diiron-oxo cluster with Mossbauer similar parameters to those of the soluble chloroplast Δ^9 desaturase.

Introduction

Eukaryotes introduce double bonds into the thermodynamically stable methylene portion of elongated acyl chains by a chemical reaction that requires molecular oxygen and an iron-containing active site. We have shown that the soluble Δ^9 desaturase has an active site containing diiron-oxo cluster using a combination of optical and Mossbauer spectroscopy (Fox *et al.* 1993). Comparison of the amino acid sequences of the soluble proteins known to contain diiron-oxo clusters yielded a consensus binding domain consisting of two copies of **(D/E) E X R H** separated by 120-170 residues.

The desaturases in all other eukaryotes, and the plant desaturases responsible for the production of polyunsaturated lipids, are integral membrane proteins that are evolutionarily distinct from the plant soluble Δ^9 desaturase. Never the less there are striking similarities between the soluble and membrane bound desaturases including: stereochemical outcome, cyanide sensitivity, CO resistance, and a requirement for exogenous reductant, and molecular oxygen. Here we report our initial

J.-C. Kader and P. Mazliak (eds.), Plant Lipid Metabolism, 18–20.

characterization of the active site components of the membrane desaturases and related hydrocarbon hydroxylases.

Results and Discussion

Since both the desaturases and hydroxylases require iron atom(s) for activity, we searched the sequences of the membrane desaturases for the soluble desaturase iron binding motif. While this motif was absent, we noted that several clusters of Histidines were conserved between the compared membrane desaturase sequences. Interestingly, similar sequences are found in two membrane bound hydroxylases from *Pseudomonas*. A comparison of divergent examples of these sequences is shown in Fig. 1.

```
                 Ia                                     Ib
         ________________________           _________________________ Dist.

RatD9   ITAGAHRLWSHRTYKARLPLRIFLIIANTMAFQNDVYEWARDHR AHHKFSETHADP 141
YstD9   ITAGYHRLWSHRSYSAHWPLRLFYAIFGCASVEGSAKWWGHSHR IHHRYTDTLRDP 137
BnF3    IFVLGHDC GHGSFSDIPLLNSVVGHILHSFILVPYHGWRISHR THHQNHGHVEND 167
RcF7    LFVLGHDC GHGSFSNNPKLNSVVGHLLHSSILVPYHGWRISHR THHQNHGHVEND 167
AtF2    IWVIAHEC GHHAFSDYQWLDDTVGLIFHSFLLVPYFSWKYSHR RHHSNTGSLERD 174
SDesA   AFVVGHDC GHRSFAKKRWVNDLVGHIAFAPLIYPFHSWRLLHD HHHLHTNKIEVD 161
SynD6   SFNVGHDA NHNAYSSNPHINRVLGMTYDFVGLSSF LWRYRHNYLHHTYTNILGHD 179
PoAlkB  ALNTGHEL GHKKETFDRWMAKIV      LAVVGYGHFFIEHNKGHHRDVATPMDP 143
PpXylM  TLPVSHEL MHRRHWLPRKMAQLL      AMFYGDPNRDIAHVNTHHLYLDTPLDS 138
                                                     *

                                                      II
                                              ________________________
RatD9   LVLNATWLVNSAAHLYGYRPYDKNIQSRENILVSLGSVGE GFHN YHHAFPYDYSASE
YstD9   VIQQATFCINSMAHYIGTQPFDDRRTPRDNWITAIVTFGE GYHN FHHEFPTDYRNAI
BnF3    HHGHDEKLPWYRGKEWSYLRGGLTTIDRDYGIFNNIHHDI GTHV IHHLFPQIPHYHL
RcF7    HHGHEDKLPWYRGKAWSYLRGGLTTLDRDYGWINNIHHDI GTHV IHHLFPQIPHYHL
Atf2    LQHTHPSLPHYDSSEWDWLRGALATVDRDYGILNKVFHNITDTHV AHHLFSTMPHYNA
SDesA   HTIPEIRFRPAADWSAAEAQLNGTVHCDYPRWVEVLCHDI NVHI PHHLSVAIPSYNL
SynD6   EFLTPDGESGAIDDEWAICQIRTTANFATNNPFWNWFCGG LNHQVTHHLFPNICHIHY
PoAlkB  YIEHYGLLRQKMEDGRYEHQKPHHSWNSNHIVSNLVLFHL QRHS DHHAHPTRSYQSL
PpXyl   VEGFNYFQHYGLVRDLDQPILLHHAWNHMGTIVRPLGCEI TNHI NHHIDGYTRFYEL
                                                    *
```

Figure 1. Dist. represents the number of residues between the starred Histidines. Residues occurring in 3 or more sequences are highlighted, and Histidines conserved in all cases are in bold case. Sequences compared are: RatD9, the rat Δ9 desaturase; YstD9, the *Saccharomyces cereviciae* Δ9 desaturase; BnF3 *Brassica napus* cytosolic omega 3 desaturase; RcF7 *Ricinus communis* chloroplastic omega 3 desaturase; SDesA, *Synechocystis* Δ12 desaturase; SynD6 *Synechocystis* Δ6 desaturase, PoAlkB, *Pseudomonas oleovorans* alkane hydroxylase; and XylM, *Pseudomonas putida* Xylene monooxygenase.

A consensus Histidine-containing motif can be described by: $\mathbf{H\ X_{(3\ or\ 4)}H\ X_{(25-31)}\ H\ X_{(2\ or\ 3)}\ HH\ X_{(137-179)}\ H\ X_{(2\ or\ 3)}\ HH}$. Region Ia and Ib are in a soluble domain separated from region II by a hydrophobic domain long enough to allow two spans of the bilayer. This would place both regions on the cytoplasmic face of the membrane and is consistent the empirically derived model for alkane hydroxylase

(Kok *et al.* 1992). Only 20 matches to this motif were found in the 97,000 entries of GenBank: 17 membrane bound fatty acid desaturases, 2 membrane bound bacterial hydroxylases, and one hemoglobin. Thus, this motif is almost exclusively limited to iron-containing membrane bound enzymes.

We have used a site directed mutagenesis approach to assess the role of these residues in catalysis. Martin had previously shown that the rat Δ^9 desaturase is capable of functionally complementing the yeast Δ^9 desaturase deletion strain L8-14C (Stukey *et al.*, 1990). We tested the effect of converting each of the 8 conserved Histidines (shown in bold type in Fig. 1) in the rat Δ^9 desaturase to Alanines. All 8 of the conserved Histidines are catalytically essential. In contrast, conversion of three adjacent non-conserved Histidines to Alanines in sites I and II had no effect on function.

To determine the structure of the active site iron center we attempted to over-express various membrane bound enzymes for spectroscopic analysis. While the membrane bound desaturases accumulated to very low levels, the alkane hydroxylase accumulated to very high levels. Membranes containing ^{57}Fe-enriched alkane hydroxylase were isolated and examined by Mossbauer spectroscopy. Preliminary spectra (Fig. 2) shows the major fraction of the iron in these preparations has Mossbauer parameters similar to those of the soluble castor Δ^9 desaturase. δ= 0.5 mm/s diamagnetic in 8 T applied field. This is the first evidence for a diiron-oxo cluster in any membrane protein.

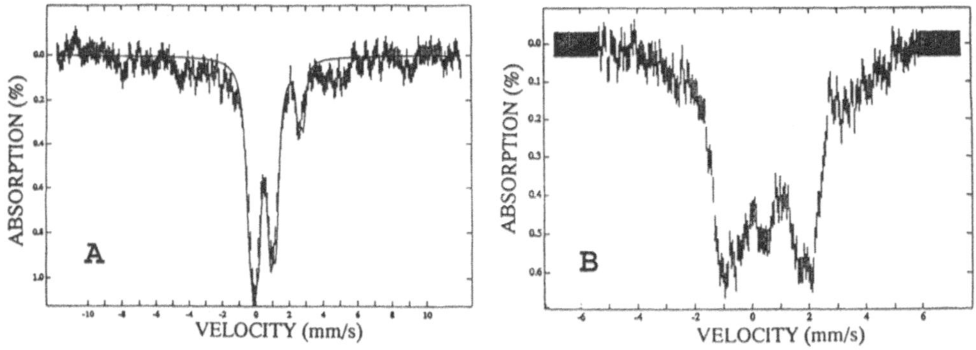

Figure 2. Mossbauer spectra of the *P. oleovorans* alkane hydroxylase at 4 K. A, zero field; B, 8 T. We thank E. Munck and K. Kaufmann for assistance with the Mossbauer spectra.

References:

Fox BG, Shanklin J, Somerville C, and Munck E. *Proc. Natl. Acad. Sci.* USA 1993;90:2486

Kok M, Oldenhuis R, van der Linden RPG, Ratjees P, Kingma J, van Lelyveld P, Widholt B. *J. Biol. Chem.* 1989; 264:5435

Stukey JE, McDonough VM, Martin CM. *J. Biol Chem.* 1989;265:16537

This work was supported by the Office of Basic Energy Research of the U.S. Department of Energy (J.S. and E.J.W).

PCR-BASED CLONING OF MEMBRANE-BOUND DESATURASES

Hermann Schmidt, Petra Sperling and Ernst Heinz
Institut für Allgemeine Botanik, Universität Hamburg
Ohnhorststr. 18, 22609 Hamburg, Germany

Introduction

The degree of unsaturation of fatty acids in membrane and reserve lipids is controlled by the action of different desaturase enzymes. Apart from scientific motivation, the prospects of genetic manipulation of plants have raised additional interest in the cloning of desaturases, particularly in view of the possibility to design plant oils of specifically adjusted unsaturation. In the last years several membrane-bound desaturases from plants and cyanobacteria were cloned by strategies based on complementation of mutants and protein purification. Comparison of the deduced amino acid sequences revealed three regions of homology with the general sequence of HXXXH that are highly conserved in membrane-bound desaturases from plants, cyanobacteria, yeast and mammals. These boxes may provide metal-chelating ligands contributing to the binding of oxygen in the reaction center. Oxygen functions as final acceptor of electrons derived from reduced cytochrome b_5 or ferredoxin and the ethylene segment of an acyl chain to be desaturated. These histidine boxes can be used to search for additional or even new desaturases from various organisms. First results of this approach are presented in the following.

Results and Discussion

Several primers of different length and degeneration were constructed from the known sequences of the first and third histidine box present in the membrane-bound desaturases. These primers were used in PCR experiments with cDNAs transcribed from polyA-mRNA of various organisms. The expected length of the resultant PCR fragments varied from 550 to 700 base pairs. These fragments were cloned and sequenced. The deduced amino acid sequences were used to construct a dendrogram together with the other presently known sequences (Fig. 1). It has to be pointed out that the attribution of our PCR-based sequences, some of which represent full-length clones, to the branches as indicated in the figure is tentative. Nevertheless, the data obtained so far suggest that this approach is useful for the attribution of homologous enzymes from different plants to the various

J.-C. Kader and P. Mazliak (eds.), Plant Lipid Metabolism, 21–23.

desaturase groups as well as for the identification of new desaturases in cases, where homology to any of the existing groups is low or missing.

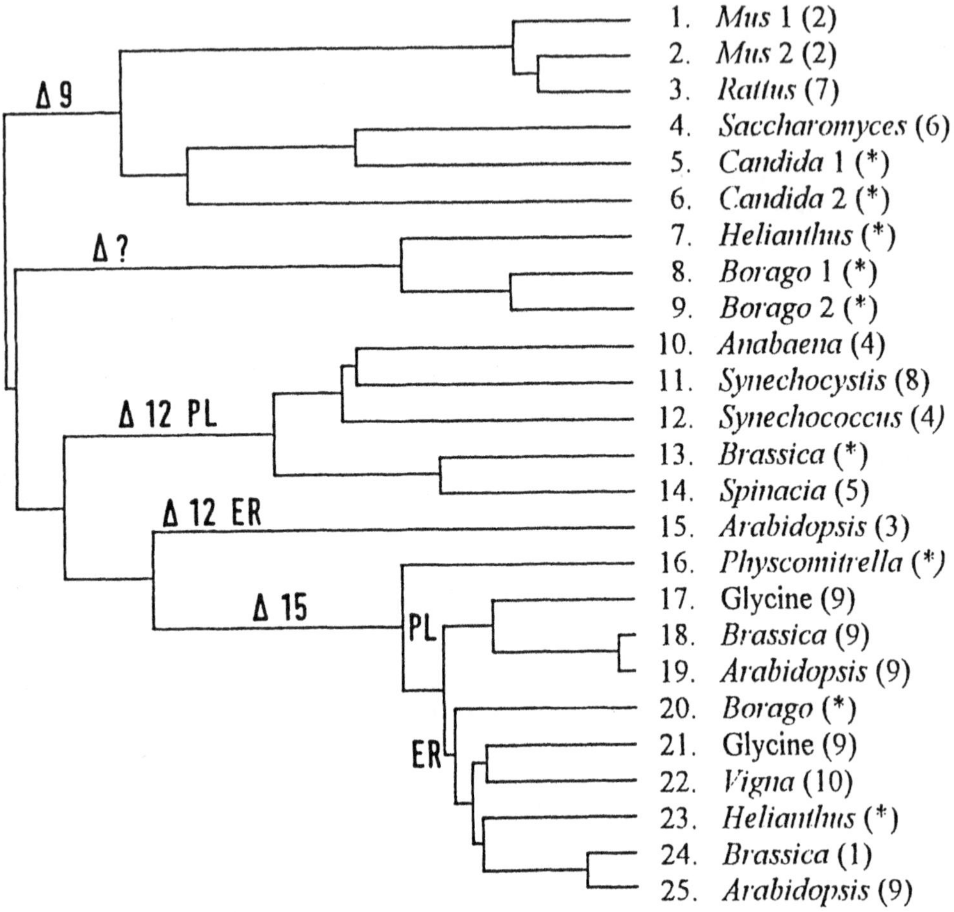

Fig. 1. Similarities of membrane-bound desaturases regarding the amino acid sequences between the first and third histidine box. The dendrogram was constructed with the CLUSTAL program using a gap weight of three. Numbers in brackets are the references, asterisks indicate enzymes sequenced in our lab.

The high homology of the Δ15(n-3)-desaturases from plastids (numbers 17-19 in the dendrogram) and ER (20-25) is obvious. A highly speculative interpretation would ascribe this to genetic kidnapping: the eucaryotic part of the cell makes use of a formerly procaryotic enzyme by targeting it to ER membranes. The sequence of a cyanobacterial n-3-desaturase will be very useful for further clarification of this point. The Δ15-desaturase from a moss (16) branches off earlier, but at present its subcellular origin cannot be assigned.

The plastidial n-6-desaturases (13,14) show homology to the corresponding cyanobacterial Δ12-desaturases (10-12) supporting their phylogenetic relationship. But in the case of the Δ12(n-6)-desaturases, the above described homology between plastids (13,14) and ER (15) does not exist. In contrast, the ER Δ12-desaturase (15) is closer to Δ 15(n-3)-desaturases, which again is open to speculation. Similarly, the cyanobacterial Δ6-desaturase (ref. 11) represents a branch of its own, which is not included in the figure.

Apart from these well-known desaturases we have found several sequences with three complete histidine boxes and the characteristic hydropathy profiles, which cannot be assigned to any of the branches identified so far. Therefore, this group (7-9) represents a separate branch to be identified yet by functional expression. The last group included in the dendrogram comprises acyl-CoA desaturases (1-6). They represent a clearly separated family, which in itself is deeply bifurcated.

Acknowledgments

Work in our lab was supported by the Bundesministerium für Forschung und Technologie (Grant BEO 22/0310 528).

References

1. Arondel V, Lemieux B, Hwang J, Gibson S, Goodman HM, Somerville CR. Science 1992; 258: 1353-1354.
2. Ntambi JM, Buhrow SA, Kaestner KH, Christy RJ, Sibley E, Kelly Jr TJ, Lane MD. J Biol Chem 1988; 263: 17291-17300.
3. Okuley J, Lightner J, Feldman K, Yadav N, Lak E, Browse J. Plant Cell 1994; 6: 147-158.
4. Sakamoto T, Wada H, Nishida I, Ohmori M, Murata N. Plant Mol Biol 1994; 24: 643-650.
5. Schmidt H, Dresselhaus T, Buck F, Heinz E. Plant Mol Biol, submitted.
6. Stukey JE, McDonough VM, Martin CE. J Biol Chem 1990; 265: 20144-20149.
7. Thiede MA, Ozols J, Strittmatter P. J Biol Chem 1986; 261: 13230-13235.
8. Wada H, Gombos Z, Murata N. Nature 1990; 347: 200-203.
9. Yadav NS, Wierzbicki A, Aegerter M, Caster CS, Perez-Grau L, Kinney AJ, Hitz WD, Booth Jr JR0, Schweiger B, Stecca KL, Allen SM, Blackwell M, Reiter RS, Carlson TJ, Russel SH, Feldman KA, Pierce J, Browse J. Plant Physiol 1993; 103: 467-476.
10. Yamamoto K, Mori H, Imaseki H. Plant Cell Physiol 1992; 33: 13-20.
11. Reddy AS, Nuccio ML, Gross LM, Thomas TL. Plant Mol Biol 1993; 27: 293-300.

Cytochrome b_5 and fatty acid desaturation

Mark A. SMITH[1,2], Johnathan A. NAPIER[2], Robert BROWNE[1,2], Peter R. Shewry[2] and A. Keith Stobart[1].
[1] Department of Biological Sciences, University of Bristol, Bristol. BS8 1UG. UK.
[2] Department of Agricultural Sciences, University of Bristol, IACR, Long Ashton Research Station, Long Ashton, Bristol. BS18 9AF. UK.

Introduction.

Cytochrome b_5 is a small haem-protein which participates in a number of oxidation/reduction reactions in the microsomal membranes of plants, animals and fungi. It has recently been shown to be involved in the microsomal Δ12 desaturation of oleate to form linoleate in Safflower (*Carthamus tinctorius*) (1,2), and the Δ12 hydroxylation of oleate to form ricinoleate in Castor bean (*Ricinus communis*) (3). Cytochrome b_5 appears to be an essential component of the acyl-complex lipid desaturase systems of the plant ER, and in the developing oil-seed may be almost totally dedicated to polyunsaturated fatty acid biosynthesis. The manipulation *in vivo* of cytochrome b_5 in oil-seed species could thus provide a means of regulating the fatty acid quality of the storage triacylglycerols. We have isolated a cDNA encoding cytochrome b_5 from a tobacco (*Nicotiana tabacum*) leaf cDNA library with the aim of altering cytochrome b_5 levels by anti-sense techniques and by over-expression of the protein in developing seeds. We have chosen tobacco for our studies as the plant is easily transformable and produces seeds rich in C18:2 fatty acids. The deduced amino acid sequence of the tobacco cytochrome shows considerable homology to mammalian cytochrome b_5. Here we describe the production of the plant cytochrome b_5 in *E.coli*, and the expression of the cytochrome b_5 message in the tobacco plant.

Cytochrome b_5 production in *E.coli*.

A cDNA (pNtcyb5) encoding cytochrome b_5 was isolated from a tobacco leaf cDNA library as described in (4). This was modified by the polymerase chain reaction (PCR) to generate a DNA sequence encoding a truncated form of the protein lacking the 30 C-terminal amino acid residues which compose the membrane binding domain. Both cDNAs were separately cloned into the vector pET-3D for expression in *E.coli.* as described in (5). Cells transformed with cDNA encoding the full length cytochrome b_5 (FL) or the truncated form (SOL) were red in colour showing absorbance spectra typical of reduced cytochrome b_5. The transformed cells accumulated cytochrome b_5 to a level of around 30% of total cell protein (fig. 1A) with the FL form accumulating in inclusion bodies and the SOL form in the cytoplasm. Being soluble the SOL form of cytochrome b_5 was easily purified and appeared to be correctly folded (as judged from CD spectra) and biologically active (being reduced by NADH in the presence of plant microsomal membranes). Absorbance spectra of the purified SOL form of tobacco cytochrome b_5 are given in figure 1B.

Expression of cytochrome b_5 message in tobacco.

RNA isolated from a number of different organs of the tobacco plant was analysed by northern blot using the cDNA encoding the full length form of the protein as a probe. The results (fig. 2A) indicated that the cytochrome b_5 message was expressed at different levels in the various tissues. Expression was highest in developing seeds and corolla tubes, weaker in leaf tissue and very low in roots and stems. A more detailed analysis of expression in the

J.-C. Kader and P. Mazliak (eds.), Plant Lipid Metabolism, 24–26.

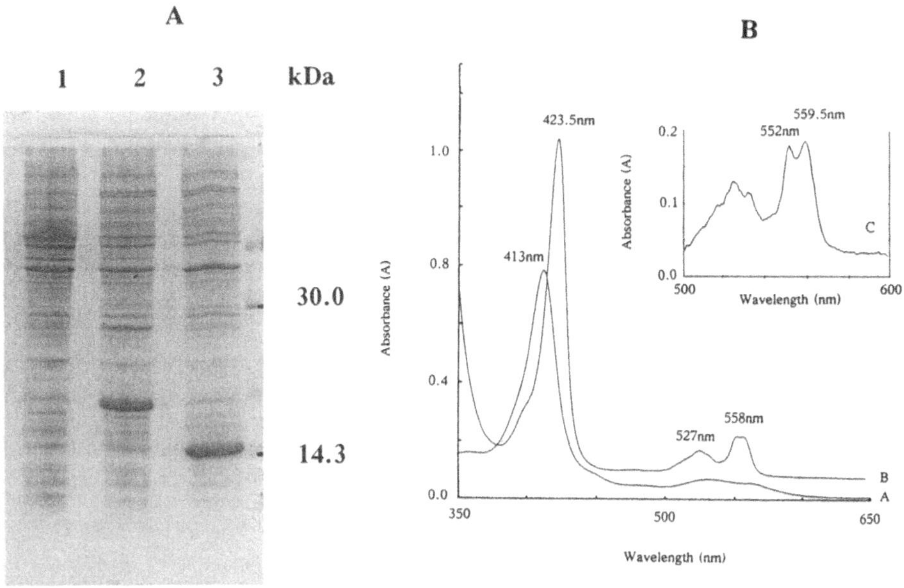

Figure 1. Expression of tobacco cytochrome b_5 cDNA in *E.coli*.
A. SDS-PAGE separation of total proteins extracted from cells transformed with (1) the vector pET-3D; or vector containing cDNA encoding the FL (2) or SOL (3) forms of cytochrome b_5. The gel was stained with Coomassie blue.
B. Absorbance spectra of the purified SOL form of cytochrome b_5; (A) oxidised spectrum, (B) reduced spectrum, (C) reduced spectrum at 77K.

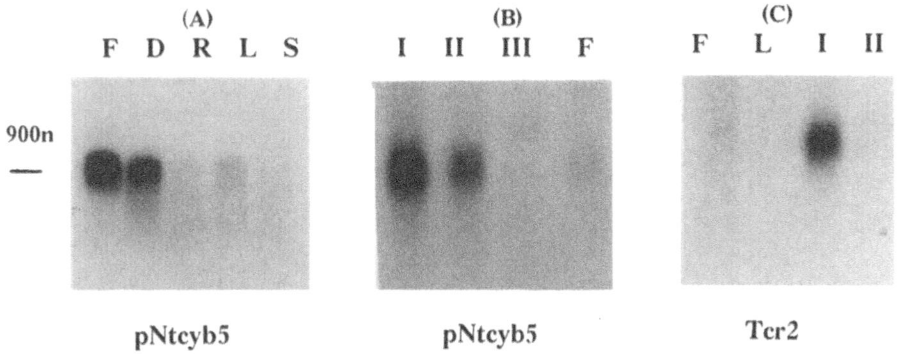

Figure 2. Northern blot analyses of cytochrome b_5 expression in tobacco.
Total RNA was isolated from tobacco tissues as indicated, 10 μg of RNA was loaded per track. Blots were probed with cDNA encoding cytochrome b_5 either isolated from the leaf cDNA library (Ntcyb5) or derived by PCR from mRNA extracted from developing seeds (Tcr2). **A**. Lane F, floral tissue; lane D, developing seeds < 15 daf; lane R, roots, lane L, leaves; lane S, stems: **B**. Lane I, developing seeds < 10 daf; lane II, developing seeds 10-15 daf; lane III, developing seeds 15-20 daf; lane F, floral tissue: **C**. Lane F, floral tissue; lane L, leaves; lanes I and II as for **B**.

developing seeds (fig. 2B) indicated that cytochrome b_5 mRNA was expressed at high levels in very early seed development, when maximum membrane biosynthesis would be expected to be occuring, and declined during seed maturation. Southern blot analysis of DNA isolated from tobacco indicated that there were several genes encoding cytochrome b_5 in the tobacco genome. RT-PCR and RACE-PCR were therefore conducted using mRNA isolated from developing seeds (stage I) and cDNAs encoding a further two forms of cytochrome b_5 were isolated. The deduced amino acid sequences of these proteins were very similar to the cytochrome b_5 initially isolated, although the 5' untranslated regions of the cDNAs appeared to differ considerably. Expression analysis of one of these cytochrome b_5 mRNAs (designated Tcr2) by northern blotting (fig. 2C) indicated that it appeared to be almost excusively expressed in the developing seed, and at an early stage of development.

Conclusions.

A cDNA encoding cytochrome b_5 has been isolated from tobacco and expressed in an active form in *E.coli*. Large amounts of this protein are therefore easily obtainable for further characterisation of the plant cytochrome, and for use in studies into the role of this haem-protein in fatty acid desaturation reactions and electron transport processes.

Analysis of cytochrome b_5 expression in tobacco indicates that there are a number of genes encoding this protein and that at least one of these may be specifically expressed in the developing seed. This form would therefore be a potential target for the manipulation of cytochrome b_5 levels in seed tissue. The isolation of the gene promotor which controls the seed specific expression of cytochrome b_5 would also be of value. A promotor confering high levels of tissue specific expression in a very early stage of seed development would be of considerable use in studies of seed development by transgenic techniques. Work is currently underway in our laboratories to isolate genes encoding cytochrome b_5 from tobacco, to alter cytochrome b_5 levels in tobacco by anti-sense techniques and over-expression, and to characterise the targeting of this protein to the ER membrane.

References.

(1) Smith M.A, Cross A.R, Jones O.T.G, Griffiths W.T, Stymne S, and Stobart A.K. (1990) Biochem. J. **272**: 23-29.

(2) Kearns E.V, Hugly S, and Somerville C.R. (1991) Arch. Biochem. Biophys. **284**: 431-436.

(3) Smith M.A, Jonsson L, Stymne S, and Stobart A.K. (1992) Biochem. J. **287**: 141-144.

(4) Smith M.A, Stobart A.K, Shewry P.R, and Napier J.A. (1994) Plant Mol. Biol. In press.

(5) Smith M.A, Stobart A.K, Tatham A, Shewry P.R, Napier J.A. (1994) Biochem. J. In press.

Acknowledgements.

This work was supported by the Science and Engineering Research Council (UK) and the Agriculture and Fisheries Research Council (UK). R.B is in receipt of an SERC Case studentship award.

PREPARATION OF 12-[^{3}H]-OLEATES FOR ASSAY OF THE Δ-12 DESATURASE

Dorota A. Abramovitch, S. Craig Tucker and Gary L. Powell
Department of Biological Sciences, Clemson University, Clemson, S. C. 29634-1903 USA

Introduction

The Δ^{12}-desaturase is an enzyme which introduces a second *cis* double bond into monounsaturated fatty acids like oleate (9-*cis*-octadecenoic acid) between the twelfth and thirteenth carbon atoms. This desaturation is the first step of polyunsaturated fatty acid biosynthesis, a process almost entirely confined to plants, some yeast and fungi, and some insects [1]. Mammals derive the essential fatty acids required in their diets from plant sources, the richest source being oil seeds. In developing oil seeds, the Δ^{12}-desaturase obtains a pair of electrons from cytochrome b_5 and molecular oxygen to abstract the two hydrogen atoms from the alkyl chain. The cytochrome b_5, the b_5 reductase, and the Δ^{12}-desaturase are normal components of the endoplasmic reticulum of developing seeds [2] although some mutants are known in which the activity of this desaturase is greatly reduced [3].

The Δ^{12}-desaturase has been assayed by observing the conversion of ^{14}C labelled oleate in oleoyl CoA, oleoyl phosphatidylcholine or a phosphatidylcholine analogue [4] to linoleate using either silver impregnated TLC [5] or HPLC to separate the methyl oleate from methyl linoleate or the corresponding lipid from the starting material [4]. The plastid Δ^9-desaturase has also been assayed by using the 9,10 tritiated oleate substrate and observing the release of tritium into water from the substrate, for example, see ref. [6]. The use of a tritiated substrate for the Δ^{12}-desaturase simplifies the separation and will make the assay more sensitive. We report procedures for synthesizing [^{13}H]-oleate, both generally labeled at the 12, 13 positions and both the R and S stereoisomer of 12-[^{3}H]oleate for the purpose of investigating the stereospecificity of the enzymic conversion of oleate to linoleate.

Results

12-[^{3}H]oleate. The desired product is oleate specifically tritiated on carbon atom 12:

$CH3(CH_2)_4CH_2CHTCH_2CH{=}CH(CH_2)_7COOH$ A

A standard method of introducing tritium is by catalytic hydrogenation, which must be done without reducing the *cis*-Δ^9 double bond. If it were possible to reduce only one of the two double bonds of linoleate (*cis,cis*-$\Delta^{9,12}$-octadienoic acid):

$CH3(CH_2)_4CH{=}CH_2CH_2CH{=}CH(CH_2)_7COOH$ B

then the desired product (A), along with an isomer with tritium at the 9,10 position and the double bond remaining at the 12 position (C), would be obtained. We needed to develop a
technique for stoichiometric introduction of one mole equivalent of tritium gas, in the presence of catalyst, on a microscale (1/2 millimole), to conserve starting product and radiolabel, and with carrier-free tritium. These requirements were met by employing commercially available linoleic acid and sodium borotritide ($NaBT_4$)(359 mCi/mmol) in ether, an aprotic solvent [7]. Each acidic proton of the free acid reacts with one hydride to produce one mole of HT, in the presence of palladium/charcoal,

J.-C. Kader and P. Mazliak (eds.), Plant Lipid Metabolism, 27–29.

to stoichiometrically reduce one double bond. The bond is reduced randomly providing a mixture of A & B. The specific activity is, in principle, half that of the borotritide. Linoleic acid in ether was added slowly to the $NaBT_4$ and Pd/C. After reduction, filtration to remove the Pd/C, methylation, and chromatography on silver impregnated TLC to separate starting material from the radiolabeled product, a peak with an Rf identical with methyl oleate and a specific activity of 50 mCi/mmol was obtained.
12-S-[^{3}H]oleate. A second method for introducing tritium into an organic compound at a specific location is by nucleophilic displacement, usually (sn2 displacements) with inversion of configuration. Methyl ricinoleate (methyl 12-(R)-hydroxy-9-octadecenoate) provides an abundant source of chiral starting material:

$CH3(CH_2)_4CH_2CH[\mathbf{OH}]CH_2CH{=}CH(CH_2)_7COOH$ D

Esterification of this hydroxy methyl ester with tosyl chloride provided a good leaving group with retention of configuration and the sn2 displacement of the tosyl group with $NaBT_4$ yields methyl 12-(S)-[^{3}H]oleate. The major product was characterized by TLC, chiral HPLC, and mass spectroscopy of the dimethylsulfide [8] and the simple product.

12-R-[^{3}H]oleate. For complete characterization of this enzymic reaction, it would be useful have available the enantiomeric specifically tritiated oleate. The Mitsunobu reaction uses triphenyl phosphine together with dimethylazodicarboxylate to form the benzoate ester of methyl ricinoleate with inversion of configuration [9]. Hydrolysis of the benzoate ester provided the 12-(R)-hydroxy-9-octadecenoate. Methylation and tosylation provided the appropriate precursor for nucleophilic displacement with $NaBT_4$ to obtain the methyl 12-(R)-[^{3}H]oleate.

Discussion

Schroepfer and Bloch [10] carried out similar syntheses, but with no requirement for unsaturation in the product, obtaining stereospecifically tritiated stearate to demonstrate the stereochemistry of the Δ^9-desaturase. Tritiated oleates were prepared using methods improved over those used earlier [12]. They have been used to synthesize the corresponding phosphatidylcholines and oleoyl-CoA. Assay of the Δ^{12}-desaturase using oleoyl-CoA will be carried just as for the Δ^9-desaturase [6] but using the corresponding cofactors. The stereospecifically tritiated substrates should permit characterizing the stereospecificity of the Δ^{12}-desaturase and the simplified assay should be of help in demonstrating expression of the recently isolated cDNA for this enzyme [11].

Experimental procedures

Methyl ricinoleate was commercially available (NU-CHEK-PREP). The corresponding methyl 12-(S)-hydroxy-9-octadecenoate was prepared using benzoic acid in the presence of triphenylphosphine and dimethyl azodicarboxylate [9]. The inverted O-benzoate was hydrolyzed and remethylated using diazomethane. The stereochemistry of these two 12-(R and S)-hydroxy-9-octadecenoates were verified by chiral HPLC of the benzoyloxy derivatives. Tosylation of the two methyl 12-(R and S)-hydroxy-9-octadecenoates was accomplished by treatment with tosyl chloride in pyridine [10] with retention of configuration. Tritium was incorporated by nucleophilic displacement of the tosyl group using NaB^3H_4 in boiling acetonitrile. The major products which comigrated with authentic methyl oleate on silver impregnated silica gel [5] were separated by preparative TLC using hexane/diethyl ether (60:40 v/v) and had identical IR spectra with methyl oleate. Using capillary GLC, two peaks were obtained with an area ratio of 1:2 with the second peak running 0.1 min faster than authentic methyl oleate. The mass spectra (molecular ion = 296 and fragmentations) of both compounds corresponded with methyl oleate. Similar results were obtained using $NaBD_4$ and treatment with dimethyl disulfide [8] verified the position of the double bond at 9,10 for the first compound but the second peak did not react with dimethyl sulfide. It may be a cyclic product. These two isomers were resolved by silver impregnated tlc. Specific activities of up to 50 mCi/mol were obtained.

Tritiated oleate was prepared by catalytic reduction of one of the two double bonds of linoleic acid. To 5 mCi of $NaBT_4$ (13.9 μmol) in 0.5 mL diethyl ether and 5 mg of 10% Pd/C was added 156 μL of 100 mg linoleic acid (55.7 μmol) (in 1 ml diethyl ether [7]). The free linoleic acid was added slowly (15 μL per min) formed sufficient tritium gas to stoichiometrically hydrogenate one of the two double bonds. The reaction mixture was taken to dryness under nitrogen, treated with 0.250 mL 10% methanolic HCl, heated to 70° for 1 hour to prepare the methyl ester. The product was separated by preparative TLC on silver impregnated silica gel as above. Three major radioactive peaks were observed, in addition to some material near the origin. The major radioactive peak corresponded with methyl oleate, the second peak with methyl vaccinate and the third peak was unidentified.

Acknowledgements
We gratefully recognize support from the EPSCoR Program of NSF (#13000-Z110) (A.G. Abbott); the Public Health Service (USA) (HL 380190) and from the USDA (NRICGP #9301064) to GLP.

References

1. Heinz E. Biosynthesis of Polyunsaturated Fatty Acids. In: Moore TS, editor. Lipid Metabolism in Plants. Boca Raton, FL: CRC Press, 1993:33-89.
2. Smith MA Cross AR Jones OTG Griffiths WT Stymne S Stobart K. Electron-transport components of the 1-acyl-3-oleoyl-sn-glycero-3-phosphocholine delta 12-desaturase (Δ^{12}-desaturase) in microsomal preparations from developing safflower (Carthamus tinctorius L.) cotyledons. Biochem. J. 1990;272:23-29.
3. Ray TK Holly SP Knauft DA Abbott AG Powell GL. The primary defect in developing seed from the high oleate variety of peanut (*Arachis hypogaea* L.) is the absence of Δ^{12}-desaturase activity. Plant Science 1993:91:15-21.
4. Sperling P Heinz E. Isomeric sn-1-octadecenyl and sn-2-octadecenyl analogues of lysophosphatidylcholine as substrates for acylation and desaturation by plant microsomal membranes. Eur J Biochem 1993;213:965-971.
5. Holloway P. Desaturation of long chain fatty acids by animal liver. Methods Enzymol. 1975;35:253-262.
6. Shanklin J Somerville C. Stearoyl-acyl-carrier-protein desaturase from higher plants is structurally unrelated to the animal and fungal homologs. Proc Natl Acad Sci USA 1991;88:2510-2514.
7. Brown HC Brown CA. A new convenient technique for the hydrogenation of unsaturated compounds. J Am Chemical Society 1962;84:1495.
8. Yamamoto K Shibahaza A Nakayama T Kajimoto G. Determination of double-bond positions in methylene-interrupted dienoic fatty acids by GC-MS as their dimethyl disulfide adducts. Chem Phys Lipids 1991;60:39-50.
9. Mitsunobu O. The use of diethyl azocarboxylate and triphenylphosphine in synthesis and transformation of natural products. Synthesis 1981;1-28.
10. Schroepfer GJ Bloch K. The stereochemical conversion of stearic acid to oleic acid. J Biol Chem 1965;240:54-63.
11. Okuley J Lightner J Feldman K Yadav N Lark E Browse J. Arabidopsis FAD2 gene encodes the enzyme that is essential for polyunsaturated lipid synthesis. J Plant Cell 1994;6:147-158.
12. Morris LJ Harris RV Kelly James AT. The stereochemistry of desaturation of long-chain fatty acids in Chlorella vulgaris. Biochem J 1968;109:673-678.

THE PHYLOGENETIC RELATIONSHIPS OF THE SPECIES WITHIN THE SECTION *Arachis* UTILIZING A FATTY ACID SYNTHESIS GENE

Patricia L. Tate, Charles Ainsworth, Gary Kochert, Albert Abbott

Department of Biological Sciences, Clemson University, Clemson,S.C. 29633, USA

Introduction

The original ancestors of the cultivated peanut (*Arachis hypogaea*), an allotetraploid, have not been conclusively determined, although several wild diploid species have been suggested. To identify the wild progenitors and to study the phylogenetic relationship of the species in the section *Arachis*, the sequence of a specific peanut gene was utilized. A cDNA of the fatty acid synthesis gene, stearoyl-ACP desaturase, was isolated and characterized. A region of this gene was then amplified from genomic DNA of nine wild species and sequenced. This sequence data, and the sequence information from the same region for the two cultivated peanut genes, were analyzed to determine the relatedness of the species. The results indicated that the cultivated peanut is derived from two separate lineages and supports the allopolyploid nature of the cultivated peanut. The amplified products were also digested with the RsaI restriction enzyme. A species-specific banding pattern unique to the DNA for some of the diploid species and one of the cultivated peanut genes was demonstrated. This technique provides a method for distinguishing diploid species and for analyzing gene expression in different tissues in the cultivated peanut.

Materials and Methods

ISOLATION OF PEANUT STEAROYL-ACP DESATURASE GENE

A cDNA library was constructed from developing peanut seed, and screened using a stearoyl-ACP desaturase clone from castor bean (a gift from J. Shanklin and C. Somerville). One clone, designated pcΔ9-1, was isolated and sequenced.

J.-C. Kader and P. Mazliak (eds.), Plant Lipid Metabolism, 30–32.

SOUTHERN HYBRIDIZATION

The DNA of nine diploid species, ten ecotypes (center of origin) tetraploid species, and eight cultivated peanut varieties was digested with Hae III and the filters were hybridized with labelled pcΔ9-1. The DNA from the parents and F2 progeny of an interspecific cross between two diploid species (A. stenosperma x A. cardenasii) was digested with DraI, Southern blotted, and hybridized with pcΔ9-1.

PCR AMPLIFICATION AND SEQUENCING

A region (350 bp) of the stearoyl-ACP desaturase gene was amplified by PCR from genomic DNA of nine wild species using primers constructed from the peanut cDNA (pcΔ9-1), and the products were sequenced. A second region (1100 bp), from the same wild diploid species, was amplified using upstream primers. The insert was sequenced and contained an intron region (504 bp). The same 1100 bp region was amplified from the cultivated peanut, and the products from both amplified genes were ligated into pUC118. One subclone was isolated and sequenced.

PARSIMONY ANALYSIS

The sequence data from the wild species and the cultivated peanut was subjected to PAUP computer analysis (Swofford 3.0), which infers phylogenies based on maximum parsimony. One phylogenetic tree was generated. The data was also subjected to bootstrap analysis to determine the confidence level of the branches.

RESTRICTION ENZYME DIGESTION OF THE 350 bp REGION

The 350 bp region was reamplified from the genomic DNA of the nine diploid species, the subclone of the cultivated peanut, and the total genomic DNA of cultivated peanut. The products were digested with a restriction enzyme, RsaI, and visualized by ethidium bromide on a 12% polyacrylamide gel.

Results

The stearoyl-ACP desaturase peanut cDNA (pcΔ9-1) was 1536 bp in length with a peptide length of 400 residues. The sequence was compared to the nucleotide sequences and amino acid sequences of other published species and was found to be similar. A highly conserved amino acid region was identified that showed 100% sequence similarity for approximately 100 bases among all species compared. Southern hybridization indicated the presence of a polymorphism in

the diploid species and not the tetraploid species. The progeny from the interspecific cross indicated the presence of a single locus that segregated in a Mendelian manner. The sequence information from the nine diploid species and the cultivated peanut was used for a phylogenetic data set and generated one parsimonious tree. The results indicated three lineages for the section *Arachis* with the two genes from the cultivated peanut representing separate lines. The two genes were most closely related to *A. ipaensis* and *A. cardenasii*.
The results of restriction digestion of the 350 bp amplified products showed that the products from three diploid species and the subclone of the cultivated peanut contained a RsaI cleavage site. The digested DNA produced two bands (180/170 bp) on the polyacrylamide gel. The products from the other diploid species did not contain the site and were represented by a single band (350 bp). Digestion of the amplified product from the cultivated peanut, an allotetraploid, produced three bands indicating one gene with the RsaI cleavage site (180/170 bp band) and one gene without the site (350 bp).

Discussion

By employing sequence information from an important fatty acid synthesis gene from the peanut, two wild diploid species have been identified, which are closely related to the cultivated peanut. Previous molecular data [1] agree that *A. ipaensis* is closely related to the cultivated peanut, while earlier reports [2,3], based on cytogenetics, suggested that *A. cardenasii* should be considered as a progenitor. The data presented here support these conclusions. In addition, the RsaI digestion pattern is a useful device for quickly identifying and categorizing the many undescribed diploid species and landrace species in the section. Also, since both genes are found in the tetraploid, this technique, coupled with RNA PCR, can be used to determine the expression of one or both genes from various tissues of the peanut plant.

References

1. Kochert G, Halward TM, Branch WD, Simpson CE. RFLP variability in peanut cultivars and wild species. Theor. Appl. Genet. 1991;81:565-570.
2. Smartt J, Gregory WC, Gregory MP. The genomes of *Arachis hypogaea* L. 1.Cytogenetic studies of putative genome donors. Euphytica 1978;27: 665-675.
3. Singh AK, Moss JP. Utilization of wild relatives in genetic improvement of *Arachis hypogaea* L. 2. Chromosome complement of species of section *Arachis*. Theor. Appl. Genet. 1982;61:305-314.

LIPID METABOLISM IN *MUCOR JAVANICUS*

Frances Jackson, Gareth Griffiths, Mark Smith, Keith Stobart.

School of Biological Sciences, Woodland Road, Bristol BS8 1UG, U.K.

Introduction

Some phycomycetes produce oil rich in linoleic acid (18:2 Δ9,12) and γ-linolenic acid (GLA; 18:3 Δ 6,9,12) but little is known in fungi of their mode of biosynthesis. Experiments following fatty acid changes in microsomal membranes from *Mucor circinelloides* in long incubations suggest that Δ6 desaturation may involve PI and may not require exogenous NAD(P)H [1]. Here, we report methods for obtaining active microsomal fractions from *Mucor javanicus* and preliminary studies on fatty acid desaturation and oil assembly.

Materials and Methods

Mucor javanicus was grown in liquid culture in a supplemented glucose medium at 25°C with shaking at 200 rpm. Fungal material was harvested after 2 days. Microsomal fractions were obtained using a method devised for yeast [2]. The fungal cells were resuspended in 0.1M phosphate buffer (pH7.2) containing 0.33M sucrose, 0.1%BSA, 1000units/ml catalase, 10mM cysteine, PVP (3%) and protease inhibitors (2mM benzamidine and 1mM PMSF) and stirred glass beads (0.2mm). The homogenate was filtered and spun at 20000xg for 15 minutes. After passing through Miracloth, the supernatant was centrifuged at 100000xg for 1 hour. The resultant pellet was resuspended in 0.1M phosphate buffer and used fresh or stored at -80°C. Lipids were separated by TLC or argentation TLC. Analysis was by GC or scintillation counting in PCS/xylene or PPO/POPOP.

Results

Lipid analysis showed that the major phospholipids in the microsomal membranes were PE and PC (48% and 39% respectively) with PI, PS and PA as minor components.PC and PE contained substantially higher quantities of GLA (over 30%) compared to PI and PS (10%). DAG was also high in GLA (25%), whereas microsomal TAG had only 10%.

FATTY ACID DESATURATION

Δ9 desaturase was assayed by following the abstration of ^3H from [9,10^{3}H] stearoyl CoA to water [3]. Following subcellular fractionation, the highest Δ9 desaturase activity was in the microsomal membranes with a specific activity of 24.1 nmol/μmolPC/min. Good desaturation rates were achieved

J.-C. Kader and P. Mazliak (eds.), Plant Lipid Metabolism, 33–35.

with NAD(P)H and 18:0-CoA. Both NADH and NADPH could serve as a source of reductant although the Vmax for NADH was double that of NADPH.The desaturase was KCN-sensitive (30% inhibition at 1mM) and CO-insensitive. The addition of soluble b5 (Smith et al, this vol.) stimulated the desaturation some two-fold.

The microsomal membranes catalysed the rapid transfer of [^{14}C]oleate in acyl-CoA to PC (Table1). LPCAT as measured spectrophotometrically with DTNB using palmitoyl lyso-PC and oleoyl CoA was extremely active (over 500 nmol/mg protein/min).

	[^{14}C] incorporation (nmol)					
time(min)	PC	PE	PA	DAG	TAG	FFA
15	11.6	1.4	0.7	0.5	0.6	1.0
30	14.2	1.7	1.0	1.0	0.7	1.6
60	12.7	1.3	0.9	0.8	0.5	1.7

Table1 Incorporation of [^{14}C]oleate into microsomal lipids. Microsomes were incubated with [^{14}C]18:1CoA (50 nmoles) and after incubation the distribution in complex lipids determined.

Desaturation of radioactive oleate to linoleate by the Δ12 desaturase occurred on the addition of reductant (200nmoles). NADH and NADPH could both serve as reductant although a two-fold increase in activity was observed with NADH compared to NADPH. Typical results from a time-course study are presented (table 2). 41% of the radioactivity incorporated in the lipid fraction accumulated in linoleate after 80 minutes incubation with NADH.

	[^{14}C] fatty acid distribution(%)		
time (min)	18:1	18:2	γ18:3
+10	83	14	4
+20	68	25	7
+40	62	34	4
+80	54	41	5
-80	88	8	4

+=+NADH

Table 2 Desaturation of [^{14}C]oleate in a time course. Microsomes were incubated with 50 nmoles [^{14}C]oleate and 200 nmoles NADH and the distribution in fatty acids of total lipid was determined.

Δ6 desaturase activity in the membranes has yet to be demonstrated although the microsomes were also active in transferring [^{14}C]linoleate in acyl-CoA to PC (table 3).

	[^{14}C] incorporation (nmol)					
time(min)	PC	PE	PA	DAG	TAG	FFA
15	11.7	1.5	2.7	0.4	2.7	2.4
30	10.0	1.5	2.6	0.4	3.5	3.2
60	9.0	1.5	2.4	0.5	2.4	4.0

Table 3 Incorporation of [^{14}C]linoleate into microsomal lipids. Microsomes were incubated with [^{14}C]18:2CoA (50 nmoles) and the distribution in complex lipids determined.

NADH and dithionite reduced minus oxidised difference spectra of microsomes showed a b-type cytochrome (cytochrome b_5) with bands at 558, 528 and 425 nm.

TAG ASSEMBLY

Microsomes were incubated with G3P and radioactive 18:1CoA. 63% of the radioactivity incorporated into membrane lipids accumulated in PC. Incubation of membranes with radioactive G3P and cold 18:1CoA, however, showed greatest incorporation of G3P in PA (table 4).

	Radioactive distribution among the complex lipids (%)				
time(min)	PC	PE	PA	DAG	TAG
30	3.1	1.2	90.6	3.1	1.0

Table 4 Microsomes were incubated with [^{14}C]G3P (400 nmoles) and 18:1CoA (50 nmoles) and the distribution of radioactivity in complex lipids determined.

The microsomes therefore efficiently acylate G3P to yield PA but little PA is converted to DAG and TAG. The transfer of oleate and linoleate in acylCoA to PC in the membranes (tables 1 & 3) was probably through the action of LPCAT rather than via a DAG-PC interconversion.

Conclusions

Microsomal membranes from *M. javanicus* catalyse Δ9 and Δ12 desaturation with the formation of oleate and linoleate respectively. The enzymes required NAD(P)H and cytochrome b_5 was identified as a possible electron transport component. The substrate for the Δ12 desaturase was acyl CoA and/or PC. No evidence was obtained that PI was involved. The membranes also catalysed the acylation of G3P to yield PA but little DAG/TAG was produced. Work is in progress on the characterisation of Δ6 desaturation and TAG assembly.

Acknowledgements

M. javanicus was supplied by Professor Herbert (Dundee University, UK). F Jackson acknowledges the receipt of a BBSRC studentship.

References

1. Kendrick A and Ratledge C. Desaturation of polyunsaturated fatty acids in *Mucor circinelloides* and the involvement of a novel membrane-bound malic enzyme. Eur. J. Biochem. 1992; 209, 667-673.
2. Murcott T, McNally T, Allen S, Fothergill-Gilmore L, Muirhead H. Purification, characterisation and mutagenesis of highly expressed recombinant yeast pyruvate kinase. Eur J Biochem 1991; 198, 513-519.
3. Shanklin J, Somerville C. Stearoyl acyl carrier protein desaturase is structurally unrelated to the animal and fungal homologues. Proc. Natl. Acad. Sci (USA).1991; 88, 2510-2514.

ARE DIACYLGLYCEROLS POSSIBLE SUBSTRATES FOR LINOLEATE DESATURATION IN SAFFLOWER COTYLEDONS (*Carthamus tinctorius* L.) ?

Kesri-Benhassaine Ghouziel and Mazliak Paul
Laboratoire de Physiologie végétale, USTBH-ISN, Alger, Algérie
Laboratoire de Physiologie Cellulaire et Moléculaire, URA 1180, Université P. et M. Curie, Paris, France

Introduction

The major lipid class in photosynthetic tissues is formed by the galactolipids rich in linolenic acid. Great advances have been realized these last years towards the elucidation of the mechanism of the biosynthesis of polyunsaturated fatty acids. Linolenate is formed by two successive desaturations of oleate. On one side oleoyl-phosphatidylcholine (PC) is converted to linoleoyl-phosphatidylcholine in microsomes [1-4]. On the other side oleoyl-monogalactosyldiacylglycerol is desaturated in the chloroplast [5]. However the cellular site and the nature of the substrate for desaturation of linoleate to linolenate are still the matter of discussions. The microsomal fraction may be involved in the last desaturation : linoleoyl-PC could be desaturated to linolenoyl-PC and transferred subsquently to the chloroplast for galactolipid synthesis [6]. Another site of desaturation has been proposed : linoleate desaturation might occur in the chloroplast and the substrate of desaturation would be the monogalactosyldiacylglycerol (MGDG) [5]. The labelling studies with [1-^{14}C] linoleate described in this paper support the view that desaturation of linoleate may involve diacylglycerols.

Material - Methods

The safflower (*Carthamus tinctorius* L.) seeds were grown for one week in total darkness and transferred into light. Cotyledons became greening and changed to leaves. Etioplasts were converted to chloroplasts. After 48 hours of illumination 50 nmoles of [1-^{14}C] linoleate (10mCi/mM) were applied on the surface of cotyledons. After different times, cotyledons were removed and fractionated by differential centrifugation. The lipids of different pellets (chloroplasts : 3000 x g, mitochondrial fraction : 20.000 x g, microsomal fraction : 100.000 x g) were analyzed by thin layer chromatography and their fatty acid methyl esters by radio gas chromatography.

J.-C. Kader and P. Mazliak (eds.), Plant Lipid Metabolism, 36–38.

Results - Discussion

INCORPORATION OF [1-^{14}C] LINOLEATE INTO THE DIFFERENT CELLULAR FRACTIONS.

After three hours of labelling, the incorporation of [1-^{14}C] linoleate was very important in the mitochondrial and microsomal fractions (respectively 42 and 40% of total incorporated radioactivity). However, after longer periods of labelling, the distribution of the radioactivity was different : a decrease in the radioactivity of the microsomal fraction was coincident with an increase in the radioactivity of the chloroplastic fraction. The radioactivity of the mitochondrial fraction did not change (table I).

Table I : Incorporation of [1-^{14}C] linoleate in the different cellular fractions of safflower cotyledons. 50 nmoles of [1-^{14}C] linoleate were applied on 5 day safflower cotyledons.

Fraction	Incubation period	pmol linoleate incorporated into total fatty acids per mg of protein	% of total fatty acid radioactivity
3000 x g	3 h	317	18
	24 h	544	35
20 000 x g	3 h	742	42
	24 h	628	40
100 000 x g	3h	673	40
	24h	395	25

DESATURATION OF [1-^{14}C] LINOLEATE IN THE DIFFERENT CELLULAR FRANCTIONS.

The distribution of the radioactivity among lipid classes showed that the amount of label increased with time in the PC of mitochondrial and microsomal fractions. The diacylglycerols were also labelled in the different fractions (from 15 to 20 p-moles of linoleate were incorporated in these lipids). In the microsomal fraction, the radioactivity of the diacylglycerols was highest after short incubation times. After longer periods, there was a loss of label in this fraction and an increase in the labelling of the diacylglycerols of the chloroplastic pellet. The desaturation of linoleate was very small in the PC of microsomal, mitochondrial and chloroplastic fractions. The percentage of total fatty acid radioactivity did not exceed 5% in the linolenate of PC. In marked contrast, the linolenate of the diacylglycerols of all fractions could contain from 9 to 28% of total fatty acid radioactivity (table II).

Table II : Percentages of total fatty acid radioactivity in the fatty acids of different cellular fractions prepared from safflower cotyledons labelled with [1-^{14}C] linoleate.

		3000 x g		100 000 x g		20 000 x g	
		3h	24h	3h	24h	3h	24h
PC	18:2	95	94	100	97	100	96
	18:3	5	6	-	3	-	4
DAG	18:2	85	72	86	84	90	84
	18:3	15	28	14	16	9	16
MGDG	18:2	70	22	72	40	72	40
	18:3	30	78	28	60	28	60

There results suggest two possibilities for linoleate desaturation :

- Linoleate phosphatidylcholine might be desaturated in the microsomal fraction to linolenate phosphatidylcholine and the diacylglycerols rich in linolenate could be rapidly transferred to chloroplasts for galactolipid synthesis. This could explain the small percent of radioactivity found in linolenate phosphatidylcholine

- Alternatively the desaturation might involve the diacylglycerols of the microsomal fraction. The linoleate diacylglycerols deriving from linoleate phosphatidylcholine might be desaturated to linolenate diacylglycerols and transferred subsquentlly to the chloroplast for galactosylation [7]. In that case, the diacylglycerols might be substrates for linoleate desaturation.

References

1. Tremolieres A, Mazliak P. Biosynthetic pathway of α-linolenic acid in developing pea leaves : in vivo and in vitro study. Plant Sci lett. 1974;2:193-201.
2. Cherif A, Dubacq JP, Mache A, Oursel A, Tremolieres A. Biosynthesis of α-linolenic acid by desaturation of oleic and linoleic acids in several organs of higher and lower plants and algae. Phytochem 1975;14:703-706.
3. Slack CR, Roughan PG, Browse S. Evidence for an oleoyl phosphatidylcholine desaturase in microsomal preparation from cotyledons of safflower (Carthamus tinctorius) seed. Biochem J 1979;179:649-656.
4. Demandre C, Tremolieres A, Justin AM, Mazliak P. Oleate desaturation in six phosphatidyl-choline molecular species from potato tuber microsomes. Biochim Biophys Acta 1986;877:380-386.
5. Heinz E. Biosynthesis of polyunsaturated fatty acids. In: Moore TS, editor. Lipid metabolism in plants: CRC Press, 1993:33-89.
6. Stobart AK, Stymne S, Appelqvist LA. Desaturation of ^{14}C-oleate in leaves of barley. Phytochem 1980;19:1397-1402.
7. Ohnishi J, Yamada M. Glycerolipid synthesis in Avena leaves during greening of etiolated seedlings. Plant cell physiol.1982;24:1553-1557.

BIOCHEMICAL AND MOLECULAR BIOLOGICAL CHARACTERIZATION OF ACETYL-CoA CARBOXYLASES

Basil J. Nikolau[1], James J. Caffrey[1], Tomás Diez[1], Ji-ying Huang[2], Xiao-min Shang[2], Lisa M. Weaver[1] & Eve Syrkin Wurtele[2]
[1]Dept. Biochem. & Biophys. and [2]Dept. Botany, Iowa State University, Ames, IA 50011.

Acetyl-CoA carboxylase (ACCase) catalyzes the ATP-dependent carboxylation of acetyl-CoA to form malonyl-CoA. Classically this reaction is considered to be the first, rate-limiting reaction of *de novo* fatty acid biosynthesis. In contrast to all other organisms, which synthesize fatty acids in the cytosol, plants are unique in that *de novo* fatty acid biosynthesis occurs almost exclusively in plastids. In addition, plants utilize malonyl-CoA for the synthesis of a number of secondary metabolites, including flavonoids, very long chain fatty acids, stilbenoids, and many malonyl derivatives; these reactions are thought to occur in the cytosol of plant cells (1, 2, introduction of reference #3). To characterize how malonyl-CoA is generated from acetyl-CoA, in at least two compartments, we have been characterizing ACCase via a comprehensive study of biotinylated proteins. Our studies, and those from other labs, indicate that plants contain at least two structurally distinct ACCases. One form of ACCase has a biotin-containing subunit of approximately 240 kDa, and a second form has a biotin-containing polypeptide of about 60 kDa.

Molecular Structure of Plant Acetyl-CoA Carboxylases

The elucidation of the molecular structure of plant ACCase has been confused by the presence of previously unrecognized two-forms of the enzyme. Following the initial identification of the enzyme in plants (4), Kanangara and Stumpf (5) suggested that it is a multi-component, "prokaryotic-like" enzyme, with the biotinylated subunit being membrane bound. Subsequently, soluble ACCases were purified from a number of plant species (2, 3 and references there in). These enzyme preparations displayed biotinylated subunits which ranged between 21 kDa to 240 kDa. These data have been interpreted to indicate that plants contain a single ACCase with a biotinylated subunit of about 240 kDa, and that smaller biotinylated polypeptides are the result of proteolytic degradation of ACCase (3, 6).

We have begun the characterization of the 240 kDa ACCase of soybean by preparing antibodies to the purified enzyme. In addition, using a PCR-based procedure we have isolated a cDNA clone that codes for the 3'-end of the mRNA. This cDNA codes for the C-terminal 320 amino acids of the 240 kDa ACCase. The soybean sequence is approximately 80% identical to the maize (7), wheat (8), alfalfa (9) and *Arabidopsis* (10) homologs, but only about 30% identical to the animal (11, 12) and yeast (13) ACCases. Recently the complete sequence of an alfalfa cDNA (9) and an *Arabidopsis* gene (10) coding for the 240 kDa ACCases have been determined. The deduced amino acid sequence of these proteins do not contain a recognizable chloroplastic transit peptide at the N-terminus. Thus, the 240 kDa ACCase appears to represent the cytosolic form of ACCase. This enzyme has a native molecular weight of approximately 500 kDa, indicating that it has a homo-dimeric quaternary structure. This enzyme appears to be similar to the yeast and animal ACCases, which also occur in the cytosol.

We have identified a second form of ACCase, which contains a 60 kDa biotinylated subunit that accumulates during embryo development. This enzyme may represent the plastidic ACCase, and may be

J.-C. Kader and P. Mazliak (eds.), Plant Lipid Metabolism, 39–42.

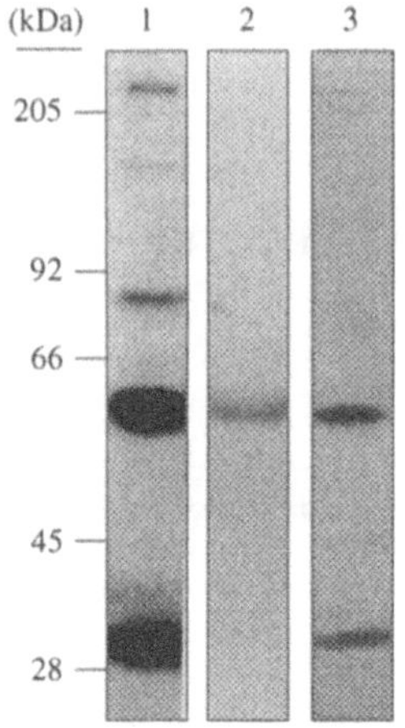

Fig. 1. Immunological characterization of the 60 kDa biotinylated protein of carrot somatic embryos. Crude extract from carrot somatic embryos was subjected to SDS-PAGE and western analysis. The blots were probed with: 1, ^{125}I-streptavidin; 2, anti-60 kDa biotinylated protein; 3, anti-30 kDa biotinylated protein. In both immunological westerns antibody-antigen complexes were detected with ^{125}I-Protein A.

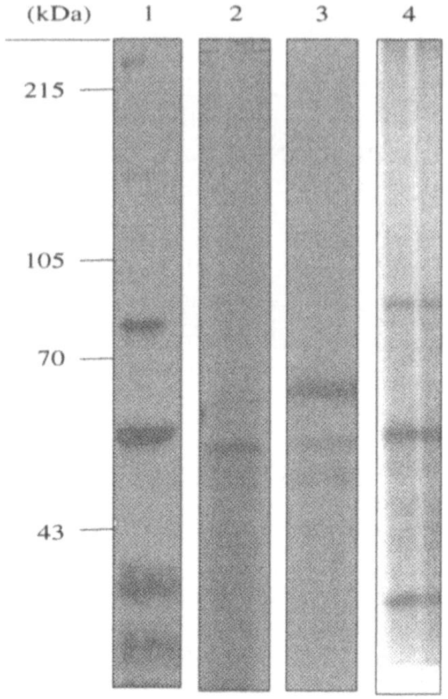

Fig. 2. Characterization of the 60 kDa ACCase of carrot. 1, streptavidin western of an extract from carrot somatic embryos. 2, *in vitro* translation of mRNA from carrot somatic embryos in wheat germ extract and immunoprecipitation with anti-60 kDa antibodies. 3, *in vitro* translation of mRNA from carrot somatic embryos in rabbit reticulocyte extract and immunoprecipitation with anti-60 kDa antibodies. 4, SDS-PAGE analysis of the immunoprecipitated complex containing the 60 kDa biotinylated protein.

analogous to the prokaryotic-type enzyme reported recently in pea chloroplasts (14). We have purified biotinylated polypeptides of 60 kDa and 30 kDa from carrot embryos and have generated antibodies to these proteins. On western blots of extracts from carrot embryos, both antibodies react with the 60 kDa biotinylated polypeptide (Fig. 1). We therefore conclude that the 60 kDa biotinylated polypeptide is an immunologically unique protein and bears no relationship to other proteins found in carrot extracts. Furthermore, these data indicate that the 30 kDa biotinylated protein purified from carrot extracts was probably a proteolytic degradation product of the 60 kDa protein. The antibodies to the 60 kDa biotinylated protein immunoprecipitate half the ACCase activity present in extracts of carrot embryos (15). This antibody-dependent precipitation of the ACCase activity requires the presence of the 60 kDa polypeptide in the extract. Thus, the removal of the ACCase activity is not due to the fortuitous precipitation of the 240 kDa ACCase. These data identify the 60 kDa biotinylated polypeptide as a subunit of a second, immunologically distinct form of ACCase.

To identify the size of the initial translation product of the 60 kDa biotinylated protein, we isolated RNA from carrot somatic embryos, *in vitro* translated the mRNA with a rabbit reticulocyte or wheat germ extracts, and immunoprecipitated the said polypeptide. With translation products from the wheat germ extracts the predominant polypeptide that was precipitated was 60 kDa, although a less abundant, 65 kDa polypeptide was also immunoprecipitated. This later polypeptide was the predominant protein immunoprecipitated from the translation products obtained with rabbit reticulocyte extracts (Fig. 2). These analyses identified the initial translation product of the protein as a 65 kDa polypeptide. This indicates that the 60 kDa biotinylated polypeptide is initially synthesized as a larger precursor, presumably due to the presence of a transit peptide. Therefore, this protein crosses a biological membrane. We are currently undertaking subcellular fractionation studies to determine if this protein is located in plastids.

Gel filtration chromatography on Sephacryl S-400 of extracts from carrot somatic embryos, indicate that the 60 kDa biotinylated polypeptide is in a very-large complex of greater than 1 MDa. This complex appears to contain additional subunits. Immunoprecipitations of the 60 kDa biotinylated protein from extracts of carrot embryos result in the precipitation of the said protein as well as nonbiotinylated proteins of 90 kDa and 35 kDa (Fig. 2). Additional analyses are required to ascertain the stoichiometry of the complex.

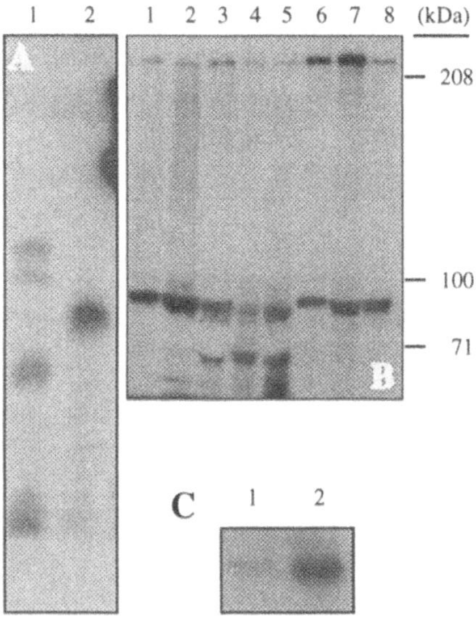

Fig. 3. Characterization of the 240 kDa ACCase of soybean. **A.** Southern analysis of soybean DNA digested with *Eco*RI (lane 1) and *Hin*d III (lane 2) and probed with a 240 kDa ACCase cDNA clone. **B.** Streptavidin western of extracts from developing seeds of soybean at 20d (1), 30d (2), 40d (3), 50d (4) and 80d (5) after flowering, and expanding (6), fully expanded (7) and mature (8) leaves of soybean. **C.** Northern analysis of RNA isolated from developing seeds (1) and expanding leaves (2) of soybean, probed with a 240 kDa ACCase cDNA.

Molecular Biology and Regulation of Plant Acetyl-CoA Carboxylases

We have available in our labs partial cDNA clones coding for the 240 kDa ACCase of soybean, maize and *Arabidopsis*. We have used these clones to determine the genetic position and number of genes that code for this enzyme in these species. In maize, we have used the recombinant inbred stocks of Burr et al. (16) to genetically map the two 240 kDa ACCase genes (*accA* and *accB*) in the maize genome. Both these genes are detectable by Southern analyses of maize genomic DNA, and *accA* maps to near the centromere of chromosome 2, while *accB* maps to chromosome 10L. It seems likely, from the work of (7), that both genes are transcriptionally active, as two mRNAs, slightly different in sequence from each other, have been cloned. Genomic Southern blots of soybean DNA probed with the cloned cDNA coding for the 240 kDa ACCase indicate that in soybeans this ACCase is coded by a small gene family, of potentially up to four copies per genome (Fig. 3A).

ACCase activity is developmentally regulated during leaf and seed development. We have investigated the relationship between ACCase activity and the accumulation of the two forms of ACCase during the development of these organs. In carrot somatic embryogenesis ACCase activity increases about 8-fold as embryos develop from embryogenic cell clusters to torpedo embryos (17). This change in ACCase activity is correlated with a dramatic increase in the accumulation of the 60 kDa ACCase, whereas the accumulation of the 240 kDa ACCase increases only slightly (16).

In soybeans, ACCase activity is higher in developing seeds than in leaves. Furthermore, ACCase activity in both organs is developmentally regulated, increasing from a basal level, to a maximum and subsequently decreasing. In comparison, although the accumulation of the 240 kDa biotinylated polypeptide is developmentally regulated during the development of seeds and leaves, it accumulates to higher levels in leaves than in seed (Fig. 3B). Consistent with this accumulation pattern, we find that the 8 kb mRNA that codes for the 240 kDa ACCase accumulates to higher levels in leaves than in seeds (Fig. 3C). In soybean seeds a 65 kDa biotinylated polypeptide accumulates late in seed development (Fig. 3B). This polypeptide accumulates to higher levels in seeds than leaves. We believe that this polypeptide is analogous to the 60 kDa biotinylated polypeptide of carrot, i.e., it may be the subunit of the plastidic ACCase.

Summary

We have identified a 60 kDa biotinylated protein of carrot that represents subunit of an ACCase that may be the plastidic form of this enzyme. We have isolated cDNA clones that code for the 240 kDa, cytosolic form of ACCase. Antibodies and clones are being used as reagents to investigate the structure and metabolic roles of these two isozymes.

References

1. Nikolau BJ, Wurtele ES, Stumpf PK. Tissue distribution of acetyl-coenzyme A carboxylase in leaves. Plant Physiol 1984;75:895-901.
2. Stumpf PK. The biosynthesis of saturated fatty acids. In: Stumpf PK, Conn EE, editors. The Biochemistry of Plants: A comprehensive treatise, vol 9. New York: Academic Press, 1987:121-157.
3. Harwood JL. Fatty acid metabolism. Annu Rev Plant Physiol Mol Biol 1988;39:101-138.
4. Hatch MD, Stumpf PK. Acetyl coenzyme A carboxylase and acyl coenzyme A-malonyl coenzyme A transcarboxylase from wheat germ. J Biol Chem 1961;236:2879-2885.
5. Kannangara CG, Stumpf PK. A procaryotic type acetyl-CoA carboxylase in spinach chloroplasts. Arch Biochem Biophys 1972;152:83-91.
6. Hellyer A, Bambridge HE, Slabas AR. Plant acetyl-CoA carboxylase. Biochem Soc Trans 1986;14:565-568.
7. Ashton AR, Jenkins CLD, Whitfeld PR. Molecular cloning of two different cDNAs for maize acetyl-CoA carboxylase. Plant Mol Biol 1994;24:35-49.
8. Elborough KM, Simon JW, Swinhoe R, Ashton AR, Slabas AR. Studies on wheat acetyl-CoA carboxylase and the cloning of a partial cDNA. Plant Mol Biol 1994;24:21-34.
9. Shorrosh BS, Dixon RA, Ohlrogge JB. Molecular cloning, characterization, and elicidation of acetyl-CoA carboxylase from alfalfa. Proc Natl Acad Sci USA 1994;91:4323-4327.
10. Roesler KR, Shorrosh BS, Ohlrogge JB. Structure and expression of an *Arabidopsis* acetyl-CoA carboxylase gene. Plant Physiol 1994;105:611-617.
11. Walid A-F, Subrahmanyan SC, Wakil SJ. Cloning of the yeast *FAS3* gene and primary structure of yeast acetyl-CoA carboxylase. Proc Natl Acad Sci USA 1992;89:4534-4538.
12. Lopez-Cassilas F, Bai D-H, Luo X, Kong I-S, Hermodson A, Kim K-H. Structure of the coding sequence and primary amino acid sequence of acetyl-CoA carboxylase. Proc Natl Acad Sci USA 1988;85:5785-5788.
13. Takai T, Yokoyama C, Wada K, Tanabe T. Primary structure of chicken liver acetyl-CoA carboxylase deduced from cDNA sequence. J Biol Chem 1988;263:2651-2657.
14. Konishi T, Sasaki Y. Compartmentalization of two forms of acetyl-CoA carboxylase in plants and the origin of their tolerance towards herbicides. Proc Natl Acad Sci USA 1994;91:3598-3601.
15. Nikolau BJ, Wurtele ES, Caffrey J, Chen Y, Crane V, Diez T, Huang J-Y, Mc Dowell MT, Shang X-M, Song J, Wang X, Weaver LM. In: Murata N, Somerville C, editors. Biochemistry and Molecular-Biology of Membrane and Storage Lipids of Plants. American Society of Plant Physiologists Press, 1993:138-149.
16. Burr B, Burr FA, Thompson KH, Albertson MC, Stubert CW. Gene mapping with recombinant inbreds in maize. Genetics 1988;118:519-526.
17. Wurtele ES, Nikolau BJ. Differential accumulation of biotin enzymes during carrot somatic embryogenesis. Plant Physiol 1992;99:1699-1703.

MAIZE ACETYL-CoENZYME A CARBOXYLASE GENES

Burle Gengenbach, Margaret Egli, Sheila Lutz, David Somers, John Gronwald+, Don Wyse
Dept. Agronomy & Plant Genetics, Univ. Minn.
+USDA-ARS, Plant Science Research Unit, St. Paul, MN

Introduction

Acetyl-coenzyme A carboxylase (ACCase) carboxylates acetyl-coenzyme A to malonyl-coenzyme A which serves as an intermediate metabolite for several diverse pathways in plant metabolism including synthesis of fatty acids, flavonoids, pigments and waxes. Plants likely have more than one ACCase isoform -- one in the plastid for fatty acid synthesis and at least one non-plastidic form to provide malonyl-coenzyme A for the other pathways. In maize, the plastid-localized ACCase required for fatty acid synthesis has been identified as a target site for several herbicide families [1] and herbicide-tolerant maize mutants have been selected [2,3].

This study was initiated to identify and characterize clones for maize plastid-localized ACCase to better understand its roles in regulation of fatty acid biosynthesis and herbicide sensitivity/tolerance. Antibodies against plastid-localized maize ACCase I [4] were used to screen a maize seedling leaf cDNA expression library. Ultimately, two related but distinct partial cDNA clones (Type A and Type B) and their corresponding genomic clones were obtained. The largest initial cDNA clone (Type A) was about 5.4 kb, so several cycles of primer extension of cDNA were required to obtain the putative full-length coding sequence of 6.9 kb. To date our studies indicate that maize has at least two functional genes for ACCase; further analysis will confirm whether one or both genes encode a polypeptide subunit targeted to the plastid.

Background

In a previous study of maize ACCase, Egli et al. [4]

J.-C. Kader and P. Mazliak (eds.), Plant Lipid Metabolism, 43–45.

isolated two isoforms by anion-exchange chromatography. ACCase I consisted of subunits of approximately 227 kDa, while ACCase II had subunits of about 219 kDa. ACCase I was present in purified chloroplasts from seedling leaves, but ACCase II could not be detected in total leaf chloroplast preparations or in mesophyll chloroplasts. ACCase I comprised most of the activity (>85%) in leaves and developing embryos and was highly sensitive to herbicides of the cyclohexanedione and aryloxyphenoxypropionate families [4,5]. ACCase II was <15% of total activity in leaves and was nearly absent in developing embryos. These observations indicate that the plastid-localized ACCase constitutes most of the enzyme activity in embryos where storage lipids accumulate [5].

Selection of maize tissue cultures for resistance to lethal herbicide concentrations resulted in at least six independent mutations for resistance to one or both herbicide families [2,3]. These mutants are inherited as single partially dominant genes and most mutations behave as alleles of the same genetic locus. ACCase I activity in mutant leaves and embryos was significantly less sensitive to herbicide inhibition indicating that the plastid-localized form had been altered [4].

Cloning Maize ACCase Coding Sequences

Polyclonal antibodies against ACCase I [4] were used to screen a maize expression library (gift of S. Gantt, U. Minn.). Several clones expressing a protein that cross-reacted with the antibody were identified, the largest of which contained a 5.45-kb cDNA insert. This clone (Type A) was sequenced and found to have an open reading frame of about 5.0 kb with high amino acid identity and similarity to the carboxytransferase sequences of rat and chicken (EMBL #A35578, #A29924). Further screening of the cDNA library with a probe from the Type A clone identified a 4.0-kb clone (Type B) with approximately 95% nucleotide identity to Type A across the 2-kb regions compared to date. These two cDNAs are nearly identical in sequence to the partial cDNAs reported recently for a different maize genotype [6]. We also screened a maize genomic library with a common probe and identified genomic clones with open reading frames identical to both Type A and Type B cDNAs, but neither genomic clone contained all of the coding sequence.

To obtain the 5′ sequences missing from the end of the Type A cDNA clone, we conducted oligonucleotide primer extension and polymerase chain reaction amplification of cDNAs synthesized from purified leaf polyA-RNA. Two amplified sequences of 1.5 kb and 0.4 kb were obtained and an additional 0.2 kb of potential coding sequence was deduced by aligning the 0.4 kb sequence with genomic clone A. The inclusive sequence data indicate an open reading frame of 6,885 nucleotides which could encode a polypeptide with an expected molecular weight of 255 kDa. The predicted polypeptide sequence includes binding domains for ATP, biotin, carboxybiotin and acetyl-coenzyme A that have about 85%, 90%, 65% and 75% identity, respectively, with alfalfa ACCase [7]. Good alignment of the complete amino acid sequence with the alfalfa sequence was evident beginning at residues #122 and #30 for maize and alfalfa, respectively.

Acknowledgements

The studies reported here were funded in part by grants from the USDA National Research Initiative Competitive Grants Program (#92-37301-7852) and from BASF AG.

References

1. Burton JD, Gronwald JW, Somers DA, Connelly JA, Gengenbach BG, Wyse DL. Inhibition of plant acetyl-coenzyme A carboxylase by the herbicides Sethoxydim and Haloxyfop. Biochem Biophys Res Comm 1987;148:1039-1044.
2. Parker WB, Marshall LC, Burton JD et al. Dominant mutations causing alterations in acetyl-coenzyme A carboxylase confer tolerance to cyclohexanedione and aryloxyphenoxypropionate herbicides in maize. Proc Natl Acad Sci USA 1990;87:7175-7179.
3. Marshall LC, Somers DA, Dotray PD, Gengenbach BG, Wyse DL, Gronwald JW. Allelic mutations in acetyl-coenzyme A carboxylase confer herbicide tolerance in maize. Theor Appl Genet 1992;83:435-442.
4. Egli MA, Gengenbach BG, Gronwald JW, Somers DA, Wyse DL. Characterization of maize acetyl-CoA carboxylase. Plant Physiol 1993;101:499-506.
5. Somers DA, Keith RA, Egli MA et al. Expression of the Acc1 gene-encoded acetyl-coenzyme A carboxylase in developing maize (Zea mays L.) kernels. Plant Physiol 1993;101:1097-1101.
6. Ashton AR, Jenkins CLD, Whitfeld PR. Molecular cloning of two different cDNAs for maize acetyl CoA carboxylase. Plant Molec Biol 1994;24:35-49.
7. Shorrosh BS, Dixon RA, Ohlrogge JB. Molecular cloning, characterization, and elicitation of acetyl-CoA carboxylase from alfalfa. Proc Natl Acad Sci USA 1994; 91:4323-4327.

EXPRESSION OF AN ALGAL ACETYL-CoA CARBOXYLASE GENE IN *E. COLI*

Paul G. Roessler, National Renewable Energy Laboratory,
Applied Biological Sciences Branch, Golden, Colorado 80401 USA

Introduction

Acetyl-CoA carboxylase (ACCase) is a biotin-containing enzyme that catalyzes the carboxylation of acetyl-CoA to form malonyl-CoA, which is one of the earliest steps in fatty acid biosynthesis. We have previously reported the isolation and characterization of the ACCase enzyme and gene from the alga *Cyclotella cryptica* [1,2]. The enzyme is large (~800 kD), and appears to be composed of four identical multifunctional subunits. The biotin carboxylase, biotin carboxylase carrier protein (BCCP), and carboxyltransferase domains of the polypeptide are in the same arrangement as in yeast and animal ACCases. The gene contains two introns (73 bp and 447 bp) and a region that encodes a putative signal sequence that could target the newly synthesized polypeptide through the endoplasmic reticulum and into the chloroplast.

ACCase is believed to catalyze a key regulatory step in fatty acid biosynthesis [3,4]. This suggests that lipid synthesis rates could be increased in certain cases by overexpressing ACCase in host organisms via genetic engineering. We are taking this approach at the National Renewable Energy Laboratory as a possible means to enhance triacylglycerol production rates in microalgae to provide a feedstock for the production of a diesel fuel substitute (biodiesel). In order to ascertain that the cloned *C. cryptica* ACCase gene produces a functional enzyme, and to determine the effects of ACCase overproduction in a bacterium, we have examined the expression of the *C. cryptica* ACCase gene in *E. coli*.

Results and Discussion

A plasmid was constructed to allow expression of the *C. cryptica* ACCase gene in *E. coli*. To make an expressible form of the ACCase gene, the 73-bp intron was removed and the 5' end of the gene (which contains the 447-bp intron and the region that encodes the putative signal sequence) was replaced with a fragment generated by the polymerase chain reaction that allows in-frame fusion to the 5' end of the *E. coli* β-galactosidase gene present in the pBluescript KS+ vector. The final construct (pACC20M9) encodes a 230 kD β-galactosidase/ACCase fusion protein that contains 44 amino acids from β-galactosidase (including polylinker-encoded amino acids) and 2066 amino acids from ACCase.

J.-C. Kader and P. Mazliak (eds.), Plant Lipid Metabolism, 46–48.

E. coli DH5αF′ cells were transformed with pACC20M9 and grown in LB medium supplemented with 50 μg/mL ampicillin, 0.5 μg/mL biotin, and 0.5 mM IPTG. Anti-ACCase immunoblots of SDS-polyacrylamide gels indicated that numerous immunoreactive polypeptides were produced by the recombinant cells (Figure 1), including an abundant polypeptide that appears to be full-length ACCase. Many truncated ACCase polypeptides were also produced in these cells. In contrast, immunoreactive polypeptides were not produced by control cells transformed with pBluescript KS+.

To determine whether the recombinant ACCase produced in *E. coli* can be biotinylated by the endogenous *E. coli* biotin ligase, gel blots were incubated with avidin conjugated to alkaline phosphatase to detect biotin-containing polypeptides. Although avidin blots are much more sensitive than anti-ACCase blots (as evidenced by the relative staining levels of purified ACCase), staining of recombinant ACCase was stronger in the anti-ACCase blot than in the avidin blot. This suggests that the endogenous *E. coli* biotin ligase is relatively inefficient at biotinylating the recombinant ACCase. Also note that the amount of endogenous biotinylated *E. coli* BCCP was much greater than the amount of biotinylated recombinant ACCase. The inefficient biotinylation of recombinant *C. cryptica* ACCase by *E. coli* biotin ligase is probably due to substantial differences in the primary (and resulting tertiary) structures of the polypeptides in the vicinity of the biotin attachment site.

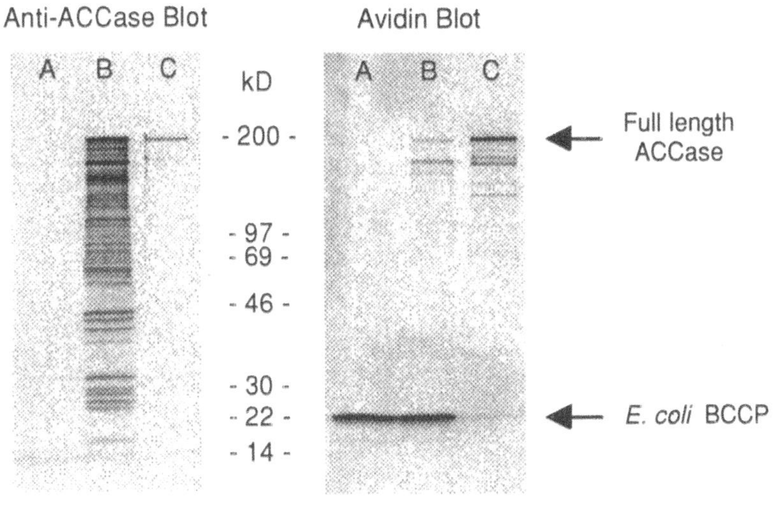

Figure 1. Immunoblots and avidin blots of recombinant *E. coli* DH5αF′ cells expressing *C. cryptica* ACCase. Lane A - Control cells containing pBluescript KS+. Lane B - pACC20M9-containing cells. Lane C - purified *C. cryptica* ACCase. Secondary antibodies and avidin were conjugated to alkaline phosphatase. The gels used for these blots were loaded with equal amounts of protein.

Expression of ACCase in various protease-deficient *E. coli* strains [ATCC 39624 (lon$^-$), ATCC 39627 (lon$^-$), and BL21 (lon$^-$, ompT$^-$)] was analyzed to assess whether the shortened ACCase polypeptides were proteolytic fragments. Because similar patterns of polypeptides were observed for each strain, proteolysis was probably not involved in generating the shortened products. It is therefore likely that these ACCase fragments are due to spurious translation initiation or termination.

Current efforts are directed at attempts to introduce the *C. cryptica* ACCase gene into yeast. Because the structures of yeast and *C. cryptica* ACCase are reasonably well conserved, it is anticipated that the endogenous biotin ligase of yeast will function more efficiently in biotinylating *C. cryptica* ACCase than was the case in *E. coli*. If so, the functionality of the recombinant ACCase can readily be evaluated.

Acknowledgements

This work was supported by the Biofuels Systems Division of the United States Department of Energy.

References

1. Roessler PG, Ohlrogge JB. Cloning and characterization of the gene that encodes acetyl-coenzyme A carboxylase in the alga *Cyclotella cryptica*. J Biol Chem 1993; 268:19254-9.

2. Roessler PG. Purification and characterization of acetyl-CoA carboxylase from the diatom *Cyclotella cryptica*. Plant Physiol 1990; 92:73-8.

3. Post-Beittenmiller D, Roughan G, Ohlrogge JB. Regulation of plant fatty acid biosynthesis: analysis of acyl-coenzyme A and acyl-acyl carrier protein substrate pools in spinach and pea chloroplasts. Plant Physiol 1992; 100:923-30.

4. Page RA, Okada S, Harwood JL. Acetyl-CoA carboxylase exerts strong flux control over lipid synthesis in plants. Biochim Biophys Acta 1994; 1210:369-72.

BIOCHEMICAL AND MOLECULAR BIOLOGICAL CHARACTERIZATION OF BIOTINYLATED PROTEINS OF PLANTS

James J. Caffrey[1], Yang Chen[2], Tomás Diez[1], Xueni Guan[2], Ji-ying Huang[2], Angela L. McKean[1], Jianping Song[1], Xiao-min Shang[2], Xun Wang[1], Lisa M. Weaver[1], Eve Syrkin Wurtele[2] & Basil J. Nikolau[1]
[1]Dept. Biochem. & Biophys. and [2]Dept. Botany, Iowa State University, Ames, IA 50011.

At least six biotinylated polypeptides can be detected in extracts from plants (1). These polypeptides are approximately 240 kDa, 120 kDa, 75 kDa, 60 kDa, 39 kDa and 34 kDa in size. Furthermore, the biotin enzymes, acetyl-CoA carboxylase (ACCase), propionyl-CoA carboxylase (PCCase), methylcrotonyl-CoA carboxylase (MCCase), pyruvate carboxylase (1) and geranoyl-CoA carboxylase (GCCase) are detectable in extracts from plants, We have undertaken biochemical and molecular biological approaches to identify and characterize the enzymes these polypeptides represent.

Acetyl-CoA Carboxylase

Plants appear to contain at least two forms of ACCase. The 240 kDa biotinylated polypeptide represents the subunit of one form of ACCase (2, 3). In soybeans this protein accumulates to higher levels in leaves than seeds. A cDNA clone that encompasses the 3'-end of the mRNA coding for this ACCase hybridizes to a single RNA species of about 8 kb, on Northern analysis of RNA isolated from soybean seeds and leaves. Consistent with the higher accumulation of the 240 kDa ACCase in leaves than in seeds, the mRNA coding for this protein accumulates to higher levels in leaves than in seeds. Southern analysis of soybean genomic DNA indicates that the 240 kDa ACCase is coded by a small gene family (4).

A 65 kDa biotinylated polypeptide accumulates specifically in developing seeds of soybean. We believe that this polypeptide is analogous to the 60 kDa biotinylated protein that accumulates in developing somatic embryos of carrot (4, 5). Antibodies to this latter protein immunoprecipitate half the ACCase activity found in extracts from somatic embryos of carrot (6). Thus, the 60 kDa biotinylated protein of carrot is a subunit of a second form of ACCase.

Propionyl-CoA Carboxylase

PCCase is a biotinylated protein that catalyzes a reaction required in the catabolism of amino acids and fatty acids of odd-numbered chain length, and in the catabolism and anabolism of branched-chain fatty acids. In order to characterize the structure of this enzyme from plants we undertook its purification. PCCase activity was purified from extracts of maize leaves by a four step scheme that included PEG precipitation, hydrophobic interaction chromatography, anion exchange chromatography and affinity chromatography. This purification scheme achieved a nearly 250-fold purification of PCCase activity. However, throughout this purification of PCCase, ACCase copurified. Indeed, SDS-PAGE analysis of the final purified PCCase preparation identified two biotinylated polypeptides of about 240 and 230 kDa. These polypeptides have previously been described as subunits of ACCase (7). Furthermore, mixed substrate kinetic studies (8) with the purified PCCase/ACCase preparation indicated that the carboxylation of propionyl-CoA and acetyl-CoA were carried out by the same enzyme. Furthermore, both PCCase and ACCase activities were similarly affected by a variety of inhibitors.

J.-C. Kader and P. Mazliak (eds.), Plant Lipid Metabolism, 49–51.

Therefore, we hypothesize that in maize, at least, all PCCase activity found in extracts is a side reaction of ACCase.

Methylcrotonyl-CoA Carboxylase

MCCase catalyzes a reaction that is required in the catabolism of leucine, and in the mevalonate shunt, which is a series of reactions by which mevalonate can be metabolized to non-isoprenoid compounds. We have purified MCCase to homogeneity from carrot embryos (9), maize leaves (10) and soybean seedlings (Fig. 1). From all these tissues MCCase is a large complex (~800,000 Da) comprising of a biotinylated subunit of about 75 kDa, and a nonbiotinylated subunit of about 60 kDa, probably in an $\alpha_6\beta_6$ configuration. Using the purified soybean MCCase we have determined that the kinetic mechanism catalyzed by this enzyme is Random Bi-Bi-Uni-Uni Ping-Pong.

We have sequenced cDNA clones coding for the biotinylated subunit of MCCase from tomato (11), soybean (12) and *Arabidopsis* (13). The deduced amino acid sequence of these biotinylated subunits show homology to biotin carboxyl carrier and to biotin carboxylase domains (11-13). Therefore, we hypothesize that this subunit catalyses the first half of the MCCase reaction.

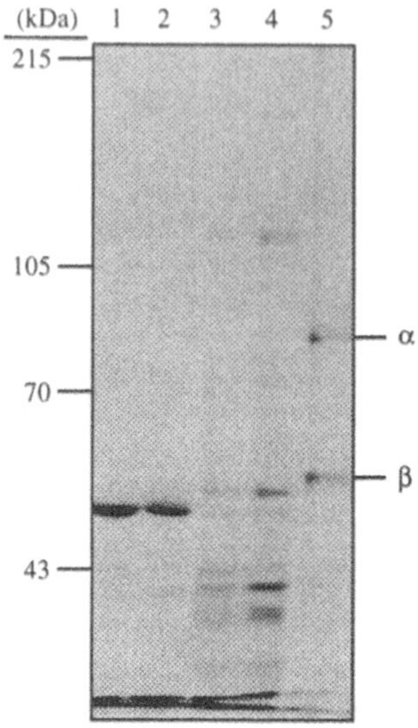

Fig. 1. Purification of MCCase from soybean seedlings. Each lane contains equal MCCase activity from each purification step. 1, crude extract; 2, PEG precipitation; 3, Cibacron Blue; 4, Q-Sepharose; 5, monomeric avidin. α=biotinylated subunit; β=nonbiotinylated subunit.

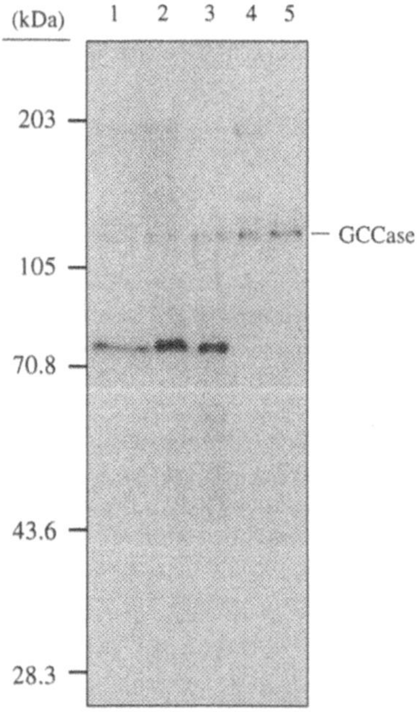

Fig. 2. Identification of the biotinylated subunit of GCCase. GCCase was purified from maize leaves and aliquots from each purification step were analyzed by SDS-PAGE and western blotting with ^{125}I-streptavidin. 1, crude extract; 2, PEG precipitate; 3, propyl-Agarose; 4, Cibacron Blue; lane 5, Q-Sepharose.

Geranoyl-CoA Carboxylase

Recently, we have found that plant extracts are able to catalyze the carboxylation of geranoyl-CoA. This activity is dependent on ATP and Mg^{2+}, and is sensitive to inhibition by avidin. These characteristics are consistent with the activity being catalyzed by the biotinylated enzyme, GCCase. This enzyme has previously only been characterized from *Pseudomonas* species (14), where it catalyzes a reaction required in the catabolism of noncyclic isoprenoid compounds.

We have purified GCCase activity nearly 200-fold from extracts of maize seedling leaves. The purification scheme included PEG precipitation, hydrophobic interaction chromatography, anion exchange chromatography and affinity chromatography. The purified GCCase preparation was examined by SDS-PAGE and western blotting with ^{125}I-streptavidin (Fig. 2). This analysis revealed that the GCCase has a biotinylated subunit of about 120 kDa. Further purification of the enzyme is required to ascertain whether GCCase contains nonbiotinylated subunit(s). Subcellular fractionations of both etiolated and green leaves of maize indicate that over 90% of the GCCase activity is localized in the plastids. We suggest that GCCase may function to degrade noncyclic isoprenoids such as carotenoids and the phytol tail of chlorophyll that are abundant in this organelle.

References

1. Wurtele ES, Nikolau BJ. Plants contain multiple biotin enzymes: Discovery of 3-methylcrotonyl-CoA carboxylase, propionyl-CoAcarboxylase and pyruvate carboxylase in the plant kingdom. Arch Biochem Biophys 1990;278: 179-186.
2. Egin-Buhler B, Ebel J. Improved purification and further characterization of acetyl-CoA carboxylase from cultured cells of parsley (*Petroselinum hortense*). Eur J Biochem 1983;133:335-339.
3. Slabas AR, Hellyer A. Rapid purification of a high molecular weight subunit polypeptide form of rape seed acetyl-CoA carboxylase. Plant Sci 1985;39:177-182.
4. Nikolau BJ, Caffrey JC, Diez T, Huang J-Y, Shang X-M, Weaver LM, Wurtele ES. Biochemical and molecular biological characterization of acetyl-coA carboxylases. This volume.
5. Wurtele ES, Nikolau BJ. Differential accumulation of biotin enzymes during carrot somatic embryogenesis. Plant Physiol 1992;99:1699-1703.
6. Nikolau BJ, Wurtele ES, Caffrey J, Chen Y, Crane V, Diez T, Huang J-Y, Mc Dowell MT, Shang X-M, Song J, Wang X, Weaver LM. In: Murata N, Somerville C, editors. Biochemistry and Molecular-Biology of Membrane and Storage Lipids of Plants. American Society of Plant Physiologists Press, 1993:138-149.
7. Egli MA, Gengenbach BG, Gronwald JW, Somers DA, Wyse DL. Characterization of maize acetyl-coenzyme A carboxylase. Plant Physiol 1993;101:499-506.
8. Dixon M, Webb EC. Enzymes. 3rd ed. New York: Academic Press, 1979: 72-75.
9. Chen Y, Wang X, Nikolau BJ, Wurtele ES. Purification and characterization of 3-methylcrotonyl-CoA carboxylase from carrot. Arch Biochem Biophys 1993;305:103-109.
10. Diez TA, Wurtele ES, Nikolau BJ. Purification and characterization of 3-methylcrotonyl-coenzyme A carboxylase from leaves of *Zea mays*. Arch Biochem Biophys 1994;310:64-75.
11. Wang X, Wurtele ES, Keller G, McKean AL, Nikolau BJ. Molecular cloning of cDNAs and genes coding for β-methylcrotonyl-CoA carboxylase of tomato. J Biol Chem 1994;269:11760-11769.
12. Song J, Wurtele ES, Nikolau BJ. Molecular cloning and characterization of the cDNA coding for the biotin-containing subunit of 3-methylcrotonyl-CoA carboxylase: Identification of the biotin carboxylase and biotin-carrier domains. Proc Nat Acad Sci USA. 1994;in press.
13. Weaver LM, Lebrun L, Hoffman N, Wurtele ES, Nikolau BJ. 3-Methylcrotonyl-CoA carboxylase of *Arabidopsis thaliana*: Isolation and characterization of cDNA coding for the biotinylated subunit. 1994;Submitted to Plant Physiol.
14. Fall RR, Hector ML. Acyl-coenzyme A carboxylases. Homologous 3-methycrotonyl-CoA and geranyl-CoA carboxylases from *Pseudomonas citronellolis*. Biochemistry 1977;16:4000-4005.

COMPARTMENTALIZATION OF TWO FORMS OF ACETYL-COA CARBOXYLASE AND PLANT TOLERANCE TOWARDS HERBICIDES

Yukiko Sasaki, Tomokazu Konishi, & Yukio Nagano
Department of Food Science and Technology, Faculty of Agriculture, Kyoto University, Kyoto 606-01, Japan.

Acetyl-CoA carboxylase (ACCase) catalyzes the synthesis of malonyl-CoA, the first intermediate in fatty acid synthesis. There are two forms of ACCase: a prokaryote form consisting of three protein components, biotin carboxylase, carboxyltransferase, and biotin carboxylase carrier protein, and a eukaryote form consisting of three functional domains on a single polypeptide. About 20 years ago, Kannangara and Stumpf reported the existence of the prokaryote form in spinach chloroplasts [1], the major site of fatty acid synthesis, but this finding has been dismissed because the prokaryote form has not yet been purified and the purified ACCases from various plants are all eukaryote form consisting of a subunit size of about 200 kDa, like that of the mammal enzyme.

Two years ago, the *acc*D gene, one of the subunit genes of carboxyl transferase, was identified in *E. coli* by protein sequencing of purified carboxyltransferase [2]. An open reading frame in the chloroplast genome has sequence similarity with this *E. coli acc*D. The open reading frame is conserved in various chloroplast genomes so far examined except for wheat and several grasses [3]. In the case of rice chloroplasts, this gene is truncated and may be a pseudo gene [4]. These previous findings prompted us to study whether the chloroplast *acc*D encodes a subunit of active acetyl-CoA carboxylase and what happened to the wheat lacking *acc*D gene.

Recently we have shown that an *acc*D gene product exists in pea chloroplasts and *acc*D encodes a subunit of active acetyl-CoA carboxylase [5]. Thus, the prokaryote form of ACCase exists in pea chloroplasts [5]. So two questions arise here. Where is the eukaryote form of ACCase in pea? Which form of ACCase exists in wheat plastids lacking the *acc*D gene in the chloroplast genome ?

Compartmentalization of two forms of ACCases.

To find the location of the eukaryote form of pea and wheat, we compared biotinyl polypeptides [6]. ACCase is a biotinyl protein and is easily detectable by streptoavidin. Since the prokaryote form of ACCase has a biotin subunit of about 35-kDa [5] and the eukaryote form has a huge biotin polypeptide of about 200-kDa, we compared the size of these biotinyl proteins in leaves and isolated chloroplasts (Fig. 1). The huge biotinyl protein was not found in pea chloroplasts, but found in wheat chloroplasts. Pea chloroplasts had a

J.-C. Kader and P. Mazliak (eds.), Plant Lipid Metabolism, 52–54.

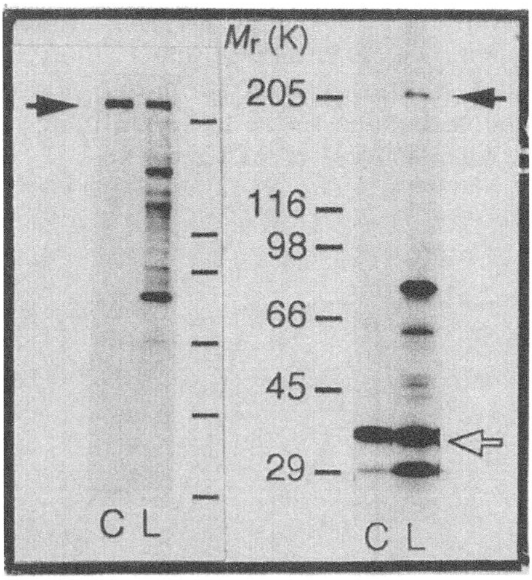

Fig. 1. Comparison of molecular mass of biotinyl polypeptides. C, chloroplast proteins; L, leaf proteins.

biotinyl protein of low molecular mass, but wheat chloroplasts did not. These results indicated that the pea eukaryote form of the enzyme existed outside the chloroplast, probably in the cytosol, as in mammals, and that the wheat eukaryote form existed in chloroplasts. These findings suggest that the two forms of ACCase are in different cell compartments, with the prokaryote form in the plastids and the eukaryote form elsewhere and that *Gramineae* lacking the *acc*D gene is an exception, with the eukaryote form in the plastids. The present findings also suggest the existence of two sites of malonyl-CoA synthesis. Malonyl-CoA is required not only for fatty acid synthesis but also for the synthesis of secondary compounds such as flavonoids in plants, so the ACCase outside of the plastids may function in such intermediate metabolism.

Plant tolerance towards herbicides

The selective grass herbicides of the phenoxypropionic acid type and the cyclohexanedione type, which were developed to kill *Gramineae* weeds in dicotyledonous crop plants, inhibit *de novo* fatty-acid synthesis [7]. Plastidic ACCase has been identified as the target of these herbicides [7], but it is not known whether the prokaryote form of the enzyme is inhibited *in vitro* by the herbicides

We examined the effects of herbicides on solubilized ACCase activity. Fenoxaprop and quizalofop did not inhibit the ACCase activity solubilized from pea plastids but inhibited the activity from wheat (Fig. 2). The ACCase activity extracted from pea leaves [6], which contained the active eukaryote form of ACCase but not the active prokaryote form, was inhibited by a higher concentration than these. These findings indicated that *in vitro* the prokaryote form of the enzyme was insensitive to these herbicides but the eukaryote form from pea was slightly sensitive. The cyclohexanedione type, sethoxydim, inhibited wheat chloroplast ACCase as already reported, but did not inhibit either form of the enzyme from the pea. The tolerance of intact dicotyledonous plants such as spinach, pea, or beans towards diphenoxypropionic acid-type and cyclohexanedione-type herbicides seems to be partly accounted for by the insensitivity of the prokaryote form of ACCase.

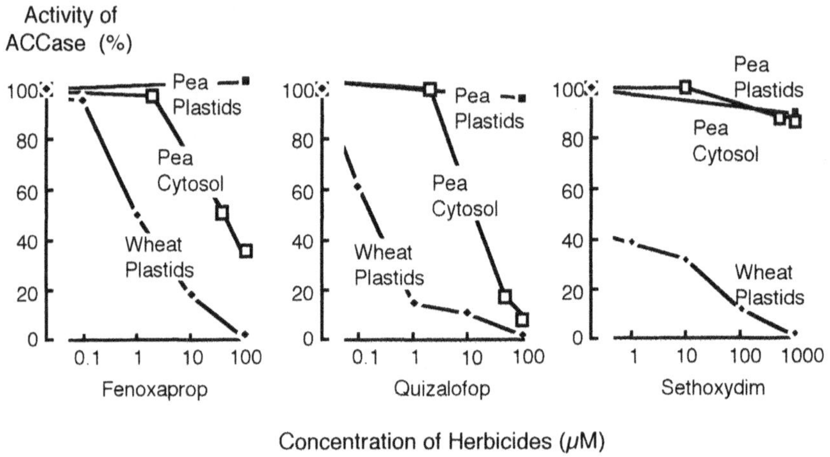

Fig. 2 Effect of graminicides on ACCase activity.
Each enzyme fraction indicated was measured as described [6].

References

1. Kannangara CG, Stumpf PK. Fat metabolism in higher plants LIV. A prokaryotic type acetyl CoA carboxylase in spinach chloroplasts. Arch. Biochem. Biophys. 1972;152:83-91.
2. Li S-J, Cronan JE Jr. The genes encoding the two carboxyltransferase subunits of *Escherichia coli* acetyl-CoA carboxylase. J. Biol. Chem. 1992: 267: 16841-16847.
3. Katayama H, Ogihara Y. Distribution of structural alterations of grass chloroplast DNA in monocots.XV International Botanical Congress, Aug, 1993 Yokohama. p 503 (abstr)
4. Hiratsuka J, Shimada H, Whittier R, Ishibashi T, Sakamoto M, Mori M, Kondo C, Honji Y, Sun C-R, Meng B-Y, Li Y-Q, Kanno A, Nishizawa Y, Hirai A, Shinozaki K, Sugiura M. The complete sequence of the rice (*Oryza sativa*) chloroplast genome. Mol. Gen. Genet. 1989; 217: 185-194.
5. Sasaki Y, Hakamada K, Suama Y, Nagano Y, Furusawa I, Matsuno R. Chloroplast-encoded protein as a subunit of acetyl-CoA carboxylase in pea plant. J. Biol. Chem. 1993; 268: 25118-25123.
6. Konishi T, Sasaki Y. Compartmentalization of two forms of acetyl-CoA carboxylase in plants and the origin of their tolerance towards herbicides. Proc. Natl. Acad. Sci. USA 1994;91:3598-3601.
7. Lichtenthaler HK. Mode of action of herbicides affecting acetyl-CoA carboxylase and fatty acid biosynthesis. Z. Naturforsch. 1990; 45c:521-528.

TOWARDS A STRUCTURAL UNDERSTANDING OF ENZYMES OF LIPID BIOSYNTHESIS

Antoni R. Slabas, Clare Brough, Johan Kroon, William Simon, Russell Swinhoe, David Rice*, John Rafferty*, Robert Winz, Antoine R. Stuitje†, Kieran Elborough

Lipid Molecular Biology Group, Department of Biological Sciences, University of Durham, U.K. Krebbs Institute of Molecular Biology, University of Sheffield, U.K.* Free University of Amsterdam†

Synopsis/Introduction

The biosynthesis of fatty acids is catalysed by two multienzyme systems (a) acetyl CoA carboxylase [ACC] and (b) fatty and synthetase [FAS]. We have purified the 220 KDa subunit form of ACC from wheat germ obtained amino and sequence and cloned a cDNA corresponding to the transcarboxylase domain of this enzyme[1]. The wheat cDNA was used to probe both an Arabidopsis genomic library and rape cDNA libraries prepared by oligo d(T) and random priming. A 19 Kb genomic clone has been isolated and the domain order ascribed following DNA sequencing[2].

Both enoyl reductase [ER] and ß-ketoreductase [ßKR] from rape have been overexpressed in E.coli and retain full biological activity. Purification strategies have been developed to obtain large mg quantities of both enzymes in homogeneous form. The two enzymes have been successfully crystallised. The ER diffracts down to 1.8Å [3] and its complete structure at near atomic resolution should be possible in the future.

Text:

(1) ACC

Figure 1 shows the domain order of ACC from Arabidopsis. The domains order from the N-terminus is CARBOXYTRANSFERASE, BIOTIN CARBOXYL CARRIER PROTEIN and TRANSCARBOXYLASE. This is the same order as the domains in animal ACC. The data has been deposited in the EMBL DATA base (Fig. 1c). Recent reports indicate that in plants two forms of ACC exist and the cloned form represents a cytoplasmic form. Anti-sense constructs to the enzyme are currently being used in transgenic studies to identify the importance of this enzyme.

(2) ER

The rape cDNA was isolated by Kater *et al.* [4], and extensive amino acid sequencing has been performed on the pure rape enzyme[5]. The cDNA

J.-C. Kader and P. Mazliak (eds.), Plant Lipid Metabolism, 55–57.

has been overexpressed in E.coli using the pET11d system. This has been purified to homogeneity and successfully crystallised. The crystals diffract down to 1.8Å (Fig. 3) which should allow new-atomic resolution to be achieved once the structure has been solved.

(3) ß-KR

This has similarly been cloned and overexpressed. Crystals have been successfully obtained and are stable to X-ray irradiation. They defract to 3Å and have a unit dimension of 80x80x300Å.

References:

[1] Elborough, K M, *et al..*, Plant Mol. Biol. 1994;24;21-34.
[2] Elborough, KM, *et al.*, Biochem. J. 1994;301;599-605.
[3] Rafferty, JB, *et al.*, J. Mol. Biol. 1994;237;240-242.
[4] Kater, MM, *et al.*, Plant Mol. Biol. 1991;17;895-909.
[5] Slabas, AR, *et al.*, Plant Mol. Biol. 1991;17;911-914.

Acknowledgements:

Biotechnology and Biological Science Research Council for financial support, Dr. Martin Kater for supplying the ER plasmid.

....

Figures:

Fig. 1 (a) Domain order of ACC in 19kb genomic clone from Arabidopsis
(b) cDNA isolated to date (c) EMBL accession numbers

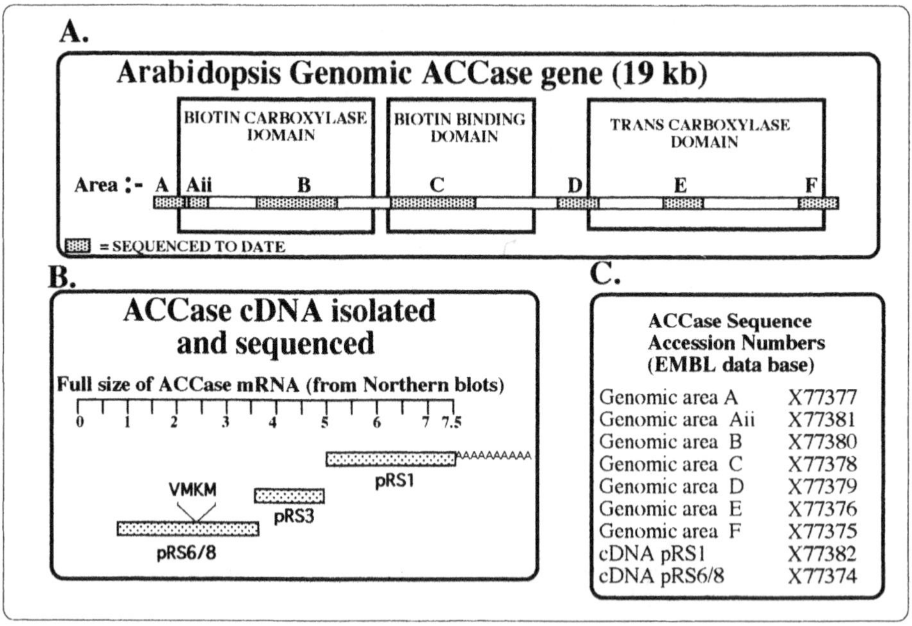

ACCase Sequence Accession Numbers (EMBL data base)	
Genomic area A	X77377
Genomic area Aii	X77381
Genomic area B	X77380
Genomic area C	X77378
Genomic area D	X77379
Genomic area E	X77376
Genomic area F	X77375
cDNA pRS1	X77382
cDNA pRS6/8	X77374

Fig.2 Crystal of ER

Fig.3 X-ray diffraction of ER -this difracts down to 1.8Å

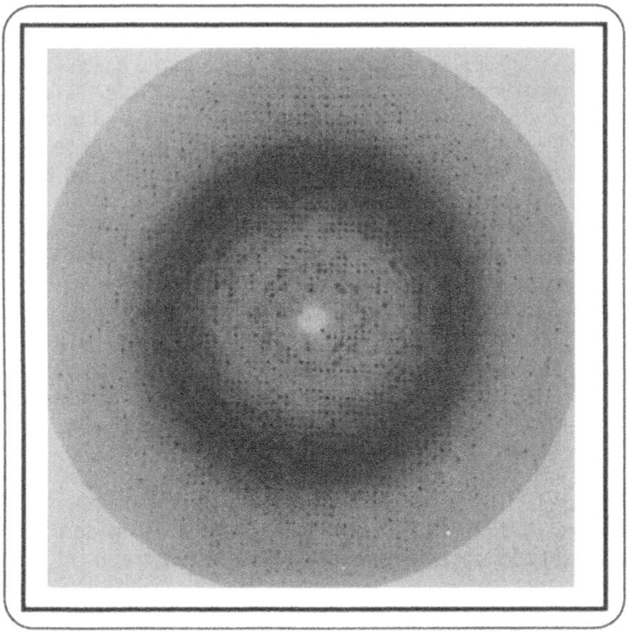

CHEMICAL REGULATION OF ACETYL-COA FORMATION AND *DE NOVO* FATTY ACID BIOSYNTHESIS IN PLANTS

Hartmut K. Lichtenthaler and Andrea Golz
Botanical Institute, University of Karlsruhe, Kaiserstraße 12, D-76128 Karlsruhe, Germany

Introduction

De novo fatty acid biosynthesis in higher plants is located in the plastids, and starts from acetyl-CoA. Acetyl-CoA can not pass biomembranes and has to be synthesized in the plastid compartment. There are two independent enzyme activities which may provide acetyl-CoA. 1) the AMP-forming acetyl-CoA synthetase (ACS), which forms acetyl-CoA from acetate, ATP and CoA and 2) the plastidic pyruvate dehydrogenase complex (pPDHC) [3]. The ACS of higher plants is localized exclusively in the plastids [1]. The substrate acetate is generated in the mitochondria and enters the plastid by diffusion [2]. The pPDHC is a large multienzyme complex, which consists of three different enzyme activities. Pyruvate can be be taken up from the cytoplasm or be generated within plastids which appear to be autonomous in forming acetyl-CoA from glucose [4].

There is still discussion which enzyme, ACS or pPDHC, provides the major acetyl-CoA pool for *de novo* fatty acid biosynthesis, and this seems to be dependent on plant species, development of plants and plastids and on the plastidic levels of pyruvate and acetate [4,6]. In order to obtain new information on the contribution of both enzymes to *de novo* fatty acid biosynthesis we looked for enzyme activities in developing chloroplasts and for specific inhibitors of the ACS and the pPDHC. The tests were performed in isolated chloroplasts and etioplasts which are capable of incorporating exogenously applied [^{14}C]-acetate as well as [2-^{14}C]-pyruvate into the total fatty acid fraction [6,7]. A specific, potent inhibitor of the plant ACS was found to be ethyl-adenylate (ethyl-AMP) (Figure 1) which mimics the tightly enzyme-bound ACS-intermediate acetyl-adenylate [7]. acetylmethylphosphinate (Figure 1) is a known inhibitor of mitochondrial PDHC [8] which also blocks the plastidic pPDHC as is shown here. We also demonstrate that etioplasts prefer pyruvate and chloroplasts acetate for fatty acid biosynthesis.

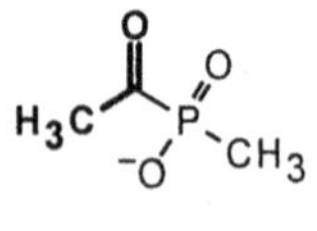

AMPI
(acetylmethylphosphinate)

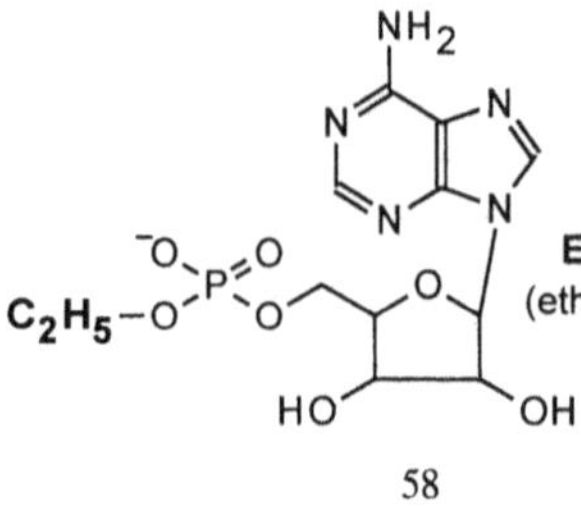

Ethyl-AMP
(ethyl-adenylate)

Figure 1: Chemical structure of the pPDHC-inhibitor AMPI and the ACS-inhibitor ethyl-AMP

J.-C. Kader and P. Mazliak (eds.), Plant Lipid Metabolism, 58–60.

Material and Methods

Intact etioplasts, etio-chloroplasts and chloroplasts were isolated from oat seedlings (*Avena sativa* L. var. Flämmingsnova) [9]. Incubation of plastids with [^{14}C]-acetate and [2-^{14}C]-pyruvate was carried out for 20 min in an iso-osmotic buffer [6], followed by saponification and extraction of total fatty acids according to [9]. During the incubation period of 20 min we did not detect a significant incorporation of radiolabel into ß-carotene or other carotenoids.

Results and Discussion

The incorporation of [2-^{14}C]-pyruvate into the total fatty acid fraction of isolated oat chloroplasts and etioplasts was inhibited by AMPI in a dose-dependent manner and increased during preincubation of plastids with the inhibitor (Figure 2). For isolated oat etioplasts the I_{50}-values were 10 µM without and 4.5 µM after a 15 min preincubation time of plastids with the inhibitor. For oat chloroplasts, where only a very low pPDHC activity remains, the I_{50}-values were much higher: 130 µM without and 80 µM after a 15 min preincubation with AMPI. The I_{50}-value for isolated etioplasts is similar to that determined for mitochondrial PDHC of pea (10 µM after 10 min preincubation) [8]. The results indicate that both mPDHC and pPDHC are inhibeted by AMPI. In contrast, the incorporation of [^{14}C]-acetate was not inhibited by AMPI, demonstrating that the other enzymes of fatty acid biosynthesis ACS, ACCase (acetyl-CoA carboxylase) and FAS (fatty acid synthetase complex) were not affected by this inhibitor.
During greening of oat seedlings and the formation of etio-chloroplasts and chloroplasts the activity of the pPDHC strongly declined as seen from the declining use of ^{14}C-pyruvate for fatty acid biosynthesis. In contrast, the activity of the ACS and the use of acetate for fatty acid biosynthesis increased. Isolated etioplasts incorporated a 9.5 fold higher amount of pyruvate than acetate into total fatty acids, whereas in isolated oat chloroplasts it was inverse, they used acetate at a 3 times higher rate than pyruvate. The incorporation of ^{14}C-pyruvate into fatty acids in etioplasts was little affected by acetate, but somewhat decreased in chloroplasts. This is in agreement with our observation that isolated oat chloroplasts prefer acetate for *de novo* fatty acid biosynthesis, whereas etioplasts prefer pyruvate. We assume that this also may be the case for etioplasts and chloroplasts of other plants. A physiological reason for this may be that etioplasts need the NADPH from the pPDHC for *de novo* fatty acid biosynthesis. In chloroplasts, in turn, ATP and NADPH is available through the photosynthetic light reactions and the acetyl-CoA is delivered via the ACS activity. With 100 µM diuron the photosynthetic electron transport was blocked and the fatty acid biosynthesis of isolated oat chloroplasts declined to 10 % of that of controls. The fatty acid biosynthesis capacity of isolated etioplasts, in turn, was not affected by diuron.

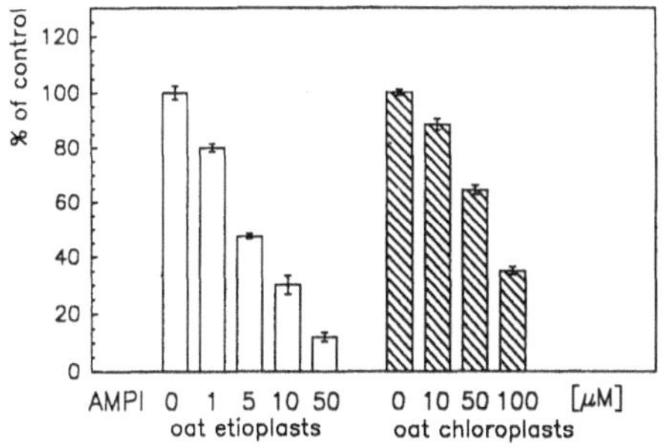

Figure 2: Dose-dependent inhibition of [^{14}C]-pyruvate incorporation into the total fatty acid fraction of isolated oat etioplasts and chloroplasts by AMPI. Time of preincubation of plastids with inhibitor: 15 min. Time of incubation 20 min.

With ethyl-AMP, we found a specific inhibitor of plant ACS which is competitive with respect to ATP (K_i-value: 42 nM) and non-competitive with respect to acetate [6]. Ethyl-AMP inhibits plant ACS preparations (I_{50}-value: 0.4 µM) but also *de novo* fatty acid biosynthesis in intact plastids starting from acetate (I_{50}-value: 10 µM in isolated radish etioplasts). The incorporation of labelled pyruvate in etioplasts was not affected by ethyl-AMP.
With ethyl-AMP and AMPI specific inhibitors of the two independent routes of acetyl-CoA formation in plastids are available. Several specific xenobiotics block efficiently *de novo* fatty acid biosynthesis at different steps and enzyme levels (Figure 3). Graminicides such as diclofop, sethoxydim or cycloxydim are specific inhibitors of acetyl-CoA carboxylase (ACCase) of grasses [10], the antibiotics cerulenin and thiolactomycin are inhibitors which affect certain of the ß-ketoacyl-ACP synthases (KAS I, II and III). With these xenobiotics one can control the metabolite flow through the fatty acid biosynthesis pathway and obtain a better understanding of the regulation of the plants' *de novo* fatty acid biosynthesis and the enzymes involved.

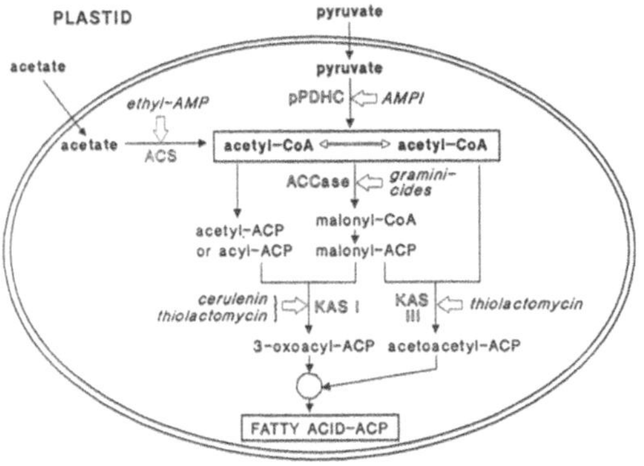

Figure 3: Scheme of *de novo* fatty acid biosynthesis pathway from acetate and pyruvate with indication of the site of action of several inhibitors.

References

1. Kuhn N, Knauf MJ, Stumpf PK. Subcellular localization of acetyl-CoA synthetase in leaf protoplasts of *Spinacia oleracea*. Arch Biochem Biophys 1981; 209:441-450.
2. Liedvogel B, Stumpf PK. Origin of acetate in spinach leaf cells. Plant Physiol 1982; 69:897-903
3. Williams M, Randall DD. Pyruvate dehydrogenase complex from chloroplasts of *Pisum sativum*. Plant Physiol 1979; 64:1099-1103.
4. Liedvogel B, Bäuerle R. Fatty acid synthesis in chloroplasts from mustard (*Sinapis alba* L.) cotyledons: Formation of acetyl coenzyme A by intraplastid glycolytic enzymes and a pyruvate dehydrogenase complex. Planta 1986; 169:481-489.
6. Golz A, Lichtenthaler HK. Inhibition of the plastidic pyruvate dehydrogenase complex in isolated plastids of oat. Z Naturforsch 1994; 49c: in press.
7. Golz A, Lichtenthaler HK. Inhibition of plant acetyl-CoA synthetase by alkyladenylates. Z Naturforsch 1992; 47c:845-850.
8. Baillie AC, Wright K, Wright BJ, Earnshaw CG. Inhibitors of pyruvate dehydrogenase as herbicides. Pest Biochem Physiol 1986; 30:103-111.
9. Kobek K, Focke M, Lichtenthaler HK, Retzlaff G, Würzer B. Inhibition of fatty acid biosynthesis in isolated chloroplasts by cycloxydim and other cyclohexane-1,3-diones. Physiol Plant 1988; 72:492-498.
10. Lichtenthaler HK. Mode of action of herbicides affecting acetyl-CoA carboxylase and fatty acid biosynthesis. Z Naturforsch 1990; 45c:521-528.

THE SECRETIVE FAMILY OF β-KETOACYL-ACP SYNTHASES

Penny von Wettstein-Knowles

Department of Physiology, Carlsberg Laboratory, Gamle Carlsberg Vej 10, DK-2500 Copenhagen, Denmark and Department of Genetics, Institute of Molecular Biology, University of Copenhagen, Øster Farimagsgade 2A, DK-1353 Copenhagen K , Denmark

Introduction

The β-ketoacyl-acyl carrier protein (ACP) synthases (KASes) or condensing enzymes create new carbon-carbon bonds by carrying out condensations. The catalysis requires activated substrates, which is achieved by a thioester linkage to the 4' phosphopantethein prosthetic group of ACP, an ACP domain or coenzyme A (CoA). As illustrated in Figure 1 a condensation reaction is tripartite. (i) An activated primer is transferred to a cysteine residue of the condensing enzyme. (ii) A nearby residue of the same enzyme provokes a decarboxylation of an activated acyl donor, generally malonate, to give a carbanion that serves as an elongating unit. (iii) A nucleophilic attack of the carbanion on the carbonyl carbon of the enzyme bound primer leads, via a presumed tetrahydral intermediate and release of an activating unit, to an elongated product with a new carbon-carbon bond and a β-keto group.

These condensing enzymes participate in quite diverse biosynthetic pathways as components of multienzyme complexes which can carry out repetitive condensations. Two types of organization have been recognized for some time; namely, complexes composed of monofunctional proteins coded for by different genes and those composed of one or two polyfunctional polypeptides determined by the corresponding number of genes. Recently a third or modular type of organization has been described. Each module consists of the complete set of genes needed to code for all activities required in a given round of elongation. Large open reading frames (ORFs) each coding for two modules are arranged in the same sequence in which they function in synthesis. Herein I will concentrate on those condensing enzymes belonging to the monofunctional enzyme complexes.

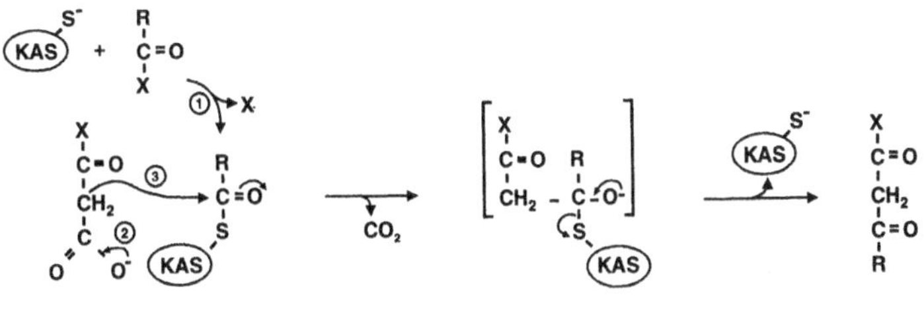

Figure 1 The condensation reaction. X represents the substrate activating group which for members of the monofunctional enzyme complexes discussed herein is ACP except that CoA activates the primer substrate for KAS III and both substrates for chalcone synthase, a polyketide synthase (PKS), and the elongases (ELSes).

J.-C. Kader and P. Mazliak (eds.), Plant Lipid Metabolism, 61–71.

Such complexes include fatty acid synthases (FASes), elongases (ELSes) and polyketide synthases (PKSes) which can function individually or in concert. FAS synthesizes the 16 and 18 carbon acyl chains of membrane lipids as well as those of the plant cutin and suberin monomers. ELSes use these acyl chains as primers to synthesize longer ones for storage lipids in some seeds and for cuticular and epicuticular waxes. PKSes participate in a wide range of secondary metabolic pathways. In plants chalcone synthase contributes to the carbon skeleton of the flavonoids, in fungi and bacteria, especially the *Actinomycetes*, PKSes are instrumental in antibiotic and toxin syntheses while in *Rhizobium* species a FAS and PKS form the acyl chains of the lipooligosaccharide Nod factors functioning in host specificity. Generally in FAS and ELS complexes the newly introduced β-keto group is immediately removed by a series of three reactions –a β-keto reduction, a β-hydroxy dehydration and an enoyl reduction– to give a completely reduced, activated substrate for another condensation. By comparison, the condensing enzymes of PKS complexes, as their name implies, generally use as a substrate the initial product of the condensation reaction or one of the intermediates formed during removal of the β-keto group. The resulting carbon skeletons are well decorated with keto groups, hydroxy groups and double bonds introduced at specified steps in the elongation. To what extent this differential use of primers reflects on the nature of the condensing enzyme rather than on compartmentalization or availability of a substrate remains to be studied. Condensing enzymes, on the other hand, can exhibit marked preferences for given chain lengths. To illustrate, three KASes participate in fatty acyl chain synthesis in plant plastids. KAS III chooses acetyl-CoA primers, KAS II prefers C_{16} and C_{14} acyl-ACP primers while KAS I is thought to be less selective and can use C_4 to C_{14} acyl-ACP primers. Another way to classify condensing enzymes is according to their sensitivity to the antibiotic cerulenin, a fatty acid analogue. Thus, the aforementioned KAS III is insensitive, KAS II somewhat sensitive while KAS I is highly sensitive.

Three condensing enzymes have been characterized in *Escherichia coli* hitherto. KAS I encoded by *fabB* is highly sensitive to cerulenin and is unique in its ability to carry out the first condensation of the unsaturated pathway. KAS II determined by *fabF* has an intermediate sensitivity to cerulenin in vivo, and is singular in accomplishing the final condensation of the unsaturated pathway. KAS III coded for by *fabH* appears to be very closely related to plant KAS III.

To delineate the contribution of the structure of a condensing enzyme to its particular function within a multienzyme complex, a store house of tools such as cloned genes, antibodies and crystal structures is needed. In the following I highlight the instrumental role of cerulenin (i) as a radiolabelled tag in enzyme purification and (ii) in activity determinations. A KAS complementation assay has been developed to the level where the more secretive members of this protein family will no longer be able to remain hidden. Initial experiments exploiting the tools as well as pertinent observations from studies on the PKS ORF I and ORF II genes from *Streptomyces* and the *nodE* and *nodF* genes of *Rhizobium* are described.

Cerulenin and the purification of KAS I

When an interest in plant condensing enzymes first manifested itself in the mid to late 1980s, all known condensing activities of FAS and PKS complexes could be inhibited by cerulenin. Considerable indirect evidence [1,2] suggested that the most sensitive enzymes formed a covalent link with the primer accepting cysteine in the active site (Figure 2A); a mechanism subsequently confirmed for the most sensitive enzymes [5,6]. Furthermore, a preliminary report had demonstrated that the ^{3}H-labelled tetrahydro derivative of cerulenin could be used to tag an apparent condensing activity during several purification steps [7]. Combining the endosymbiont theory for the origin of plant plastids, the ability to obtain and process large amounts of *E. coli* in which KAS I was reported to represent 0.6% of total proteins [8], the availability of mutants in the gene encoding this *E. coli* condensing enzyme [9] and the extensive literature available on fatty acid synthesis in *E. coli* suggested that this organism would be a

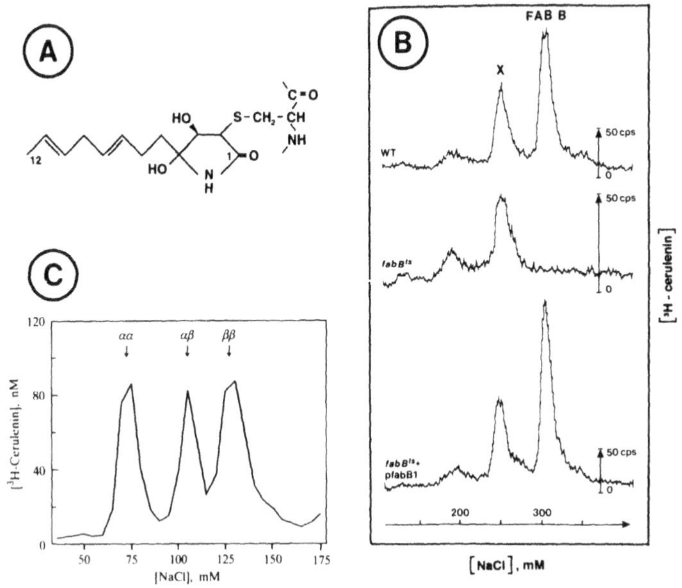

Figure 2 Cerulenin binds covalently to the active site cysteine of KAS I condensing enzymes (A). ^{3}H-cerulenin labelled or tagged proteins are resolved by anion exchange chromatography as illustrated in (B) for *E. coli* [3] and in (C) for barley [4]. While only three KAS I proteins are tagged in enriched barley chloroplast preparations, variable numbers of proteins in addition to X and B (KAS I) can be tagged in crude *E. coli* preparations depending on the experimental conditions.

suitable system for establishing the tagging technique. ^{3}H-cerulenin, which has a greater affinity for condensing enzymes than the tetrahydro derivative [10] was available. Thus, ^{3}H-cerulenin was added to a crude *E. coli* extract, the non bound antibiotic plus membranes removed by gel filtration chromatography and the result analyzed by sodium dodecyl sulphate - gel electrophoresis. When the fractions corresponding to the labelled band detected by fluorography were subjected to anion exchange chromatography, two major peaks were revealed (Figure 2B). The peak absent in a *fabB^{ts}* mutant strain was further purified and amino acid sequences obtained. The latter were present in the sequence deduced from an ORF, cloned by gene rescue of the *fabB^{ts}* mutant. Activity measurements supported the deduction that the tagged, purified protein of 406 residues was the condensing enzyme, KAS I [3,5]. Its amino acid composition is very similar to that reported previously for KAS I purified by activity assays [8].

Adding ^{3}H-cerulenin to crude extracts of barley greening seedling leaves labels a very large number of proteins. When isolated barley chloroplasts are allowed to take up the antibiotic, however, specific labelling ensues. The anion exchange chromatography step separates three proteins (Figure 2C). Further purification and structural characterization revealed that the peaks contained dimers of two closely related polypeptides, αα, αβ and ββ. By comparison only a single dimeric protein is tagged in spinach. The obtained amino acid sequence data of the β-polypeptide was exploited to obtain a probe for screening cDNA libraries. Sequence analyses of the resulting cDNA clones revealed that they coded for the β-polypeptide which was 427 residues in length and preceded by a 35 residue transit peptide. That *Kas12* coded for a KAS I enzyme was deduced from the marked sensitivity to cerulenin, 35% identity to the *E. coli* KAS I sequence and that the pertinent column fractions not exposed to cerulenin were able to synthesize major amounts of C_{16} fatty acids [4].

Isolation of plant genes coding for KAS ?

A major difference in the described barley and *E. coli* systems is that by altering the experimental conditions proteins other than KAS Is can be tagged in the latter, but not in the former system. This observation precludes the use of cerulenin tagging for isolating plant KAS II, even though it is inhibited by the antibiotic, albeit at much higher concentrations than KAS I. Other avenues, such as homology to known nucleotide sequences, activity assays or use of antibodies for monitoring purification have to be taken to obtain KAS II.

Thus, two proteins were purified from developing castor bean endosperms [11] using the Garwin assay [8]. The first was a homodimer of a 50 kDa polypeptide with KAS I activity and the second a heterodimer of the 50 kDa polypeptide and a 46 kDa polypeptide with some KAS II activity. Primary sequence data were used as a starting point to obtain cDNA clones coding for both castor bean polypeptides. At the deduced amino acid level, the 50 kDa encoding ORF is 88% similar to KAS 12. To seek for the barley analogue of the 46 kDa encoded polypeptide, PCR was carried out using primers constructed to the highly conserved region encompassing the active site cysteine and to the most different region of the 46 kDa ORF. The template was an appropriate first strand cDNA population. Two barley cDNAs, ORFs 22 and 25, active in greening seedling leaves of barley were obtained having 89% similarity to the 46 kDa castor bean ORF [12]. Cloning of ORFs 22 and 25 into expression vectors yielded large amounts of protein in inclusion bodies. Although activity assays could therefore not be carried out on the barley ORF 22 and 25 recombinant proteins, they were used as antigens to prepare highly cross reactive antibodies [13]. Since a homodimer of the 46 kDa castor bean ORF and the barley 22 and 25 kDa ORFs have not been shown to have KAS II activity, they will be referred to as KAS ? herein. Two cDNAs from an embryo cDNA library of *Cuphea wrightii* [14] and one from *Glycine max* [15] also code for KAS ? on the basis of sequence analyses [15,16]. They were isolated with the aid of PCR and by exploiting sequences for other *Kas* genes.

That the castor bean protein with KAS II activity is a heterodimer is intriguing with respect to recent deductions drawn from analyses of *Streptomyces* monofunctional PKSes. Such PKS gene clusters have two ORFs with homology to KASes; ORF I with an active site cysteine and ORF II without. Complementation experiments in which plasmids containing either ORF I or ORF II from one species were transformed into another species with mutations in either ORF I or ORF II demonstrated that ORF II is required for synthesis of the antibiotic carbon skeletons [17]. The length of the latter depends on the species. Results from deleting a PKS gene cluster and replacing it with selected ORFs intimate that the function of ORF II is to determine the chain length [18]. This insinuates that a homodimer of plant KAS ? will function as a KAS II.

Purification and cloning of a new *E. coli* condensing enzyme using cerulenin

KAS IIs of plants and *E. coli* have in common some tolerance to cerulenin and a preference for extending 16 carbon chains by a C_2-unit. Could one of the cerulenin taggable proteins in *E. coli* (Figure 2B) be KAS II? A systematic approach [19] was undertaken to purify each tagged protein to the extent that a single well resolved band was obtained upon gel electrophoresis. The band was then transferred to a membrane for amino terminal sequencing. With the caveat that each analyzed sample contained only a single protein, the potential candidates for other cerulenin binding members of the *E. coli* KAS family were limited to a 45 kDa protein in peak X of Figure 2B, and a 55 kDa protein, potentially corresponding to the unmarked peak preceding X in Figure 2B. Results of earlier work gave an estimate of 44-45 kDa for KAS II [8], which would make X a more likely candidate than the 55 kDa protein for this enzyme.

Further purification of X was initially thwarted by its disappearance, eventually attributed to its binding to the Millex-GV filter used for concentration between chromatographic steps [19]. Subsequent work led to amino terminal sequence data and the observations that X was very cross reactive with antibodies against the recombinant ORF 22 and ORF 25 proteins from barley, but not with antibodies against native KAS I protein from *E. coli* [13,20]. The map position of *fabF* coding for KAS II was presumed to be just downstream from *acpP* determining acyl carrier protein. λ clones containing the just mentioned region of the *E. coli* genome were used as templates in PCR reactions with primers homologous to *acpP* sequences and the deduced 5' sequence of X. The results revealed that X was coded for by the adjacent downstream ORF which was then sequenced and found to be more similar to plant *Kas1* and *Kas?* genes than to the *fabB* gene of *E. coli* [21]. The deduced amino acid composition of X does not match very well to that reported earlier for *E. coli* KAS II [8], and furthermore the activity of X was stable at 43°C, contrary to expectation [22]. The identity of X had therefore to be established by activity assays.

Development of the KAS complementation assay

Several assays for condensing activity were developed during early studies of fatty acid synthesis in *E. coli* in the 1960s. The first was the malonyl-CoA : CO_2 exchange reaction [23] which measures the transfer of radio label from $^{14}CO_2$ into acid precipitable protein, a partial reverse reaction of the condensation (Figure 1). Two were spectrophotometric methods [24] in which either the formation of the β-keto group was measured directly at 303 nm or indirectly at 340 nm by including a source of β-ketoacyl-ACP reductase in the reaction mix which couples β-keto formation to oxidation of NADH. The Garwin assay [8] developed more recently requires both radio and nonradio labelled acyl-ACP substrates. The addition of a single donor unit to the primer substrate is followed by reducing the β-keto group with $NaBH_4$ to a β-hydroxy group. This converts the condensation product into an organic soluble compound that is readily extractable for counting. These assays are either relatively insensitive or require synthesis of the acyl-ACP substrates.

While crude extracts of wild type *E. coli* readily incorporate radio labelled C_2-units from ^{14}C-malonyl-CoA into long chain fatty acids (Figure 3 left), extracts from fatty acid requiring mutants do not do so. This was the basis for the first FAS complementation tests in *E. coli* [25] in which equal amounts of crude extracts from two such auxotrophic mutants were mixed and assayed. Positive results indicated that a FAS moiety defective in one mutant strain was functional and present in the other mutant strain, and vice versa. In cerulenin based complementation assays, the antibiotic replaces one of the mutant strains by selectively inactivating one or more of the condensing enzymes. Low concentrations eliminate

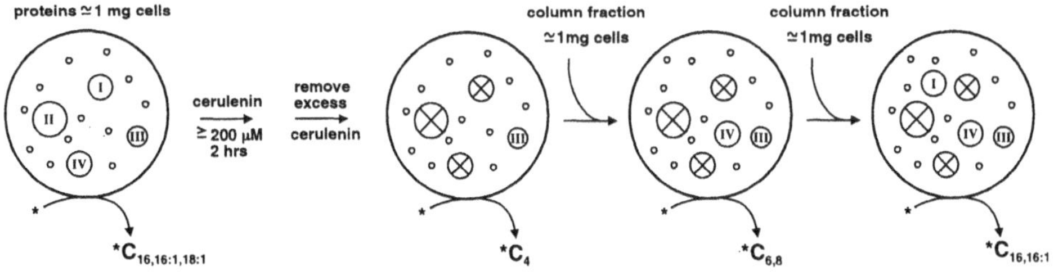

Figure 3 The KAS complementation assay uses crude *E. coli* extracts. Condensing enzymes in varying stages of purification and in circa the same concentration as those inactivated by cerulenin treatment are added to the extract. The ability of the extract to incorporate radio label from ^{14}C-malonyl-CoA (*) into elongated acyl chains is assayed. Acyl products are shown at bottom. Left to right: before cerulenin treatment all four KASes (I-IV) are active; after cerulenin treatment only KAS III is active; after cerulenin treatment KAS IV is added; after cerulenin treatment both KAS I and IV are added. In the latter case if a plant protein fraction containing KAS II had been added simultaneously C_{18} chains would also be synthesized.

KAS I, higher ones KAS II. Since all the other FAS components are present and apparently functional the need to add acyl-ACP substrates is obviated. The cerulenin extract is readily prepared and stored. Together with ^{14}C-malonyl-CoA, a protein fraction of interest, crude to pure, is added to replace the second mutant strain. Optimized parameters for this assay are described elsewhere [20]. Radio-gas chromatography (GC) which is very sensitive has been used to monitor incorporation of radio label into the fatty acids [3,4,26]. Hydrolysis of the synthesized acyl-ACPs, preparation of derivatives and radio-GC, however, are very time consuming. The acyl-ACP products can also be separated by electrophoresis on mini polyacrylamide urea gels, and transferred directly to a PVDF membrane for phosphorimaging, which not only visualizes the result but allows very accurate quantitation. This is a very much faster method than radio-GC even though a series of gels varying in urea concentration must be run for each sample to resolve all chain lengths [27].

An alternative method has been described for determining if a protein fraction can restore condensing activity to a cerulenin inhibited crude *E.* coli extract based on the original malonyl-CoA : CO_2 exchange assay. C_6- or C_{18}-CoAs were supplied as substrates to assay for KAS I and KAS II, respectively. This approach lead to the first purification of a plant KAS, namely from rape [28].

As illustrated in Figure 3, examples given in [21,28], addition of ^{14}C-malonyl-CoA to a cerulenin treated extract results in the synthesis of C_4-acyl chains. This is due to the presence of the uninhibited KAS III. When protein X was added to the cerulenin treated extract, C_6- and C_8-acyl chains were synthesized, revealing that X can elongate C_4-ACP. By comparison when KAS I or KAS I + X were added, C_{16}- and $C_{16:1}$- but not $C_{18:1}$-acyl chains were produced demonstrating that X can not use $C_{16:1}$ as a substrate. Further results demonstrated that the lengths of available substrates in an assay can be varied by intentionally adding known condensing enzymes. In other words, two or more fractions can be added simultaneously to determine whether an acyl-ACP synthesized by a condensing activity present in one fraction can be further elongated by the activity in a second fraction. In this way *E. coli* KAS II activity was found in another fraction from the anion exchange column, circa in the same position as the unmarked peak preceeding X in Figure 2B. Together the results imply that X is a hitherto undescribed condensing enzyme, specific for short chain primers. It has been denoted KAS IV and is encoded by *fabJ* [21]. Preliminary experiments have demonstrated that this technique can be used to follow plant condensing enzymes during purification [29].

Facets of the *E. coli* and plant KAS families

In Figure 4A the condensing enzymes of *E. coli* are placed in the fatty acid biosynthetic pathway according to our present concept of their preferred primer substrates [29]. A potential role for the newest family member (KAS IV) is in synthesis of C_8-acyl chains which are precursors of lipoic acid and the lipoamide cofactor [30]. Interesting questions can be posed such as when in evolution did KAS IV appear in bacteria and does a KAS IV occur in plants? Older observations in the literature can be interpreted in support of the latter notion [31], as can the endosymbiont theory for the origin of plastids and the many similarities in fatty acid biosynthetic systems of plastids and *E. coli*.

Very recently the existence of a fourth condensing in plants has been deduced based on the results of studies with plant KAS III [32]. Assays of cerulenin treated extracts for KAS III activity revealed elongation up to 6 carbons in spinach [33] and up to 10 carbons in *Cuphea lanceolata* [32]. The ability to extend beyond 4 carbons was lost in the later stages of purification in spinach, however, and purified KAS III was highly specific for acetyl-CoA [34]. The presence of a new condensing enzyme was proposed to explain the origin of the cerulenin insensitive products longer than C_4. While the former attribute suggests that KAS III encoded by *fabH* was the progenitor of this deduced enzyme, the latter characteristic infers that KAS IV encoded by *fabJ* was. I predict that an analogue of the *E. coli* KAS IV with the same role occurs in plants, and that the complementation assay described above can be used for its isolation. Interestingly, the glycine decarboxylase complex with its lipoamide cofactors which participates in photorespiration is localized in mitochondria and not in plastids. This complex has been

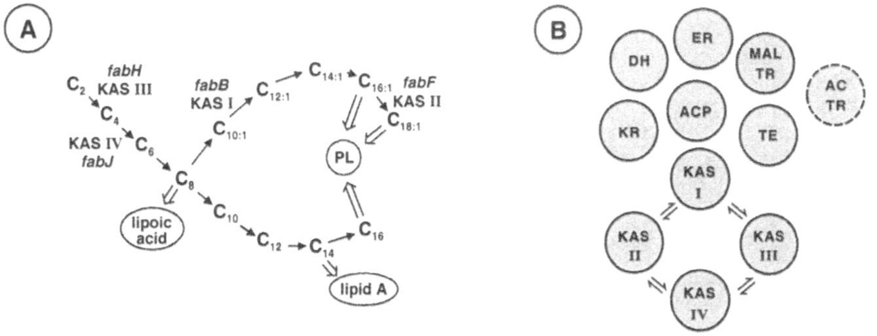

Figure 4 (A) Coordinated interaction of members of the condensing (KAS) enzyme family results in biosynthesis of fatty acids in *E. coli*. The genes encoding each KAS as well as major destinations of the fatty acyl products are shown. Lipoic acid is a precursor of the coenzyme lipoamide. Lipid A consists of β-hydroxymyristate linked to saccharides in the cell membrane. PL are the membrane phospholipids. (B) How do KASes interact with one another and the other members of the FAS complex? ACP, acyl carrier protein; KR, β-ketoacyl-ACP reductase; DH, β-hydroxyacyl-ACP dehydrase; ER, enoylacyl-ACP reductase; MAL TR, malonyl-CoA:ACP transacylase; TE, thioesterase; AC TR acetyl-CoA:ACP transacylase whose contribution to fatty acid synthesis is uncertain since the discovery and characterization of KAS III [33,38].

estimated to comprise one third of the total soluble mitochondrial proteins in green pea leaves [35] thereby implying the presence of a large number of C_8-acyl chains. Perhaps this scenario explains the presence of the recently reported ACP in plant mitochondria [36,37] which was unforeseen. Will another KAS IV analogue occur in plastids? Experiments to determine the subcellular compartmentalization of plant condensing enzymes have been initiated [29].

Considering that the condensing enzymes shown in Figure 4A are part of monofunctional enzyme complexes, what is the relationship of the KASes to one another and to the other members of the active unit(s). For example, do all the KAS isoforms exist in a single complex, is there only one KAS per complex or do the KASes rotate in and out of a functional unit as cartooned in Figure 4B? Do specific isoforms of the other complex members occur and if so do they interact with designated KASes? While answers to such questions are for the future, the first steps can be taken now.

Structure-function

The most direct approach to structure-function correlations necessitates an x-ray crystal structure. While crystals of *E. coli* KAS I were prepared in 1970 [39], they were not studied. In the past year new crystals have been made from KAS I that have a different structure [20]. A small number of highly or completely conserved residues potentially functioning in catalysis have been identified from analyses of primary sequences of enzymes carrying out Claisen condensations via various mechanisms (KASes in FAS and PKS plus biosynthetic thiolases) [16]. The bases coding for these residues in *fabB* can be mutagenized in vitro, and the resulting constructs transformed into the *fabB*^{ts} mutant where their ability to complement the mutation can be assessed. A different, supplementary avenue which can be pursued is transfering genes from one species to another and looking for changes in fatty acyl composition. This is illustrated below with three KASes using long chain, unsaturated primer substrates. The double bonds are close to the end of the chain where the condensation takes place (trans Δ2, cis Δ3, cis Δ4).

KAS I encoded by *fabB* in *E. coli* is peculiar in using cis Δ3-C_{10} as a substrate. Can a plant KAS use this primer? Overexpression in *E. coli* of barley KAS I and KAS ? as well as of *E.* coli KAS I yields

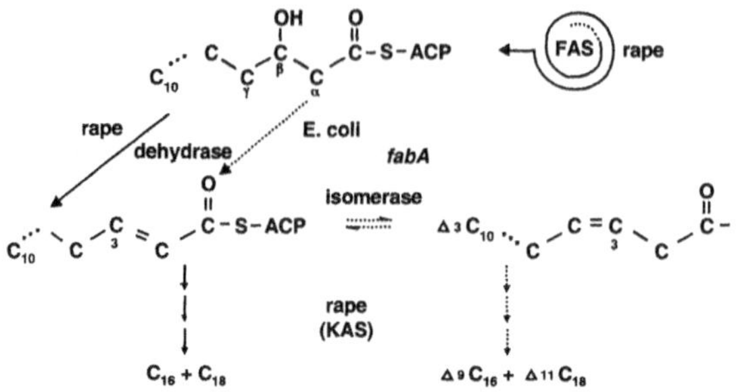

Figure 5 Can a plant KAS use cis Δ3-C_{10} as a substrate for unsaturated fatty acyl chains? This compound is formed in *E. coli* by the isomerase activity of the *fabA* encoded dehydrase, which as the plant dehydrase synthesizes trans Δ2-C_{10} used in synthesis of saturated fatty acids. If both *fabA* activities (broken arrows) function in transgenic rape seeds, and the rape KAS can use the cis Δ3-C_{10} substrate then the *E. coli* specific unsaturated fatty acyl chains, cis Δ9-C_{16} and cis Δ11-C_{18} (right side), as well as the plant unsaturated fatty acids derived from C_{16}- and C_{18}-acyl chains (left side) will be synthesized.

inactive, inclusion body localized proteins [12,40]. Thus the answer is being sought by producing rape seeds transgenic for *fabB*, *fabA* (synthesizing the substrate) or *fabA* + *fabB* and analyzing their fatty acid compositions (see Figure 5). In *Coriander* endosperms an acyl-ACP desaturase has recently been described acting primarily on 16 carbon chains to give cis Δ4-C_{16}-ACP [41]. This is elongated by one C_2-unit forming petroselinoyl-ACP a substrate for another specialized enzyme, a cis Δ6 thioesterase [42]. The condensation carried out in this pathway for petroselinic acid production is conceivably achieved either by a modified KAS II or by an additional KAS. In the latter case one acts in synthesis of the petroselinic acyl moieties for triacyl glycerols and the second yields those for membrane lipids. An analogous situation may occur in epidermal cells at the same stage of elongation (C_{16}-C_{18}): a specialized KAS for acyl chains destined for waxes, cutin and suberin and a KAS II for membrane lipids.

Rhizobium species synthesize Nod factors with an array of substitutions which participate in selecting specific plants as hosts. The group of interest here comprises unsaturated fatty acyl chains, specifically, $C_{16:2}$, $C_{16:3}$ and $C_{18:4}$ (Figure 6A). The origin of all three is dependent upon *nodE* and *nodF* with homology to genes coding for KAS and ACP, respectively (Figure 6B). Presence of *nodE* and *nodF* determined proteins simulates a PKS which bypasses the third reductive step in an elongation cycle and uses trans Δ2 substrates as primers (Figure 6C) for $C_{16:3}$ and $C_{18:4}$ synthesis. The prominent amounts of $C_{16:2}$ products implies that acyl chains lacking a trans Δ2 double bond are also used as substrates by the *nodE* encoded KAS, just as *E. coli* KAS I can employ saturated as well as cis Δ3-C_{10} chains as substrates. In the absence of functional *nodE* and *nodF* coded proteins in *R. meliloti*, a single extension takes place yielding cis Δ11-C_{18}, a characteristic FAS end product, which is acylated to the Nod factor backbone in place of $C_{16:2}$ and $C_{16:3}$. This intimates that the *Rhizobium* FAS components for the $C_{16:1}$ to $C_{18:1}$ elongation cycle are in the same compartment as the two Nod enzymes (Figure 6c). That the latter contain sufficient information to specify given unsaturated acyl chains was shown by deleting both *nodE* and *nodF* from *R. meliloti* and replacing them in trans with their *R. leguminosarum* analogues. The transformed *R. meliloti* synthesized $C_{18:4}$ peculiar to *R. leguminosarum* as well as $C_{18:2}$, $C_{18:3}$, $C_{16:2}$ and $C_{16:3}$ [43]. The *nodE* determined KAS differs in three respects from that encoded by *E. coli fabB* and the petroselinic KAS in (i) requiring a special ACP, namely that coded for by *nodF*, (ii) functioning in more than a single elongation step and (iii) using trans rather than cis unsaturated substrates. Among the experiments that could conceivably be carried out are expanding those described above with transgenic rape to include *nodE*, replacing *nodE* with *fabB* in *Rhizobium* and vice versa as well as constructing hybrid KAS genes to see if regions specifying use of unsaturated substrates can be pinpointed.

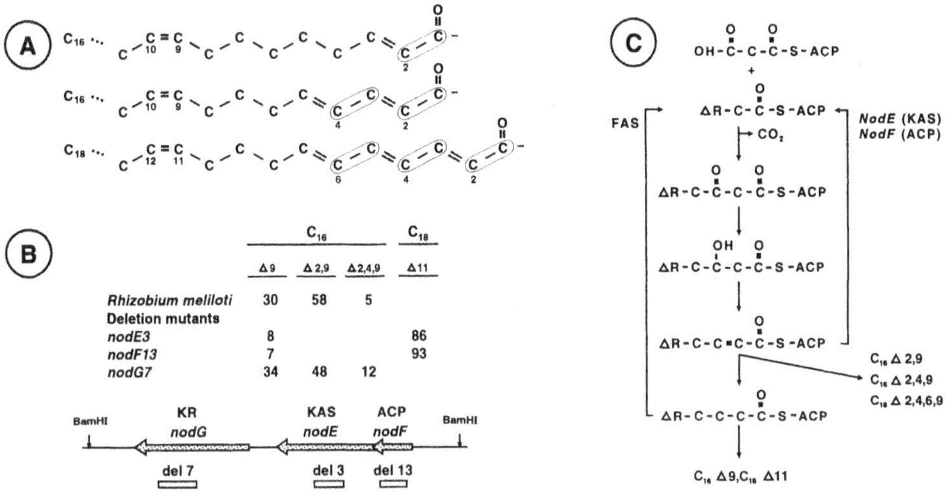

Figure 6 The unsaturated acyl chains present on the chitooligosaccharide backbones of Nod factors contribute to host specificity of *Rhizobium* species. (A) In *R. meliloti* these are $C_{16:2}$ (trans Δ2, cis Δ9) and $C_{16:3}$ (trans Δ2, trans Δ4, cis Δ9) and in *R. leguminosarum* $C_{18:4}$(trans Δ2, trans Δ4, trans Δ6, cis Δ9). (B) Three genes with homology to FAS components occur on a *R. meliloti* Bam HI fragment. Deletion analyses have established that the *nodE* and *nodF* genes are essential for synthesis of the unsaturated fatty acyl chains, but that *nodG* is not [43]. (C) Relationship between FAS (left side) and the PKS Nod reactions (right side). While formation of $C_{16:2}$ requires only a premature termination of fatty acid synthesis, formation of $C_{16:3}$ and $C_{18:4}$ also necessitates either one or two rounds of elongation before premature termination.

References

1. Funabashi H, Iwasaki S, Okuda S. A model study on the mechanism of fatty acid synthetase inhibition by antibiotic cerulenin. Tetrahedron Lett 1983;24:2673-2676.
2. Omura S. Cerulenin. Meth Enzymol 1981;72:520-532.
3. Siggaard-Andersen, M. Role of Escherichia coli β-ketoacyl-ACP synthase I in unsaturated fatty acid synthesis. Carlsberg Res Commun 1988;53:371-379.
4. Siggaard-Andersen M, Kauppinen S, von Wettstein-Knowles P. Primary structure of a cerulenin-binding ß-ketoacyl-[acyl carrier protein] synthase from barley chloroplasts. Proc Natl Acad Sci USA 1991;88:4114-4118.
5. Kauppinen S, Siggaard-Andersen M, von Wettstein-Knowles P. ß-Ketoacyl synthase I of Escherichia coli: nucleotide sequence of the fabB gene and identification of the cerulenin binding residue. Carlsberg Res Commun 1988;53:357-370.
6. Funabashi H, Kawaguchi A, Tomoda H, Omura S, Okuda S, Iwasaki S. Binding site of cerulenin in fatty acid synthetase. J Biochem 1989;105:751-755.
7. Roberts G, Leadlay PF. Use of [^{3}H]tetrahydrocerulenin to assay condensing enzyme activity in Streptomyces erythreus. Biochem Soc Trans 1984;12:642-643.
8. Garwin JL, Klages AL, Cronan JE. Structural, enzymatic, and genetic studies of ß-ketoacyl-acyl carrier protein synthases I and II of Escherichia coli. J Biol Chem 1980;255:11949-11956.
9. Ulrich AK, de Mendoza D, Garwin JL, Cronan JE. Genetic and biochemical analyses of Escherichia coli mutants altered in the temperature-dependent regulation of membrane lipid composition. J Bacteriol 1983;154:221-230.
10. D'Agnolo G, Rosenfeld IS, Awaya J, Omura S, Vagelos PR. Inhibition of fatty acid synthesis by the antibiotic cerulenin. Specific inactivation of ß-ketoacyl-acyl carrier protein synthetase. Biochim Biophys Acta 1973;326:155-166.

11. Genez A, McCarter D, Nelson J, Bleibaum J, Thompson GA, Knauf VC, Stalker D. Molecular cloning and characterization of ß-ketoacyl-ACP synthase genes from Ricinus communis. Third Intl Congr Plant Mol Biol; 1991 Oct 6-12; Tucson, AZ; Abst. 719:(Abstract)
12. Wissenbach M. New members of the barley Kas gene family encoding ß-ketoacyl-acyl carrier protein (ACP) synthases (KASes). Plant Physiol 1994;(In Press)
13. Wissenbach M, Chuck J-A, Siggaard-Andersen M, Olsen JG, von Wettstein-Knowles P. Powerful new antibodies to ß-ketoacyl-ACP synthases. In: These proceedings.
14. Slabaugh M, Leonard J, Tai H, Jaworski JG, Knapp S. Condensing enzymes and thioesterases expressed in immature embryos of Cuphea wrightii. Plant Lipid Symposium; 1993 July 29-31; Minneapolis, MN; Abst. A9:(Abstract)
15. Kinney AJ, Hitz WD, Yadav NS, Perez-Grau L. Genes of fatty acid biosynthesis in developing oilseeds. Plant Physiol 1993;102:1(Abstract)
16. Siggaard-Andersen M. Conserved residues in condensing enzyme domains of fatty acid synthases and related sequences. Protein Seq Data Anal 1993;5:325-335.
17. Sherman DH, Kim E-S, Bibb MJ, Hopwood DA. Functional replacement of genes for individual polyketide synthase components in Streptomyces coelicolor A3(2) by heterologous genes from a different polyketide pathway. J Bacteriol 1992;174:6184-6190.
18. McDaniel R, Ebert-Khosla S, Hopwood DA, Khosla C. Engineered biosynthesis of novel polyketides. Science 1993;262:1546-1550.
19. Siggaard-Andersen M. A study of ß-ketoacyl-ACP synthases. [dissertation]. Copenhagen, DK: Univ of Copenhagen, 1993.
20. Olsen JG, Siggaard-Andersen M, Kadziola A, Wissenbach M, Chuck J-A, Larsen S, von Wettstein-Knowles P. Purification, crystallization, and preliminary x-ray diffraction analysis of a condensing enzyme. In: These proceedings.
21. Siggaard-Andersen M, Wissenbach M, Chuck J-A, Svendsen I, Olsen JG, von Wettstein-Knowles P. The *fabB* encoded ß-ketoacyl-[acyl carrier protein] synthase IV from Escherichia coli is sensitive to cerulenin and specific for short chain substrates. Proc Natl Acad Sci USA 1994;submitted.
22. Garwin JL, Cronan JE. Thermal modulation of fatty acid synthesis in Escherichia coli does not involve de novo enzyme synthesis. J Bacteriol 1980;141:1457-1459.
23. Vagelos PR, Alberts AW. Malonyl coenzyme A-CO_2 exchange reaction. J Biol Chem 1960;235:2786-2791.
24. Alberts AW, Majerus PW, Vagelos PR. ß-Ketoacyl carrier protein synthase. Meth Enzymol 1969;14:57-60.
25. Cronan JE, Birge CH, Vagelos PR. Evidence for two genes specifically involved in unsaturated fatty acid biosynthesis in Escherichia coli. J Bacteriol 1969;100:601-604.
26. Siggaard-Andersen M. Role of Escherichia coli ß-ketoacyl-ACP synthase I in unsaturated fatty acid synthesis. Carlsberg Res Commun 1988;53:371-379.
27. Post-Beittenmiller D, Roughan G, Ohlrogge JB. Regulation of plant fatty acid biosynthesis. Analysis of acyl-coenzyme A and acyl-acyl carrier protein substrate pools in spinach and pea chloroplasts. Plant Physiol 1992;100:923-930.
28. MacKintosh RW, Hardie DG, Slabas AR. A new assay procedure to study the induction of ß-ketoacyl-ACP synthase I and II, and the complete purification of ß-ketoacyl-ACP synthase I from developing seeds of oilseed rape (Brassica napus). Biochim Biophys Acta 1989;1002:114-124.
29. Siggaard-Andersen M, Wissenbach M, Chuck J-A, Olsen JG, von Wettstein-Knowles P. Screening barley for condensing enzyme activities. In: These proceedings.
30. Parry RJ. Biosynthesis of some sulfur-containing natural products. Investigations of the mechanism of carbon-sulfur bond formation. Tetrahedron 1983;39:1215-1238.
31. Packter NM, Stumpf PK. Fat metabolism in higher plants. Production of short- and medium-chain acyl-acyl carrier protein by spinach stroma preparations treated with cerulenin. Biochim Biophys Acta 1975;409:274-282.
32. Schuch R, Brummel M, Spener F. ß-Ketoacyl-acyl carrier protein (ACP) synthase III in Cuphea lanceolata seeds: Identification and analysis of reaction products. J Plant Physiol 1994;143:556-560.
33. Jaworski JG, Clough RC, Barnum SR. A cerulenin insensitive short chain 3-ketoacyl-acyl carrier protein synthase in Spinacia oleracea leaves. Plant Physiol 1989;90:41-44.
34. Clough RC, Matthis AL, Barnum SR, Jaworski JG. Purification and characterization of 3-ketoacyl-acyl carrier protein synthase III from spinach. A condensing enzyme utilizing acetyl-coenzyme A to initiate fatty acid synthesis. J Biol Chem 1992;267:20992-20998.

35. Oliver DJ, Neuburger M, Bourguignon J, Douce R. Interaction between the component enzymes of the glycine decarboxylase multienzyme complex. Plant Physiol 1990;94:833-839.
36. Chuman L, Brody S. Acyl carrier protein is present in the mitochondria of plants and eucaryotic micro-organisms. Eur J Biochem 1989;184:643-649.
37. Shintani DK, Ohlrogge JO. The characterization of a mitochondrial acyl carrier protein isoform isolated from Arabidopsis thaliana. Plant Physiol 1994;104:1221-1229.
38. Jaworski JG, Post-Beittenmiller D, Ohlrogge JB. Acetyl-acyl carrier protein is not a major intermediate in fatty acid biosynthesis in spinach. Eur J Biochem 1993;213:981-987.
39. Prescott DJ, Vagelos PR. Acyl carrier protein. XIV. Further studies on ß-ketoacyl acyl carrier protein synthetase from Escherichia coli. J Biol Chem 1970;245:5484-5490.
40. Chuck J-A, Wissenbach M, Siggaard-Andersen M, Olsen JG, von Wettstein-Knowles P. Studies in the optimization of the overexpression of ß-ketoacyl-ACP synthase I from barley and Escherichia coli. In: These proceedings.
41. Cahoon EB, Ohlrogge JB. Metabolic evidence for the involvement of a Δ^4-palmitoyl-acyl carrier protein desaturase in petroselinic acid synthesis in coriander endosperm and transgenic tobacco cells. Plant Physiol 1994;104:827-837.
42. Dörmann P, Frentzen M, Ohlrogge JB. Specificities of the acyl-acyl carrier protein (ACP) thioesterase and glycerol-3-phosphate actyltransferase for octadecenoyl-ACP isomers. Plant Physiol 1994;104:839-844.
43. Demont N, Debellé F, Aurelle H, Dénarié J, Promé JC. Role of the Rhizobium meliloti *nodF* and *nodE* genes in the biosynthesis of lipo-oligosaccharidic nodulation factors. J Biol Chem 1993;268:20134-20142.

IN PLANTA EXPRESSION OF cDNA ENCODING 3-KETOACYL-ACYL CARRIER PROTEIN SYNTHASE III (KAS III) FROM SPINACH

Heeyoung Tai and Jan G. Jaworski
Department of Chemistry, Miami University, Oxford, Ohio, U.S. A.

The 3-ketoacyl-ACP synthase III (KAS III) catalyzes the condensation reaction of acetyl-CoA with malonyl-ACP during the initial cycles of fatty acid synthesis. Although KAS III has been implicated as a site of regulation for de novo fatty acid synthesis in *E. coli* [1, 2], its regulatory role in plants has not been clearly delineated. Alteration of KAS III expression in plants will help elucidate its function in vivo.

To facilitate manipulation of KAS III level in plant, we have isolated the cDNA for KAS III from spinach [3] and, more recently, from *Arabidopsis* [4]. The KAS IIIs from these two plants share a large degree of homology (71.8% identity at amino acid level), but show no similarity to the clones of other KAS isozymes from either plants or *E. coli*.

Using *Agrobacterium* mediated transformation with a construct containing spinach KAS III sequence under control of the double cauliflower mosaic virus (CaMV) 35S promoter, we were able to achieve up to 300-fold overexpression of KAS III activity in transgenic tobacco leaves.

Materials and Methods

CONSTRUCTION OF TRANSGENIC TOBACCO

A coding region of spinach KAS III cDNA, including leader sequence, was generated by a PCR using 26 nucleotide long primers containing *Sal*I site. The resulting 1232bp PCR fragment was subcloned into a cassette vector p1079 (Fig. 1) which contained a repeat of the CaMV 35S promoter and the 3' end of nopaline synthase gene (*nos*). The promoter/KAS III/*nos* sequence was cleaved from p1079-KAS and ligated into a binary vector pCGN 1557, which carries T-DNA border sequences (LB, RB: Fig. 1) as well as neomycin phosphotransferase (*npt*) and gentamycin resistant sequences for antibiotic selections. The construct was introduced into *Agrobacterium tumefaciens* strain EHA 105 by direct transformation and the transformed *Agrobacterium* was used to infect tobacco leaf discs grown under sterile conditions.

IDENTIFICATION OF TRANSGENIC PLANTS

Transformed tobacco plants were initially screened for their resistance to kanamycin (300mg/l). Once plants generated roots in the selection media, they were transferred to soil. Genomic DNA was prepared from leaves of these putative transformants and subjected to a PCR reaction using the primers that generated the KAS III coding fragment

J.-C. Kader and P. Mazliak (eds.), Plant Lipid Metabolism, 72–74.

from spinach cDNA clone (Fig 1).

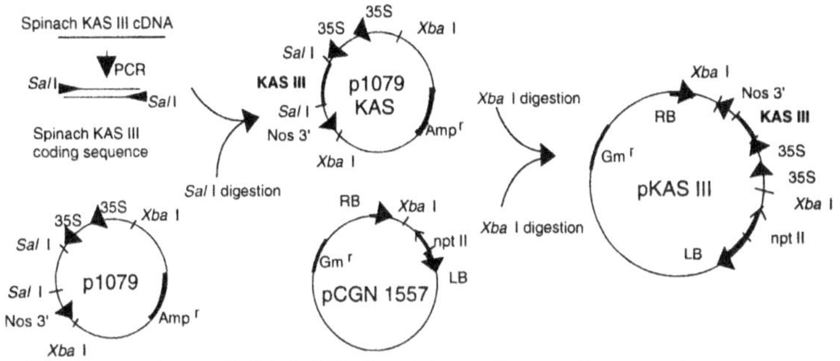

Fig. 1. Construction of KAS III plant transformation vector

In order to determine the presence of inserted KAS III gene product, antibody to the spinach KAS III was obtained using overexpressed KAS III fusion protein from *E. coli* [3]. The KAS III part of the *E. coli* fusion protein was purified with an affinity matrix, followed by a monoQ column for antibody production. Protein extracts from the transgenic tobacco plants were subjected to 12% SDS-PAGE for immunoblot.

KAS III ASSAY OF LEAF EXTRACTS

Leaf samples from transgenic and control plants were frozen in liquid nitrogen and homogenized in extraction buffer (50mM Tris.Cl (pH 8), 10mM EDTA, 10mM Dithiothreitol, 1mM Phenylmethylsulfonyl fluoride, 2mM ε-Amino-n-caproic acid, and 10% (w/v) Polyvinylpolypyrrolidone). The homogenates were centrifuged at 4°C at 13000 g. In order to determine the amout of extract at which enzyme was limiting, four different amounts of the supernatant (0.5, 1.0, 2.5, 5.0 μl) from each preparation were assayed for KAS III activity as described [5].

Results and Discussion

Six transgenic tobacco plants were confirmed to carry the spinach KAS III sequence by PCR analysis (Fig 2). The primers used in the PCR were specific for the spinach KAS

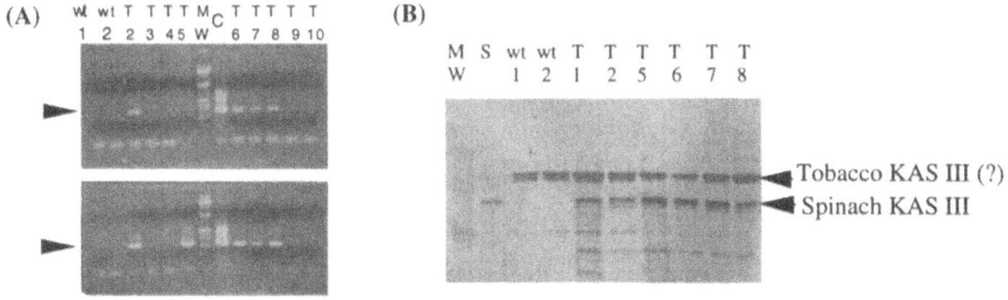

Figure. 2. (A) PCR results and (B) Immunoblot of transgenic tobacco plants. Two different pairs of primers were used (top and bottom row of the gel) for PCR and spinach KAS III antibody was used for the immunoblot. MW: molecular weight markers; wt: wild type; T: putative transgenic plants; C: control using spinach KAS III cDNA as template; S: spinach extract.

III, since they were derived from divergent regions of the clone. The absence of the PCR product from wild type tobacco DNA confirmed the specificity of the primers. All six PCR-positive plants showed a unique spinach KAS III gene product on an immunoblot (Fig. 2). The antibody raised against spinach KAS III seemed to cross react with KAS III of other plants, and therefore detected both spinach and tobacco KAS III. However, the endogenous tobacco KAS III was easily distinguishable from the inserted spinach KAS III due to its different mobility on the gel.

KAS III activity was measured and found to be as much as 297-fold higher in the transgenic tobacco than in the control plants (Table 1). Considering that transformation frequency was low and other *Agrobacterium* strains failed to generate transgenic plants, this high level of KAS III may be the maximum activity that the tobacco can tolerate. However, the appearance of the transgenic plants were not distinguishable from wild type tobaccos.

Table 1. KAS III activity in leaves of control and transgenic tobacco plants.

	KAS III Activity (pmol/min/mg)
Wild type1	26.49
2	185.70
Transgenic 1	5008.70
5	7877.20
6	3947.03
7	6625.42

We are in the process of determining the fatty acid composition and acyl-ACP pools in these transgenic plants. The requirement for KAS III in plant fatty acid biosynthesis can also be investigated by using antisense construct in combination with inducible plant promoters. Information obtained from these transgenic plants will help define the influence of KAS III on regulation of fatty acid biosynthesis in plants.

References

1. Jackowski S, Rock CO. Altered molecular form of acyl carrier protein associated with β-ketoacyl-acyl carrier protein synthase II (*fabF*) mutants. J Bacteriol 1987;169:1469-1473.

2. Jackowski S, Murphy CM, Cronan JE, Rock CO. Acetoacetyl-acyl carrier protein synthase. A target for the antibiotic thiolactomycin. J Biol Chem 1989;264(13):7624-7629.

3. Tai H, Jaworski JG. 3-ketoacyl-acyl carrier protein synthase III from spinach (*Spinacea oleracea*) is not similar to other condensing enzymes of fatty acid synthase. Plant Physiol 1993;103:1361-1367.

4. Tai H, Post-Beittenmiller D, Jaworski JG. Cloning of cDNA encoding 3-ketoacyl-acyl carrier protein synthase III from *Arabidopsis*. Plant Physiol 1994;in press.

5. Clough RC, Matthis AL, Barnum SR, Jaworski JG. Purification and characterization of 3-ketoacyl-acyl carrier protein synthase III from spinach. J Biol Chem 1992;267(29):20992-20998.

Purification, Crystallization, and Preliminary X-ray Diffraction Analysis of a Condensing Enzyme

Johan Gotthardt Olsen[2], Mads Siggaard-Andersen, Anders Kadziola[3], Margit Wissenbach[1], Jo-Anne Chuck[1], Sine Larsen[3], and Penny von Wettstein-Knowles[2].

[1]Department of Physiology, Carlsberg Laboratory, Gamle Carlsberg Vej 10, DK-2500 Valby, [2]Department of Genetics, Institute of molecular biology, University of Copenhagen and [3]Center for Crystallographic Studies, H.C. Ørsted Institute, Universitetsparken 3, University of Copenhagen, Denmark.

Introduction

The β-ketoacyl-ACP synthases (KAS's), or condensing enzymes, occupy a key position in the fatty acid biosynthetic pathway (FAS) as they catalyse the elongation of the acyl chain. Much interest has developed in the regulation of production of fatty acids with respect to amount, chain length and degree of saturation [1,2]. In large part, this is due to the obvious industrial interests in controlling the synthesis of specific oils for uses such as nutrients, lubricants and fuels [3]. Indeed, the fatty acid biosynthetic pathway is an extremely interesting biochemical and physiological system of considerable complexity. Our objective is to improve our understanding of some of the molecular mechanisms of regulation, catalysis, substrate specificity and intermolecular interactions with other fatty acid synthase components. We hope that knowledge of the crystal structure of a condensing enzyme will participate in the elucidation of some of these questions as well as guide future experiments. Overexpression of plant (and bacterial) condensing enzymes in bacteria, results largely in inclusion bodies which renders the protein unsuitable for crystallization. The *fabB* gene product, KAS I from *Escherichia coli* was chosen as a model system for our experiments because of its high degree of homology to the barley condensing enzyme I [4] because it has been crystallized previously [5] and it can be purified in sufficient amounts from wild type *E. coli*. We hereby report the purification and crystallization of the *fabB* product.

Complementation assay

A crude protein extract from *E. coli* is capable of incorporating radioactively labelled malonyl-CoA into long chain fatty acids, demonstrating that components of the FAS are present and can function [6]. Condensing enzymes I, II and IV from *E. coli* are inhibited by the antibiotic cerulenin by covalent attachment to the active site cysteine [7]. In a cerulenin treated protein extract, virtually no fatty acid synthesis takes place due to the specified inhibitions. After removal of excess antibiotic, the cerulenin treated preparation can be used to determine if a protein fraction contains condensing enzyme(s) by monitoring the incorporation of radioactively labelled carbon from malonyl-CoA into fatty acids. The assay was used in the isolation of condensing enzymes I and IV from *E. coli*.

The optimized assay mix (500μL) contains 1 mM NADH, 1 mM NADPH, 2 mM DTT, 10 μM acyl carrier protein (ACP), 10 μM acetyl-CoA, 59.4 μM malonyl-CoA, 0.51 μM (2-^{14}C)-malonyl-CoA (59 mCi/mmol purchased from New England Nuclear - all other chemicals and ACP from Sigma), and ≈3 μg/assay of cerulenin treated protein extract. Circa 15 ng/assay of purified KAS I is appropriate to give a clear signal. The specific activity in this assay is in the range of 5 μmol/min/mg. The assay reaction runs for 1 h at 37°C and the assay was linear up to 25% incorporation of the malonyl-CoA, corresponding to circa 8000 dpm. The hydrolysed fatty acids are extracted in hexane and the radioactivity is determined.'

J.-C. Kader and P. Mazliak (eds.), Plant Lipid Metabolism, 75–77.

Purification

260 g *E. coli* K12 (wet weight) was thawed and resuspended in a minimum volume of buffer A (20 mM Tris-HCl, pH 8.0, 1 mM EDTA, 2 mM DTT) with 0.1 M NaCl. Cells were ruptured in a French press, polytron homogenised and ultracentrifuged (step 1). The supernatant was fractionated on a 530 mL Q-sepharose column (anion exchange) in an 11 L 0-1 M NaCl gradient (step 2). The fractions were assayed according to procedures described above. Two distinct activity peaks were detected of which the early one constitutes the *fabJ* product, β-ketoacyl-ACP synthase IV [8] and the late one the *fabB* product, β-ketoacyl-ACP synthase I (KAS I) [6].

The late peak was dialysed against buffer A and loaded on a 91 mL Q-sepharose column. The protein was eluted in 20 mM piperazin-HCl, pH 6.0, 1 mM EDTA, 2 mM DTT in a 3 L 0-1 M NaCl gradient (step 3). The activity peak was concentrated, equilibrated with buffer A using Amicon centriprep 30 concentrators, size fractionated on a 110 mL sephacryl S-300 column (step 4). To the pooled activity peak fractions, a solution of $(NH_4)_2SO_4$ was added to a final concentration of 2 M. Precipitate was removed and the supernatant fractionated on an alkyl superose 10/10 column (hydrophobic interaction) in a 240 mL 2-0 M $(NH_4)_2SO_4$ gradient (step 5). The activity peak from this column constitutes the end product of the purification. All purification steps were carried out at 4°C. Sodium dodecyl sulfate-polyacrylamide gel electrophoresis (SDS-PAGE) was used to monitor the degree of purity (figure 1). The purified protein was concentrated to approximately 20 mg/mL.

Crystallization

A fast screen for protein crystallization was carried out by use of the hanging drop vapour diffusion method in which a drop, typically containing equal amounts of protein- and precipitant solution (eg 5 μL + 5 μL), is placed on a cover slip which is positioned (airtight) over a reservoir containing 1 ml of the precipitant solution. By vapour diffusion, the water in the drop will eventually reach the same chemical potential as that of the reservoir, thereby causing supersaturation of the protein which at some point is forced out of solution.

Figure 1

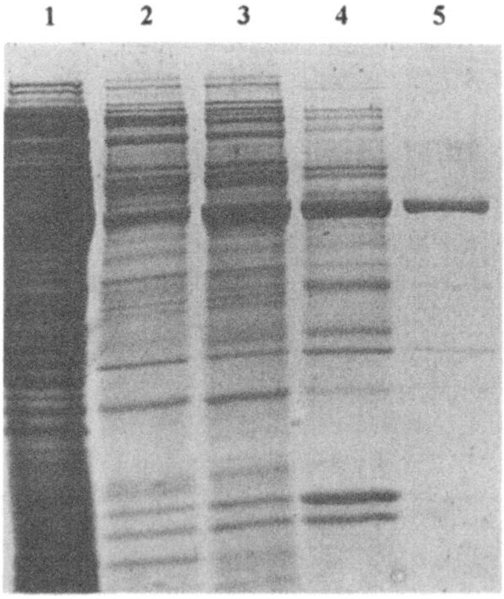

Figure 2

Figure 1. Purification of KAS I from *E. coli* as revealed by SDS-PAGE and coomassie staining. Numbers denote steps in purification as defined in text.
Figure 2. Crystal of β-ketoacyl-ACP synthase I.

Needle-shaped crystals appeared immediately over a reservoir containing 0.1 M Hepes buffer, pH 7.5, 2% polyethylene glycol, MW 400 kDa and 2.0 M $(NH_4)_2SO_4$. Thicker, more well defined crystals were obtained using higher precipitant concentrations (3% PEG400 and 2.2M $(NH_4)_2SO_4$) (fig. 2). Crystals typically appear within 24 hours. Attempts to slow down crystal growth have as yet proven futile but macroseeding to conditions with lower precipitant concentration, at which no nucleation is detected, gives larger crystals. To confirm that the crystals were indeed KAS I, some crystals were dissolved and analysed by SDS-PAGE (results not shown).

Preliminary diffraction data to 3.2Å reveal that the enzyme crystallizes in the orthorhombic system and the cell dimensions are: a =59.6 Å, b = 142.1 Å and c = 214.2 Å. The molecular structure is not known and to our knowledge, neither is that of any protein of homologous sequence. Therefore, the structure will have to be solved by application of the multiple isomorphous replacement method. Currently, we are in the process of screening for appropriate heavy atom derivatives.

ACKNOWLEDGEMENTS

The bacteria were a generous and much appreciated gift from Dr. Anthony Kinney.

REFERENCES

[1]Magnuson K, Jackowski S, Rock CO and Cronan JE. Regulation of Fatty Acid Biosynthesis in *Escherichia coli*. Microbiol. Rev. 1993;57:522-542.

[2]Slabas AR and Fawcett T. The biochemistry and molecular biology of plant lipid biosynthesis. Plant Mol. Biol. 1992;19:169-191.

[3]Ohlrogge JB. Design of new plant products: Engineering of fatty acid metabolism. Plant Physiol. 1994;104:821-826.

[4]Siggaard-Andersen M, Kauppinen S, and von Wettstein-Knowles P. Primary structure of a cerulenin-binding β-ketoacyl-[acyl carrier protein] synthase from barley chloroplasts. Proc. Natl. Acad. Sci. USA 1991;88:4114-4118.

[5]Prescott DJ and Vagelos PR. Acyl Carrier Protein XIV. Further studies on β-ketoacyl acyl carrier protein synthetase from *Escherichia coli*. J. Biol. Chem. 1970;244:5484-5490.

[6]Siggaard-Andersen M. Role of Escherichia coli β-ketoacyl-ACP synthase I in unsaturated fatty acid synthesis. Carlsberg Res. Commun. 1988;53:371-379.

[7]Kauppinen S, Siggaard-Andersen M, and von Wettstein-Knowles P. β-ketoacyl-ACP synthase I of Eschericia coli: Nucleotide sequence of the *fabB* gene and identification of the cerulenin binding residue. Carlsberg Res. Commun. 1988;53:357-370

[8]Siggaard-Andersen M, Wissenbach MH, Chuck J, Svendsen I, Olsen JG and von Wettstein-Knowles P. The *fabJ* encoded β-ketoacyl-[acyl carrier protein] synthase IV from *Escherichia coli* is sensitive to cerulenin and specific for short chain substrates. Proc. Natl. Acad. Sci. USA 1994. Submitted.

INHIBITION OF FATTY ACID CONDENSING ENZYMES IN PLANTS

A. Lesley Jones, Jane E. Dancer and John L. Harwood

Institute of Medical Genetics
Biochemistry Department
University of Wales, College of Cardiff,
PO Box 903, Cardiff, CF1 1ST, U.K.

Introduction

Fatty acid synthesis in plants is carried out by a type I dissociable fatty acid synthase (FAS). There are three β-ketoacyl-ACP synthases (KAS) associated with FAS in plants. The short-chain condensing enzyme (KAS III) catalyses the initial condensation of acetyl-CoA with malonyl-ACP [1,2] to form acetoacetyl-ACP (4C) and may catalyse further rounds of condensation *in vivo* [3]. Further rounds of condensation are carried out by KAS I which catalyses the condensation of malonyl-ACP with intermediate length acyl-ACPs from acetoacetyl-ACP to myristoyl-ACP (14C) to give palmitoyl-ACP (16C). KAS II elongates palmitoyl-ACP (16C) to stearoyl-ACP (18C) [4].

There are some known inhibitors of fatty acid synthesis. Cerulenin inhibits KAS I completely and irreversibly at 20μM [5] and inhibits KAS II but at higher concentrations. Cerulenin also inhibits the condensing enzyme function of multifunctional protein type I fatty acid synthases [6]. For all these enzymes the mechanism of action is similar, involving the covalent binding of the inhibitor to the cysteine of the active site of the protein [7]. Arsenite, an inhibitor of enzymes containing vicinal thiol groups, inhibits KAS II but no other condensing enzyme. KAS III is insensitive to inhibition by either cerulenin or arsenite but its activity is inhibited by thiolactomycin (TLM). This inhibition has been seen in bacteria [8] and plants [9] but is rather variable as reported in the literature [10-14].

Results and Discussion

Table 1 shows that TLM does indeed inhibit fatty acid synthesis in intact pea chloroplasts, but is not a particularly good inhibitor of KAS III *in vitro*, even at relatively high concentrations. Jaworski [10] reported 60% inhibition of fatty acid synthesis in spinach chloroplasts by 50μM TLM but Yamada *et al.* [11] found much greater inhibition in spinach chloroplasts (I_{50}=3.8μM). Why there should be such a range of inhibition is unclear. Yamada *et al.* [11] used the naturally occurring TLM (derived from *Nocardia* sp. 2-200). We used a racemic mixture and the I_{50} was between 10-100μM [12]. Variability is also apparent in levels of inhibition reported for *E. coli* fatty acid synthesis where Jackowski *et al.* [13] observed an I_{50} of 75μM and Hayashi *et al.* [14] one of 2μM. In his study of the effect of TLM on KAS III, Jaworski *et al.* [10] presented results similar to those presented here - that is, a major inhibition of overall fatty acid synthesis but little inhibition of KAS III. However, Jaworski *et al.* [11] did observe strong inhibition of KAS III if the compound was exposed to air for 5-10mins before assaying. It is possible that KAS III is inhibited by an oxidation product of TLM and that TLM is also converted by chloroplasts to a more effective inhibitor. However we did not see this effect in peas [Jones, A.L. and Harwood, J.L., unpublished results].

J.-C. Kader and P. Mazliak (eds.), Plant Lipid Metabolism, 78–80.

Figure 1 Structure of thiolactomycin and its analogues

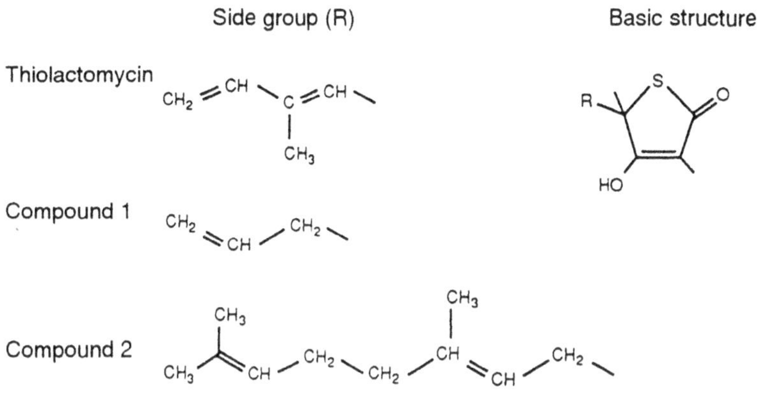

Table 1 also shows that compound 1 (Fig. 1), which has only a 3C side chain, has no inhibitory effect at all, whereas compound 2, which has a 8C side chain plus two substituent methyl groups completely inhibits *de novo* fatty acid synthesis. However, although it inhibits KAS III activity more than TLM 60% of the KAS III activity remains (Table 1). Again KAS III activity is less affected than overall fatty acid synthase.

Table 1 Effect of 1mM TLM and analogues on total fatty acid synthesis in intact pea chloroplasts and on KAS III activity *in vitro*

Inhibitor	Fatty acid synthesis (% control)	KAS III activity (% control)
Control	100±3.1	100±2.7
TLM	15.9±2.2	82.8±1.3
Compound 1	100±2.6	104.5±1.1
Compound 2	0	60±2.9

Inhibitors were dried down under N_2 in the incubation tube and preincubated with chloroplasts for 10mins. before addition of ^{14}C-acetate. Chloroplasts were isolated and incubated with [1-^{14}C] acetate as in [15]. Means ± S.D. (n=2) shown. KAS III assays were performed using a 40-80% $(NH_4)_2SO_4$ cut from pea leaves. Final concentrations in the assay were 10μM ^{14}C-acetyl-CoA (55μCi/mmol), 5μg ACP (*E. coli*), 0.1M Tris/HCl pH 7.5, 1mM NADH, 2mM NADPH and 10μM malonyl CoA. Activity was measured as acid precipitable counts. Means ± S.D. (n=3) shown.

To clarify these results by examining the effect of these compounds on the other condensing enzymes we used ^{14}C-laurate as a substrate for FAS. Laurate should be a substrate for KAS I but not KAS III, nor directly for KAS II. Initial work indicated that ^{14}C-laurate was poorly taken up by chloroplasts, so a leaf disc system was used to analyse ^{14}C-laurate metabolism. Table 2 shows that 100μM TLM inhibited overall labelling indicating an effect on KAS I but also that KAS II activity was completely abolished. Compound 2 gave greater inhibition of *de novo* fatty acid synthesis than TLM and abolished any further metabolism after palmitate - again KAS II was more sensitive to the inhibitor than KAS I. Compound 1 produced a similar radiolabelled fatty acid pattern to that of the control but actually increased the amount of label taken into fatty acids.

Table 2 Distribution of radioactivity in fatty acids of pea leaf discs incubated with [1-^{14}C]laurate and 100μM TLM or analogue

Treatment	% control labelling	% total ^{14}C-labelled fatty acids				
		16:0	16:1	18:1	18:2	others
Control	100±3	43±2	11±4	27±6	17±2	n.d.
TLM	41±3	72±2	28±2	n.d.	n.d.	n.d.
Compound 1	151±10	54±tr.	7±3	22±1	15±1	3±4
Compound 2	24±2	100±tr.	n.d.	n.d.	n.d.	n.d.

Others includes 18:0 and very long chain saturated fatty acids. Experiments were carried out as in [15], with leaf discs from 14 day Onward pea plants incubated with 100μM inhibitor which was dried down under N_2 before the addition of leaf discs, 1μCi of [1-^{14}C] laurate and water to a final vol. of 5ml. Means ± S.D. shown (n=2), n.d.=not detected, tr. < 0.5%

From this study we can conclude that TLM is indeed a better inhibitor of KAS I than KAS III, and that KAS II is more sensitive to TLM than either KAS I or III. The increased side-chain length of compound 2 appears to make it a more effective inhibitor of condensing enzymes than TLM, but the order of sensitivity is the same. Compound 1, with a shorter side-chain is not inhibitory to any of the condensing enzymes. Cerulenin has a C8 hydrophobic side-chain attached to a polar head group and length of side chain appears to be important. Neither cerulenin, nor compound 1 with a shorter side chain affect KAS III (which has two short chain length substrates) although TLM and compound 2 do, despite their longer side chains. This supports other data indicating that the active site of this enzyme is markedly different to those of KAS I and II [7] although the mechanism of reaction is probably similar as the putative active site cysteine is conserved [Siggaard-Andersen, M. pers comm.].

This work was carried out as a LINK project funded by SERC, AFRC and AgrEvo UK Ltd.

1. Jackowski, S. & Rock, C.O. J. Biol. Chem. 1987; 262: 7927-31
2. Jaworski, J.G., Clough, R.C., Barnum, S.R. Plant Physiol. 1989; 90: 41-4
3. Jaworski, J.G., Tai, H., Ohlrogge, J.B. & Post-Beittenmiller, D. Prog. Lipid Res. 1994; 33:47-54
4. Harwood, J.L. Ann. Rev. Plant Physiol. Plant Mol. Biol. 1988; 39: 101-147
5. Siggaard-Andersen, M. Carlsberg Res. Commun. 1988; 53: 371-9
6. Wakil, S.J. Biochemistry 1989; 28: 4523-30
7. Kauppinen, S., Siggaard-Andersen, M. & von Wettstein-Knowles, P. Carlsberg Res. Commun.; 53: 357-70
8. Nishida, I., Kawaguchi, A. & Yamada, M. J. Biochem. 1984; 25: 265-
9. Nishida, I., Kawaguchi, A. & Yamada, M. J. Biochem. 1986; 99: 1447-54
10. Jaworski, J.G., Clough, R.C., Barnum, S.R. *et al.* In: Quinn, P.J. & Harwood, J.L. eds. Plant lipid biochemistry, structure and utilization. London:Portland Press, 1990: 97-104
11. Yamada, M. Kato, M., Nishida, I. *et al.* In Stumpf, P.K., Mudd, J.B. & Nes, W.D., eds. The metabolism structure and function of plant lipids. New York: Plemun Press, 1987:447-54
12. Jones, A.L., Dancer, J.E. & Harwood, J.L. Biochem. Soc. Trans. 1994; 22:258S
13. Jackowski, S., Murphy, C.M., Cronan, J.E. & Rock, C.O. J. Biol. Chem. 1989; 264: 7624-9
14. Hayashi, T., Yammamoto, O. *et al.* Biochem. Biophys. Res. Comm. 1983; 115, 1108-13
15. Walker, K.A., Ridley, S.M. & Harwood, J.L. Biochem. J. 1988; 254: 811-7

POWERFUL NEW ANTIBODIES TO β-KETOACYL-ACP SYNTHASES

Margit Wissenbach[1], Jo-Anne Chuck[1], Mads Siggaard-Andersen[1], Johan Gotthardt Olsen[1,2] and Penny von Wettstein-Knowles[1,2]
[1]Department of Physiology, Carlsberg Laboratory, Gamle Carlsberg Vej 10, DK- 2500 Copenhagen, Denmark and [2]Department of Genetics, Institute of Molecular Biology, University of Copenhagen, Denmark

Introduction

In higher plants fatty acid biosynthesis takes place primarily in plastids. Three condensing enzymes (β-ketoacyl-ACP synthases, KASes), namely KAS I, KAS II and KAS III, have been described. They are different with respect to the preferred length of fatty acyl substrates as well as in their sensitivity to the inhibitor cerulenin.
Three barley cDNAs representing two ORFs encoding two proteins with high homology to a 46 kDa β-ketoacyl-ACP synthase from castor bean [1] have been cloned and characterized [2]. The two novel barley KASes were expressed in *Escherichia coli* using a T7 expression system (pET5a). Both recombinant proteins were completely insoluble leading to the formation of inclusion bodies precluding activity studies. To get a specific tool to isolate the natural proteins for characterization, antibodies against the two inclusion body proteins were raised in prescreened rabbits. The antibody specificity was tested in western blots using the two new recombinant barley KASes, barley KAS I (KAS 12) and *E. coli* KAS I (FAB B) proteins. All KAS proteins tested showed strong cross reactivity with both types of antibodies. This fact enabled us to isolate a hitherto undescribed KAS enzyme from *E. coli* (KAS IV) and clone its gene [3].

Two new barley KASes

The polymerase chain reaction (PCR) was used to generate a probe with specificity for barley cDNAs analagous to the castor bean 46 kDa synthase clone [4]. Screening a barley greening leaf cDNA library (λZAP II) lead to the isolation of three clones (pMaW20, pMaW21, pMaW25; [2]) which were sequenced. The inserts of pMaW20 and pMaW21 are overlapping, showing complete sequence identity in the 940 bp overlap (figure 1). Therefore the two inserts were spliced at a common Pst I site, resulting in clone pMaW22, which encodes open reading frame (ORF) 22. pMaW25 contains a complete open reading frame (ORF 25). Both ORF proteins start with a putative transit peptide region of about 70 amino acids followed by a mature protein region of around 420 amino acids. Within this region ORFs 22 and 25 show 93% sequence similarity. Compared to

J.-C. Kader and P. Mazliak (eds.), Plant Lipid Metabolism, 81–83.

the 46 kDa synthase sequence, the same degree of similarity (around 90%) is found. The similariry to the 50 kDa synthase and KAS 12 is only in the range of 75%.

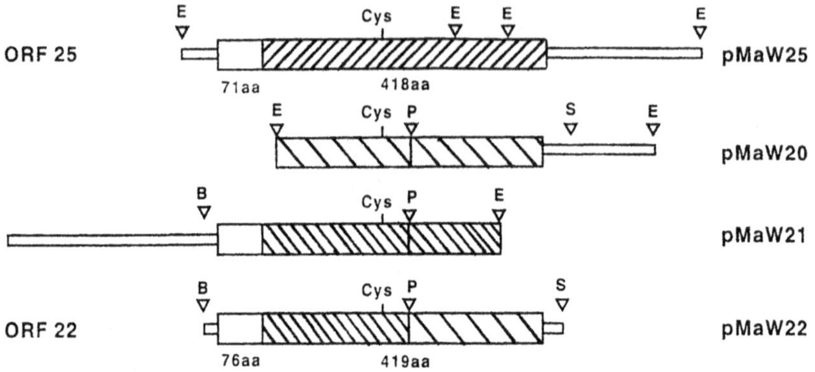

Figure 1: cDNA clones and ORFs. Coding regions are drawn as big boxes, the white parts represent putative transit peptide encoding sequences, the grey parts represent the putative mature coding regions. 5' and 3' non-coding regions are represented by narrow boxes.
B = BamH I, E = EcoR I, P = Pst I, S = Spe I, Cys = active site cysteine, aa = amino acids.

Expression of ORF 22 and ORF 25 proteins in *E. coli*

After introduction of appropriate restriction sites the putative mature coding parts of ORF 22 (encoding 419 amino acids) and ORF 25 (encoding 418 amino acids) were subcloned into the pET5a vector and expressed in *E.coli* strain BL21(DE3)pLysS. Both recombinant ORF proteins were produced in relatively large amounts, however in a totally insoluble form in inclusion bodies. These were isolated and further purified by preparative SDS-PAGE. Amino-terminal sequencing was performed to confirm the correct translation start. The purified proteins were injected into rabbits for antibody production.

Cross reactivity of KASes with serum 22 and 25

Western blotting using KAS 12 and FAB B inclusion bodies as well as total extracts of BL21(DE3)pLysS expressing the putative mature ORF 22 and 25 proteins showed that all four proteins cross reacted with serum 22 (directed against ORF 22 protein) and 25 (directed against ORF 25 protein). Using crude barley chloroplast extracts one major band with a molecular mass of ca. 45 kDa was detected. This band represents several KASes (at least KAS 12 and ORF 22 and 25 proteins) of the same molecular mass. Only very few other chloroplast and *E. coli* proteins showed cross reactivity.

Identification of a novel *E. coli* KAS enzyme

In the course of purifying an elusive *E. coli* KAS, serum 22 was used to identify the protein by western blotting. Activity measurements revealed a short chain condensing activity associated with a hitherto undescribed enzyme, which has been designated KAS IV [3].

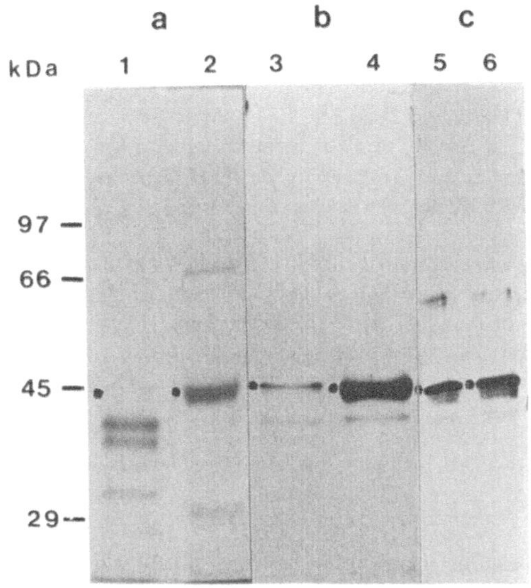

Figure 2: Cross reactivity of *E. coli* KASes with serum 22. Coomassie stained gel (a) and western blots of the same fractions (b), plus fractions from another purification (c) containing *E. coli* KAS IV (FAB J) and KAS I (FAB B) proteins using serum 22, 1:10,000 dilution.
Lanes 1,3,5,6 fractions containing KAS IV protein; lanes 2 and 4, fractions containing KAS I protein; proteins were fractionated using ion exchange and hydrophobic interaction chromatography. The amount of KAS IV loaded (lanes 1 and 3) is too low to be detected by coomassie staining. Lanes 1-4 contain ca. 1.25 μg, lane 5 4.0 μg and lane 6 2.8 μg total protein. The positions of the KAS proteins are indicated.

Conclusions

Two new barley cDNAs encoding β-ketoacyl-ACP synthases have been characterized. Antibodies against recombinant proteins, coded for by the cDNAs recognize condensing enzymes from *E. coli*. This high cross reactivity of condensing enzymes from barley and *E. coli* with the new antibodies suggests that these will be powerful tools for isolating e.g. other plant condensing enzymes, such as those in elongase complexes.

References

[1] Genez A, McCarter D, Nelsen J, Bleibaum JL, Thompson GA, Knauf VC, Stalker D. Molecular cloning and characterization of β-ketoacyl-ACP synthase genes from *Ricinus communis*. Int Soc Plant Mol Biol Congress, Tucson, Arizona 1991; Abst. 719.

[2] Wissenbach, M. New members of the barley *KAS* gene family encoding β-ketoacyl-acyl carrier protein (ACP) synthases (KASes). Plant Physiol 1994, submitted.

[3] Siggaard-Andersen M, Wissenbach M, Chuck J, Svendsen I, Olsen JG, Wettstein-Knowles P von. The *fabJ* encoded β-ketoacyl-[acyl carrier protein] synthase IV from *Escherichia coli* is sensitive to cerulenin and specific for short chain substrates. Proc Natl Acad Sci USA 1994, submitted.

[4] Wissenbach M, Siggaard-Andersen M, Kauppinen S, Wettstein-Knowles P von. Condensing enzymes of barley. In: Cherif A, Miled-Daoud DB, Marzouk B, Smaoui A, Zarrouk M, editors. Metabolism, Structure and Utilization of Plant Lipids. 1992: 393-396.

SCREENING BARLEY FOR CONDENSING ENZYMES OF THE FATTY ACID SYNTHASE

Mads Siggaard-Andersen[1], Margit Wissenbach[1], Jo-Anne Chuck[1], Johan Gotthardt Olsen[1,2] and Penny von Wettstein-Knowles[1,2].

[1]Carlsberg Laboratory, Department of Physiology, Gamle Carlsberg Vej 10, DK-2500 Copenhagen, Denmark, and
[2]Institute of Molecular Biology, Department of Genetics, University of Copenhagen, Øster Farimagsgade 2A, DK-1353 Copenhagen K, Denmark.

Three condensing enzymes (KASes) have been characterized from the fatty acid synthases of plants and *E. coli*. Recently we have found an *E. coli* KAS IV with specificity for short chain acyl-ACPs and a high sensitivity to cerulenin. We are now investigating the possibility of a similar enzyme existing in plants. Here we describe both our basic KAS assay, as well as some modifications that expand the substrate specificity range over which the assay is useful. In this initial work we have also observed a possible compartmentalization of KASes which may further add to the complexity of plant FASes.

Introduction

β-Ketoacyl-ACP synthases (KASes) carry out the chain elongation steps in the course of fatty acid synthesis. In plants and most bacteria the enzyme is part of a separable fatty acid synthase (FAS), and in both cases three enzymes with different substrate specificities and inhibitor sensitivities have been characterized [1,2]. After finding a KAS IV in *E. coli* [3] we are now searching for a similar activity in barley; specific for short chain elongations of acyl-ACPs ($C_4 \rightarrow C_6 \rightarrow C_8$) and highly sensitive to cerulenin.

Complementation assay

We use a complementation assay, where fractions are tested for their ability to restore fatty acid synthesis to an *E. coli* protein extract previously treated with cerulenin. The latter specifically inhibits the condensing enzyme component of the fatty acid synthase through covalent attachment at the active site cysteine [4]. Cerulenin (Sigma) is dissolved in acetone, aliquoted to the incubation tubes and dried under a stream of nitrogen. Crude *E. coli* proteins at 10 mg/mL in 50 mM Tris-HCl pH 8.0 [4] are then added to give a final cerulenin concentration of 200 μM. The reaction proceeds for 1 hour at 4°C and excess cerulenin is then removed on a small gel-filtration column (G25-SF, Pharmacia) using 50 mM Tris-HCl pH 8.0 containing 1 mM EDTA and 2 mM DTT. Such preparations are completely devoid of FAS activity beyond C_4-ACP, the product of the cerulenin resistant KAS III. Since only the condensing component is inhibited, FAS activity beyond C_4-ACP can be restored by addition of KASes using C_4-ACP as substrate: Assays contain 20 μg cerulenin treated *E. coli* protein extract, 10 μM *E. coli* ACP (Sigma), 10 μM acetyl-CoA, 60 μM [^{14}C]malonyl-CoA (18-122 GBq/mol), 1 mM NADH, 1 mM NADPH, 2 mM DTT, 1 mM EDTA, 50 mM K-PO_4 pH 6.8 buffer and condensing enzyme to give an approximate activity of 20 pmol/min in 100 uL assays. Formation of ^{14}C labelled fatty acids is allowed to proceed for 60 minutes at 37°C. Assays are then stopped by adding NaOH to a final concentration of 2 M followed by hydrolysis for 30 minutes at 80°C to yield free fatty acids. After acidification using H_2SO_4, fatty acids are extracted using 3×2 mL *n*-hexane and the ^{14}C radioactivity

J.-C. Kader and P. Mazliak (eds.), Plant Lipid Metabolism, 84–86.

label directly counted in a liquid scintillant. Of the 6400 dpm in each assay (18 GBq/mol malonyl-CoA), up to 1600 dpm are recovered in the hexane phase within the linear range of the assay. Interestingly, the background is only 100 dpm, indicating that the C_4 synthesized by the cerulenin resistant *E. coli* KAS III is not recovered quantitatively. Carried out this way, the assay detects KAS activity in any fraction capable of elongating C_4-ACP. As the incorporation of ^{14}C label is also a function of the KAS activity in elongations following the $C_4 \rightarrow C_6$ step, the assay is not suited to do quantitative comparisons of individual condensing enzyme activities.

BARLEY KAS ACTIVITIES

Barley leaves were broken in a blender equipped with replaceable razor blades, using a 100 mM Tris-HCl pH 8.5 buffer containing 600 mM glycerol [5]. To isolate KASes from the chloroplast located fatty acid synthase, chloroplasts were recovered by filtration through two layers of nylon mesh and centrifugation at 1500*g for 20 min. They were broken in a French pressure cell at 150 MPa and soluble proteins recovered in a 60 min centrifugation at 135,000*g in the presence of 25 mM $MgCl_2$. When these proteins are fractionated by anion-exchange chromatography and analyzed for KAS activities in the complementation assay, a pattern of poorly resolved activities is obtained (Fig. 1, Panel A). We have previously analyzed the produced fatty acids by radio-GC with little or no improvement on the identification of individual activities [5]. This probably reflects a large number of KASes in the plant, as already suggested by the finding of three KAS I isoforms (αα, αβ, and ββ) [5].

COMPARTMENTALIZATION OF BARLEY KAS

Upon careful washing of the chloroplasts by repetitive resuspension and centrifugation at 1500×g for 1 min until a clear supernatant is obtained, the KAS profile obtained was different (Fig. 1, Panel B), and at least one isoform of KAS I (αα) is now absent from the preparation. In a control experiment, all material precipitating at 1500×g in 20 min centrifugations was removed, and a 12000×g pellet collected and thoroughly washed. The sample was otherwise treated as the chloroplast preparations, and found to contain several KAS activities which eluted with a profile different from that of the above preparations (Fig. 1, Panel C). For an activity eluting at the position of the αα KAS I, the acyl-ACP products were analyzed (see below) and found to be the same as originally identified for the αα KAS I by radio-GC, namely C_{14} and C_{16}-ACP [5]. These results suggest that condensing enzymes may also be found at locations outside the chloroplast in a fraction precipitating at 12000×g. One such location could be the mitochondrion, where previously ACP has been demonstrated in *Arabidopsis* [6], and a specialized condensing enzyme implicated in maintaining respiratory function in yeast [7].

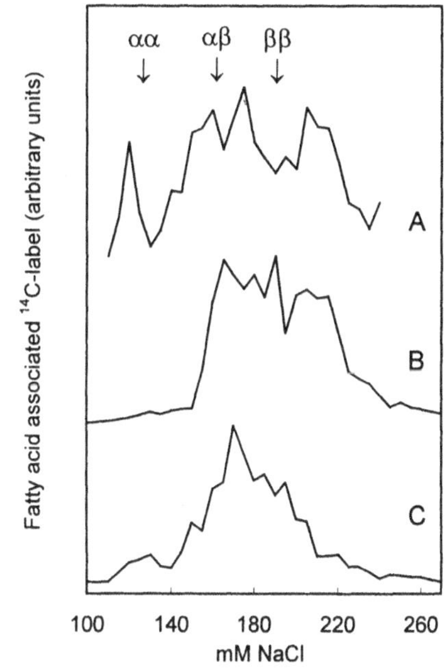

Figure 1. Complementation assays of KAS activities from barley. Proteins were separated by anion exchange on a 5×50 mm column of mono Q using a 20 mM Tris-HCl buffer containing 1 mM EDTA, 2 mM DTT and a linear gradient from 0 to 1 M NaCl in 100 mL. Produced fatty acids were extracted and quantitated using scintillation counting. Proteins are from 1500×g 20 minute (A), 1500×g 1 minute (B) and 12000×g 20 minute (C) pellets as described. The elution positions of previously characterized KAS Is (αα, αβ, and ββ) [5] are indicated.

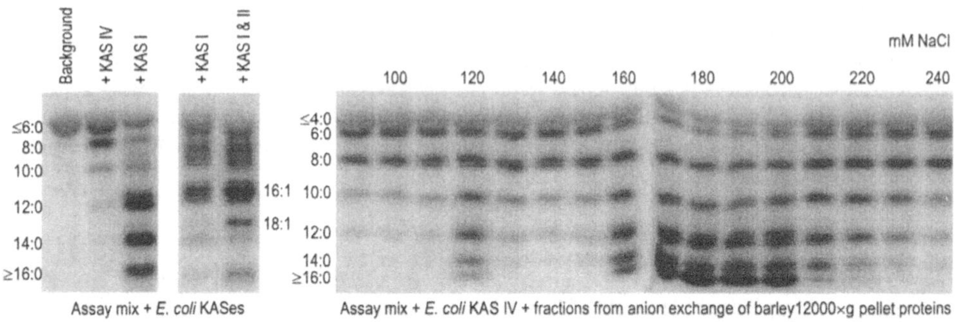

Figure 2. [^{14}C]Acyl-ACP products formed in the complementation assay separated by 3M urea-PAGE [8] and visualized using a phosphorimager. The left panel shows results obtained with different *E. coli* condensing enzymes, and illustrates the power of combining complementation assays, purified KASes and analysis by urea-PAGE. The right panel shows products formed when barley KASes obtained from the 12000×g pellet were fractionated by anion-exchange and analyzed in complementation assays including *E. coli* KAS IV (that is, assaying for elongation of C_8-ACP).

Modified complementation assays

Rather than hydrolyzing the acyl-ACPs, they can be directly identified using conformationally sensitive gels [8]. This yields information not only about the ability to elongate C_4-ACP, but also characterizes the enzyme with respect to elongation of a range of contiguous substrates (Fig. 2). The above complementation assays are only useful for detecting condensing enzymes capable of elongating C_4-ACP. However, after supplementing the assay mix with purified KASes capable of elongating C_4-ACP the produced acyl-ACPs can serve as substrates for the condensing enzyme of interest. For example, addition of purified *E. coli* KAS IV results in a background of C_8-ACP suitable for assaying condensing enzymes utilizing this substrate (Fig. 2, right panel). Similarly, supplementation with *E. coli* KAS I results in synthesis of longer chain products such as $C_{16:0}$ and $C_{16:1}$, providing substrates for plant and *E. coli* KAS II, respectively (Fig. 2, left panel). Such modified assays are simple, fast, sensitive and relatively specific. Using this approach in the course of purifying a plant KAS IV, we expect to identify fractions elongating C_4-ACP but not C_8-ACP.

1. Slabas AR, Fawcett T. The biochemistry and molecular biology of plant lipid biosynthesis. Plant Mol Biol. 1992;19:169-91.
2. Magnuson K, Jackowski S, Rock CO, Cronan JE, Jr. Regulation of fatty acid biosynthesis in *Escherichia coli*. Microbiol Rev. 1993;57:522-42.
3. Siggaard-Andersen M, Wissenbach M, Chuck J-A, Svendsen I, Olsen JG. The *fabJ* encoded β-ketoacyl-[acyl carrier protein] synthase IV from *Escherichia coli* is sensitive to cerulenin and specific for short chain substrates. Proc Natl Acad Sci USA. 1994; in press
4. Kauppinen S, Siggaard-Andersen M, Wettstein-Knowles P von. ß-Ketoacyl-ACP synthase I of *Escherichia coli*: Nucleotide sequence of the *fabB* gene and identification of the cerulenin binding residue. Carlsberg Res Commun. 1988;53:357-70.
5. Siggaard-Andersen M, Kauppinen S, Wettstein-Knowles P von. Primary structure of a cerulenin binding β-ketoacyl-[acyl carrier protein] synthase from barley chloroplasts. Proc Natl Acad Sci USA. 1991;88:4114-8.
6. Shintani DK, Ohlrogge JB. The characterization of a mitochondrial acyl carrier protein isoform from *Arabidopsis thaliana*. Plant Physiol. 1994;104:1221-9.
7. Harington A, Herbert CJ, Tung B, Getz GS, Slonimski PP. Identification of a new nuclear gene (CEM1) encoding a protein homologous to a β-keto-acyl synthase which is essential for mitochondrial respiration in Saccharomyces cerevisae. Mol Microbiol. 1993;9:545-55.
8. Post-Beittenmiller D, Jaworski JG, Ohlrogge JB. *In vivo* pools of free and acylated acyl carrier proteins in spinach. J Biol Chem. 1991;266:1858-65.

A MOLECULAR STUDY ON THE FUNCTIONAL RELATIONSHIP BETWEEN PROKARYOTIC AND PLANT ENOYL-ACP REDUCTASES.

Antoine R. Stuitje, Martin M. Kater and H. John J. Nijkamp.

Department of Genetics, Institute for Molecular Biological Sciences, BioCentrum Amsterdam, Vrije Universiteit,De Boelelan 1087,1081 HV Amsterdam, The Netherlands

Introduction

From an evolutionary point of view the fatty acid biosynthetic pathways of plants and prokaryotes are considered to be closely related. In these organisms, *de novo* fatty acid synthesis is catalysed by a so-called type II fatty acid synthetase complex (FAS II), which consists of eight structurally independent and monofunctional components. In this study we present a comparative molecular genetic and biochemical study of the enoyl-ACP reductase FAS components of plant and bacterial origin. The putative bacterial enoyl-ACP reductase gene (*envM*) was identified on the basis of amino acid sequence similarities with the recently cloned plant enoyl-ACP reductase. Subsequently, it was unambiguously demonstrated by over-expression studies, that the *envM* gene encodes the bacterial enoyl-ACP reductase. An anti-bacterial agent called diazaborine was shown to be a specific inhibitor of the bacterial enoyl-ACP reductase, whereas the plant enzyme was insensitive to this synthetic antibiotic. The close functional relationship between the plant and bacterial enoyl-ACP reductases was inferred from genetic complementation of an *envM* mutant of *Escherichia coli* . *envM* gene-replacement studies, facilitated by the use of diazaborine, demonstrated for the first time that a single component of the plant FAS system can functionally replace its counterpart within the bacterial multi-enzyme complex. Finally, lipid analysis of recombinant *Escherichia coli* strains with the hybrid FAS system unexpectedly revealed that enoyl-ACP reductase catalyses a rate limiting step in the elongation of unsaturated fatty acids.

Results

THE *E. COLI ENVM* GENE ENCODES THE BACTERIAL COUNTERPART OF PLANT ENOYL-ACP REDUCTASE.

There are several lines of evidence indicating that the protein encoded by the *envM* gene of *Salomonella* and *E. coli* is involved in core fatty acid biosynthesis and encodes the bacterial enoyl-ACP reductase. First, amino acid sequence comparison reveals significant homology between plant enoyl-ACP reductase and the protein encoded by the *envM* gene.[1].
Secondly, in both *Salmonella* and *E. coli* allelic forms of the *envM* genes have been isolated which result in resistance against a group of antibacterial drugs called diazaborines (Fig. 2) [2, 3]. As diazaborine inhibits the synthesis of both, phospholipids and lipopolysaccharides [3], these studies have pointed to a role of the envM protein in either

J.-C. Kader and P. Mazliak (eds.), Plant Lipid Metabolism, 87–89.

fatty acid or lipid biosynthesis.

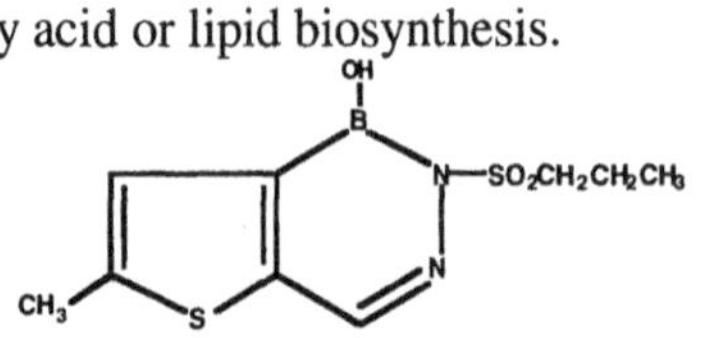

Figure.1, structure of diazaborine.

In fact, it was recently demonstrated that over-expression of the plastidial mature form of rapeseed enoyl-ACP reductase in *E. coli* results in a high level of resistance towards diazaborine [1]. Thirdly, genetic complementation experiments in an *E. coli* strain with a temperature-sensitive growth phenotype, due to the presence of the *envM392* (ts) allele, revealed that expression of the rapeseed enoyl-ACP reductase coding sequence resulted in complementation of the mutation [1]. Finally, by over-expression of the *envM* gene in *E. coli* it was demonstrated that the *envM* protein catalyses the NADH-dependent reduction of crotonyl-CoA [1]. Enzyme assays on bacterial and plant enoyl-ACP reductases, overproduced in *E. coli* , demonstrated that the wild-type *E. coli* enoyl-ACP reductase (the *envM* gene product) is inhibited by diazaborine in a NAD^+ dependent manner, whereas both the rapeseed reductase and the reductase encoded by the *envM* allele of the diazaborine-resistant *E. coli* are relatively insensitive to the drug (not shown, [1]).

REPLACEMENT OF THE BACTERIAL *ENVM* GENE BY A PLANT COUNTERPART

In order to demonstrate unambiguously that the plant enoyl-ACP reductase is functionally related to its counterpart in the bacterial FAS system, we set up an experiment to replace the protein encoded by the envM gene with that encoded by the rapeseed enoyl-ACP reductase c-DNA clone [4]. For this purpose the rapeseed enoyl-ACP reductase was cloned between the promoter and terminator region of the *envM* gene. We made use of the diazaborine-resistant character of the rapeseed enoyl-ACP reductase and a M13 amber phage vehicle to be able to efficiently select for homologous recombination events in *E. coli.* Acquisition of diazaborine resistance after infection of a non-suppressor host by the recombinant amber phage vehicle, will in most cases be the result of a single cross-over event near the *envM* locus which leaves the *envM* locus intact (see Fig. 2A). Next, selection for a second cross-over event which removes the prophage is facilitated by the deoxycholate (DOC) sensitivity of M13 lysogens. Molecular analysis of the recombinants (Fig. 2B) unambiguously demonstrated that replacement of the bacterial gene by the plant counterpart is possible [1].

EFFECT OF A HYBRID PLANT-BACTERIAL FAS COMPLEX ON BACTERIAL GROWTH AND FATTY ACID COMPOSITION OF THE PHOSPHOLIPIDS

Although there are no obvious differences between the recombinant strain and the parental *E. coli* strain at growth temperatures below 40°C, bacterial growth was found to be severely inhibited at higher temperatures. While the optical density still increases exponentially during incubation at 42°C, the doubling time increased from about 55 minutes in the parental strain, to about 145 minutes in the strain with the recombinant bacterial-plant FAS system.

Enzyme studies on the enzyme over-expressed in *E. coli* revealed that the rapeseed enzyme is quite unstable at temperatures above 30 °C, although it is not as rapidly heat inactivated as the enzyme encoded by the temperature sensitive *E. coli* strain with the *envM392* (ts) allele (not shown, [1]). These data revealed that the growth defect at 42°C is most likely the result of the instability of the plant enoyl-ACP reductase at this temperature.

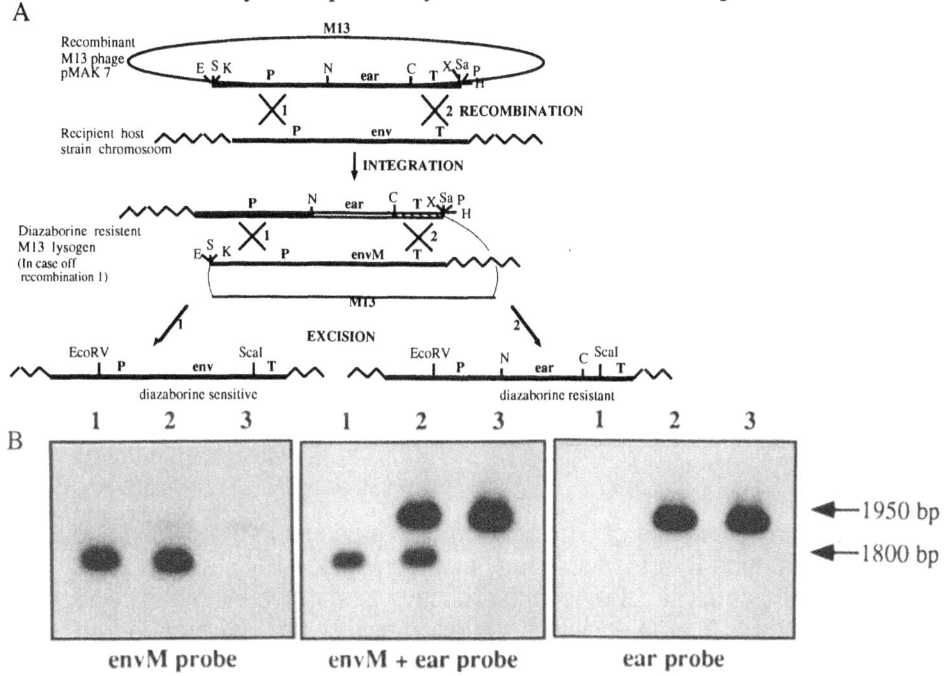

Figure 2. Gene replacement by homologous recombination. (A) Schematic representation of the formation of a M13 lysogen by a single cross-over event between the promoter sequences of the envM gene on the E.coli chromosome and homologous sequences on the M13 clone (recombination 1). Excision of the M13 prophage may occur either by a single cross-over between homologous sequences of the promoter (1) or the 3' downstream region of the gene (2), which will respectively result in a restoration of the normal gene configuration (1, diazaborine sensitive) or gene replacement (2, diazaborine resistant). (B) Southern blots showing M13 mediated replacement of the E.coli envM gene by the enoyl-ACP reductase gene of Brassica napus. 1=host strain, 2=lysogen, 3 =final recombinant

Moreover, lipid analyses of recombinant *E. coli* strains with the hybrid FAS system strongly suggested that in *E. coli* enoyl-ACP reductase catalyses a rate limiting step in the elongation of palmitoleic acid (C16:1) to cis-vaccenic acid (18:1), [1] .

References

1 Kater MM, Koningstein GM, Nijkamp HJJ and Stuitje AR. The use of a Hybrid Genetic System to Study the Functional Relationship between Prokaryotic and Plant Multi-Enzyme Fatty Acid Synthetase Complexes. Plant Mol Biol 1994: in press.

2 Bergler H, Högenauer G and Turnowsky F. Sequences of the *envM* gene and of two mutated alleles in *Escherichia coli.* J Gen Microbiol 1992;138: 2093-2100.

3 Turnowsky F, Fuchs K, Jeschek C and Högenauer G. *envM* Genes of *Salmonella typhimurium* and *Escherichia coli.* J Bacteriol 1989;171: 6555-6565.

4 Kater MM, Koningstein GM, Nijkamp HJJ and Stuitje AR. cDNA cloning and expression of *Brassica napus* enoyl-acyl carrier protein reductase in *Escherichia coli.* Plant Mol Biol 1991; 17: 895-909.

EXPRESSION OF mRNA AND STEADY-STATE LEVELS OF PROTEIN ISOFORMS OF ENOYL-ACP REDUCTASE FROM *Brassica napus*.

Tony Fawcett, William J. Simon, John Shanklin* and Antoni R. Slabas.

Lipid Molecular Biology Group, Department of Biological Sciences
University of Durham, South Road, Durham, DH1 3LE, U.K.
*Dept of Biology, Brookhavn National Lab, Upton, NY 11973 USA

Introduction.

The *de novo* biosynthesis of fatty acids in plants occurs in the plastid and is catalysed by a type II, dissociable fatty acid synthetase(FAS). Enoyl-ACP reductase (ER) catalyses the second reductive step in the synthetic cycle, converting a *trans* 2,3 moiety into a saturated acyl chain. The protein from *B. napus* has been purified[1], extensively sequenced[2] and the absolute mass of the enzyme in developing rape seed determined[3]. The cDNA which encodes ER has been cloned and fully sequenced; Southern hybrisization with this cDNA shows the presence of four genes in *B. napus*[4]. This study investigates the expression of ER mRNA and the steady-state levels of ER protein isoforms by Northern and two-dimensional Western analyses.

The maximum level of ER mRNA in rape seed is detected before the maximum level of corresponding protein, which in turn precedes the maximum level of fatty acids synthesized. A second component of plant FAS, stearoyl-ACP desaturase(SD) has also been analysed using Northern and 1-D Western blots. The two cDNA's show coordinated expression and the proteins that they encode have maximum steady-state levels at the same point in embryogenesis, indicating that some overall control of FAS gene expression is taking place.

Four isoforms of ER are shown to be present in both seed and leaf tissue; the most abundant two forms present in seed are also the most abundant in leaf. This data shows that the greater demand for ER in seeds is met by an increased expression of all four genes rather than by switching on a seed specific gene.

Results.

Biological material. B. napus var. Jet Neuf was grown in a greenhouse and

J.-C. Kader and P. Mazliak (eds.), Plant Lipid Metabolism, 90–92.

flowers tagged to establish an accurate seed developmental time course. mRNA and protein was extracted from seeds of various ages(days after flowering; DAF) for Northern and Western analyses. The total amount of fatty acids per seed was also determined.

Northern Analysis. Poly(A)$^+$ enriched RNA (1μg) from seeds, of various ages, and leaf was subjected to Northern analysis after gel electrophoresis and transfer to Hybond N(Amersham) membrane. When probed with a radiolabelled enoyl-ACP reductase cDNA a single hybridising band of approximately 1.6kb was seen. The maximum level of hybridisation in seed was seen at 29 DAF and the amount of message fell off rapidly in older seeds(Fig. 1); the levels in seed are approximately 15-fold greater than in young leaf tissue.

Following washing the blot to remove hybridisation, it was reprobed with a radiolabelled SD cDNA. A single band of approximately 1.9kb was seen; again the maximum hybridisation was in the 29 DAF seed sample, however the message was longer-lived than that of ER. The levels of message in seed was shown to be approximately 30-fold greater than in leaf.

Western Analysis. Western analysis of protein extracts, detected using specific antibodies[3,5] showed that the maximum steady-state levels of both ER(Fig. 1) and SD was in the 35 DAF seed sample.The levels of both proteins showed very similar kinetics and their levels remained at about 70% of the maximum even after fatty acid levels had stopped increasing.

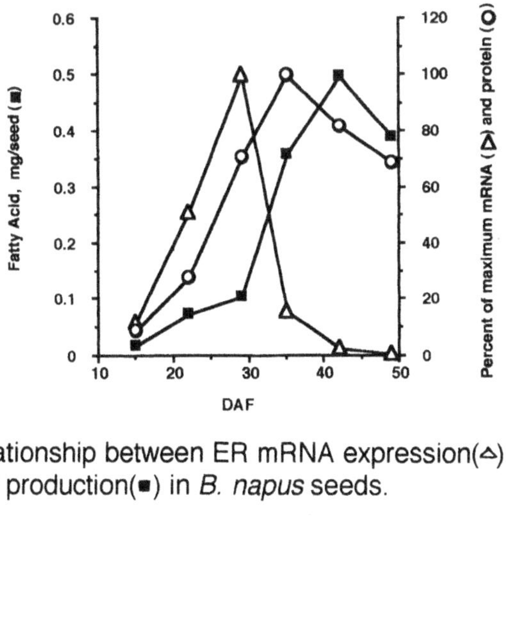

Figure 1: Relationship between ER mRNA expression(△); protein synthesis(○) and fatty acid production(■) in *B. napus* seeds.

Isoforms of Enoyl-ACP Reductase. Protein extracts from seeds and leaf were subjected to two-dimensional electrophoresis and Western analysis. The same four isoforms of ER were present in the leaf and seed material(Fig. 2), indicating that all four genes in *B. napus* show increased expression in seeds , rather than switching on a seed specific gene. The denatured isoelectric point of the four isoforms was determined by a comparison with carbamylated creatine kinase(BDH) markers. The relative pI's of the polypeptides are ER I, 5.9; ER II, 6.1; ER III, 6.4 and ER IV, 6.8.

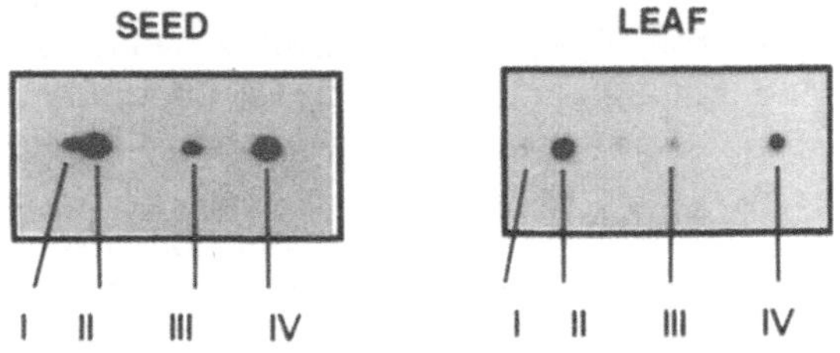

Figure 2: Comparison of seed and leaf isoforms of ER.

Acknowledgements.

This work was supported by AFRC grant number PG 12/523.

References.

1. Slabas AR, Sidebottom CM, Hellyer A, Kessell RMJ, Tombs MP. Induction, purification and characterisation of NADH-specific enoyl-acyl carrier protein reductase from seeds of oil seed rape (*Brassica napus*). Biochem Biophys Acta 1986;877:271-280.
2. Slabas AR, Cottingham IR, Austin A, Fawcett T, Sidebottom CM. Amino acid sequence analysis of rape seed (*Brassica napus*) NADH-enoyl-ACP reductase. Plant Mol Biol 1991; 17: 911-914.
3. Slabas AR, Cottingham IR, Austin A, Safford R, Smith CG. Immunological detection of NADH-specific enoyl-ACP reductase from rape seed (*Brassica napus*) - induction, relationship of α and β polypeptides, mRNA translation and interaction with ACP. Biochem Biophys Acta 1990;1039:181-188.
4. Kater MM, Koningenstein GM, Nijkamp JJ, Stuitje AR. cDNA cloning and expression of *Brassica napus* enoyl-acyl carrier protein reductase in *Escherichia coli.* Plant Mol Biol 1991;17:895-909.
5. Shanklin J, Sommerville C. Stearoyl-acyl-carrier protein desaturase from higher plants is structurally unrelated to the animal and fungal homologs. Proc Natl Acad Sci USA 1991;88:2510-2514.

SOLUBLE AND MEMBRANE ASSOCIATED ENOYL REDUCTASES IN *BRASSICA NAPUS*.

Anne Fowler, J William Simon, Tony Fawcett and Antoni R Slabas.

Lipid Molecular Biology Group, Department of Biological Sciences
University of Durham, South Road, Durham, DH1 3LE, U.K.

Introduction.

Enoyl reductase, a component of fatty acid synthase (FAS) catalyses the second reductive step of fatty acid synthesis.

Soluble enoyl reductase is involved in the synthesis of fatty acids from C4:1 to C18:1. It is located in the plastid and uses acyl-ACP's as substrates. However, acyl-CoA's can also be used as substrates *in vitro* [1].

The elongation of fatty acids from C18:1 involves membrane associated enoyl reductase, which uses acyl-CoA substrates [2]. Whilst soluble enoyl reductase has been well studied, there is little known about membrane associated enoyl reductase.

To assay membrane associated enoyl reductase, the substrate (*trans*)-2,(*cis*)-11-eicosadienoyl-CoA [C20:2(*t*2,c11)-CoA] has been synthesised. The properties of membrane associated and soluble enoyl reductase have been compared. The enzymes have different substrate specificities, distinct pH optima and differential Triton X-100 and NaCl affects. The data clearly demonstrates three forms of enoyl reductase - one soluble and two membrane associated forms.

Results.

BIOLOGICAL MATERIAL.

To study the membrane associated enzymes involved in elongation in oil seed rape (*Brassica napus*), a high erucic acid rape (HEAR) variety was used. Low erucic acid rape variety contains a mutation which prevents elongation of C18:1 fatty acids. As the nature of the mutation is not known, and could be in the elongating enoyl reductase, HEAR was used throughout these studies.

Membranes used were a 0.5M NaCl washed, 40 000g - 200 000g membrane pellet.

The soluble enoyl reductase used was purified homogeneous recombinant protein, that had been expressed in *E. coli* [3].

J.-C. Kader and P. Mazliak (eds.), Plant Lipid Metabolism, 93–95.

ACYL-CoA SUBSTRATES.

(*Trans*)-2-(*cis*)-11-eicosadienoyl-CoA [C20:2(*t*2,c11)-CoA] was synthesised from (*cis*)-11-eicosaenoyl-CoA [C20:1(*c*11)-CoA] by the action of acyl-CoA oxidase [4].

(*Trans*)-2-crotonyl-CoA [C4:1(*t*2)-CoA] was obtained commercially.

WESTERN ANALYSIS.

A blot containing salt washed, membrane pellet showed a band of 32kDa by western analysis. The antibody used was raised against native, soluble enoyl reductase. This may indicate that membrane associated enoyl reductase had the same molecular weight as soluble enoyl reductase, or that some contamination of soluble enoyl reductase remained in the pellet even after stringent washing. At present, this cannot be resolved.

A COMPARISON OF MEMBRANE ASSOCIATED AND SOLUBLE ENOYL REDUCTASE ACTIVITIES.

Substrate specificity. The substrate specificity of both types of enoyl reductase was compared.

Soluble enoyl reductase could utilise either C4:1(*t*2)-CoA or C20:2(*t*2,*c*11)-CoA with NADH; NADPH could not be used.

Membrane associated enoyl reductase could utilise C20:2(*t*2,*c*11)-CoA with either NADH or NADPH. The data suggested two alternatives for membrane associated enoyl reductase: i) either there was one form that prefers NADH, but would use NADPH, ii) there were two separate enzymes.

C4:1(*t*2)-CoA could only be used by the membranes in the presence of NADH. This activity may have been due to the membrane associated enoyl reductase, or contaminating soluble enoyl reductase.

pH effect. The effect of pH on the soluble and membrane associated forms of enoyl reductase was studied (see table 1). There was a clear difference between the three pH optima. This suggested the existence of three types of enoyl reductase.

The effect of Triton X-100 and NaCl. In order to further differentiate the enzymes, the sensitivity of soluble and membrane associated enoyl reductases to Triton X-100 and NaCl was looked at.

At 0.04% Triton X-100, soluble enoyl reductase was inhibited 53%, but membrane associated enoyl reductase was more severely affected (83% inhibition with NADH, 78% with NADPH).

NaCl also affected the enzyme activities. To obtain maximal activity, soluble enoyl reductase required 0.08M NaCl. Membrane associated enoyl reductase required no NaCl with NADH; 0.4M with NADPH.

Enzyme (substrates)	pH	Buffer (80mM)
Soluble enoyl reductase (C20:2(*t*2,*c*11)-CoA + NADH)	8.0	Tris
Membrane associated enoyl reductase (C20:2(*t*2,*c*11)-CoA + NADH)	6.0	MOPS
Membrane associated enoyl reductase (C20:2(*t*2,*c*11)-CoA + NADPH)	5.0	MES

Table 1. pH optima for soluble and membrane associated enoyl reductases.

The data clearly demonstrated three types of enoyl reductase in *Brassica napus*; one soluble and two membrane associated forms. More information will be obtained on the membrane associated forms as they are purified (in progress).

Acknowledgements.

This work is funded by BBSRC.

References.

1 Slabas AR, Sidebottom CM, Hellyer , Kessell RMJ, Tombs MP. Induction, purification and characterisation of NADH-specific enoyl acyl carrier protein reductase from developing seeds of oil seed rape (*Brassica napus*). Biochem Biophys Acta 1986; 877: 271-280.

2 Slack CR, Browse JA. In: Murray DR, editor. Seed physiology Vol 1. Academic Press, Australia, 1984: 209-243.

3 Kater MM, Koningstein GM, Nijkamp HJJ, Stuitje AR. cDNA cloning and expression of *Brassica napus* enoyl-acyl carrier protein reductase in *Escherichia coli*. Plant Mol Biol 1991;17: 895-909.

4 Broadway NM, Dickinson FM, Ratledge C. Long chain acyl-CoA ester intermediates of ß-oxidation of mono- and di- carboxylic fatty acids by extracts of *Corynebacterium* spp. strain 7E1C. Biochem. J. 1992; 285: 117-122.

THE PURIFICATION OF ACETOACYL CARRIER PROTEIN SYNTHASE FROM AVOCADO AND IDENTIFICATION OF A SEPARATE ACETYL COA:ACP TRANSACYLASE ACTIVITY

Bethan S. Gulliver & Antoni R. Slabas.

Lipid Molecular Biology Group, Department of Biological Sciences
University of Durham, South Road, Durham, DH1 3LE, U.K.

Introduction

The biosynthesis of fatty acids in plants is catalysed by a type II, dissociable, fatty acid synthetase (FAS) made up of at least six catalytic polypeptides and a central, acyl carrier protein (ACP). There are three β ketoacyl-ACP synthetases (KAS): KAS 1, catalysing the initial reaction between acetyl ACP and malonyl ACP; KAS 2, catalysing the addition of acetyl ACP to the elongating fatty acid chain and the recently identified KAS 3. A further enzyme implicated in the initiation of FAS is acetyl CoA:ACP transacylase (ACAT) [3]. ACAT catalyses the formation of acetyl ACP. KAS 3, however does not utilise acetyl ACP, instead the condensation occurs between malonyl-ACP and acetyl CoA. There has recently been much discussion about the possible roles of KAS 3 and ACAT which appear to have partially overlapping functions

KAS 3 activity has been purified to homogeneity from spinach [1] and *E. coli* [4]. In both the purified enzyme has the ability to load an acetyl group from acetyl CoA, 'ACAT' activity. Purification of ACAT from *E. coli* has been reported [3], but as this work predates the discovery of KAS 3 it is not clear whether this activity is resolvable from KAS 3. There is still confusion as to whether plants posses distinct ACAT activity since the assays that have been used for both enzymes contain ACP.

We designed two novel assay systems, requiring the purification of *E. coli* ACP and recombinant *E. coli* malonyl ACP acyl transferase (MCAT) and also the preparation of malonyl ACP. KAS 3 was purified and characterised from avocado tissue. Following anion exchange a peak of ACP acetyl loading activity not associated with KAS 3 activity was observed. The absolute dependence of ACAT on the presence of ACP was used as the basis for a radio metric assay to demonstrate the existance of ACAT

In both avocado (*Persea americana*) and rape (*Brassica napus*) tissue we have: demonstrated the presence of KAS 3 and a distinct ACAT, obtained clear resolution of KAS 3 and ACAT activities following anion exchange chromatography and determined the ratio of activities of these two enzymes.

J.-C. Kader and P. Mazliak (eds.), Plant Lipid Metabolism, 96–98.

Results

KAS 3 was prepared from avocado tissue using a 6-step purification procedure. KAS 3 activity was distinguished from ACAT activity on the basis that KAS 3 is highly stimulated by the addition of malonyl CoA in the presence of MCAT [2].

As shown in figure 1, KAS 3 will itself bind acetyl CoA, forming an acetyl CoA-KAS 3 intermediate. However, in order to convert acetyl CoA into acetyl ACP (an acid precipitatable product) ACAT requires ACP. We used the binding of [1-^{14}C] from [1-^{14}C] acetyl CoA to acid precipitatable protein, in the presence and absence of ACP, to distinguish between ACAT and KAS 3 activities. The action of ACAT in the presence of ACP is catalytic, whereas KAS 3, in the absence of malonyl ACP, is simply binding acetyl CoA.

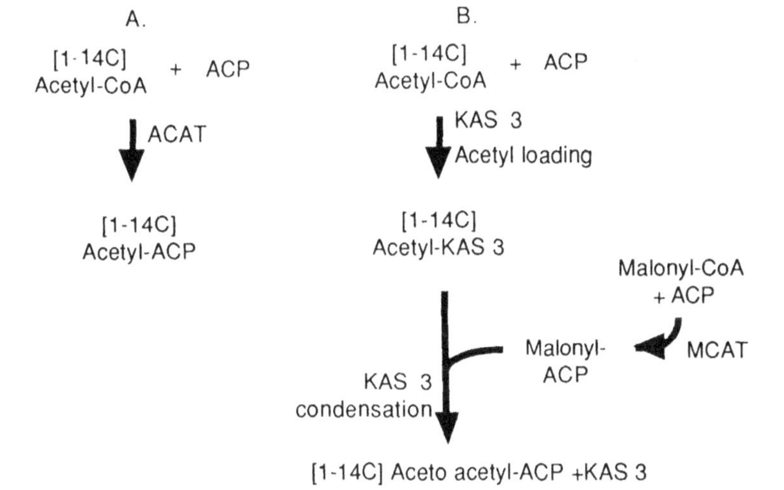

Figure 1. Proposed scheme for the reactions catalysed by ACAT and KAS 3 A. ACAT catalyses the formation of acetyl ACP. This is detected in the assay by the use of [1-14C].

B. KAS 3 first forms a [1-14C] labelled acetyl-KAS 3 complex and then condenses with malonyl-ACP, generated in situ and catalysed by MCAT, resulting in the synthesis of aceto acetyl-ACP, the final product of the KAS 3 reaction.

DETERMINATION OF THE PRESENCE OF ACAT

A PEG precipitated homogenate of avocado was prepared and subjected to anion exchange on a Pharmacia Mono Q column. 0.5ml fractions were collected and assayed for binding of [1-^{14}C] from [1-^{14}C] acetyl CoA in the presence and absence of *E.coli* ACP. The first peak was absolutely dependent on the presence of ACP for incorporation of [1-^{14}C], confirming this peak as ACAT. The second peak gave a similar, small, incorporation of [1-^{14}C] in the presence and absence of ACP as expected for KAS 3. This work was repeated with a crude Triton-X-100 extract of rape seed and gave similar results.

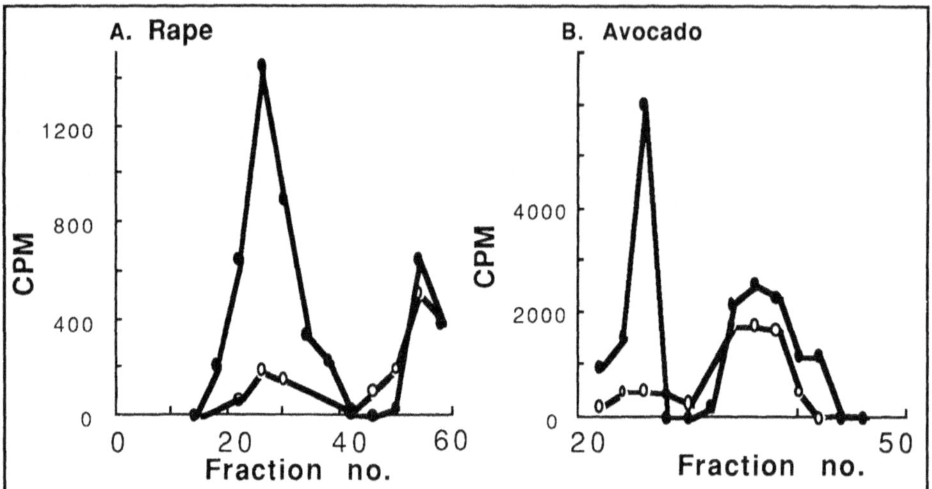

Figure 2. Separation of KAS 3 & ACAT by anion exchange. **A**. Rape seed extract and **B**. Avocado extract were run on anion exchange and the fractions assayed for [1-^{14}]C acetyl binding in the presence (●) and absence (O) of ACP.

Acknowledgements

We acknowledge the financial support of the AFRC for this work under Grant No. PG 12/534 to ARS.

References

1. Clough RC, Matthis AJ, Barnum SR, Jaworski JG: Purification and characterisation of 3-ketoacyl carrier protein synthase iii from spinach/acondensing enzyme utilising acetyl-coenzyme-A to initiate fatty acid synthesis. J Biol Chem 267: 20992-20998 (1992)
2. Gulliver BS & Slabas AR: Acetoacyl-acyl carrier protein synthase from Avocado-purification, characterisation and clear resolution from acetyl CoA:ACP transacylase. Plant mol Biol: in press.
3. Lowe PN & Rhodes S: Purification and characterisation of [acyl-carrier-protein] acetyl transferase from *Escherichia coli.* Biochem J 250: 789-796 (1988)
4. Tsay J-T, Oh W, Larson TJ, Jackowski S, Rock CO, isolation and characterisation of the b-ketoacyl carrier protein synthase 3 gene (fab H) from *Esherichia coli* K-12.. J Biol Chem 267: 6807-6814 (1992).
5. Verwoert IIGS, Verbree EC, Van Der Linden KH, Nijamb HJJ, Stuitje AR, Cloning nucleotide sequence and expression of the *Escherichia coli* Fab D gene encoding malonyl coenzyme A-Acyl carrier protein transacylase: J Bact 174 (1992).

OVER-EXPRESSION AND STUDY OF ß-KETOACYL-[ACP] REDUCTASE FROM *BRASSICA NAPUS* SEED

Neil C Thomas, Tony Fawcett, Jose M Martinez-Rivas, Antoni R Slabas.

Lipid Molecular Biology Group, Department of Biological Sciences University of Durham, South Road, Durham, DH1 3LE, U.K.

Introduction

A cDNA clone has been isolated from a Brassica *napus* var. Jet Neuf seed library, that encodes ß-ketoacyl-[ACP] reductase (3-oxoacyl-ACP reductase) (ßKR) [1].

This has been used to construct an over-expression vector that produces large amounts of the protein in *E. coli* strain BL21(DE3).

The enzyme produced is in an active form and accounts for approximately 5% of the total protein in a lysozymal extract.

Antibodies against the purified ßKR protein have been raised in mice, enabling detection of ßKR on western blots of crude seed extracts.

It is now possible to produce and purify milligram quantities of active ßKR with ease, thus providing enhanced opportunities for the study of the enzyme.

Experimental

OVER-EXPRESSION VECTOR CONSTRUCTION

The plasmid containing the full length ßKR clone was used as a PCR template, with primers that would amplify the mature protein coding sequence.

The primers were also designed to flank the sequence with *Nco*I and *Bam*HI restriction sites, suitable for insertion into a pET-11d vector, without altering the amino acid sequence of the encoded ßKR protein.

Digestion and ligation of vector and insert produced pETJRS10.1 which was used to transform expression strain BL21(DE3).

Lysozymal extracts of transformed clones showed expression of ßKR upon SDS-PAGE analysis. See Figure 1.

J.-C. Kader and P. Mazliak (eds.), Plant Lipid Metabolism, 99–101.

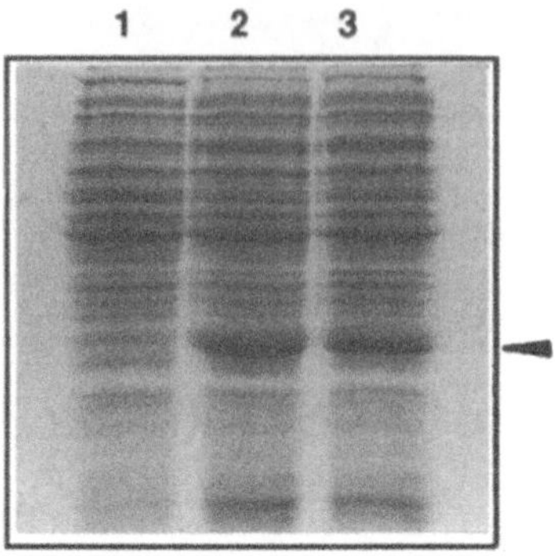

Figure 1. 10% SDS-PAGE gel. Lane 1: BL21(DE3) untransformed cell extract; Lanes 2,3: Extracts from BL21(DE3) cells transformed with pETJRS10.1

ASSAY AND PURIFICATION OF ßKR

Assays for ßKR activity, using the substrate analogue Acetoacetyl-CoA, showed transformant crude extracts to have a specific activity 100 fold higher than a developing rape seed extract. The activity detected in untransformed extracts was negligible.

The protein was purified in two steps, both on FAST S ion-exchange matrix. The first, at pH 6.0, purified ßKR to 85%. The second, at pH7.0 eluted the protein at >95% purity.

The pure enzyme has a specific activity in the region of 20-30 μmole $min^{-1} \mu g^{-1}$. This compares with a figure of 7.34 μmole min^{-1} μg^{-1} for the purified rape seed enzyme [2] (when amino acid composition was used to measure [ßKR]). This data suggests that the enzyme is fully active before purification.

Amino acid composition and spectral analyses of the purified ßKR provided an absolute protein standard and an extinction coefficient. See Figure 2

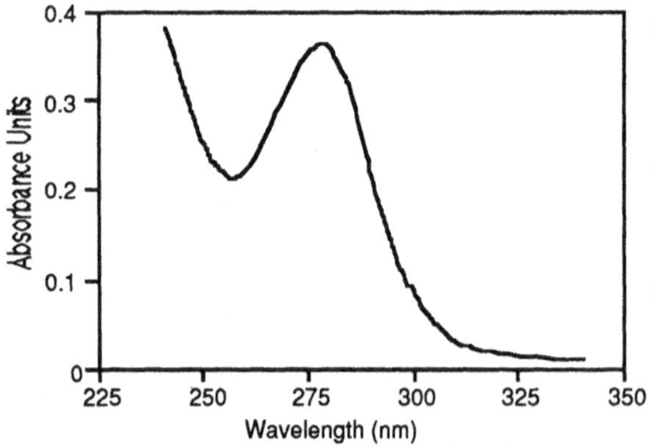

Spectral Data:

280nm peak = 0.358 AU

Amino Acid Analysis Data:

[ßKR] = 0.346 mg / ml

Extinction Coefficient:

$\varepsilon_{\beta KR} = 1.12 \times 10^{5}\ M^{-1} cm^{-1}$

Figure 2. Spectral anaylsis of purified ßKR and extinction coefficient obtained in conjunction with amino acid analysis.

RAISING POLYCLONAL ANTIBODIES TO ßKR

Antibodies were raised in mice using immunisations of 45µg and 6µg ßKR per animal with equal success. This preparation is capable of detecting ßKR in crude seed extracts. See Figure 3. Used in conjunction with the ßKR absolute standard, this antibody preparation allows the quantification of ßKR in seeds at given time points in a developing seed series.

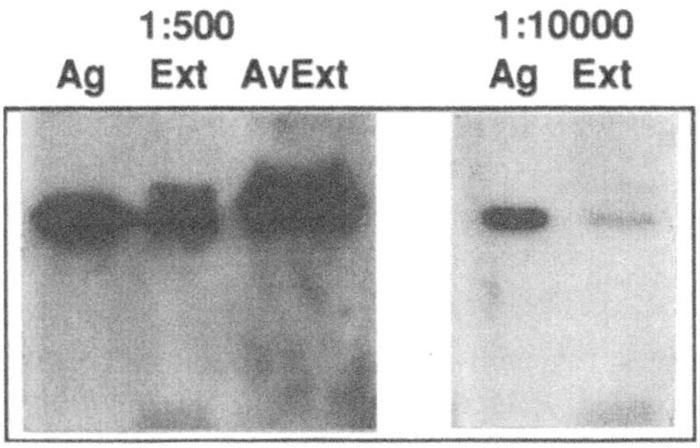

Figure 3. Titre of Anti-ßKR antibodies. Key: "1:500" = dilution of antibody; "Ag"= 100ng ßKR; "Ext" = 50µg Rape seed protein; "AvExt"= 50µg of Avocado protein.

Conclusion

The ease of production and purification of milligram quantities of ßKR facilitates detailed structural and mechanistic studies.

The availability of antibodies to ßKR, and a ßKR standard, allows the quantification of the absolute amount of ßKR in developing seeds.

Acknowledgements

N.C.T. would like to thank the BBSRC (UK) for a research studentship.

References

1. Martinez-Rivas JM, Thomas NC, Chase D, Fawcett T, Slabas AR. cDNA cloning and overexpression of 3-oxoacyl-ACP reductase from *Brassica napus* seed. Grasas y Aceites 1993;44:119-120

2. Sheldon PS, Keckwick RGO, Smith CG, Sidebottom C, Slabas AR. 3-Oxoacyl-[ACP] reductase from oilseed rape (*Brassica napus*). Biochim Biophys Acta 1992;1130:151-159

ISOLATION AND CHARACTERIZATION OF STEAROYL-ACP THIOESTERASE

Dehua Liu and Dusty Post-Beittenmiller
Plant Genetics Section, The Samuel Roberts Noble Foundation, Inc.
P. O. Box 2180, Ardmore, Oklahoma 73402 USA

SYNOPSIS

Leek extracts possess at least two acyl-acyl carrier protein (ACP) thioesterase (acyl-ACP hydrolase) activities. One is the well characterized oleoyl-ACP thioesterase (OTE) and the other is a stearoyl-ACP thioesterase (STE) with high specificity for its substrate. The relative activities of these two thioesterases differed between mesophyll and epidermis. STE has been separated from OTE and purified to near homogeneity by a seven-step procedure that included an ACP-affinity column. The purified STE was found to primarily hydrolyze stearoyl-ACP.

INTRODUCTION

Fatty acid biosynthesis in higher plant cells occurs in the plastids of non-photosynthetic tissues and in chloroplasts of green tissue. The primary products of de novo synthesis are the saturated fatty acids, palmitate (C16:0), stearate (C18:0), and the monounsaturated fatty acid, oleate (C18:1). The 16- and 18-carbon fatty acids act as substrates for additional enzymes in the synthesis of membrane and storage lipids or are further elongated to very long-chain fatty acids for wax biosynthesis. In all plants analyzed to date, OTE prefers oleoyl-ACP, but also has activities with palmitoyl-ACP (16:0-ACP) and stearoyl-ACP (18:0-ACP). Generally, OTE activity is 5- to 10-fold higher with 18:1-ACP than with 18:0-ACP and 16:0-ACP [1,2]. It seems unlikely however, that the nonspecific OTE activity with these acyl-ACPs would provide an adequate supply of saturated fatty acids in some plant tissues where oleic acid is not the major product of plastid fatty acid biosynthesis, i.e. in epidermal cells, where saturated fatty acids are required for the epicutiucular wax biosynthesis.

In the course of our study, we identified two active thioesterase activities: 16:0-ACP and a 18:0-ACP thioesterase in leek epidermis. Here we report the isolation, purification and characterization of one thioesterase from leek epidermis.

MATERIALS AND METHODS

Fresh leeks (Allium porrum L.) were regenerated in a growth chamber (50% humidity and 18 h light). After 10 days, epidermis and mesophyll of the regenerated leaf were harvested and stored at -70°C until used.

J.-C. Kader and P. Mazliak (eds.), Plant Lipid Metabolism, 102–104.

Acyl-ACP thioesterase activity was assayed as described [1]. All purification steps were carried out at 4°C. Two hundred grams of frozen leek epidermis were ground to a powder in liquid nitrogen using a prechilled mortar and pestle. The protein was precipitated with 75% (w/v) ammonium sulfate, and then subjected to hydroxyapatite, Mono-S, Mono-Q, and ACP affinity chromatography. Proteins from each purification step were separated by SDS polyacrylamide gel electrophoresis according to Laemmli (1978)[3] and silver stained (Bio-Rad).

RESULT

Acyl-ACP thioesterase activities with 16:0-ACP, 18:0-ACP and 18:1-ACP were assayed under standard conditions with different plant species. Total leaf extracts from leek showed high activity for all three substrates. The significance of this comparison is the activity ratio. i.e. the activity with both 16:0- and 18:0-ACP was close to the activity with 18:1-ACP (Fig.1) indicating leek may posses a novel thioesterase with specific activity for 16:0-ACP or 18:0-ACP.

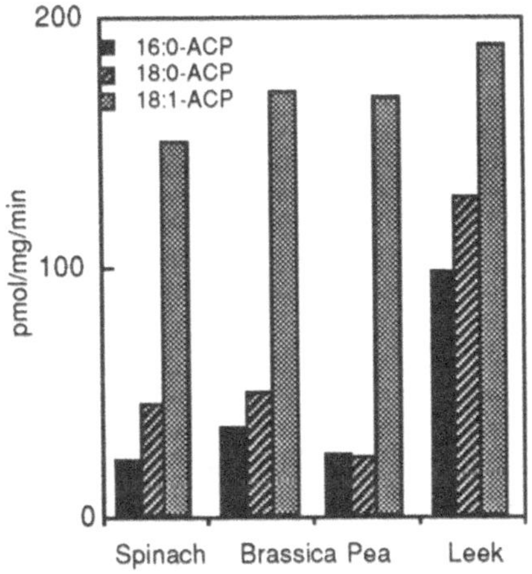

Fig. 1. Acyl-ACP thioesterase activities in total leaf extracts of different plant species. Enzyme activity was assayed under standard conditions.

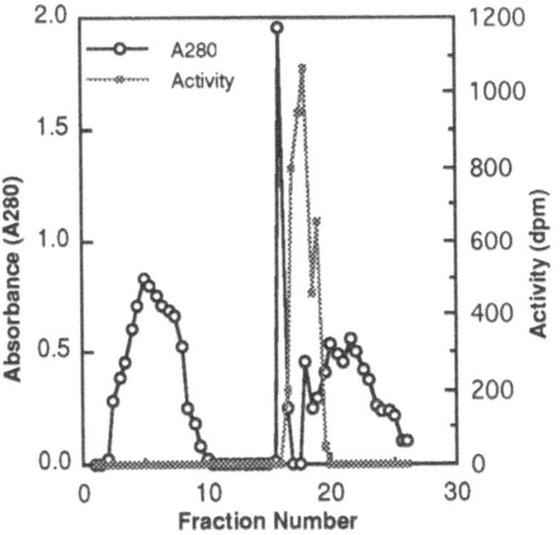

Fig.2. FPLC Mono-Q chromatography of STE. The partially purified STE was applied to Mono-Q column and eluted as described in the text.

The next step was to further isolate STE . This was accomplished by hydroxyapatite chromatography with a stepwise elution. The partially-purified STE almost exclusively hydrolyzed stearoyl-ACP. The STE was purified from other proteins by FPLC Mono S. The active STE was eluted by a succinate buffer (20 mM, pH 4.5) which proved to be an efficient approach to obtain the high recovery. The STE was further purified by a Mono Q chromatography (Fig. 2) using Tris buffer (20 mM pH 8.5) with a 0 - 0.4 M NaCl gradient. The active fractions were then pooled and applied to an ACP-affinity column; STE was eluted in two steps.

STE activity correlated with a subunit molecular mass of 40 kDa, and was purified to near homogeneity as judged by SDS-PAGE silver staining (Fig. 3). The native molecular mass of the purified STE was estimated by gel filtration (Superdex 200 HR

10/30). STE activity eluted as a single peak with an estimated molecular mass of 84 kDa (data not shown) indicating it was a dimer.

The specificity of the acyl-ACP thioesterase to acyl-ACP substrates with different acyl chains was evaluated. The purified STE has a high preference for 18:0-ACP (Fig.4). To eliminate the possibility that STE is a acyl-CoA thioesterase, 18:0-CoA was also tested as substrate. and no reaction was observed. These data confirmed that the highly purified STE prefers 18:0-ACP as its substrate.

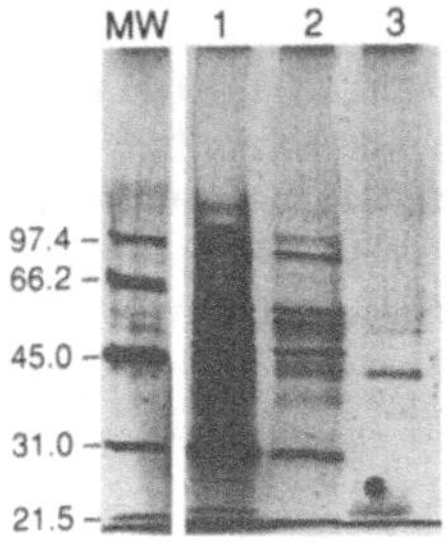

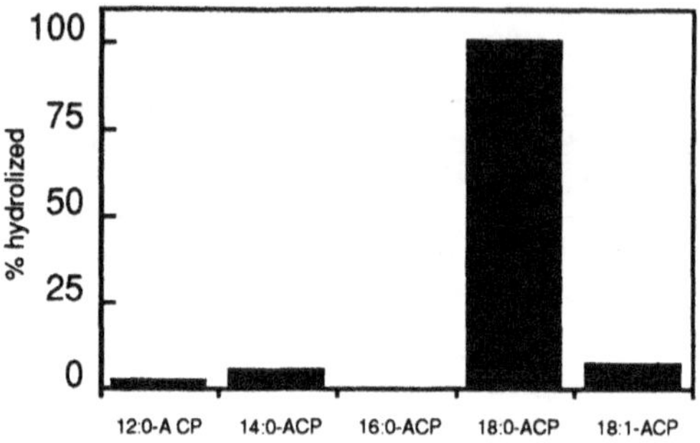

Fig. 3. SDS-PAGE of various STE fractions. Fractions were separated on a 10% acrylamide gel followed by silver staining. Lane 1: HA fraction. Lane 2: 0.2 M NaCl elution(ACP column) and lane 3: 1.0 M NaCl elution. (ACP column).

Fig. 4. Substrate specificity of the purified STE. Acyl-ACPs at a concentration of 1 μM were assayed with the purified STE.

DISCUSSION

We have developed an isolation and purification protocol for STE. The purification has been difficult both due to the cross activity of OTE with 18:0-ACP and the low abundance of STE in epidermal tissues. In sharp contrast to OTE, the purified STE showed high substrate specificity with 18:0-ACP. STE essentially resides in epidermal cells suggesting its role may related to a specific function of epidermal cells. Since wax biosynthesis uses only saturated fatty acids as substrate, the fatty acid (18:0) produced by the hydrolysis of 18:0-ACP by STE can serve as the substrate. Therefore, the role of STE may be to terminate fatty acid biosynthesis in the plastid and supply saturated fatty acids for wax biosynthesis.

Reference

1. Ohlrogge JB, Shine WE, Stumpf PK. Fat metabolism in higher plants - Characterization of plant acyl-ACP and acyl-CoA hydrolases. Arch Biochem Biophys 1979;189:382-391.

2. Mckeon T A, Stumpf PK. Purification and characterization of the stearoyl-acyl carrier protein desaturase and the acyl-acyl carrier protein thioesterase from maturing seeds of safflower. J Biol Chem 1982; 257:12141-12147.

3. Laemmli UK. Cleavage of structural proteins during the assembly of the head of bacteriophage T4. Nature 1970;227:680-685.

A NOVEL ACYLTRANSFERASE ACTIVITY IN AN OLEAGINOUS ALGA

Jane C. Schneider and Paul G. Roessler
National Renewable Energy Laboratory
1617 Cole Boulevard
Golden, CO 80401 USA

Introduction

Nannochloropsis sp. is a unicellular marine alga that accumulates fatty acids in triacylglycerols (TAGs) as well as in membrane lipids [1]. Whereas TAG contains mostly 16:0 and 16:1 (n-7) fatty acids, the phospholipids and galactolipids contain a large proportion of eicosapentaenoic acid [20:5 (n-3)]. The accumulation of this fatty acid is regulated by environmental conditions such as light intensity [2]. We have previously developed a preliminary model of polyunsaturated fatty acid (PUFA) synthesis based on pulse-chase radiolabeling studies [3] and analysis of a mutant strain deficient in 20:5 (n-3) (data not shown). The model proposes that C_{18} fatty acids are desaturated while attached to phosphatidylcholine (PC), whereas C_{20} fatty acids are desaturated when attached to phosphatidylethanolamine (PE). The PUFAs formed on these extrachloroplastic lipids are the sole source for PUFAs that accumulate in monogalactosyldiacylglycerol (MGDG).

In higher plants, acyl exchange is considered to be a major contributor to the formation of PUFAs. In this reaction, 18:1 is transferred to the *sn*-2 position of PC by lyso-PC acyltransferase and subsequently desaturated [4]. Radiolabeling studies and positional analysis of the labeled lipids by digestion with phospholipase A_2 indicated that *Nannochloropsis* also contains this lyso-PC acyltransferase activity. The enzyme incorporates 18:1 and 20:4 equally well into the *sn*-2 position of PC. In addition, a novel acyltransferase activity was found that preferentially incorporates 20:4 (n-6) into both the *sn*-1 and *sn*-2 positions of PE.

Results and Discussion

Incubation of *Nannochloropsis* membrane preparations with [^{14}C]-acyl-CoA results in the accumulation of label in lipids. The distribution of label among lipid classes varied depending on the nature of the acyl-CoA. When 18:1-CoA was used, PC was the most strongly labeled lipid, whereas in the presence of 20:4-CoA, PE was the most strongly labeled. A few unidentified lipids, diacylglyceroltrimethylhomoserine (DGTS), and neutral lipids accumulated label in smaller amounts.

Three [^{14}C]-acyl-CoA substrates were used to measure the specificity of enzymes that incorporate label into the three most highly labeled lipids (PC, PE and DGTS) (Table 1). PC incorporated 18:1-CoA and 20:4-CoA to the same extent, whereas DGTS and PE

J.-C. Kader and P. Mazliak (eds.), Plant Lipid Metabolism, 105–107.

were more readily labeled with 20:4-CoA than with 18:1-CoA or 16:0-CoA. The incorporation of 20:4 into PE was the highest activity observed, more than six times higher than any other reaction.

	nmol incorporated•mg^{-1}•h^{-1} (±SD)					
[^{14}C]-acyl-CoA	PC		PE		DGTS	
16:0	2.2	(0.2)	1.3	(0.7)	2.5	(0.7)
18:1	20.9	(2.7)	1.6	(0.2)	1.2	(0.2)
20:4	16.6	(1.8)	117.6	(11.0)	14.3	(0.9)

Table 1. Specificity of lyso-PC and lyso-PE acyltransferases for different acyl-CoAs.

The position into which the labeled fatty acid was incorporated was examined by digestion of the *in vitro* labeled lipids with phospholipase A_2, which hydrolyzes the *sn*-2 position of phospholipids, resulting in a free fatty acid and lyso-phospholipid. About 95% of the label from 18:1-CoA or 20:4-CoA in PC was found at the *sn*-2 position (Table 2). Thus, this reaction is similar to the lyso-PC acyltransferase activity described in safflower [5]. Surprisingly, incorporation of 20:4-CoA occurred to approximately the same extent at both positions of PE. This incorporation is unlikely to be the result of *de novo* lipid synthesis, since the soluble substrates for such reactions were not provided in the assay, and label in phosphatidic acid was not observed. Further experiments will be conducted to examine whether 20:5 (n-3)-CoA is also a good substrate for this activity. Our present model of PUFA synthesis in *Nannochloropsis* proposes that fatty acids of MGDG are desaturated while attached to a phosphoglycerol backbone. Since most molecular species of MGDG contain 20:5 in the *sn*-1 position [2], this novel lyso-PE acyltransferase activity may be responsible for the formation of the extrachloroplastic glycerolipid backbone used for MGDG synthesis.

	% of dpm in all lipids after phospholipase A_2 digestion					
Lipid:	PC		PC		PE	
[^{14}C]-acyl-CoA provided:	18:1-CoA		20:4-CoA		20:4-CoA	
phospholipase A_2:	-	+	-	+	-	+
Fatty acid(*sn*-2)	0.8	93.0	0.9	88.5	1.1	42.9
Phospholipid (substrate)	95.5	4.0	93.5	7.6	96.5	2.2
lyso-phospholipid (*sn*-1)	3.5	3.1	5.7	4.0	2.5	55.0

Table 2. Positional analysis of PC and PE labeled *in vitro* with [^{14}C]-acyl-CoA.

Methods

Membranes were prepared from mid-log phase cultures as follows: cell pellets were disrupted with a teflon homogenizer and approximately two volumes of glass beads

(<106 micron diameter). Buffer (150 mM Tricine KOH pH 7.5, 1 mM EDTA, 1 mM EGTA, 5 mM DTT, 0.33 M sorbitol) was added and cell debris removed by centrifugation at 6000 x *g*. The supernatant was centrifuged at 100,000 x *g* to pellet membranes, which were resuspended with a ground-glass homogenizer in a small amount of buffer without DTT. Protein concentration was measured using the Bio-Rad DC assay.

Membrane preparations from *Nannochloropsis* were provided with the specified [^{14}C]-acyl-CoA (0.03 uCi), 100 mM Tris-HCl pH 8.5, 1 mM CoA, 2 mM ATP and 2 mM $MgCl_2$ for 4 to 10 minutes at 30°C. Rates of acyl-CoA incorporation into lipids were linear and proportional to protein concentration in the ranges used. Lipids were separated by silica gel TLC using chloroform:methanol:water (65:25:4) and scraped into tubes for scintillation counting. For positional analyses, lipids were eluted after separation on TLC, and digested with phospholipase A_2 from *Crotalus adamanteus* in the presence of 100 μg of dioleoyl-PC or dioleoyl-PE carrier in 0.9 M borate buffer, pH 8.0 and ether. The digested lipids were separated by TLC as above and scraped into tubes for scintillation counting.

Acknowledgements

This work was supported by the United States - Israel Binational Agricultural Research and Development Fund (No. IS-1819-90C) and the Biofuels Systems Division of the United States Department of Energy.

References

1. Sukenik A, Carmeli Y, Berner T. Regulation of fatty acid composition by irradiance level in the Eustigmatophyte *Nannochloropsis* sp. J Phycol 1989;25:686-92.
2. Sukenik A, Yamaguchi Y, Livne A. Alterations in lipid molecular species of the marine Eustigmatophyte *Nannochloropsis* sp. J Phycol 1993;29:620-6.
3. Schneider JC, Roessler P. Radiolabeling studies of lipids and fatty acids in *Nannochloropsis* (Eustigmatophyceae), an oleaginous marine alga. J Phycol 1994;30:594-8.
4. Heinz E. Biosynthesis of polyunsaturated fatty acids. In: T S Moorc J, editor. Lipid Metabolism in Plants. Ann Arbor: CRC Press, 1993: 33-89.
5. Stymne S, Appelqvist LA. The biosynthesis of linoleate from oleoyl-CoA via oleoyl-phosphatidylcholine in microsomes of developing safflower seeds. Eur J Biochem 1978;90:223-9.

DIHYDROSTERCULATE IN TOBACCO TRANSFORMED WITH BACTERIAL CYCLOPROPANE FATTY ACID SYNTHASE

Katherine M. Schmid
Dept. of Biological Sciences, Butler University, Indianapolis, IN 46208 U.S.A.

Many gram negative bacteria accumulate cyclopropane fatty acids (CPFAs) in their membranes during stationary phase. In plants, on the other hand, CPFAs are best known as constituents of certain seed oils. CPFAs make up around 40% of *Litchi chinensis* seed oil [1] and a significant proportion of oils from other Sapindales. In order Malvales, small amounts of CPFA typically accompany cyclopropene fatty acids in seed oils, although up to 5% CPFA has been observed in the polar lipid fraction from Malvaceous roots [2]. Genetic engineering of oilseeds for unusual fatty acid production requires that the introduced fatty acids accumulate in triacylglycerol without compromising membrane structure or function. A major purpose of this study has been to determine whether a higher plant would express a bacterial CPFA synthase, and, if so, how the resulting CPFA would be partitioned.

The native substrates of bacterial CPFA synthases are the membrane lipids themselves. The enzyme from *Escherichia coli* adds a methylene group from S-adenosyl-methionine across the double bond of either vaccenate (18:1 Δ^{11}) or palmitoleate (16:1 Δ^{9}) esterified to phosphatidylethanolamine, phosphatidylglycerol or cardiolipin [3]. However, unsaturated fatty acid auxotrophs grown on oleate (18:1 Δ^{9}) have been shown to accumulate its cyclopropane analog, dihydrosterculate[4,5]. This suggests that plants should contain appropriate substrates for the bacterial enzyme, and that dihydrosterculate esterified to phospholipid would be the most likely primary product. In this paper, it is shown that tobacco transformed with the *Escherichia coli* CPFA synthase under the cauliflower mosaic virus 35S promoter does produce dihydrosterculate. Most of the product remains associated with polar lipids, but the triacylglycerol fraction from tobacco callus also contains CPFA.

Methods and Materials

Both a *Nicotiana tabacum* cell line and tobacco leaf disks were transformed with *Agrobacterium tumefaciens* strain LBA4404 carrying either pCP4, a construct in which the *Escherichia coli* CPFA coding sequence [6] had been inserted into the binary vector pGA748 [G. An], or the vector alone, which included the cauliflower mosaic virus 35S promoter, the octopine synthase polyadenylation site, and the neomycin phosphotransferase gene. The CPFA synthase coding sequence was prepared by polymerase chain reaction using a derivative of pGI22 [7] as template. *A. tumefaciens* was killed with 500 μg/ml carbenicillin, and transformants were selected on kanamycin.

Portions of kanamycin-resistant calli and plants were killed in boiling isopropanol and extracted with hexane/isopropanol (3:2). Lipids were fractionated on BioSil A, and

J.-C. Kader and P. Mazliak (eds.), Plant Lipid Metabolism, 108–110.

triacylglycerols were isolated by thin layer chromatography of the neutral lipids on silica gel G with hexane/diethyl ether/acetic acid (80:20:2). Lipid fractions were transesterified with sodium methoxide. Fatty acid methyl esters were separated by gas-liquid chromatography on a packed 12 ft DEGS column. *Litchi chinensis* oil and hydrogenated *Sterculia foetida* oil were used as qualitative standards. Gas chromatography/mass spectrometry was performed on an HP5890 gas chromatograph with a 30 m DB-23 column linked to an HP5970 mass selective detector.

Results

As exemplified in Figure 1, a peak comigrating with methyl dihydrosterculate was identified in gas chromatograms of transesterified leaf lipids from pCP4 transformants, but not in leaf lipids of plants transformed with vector. Putative dihydrosterculate in total leaf lipids ranged from 0.9 to 2.5%. The apparent CPFA peak was also observed in 23 of 30 pCP4 tobacco cell line transformants. Proportions of CPFA in total lipids reached 3% in about half of the positive calli, with a record level of 9.9%. Analyses of more than one callus from the same line gave somewhat variable proportions of CPFA. Size, density, and age of callus could have contributed to this variability.

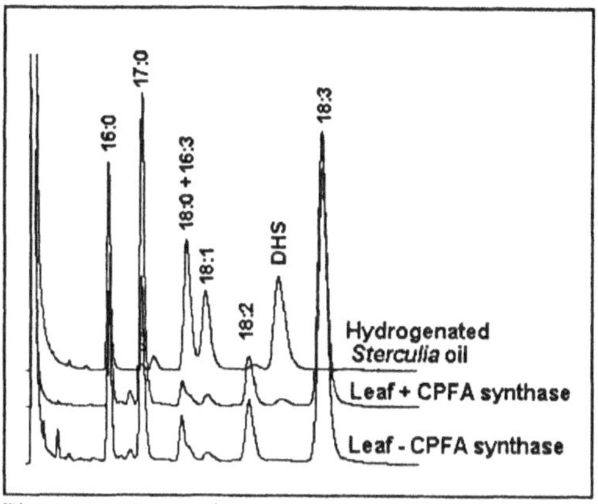

Figure 1. Fatty acid composition of leaf lipids from transformed tobacco. Comigration of a peak with dihydrosterculate (DHS). 17:0 = internal standard.

Identification of the putative dihydrosterculate as a nineteen carbon cyclopropane fatty acid was verified by gas chromatography/mass spectrometry. The mass spectrum showed a molecular ion (Z/e = 310) and other fragments (Z/e = 184, 136, 278) characteristic of methyl dihydrosterculate [8]. Note that ring position could not be determined by this method.

When callus lipids were fractionated, CPFAs were observed in both polar and nonpolar lipids. Quantitatively, phospholipids, the primary lipid fraction in the calli, retained the bulk of the dihydrosterculate produced. Triacylglycerols, which comprised about to 7% of total lipid, were somewhat enriched in CPFAs (Figure 2). The callus data suggested that oil-rich tissues would concentrate CPFAs in triacylglycerols. However, seed analyses to date have been disappointing. CPFAs are undetectable in total lipid from first generation seed of plants selected for leaf dihydrosterculate content, although a large proportion of second generation seedlings are kanamycin resistant.

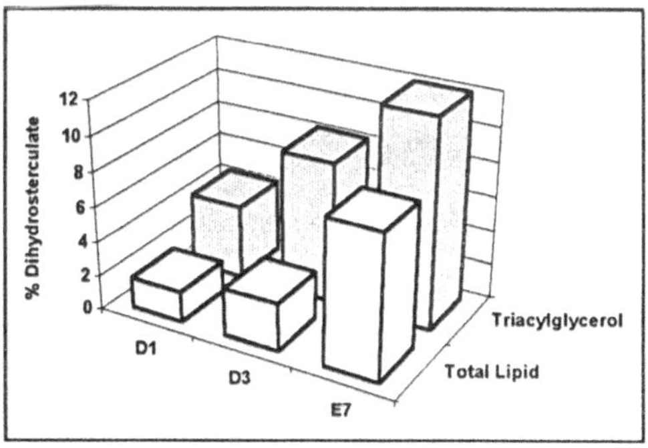

Figure 2. Proportion of dihydrosterculate (weight %) in lipids from three callus isolates.

Discussion

The results presented above show that active *E. coli* CPFA synthase can be expressed in plants. Since no specific targeting sequence was used, the enzyme is presumed to have interacted with oleate moieties in membranes accessible to the cytoplasm. Although not an integral membrane protein, the synthase is normally associated with membranes in the bacterium and is active on monounsaturated vesicles *in vitro* [3].

The identification of CPFA in callus triacylglycerol indicates that enzymes capable of dihydrosterculate transfer to that fraction are available in tobacco. The failure of the cauliflower mosaic virus 35S promoter to drive significant CPFA accumulation in seeds may therefore be unrelated to lipid metabolism *per se*. The 35S promoter is relatively weak during the latter part of seed development [9], and the relatively protease-sensitive CPFA synthase [10] may turn over rapidly. It is hoped that manipulation of promoters, enzyme localization, and substrate concentration will result in both dihydrosterculate-rich oils and a better understanding of fatty acid flow between phospholipid and triacylglycerol.

Acknowledgements

This research was supported by USDA research award INDR-9204324. Thanks are also due to Edgar B. Cahoon and Betsy Murphy for technical assistance, David K. Shintani for tobacco cell culture expertise, and John E. Cronan, Jr. for the CPFA synthase construct pGI22.

References

1. Gaydou EM, Ralaimanarivo A, Bianchini J-P. Cyclopropanoic fatty acids of Litchi (Litchi chinensis) seed oil. A reinvestigation. J Agric Food Chem 1993; 41:886-890.
2. Schmid KM, Patterson GW. Distribution of cyclopropenoid fatty acids in Malvaceous plant parts. Phytochemistry 1988; 27:2831-2834.
3. Taylor FR, Cronan JE Jr. Cyclopropane fatty acid synthase of Escherichia coli. Stabilization, purification, and interaction with phospholipid vesicles. Biochemistry 1979; 18:3292-3300
4. Marinari LA, Goldfine H, Panos C. Specificity of cyclopropane fatty acid synthesis in Escherichia coli. Utilization of isomers of monounsaturated fatty acids. Biochemistry 1974; 13:1978-1983
5. Ohlrogge JB, Gunstone FD, Ismail IA, Lands WEM. Positional specificity of cyclopropane ring formation from cis-octadecanoic acid isomers in Escherichia coli. Biochim Biophys Acta 1976; 431:257-267
6. Wang A-Y, Grogan DW, Cronan JE Jr. Cyclopropane fatty acid synthase of Escherichia coli: Deduced amino acid sequence, purification, and studies of the enzyme active sites. Biochemistry 1992; 31:11020-11028.
7. Grogan DW, Cronan JE 1984. Cloning and manipulation of the Escherichia coli cyclopropane fatty acid synthase gene: Physiological aspects of enzyme overproduction. J Bacteriol 1984; 158:286-295.
8. Christie WW, Holman RT. Mass spectrometry of lipids. 1. Cyclopropane fatty acid esters. Lipids 1966; 1:176-182.
9. Williamson JD, Hirsch-Wyncott ME, Larkins BA, Gelvin SB. 1989. Differential accumulation of a transcript driven by the CaMV 35S promoter in transgenic tobacco. Plant Physiol 1989; 90:1570-1576
10. Wang A-Y, Cronan JE Jr. The growth phase-dependent synthesis of cyclopropane fatty acids in Escherichia coli is the result of an RpoS(KatF)-dependent promoter plus enzyme instability. Molecular Microbiol 1994; 11:1009-1017.

MODULATION OF THE VERY-LONG-CHAIN FATTY ACID (VLCFA) FORMATION IN LEEK

Cassagne Claude, Bessoule Jean-Jacques, Schneider Florence, Lessire René, Sturbois Bénédicte, Moreau Patrick, Spinner Christel.

Laboratoire de Biogenèse Membranaire, URA 1811-CNRS, Université de Bordeaux II, 1 rue Camille Saint Säens, 33077 Bordeaux Cedex, France

Introduction

Acyl-CoA elongases (ACE) are membrane-bound multienzymes which have been purified in a very few cases (for review:1). The purified enzymes embedded in Triton X100 micelles have an apparent Mw of 300 kDa and, upon SDS treatment can be resolved in 3-4 bands of 59, 61 and 65 kDa (2). Three of the four postulated steps have been demonstrated (3). From these studies, it appears that the condensation (i.e. the synthesis of 3-ketoacyl-CoA) is the rate limiting step. The regulation of the overall process of elongation could therefore firstly affect this step. Hence, molecules allowing an increase or a decrease of the VLCFA synthesis in plants must be firstly identified among molecules interacting with ketoacyl-CoA synthase(s) (KAS). The acyl-CoA elongases are unique among fatty acid synthesizing enzymes in that there is apparently no covalent link between the enzymes and the substrates, so that no acyl transferases are involved in the elongation process.

Owing to the nature of the substrates, it has been proposed that acyl-CoAs are attached by partly ionic, partly hydrophobic, interactions to the condensing enzyme, whereas the malonyl-CoA is attached chiefly by ionic interactions (1). The condensation creates a malonyl-CoA like structure which disappears progressively as the reaction proceeds, so that the chain under formation leaves the malonyl-CoA site for the acyl-CoA site after the dehydration. A study of the modulation of the activity of the acyl-CoA elongases requires that these properties be taken into account.

What makes that an acyl-CoA elongase from Simmondsia chinensis forms only monounsaturated VLCFA whereas those from leek leaves form exclusively the saturated ones ? It has been postulated that this could result from either the specificity of the elongases towards saturated or unsaturated acyl-CoAs, or from the competition of other enzymes for the same substrates. The first hypothesis was ruled out in leek when it was demonstrated that the purified elongases had no specificity towards the acyl chain-length or the unsaturation (3).

It was proposed that the specificity was due to the actual concentration of substrate in the vicinity of the elongase, though no clear-cut link could be established between the latter parameter and the activity of other acyl-CoA-using enzymes towards e. g. C18-CoA and C18:1-CoA (Bessoule, unpublished observations).

J.-C. Kader and P. Mazliak (eds.), Plant Lipid Metabolism, 111–114.

Modulation of the local concentration :

PARTITION OF ACYL-CoAS

The above mentioned observation prompted some authors to further analyze both theoretically and experimentally the parameters governing the concentration of the subtrates in the vicinity of the enzyme. Whereas the water-soluble substrates (malonyl-CoA, NADPH-NADH) did not exhibit unexpected properties, it was observed that not only the acyl-CoA partitions between water and the biological membranes (as had been known for years), but that, unexpectedly, the partition coefficient seemed to vary as a function of the surface density (4). More surprising was the observation that the partition of a given amount of acyl-CoA between water and a given amount of membranes was independent of the water volume, i.e. was in marked contrast to the definition itself of a partition coefficient which involves the ratios of concentrations of a given amphiphilic molecule in membrane and in water (5).

PARTITION OF ACYL-CoAS AND METABOLISM

It was also demonstrated that the binding of acyl-CoA to the membrane allowed its metabolism (4) which in turn allowed a further binding. This was established in the case of a mammalian microsomal acyl-CoA thioesterase (ACTE) (4). The free fatty acids (FFA) formed in the membrane upon acyl-CoA hydrolysis were partly released from the membrane, and the FFA partition coefficient (from membrane to water) increased as a function of time, on the other hand, the partition of the total aliphatic chains remained constant irrespective of their form (acyl-CoAs or FFA), explaining an increased binding of acyl-CoAs to the membranes (4).

Modulation of the elongating activity by the membrane :

Acyl-CoA elongases use exclusively membrane-bound acyl-CoAs, and a prominent role of the membrane, besides insuring a proper environment to the enzyme(s), is to increase the actual concentration of the acyl-CoAs in the vicinity of the elongases by at least two orders of magnitude ; this implies a dramatic decrease of the apparent K_m of the substrate for the catalytic site. A theoretical analysis of the activity of the native microsomal elongase allowed to interpret why the curve of elongating activity as a function of the stearoyl-CoA concentration (from a 0.9 μM to 9 μM) was sigmoidal : these results could be explained by the competition of various enzymes (chiefly acyl-CoA elongase:ACE, acyl-CoA thioesterase ACTE and acyl-CoA transacylase ACTA) for the same substrate with K'_m (ACE) >>K'_m (ACTE) and K'_m (ACTA) (6). In excellent agreement with this interpretation, inserting the purified ACE into phospholipid (PL) vesicles yielded a michaelian curve of overall elongation vs acyl-CoA concentration, indicating that the ACE is not an allosteric multienzyme (6).

By analyzing the effect of higher stearoyl-CoA concentrations on the activity of native ACE, it was shown that, under standard experimental conditions, a maximal activity was reached at *ca* 10 μM stearoyl-CoA (i. e. at a 3 mM intramembrane substrate concentration), and then the activity decreased regularly up to 90 μM stearoyl-CoA(5). The variation of the activity was shown to be typical of an inhibition by substrate excess. The results were best explained by the existence of two sites for the binding of stearoyl-CoA on the elongases : the catalytic site (Site I), and a second site that binds the substrate, thus preventing a further elongation of the substrate. The calculated apparent affinities of the acyl-CoAs were 6 μM (Site I), and 18 μM (Site II) (5).

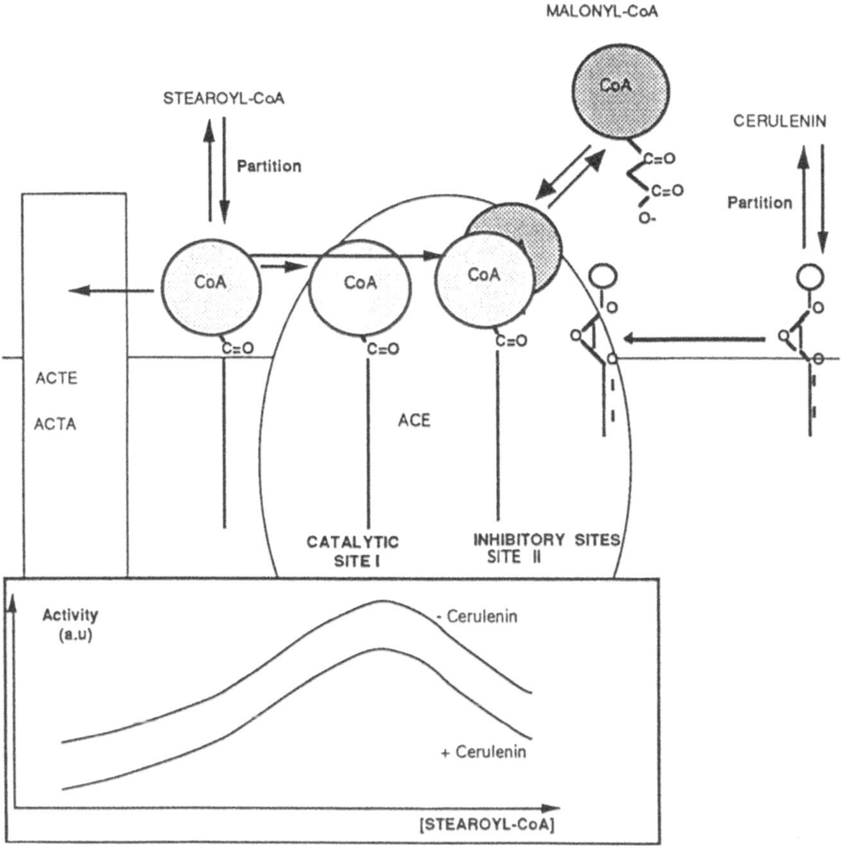

Scheme I : The main parameters modulating the ACE are presented. The location of the inhibitory site II is not demonstrated. The curves show a simulation of the ACE activity (a.u. = arbitrary units) as a function of the stearoyl-CoA concentration and takes into account the partition of the acyl-CoAs, the presence of ACTE and ACTA, the inhibitory site II and the cerulenin site.

The nature of the site II, and, in particular, its analogies with the site of malonyl-CoA binding are not known : the importance of such an inhibitory site in terms of overall lipid metabolism is not understood and will require the study of the origin (and concentration) of cytoplasmic malonyl-CoA, the *in vivo* concentration of acyl-CoAs, etc... On the other hand, the specificity of Site II for stearoyl-CoA remains to be demonstrated (and for example the effect of CoASH should be tested). The competition of ACE at low (ca 5μM) substrate concentration with ACTE or ACTA, the inhibition of ACE at high concentration delineate a narrow concentration range allowing full elongating activity.

Modulation of the VLCFA synthesis : Specific inhibitors ?

Modulating the VLCFA synthesis is a key to wax modulation. As seen above, the condensation is the limiting step of the whole elongation process so that any attempt to modulate chemically the wax biosynthesis, (i. e.without using the tools of molecular biology), should consider the condensing enzyme as the privileged target. As a first attempt, the effect of cerulenin on the elongation was studied. Cerulenin is known to inhibit the *de novo* fatty acid synthesis in E. Coli, yeast, mammals and plants by inactivating the condensing enzyme after a covalent binding at the catalytic site. In our hands, a partition of cerulenin between the water and the microsomes was observed, and the membrane-bound cerulenin was suspected to be active on the ACE (5). No evidence of a covalent binding of cerulenin to the ACE was gathered. As in the soluble FAS, the cerulenin inhibited almost uniquely the condensing enzyme(s). After a preincubation, the inhibition could be quantitative. From the kinetic analysis it was concluded that the inhibition was non-competitive and the inhibition constant was *ca* 15 μM, very close to that reported for site II of stearoyl-CoA binding. These sites however are probably not identical. Recently it was shown in yeast, by using cerulenin analogs of various chain lengthes that the size of the hydrophobic cavity of the condensing enzyme may be a factor by which chain extension is controlled (7). This prompted us to develop a new strategy aiming to affect specifically the VLCFA content in plants, without affecting the *de novo* synthesis and the overall fatty acid metabolism up to C18 (see Schneider et al., this volume).

The recent results on the activity of the condensing enzyme are summarized in scheme I.

Acknowledgements
The work of the authors discussed above was supported by CNRS and Université de Bordeaux. Part of the study was conducted under the Bioavenir Programme/groupe de Recherches "Barrières Cuticulaires" and supported by a grant from Rhône-Poulenc with the contribution of the Ministère de l'Enseignement Supérieur et de la Recherche, and the Ministère de l'Industrie.

References

1. Cassagne C, Lessire R, Bessoule JJ, Moreau P, Créach A, Schneider F, Sturbois B. Biosynthesis of very long chain fatty acids in higher plants. Prog Lipid Res 1994;33:55-69.
2. Bessoule JJ, Lessire R, Cassagne C. Partial purification of the acyl-CoA elongase of Allium porrum leaves. Arch Biochem Biophys 1989;268:475-484.
3. Lessire R, Bessoule JJ, Cassagne C. Involvement of a 3-ketoacyl-CoA intermediate in acyl-CoA elongation by an acyl-CoA elongase purified from leek epidermal cells. Biochim Biophys Acta 1989;1006:35-40.
4. Juguelin H, Bessoule JJ, Cassagne C. Ineraction of amphiphilic substrates and their metabolites with microsomes from mouse sciatic nerves. Biochim Biophys Acta 1991;1068:41-51.
5. Schneider F, Lessire R, Bessoule JJ, Juguelin H, Cassagne C. Effect of cerulenin on the synthesis of very-long-chain fatty acids in microsomes from leek seedlings. Biochim Biophys Acta 1993;1152:243-252.
6. Bessoule JJ, Lessire R, Cassagne C. Theoretical analysis of the activity of membrane-bound enzymes using amphiphilic or hydrophobic substrates. Biochim Biophys Acta 1989;983:35-41.
7. Morisaki N, Funabashi H, Shimazawa R, Furukawa J, Kawaguchi A, Okuda S, Iwasaki S. Effect of side-chain structure on inhibition of yeast fatty acid synthase by cerulenin analogues. Eur J Biochem 1993;211:111-115.

THIOCARBAMATE ACTION ON VERY LONG CHAIN FATTY ACID SYNTHESIS IN PLANTS

Philippa B. Barrett and John L. Harwood

Biochemistry Department, University of Wales, College of Cardiff, PO Box 903, Cardiff, CF1 1ST, U.K.

Introduction

Thiocarbamate herbicides have been reported to affect the biosynthesis of epicuticular lipids of leaves from pea [1]. These alterations may result in an increased vulnerability to environmental factors and enhanced levels of transpiration. Ultimately the death of susceptible plants may ensue. It has been postulated that, once transported into the plant, the thiocarbamate compound is oxidised by a cytochrome P450 mixed function oxidase to its sulphoxide derivative [2]. It is the latter which is believed to be the active herbicidal form. This sulphoxide metabolite is believed to have an inhibitory affect on the various elongases involved in very long chain fatty acid synthesis (precursors of wax and suberin). This inhibition is thought to be selective, no effect having been reported on *de novo* synthesis of palmitate and stearate [3]. The effects of the thiocarbamate herbicide, pebulate (S-propyl butylethylthiocarbamate) and its sulphoxide on very long chain fatty acid (VLCFA) synthesis were investigated. Experimental data showed, both *in vivo* and in microsomal fractions from germinating pea seeds, that the sulphoxide caused a much greater inhibition of VLCFA synthesis, than the parent molecule. This suggested that the sulphoxide was the active herbicidal form. Subcellular fractionation studies indicated that the elongases are associated with the endoplasmic reticulum. The secondary effects on the wax quantity and quality due to this direct inhibition of the various elongases were confirmed by scanning electron microscopy.

Materials and Methods

Pebulate and its sulphoxide, a kind gift from C.J. Spillner (ICI Americas), were dissolved in ethanol (2.5% (v/v) final concentration). Systems which had been previously shown to give good rates of VLCFA synthesis were used - germinating pea seeds and young barley leaves. Barley leaves were pre-incubated for 8h with herbicide followed by an incubation of 15h in the presence of 2μCi of the radiolabelled precursor [1-^{14}C]acetate for 24h. Radiolabelled fatty acids were extracted and analysed by radio-g.l.c. in glass columns packed with SP-2100, using a temperature programme (210°C for the first 10 min, then 4°C/min up to 250°C)

Microsomal fractions were prepared and incubated for 3h with 0.05μCi [2-^{14}C]malonyl-CoA and 9μM stearoyl-CoA or arachidoyl-CoA, in the presence of the appropriate precursors [4]. Reactions were stopped by addition of 60% KOH followed by heating for 30min at 70°C. Fatty acids were extracted and bromophenacyl derivatives prepared. These were

J.-C. Kader and P. Mazliak (eds.), Plant Lipid Metabolism, 115–117.

analysed by reverse-phase HPLC at 242nm using a gradient of 80/20% acetonitrile /water increasing linearly to 100% acetonitrile over 20min (flow rate 2.0ml/min, column temp. 25°C).

Microsomal fractions were loaded onto a discontinuous gradient of 5/15% Ficoll suspended in 0.32M sucrose - 2mM Tris-HCl, pH 7.4 and centrifuged at 23,000rpm for 15min (4°C). NADH:cyt c reductase was assayed spectrophotometrically at 550nm [5]. TPPase and AMPase were assayed as described [5].

Results and Discussion

The effects of pebulate on the % fatty acid radiolabelling in germinating pea seeds and young barley leaves may be seen in Table 1. It can be surmised, assuming that separate enzyme systems are involved in VLCFA synthesis, that stearate elongation is not as sensitive to pebulate as are the elongation of arachidic and behenic acids. When pebulate sulphoxide was incubated with both systems (Table 1), at comparable concentrations to its parent compound, it appeared to be a more potent inhibitor of VLCFA synthesis. It was further noticeable that neither pebulate or its sulphoxide had an effect upon *de novo* syntheis of palmitate or stearate, in keeping with the specific effects reported for other thiocarbamate herbicides i.e. no effect upon *de novo* synthesis but inhibition of VLCFA synthesis.

This inhibition of the VLCFAs was found to have secondary effects upon the quantity and quality of the epicuticular wax structure when electron micrographs (not shown here) of both barley and pea leaf controls were compared with treated samples.

Table 1 - The effect of Pebulate and its sulphoxide on % fatty acid labelling in germinating pea seeds and young barley leaves
P = Pebulate ($1x10^{-5}$M), PS = Pebulate sulphoxide ($1x10^{-5}$M), n.d. = none detected

	≤18:0	20:0	22:0	24:0
Pea				
control	82±4	7±1	7±2	5±2
P	88±4	8±1	4±1	nd
PS	98±4	2±2	nd	nd
Barley				
control	62±7	5±1	19±3	14±4
P	77±2	5±0	12±1	6±1
PS	88±6	2±1	5±4	4±4

The effects of these Pebulate and its sulphoxide on the individual elongases involved in VLCFA synthesis were investigated further in microsomal fractions from germinating pea seeds. It was evident (Table 2), that pebulate sulphoxide caused greater inhibition of both arachidic and behenic acid synthesis *in vitro* than its parent compound. This isin keeping with the hypothesis that the sulphoxide metabolite is the active form of the herbicide.

Subcellular fractionation of the germinating pea seed was performed to enable us to elucidate the location of these enzymes involved in VLCFA synthesis (Fig. 1). It was concluded that the elongases responsible for the synthesis of arachidate and behenate respectively were associated with the endoplasmic reticulum.

Table 2 - Inhibition fatty acid synthesis by Pebulate or Pebulate sulphoxide *in vitro*
P = Pebulate ($2.5x10^{-5}M$), PS = Pebulate sulphoxide ($2.5x10^{-5}M$), VLCFAs = very long chain fatty acids

Elongase substrate		Radioactivity (d.p.m)			
		20:0	22:0	24:0	total VLCFAs
18:0-CoA	control	976	425	226	1627
	P	1363	384	128	1875
	PS	451	92	41	584
20:0-CoA	control	166	823	64	1053
	P	141	1232	94	1467
	PS	69	383	74	526

Figure 1 - Subcellular fractionation studies to show A) stearoyl-CoA elongase activity and B) arachidoyl-CoA elongase activity
ER = endolpasmic reticulum, GA = Golgi apparatus, PM = plasma membrane

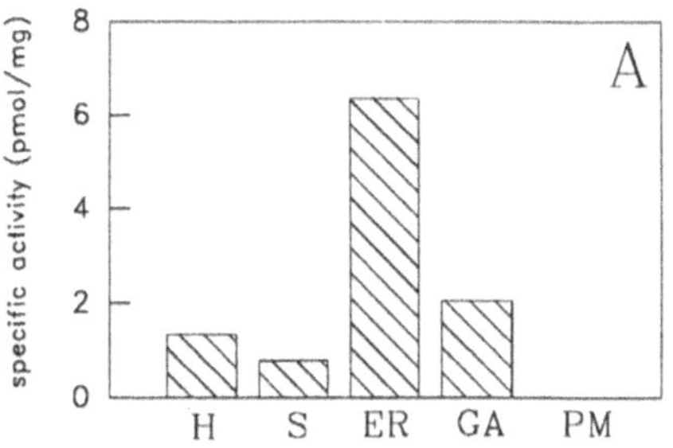

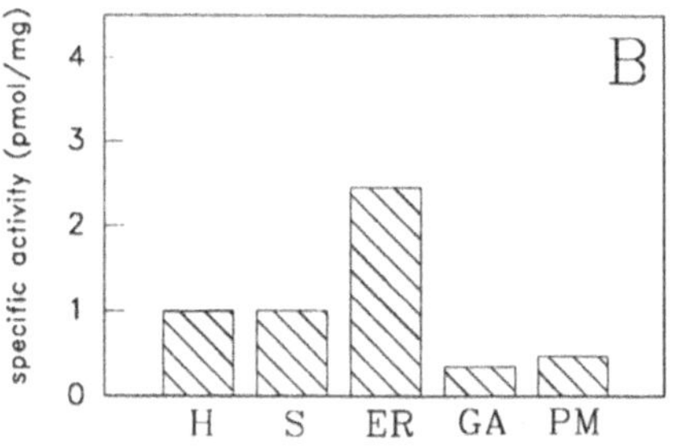

Acknowledgement

We are grateful to the SERC for a studentship to PBB.

References

1. Still GG, Davis DG, Zander GL. Plant epicuticular lipids: alteration by herbicidal carbamates. Plant Physiol 1970;46:307-314.
2. Hathaway DE. Molecular mechanisms of herbicide selectivity. Oxford and New York: Oxford University Press, 1989.
3. Abulnaja KO, Tighe CR, Harwood JL. Inhibition of fatty acid elongation provides a basis for the action of the herbicide, ethofumesate, on surface wax formation. Phytochem 1992;31:1155-1159.
4. Barrett PB, Harwood JL. The basis of thiocarbamate action on surface lipid synthesis in plants. Brighton Crop Protection Conference - Weeds 1993:183-188.
5. Hobbs MC, DeLarge MHP, Baydoun EA-H, Brett CT. Differential distribution of a glucuronyltransferase, involved in glucuronoxylan synthesis, within the Golgi apparatus of pea (*Pisum sativum* var. Alaska).Biochem J 1991;277:653-658.

ELONGATION SYSTEM INVOLVED IN THE BIOSYNTHESIS OF VERY LONG CHAIN FATTY ACIDS IN *BRASSICA NAPUS* SEEDS: CHARACTERIZATION AND SOLUBILIZATION

Hiroyuki Imai, Alenka Hlousek-Radojcic, Andrea Matthis and Jan Jaworski
Department of Chemistry, Miami University, Oxford, Ohio 45056 USA

Erucic acid is synthesized by extending oleic acid, presumably as oleoyl-CoA, through two rounds of elongase reactions involving the addition of two carbons each cycle. This elongase catalyzes at least four reactions required for the extension, reduction, and dehydration of the acyl-CoA. In this communication, we report a preliminary characterization and solubilization of elongase from developing seeds of high erucic acid variety of *Brassica napus*.

We were able to obtain comparable *in vitro* elongation activities (5 nmol/min/g fresh weight) to *in vivo* rate of the accumulation of erucic acid (6 nmol/min/g fresh weight (as calculated from data in [1])). The distribution of the total elongation activity between the oil body fraction and microsomal pellet was analyzed (Table 1). We detected 50-70% of the total elongation activity in the floating oil body fraction. We measured the activity of the cholinephosphotransferase (CPT) activity, a microsomal marker enzyme, to determine the extent to which endolpasmic reticulum contaminated the oil body fraction. The CPT

Table 1. Distribution of elongase and cholinephosphotransferase (CPT) activities in homogenates from 1 g of developing seeds of *Brassica napus*

The elongase activity was measured in the reaction mixture (25 µl) containing 80 mM HEPES-KOH (pH 7.2), 20 mM $MgCl_2$, 1 mM ATP, 500 µM NADPH, 15 µM oleoyl-CoA and 100 µM [2-^{14}C]malonyl-CoA, and usually 5 µl of enzyme. The CPT activity was measured as described in [2]. Protein concentrations were determined using the BCA protein assay (PIERCE) with bovine serum albumin as a standard.

Fraction	Elongase activity (nmol/min)	CPT activity (nmol/min)	Protein (mg)	Specific activity of elongase (nmol/min·mg protein)
Microsome	2.6	3.1	7.5	0.35
Oil body fraction	5.0	0.75	0.78	6.4

J.-C. Kader and P. Mazliak (eds.), Plant Lipid Metabolism, 118–120.

activity was four fold higher in the microsomal fraction compared to the oil body fraction. In addition, the specific activity of the elongase reaction in the oil body fraction (6.4 nmol/min/mg protein) was eighteen fold greater than the specific activity of the microsomal pellet (0.35 nmol/min/mg protein). This suggests that most of the elongase activity in the oil body fraction was not due to contamination by the endoplasmic reticulum.

The elongation activity was solubilized from the oil body fraction with zwittergent 3-10. Solubilization was confirmed both by high speed centrifugation and by gel filtration chromatography. At the zwittergent concentration of 25 mM, about 85% of the elongation activity remained in the supernatant after a 100,000 *g* spin for 1 hour (Fig. 1), and the elongase activity eluted as a major peak at approximately 60 kDa from a 15 x 450 mm column of Superose 6 gel filtration (Fig. 2).

We analyzed the products of elongase by using $NaBH_4$ reduction (Fig. 3). The enzyme reaction was terminated by addition of 0.5 ml of 0.1 M K_2HPO_4, 0.4 M KCl, 30% tetrahydrofuran and 5 mg/ml $NaBH_4$ as described by Garwin et al. [3]. After addition of toluene, the upper phase of the resulting two phase mixture was taken and applied to a silica gel thin layer plate. The plate was developed in a mixture of diethyl ether-concentrated NH_4OH (100/1, v/v). A major product of elongase in oil body fraction was a monol produced by the reduction of a saturated acyl-thioester (lane 1). Omission of NADPH from the reaction mixture resulted in the formation of diol which is the expected product from the reduction of 3-ketoacyl thioester (lane 2). The major product of elongase in the activity peak at 60 kDa from the gel filtration column was the diol, whether the enzyme was analyzed in the presence (lane 3) or in the absence (lane 4) of NADPH. A similar result was obtained by analyzing the products of elongase before gel filtration. This result suggested that an acyl condensing enzyme was solubilized with the detergent and separated from the other component of the multicomponent elongase. These findings are in contrast to an earlier study in which elongase activity solubilized from a microsomes with Triton X-100 retained all the component reactions [4].

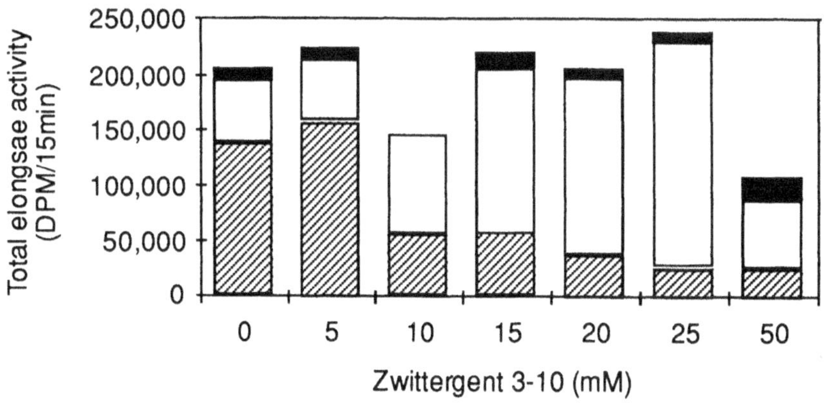

Fig. 1 Effect of increasing zwittergent 3-10 concentration on the solubilization of the elongase activity from the oil body fraction. The elongase activity was measured after a 100,000 *g* spin. Legend: floating oil body fraction, hatched bar (▨); supernatant, hollow bar (□); pellet, solid bar (■).

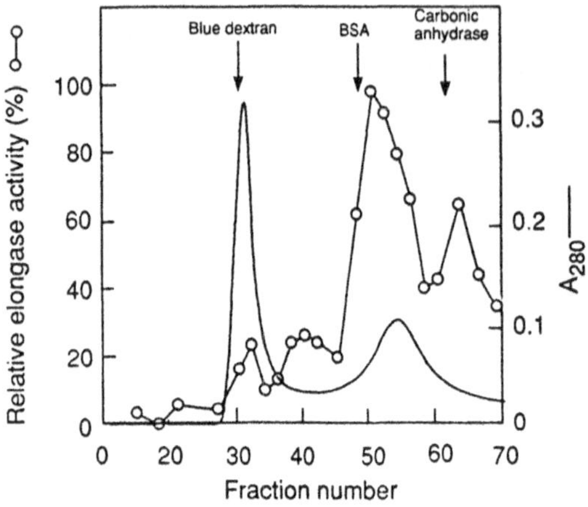

Fig. 2 Gel filtration chromatography on Superose 6 of the elongase in the oil body fraction.

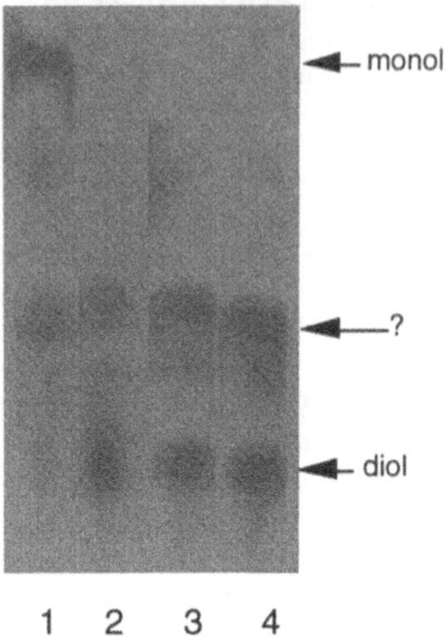

Fig. 3. Analysis of the product of elongase in oil body fraction and in the activity peak at 60 kDa from a Superose 6 gel filtration column. Lane1,2, oil body fraction, plus or minus NADPH. Lane 3, 4, elongase activity after gel filtration, plus or minus NADPH.

References

1. Slabas AR, Sidebottom CM, Hellyer A, Kessel RNJ, Tombs MP. Induction, purification and characterization of NADPH-specific enoyl acyl carrier protein reductase from developing seeds of oil seed rape (*Brassica napus*). Biochim Biophys Acta 1986, 877; 271-280.

2. Thomas S, Moore J. Phosphoglyceride synthesis in endoplasmic reticulum. In: Packer L, Douce R editors. Methods in Enzymology, vol 148. Academic Press, 1987; 585-596.

3 Garwin JL, Klages AL, Cronan JJE. Structural, enzymatic, and genetic studies of ß-ketoacyl-acyl carrier protein synthases I and II of *Escherichia coli*.. J Biol Chem. 1980, 255; 11949-11956.

4 Créach A, Lessire R. Solubilization of acyl-CoA elongases from developing rapeseed (*Brassica napus* L.). J. Amer Oil Chem Soc. 1993, 70; 1129-1133.

Study of the partially purified C18:1-CoA elongase from developing rapeseeds *(Brassica napus L.)*

Créach Anne, Domergue Frédéric and Lessire René

*Laboratoire de Biogenèse Membranaire, URA 1811-CNRS, Université de Bordeaux II, 1 Rue Camille Saint Säens, 33077 Bordeaux Cedex

Introduction :

Erucic acid is the main constituent of the triacylglycerol from rapeseed oil, it results from the elongation of oleoyl-CoA by malonyl-CoA (1). The acyl-CoA elongase, the enzyme responsible for the erucic acid synthesis, is mainly located in the ER (2). Since these intermediates have been evidenced in L. annua (3), these results supporte that the elongation process could involve the following reactions : i) condensation leading to the formation of a 3-keto acyl-CoA, ii) reduction of the keto acyl-CoA into a 3-OH acyl-CoA, iii) dehydratation of the 3-OH acyl-CoA and, finally, reduction of the trans 2-3 enoyl-CoA. Nevertheless, the organization, the structure and the different reactions involved in the elongation process remain largely unknown. This is due for a large part to the difficulties encountered during the purification of the membrane-bound enzymes. The solubilization of acyl-CoA elongases from oil seeds has been reported (4,5), and only one purification has been attempted leading to the obtention of a partially purified enzyme (4).

Here, is reported a purification protocol leading to the obtention of an acyl-CoA complex from developing rapeseeds highly delipidated and largely purified.

Materials and Methods :

Eight weeks old rapeseeds were from INRA (Le Rheu, France). 15000g particulate fraction was prepared as described previously (6).The elongation activity was measured using [2-^{14}C]malonyl-CoA [17 µM] and oleoyl-CoA [9µM] in the presence of NADH and NADPH (6).

The membrane proteins, solubilized, using Triton X-100 protein ratio of 2.5 according to Créach and Lessire (5) were subjected to a C_{16}-CoA-agarose chromatography using a Hepes 0.08 M pH 7.2 containing 0.02% Triton X-100, 10 mM ß mercaptoethanol and 10% ethylene glycol (Buffer A). The acyl-CoA elongase was eluted using a 0.5 M NaCl concentration. Then, the acyl-CoA elongase enriched fractions were loaded onto an Ultrogel AcA34 column previously

J.-C. Kader and P. Mazliak (eds.), Plant Lipid Metabolism, 121–123.

equilibrated with the buffer A and calibrated with molecular mass markers. Finally, the proteins were submitted to a DEAE 52 chromatography. The acyl-CoA elongase was eluted using a 0-1M NaCl gradient in the buffer A.

The lipids were analyzed by tlc and quantified using the method reported by Heape et al. (7). The acyl-CoA fraction and intermediates were reported and characterized by tlc as reported (6).

Electrophoresis analysis and immunoblotting experiments have been performed in the conditions described in reference (8). Proteins were estimated using Bradford's protocol (9).

Table 1: Properties of the purified acyl-CoA elongase from developing rapeseeds.

Molecular mass	270kDa
Multienzyme complex	3 subunits
Lipid content	<0.1µg
Substrates	C18:1-CoA>C18:0-CoA>>C16:0-CoA
Reaction Products	3-OH acyl-CoA (92%) and acyl-CoA (8%)
Phospholipid activation	PC (20µg) +PE (5µg)
Antibody reactivity	Recognition by anti leek elongase

Results :

The elution profile of the C16-CoA agarose chromatography shows that the majority of proteins was not retained on the column and that no acyl-CoA elongase activity was associated with these proteins. The acyl-CoA elongase activity was eluted as a sharp peak using a 0.5 M NaCl concentration. During the second chromatography, the proteins associated with the acyl-CoA activity were eluted close to the void volume. In our conditions, the estimated molecular mass was about 270 kDa. This step is very important because the enrichment of the activity was 59 fold higher compared to that measured using native membranes. But the presence of contaminant activities like acyl-CoA thioesterase led to the addition of an other step of purification. The proteins non retained onto the DEAE column did not present acyl-CoA activity, and the elongase necessited a 0.3-0.4 M NaCl concentration to be eluted. The activity presents 2 peaks which were considered as the purified enzyme.

The purified enzyme has been particularly studied and its principal properties are summarized in Table 1.

The lipid amount of the purified acyl-CoA elongase has been determined. Assuming, that the lipid measurement was able to detect a minimum of 1µg, no lipid was evidenced in the purified enzyme fraction reflecting a strong delipidation. Preliminary reconstitution experiments of the acyl-CoA elongase have been carried out. The addition of PC (20µg) led to an increase of the activity of 38%. The acyl-CoA elongation was doubled when PC (20µg) and PE (5µg) were both added.

The acyl-CoA (purified) elongase was analyzed by SDS-electrophoresis. The Coomassie blue staining showed the presence of only 3 major protein bands of 61,62 and 67 kDa molecular mass. These data were compatible with the molecular mass estimated by the gel filtration for the acyl-CoA elongase embedded in a Triton micelle and suggest that the acyl-CoA elongase is a multienzyme complex.

Moreover, the anti-leek acyl-CoA elongase antibodies were able to recognize the enzyme from rapeseeds. Immuno-blotting experiments showed that the proteins constitutive of the purified fraction were able to bind the anti-leek acyl-CoA elongase. These results strongly suggest that common epitopes could exist between both elongases.

Acknowledgements:

This work was supported by ONIDOL and CETIOM. We thanks very much Dr M. Renard and H. Picault for providing the rapeseeds.

References:

1. Agrawal V.P., Stumpf P.K., Elongation systems involved in the biosynthesis of erucic acid from oleic acid in developing Brassica juncea seeds,Lipids,1985;20:361-366.

2. Whithfield H.V., Murphy D.J., Hills M.J., Sub-cellular localization of fatty acid elongase in developing seeds of Lunaria annua and Brassica napus, Phytochem.,1993;32:255-258.

3. Fehling E., Mukherjee K.D., Acyl-CoA elongase from higher plant (Lunaria annua): Metabolic intermediates of very long chain acyl-CoA products and substrate specificity, Biochim. Biophys. Acta,1991;1082:239-247.

4. Fehling E., Lessire R., Cassagne C., Mukherjee K.D., Solubilization and partial purification of constituents of acyl-CoA elongase from Lunaria annua, Biochim. Biophys. Acta,1992;1126:88-94.

5. Créach A., Lessire R., Solubilization of acyl-CoA elongases from developing rapeseed (Brassica napus), J. Am. Oil. Chem. Soc.,1993;70:1129-1133.

6. Créach A., Lessire R., Cassagne C., Kinetics of C18:1-CoA elongation and transacylation in rapeseeds, Plant Physiol. Biochem.,1993;31:923-930.

7. Heape A.M., Juguelin H., Boiron F., Cassagne C., Improved one dimensional thin layer chromatographic technic for polar lipids, J. Chromatogr.,1985;322:391-395.

8. Bessoule J.J., Lessire R., Cassagne C., Partial purification of the acyl-CoA elongase of Allium porrum leaves, Arch. Biochem. Biophys.,1989;268:475-484.

9. Bradford M.M., A rapid and sensitive method for the quantification of microgram quantities of protein utilizing the principle of protein-dye binding, Anal.Biochem., 1976;72:248-254.

Characterization of the trans 2-3 enoyl-CoA reductase of the acyl-CoA elongase from leek (Allium porrum L.).

Spinner Christel*, Levannier Karine+, Lessire René*, Lellouche Jean-Paul+, Mioskovski Charles+ and Cassagne Claude*.

*Laboratoire de Biogenèse Membranaire, URA 1811-CNRS, Université de Bordeaux II, 1 Rue Camille Saint Säens, 33077 Bordeaux Cedex and +Bat 574, DBMCM/SMM, CEN Saclay, 91191 Gif sur Yvette Cedex, France.

Introduction :

The acyl-CoA elongase, responsible for the synthesis of very long chain fatty acids (VLCFA), has been purified from leek (1) Lunaria annua (2) and rapeseeds (3). In both cases the molecular mass of the Triton micelle containing the acyl-CoA elongase was estimated by gel filtration chromatography as about 300kDa. The SDS-Page electrophoresis analysis has shown the presence of 3 to 4 different proteins, having 59 to 65 kDa as molecular mass (1,2), strongly suggesting that the acyl-CoA could be a multienzyme complex. It was generally admitted that the elongation process involved the following reactions: condensation of the acyl-CoA to malonyl CoA to form a 3-ketoacyl-CoA, reduction of this 3-ketoacyl-CoA, dehydratation of the resulting 3-OH acyl-CoA and finally reduction of the trans 2-3 enoyl-CoA (4). An important point concerning the knowledge of the elongation mechanism and the structure of the acyl-CoA elongase is to determine the nature of the intermediate reaction catalyzed by each protein constituting the complex. The difficulty was the lack of acyl-CoAs susceptible to be the substrates of the intermediate reactions. The enoyl-CoA reductase from etiolated leek seedlings has been characterized using trans 2-3 C20-CoA synthesized chemically.

Materials and Methods :

The trans 2-3 enoyl-CoA has been synthesized chemically by K. Levannier and J.P. Lellouche (CEA Saclay). The microsomes, used as enzyme source, were prepared from 7-day old leek seedlings as described earlier (5).The elongation activity was measured using [2-^{14}C]malonyl-CoA [17 µM] and acyl-CoA [9µM] in the presence of DTT, MgCl2, NADH, NADPH (1) and 60µg of proteins. The fatty acids were extracted and the radioactivity was measured.

J.-C. Kader and P. Mazliak (eds.), Plant Lipid Metabolism, 124–126.

A spectrophotometric measurement of the reductase activity based on the decrease of the A340 due to the oxydation of NADPH or NADH was also developed. A typical assay contained: 100µg of proteins, NADH or NADPH (50µM), Triton X-100 (50µM), MgCl2 (1mM) and trans 2-3 enoyl-CoA (9µM).

Results :

Trans 2-3 C18:1-CoA elongation: The elongation of the trans 2-3 C18:1-CoA by the [2-14C] malonyl-CoA in the presence of NADH and NADPH was compared to the VLCFA synthesis obtained using C18:0-CoA as precursor. The results reported in the Table 1, show that the elongation of the unsaturated acyl-CoA was about half that observed with C18:0-CoA. A first improvement was obtained by the addition of Triton X-100 (50µM) in the reaction mixture, the elongation increased by about 3 fold and was of the same order of magnitude as with C18:0-CoA which is the best precursor of the saturated VLCFA (4). Moreover, after a 10 min pre-incubation of the trans 2-3 enoyl-CoA with the microsomes, the labelled malonyl-CoA was added and the reaction was running for one hour. Under these conditions, the level of the incorporation of [2-14C] malonyl-CoA in the VLCFA was the same whatever the precursor used. The tlc analysis showed that the reaction product was, as expected, a C22:0-CoA. This result demonstrates the involvement of the trans 2-3 enoyl-CoA in the elongation process.

Table 1: Elongation of C18:0-CoA and trans 2-3 C20:1-CoA by microsomes from leek seedlings.

Substrates	VLCFA synthesis (nmoles/mg/h)		
		+triton	+pre-incubation
C18:0-CoA	0.42	0.81	nd
trans 2-3 C20:1-CoA	0.25	0.72	1.15
C20:0-CoA	nd	nd	1.14

nd: not determined

The specificity towards the trans 2-3 enoyl-CoAs has been determined using a 9µM concentration of acyl-CoA. The incorporation of [2-14C] malonyl-CoA into VLCFA shows that the trans 2-3 C12:1 and trans 2-3 C16:1-CoA were poorly elongated compared to the trans 2-3 C20:1-CoA. The influence of NADH and NADPH on the elongation of trans 2-3 eïcosanoyl-CoA has been also studied. In the range of concentration studied (0 to 500µM), the elongase activity was always higher in the presence of NADPH than with NADH. The apparent Km calculated from these concentration curves were 9 and 35µM for the NADPH and NADH respectively, indicating that NADPH could be the preferred reductant of the enoyl-CoA reductase.

Table 2: Trans 2-3 enoyl-CoA elongation by etiolated leek seedling microsomes.

Acyl-CoAs	VLCFA synthesis (nmoles/mg/h)
trans 2-3 C12:1-CoA	nd
trans 2-3 C16:1-CoA	0.22
trans 2-3 C18:1-CoA	0.39
trans 2-3 C20-:1CoA	0.60

Spectrometric measurements: The trans 2-3 enoyl-CoA activity has been directly measured by the decrease of the absorbance due to NADH and NADPH oxidation at 340nm. Under these conditions, it has been shown than the NADPH disappearance was more rapid that of NADH. The rate of the reaction was the same for 15 min and depended on the amount of the proteins in the incubation medium. As previously demonstrated the trans 2-3 C20:1-CoA appeared also to be the best substrate for the reduction.

In conclusion, the trans 2-3 enoyl-CoA reductase, one of the proteins constituting the acyl-CoA elongase, has been evidenced in the leek microsomes. The preferred sustrate for this enzyme is the trans 2-3 C20:1-CoA and NADPH is the reductant the more efficient.

Acknowledgements:

This work was a part of the Bio-avenir program supported by Rhône Poulenc Agrochimie and CNRS. We thank very much Dr Cinti (Farmington State University) for providing the trans 2-3 C12 and C16 acyl-CoAs.

References:

1. Bessoule J.J., Lessire R., Cassagne C., Partial purification of the acyl-CoA elongase of Allium porrum leaves, Arch. Biochem. Biophys.,1989;268:475-484.
2. Fehling E., Lessire R., Cassagne C., Mukherjee K.D., Solubilization and partial purification of constituents of acyl-CoA elongase from Lunaria annua, Biochim. Biopys. Acta,1992;1126:88-94.
3. Créach A., Lessire R. ,Solubilization and partial purification of the acyl-CoA elongase from developing rapeseeds (Brassica napus L.), Grasas y Aceites,1994;44:120-122.
4. Fehling E., Mukherjee K.D., Acyl-CoA elongase from higher plant (Lunaria annua): Metabolic intermediates of very long chain acyl-CoA products and substrate specificity, Biochim. Biophys. Acta,1991;1082:239-247.
5. Lessire R., Juguelin H., Moreau P, Cassagne C., Elongation of acyl-CoAs by microsomes from etiolated leek seedlings, Phytochem.,1985;24:1187-1192.

MOLECULAR CLONING AND CHARACTERIZATION OF GENES INVOLVED IN CUTICULAR WAX BIOSYNTHESIS

Basil J. Nikolau[1], Xiaojie Xu[3], Yiji Xia[3], Joel D. Hansen[1], Shane Heinen[3], Tsui-Jung Wen[3], Massimo Delledonne[3] & Patrick S. Schnable[2,3].
[1]Dept. Biochem. & Biophys., [2]Dept. Agronomy and [3]Dept. Zoology & Genetics, Iowa State University, Ames, IA 50011.

The surfaces of the aerial parts of plants are coated with a layer of lipids collectively known as the cuticular waxes (1, 2). This first, outer barrier to the environment is thought to play an important, but as yet undefined role in the survival of the organism. The cuticular waxes are a complex mixture of very long chain (20 - 36 carbons) aliphatic compounds, including fatty acids, alcohols, esters, aldehydes, ketones, and alkanes, which are biosynthetically derived from palmitic acid (2). We are taking a molecular genetic approach to clone genes involved in the biosynthesis and/or deposition of these compounds. In maize, mutants have defined 17 loci that affect the appearance of seedling cuticular waxes (the *glossy* loci) (3). In *Arabidopsis,* a total of 21 loci that affect cuticular wax accumulation on the stems and siliques have been defined genetically (the *cer* loci) (4).

Molecular cloning and characterization of *glossy* loci

Using the maize *Mutator* system we have generated a large collection of transposon-induced mutations at seven known *glossy* loci: *gl1* (14 alleles), *gl2* (13 alleles), *gl3* (2 allele), *gl4* (5 alleles), *gl6* (3 alleles) *gl8* (60 alleles), and *gl11* (9 alleles). In addition, two new mutants not allelic to any previously described *gl* locus have been isolated (*gl25* and *gl26*). This collection of alleles is being analyzed for the purpose of identifying and cloning the specific *Mutator* element that

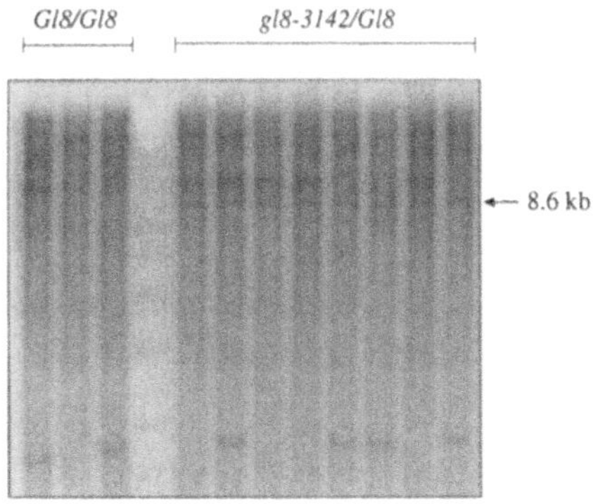

Fig. 1. Cosegregation of an 8.6-kb, *Mu8*-containing *Hin*dIII fragment with the *gl8-3142* allele in progeny from cross: *gl8-3142/Gl8* X *Gl8/Gl8*.

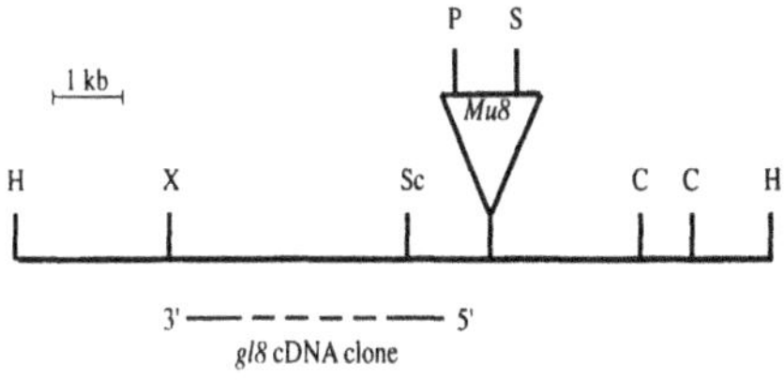

Fig. 2. Restriction enzyme map of the cloned 8.6-kb *Hin*dIII fragment containing the *gl8* locus and alignment of the sequence with the 0.8 kb *gl8* cDNA. C=*Cla*I, H=*Hin*dIII, P=*Pst*I, S=*Sal*I, Sc=*Sac*I, X=*Xho*I.

J.-C. Kader and P. Mazliak (eds.), Plant Lipid Metabolism, 127–130.

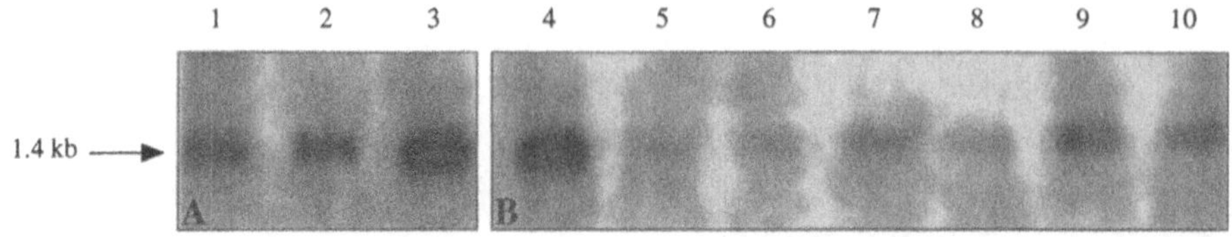

Fig. 3. Northern analysis of the accumulation of *gl8* mRNA. A. RNAs were isolated from seedlings with the following genotype: *gl8-3142* (1), *gl8-ref* (2), Q60, F2 plants (3). B. RNAs were isolated from the following organs of F2 plants of Q60: seedlings (4), mature leaf (5), mature husk (6), mature ear (7), endosperm (8), mature root (9) and seedling root (10).

generated the mutation.

These analyses have progressed furthest with the collection of 60 independent Mut*ator*-induced mutations at the *gl8* locus and 14 independent alleles of the *gl1 locus*. Southern blot analyses of DNA isolated from sibling plants carrying either the *Mutator*-induced *gl1* mutant allele, or the wild-type *Gl1* allele, identified a 12.5 kb, *Mu1*-containing *Hin*dIII fragment, which co-segregates with the mutant phenotype (5). This fragment is currently being further analyzed.

Southern blot analyses of DNA isolated from sibling plants carrying either the *Mutator*-induced *gl8-3142* mutant allele, or the wild-type *Gl8* allele, identified an 8.6 kb *Hin*dIII fragment, which contains the transposon *Mu8*, that cosegregates with the mutant allele (Fig. 1). This DNA fragment, putatively identified as containing at least a portion of the *gl8* locus, was cloned in a la*mbda* vector. Figure 2 presents a restriction map of the cloned fragment.

A DNA fragment flanking the *Mu8* transposon was isolated from this clone and used to establish that the cloned sequence represents the *gl8* locus. Southern blot analyses revealed that 9 of 12 *gl8 -Mu* alleles exhibit size polymorphisms relative to their corresponding wild-type progenitor alleles. The cloned *gl8* fragment was used as a probe in northern analyses of RNA isolated from maize seedlings. These analyses and sequencing of genomic and cDNA clones enabled us to map the *gl8* gene on the cloned fragment (Fig. 2).

Northern analyses of RNA isolated from various organs of maize identified the *gl8* mRNA as a 1.4 kb band (Fig. 3). In homozygous *gl8/gl8* seedlings the accumulation of the *gl8* mRNA is reduced to between 1/3 and 1/5 of the level in found in wild-type seedlings (Fig. 3). In wild-type plants the *gl8* mRNA is detectable in all organs examined, however, maximum accumulation occurs in seedling tissue (Fig. 3).

A 0.8-kb *gl8* cDNA clone was isolated from a library prepared by Dr. Alice Barkan, from two-week-old B73 maize seedlings. The isolated cDNA was completely sequenced and was compared to previously sequenced genes. This analysis revealed that the deduced amino acid sequence of the *gl8* protein shows significant similarity to 3-ketoacyl reductases. For example, the *gl8* sequence shows similarity to 3-ketoacyl-ACP reductase of *Escherichia coli* (60% similarity). We speculate that the *gl8* gene codes for a ketoacyl reductase that is involved in the elongation reactions required for the biosynthesis of very long chain fatty acids. Additional research is required to elucidate the structure and regulation of the enzyme encoded by the *gl8* gene. Furthermore, the cloned gene will provide us with the means of identifying the substrate specificity of the enzyme, in particular, the preferred chain length of the of the acyl substrate, and the thioester form of the substrate.

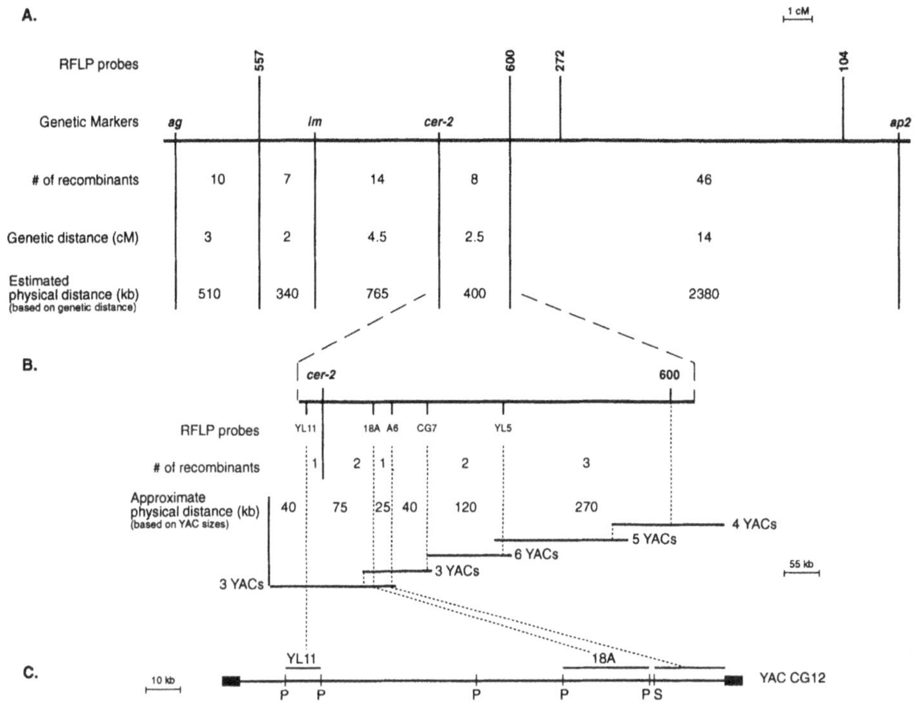

Fig. 4. A chromosome walk to the *cer2* locus of *Arabidopsis*.

Progress towards cloning the *cer2* locus of *Arabidopsis*.

We have carried out a "chromosome walk" on chromosome 4 of *Arabidopsis* to the cer2 locus. This walk encompasses five steps covering a genetic distance of 2.5 cM from the RFLP probe λbAt600 to *cer2*. Figure 4 **schematically illustrates the walk.** The estimated genetic and physical distances between RFLP and visible genetic markers and the numbers of genetic recombinants isolated within each interval are shown in panel A. A summary of the chromosome walk is shown in panel B, in which vertical dotted lines connecting YACs and chromosome 4 depict cross-hybridization. The probe 18A maps to the right of two of the eight recombination breakpoints that were recovered between λbAt600 and *cer2*. Similarly, probe YL11 maps to the left of one of the 14 recombination breakpoints between *cer2* and *im*. These results establish that the *cer2* locus is located within a 75 kb interval of YAC CG12. This 75 kb cloned segment has been subcloned into the cosmid vector pLZO3, which is suitable for *Agrobacterium*-mediated transformation of *Arabidopsis*. Seventeen overlapping subclones encompassing the 75 kb interval containing the *cer2* locus have been transformed into homozygous *cer2* plants. Transgenic plants are being regenerated for the purpose of screening to identify the DNA fragment(s) that are able to complement the *cer2* phenotype; such fragments must include the *cer2* locus.

References

1. Tulloch AP. Chemistry of waxes of higher plants. In: Kolattukudy PE, editor. Chemistry and biochemistry of natural waxes. New York: Elsevier, 1976:235-287.
2. Kolattukudy PE. Biochemistry of plant waxes. In: Kolattukudy PE, editor. Chemistry and biochemistry of natural waxes. New York: Elsevier, 1976:289-347.
3. Coe EH, Neuffer MG, Hoisington DA. The genetics of corn. In: Sprague GF, Dudley JW, editors. Corn and corn improvement, 3rd edition. Madison: American Society of Agronomy, 1988:81-258.
4. Koornneef CJ, Hanhart J, Thiel F. A genetic and phenotypic description of *Eceriferum* (*cer*) mutants in *Arabidopsis thaliana*. J. Heredity 1989;80:118-122.
5. Schnable PS, Nikolau BJ, Stinard PS, Xia Y, Hansen JD, Xu X, Heinen S, Delledonne M, Myszewski, ME. Genetic Approaches to isolating maize and *Arabidopsis* genes involved in cuticular wax biosynthesis. In: Murata N, Somerville C, editors. Biochemistry and molecular biology of membrane and storage lipids of plants. American Society of Plant Physiologists Press, 1993:196-206.

DESIGN OF NEWS MODULATORS OF THE VERY-LONG-CHAIN FATTY ACIDS BIOSYNTHESIS.

Schneider F[1], Bessoule J.J[1], Mioskowski C[2], Lellouche J.P[2], Levannier K[2], Cassagne C[1].
[1]Laboratoire de biogenèse membranaire, 1 rue Camille Saint-Saëns, 33077 Bordeaux Cedex.
2 Centre d'études nucléaires de Saclay. SMM/DBCM/DSV. Batiment 547, 91191 Gif-sur-Yvette Cedex.

Introduction

The natural waxes play an important role, not only in controlling the water balance of the higher plants, but also in the penetration of xenobiotics. Attempts to improve the drug penetration rely on the control of waxes production As elongases are key enzymes in the synthesis of the aliphatic components of natural waxes, modulating very-long-chain fatty acids (VLCFA) production will have repercussions on the rate of wax synthesis.
VLCFA synthesis is sensitive to cerulenin (1), but it is well-known that cerulenin is also a potent inhibitor of the fatty-acid"de novo" synthesis. As a consequence, a new strategy is being developed to elaborate specific inhibitors of VLCFA synthesis. To study the structural factors affecting the inhibition, we analyse two types of inhibitors : those with a variable chain length, but having a common polar head (e.g cerulenin and higher homologs) and those with fixed chain length and variable polar heads.

Morisaki et al synthesized cerulenin analogs having different chain lengthes and tested them on the Yeast Fatty Acid Synthase (FAS). They showed that cerulenin with 12 or 16 carbon atoms (respectively Cer and Cer 16) could inhibit the FAS but an analog with 18 carbon atoms (Cer 18) was uneffective .They proposed that the site of Yeast FAS could not accept the higher analog (2).

Results

For our part, we tested the effect of these cerulenin analogs on the three elongation systems present in leek seedlings microsomes (3). The different homologs of cerulenin are effective on C_{18}-CoA elongation in the microsomes of leek seedlings. The C_{20}-CoA elongase was sensitive to those three molecules, but to a lesser extent to cer 18. As the ATP-dependent elongation activity is highly reduced in presence of cer, cer 16 and cer 18, it appears that in presence of controlled amounts of these molecules, it is possible to modulate all the elongase activities of leek seedlings microsomes. We observed that for cerulenin, a preincubation time of 20 minutes was necessary to obtain maximal inhibition of the VLCFA synthesis (1). For analogs with 16 and 18 carbon atoms, a preincubation time of 45 minutes was require to reach a maximal inhibition.
Since the cerulenin analogs have the same polar head as cerulenin, they are likely to affect chiefly the condensation step. Accordingly, cer 16 and cer 18, exactly as

J.-C. Kader and P. Mazliak (eds.), Plant Lipid Metabolism, 131–132.

cerulenin, inhibit the first reaction involved in the elongation : the 3-ketoacyl-CoA synthase.
The sensitivity of the leek seedling FAS to these analogs was tested : whereas the FAS activity was, as expected inhibited by 90% with 50 μM control cerulenin, the replacement by the Cer 16 analog led to a lesser inhibition (60% at 50 μM). But the most interesting- and promising - result was that cerulenin with 18 carbon atoms had no inhibitory effect on the "de novo" synthesis.

Another strategy to design specific inhibitors of VLCFA synthesis was developed. Different molecules having the same length chain (20 carbon atoms) were synthesized chemically. Three structures were studied : dioxolane amide (7III), dioxolane acid (177II), dioxolane-CoA.
Of the three molecules tested, only the CoASH ester of dioxolane was effective on the elongation of C_{18}-CoA (50% inhibition at 90 mM). As for cerulenin, a preincubation of microsomes with the effectors raised the inhibition level. At 15 μM, in absence of preincubation, only 8% of inhibition was obtained, whereas after a 20 minutes preincubation of the same molecule at the same concentration, elongation was inhibited by 50%. Addition of Triton X100 did not increase inhibition level.
The ATP-dependent elongation was also sensitive to dioxolane-CoA.
The analysis of reaction products by thin layer chromatography suggested that this molecule could act on the condensation reaction. This was demonstrated directly by measuring only the 3-ketoacyl-CoA synthase in presence of increasing concentrations of dioxolane-CoA : the level of inhibition was similar to that observed on the overall elongation process.
Moreover, the leek seedling FAS activity was not inhibited by this molecule. On the contrary, this activity seemed to increase as a function of dioxolane-CoA concentration.

In conclusion, we demonstrate that "de novo" fatty acid synthesis and VLCFA synthesis could be distinguished using different effectors. Cerulenin with 18 carbon atoms and dioxolane-CoA with 20 carbon atoms appear to be specific inhibitors for all elongation systems in the microsomes of leek seedlings and specifically on the condensation enzymes of these elongases. This differential effect may be due to the chain length of the effectors for cerulenin 18 as for dioxolane-CoA. In addition, the preliminary results reported above strongly suggest that a CoA-like structure is required for the inhibition in the case of the homologs of dioxolane.

(1) Schneider F, Lessire R, Bessoule J-J, Juguelin H, Cassagne C. Effect of cerulenin on the synthesis of very-long-chain fatty acids in microsomes from leek seedlings. Biochem Biophys Acta 1993;1152:243-252.

(2) Morisaki N, Funabashi H, Shimazawa J, Kawaguchi A, Okuda S, Iwasaki S. Effect of side-chain structure on inhibition of Yeast fatty-acid synthase by cerulenin analogues. Eur J Biochem 1993;211:111-115.

(3) Moreau P, Bertho P, Juguelin H, Lessire R. Intracellular transport of very-long-chain fatty acids in etiolated leek seedlings. Plant Pysiol Biochem 1988;26(2):173-178.

Acknowledgements : We are indebted to Dr Iwasaki for the gift of cerulenin analogs. This study has been conducted under the BIOAVENIR programme / Groupe de Recherche "Barrières Cuticulaires" financed by Rhône-Poulenc.

Section 2

Membranes

ULTRASTRUCTURAL FEATURES OF CHLOROPLASTS FROM FATTY ACID DESATURASE MUTANTS OF *Arabidopsis thaliana*.

Peter J. Quinn

Biochemistry Section, Division of Life Sciences, King's College London, Campden Hill, London W8 7AH, U.K.

Introduction

The chloroplasts of higher plant leaves are characterised by their ultrastructural morphology as revealed by thin-section electron microscopy. Three membranes can be distinguished by their location within the organelle, namely, inner and outer envelope membranes and the thylakoid membrane. Biochemical analyses of each of these membranes has revealed differences related to their particular functions but our knowledge of how the components are assembled in the membranes and the factors that regulate their interactions is still rudimentry. In this chapter the influence of polyunsaturated membrane lipids on the ultrastructure of higher plant chloroplasts will be described in an attempt to extrapolate the effects of interactions at a molecular level to assembly and morphology of the photosynthetic membrane.

Functions of Polyunsaturated Lipids

One of the characteristic features of membrane lipids of higher plant chloroplasts is their content of polyunsaturated fatty acids. In some plant species, such as *Vicia faba*, trienoic fatty acids may comprise 80% of more of the total fatty acyl residues of the chloroplast membrane lipids [1]. The presence of unsaturated membrane lipids is believed to be required to achieve a certain degree of fluidity in the membrane such that interactions between proteins that are required for processes like electron transport can proceed at meaningful rates. Other functions that have been suggested is that polyunsaturation of the nonbilayer-forming lipids such as the mono-galactosyldiacylglycerols is necessary to preserve their phase behaviour. Nonbilayer-forming lipids could function to promote oligomeric protein assemblies in formation of

J.-C. Kader and P. Mazliak (eds.), Plant Lipid Metabolism, 135–140.

complexes of protein through lateral interactions and to seal the irregular interface between the hydrophobic domains of intrinsic proteins and lipids thereby preventing the leakage of solutes across the membrane. In these functions the nonbilayer-forming lipids are constrained to adopt a bilayer configuration by their interaction with other membrane components.

CHARACTERISATION OF LIPID FUNCTION

Several strategies have been developed to characterise the structural and functional role of particular molecular species of membrane lipid. The use of phospholipid exchange or transfer proteins have been described to exchange one molecular species of a lipid class for another or enrich membranes in specific molecular species, respectivly. Another method has been to hydrogenate the double bonds of polyunsaturated membrane lipids *in situ* using homogeneous catalysts and to study the consequences of lipid saturation on membrane functions [2]. With the isolation of mutant strains of *Arabidopsis* defective in desaturation of chloroplast fatty acids [3,4] the question can be addressed directly without the problems associated with the use of exchange or transfer proteins or introduction of transition metal catalysts.

FATTY ACID DESATURASE MUTANTS OF *ARABIDOPSIS*

The JB67 mutant of *Arabidopsis* is characterised by an increased proportion of palmitic acid (16:0) and the unsaturated species of palmitic acid are correspondingly reduced in the thylakoid membrane lipids [3]. The defect has been shown to be the result of a single mutation of a nuclear gene. The reduced level of unsaturation of the 16-carbon acyl chains results in a slight decrease in CO_2 fixation and reduction is the rate of photosynthetic electron transport but there in an increased thermal stability of the photosynthetic apparatus at elevated temperatures.

Another mutant, LK3, is deficient in activity of ω-6 desaturase and palmitoleic and oleic fatty acyl residues dominate the composition of thylakoid galactolipids [4]. Biochemical studies of the mutant have shown that the reduced level of lipid unsaturation is associated with marked changes in protein and chlorophyll content of the thylakoid membranes and a reduction in photosynthetic electron transport rates. Like the JB67 mutant, photosynthetic electron transport of the LK3 mutant is more heat stable compared to the wild-type [6].

Characterisation of Membrane Morphology

While the effect of reduced levels of lipid unsaturation on chloroplast functions have been well characterised, less information is available regarding the molecular basis underlying these observed effects. Freeze-fracture electron microscopy has revealed differences in the size and distribution of membrane-associated particles in both exoplasmic (E) and protoplasmic (P) surfaces of membranes fractured laterally in stacked and unstacked regions of the thylakoid [5]. Biochemical studies are consistent with the lateral heterogeneity in the distribution of chlorophyll-protein complexes of photosystem-I and photosystem-II with photosystem-II localized in the appressed regions of the grana and photosystem-I confined predominantly to the non-appressed grana-end membranes and margines and stroma lamellae.

A detailed analysis of membrane-associated particle size and density in broad bean chloroplasts subjected to heat stress has been intepreted as an initial dissociation of the core particles of the photosynthetic complex of photosystem-II from their associated light-harvesting chlorophyll-protein complexes. Combined with data that showed phase separation of non-bilayer lipids in the membrane and that the tendency of MGDG to form non-bilayer structures was critically dependent on polyunsaturation of the hydrocarbon chains, it was suggested that polyunsaturated molecular species of monogalactolipid could function to "package" the complexes of the photosynthesic apparatus into the planar thlakoid membrane [6].

Thin-section Electron Microscopy

The ultrastructural features of chloroplasts in leaves of Columbia wild-type *Arabidopsis thaliana* (L) Heyn and mutant strains JB67 and LK3 isolated therefrom have been examined by thin section electron microscopy. Leaves from 3 week old plants grown at 22°C under standard conditions were processed in tendem. Tissue was fixed in 4% gluterdialdehyde, sectioned and stained with uranyl acetate and lead citrate. The sections were examined by transmission electron microscopy and morphometric features of thin sections through chloroplasts were analysed (see Table 1).

One of the most conspicuous differences between the wild-type and the mutants was the presence of large starch granules within the chloroplasts of the mutants and which were not observed in sections through chloroplasts of the wild-type. The most striking feature is a decrease in the length of the appressed membranes per plastid primarily due to a decrease in the number of thylakoids per granum and a

Table 1
Morphometric analysis of chloroplasts from *Arabidopsis thaliana* wild-type (w.t.) and Mutant strains deficient in fatty acid desaturation.

Measurement	w.t.	JB67	LK3
grana/plastid	58 ± 6.2	52 ± 7.1	64 ± 7.2
thylakoids/granum	5.8 ± 2.1	4.1 ± 1.8	3.9 ± 2.1
granal width (μm)	0.5 ± 0.09	0.37 ± 0.08	0.44 ± 0.07
stromal thylakoids/ plastid	97.7 ± 23.5	94.3 ± 13.1	101 ± 11.5
stromal thylakoid length (μm)	0.21 ± 0.07	0.17 ± 0.08	0.16 ± 0.03
appressed thylakoids/ plastid (μm)	118.3	64.6	71.2
non-appressed thylakoids/plastid (μm)	47.4	36.9	37.5
total thylakoids (μm)	165.7	101.5	108.7
appressed/non-appressed thylakoids	2.50	1.75	1.89
average cross-sectional area of chloroplast (μm^2)	11.7 ± 2.0	11.2 ± 2.7	9.8 ± 3.9

decrease in the granal width. The decrease in the granal size in LK3 is partially compensated by an increase in the number of grana per plastid.

The mutations also produce a decrease in the stromal membrane length. This results in a reduction of the average amount of the non-appressed thylakoid membranes leading to a decrease in the appressed-to-non-appressed membrane ratio. The reduction in the amount of the appressed and non-appressed membranes in JB67 and LK3 mutants corresponds to a decrease of 39% in the total membrane length per chloroplast in JB67 and 34% in LK3. Since there was no significant difference in the number of plastids per section for the three plants, the lower thylakoid membrane length reflects a decrease in the amount of photosynthetic membrane in the cell.

Freeze-fracture Electron Microscopy

The ultrastructural features, visualized by freeze-fracture electron microscopy, of chloroplasts isolated from leaves of *Arabidopsis* mutants harvested at the rosette stage [18] have also been compared. A striking feature of the fracture faces of stacked thylakoids in the two mutants is the presence of aligned membrane-associated particles together with regions of randomly distributed particles.

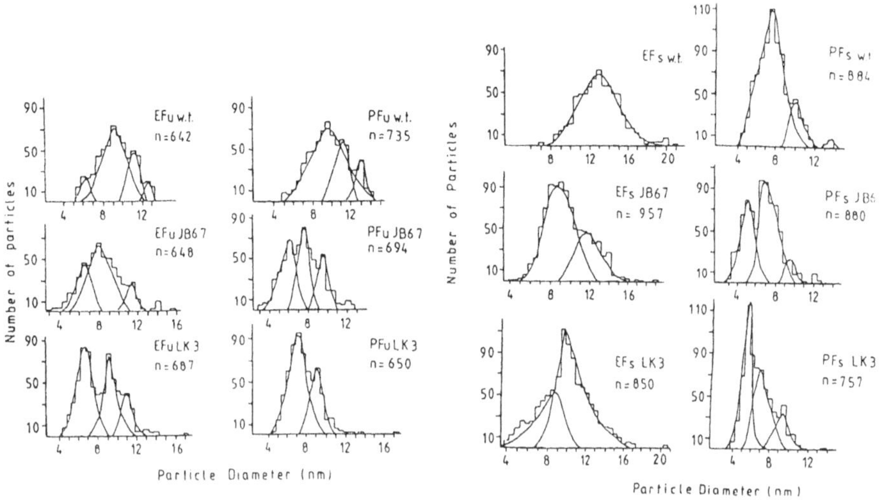

Figure 1. Histograms of particle size distribution in photosynthetic membranes of *Arabidopsis thaliana*

Aligned particles were never observed in wild-type thylakoids thermally quenched from 20°C but a tendency to form arrays was seen in suspensions thermally quenched from 0°C. In JB67 and LK3, the aligned particles on the EF_s fracture face were organized in a rectangular lattice with repeat distance along the two axes of 29 nm and 34 nm, and 31 nm and 36 nm, in the respective mutants. The complementary PF_s fracture face contained areas with aligned and randomly distributed particles. In JB67 the aligned particles were organized in single rows separated by a distance of 25 nm, while in LK3 they were arranged in double rows separated by equally spaced single depressions of 27 nm.

Histograms of particle size distribution observed in the wild-type and mutant fracture faces of thylakoid membranes are presented in Figure 1. A summary of the average particle diameters in the three plants is shown in Table 2 and respective particle densities are presented in Table 3. The general trend with this data is that the particle size in chloroplasts from the two mutant strains is lower and the density of particles correspondingly greater in the two mutant strains compared to the wild type. Because the protein composition remains relatively unaffected by the mutation one intepretation of the results is that when the level of unsaturation is decreased there is a tendency for the oligomeric complexes to dissociate into their composite parts. This would support the notion that a role of poly-

Table 2
Average particle diameter in wild type (w.t.) and mutant chloroplast membranes of *Arabidopsis thaliana*

Fracture Face	Average Diameter (nm) (+S.D.)		
	w.t.	JB67	LK3
EFs	13.2 ± 2.3	8.9 ± 1.2 aligned 12.3 ± 1.9 random	8.0 ± 1.6 aligned 10.8 ± 1.2 random
PFs	7.6 ± 2.7	7.2 ± 0.9 aligned 6.7 ± 2.4 random	7.5 ± 0.8 aligned 7.4 ± 2.1 random
EFu	8.8 ± 3.1	8.9 ± 2.5	8.3 ± 1.9
PFu	9.8 ± 2.5	7.4 ± 2.4	7.6 ± 1.5

Table 3
Average particle density in wild type (w.t.) and mutant chloroplast membranes of *Arabidopsis thaliana*

Fracture Face	Particle Density (particles/μm^2) (+S.D.)		
	w.t.	JB67	LK3
EFs	1527 ± 70	1469 ± 49 random 2262 ± 5 aligned	1281 ± 78 random 1818 ± 10 aligned
PFs	6234 ± 163	7583 ± 184 random 4374 ± 43 aligned	7229 ± 239 random 3586 ± 32 aligned
EFu	381 ± 64	449 ± 23	459 ± 28
PFu	4219 ± 98	6975 ± 217	6062 ± 49

unsaturated membrane lipids might be to preserve the interactions that take place between components of the photosynthetic complexes. The formation of particle arrays may result from failure of the components to interact in a manner similar to their behaviour in the wild type strain.

References

1. Murphy DJ. The molecular organisation of the photosynthetic membranes of higher plants. Biochim Biophys Acta 1986;864:33-94.
2. Quinn PJ, Joo F, Vigh L. Role of unsaturated lipids in membrane structure and stability. Prog Biophys Mol Biol 1989;53:71-103.
3. Kunst L, Browse J, Somerville CR. Enhanced thermal tolerance in a mutant of *Arabidopsis* deficient in palmitic acid unsaturation. Plant Physiol 1989;91:401-408.
4. Hugly S, Kunst L, Browse J, Somerville CR. Enhanced thermal tolerance and altered chloroplast ultrastructure in a mutant of *Arabidopsis* deficient in lipid unsaturation. Plant Physiol 1989;90:1134-1142.
5. Arntzen CJ. Dynamic structural features of chloroplast lamellae. Curr Top Bioenerg 1978;8:111-160.
6. Quinn PJ. Lipid phases and lipid-protein interactions in the chloroplast photosynthetic membrane. Biochem Soc Trans 1987;15:86-91.

ELECTRON MICROSCOPY AND X-RAY SCATTERING STUDIES OF THE STRUCTURAL ORGANISATION OF PROLAMELLAR BODIES ISOLATED FROM *Zea mays.*

Eva Selstam[a], Patrick Williams[b], Tony Brain[b] and Wim Bras[c].

University of Umeå[a], King's College London[b], Daresbury Laboratory[c], Warrington, UK.

Introduction

PLB of the type found in maize are thought to be built up of lipid bilayers organised into a bicontinuous cubic phase based on a set of intersecting tubes following the geometry of the diamond lattice. It has been suggested that this type of structure constitutes a Schwartz minimal energy surface commonly referred to as a D-surface [1]. We have performed parallel electron microscopy and X-ray diffraction methods to test this hypothesis.

Results and Discussion

PLB were isolated from *Zea mays* using techniques based on earlier studies on wheat [2]. Samples were either prepared for thin-section electron microscopy or subjected to freeze-fracture. Typical thin-section electronmicrographs showing two characteristic staining patterns are presented in Fig. 1. The relationship between the images seen in the electronmicrographs and the diamond lattice can be readily appreciated by reference to the accompanying photographs. The square pattern arises from the intersection of orthogonal (2 2 0) planes of the diamond lattice; the diamond pattern from the intersection of two sets of (1 1 1) planes. Unit cell sizes corresponding to these images are readily calculated from the spacing of the planes. The values obtained varied considerably, probably reflecting the different lipid composition of the individual PLB, but were normally in the range 45-65 nm.

Similar measurements were performed using freeze-fracture electron microscopy. Two major sets of fracture planes were identified. One set, parallel to the (2 2 0) planes, yielded a square lattice similar to that seen in the thin-section study. The other set, parallel to the (1 1 1) planes, yielded an hexagonal lattice. Unit cell sizes, calculated on the basis of the diamond lattice, were very similar to those obtained from the thin-section electronmicrographs.

We have measured the X-ray scattering profile of maize PLB using the 3 metre, small-angle diffraction camera of Station 8.2 of the BBSRC Synchrotron facility at Daresbury (UK). A typical scattering pattern as measured and the corresponding pattern after

J.-C. Kader and P. Mazliak (eds.), Plant Lipid Metabolism, 141–143.

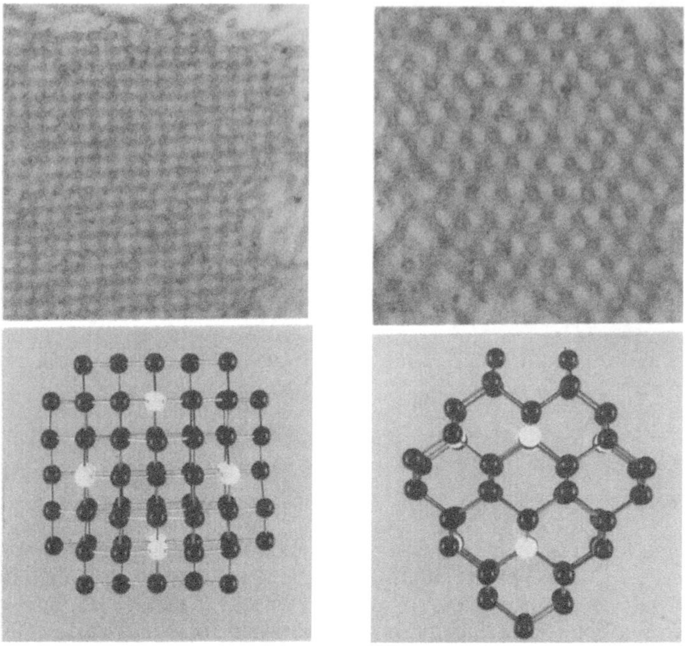

Figure 1. Thin-section electronmicrographs and corresponding photographs of a model of the diamond structure showing the origins of their staining patterns.

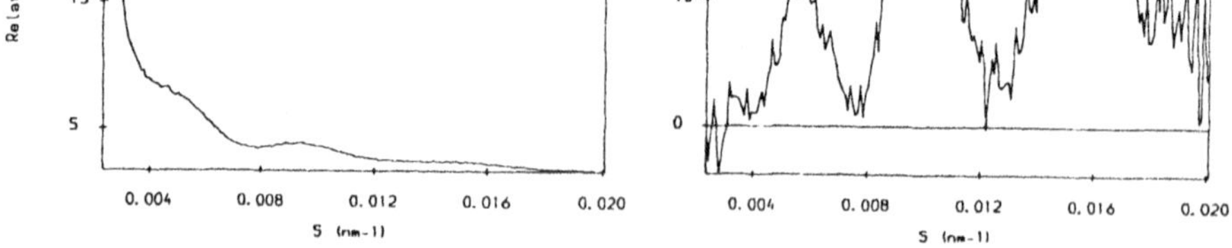

Figure 2. Typical X-ray scattering pattern obtained for maize PLB (a) original pattern (b) after correction for background scattering and geometric factors.

allowance for background scattering and correction using the Lorenz polarisation factor, are presented in Fig. 2.

Longley and McIntosh [3] have shown that glycerol mono-oleate dispersed in excess water forms a bicontinuous cubic phase based on a D-surface, directly analogous to that proposed for maize PLB. They pointed out that in terms of X-ray diffraction, the symmetry properties of the D-surface are equivalent to those of two interlocking diamond lattices with a unit cell of half the dimensions of the original diamond lattices. The patterns shown in Fig. 2 consist of three main peaks which appear to index on the series $1/\sqrt{3}$, $1/\sqrt{8}$, $1/\sqrt{21}$. On this basis, the unit cell length is 29 - 30 nm (i.e. about half the size of the diamond unit cell corresponding to the optical image). This observation adds considerable support to the view that the description of the organisation of the maize PLB in terms of the D-surface is indeed correct.

References

[1] Lindstedt, I. and Liljenberg, C. (1990) Physiol. Plant. 80, 1-4.
[2] Selstam, E. and Widell-Wigge, A. (1989) Physiol. Plant. 77, 401-406.
[3] Longley, W. and McIntosh, T.J. (1983) Nature, 303, 612-614.

MGDG SYNTHASE OF SPINACH CHLOROPLAST ENVELOPE: PROPERTIES OF THE SUBSTRATE BINDING SITES

Eric MARECHAL, Maryse A. BLOCK, Jacques JOYARD and Roland DOUCE
CEA-Laboratoire de Physiologie Cellulaire Végétale,
CNRS-URA n° 576,
DBMS, CENG
17, Rue des Martyrs, 38054 GRENOBLE cedex 9, France.

Synopsis

Kinetic properties of the delipidated monogalactosyldiacylglycerol (MGDG) synthase with respect to the various diacylglycerol molecules were first investigated in mixed micelles containing the partially purified enzyme, the substrate (diacylglycerol) and the detergent, according to the "surface dilution" kinetic model proposed by [1]. We concluded that MGDG synthase has different affinities for the diacylglycerol species tested: a lower affinity for C18:0/C18:0, C16:0/C16:0, C16:0/C18:1, C18:3/C18:3 (Km from 0.040 to 0.066 mol-fraction), a higher affinity for C16:0/C18:2, C18:1/C16:0, C18:0/C18:1, C18:1/C18:1 (Km values from 0.024 to 0.031 mol-fraction), and an even higher affinity for C18:2/C18:2 with Km of 0.0089 mol-fraction. To provide further evidence for a physiological significance of these data, enzymatic assays were performed within envelope vesicles prepared from thermolysin-treated intact chloroplasts. Kinetic experiments suggested that MGDG synthase possesses distinct and independant binding sites for each substrate, UDP-galactose or diacylglycerol. This was verified by studying pattern of inhibition of MGDG synthase activity by UDP relatively to each substrate. Moreover, protection of MGDG synthase activity against citraconic anhydride (a specific reagent of lysines) by incubating with either UDP-gal or 1,2-diacylglycerol demonstrates that substrate binding is not ordered.

Introduction

Monogalactosyldiacylglycerol (MGDG), the major polar lipid in the plastid is synthesized by a UDP-galactose:1,2-diacylglycerol 3-β-galactosyltransferase (or MGDG synthase; E.C. 2.4.1.46) which transfers a galactose from a water soluble donor, UDP-galactose, to a hydrophobic acceptor, 1,2-diacylglycerol, to synthesize MGDG with the release of UDP. This enzyme is located in the inner envelope membrane of the plastid [2]. Purification of MGDG Synthase conducted by [3,4] led to the conclusion that,

J.-C. Kader and P. Mazliak (eds.), Plant Lipid Metabolism, 144–151.

despite its importance for chloroplast membrane biogenesis, MGDG synthase is only a minor protein. This conclusion raises the questions of the mechanism of MGDG synthase activity and of its regulation. In this presentation, we analyse the functioning of MGDG synthase relatively to its substrates.

Affinity of MGDG Synthase Towards Diacylglycerol

KINETIC ANALYSIS IN MIXED MICELLES

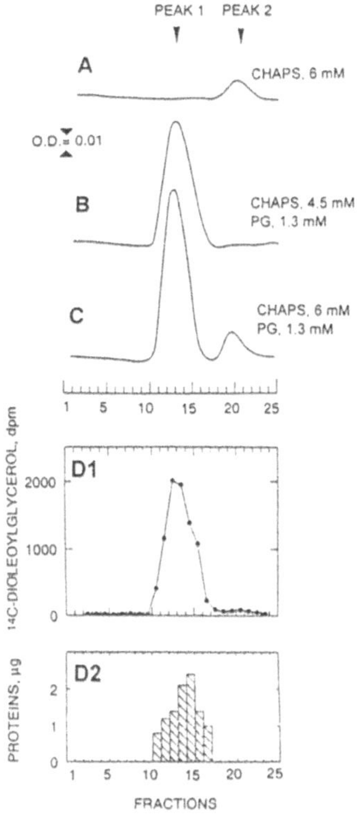

Figure 1: Sephacryl S-200 chromatography of micelles. Panel **A**: micelles containing 6 mM CHAPS; Panel **B**: micelles containing 4.5 mM CHAPS and 1.3 mM PG; Panel **C**: micelles containing 6 mM CHAPS and 1.3 mM PG; Panel D: micelles containing 6 mM CHAPS, 1.3 mM PG, 160 mM ^{14}C-dioleoylglycerol and partially purified MGDG synthase (corresponding to 8 mg protein). Samples were applied and eluted from a Sephacryl S-200 column as described under [5]. The two separated peaks, peak 1 and peak 2, correspond to micelles of 90,000 and 18,000 daltons, respectively. The same elution profile as curve in panel **C** was obtained with the complete mixture (i.e. for micelles containing 6 mM CHAPS, PG, ^{14}C-dioleoylglycerol and partially purified MGDG synthase), but diacylglycerol (panel **D1**) and proteins (panel **D2**) were present only in peak 1. Pure CHAPS micelles were eluted in peak 2.

We have investigated MGDG synthase activity in partially purified preparation, using mixed micelles containing 1,2-diacylglycerol, CHAPS and phosphatidylglycerol (PG). MGDG synthase was purified on a hydroxyapatite-ultrogel (IBF, France) according to [5]. We have previously shown that this fraction is delipidated [6]. The presence of PG was essential for maximal galactosylation activity because it strongly improves 1,2-diacylglycerol solubilization by CHAPS [5,6]. In these conditions, all diacylglycerol molecules were solubilised above 3.5 mM of CHAPS. The homogeneity of the micelles present in the reaction medium was studied by gel filtration on a Sephacryl S-200 column. Figure 1 shows the elution patterns of the different types of micelles that could be encountered. Pure CHAPS micelles are rather small, their size was estimated by gel filtration to be arround 18-20 kDa, whatever the CHAPS concentration was (up to 9 mM). When mixed CHAPS/PG micelles were analyzed, larger micelles were formed whose size was estimated to be arround 80-90 kDa. However, at 6 mM CHAPS, an additional population of vesicles was obtained with a size of 17-19 kDa. Finally, when PG, dioleoylglycerol and delipidated MGDG synthase were mixed with CHAPS (below 6 mM), gel filtration analyses demonstrated that only large micelles of 80 to 90 kDa were present and that they contained dioleoylglycerol molecules and proteins besides CHAPS and PG. Small micelles containing only CHAPS were formed when CHAPS concentration was raised to 6 mM. The resulting mixed micelles rather homogenous in size between 3.5 mM and 6 mM CHAPS could therefore be used to determine whether the surface dilution kinetic model [1] was suitable for MGDG synthase. Figure 2 shows MGDG synthase activity (at saturating UDP-galactose concentration) represented as a function of the total molar concentration of the three amphiphiles: CHAPS, PG and diacylglycerol for a series of given mole fractions of diacylglycerol (CHAPS and PG considered as a single complex detergent being kept at a given molar ratio CHAPS/PG (1:0.29)). We concluded from this figure that, between 3.5 mM and 6 mM of CHAPS, MGDG synthase activity was dependent on diacylglycerol surface concentration and independant of diacylglycerol molar concentration. In addition, we verified that in this range of CHAPS concentration, the apparent Km for diacylglycerol expressed in mol fraction remains constant [5]. Altogether these results demonstrated that under our experimental conditions, MGDG synthase follows the "surface dilution" kinetic model.

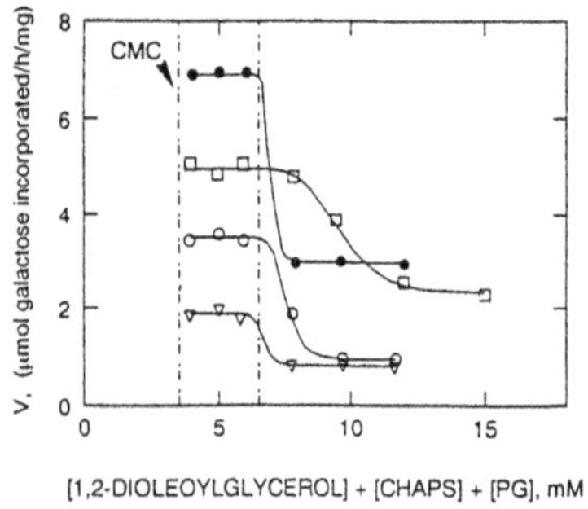

Figure 2: Activity of MGDG synthase in mixed micelles, according to the "surface dilution" kinetic model. MGDG synthase activity was measured as a function of the sum of the molar concentrations of CHAPS, PG and diacylglycerol at given mole fractions of diacylglycerol: —▽—, 0.005; —○—, 0.01; —□—, 0.02; —●—, 0.05 as described under [5].

Kinetic parameters of MGDG synthase for 9 different diacylglycerol molecular species were determined using the experimental conditions determined above. Double-reciprocal plots of the results were drawn for each diacylglycerol molecular species. In all cases, plots of the data were linear (Figure 3A and C) and were used to calculate kinetic constants. Apparent Km values (expressed as mol fractions) for all molecular species analyzed are presented in Table 1. On the basis of these values, diacylglycerol molecular species can be divided in 3 groups. The envelope MGDG synthase shows the lowest affinity (Km values higher than 0.04-mol fraction) for diacylglycerol molecular species comprising 18:0/18:0, 16:0/16:0, 16:0/18:1 and 18:3/18:3. Most other diacylglycerol molecular species (16:0/18:2, 18:1/16:0, 18:1/18:1, 18:0/18:1) presented apparent Km values between 0.025- and 0.033-mol fraction. The third group comprised only dilinoleoylglycerol (18:2/18:2), because the affinity of MGDG synthase for substrate was much higher than for any other diacylglycerol molecular species. Taking into account the size of the mixed micelles (90 kDa, corresponding to 140 molecules, see above), this Km value represents only 1.25 molecule of dilinoleoylglycerol per micelle, whereas Km values for dioleoylglycerol, dipalmitoylglycerol and distearoylglycerol represent respectively 4, 7 or 9.3 molecules per micelle.

Table 1: Kinetic parameters of MGDG synthase in mixed micelles. Mixed micelles were prepared as described by [5]. MGDG synthase activity was determined with various diacylglycerol molecular species as the varied substrate. Diacylglycerol concentration was expressed as mole fraction, i.e. [diacylglycerol]/([CHAPS] + [PG] + [diacylglycerol]). The Vmax and app Km values were respectively calculated from the 1/V intercept and the 1/[substrate], expressed as mole fraction, intercept axes of double reciprocal plots. Diacylglycerol concentration within a micelle was also expressed as the number of diacylglycerol molecules per micelle, assuming that each 90 kDa-mixed micelle contains about 140 molecules of CHAPS, PG and diacylglycerol.

diacylglycerol molecular species	Km (mol fraction)	diacylglycerol concentration per micelle corresponding to Km values (molecule/ micelle)	diacylglycerol concentration in the micelles corresponding to Km values (mmol/nm^2)	Vmax (mmol galactose incorporated/h/ mg protein)	Vmax /Km
18:0/18:0	0.066	9.3	52	12.5	187
16:0/16:0	0.050	7	39	12.5	250
16:0/18:1	0.042	6	33	11.7	280
18:3/18:3	0.040	5.6	31	10.5	262
16:0/18:2	0.031	4.3	24	11.3	361
18:1/16:0	0.029	4	22	10.8	378
18:1/18:1	0.028	4	22	11.1	399
18:0/18:1	0.024	3.3	18	10.6	445
18:2/18:2	0.0089	1.25	7	10.5	1176

KINETIC ANALYSIS IN ENVELOPE VESICLES

To analyse the physiological significance of the kinetic data obtained with mixed micelles, we determined the kinetic parameters of MGDG synthase for diacylglycerol in isolated envelope membranes. These membranes contain high levels of diacylglycerol that are not normal envelope constituents [7] but are produced from galactolipids within the envelope membranes during the course of their purification owing to a galactolipid:galactolipid galactosyltransferase [8]. However, it is possible to prepare diacylglycerol-free envelope membranes by using thermolysin-treated chloroplasts for envelope purification. We therefore assayed MGDG synthase activity in these diacylglycerol-free envelope membranes supplied with exogenous diacylglycerol. Our previous results with mixed micelles led us to use for these experiments dioleoylglycerol, for which MGDG synthase had an affinity representative of those of most diacylglycerol molecular species analyzed, and dilinoleoylglycerol which was the best substrate for the enzyme. In contrast to the data obtained in mixed micelles (Figure 3A and C), rather complex double-reciprocal plots of the results were obtained with envelope vesicles for each diacylglycerol molecular species (Figure 3B and D). At very low diacylglycerol concentration, the velocity of the reaction was not dependent on diacylglycerol concentration and remained almost constant. Therefore, reciprocal plots of the data were almost parallel to the 1/[substrate]-axis. In contrast, at high diacylglycerol concentration, MGDG synthase was inhibited and the curve was therefore asymptotic to the 1/V-axis. Kinetic constants were determined from the linear part of the curves (Figure 3B and D).

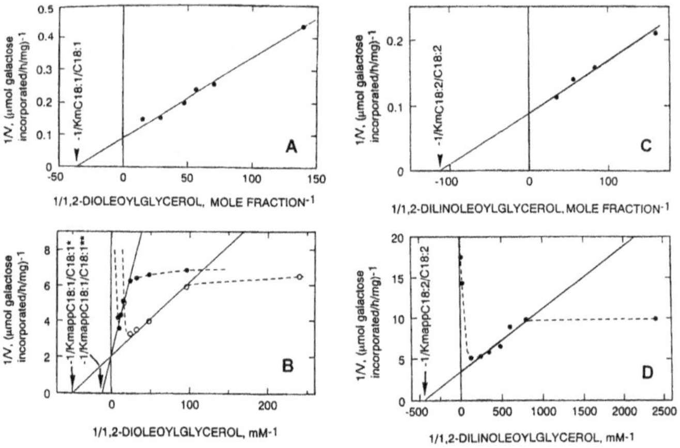

Figure 3: Kinetic analyses of MGDG synthase activity in mixed micelles, as a function of diacylglycerol concentration, and in envelope membranes. MGDG synthase activity was determined with dioleoylglycerol (**A** and **B**) and dilinoleoylglycerol (**C** and **D**) as the varied substrate. Diacylglycerol concentration is expressed as mole fraction, [diacylglycerol]/([CHAPS] + [PG] + [diacylglycerol]), in mixed micelles (**A** and **C**) and in molar values for experiments with envelope membranes (**B** and **D**). With dioleoylglycerol as the varied substrate, MGDG synthase activity was assayed in presence of different amounts of envelope membranes: —○—, 170 mM envelope glycerolipids (Km 18:1/18:1* = 20 mM or 0.1 mol fraction); —●—, 500 mM envelope glycerolipids (Km 18:1/18:1** = 90 mM or 0.14 mol fraction). With dilinoleoylglycerol as the varied substrate, MGDG synthase activity was assayed with envelope membranes corresponding to 170 mM glycerolipids (Km 18:2/18:2 = 2.2 mM or 0.0093 mol fraction). In mixed micelles, Km values were respectively 0.028 and 0.089 mol fractions for dioleoylglycerol and dilinoleoylglycerol.

The data led to a series of observations. Firstly, the appVmax was much lower in isolated vesicles (≈ 0.5 mmol/h/mg protein) than in mixed micelles (10 to 12 mmol/h/mg protein). This was expected since mixed micelles contained partially purified MGDG synthase, i.e. there is more MGDG synthase molecules per mg protein in mixed micelles than in envelope vesicles. Secondly, like in mixed micelles, dilinoleoylglycerol was within isolated membranes a much better substrate for MGDG synthase than dioleoylglycerol. Figure 3 also demonstrates that apparent Km values for diacylglycerol determined in envelope vesicles are dependent on the amount of membranes. Analysis of the data according to the "zero order membrane partition" kinetic model proposed by

[9,10] led to the conclusion that, in isolated membrane, the effective Km constant for dioleoylglycerol was of the same order of magnitude as the Km calculated for dioleoylglycerol in mixed micelles.

Independence of Diacylglycerol and UDP-gal Binding Sites

When two-substrate kinetics were performed, data showed that the affinity of MGDG synthase for each of its substrates (UDP-gal and 1,2-diacylglycerol) did not vary when the cosubstrate supply was varied [5]. This observation suggested that MGDG synthase possesses distinct and independent binding sites for each substrate and that the enzyme obeys a sequential bireactant mechanism. We further investigated the structure and properties of these binding sites inside the catalytic site using various inhibitors of the enzyme and protection from inhibition by substrates.

INHIBITION BY UDP

UDP is a well known inhibitor of MGDG synthase, competitive relatively to UDP-gal, at fixed diacylglycerol supply [11,6], see Fig. 4A.

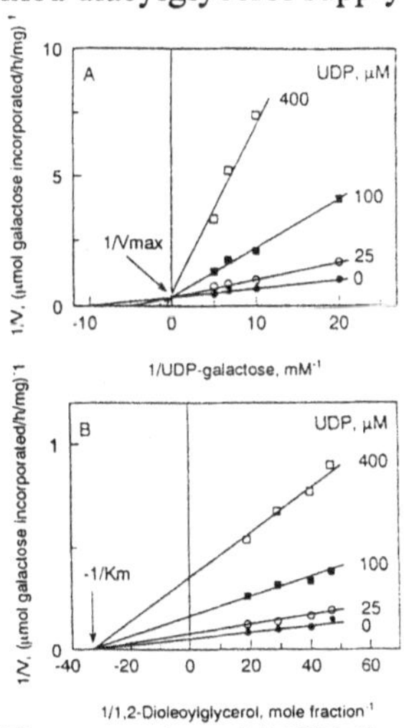

Figure 4: MGDG synthase inhibition by UDP. A: UDP-galactose is the varied substrate and dioleoylglycerol mol fraction is kept constant (0.03-mol fraction). The family of curves exhibits a competitive inhibition pattern : Ki_{UDP} = 23.5 ± 1.5 mM. **B:** dioleoylglycerol is the varied substrate, and UDP-gal concentration is kept constant (1 mM). 1 / dioleoylglycerol is expressed in mol fraction according to the surface dilution model. MGDG synthase inhibition by UDP shows a non-competitive pattern which calculated Ki for UDP is 52 ± 16 mM.

When we plotted MGDG synthase inhibition by UDP at fixed UDP-gal molarity (1 mM), relatively to diacylglycerol surface concentration according to the "surface dilution kinetic model", we obtained fairly simple curves, coherent with a non-competitive inhibition (Fig. 4B). Inhibition constant (Ki_{UDP}), calculated from experiment at fixed diacylglycerol mole fraction (Fig. 1A), is 23.5 ± 1.5 mM, whereas Ki_{UDP} value calculated from experiment at fixed UDP-gal molarity is slightly higher : 52 ± 16 mM UDP binds competitively to the UDP-gal binding site, while it does not interfere with diacylglycerol binding. This result demonstrates that UDP-gal binds to MGDG synthase at least by part of its nucleotidic side, on a topologicaly distinct site than 1,2-dioleoylglycerol.

INHIBITION BY CITRACONIC ANHYDRIDE

Table 2: MGDG synthase inactivation by citraconic anhydride. Partially purified MGDG synthase was preincubated 30 min at 25°C as described, then subsequently incubated 30 min at 25°C with citraconic anhydride. Galactosylation activity was measured as in [5].

Preincubation	Incubation with citraconic anhydride	Activity (%)
-	-	100
+PG (1.3 mM)	10 mM	13
+1,2-DAG (500 μM) +PG (1.3 mM)	10 mM	65
+UDP-gal (1mM)	10 mM	77

Table 2 shows MGDG synthase inactivation by a specific reagent of primary amines, i.e. citraconic anhydride which blocks lysines at alcaline pH. MGDG synthase was partially inactivated after a 30-minutes incubation with this reagent. Preliminary 30-minutes incubations with either MGDG synthase substrate protected the enzyme against inactivation by citraconic anhydride. This protection indicates that primary amines are essential for binding of both substrates. Furthermore, it shows that substrate binding is random.

Conclusion

From these observations, we can conclude that MGDG synthase from spinach chloroplast envelope contains two functionally independent and topologicaly distinct binding sites. MGDG synthase bireactant mechanism occurs by a sequential binding of each substrate, followed by the release of each product (UDP and MGDG). Since MGDG synthase inactivation by citraconic anhydride was affected when the enzyme was preincubated by either substrate, substrate binding is therefore random. MGDG synthase inhibition by UDP was strictly competitive relatively to UDP-gal, and noncompetitive relatively to 1,2-dioleoylglycerol. Therefore it can be assumed that UDP-gal binds to MGDG synthase at least by part of its nucleotidic side, on a topologicaly distinct site than 1,2-dioleoylglycerol. MGDG synthase affinity for 1,2-diacylglycerol was dependent on the fatty acid length (16 or 18 carbones), position on glycerol backbone (*sn*-1 or *sn*-2), and desaturation degree (0 to 3 double bonds). Therefore, 1,2-diacylglycerol binding on MGDG synthase involves partly its acylated side. Future work aims to explain why in spite of the low molecular weight (the predicted Stockes radius for a 19 kDa globular protein should be ~2 nm), MGDG synthase is capable of binding large portions of both substrates in separate binding domains and how catalysis occurs.

References

1. Deems R.A., Eaton, B.R., and Dennis, E.A. Kinetic analysis of phospholipase A2 activity towards mixed micelles and its implication for the study of lipolytic enzymes. J. Biol. Chem. 1975; 250, 9013-9020.
2. Block, M. A., Dorne, A.-J., Joyard, J., and Douce, R. Preparation and characterization of membrane fractions enriched in outer and inner envelope membranes from spinach chloroplasts. II-Biochemical characterization. J. Biol. Chem. 1983; 258, 13281-13286
3. Maréchal, E., Block, M. A., Joyard, J., and Douce, R. Purification de l'UDP-galactose:1,2-diacylglycérol galactosyltransférase de l'enveloppe des chloroplastes d'épinard. C. R. Acad. Sci. Paris 1991; 313, 521-528
4. Teucher, T., and Heinz, E. Purification of UDP-galactose:diacylglycerol from chloroplast envelopes of spinach (Spinacia oleracea L.) Planta 1991; 184, 319-326
5. Maréchal, E., Block, M.A., Joyard, J., and Douce, R. Kinetic properties of monogalactosyldiacylglycerol synthase from spinach chloroplast envelope membranes. J. Biol. Chem. 1994; 269, 5788-5798
6. Covès, J., Joyard, J., and Douce R. Lipid requirement and kinetic studies of solubilized UDP-galactose:diacylglycerol galactosyltransferase activity from spinach chloroplast envelope membranes. Proc. Natl. Acad. Sci. USA 1988; 85, 4966-4970
7. Dorne, A.-J., Block, M.A., Joyard, J., and Douce, R. The galactolipid:galactolipid galactosyltransferase is located on the outer membrane of the chloroplast envelope. FEBS Lett. 1982; 145, 30-34
8. Van Besouw, A., and Wintermans, J.F.G.M. Galactolipid formation in chloroplast envelopes. I. Evidence for two mechanisms in galactosylation. Biochim. Biophys. Acta 1978; 529, 44-53
9. Heirwegh, K.P.M., Meuwissen, J.A.T.P., Vermeir, M., and De Smedt, H. Liposomes as carriers of poorly water-soluble substrates: linear modelling of membrane systems with catalytic or binding sites of different facedness. Significance of experimental membrane partition coefficient and of kinetic and equilibrium parameters. Biochem. J. 1988; 254, 101-108.
10. Heirwegh, K.P.M., Meuwissen, J.A.T.P., Van den Steen P., and De Smedt, H. Modelling of chemical reactions catalysed by membrane-bound enzymes. Determination and significance of the kinetic constants. Biochim. Biophys. Acta 1989; 995, 151-159.
11. Van Besouw, A., and Wintermans, J.F.G.M. The synthesis of galactolipids by chloroplast envelopes. FEBS Lett. 1979; 102, 33-37

UDP-GALACTOSE: DIACYLGLYCEROL GALACTOSYLTRANSFERASE IN CUCUMBER SEEDLINGS: PURIFICATION OF THE ENZYME AND THE ACTIVATION BY PHOSPHATIDIC ACID

Hiroyuki Ohta, Mie Shimojima, Takeshi Arai, Tatsuru Masuda, Yuzo Shioi and Ken-ichiro Takamiya, Faculty of Bioscience & Biotechnology, Tokyo Institute of Technology, Nagatsuta, Yokohama 227, Japan

Introduction

UDP-galactose:diacylglycerol galactosyltransferase (UDGT) catalyzes the final step of the synthesis of monogalactosyl diacylglycerol (MGDG), a major glycolipid in plastid membrane. We found that the enzyme activity was strongly activated coincided with accumulation of galactolipids during germination of cucumber seedlings, suggesting that activation step of the UDGT was important in the development of chloroplast membrane [1]. However, mechanism of regulation of the UDGT activity during chloroplast formation remains unclear.

Purification of the UDGT from spinach envelopes was reported by Teucher and Heinz [2] and Maréchal et al [3]. According to their results, molecular mass of the purified proteins and the properties of the solubilized enzymes were different from each other. Teucher and Heinz [2] proposed that a polypeptide of 22 kDa was associated with the activity. In contrast, Maréchal et al. [3] reported that they led to 90% enrichment of a 19 kDa polypeptide as the enzyme, and furthermore, no 22 kDa polypeptide could be visualized in the purest fraction. In addition to such discrepancy of the molecular weight, Covès et al [4] reported lipid requirement of the CHAPS-solubilized UDGT, although cholate-solubilized enzyme did not show such requirement for the enzyme assay [2].

In the present study, we purified UDGT from cucumber cotyledons and also investigated the enzyme properties.

Materials and Methods

PLANT MATERIALS

Cucumber seeds (Cucumis sativus L. cv. Aonagajibai) were sown on wet vermiculite, and were grown in the dark at 27°C. After illumination for 29 h, cotyledons without hypocotyl hook were collected and stored at -80°C.

ASSAY OF UDGT ACTIVITY

UDGT activity was assayed using 1,2-dioleoyl-*sn*-glycerol (200 μg/ml) and UDP-[^{3}H]-galactose (400 μM) as substrates according to Teucher and Heinz [2] .

PREPARATION OF MICROSOME AND SOLUBILIZATION OF THE ENZYME

J.-C. Kader and P. Mazliak (eds.), Plant Lipid Metabolism, 152–155.

Cucumber cotyledons were homogenized by a Warning blender with suspension buffer: 50 mM Tris-HCl, pH 7.8 / 0.8 M sodium acetate / 10 mM dithiothreitol (DTT) / 1 mM phenylmethanesulfonyl fluoride (PMSF) / 0.02% NaN_3 / 20% glycerol (TSPB). The homogenate was filtrated through four layers of gauze and centrifuged at 15,000 x*g* at 4°C for 15 min. The supernatant was centrifuged at 165,000 x*g* for 1 h, and then, the sediments (microsomes) were resuspended in the same buffer and stored at -30°C.

Microsomes in TSPB were mixed with 12 mM lauryldimethylamine oxide (LDAO) / TSPB to result in 3 mg protein/ml, and stirred on ice for 10 min. Subsequent centrifugation for 1.5 h at 165,000 x*g* yielded the supernatant fraction containing the solubilized enzyme.

Results and discussion

SOLUBILIZATION OF CUCUMBER UDGT

For the solubilization of UDGT, we investigated the effect of the several detergents having anionic (sodium cholate), nonionic (Triton X-100, TX-100; Octylglucoside, OG, Brij-35, LDAO) or Zwitterionic (CHAPS) properties. Forty % of the enzyme activity was solubilized with OG, 18% with CHAPS, and 12% with sodium cholate (Fig. 1). Interestingly, Fig. 1 also indicates that solubilized cucumber UDGT required lipids for the enzyme assay, similarly to the CHAPS-solubilized enzyme in spinach [4]. The UDGT activity solubilized with these detergents was markedly increased by the addition of microsome lipids (ML) to the reaction mixture prepared from the microsome fraction.

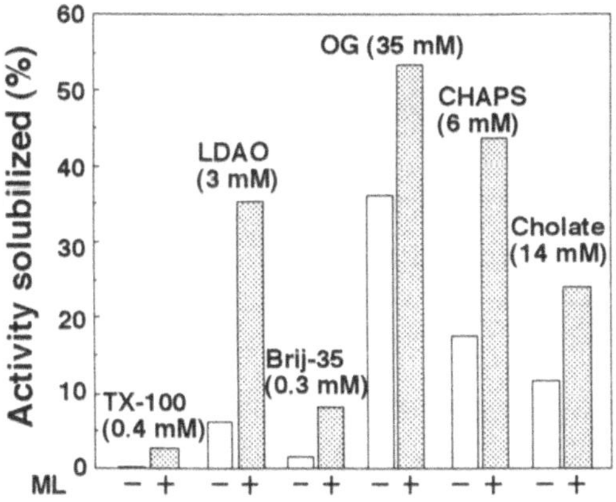

Fig. 1 Solubilization of UDGT by detergents. Microsome protein (3 mg/ml) was stirred with 1.5 times of critical micelle concentration of each detergent for 10 min. The mixture was centrifuged at 165,000 x*g* for 1 h, and the supernatant was assayed under two different conditions with (+) and without (-) the addition of microsome lipids (ML, 225 μg) .

The activation of LDAO-solubilized UDGT was the most drastic of these solubilized fractions. When solubilized activity was compared after addition of ML, LDAO showed 35% of solubilization, similarly to CHAPS (44%) and OG (52%). Among these detergents, LDAO, a nonionic detergent, was employed on solubilization of the enzyme, because of the advantage in chromatographic procedure.

PURIFICATION OF CUCUMBER UDGT

UDGT was solubilized with LDAO from chloroplast envelope-enriched microsome fraction of cucumber cotyledons, and purified by several chromatographic procedures (Table 1). Metal chelate affinity chromatography (MCAC) and CM-Toyopearl were the most effective steps on the purification. After the final step of 2nd CM-Toyopearl, cucumber UDGT was purified more than 7,400-fold from solubilized fraction with a specific activity more than 370 nmol/min/mg (Table 1). SDS-PAGE of the purified protein gave a major band of 47 kDa with a few minor components. Throughout our experiments, UDGT activity was tightly associated with this 47 kDa protein, suggesting that the protein is UDGT. However, the molecular weight of the purified protein is quite different from spinach enzyme [2,3]. Currently, we are preparing a polyclonal antibody against the 47 kDa protein. Further investigation of the protein using immunochemical and molecular biological techniques is necessary to determine whether the protein is UDGT or not.

Table 1. Summary of Purification of UDGT from Cucumber Cotyledons

Purification step	Protein (μg)	Activity (nmol/min)	Specific act. (nmol/min/mg)	Yield (%)	Purification (-fold)
Solubilized Supernatant	15,529,000	531	0.05	100	1
MCAC	1,292,000	271	0.21	51.0	4.2
1st Hydroxyapatite	225,000	127	0.56	24.0	11.2
Toyopearl HW-55	26,300	48.2	1.8	9.1	36.0
2nd Hydroxyapatite	4,100	36.7	9.0	6.9	180
1st CM-Toyopearl[1)]	104	13.4	128.8	2.5	2,576
2nd CM-Toyopearl[2)]	<1	0.37	>370	0.07	>7,400

1) Half of the total activity obtained by 2nd hydroxyapatite chromatography was applied.
2) Most active fraction of 1st CM-Toyopearl chromatography was applied.

LIPID REQUIREMENT OF CUCUMBER UDGT

As shown in Fig. 1, the solubilized UDGT was activated by the addition of microsome lipids to the assay mixture. However, more than 200 μg of lipids was required to exhibit maximal activity. Major phospholipids and galactolipids of chloroplast membrane, a localization site of UDGT, were prepared from cucumber cotyledons, and the effect on the UDGT activity was examined to identify component(s) which contributes to the stimulation of UDGT (Fig. 2). In these lipids, phosphatidylglycerol (PG) showed the highest activation of UDGT and it reached to 15-fold of that without lipids. In addition to

PG, stimulations of UDGT by phosphatidylinositol (PI) and sulfoquinovosyl diacylglycerol (SQDG) were also observed. In contrast, MGDG, digalactosyl diacylglycerol (DGDG), phosphatidylcholine (PC) and phosphatidylethanolamine (PE) had little effect on the UDGT activity. All of the effective lipids (PG, PI, and SQDG) were acidic lipids, suggesting that a negative charge in lipids was necessary to stimulate the activity. However, similarly to ML, at least 100 μg of the acidic lipids in the reaction mixture (=1.3 mM) was required to show maximal activity, although each content of these acidic lipids was less than 5% of total microsome lipids. Thus, stimulation of the UDGT activity by minor components in chloroplast was also investigated. In addition to the major acidic lipids, phosphatidic acid (PA), lyso PA and lyso PG also showed high stimulation of the UDGT (Fig. 3). In particular, at 60 μM, only PA exhibited 30-fold activation of the UDGT. Furthermore, the ratio of the activation of the UDGT with the microsome lipids corresponded to the content of PA in the microsome fraction (data not shown). These results suggest that the activity of UDGT is regulated by the level of PA in chloroplast envelope, a localization site of the enzyme.

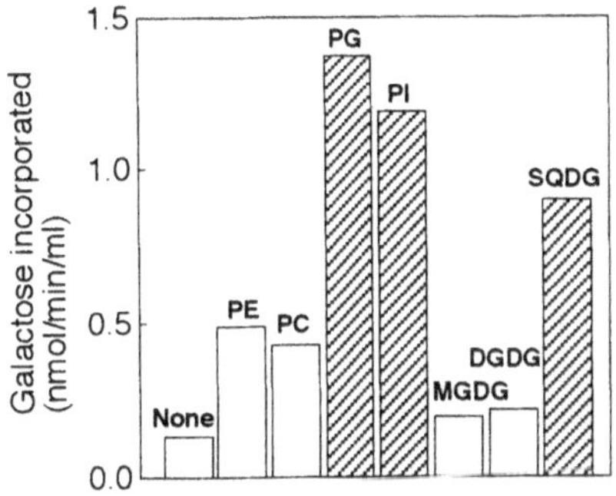

Fig. 2 Effects of chloroplast lipids on UDGT activity. UDGT activity was measured in the presence of 200 μg of each lipid.

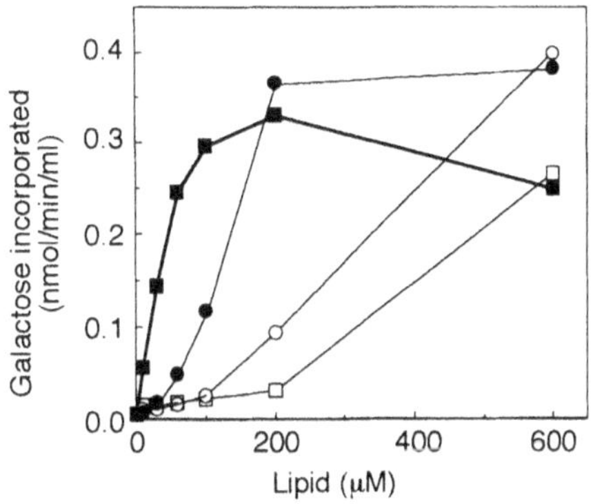

Fig. 3 Effect of PA on UDGT activity. ■, PA; ●, Lyso PA; ○, Lyso PG; □, PG.

References

[1] Ohta H, Shimojima M, Ookata K, Masuda T, Shioi Y, Takamiya K. A galactosyltransferase is activated in light-mediated accumulation of galactolipids during development in cucumber seedlings. 1994;submitted.

[2] Teucher, T and Heinz, E. Purification of UDP-galactose:diacylglycerol galactosyltransferase from chloroplast envelopes of spinach (*Spinacia oleracea* L.) *Planta* 1991;184: 319-326.

[3] Maréchal É, Block MA, Joyard J, Douce R. Purification of UDP:galactose 1.2-diacylglycerol galactosyltransferase from spinach chloroplast envelope membranes.*C. R. Acad. Sci. Paris*, t. 313, Série III, 1991;521-528.

[4] Covès J, Joyard J, Douce R. Lipid requirement and kinetic studies of solubilized UDP-galactose:diacylglycerol galactosyltransferase activity from spinach chloroplast envelope membranes. *Proc. Natl .Acad .Sci. USA* 1988;85:4966-4970.

DETECTION OF PHOSPHATIDYLGLYCEROL AND MONOGALACTOSYL-DIGLYCERIDE ON PEPTIDES OF PHOTOSYSTEM I IN *NICOTIANA TABACUM* SPECIES

Anja Makewicz, Alfons Radunz and Georg H. Schmid
Universität Bielefeld, Fakultät für Biologie, Lehrstuhl für Zellphysiologie, 33501 Bielefeld, Germany

Introduction

Lipids play an important role in the architecture of photosynthetic membranes and probably in their functioning. Four major lipid classes comprise the lipid content of the thylakoids of higher plant chloroplasts, eucaryotic algae and cyanobacteria. These are the galactolipids, monogalactosyldiglyceride (**MGDG**) and digalactosyldiglyceride (**DGDG**), the glycolipid sulfoquinovosyldiglyceride (**SQDG**) and the phospholipid phosphatidylglycerol (**PG**).

The major role of lipids is not only the formation of a lipid bilayer which will act as a structural matrix for proteins and protein complexes but also the association with proteins and protein complexes to build up lipid-protein-complexes. The function of lipids in lipid-protein-complexes is probably to stabilize the activity and conformation of proteins and protein complexes on their own and in the membrane. That shows that lipids provide a hydrophobic environment in which proteins and protein complexes are embedded, in which they can interact with each other and with other membrane components and in which they can change their conformation. So the surrounding lipid phase is very important for the structure, function and activity of membrane proteins [1,2].

By means of the Western blot technique and by using monospecific polyclonal antisera to lipids and carotenoids we have shown that certain lipids and carotenoids are bound onto proteins and functional protein complexes of the thylakoid membrane like the PS I and PS II complex, the light-harvesting pigment-protein complex (LHCP) and the CF_1 factor of the ATPase synthetase complex [3-7]. Furthermore our own studies have shown that antisera to lipids and carotenoids can inhibit photosynthetic electron transport reactions in the region of PS II as well as in the region of PS I. The inhibitory effect on electron transport reactions can be interpreted by assuming that the binding of antibodies induces a conformational change in certain electron transport peptides, thus leading to an inhibition of electron flux. This means that lipids and carotenoids as antigenic determinants are bound on these components.

In the present publication we report on the binding of phosphatidylglycerol and monogalactosyldiglyceride on peptides of PS I in *Nicotiana tabacum* var. JWB and the three

J.-C. Kader and P. Mazliak (eds.), Plant Lipid Metabolism, 156–160.

mutants Su/su, Su/su var. Aurea and NC 95. The binding of these lipids onto PS I peptides is demonstrated by monospecific polyclonal antisera with the Western blot technique and by lipid analysis by HPLC.
The mutants differ from the wild type by their chloroplast structure [6]. The yellow-green mutant Su/su has lower grana stacks and more extended intergrana regions whereas the structure of the lamellar system of the yellow mutant Su/su var. Aurea consists just of membrane doublings and very extended intergrana lamellae. The mutant NC 95 differs from the other *Nicotiana tabacum* mutants by the phenomenon that it has variegated leaves with green and yellow-green leaf patches. Chloroplasts from the green leaf areas have a normal lamellar system with grana and intergrana regions whereas chloroplasts from the yellow-green leaf areas have only single stranded isolated thylakoids.

Materials and Methods

Photosystem I was isolated from *Nicotiana tabacum* var. JWB and the three mutants Su/su, Su/su var. Aurea and NC 95. PS I-200-preparations from the four tobacco varieties were prepared according to Wynn and Malkin [8]. Solubilized PS I and LHCP complexes were loaded onto a 0,4-1 M sucrose gradient that contained 0.02% Triton X-100 with a 2 M sucrose cushion and centrifuged at 100.000 x g for 18 h at 4°C. The dark green band at the top of the 2 M cushion contained the native PS I-200 particles.
Chlorophyll was determined according to Schmid [9] in methanol/water 90/10 (v/v) and protein determinations were done according to Bradford [10].
PS I polypeptides were modified in the SDS polyacrylamide gel electrophoresis according to the methods of Weber and Osborn [11] and Laemmli [12]. A 1,5 mm gel with a 10-20% gradient separation gel and a 3% collection gel was used. Prior to electrophoresis, samples were solubilized with 200 mM DTT, 2% SDS and 20 mM Tris-HCl buffer (pH 8,4) at 50°C for 20 min. Electrophoresis was carried out at a constant current of 25 mA for 6 h at 4°C. Following electrophoresis the gel was stained with Coomassie Brilliant Blue (Fig. 1).
Western-Blotting was performed as described by Renart et al. [13]. The proteins were transferred by pressure from SDS gels to nitrocellulose membranes for 20 h at room temperature. The membranes were blocked with 2,5% fish gelatine. The first antibody (against MGDG and PG) was diluted 200 or 500 fold, depending on the antisera. The second antibody, peroxidase-conjugated pig immunglobulins to rabbit immunglobulins (anti-rabbit IgG, DAKO), was diluted 1000-fold. Specifically bound antibodies were stained by the reaction of peroxidase with H_2O_2 and 4-chloro-1-naphtol (Sigma).
In order to determine the type of lipid binding the peptides were transferred to nitrocellulose membranes and were incubated before antibody incubation for 60 min with phospholipase A_2 (Sigma, lipase concentration: 1 mg/mg chlorophyll) or the PS I particles were incubated before SDS-PAGE for 60 min with phospholipase A_2. Furthermore, the PS I particles were washed with acetone or methanol before SDS-PAGE in order to extract lipids that were only weakly bound by intermolecular interaction. Therefore, PS I particles were incubated with methanol or acetone for 30 min at 4°C, centrifuged at 5.000 x g for 20 min and washed two times with water. The sediments

were resuspended in buffer (0,05 M Tris + 0.05% Triton X-100, pH: 7.8).
For isolation of the core peptides of PS I (**CPI**) the 66 kDa region was cut out after SDS-PAGE and was eluted from the gel by electroelution at 200 V for 8 h. The purity test of the obtained CPI was carried out by SDS-PAGE and by Western blot procedure.
For analysis of lipids the PS I particles and the isolated CPI (PS I core peptides) were dialysed against water, precepitated with acetone, centrifuged at 5.000 x g for 20 min and washed two times with water. The sediments were extracted with boiling methanol and subsequently extracted with methanol/chloroform (1:2 v/v). For the HPLC analysis a lipid solution in methanol was used. The analysis was carried out with a RP-18 column with acetonitrile/water/phosphoric acid (85%) 60:39, 9:0, 1 (v/v/v). The column pressure was 5,5 bar and the flow-rate 0.5 ml/min. Detection of eluted lipids was obtained with a spectral-detector at 205 nm.

Results and Discussion

1. PS I particles were isolated from the wild type tobacco *Nicotiana tabacum* var. JWB and the three mutants Su/su, Su/su var. Aurea and NC 95 by sucrose density centrifugation. In Fig. 1 are shown the polypeptide compositions of all four species analyzed by SDS-PAGE. The PS I preparations of all four species can be fractionated into the PS I core complex and the LHCP I antenna complex. The PS I core complex contains two high molecular mass polypeptides (66 kDa from SDS-PAGE analysis) and several low molecular mass subunits (22, 19, 17, 16, 10 and 9 kDa). The LHCP I fraction is composed of 4 subunits (28, 26, 25 and 24 kDa). PS I preparations of the mutants Su/su and Su/su var. Aurea show impurities of PS II (D1/D2 peptides and chlorophyll-binding peptides).

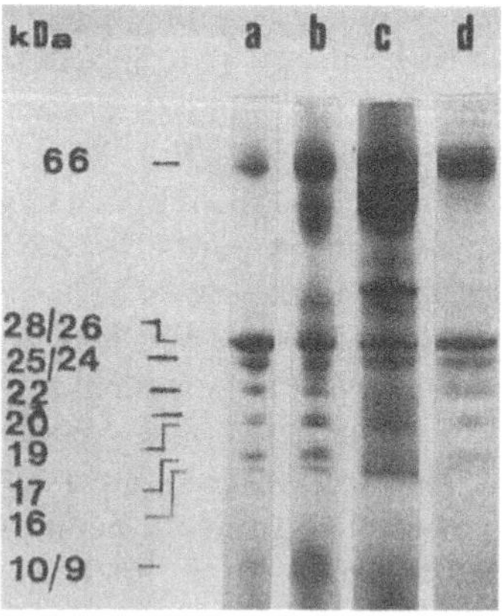

Fig. 1: Coomassie blue stained 10-20% SDS polyacrylamide gel showing the polypeptide compositions of PS I complexes from the four *Nicotiana tabacum* species; **a**) Wild type tobacco **JWB** (10 μg chlorophyll); **b**) Mutant **Su/su** (10 μg chlorophyll); **c**) Mutant **Su/su var. Aurea** (10 μg chlorophyll); **d**) Mutant **NC 95**, yellow leaf patches (10 μg chlorophyll)

2. In the Western blot procedure the polypeptides of PS I from all four species were transferred by diffusion to nitrocellulose membranes and immunostained with different antisera against lipids. But there was only a positive reaction with the antisera against phosphatidylglycerol (PG) and monogalactosyldiglyceride (MGDG) but no reaction with other glycolipids like digalactosyldiglyceride or sulfoquinovosyldiglyceride. The control sera did not react at all. In Fig. 2 and Fig. 3 is shown that antibodies to PG only marked the core peptides of PS I whereas antibodies to MGDG marked the core peptides as well as the peptides of the light-harvesting complex (LHCP) in all for species.

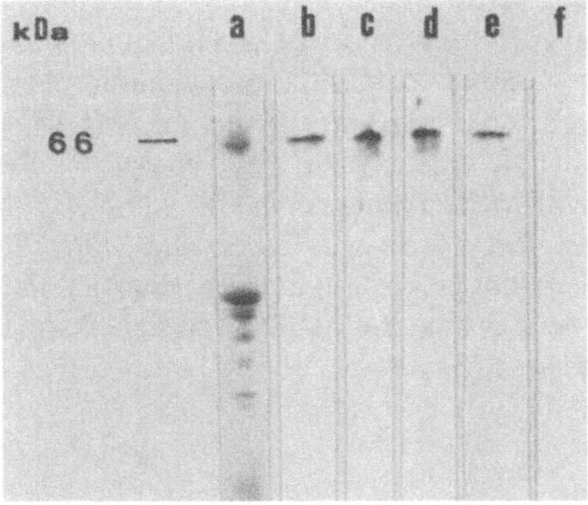

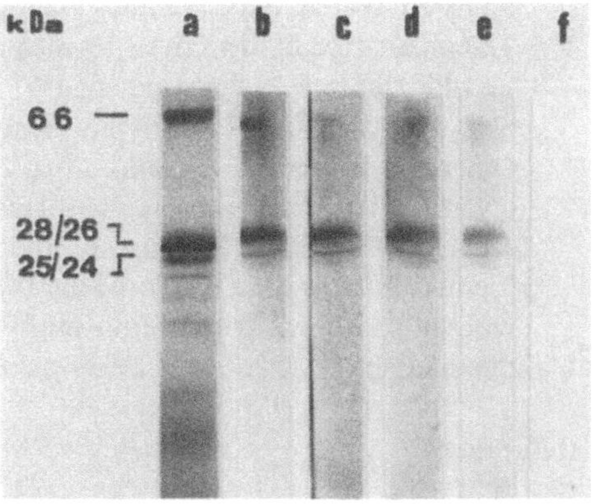

Fig. 2: Immunological determination of phosphatidylglycerol (**PG**) by Western blot analysis of PS I polypeptides from the four *Nicotiana tabacum* species; a) SDS-PAGE analysis of PS I polypeptides from **JWB** (10 µg chlorophyll); b-e) Nitrocellulose membranes after the reaction with PG antiserum (8 $P_{2/4}$, dilution: 1:200) with PS I polypeptides from b) Wild type tobacco **JWB** (10 µg chlorophyll), c) Mutant **Su/su** (10 µg chlorophyll), d) Mutant **Su/su var. Aurea** (10 µg chlorophyll), e) Mutant **NC 95**, yellow leaf patches (10 µg chlorophyll), f) Nitrocellulose membrane with PS I polypeptides from **JWB** (10 µg chlorophyll) after the reaction with control antiserum (8 P_0, dilution: 1:200)

Fig. 3: Immunological determination of monogalactosyldiglyceride (**MGDG**) by Western blot analysis of PS I polypeptides from the four *Nicotiana tabacum* species; a) SDS-PAGE analysis of PS I polypeptides from **JWB** (10 µg chlorophyll); b-e) Nitrocellulose membranes after the reaction with MGDG antiserum (83 H_4, dilution: 1:500) with PS I polypeptides from b) Wild type tobacco **JWB** (10 µg chlorophyll), c) Mutant **Su/su** (10 µg chlorophyll), d) Mutant **Su/su var. Aurea** (10 µg chlorophyll), e) Mutant **NC 95**, yellow leaf patches (10 µg chlorophyll), f) Nitrocellulose membrane with PS I polypeptides from **JWB** (10 µg chlorophyll) after the reaction with control antiserum (83 H_0, dilution: 1:500)

3. In order to extract lipids that are only weakly bound and to see how strong the binding of lipids onto PS I peptides is, the PS I particles were washed with solvents like methanol or acetone. But the washing had no influence on the immunoreactions. That means that the binding must be very strong.
   The positive antibody reaction was at first lost after treatment of the PS I particles with phospholipase A_2. If the nitrocellulose membranes were incubated with phospholipase A_2 before antibody incubation the antibody reaction is positive but the intensity of the immunostaining is diminished. This shows that on the nitrocellulose membrane not all of the lipid molecules are accessible for the phospholipase A_2. Phospholipase A_2 splits fatty acid in position 2 of the diglyceride. From the reaction of the antibodies with the bound lipids it can be concluded that the lipids are bound to the peptides via the fatty acids.
4. The qualitative lipid analysis by means of HPLC has led to the result that the purified PS I complexes and purified CPI complexes contain the phospholipid phosphatidylglycerol (**PG**) and the glycolipid monogalactosyldiglyceride (**MGDG**). Other lipids were not identified in both lipid fractions. In PS I complexes the quantity of PG is higher as the quantity of MGDG whereas in CPI complexes it seems to be equal. Because of the fact that PG is only associated with CPI (demonstrated by the results of the Western blot procedure) it can be assumed that with an additional purification step PG gets lost but not MGDG. This shows that the binding of MGDG is stronger than the binding of PG.

References

1. Webb MS, Green BR. Biochim Biophys Acta 1991; 1060: 133-58.
2. Sandermann jr H. Biochim Biophys Acta 1978; 515: 209-37.
3. Bednarz J, Radunz A and Schmid GH. Z Naturforsch 1988; 43c: 423-30.
4. Voß R, Radunz A and Schmid GH. Z Naturforsch 1992; 47c: 406-15.
5. Kruse O, Radunz A and Schmid GH. Z Natuforsch 1993; 49c: 115-24.
6. Schmid GH, Radunz A and Gröschel-Stewart UG. Immunologie und ihre Anwendung in der Biologie. Georg Thieme Verlag 1993, Stuttgart, New York.
7. Makewicz A, Radunz A and Schmid GH. Z Naturforsch 1994; 49c: (in press).
8. Wynn RM and Malkin R (1988) Biochemistry 1988; Vol 27, No 16: 5863-69.
9. Schmid GH. In: San Pietro A, editor. Methods in Enzymology. Academic Press, London, Vol 23, 1971: 171-94.
10. Bradford MM. Anal Biochem 1976; 72: 248-54.
11. Weber K and Osborn M. J Biol Chem 1969; 244: 4406-12.
12. Laemmli UK. Nature 1970; 227: 680-85.
13. Renart J, Reiser J and Stark GR (1979). Proc Natl Acad Sci USA 76: 3116-20.

INTRACHLOROPLASTIC LIPID EXPORT : A GENERAL PROCESS WITH A KEY ROLE FOR THE CHLOROPLAST ENVELOPE IN THE ESTABLISHMENT OF LIPID ASYMMETRY IN THYLAKOID MEMBRANES

André RAWYLER, Marlyse MEYLAN-BETTEX and Paul-André SIEGENTHALER, Laboratoire de Physiologie Végétale, Université de Neuchâtel

Introduction

The final steps leading to the synthesis of chloroplast (CP) lipids take place in the envelope membranes. Over 90% of the CP lipids belong to thylakoid membranes (TM) [**1**]. Since TM are deprived of lipid synthesis ability [**2**], they must rely entirely on the envelope for their lipid supply. Lipid export from the inner envelope membrane (IEM) to TM is thus implicitly assumed, although this process is still largely unknown. Here, we extend our earlier study [**3**], now using different labelled precursors and plant species, in order to answer the following questions: (1) What are the mechanism(s) of lipid export? (2) Are all lipid classes and molecular species exported in the same way? (3) Is the export mechanism identical among plants of different species and age?

Material & Methods

Intact CP [**4**] were incubated with a labelled precursor (UDP-[^{14}C]galactose [**3**], [^{14}C]glycerol-3-phosphate [**5**] or [^{14}C]acetate [**6**]) and repurified [**7**] after labelling. Purified, envelope-free TM were then prepared [**3**] and the radioactivity of each lipid measured in both CP and TM. In addition, TM were treated with the lipase from Rhizopus arrhizus (LRa) [**3**] or with Vipera russelli phospholipase A_2 (PLA_2) [**8**] in order to compare mass & label transbilayer distributions of mono-, di-galactosyldiacylglycerol (MGDG, DGDG) and phosphatidylglycerol (PG) (see Rationale in [**3**]). Analysis of sugar lipid, lipid phosphorus, pigments and radioactivity were as in [**3**].

Results & Discussion

The intrachloroplastic export of MGDG, DGDG and PG depended linearly on their synthesis (both in extent & rate)

J.-C. Kader and P. Mazliak (eds.), Plant Lipid Metabolism, 161–163.

between 5 and 60 min at 25°C, regardless of the precursor used and with regression coefficients always > 0.950.

The export efficiency (= percentage of the synthesized lipid used for export) was class-selective: PG (87%), MGDG (48-82%), DGDG (19-32%). The label incorporation ratio between TM and CP was 1.1-1.3 for MGDG and PG but only 0.5 for DGDG, indicating that the class selectivity of lipid export is mainly attributable to a marked restriction of DGDG export.

The export efficiency for MGDG was significantly higher in CP from young spinach (82%) than in those from mature spinach (53%), whereas the export efficiency of DGDG was identical (32%). Both galactolipids were also exported more efficiently in CP from spinach ("16:3-plant") than in those from pea or lettuce ("18:3-plants").

When spinach CP were incubated with [^{14}C]acetate (+ light), all molecular MGDG species synthesized under our conditions (monoene to tetraene) were exported with an efficiency similar to that of the hexaene MGDG species obtained (in the dark) with UDP-[^{14}C]galactose as precursor.

From the determination of galactolipid synthesis and export rates at 5 and 25°C, we obtained similar Q_{10} values for MGDG synthesis and export (3.9), suggesting that both processes are influenced by temperature in a similar way. On the other hand, the Q_{10} value of DGDG export (4.5) was higher than that of its synthesis (3.8), which thus differ markedly with respect to their temperature dependence.

The sidedness analysis of lipid mass and label in TM treated by lipolytic enzymes revealed the following features. In most cases, the transmembrane distribution of labelled lipids in TM matched closely the corresponding distribution of mass **[3, 9]**, regardless of plant age and species. However, when significative differences occurred, they were found to depend exclusively on the lipid class involved: labelled MGDG molecules were always more inwardly oriented than the bulk of MGDG, whereas labelled DGDG and PG molecules were always more outwardly oriented than the bulk of DGDG and PG.

The transbilayer distribution of [^{14}C]acetate-labelled MGDG molecular species in TM was also investigated. Mono- and diene species, which together accounted for 72% of the total MGDG label in TM, showed an outside/inside molar ratio of 27/73. On the other hand, tri- and tetraene species, which together made up the remaining 28% of the total MGDG label, adopted a outside/inside molar ratio of 65/35. This latter value is equal to the native (mass) transmembrane distribution of MGDG **[3, 9]**. Consequently, the transbilayer distribution of molecular MGDG species synthesized de novo depends on their unsaturation level, with a critical double bond number of 3 per molecule. A functional link between lipid asymmetry in TM and lipid desaturation (which takes place in the IEM **[10]**) is thus established, via the lipid export mechanism.

How to account for these facts ? Intrachloroplastic

lipid export from IEM to TM would be achieved only by those mechanisms which allow the appearance of new lipids in both thylakoid monolayers simultaneously, i.e. a vesicular transfer and/or a non-vesicular process (direct, transient fusional interaction between IEM and TM), both involving complete fusion events and being ATP-independent. Both mechanisms might coexist or occur sequentially during development.

We propose to endow the envelope with a new competence, namely the biogenesis of lipid asymmetry in TM (see Figure). The localization of the various lipid-synthesizing (1, 2, 3) and -desaturating enzymes, the occurrence of contact sites (4) and the relative rates of transbilayer movement (short arrows) of each lipid molecule would play key roles in establishing the proper transbilayer distribution of each lipid in IEM. Transient fusions (5) with TM (connecting stroma-facing sides together and lumen-facing sides together) would allow the newly synthesized molecules to diffuse laterally from each monolayer (a, b, d, e) into TM, where adjustments in lipid asymmetry might later be made. The model requires a few sensible assumptions on the relative rates of transbilayer movement of nonbilayer- and bilayer forming lipids (DAG > MGDG >> DGDG, PG). It allows a single fusion mechanism to account for both lipid export to and lipid asymmetry in TM.

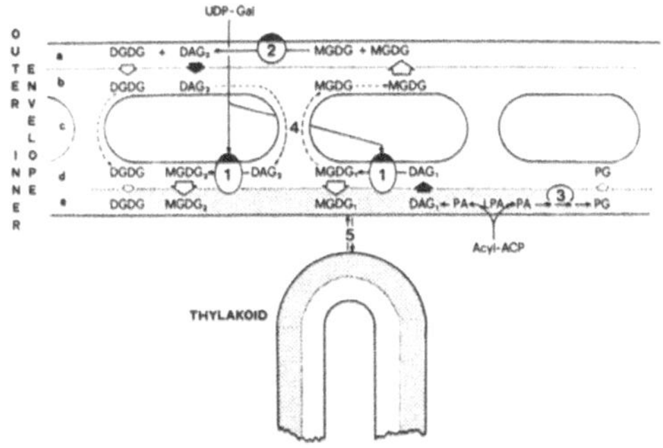

(Supported in part by grants 31.263 86.89 and 3100.33693.92 of the Swiss National Science Foundation)

References

[1] Joyard J, Douce R. Physiol. Vég. 1976;14:31-48
[2] Dorne AJ, Joyard J, Douce R. Proc. Natl. Acad. Sci. USA 1990;87:71-74
[3] Rawyler A, Meylan-Bettex M, Siegenthaler PA. Biochim. Biophys. Acta 1992;1104:331-341
[4] Mourioux G, Douce R. Plant Physiol. 1981;67:470-473
[5] Mudd JB, DeZacks R. Plant Physiol. Arch. Biochem. Biophys. 1981;209:584-591
[6] Heemskerk JWM, Bögemann G, Helsper JPFG, Wintermans, JFGM. Plant Physiol. 1988;86:971-977
[7] Mills JR, Joy KW. Planta 1980;148:75-83
[8] Siegenthaler PA, Smutny J, Rawyler A. Biochim. Biophys. Acta 1987;891:85-93
[9] Rawyler A, Unitt MD, Giroud C, Davies H, Mayor JP, Harwood JL, Siegenthaler PA. Photosynthesis Res. 1987;11:3-13
[10] Schmidt H, Heinz E. Proc. Natl. Acad. Sci. USA 1990;87: 9477-9480

LOCALIZATION OF LIPIDS AND XANTHOPHYLLS IN INSIDE-OUT VESICLES FROM THYLAKOIDS OF *NICOTIANA TABACUM*

J. Dädelow, A. Radunz and G. H. Schmid;Universität Bielefeld, Fakultät für Biologie, Lehrstuhl für Zellphysiologie, Postfach 10 01 31, 33501 Bielefeld, Germany

Introduction

Lipids of the thylakoid membrane show, as demonstrated by chemical, enzymatic, immunological and extraction analyses with solvents of differing polarities, an asymetric distribution in the transversal direction of the membrane [1-6]. Moreover, as functional complexes as PSI, PSII, the LHCP-complex, cytochrome b_6 and the ATPase complex of photophosphorylation differ with respect to their lipid composition, and as these complexes show lateral heterogenety in their distribution between grana and intergrana, the lipid distribution differs also in the grana and intergrana regions of the membrane system [1, 5, 7-9]. The lipids are adsorbed to these functional complexes by intermolecular interactions of differing strength. Corresponding to these different binding affinities, different functions might be attributed to these lipids. The major portion of these lipids is extractable with acetone and other polar organic solvents. Only a few lipid molecules are tighter bound onto the peptide subunits of these complexes and escape the extraction procedure. Even after SDS-polyacrylamide electrophoretic analyses the different galactolipids, anionic lipids, phospholipids and also carotenoids may still be detected on peptides of these complexes by means of the Western Blot-procedure and the use of monospecific antisera [10-13].

In context with structure-function analyses of operational photosynthetic complexes so-called inside-out vesicles from thylakoids of *N. tabacum* var. JWB were used [14, 15]. These vesicles have the advantage to originate from the partition region of the thylakoid membrane and to occur as particles in which by vesicle formation a defined membrane side, namely the originally towards the lumen directed surface becomes surface exposed. This is obtained by an isolation procedure in which distribution between the two-phase gradient polyethylene glycol dextran is used [15]. These inside-out vesicles represent an ideal material for immunological analyses of structure-function relationships. In a first paper we report here on the reaction of monospecific antisera towards the main components of lipids and xanthophylls with these vesicles.

J.-C. Kader and P. Mazliak (eds.), Plant Lipid Metabolism, 164–169.

Materials and Methods

Isolation of inside-out vesicles

Inside-out vesicles were isolated from *Nicotiana tabacum* var. JWB according to the procedure of Andersson and Åkerlund [15]. The breaking of the cells was carried out in a French press by two subsequent pressings at 560 *psi*. The mixture of the resulting inside-out and right-side out vesicles, having differing surface structure, was separated in a two-phase gradient of polyethylene glycol and dextran. Whereas the right-side vesicles accumulate in the upper polyethylene glycol phase, the inside-out vesicles occur in the lower dextran-phase. From this phase; the inside out vesicles are centrifuged at 100.000 xg according to Åkerlund et al. [16] and are resuspended in 5 mM Na_2HPO_4/NaH_2PO_4, 2.5 mM NaCl and 0.5 M sucrose (pH 7.4).

Measurement of photosynthetic electron transport reactions

Photosynthetic electron transport reactions were measured with an oxygen electrode of Rank Brothers (England). The conditions are summarized in Table 1. In order to determine optimal conditions; the reduction of 2.6 dichlorophenol indophenol was measured in dependance on its pH and the KCl-concentration. These experiments show that the maximal reaction was obtained in 30 mM tricine buffer at pH 6.5 and a KCl addition of 20 mM.

Antisera

The antisera to lipids and carotenoids have been characterized in an earlier publication [4, 5, 10-11]. The protein antisera to the D-proteins, the LHCP-complex and the OEE_1-protein were obtained by immunization of rabbits. Monospecificity of the antisera to the core-peptides and to the LHCP was demonstrated by means of the Western-Blot procedure.

Electronmicroscopy

The preparations (Fig. 1) were fixed in 5% glutaraldehyde and post-fixed with a solution of OsO_4. Dehydratation was carried out with an acetone water mixture with acetone increasing from 15 to 100%. The preparation was embedded in a Spurr mixture. The 60 nm thick sections were stained with 1% permanganate and post-stained with 2% lead citrate and viewed under the electron microscope.

Results and Discussion

Figure 1 shows an inside-out vesicle preparation [15,16] in comparison to the photosystem II membrane preparation [17] from *Nicotiana tabacum* var. JWB. The vesicle character of the preparation is clearly seen. From the strong agglutination reaction of the antiserum to the OEE_1 protein (with a molecular mass of 33 kDa [18]) and the strong inhibitory effect of this antiserum on photosynthetic electron transport in the region of photosystem II, it is clearly seen that the used vesicles are indeed inside-out vesicles, in which this protein, which in native chloroplasts is located on the luminal side of the membrane, is now directly surface exposed and thus accessible to antibodies. By washing with 1.5 M $CaCl_2$ this protein as well as the other two extrinsic proteins OEE_2 and OEE_3 were removed to a large extent.

The protein/chlorophyll ratio of these particles obtained from grana thylakoids is 4.6 +/- 0.4 and corresponds to that of vesicles from spinach [15]. With PSII-membrane preparations of *Nicotiana tabacum*, which have been obtained according to the method

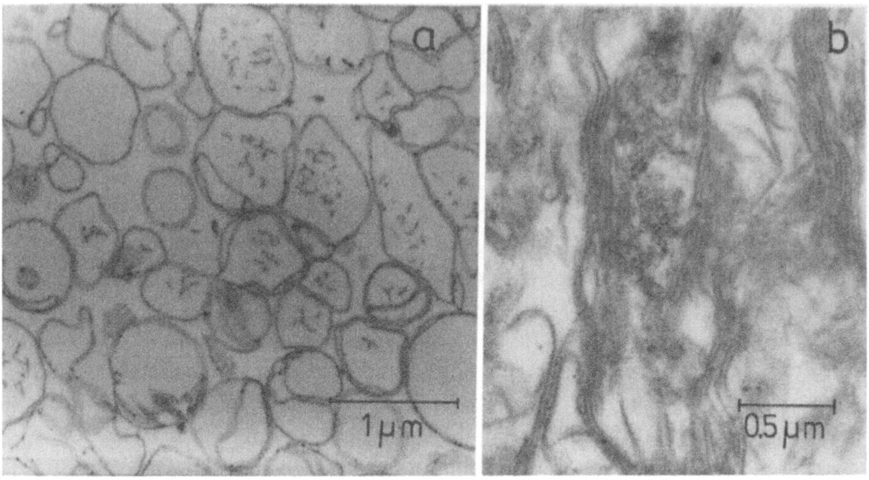

Fig. 1. EM-picture of an inside-out vesicle preparation (a) and a photosystem II-particle preparation (b) of *N. tabacum* var. JWB

of Berthold et al. [17] also from grana thylakoids, the chlorophyll content equally makes up for 25% of the amount of protein present. Measurement of photosynthetic electron tranpsort in the region of photosystem II shows that the rate corresponds to 200 µMoles DCPiP reduced mg $Chl^{-1} \cdot h^{-1}$. The PSI-activity measured as electron transport from DCPiP to methylviologen is about half the former rate. This shows that these preparations contain in addition PSI-membrane regions. The activity measurements give no information on the presence of potential other impurities.
The agglutination reactions of the inside-out vesicles with the antisera to lipids, xanthophylls and peptides are compared in table 1 with agglutination reactions of stroma-freed chloroplasts in which the thylakoid surface directed to the chloroplast stroma is accessible. Whereas antibodies to the anionic lipids are only monovalently bound to the outer surface, and therefore do not yield agglutination, inside-out vesicles are directly agglutinated. Antibodies to phosphatidylglycerol cause a considerably stronger agglutination than antibodies to sulfolipids. For the xanthophyll antibodies with both chloroplast preparations the same reactions are observed as with antibodies to anionic lipids. A comparison between the antibodies to violaxanthin and lutein shows that the antibodies to violaxanthin cause a considerably higher degree of agglutination than the lutein antibodies. With the neutral galactolipids the differences in the reaction of the antibodies with the outer and the inner surface of the thylakoid membrane are not so high. Thus both, inside-out vesicles as well as stroma-freed chloroplasts are agglutinated by the two galactolipid antisera. However, antibodies to digalactolipid react with inside-out vesicles considerably weaker than those to monogalactolipid. These results clearly show that monogalactolipid, the negatively charged phosphatidylglycerol and violaxanthin are stronger exposed in the surface directed to the thylakoid lumen than digalactolipid, lutein and the anionic

Table 1: Comparison of Agglutination Reactions of Antisera to Lipids, Xanthophylls and Proteins with Inside-out Vesicles and with Stroma-freed Chloroplasts of *N. tabacum* var. J.W.B. and Comparison of the Inhibitory Effect of these Antisera on Electron Transport Reactions of Photosystem II.

Antiserum	Agglutination of		Inhibition of PSII-reaction in % of the control reaction	
	inside-out vesicles	stroma-freed chloroplasts	inside-out vesicles[a]	stroma-freed chloroplasts[b]
MGDG	3+	4+	0	17-54
DGDG	+	4+	not determined	13-49
SQDG	+	only monovalent antibody binding	16	38
PG	4+	only monovalent antibody binding	not determined	48[c]
Violaxanthin	4+	only monovalent antibody binding	8	25-37
Lutein	2+	only monovalent antibody binding	not determined	35
D1-Corepeptide	2+	+	25	11-22
D2-Corepeptide	+	+	13	11-20
LHCP-complex	2+	+	27	18
33 kDa (OEE_1)	4+	+	30	38

a. The electron transport reaction was measured with inside-out vesciles as DCPiP-reduction with water as the electron donor (DCPiP-concentration 8.3 x 10^{-5} M/assay). **b.** With stroma-freed chloroplasts in addition to reaction a) also the reduction of ferricyanide or silicomolybdate with water as the electron donor was measured. (Electron acceptor concentration in the assay was: 34 x 10^{-5} M or 2.3 x 10^{-4} M). **c.** The antiserum to phosphatidylglycerol inhibits electron transport in the region of photosystem II only after washing the chloroplasts with 0.8 M Tris buffer. Electron transport was measured with an oxygen electrode from Rank Brothers (England).

With Inside-out vesicles only the surface directed to the thylakoid lumen is accessible to antibodies. With stroma-freed chloroplasts only the surface directed towards the chloroplast stroma is accessible.

sulfolipid. These conclusions can at least be drawn for the partition regions of the thylakoid membrane, from which the inside-out vesicles have been obtained. On the surface directed to the stroma at least the accessibilities of the monogalactolipid and sulfolipid to antibodies are hampered by the CF1-complex of the ATPase.

Measurement of photosynthetic electron transport revealed that antibodies to sulfolipid inhibit the reduction of DCPiP with water as the electron donor strongly although inside-out vesicles are only weakly agglutinated by this antiserum. On the other hand, the antiserum to violaxanthin, which strongly agglutinates these vesicles, affects only weakly electron transport reactions. The monogalactolipid antiserum which also strongly agglutinates these vesicles has apparently no influence on electron transport reactions in the region of photsystem II. Part of these results are in contrast to the effect of the corresponding antisera on electron transport reactions in the region of photosystem II in intact stroma freed chloroplasts (table 1). Binding of antibodies to all investigated lipids located on the surface directed towards the stroma, inhibited photosystem II-mediated electron transport reactions independent on whether DCPiP, $K_3Fe(CN)_6$ or $SiMoO_4^{-2}$ were the electron acceptors. The degree of inhibition was between 17 and 54% of the control reaction. These different inhibitory effects, caused by antibody binding are apparently caused by the different binding affinities of the antibodies onto the lipid antigens of the thylakoid membrane surface directed towards the outside or towards the membrane lumen. On the other hand, it is conceavable that due to the binding of antibodies onto lipid antigens the protein molecules of the different thylakoid membrane surfaces undergo different conformational changes which in turn lead to different inhibitions of photosynthetic electron transport.

A study on the binding of lipids on peptides of inside-out vesciles by means of the Western-Blot procedure led to the same results as the study on binding of lipids onto peptides of photosystem II-particles [10]. Sulfolipid is detectable on the D_1-core peptide, if the peptide occurs in the dimeric form. The violaxanthin antiserum also reacts with the D_1-peptide. The monogalactolipid antiserum labels besides the peptides of the LHCP-complex with molecular masses 28, 26 and 25 kDa and the chlorophyll-binding peptides of 43 and 47 kDa also peptides with the molecular masses 66 kDa. The labelling of the 66 kDa peptides seems to be due to the reaction of the monogalactolipid antiserum with the core peptides of photosystem I, since these vesicles, as shown above, contain impurities from photosystem I [19].

In this context the behaviour of the antisera to the core peptides D_1 and D_2 is of particular interest. Both antisera cause in inside-out vesicles or in stroma-freed chloroplasts only a small inhibition of photosynthetic electron transport. The small effect is remarkable not only in the sense that here monospecific antisera have reacted but also in the sense that antibodies bind onto antigenic determinants which on both

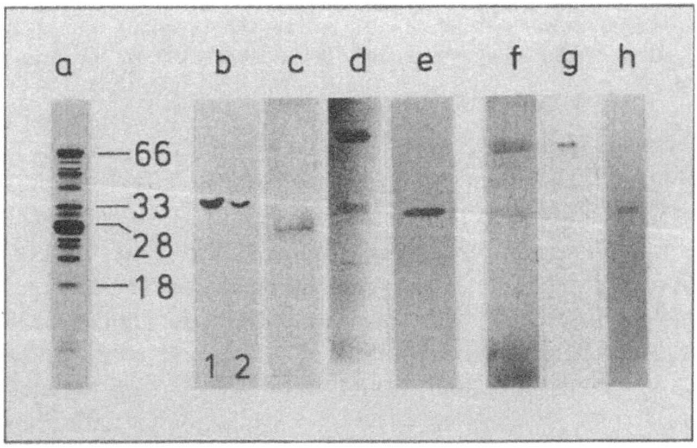

Fig. 2: Peptide analysis of inside-out vesicles in the SDS-polyacrylamide gel electrophoresis and result of Western-Blot analyses. a) Peptide analysis in the SDS-gel; b-h) Nitrocellulose membranes with peptides of inside-out vesicles after reaction of the antisera to b_1) 33 kDa peptide (OEE_1); b_2) 33 kDa peptide obtained by $CaCl_2$-washing of the reaction with the homologous antiserum; c) LHCP-complex; d) D_1-core peptide (the antiserum labels the 32 kDa D_1-peptide and a 66 kDa aggregate formed with a 17 kDa fragment of this peptide); e) D_2-core peptide; f) Monogalactolipid; g) Sulfolipid; h) Violaxanthin

sides reach out of the membrane. That means that the antibody binding does not affect or reach reactive regions of the electron transport components. On the other hand in the case of the OEE_1 peptide, which is located on the luminal surface of the thylakoid membrane and which is directly accessible to antibodies in inside-out vesicles, antibody binding causes a high inhibitory effect of electron transport reactions in the region of photosystem II.

References

1. Gounaris K, Barber J, Harwood JL. Biochem J 1986; 237: 313-326.
2. Rawyler H, Siegenthaler A. Eur J Biochem 1980; 110: 979.
3. Unitt MB, Harwood JL. In: Wintermans GM, Kuiper PJC, editors. Biochemistry and Metabolism of Plant Lipids. Elsevier/North Holland, Biomed. Press, Amsterdam, 1982: 359-362.
4. Radunz A. Ber Deutsch Bot Ges 1981; 94: 477-489.
5 Schmid GH, Radunz A, Gröschel-Stewart U. Immunologie und ihre Anwendung in der Biologie. Georg Thieme Verlag, 1993.
6. Costes C, Bazier R, Lechevallier D. Physiol Veg 1972; 10: 291-297.
7. Bednarz J, Radunz A, Schmid GH. Z Naturforsch 1988; 43c: 423-430.
8. Ryrie IJ, Andersson JM, Goodshild DJ. Eur J Biochem 1980; 137: 205-213.
9. Murphy DJ, Woodrow JE. Biochim Biophys Acta 1983; 725: 104.
10. Voß R, Radunz A, Schmid GH. Z Naturforsch 1992; 47c: 406-415.
11. Kruse O, Radunz A, Schmid GH. Z Naturforsch 1994; 49c: 115-124.
12. Haase R, Unthan M, Coutourier P, Radunz A, Schmid GH. Z Naturforsch 1993; 48c: 623-631.
13. Makewicz A, Radunz A, Schmid GH. Z Naturforsch 1994; 49c (in press).
14. Schmid GH, Price JM, Gaffron H. J de Microskop 1966; 5: 205-216.
15. Andersson B, Åkerlund HE. Biochim Biophys Acta 1978; 503: 462-472.
16. Åkerlund HE, Jansson C, Andersson B. Biochim Biophys Acta 1982; 681: 1-10.
17. Berthold DA, Babcock GT, Yocum CF. Febs Lett 1981; 134: 231-234.
18. Menke W, Koenig F, Radunz A, Schmid GH. Febs Lett 1975; 49: 372-375.
19. Makewicz A, Radunz A, Schmid GH. (this volume)

THOUGHTS CONCERNING A NEW PARADIGM OF THE PHOTOSYSTEM II REGION OF THE THYLAKOID MEMBRANE BASED ON LIPID STRUCTURE AND FUNCTION

Paul-André Siegenthaler, Yinong Xu, Jana Smutny, Marlyse Meylan Bettex, Jarmila Vallino and André Rawyler

Laboratoire de Physiologie végétale, Université de Neuchâtel, rue Emile-Argand 11, CH-2007 Neuchâtel, Switzerland

Each type of membrane is characterized by the amount and kind of its proteins and lipids. When considering the numerous models which have been published in the literature on the molecular organization of the thylakoid membrane (TM), it is surprising that almost nobody has attempted to represent the acyl lipids according to their topological distribution and associated functions. This failure is probably due to the fact that, contrarily to their counterpart protein moiety, lipids have no catalytic properties by themselves. The purpose of this communication is to propose a tentative paradigm shedding light on the molecular architecture of the TM in which not only proteins but also acyl lipids are considered. The concept of this model is based on recent studies [1-5].

Murata et al. [1] have studied three different photosystem II (PSII) preparations and compared their polypeptide and glycerolipid components as well as their photosynthetic activities. Several of their findings can bring interesting informations for the construction of a new paradigm of the molecular organization of acyl lipids in the TM : (1) The reaction center (RCII) which contains five proteins (the D1 and D2, the large and small subunits of cytochrome b_{559} and the 4.8 kDa hydrophobic protein) is tightly associated with 1 mol of MGDG (monogalactosyldiacylglycerol) and 0.1 mol PG (phosphatidylglycerol) per mol P_{680}. The presence of these two lipids is probably necessary for the photochemical charge separation; (2) the PSII core complex contains two additional antenna core proteins (43 and 47 kDa), the 33 kDa extrinsic protein and one hydrophobic protein. These proteins which bound firmly MGDG, DGDG (digalactosyldiacylglycerol) and PG (3-4 mol / mol P_{680} for each lipid) confer to this complex the capacity of performing O_2 evolution; (3) the molar ratio of glycerolipids to P_{680} reveals that PSII membranes (i.e. PSII core complex + LHCII + the 23 and 18 kDa extrinsic proteins) can loose about 97, 90 and 99% of their MGDG, DGDG and PG molecules, respectively, without affecting the ability to support charge separation and O_2 evolution. This is an indication that most lipids are in the bulk phase and only a few of them are involved in protein-lipid interactions and consequently may play particular structural and/or functional role; (4) when compared to

J.-C. Kader and P. Mazliak (eds.), Plant Lipid Metabolism, 170–172.

the bulk phase lipids the relative content of saturated fatty acids (16:0; 18:0) increases and that of trienoic acids (18:3) decreases in the PSII core complex and even more in the RCII complex.

Trémolières et al. [2] have determined the lipid and fatty acid composition in various purified light-harvesting complexes II, i.e. in the major (Lhcb1 : 28 kDa; Lhcb2 : 27 kDa) and the minor LHCII (Lhcb3 : 25 kDa; Lhcb4 : CP 29; Lhcb5 : CP26; Lhcb6 : CP 24). Their data support the view that a great heterogeneity exists in the lipid class distribution and in the degree of unsaturation in this region of the TM: (1) the minor LHCII complexes retain high amounts of lipids in the decreasing order MGDG > DGDG >> PG. The estimated lipid/polypeptide molar ratio varies from 15 to 25 for MGDG, 3 to 10 for DGDG and about 1 for PG; (2) in contrast, in the major LHCII complexes this ratio drops to about 1 for all the above lipids; (3) the relatively high amount of MGDG and to some extent of DGDG molecules which are associated with the minor complexes suggests that these lipids have linking properties between the PSII core complex and the antennae (Lhcb 3-6). Concerning the PG molecule, it might be anchored in the apoprotein-chlorophyll domains of these minor complexes; (4) in the major LHCII complexes, the low lipid/polypeptide ratio excludes the view that lipids can play a cohesion role in the organization of these antennae but rather a specific role in the function of LHCII; (5) in LHCII complexes, the galactolipids are slightly unsaturated and PG much more saturated than in the corresponding bulk lipids.

Siegenthaler's group [3-5] has studied the bulk lipid phase of the TM : (1) when acyl lipids are expressed in mol % of total lipids, MGDG (35 mol %) + DGDG (4 mol %) in the outer monolayer and DGDG (21 mol %) + MGDG (19 mol %) in the inner monolayer are prominent - both galactolipids representing about 40 mol % of the total lipid in each monolayer; (2) the minor lipid classes SQDG (sulfoquinovosyldiacyl-glycerol), PG and PC (phosphatidylcholine) represent not more than 10% in each monolayer. (3) Altogether, these results strongly suggest that most galactolipids have a structural role in the bulk lipid phase whereas the minor lipids and a few saturated galactolipid molecules display a particular functional role in the TM.

In addition to these topological studies, we have shown [5] - using a lipid depletion technique, that the population of PG molecules (70 mol % of the lipid class) localized in the outer monolayer is likely to have a structural role in the bulk phase. In contrast, the inner PG population includes several yet undefined subpopulations, some of them, very discrete (3 mol %) being highly efficient in supporting electron flow and O_2 evolution activities. Thus, these subpopulations of PG have a particular role to play in the TM function.

Based on these results [1-5], we have built a new paradigm of the molecular organization of the PSII region of the TM (Fig. 1). This model inspired from that of Jansson [6] underlines the topological distribution of acyl lipids. All big circles, squares and triangles represent lipids which are firmly bound to proteins and thus are likely

to play a special structural and/or functional role. Small circles and stars represent MGDG and DGDG which are located in the bulk lipid phase of TM. All open symbols underline the fact that lipids closely bound to proteins, especially in the RCII, are saturated. Though we have tried to respect the stoichiometry between lipids and polypeptides, one has to keep in mind that the use of detergents for the isolation of PSII preparations may cause the loss of lipids and the perturbation in their topological distribution. Anyhow, this model shows clearly a great heterogeneity of lipid classes and of their molecular species.
Supported in part by the Swiss National Science Foundation.

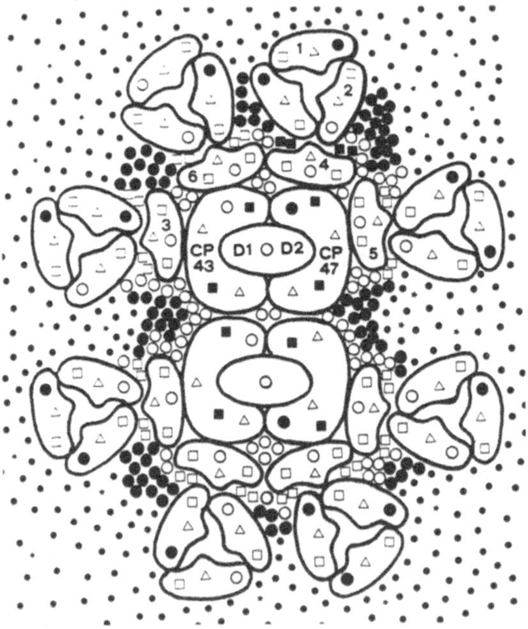

Fig. 1. A new paradigm of the PSII region of the thylakoid membrane (top view). ○●, saturated and unsaturated MGDG; □ ■, saturated and unsaturated DGDG; △ ▲, saturated and unsaturated PG; ● ✶, MGDG and DGDG of the bulk lipid phase of TM. The reaction center is represented only by the D1 and D2 proteins, the core complex by additional proteins CP43 and CP47, and the different LHCII complex by 1 and 2 (Lhcb1,2), 3,4,5 and 6 (Lhcb3-6), according to Jansson [6].

References
[1] Murata N, Higashi SI, Fujimura Y. Glycerolipids in various preparations of photosystem II from spinach chloroplasts. Biochim Biophys Acta 1990;1019:261-268.
[2] Trémolières A, Dainese P, Bassi R. Heterogenous lipid distribution among chlorophyll-binding proteins of photosystem II in maize mesophyll chloroplasts. Eur J Biochem 1994;221:721-730.
[3] Rawyler A, Siegenthaler PA. Transversal localization of monogalactosyldiacylglycerol and digalactosyldiacylglycerol in spinach thylakoid membranes. Biochim Biophys Acta 1985;815:287-298.
[4] Siegenthaler PA, Rawyler A, Smutny J. The phospholipid population which sustains the uncoupled non-cyclic electron flow activity is localized in the inner monolayer of the thylakoid membrane. Biochim Biophys Acta 1989;975:104-111.
[5] Jansson S. The light-harvesting chlorophyll a/b-binding proteins. Biochim Biophys Acta 1994;1184:1-19.

LIPID DEPENDENCE OF PROTEIN KINASE ACTIVITIES IN SPINACH CHLOROPLAST ENVELOPE MEMBRANES

Lucien BOVET, Jana SMUTNY, Paul-André SIEGENTHALER
Laboratoire de Physiologie végétale, Université de Neuchâtel

INTRODUCTION

Spinach chloroplast envelope membranes, incubated *in vitro* in the presence of [γ^{32}P]ATP, incorporated labelled phosphate mainly in three proteins exhibiting Mr of 67, 26 and 14 kDa. Two of them, the 26 and 14 kDa phosphoproteins, are pools of LHCIIb and Rubisco small subunit tightly bound to envelope membranes [1] and are phosphorylated by a unique Ca^{2+}-dependent serine protein kinase [2]. The 67 kDa phosphoprotein is likely to be autophosphorylated; its ^{32}P-labelling would take place *via* its own phosphoglucomutase activity [2-4]. The aim of this investigation is to modulate envelope protein kinase activities following incubation for various times with the phospholipase C (PLC) from Bacillus cereus or the lipase from Rhizopus arrhizus (LRa).

METHODS

Spinach chloroplast envelope membranes were prepared according to [5]. Envelope protein samples (30 µg) were phosphorylated in the presence of 33 nM [γ^{32}P]ATP as described in [2]. The envelope protein kinase activities were stopped after incubation at 23°C for 1.5 min by adding one volume of SDS-PAGE sample buffer. The incorporated [^{32}P]phosphate into proteins was quantified by the Cerenkov procedure. Lipid extraction and thin layer chromatography (TLC) were carried out as described in [2] and [6]. Immediately after charring, the separated lipids on TLC plates were quantified by laser densitometry (Bioimage Millipore).

RESULTS

Figure 1A-B shows that in the presence of a small amount of PLC, phosphatidylcholine (PC) was rapidly hydrolysed (>90% in 5 min) whilst the content of phosphatidylglycerol (PG) underwent a decrease of less than 20%. An addition of a large amount of PLC induced the complete hydrolysis of the remnant pool of PG. In the presence of a small amount of LRa, monogalactosyldiacylglycerol (MGDG) was rapidly hydrolysed (>95% in 20 min) whilst the content of digalactosyldiacylglycerol (DGDG) remained constant. The addition of a large excess of LRa induced the

J.-C. Kader and P. Mazliak (eds.), Plant Lipid Metabolism, 173–175.

complete hydrolysis of DGDG. During the hydrolysis, the content in free fatty acids (FFA), lyso-MGDG (LMG) and lyso-DGDG (LDG) increased in a concomitant fashion.

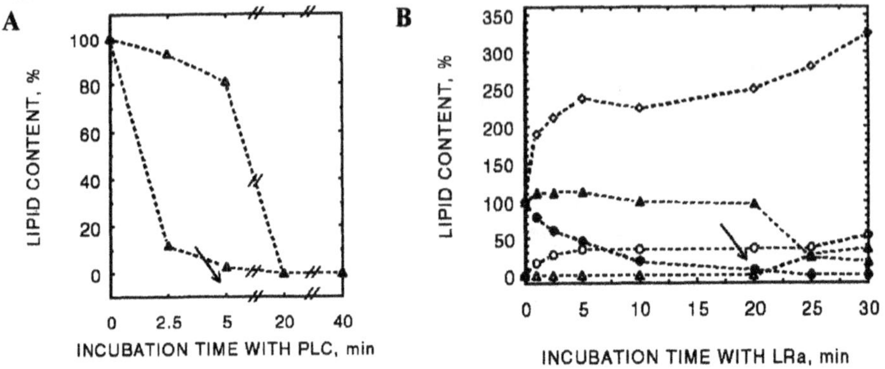

Figure 1A-B. Envelope membrane vesicules (150 µg proteins for phospholipase C from Bacillus cereus: PLC and 240 µg proteins for the lipase from Rhizopus arrhizus: LRa) were incubated for various times with **(A)** PLC (~0.6U for 210 µl of reaction medium) → low [ ]; > 8U for the remaining volume of reaction medium → high [ ] (see arrow) or **(B)** LRa (~7U for 480 µl of reaction medium) → low [ ]; > 50U for the remaining volume of reaction medium → high [ ] (see arrow) before lipid extraction. **A** Time course of the PC (▲) and PG (△) hydrolysis with PLC. **B** Time course of the MGDG (•) and DGDG (▲) hydrolysis with the LRa. The hydrolysis products (△) LDG (○) LMG (◊) FFA are also shown in the graph.

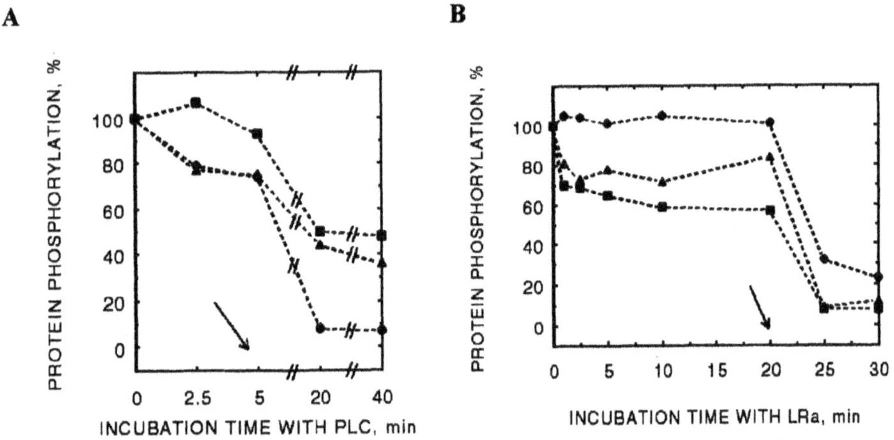

Figure 2A-B. Envelope protein phosphorylation following lipid hydrolysis with PLC **(A)** or LRa **(B)**; graphic representation showing the ^{32}P labelling kinetics of the 67 (•), 26 (▲) and 14 (■) kDa phosphoproteins. The additions of high [ ] of PLC or LRa are indicated by arrows.

Figure 2A-B shows that the complete hydrolysis of PC (>95%) affected only slightly the [^{32}P] incorporation in the envelope proteins. On the other hand, the complete hydrolysis of PG (>95%) inpaired markedly the [^{32}P] incorporation in the envelope proteins. Interestingly the [^{32}P] incorporation in the 67 kDa phosphoprotein was completely abolished. This was not the case for the 26 and 14 kDa phosphoproteins since both of them kept notwithstanding about 40% of their initial labelling. The hydrolysis of MGDG affected moderately the [^{32}P] incorporation in the three envelope proteins, whereas the hydrolysis of DGDG inhibited significantly their phosphorylation.

The complete hydrolysis of PG and PC by PLC had almost no effects on the phosphorylation of an exogenous substrate such as myelin (data not shown). However, the hydrolysis of galactolipids by LRa (particularly the DGDG pool) affected significantly the myelin phosphorylation, since more than 60% [^{32}P]-labelling inhibition was observed following the addition of high LRa concentrations (data not shown).

CONCLUSION

These results indicate that the incorporation of phosphate in the lumenal phosphoglucomutase (67 kDa), likely occuring *via* its own autophosphorylation [2-4], is dependent on the presence of envelope lipids (particularly phospholipids). Concerning the Ca^{2+}-dependent serine protein kinase responsible for the phosphorylation of LHCIIb (26 kDa) and the Rubisco small subunit (14 kDa) bound to envelope membranes [1], the PLC and LRa treatments reveal that the protein kinase needs envelope lipids to be activated. The comparison of the results obtained with or without the addition of an exogenous substrate (myelin) in the reaction medium suggests that PG and PC (mainly PG) are necessary to present the phosphorylation site(s) of LHCIIb and Rubisco small subunit to the protein kinase, since PLC has no effect on the myelin phosphorylation. On the contrary, galactolipids (more specifically DGDG) seem essential for the activation of the protein kinase itself, since the phosphorylation of endogenous envelope substrates (LHCIIb (26 kDa) and Rubisco small subunit (14 kDa) as well as exogenous substrates (myelin; histone III-S, data not shown) is indiscriminately sensitive to LRa treatments.

REFERENCES

[1] Bovet L, Müller MO, Siegenthaler PA. The 26 and 14 kDa phosphoproteins associated with spinach chloroplast envelope membranes are distinct membrane-bound pools of LHCII and of Rubisco small subunit. Planta 1994 (in press)

[2] Siegenthaler PA, Bovet L. A unique protein kinase activity is responsible for the phosphorylation of the 26 and 14 kDa proteins but not of the 67 kDa protein in the chloroplast envelope membranes of spinach. Planta 1993;190:231-240.

[3] Salvucci ME, Drake RR, Broadbent KP, Haley BE, Hanson KR, McHale NA. Identification of the 64 kDa chloroplast stromal phosphoprotein as phosphoglucomutase. Plant Physiol 1990;93:105-109.

[4] Soll J, Berger, V, Bennett J. Adenylate effects on protein phosphorylation in the interenvelope lumen of pea chloroplasts. Planta 1989;177: 393-400.

[5] Siegenthaler PA, Dumont N. Characteristics of spinach chloroplast envelope, Thylakoid and stroma polypeptides as revealed by Triton X-114 phase partition. Plant Cell Physiol 1990;31:1101-1108.

[6] Siegenthaler PA, Rawyler A, Smutny J. The phospholipid population which sustains the uncoupled non-cyclic electron flow activity is localized in the inner monolayer of thylakoid membrane. Biochim Biophys Acta 1989;975:104-111.

POLAR LIPID COMPOSITION, PHOTOSYNTHETIC ACTIVITY AND CHLOROPHYLL-PROTEIN COMPLEXES OF CHLOROPLASTS FROM HEMIPARASITIC ANGIOSPERMS ORIGINATING FROM TEMPERATE AND TROPICAL AREAS

Christiane Tuquet[1] and Georges Sallé[2]

Laboratoire de Biologie Cellulaire Végétale[1] and Laboratoire de Cytologie Expérimentale et Morphogenèse Végétale[2], Université Pierre et Marie Curie Paris, France.

Introduction

Parasitic flowering plants are usually named hemiparasites or holoparasites according to the presence or the absence of chlorophyll. Hemiparasitic plants are capable of independent photosynthesis but the rates are low. In a previous work [1], a study of chloroplasts isolated from the witchweed *Striga* (Scrophulariaceae), a root parasite in semi-arid tropics, showed some specific changes concerning more particularly the polar lipid composition. However, it was difficult to relate these changes to parasitism only, or both to parasitism and adaptation to drought since *Striga* sustains highly dry conditions in its natural environment. Therefore, the study of the biochemical composition and photosynthetic activity of chloroplasts, isolated from two different mistletoes growing in temperate (*Viscum album*, Viscaceae) and in semi-arid (*Tapinanthus dodoneifolius,* Loranthaceae) countries, seemed to be interesting to perform.
This study showed that chloroplasts from hemiparasitic plants reveal a trend for low values of chlorophyll and polar lipids while the protein content is less affected. "Tropical chlroroplasts" present marked changes in polar lipid composition along with a decrease in unsaturation of fatty acids. Chloroplasts from all hemiparasitic plants are capable of a more or less limited photosynthesis, probably reflecting the degree of dependence on the host.

Materials and methods

MATERIALS

Striga hermonthica plants parasitizing sorghum (*Sorghum bicolor)* were collected in their natural surroundings in Mali. *V album* was collected on apple trees in the countryside around Paris. *T. dodoneifolius* growing on shea trees was harvested in Mali.

METHODS

Plastid isolation. Chloroplasts were isolated from green leaves of *S. hermonthica* plants according to [1]. The technique of plastid isolation was modified for mistletoes as the large amounts of lipids present in the thick leaves made the isolation very difficult.

J.-C. Kader and P. Mazliak (eds.), Plant Lipid Metabolism, 176–178.

Pigment, lipid and protein analyses. Chlorophyll, protein, polar lipids and fatty acids were analyzed as in [1]. Measurement of photosynthetic activities and separation of chlorophyll-protein complexes were performed according to [1].

Results

For all quantitative determinations, a range of values corresponding to those usually found in non-parasitic higher plants is reported for comparison [2,3].

CHLOROPHYLL AND PROTEIN CONTENT.

Chloroplasts from the three hemiparasitic plants display a normal Chl*a* /Chl *b* ratio but contain very low amounts of chlorophyll, specially those from *T. dodoneifolius*. On the other hand, the protein content is less reduced (Table 1).

Table 1. Chlorophyll, protein and polar lipid contents.

Material	Chlorophyll (molecules plastid^{-1})	Protein/Chl (w/w)	Polar lipids (molecules plastid^{-1})
S. hermonthica	465 ± 95(x10^6)	19.6 ± 15	470 ± 120(x10^6)
V. album	451 ± 88(x10^6)	16.1 ± 45	1402 ± 98(x10^6)
T. dodoneifolius	267 ± 97(x10^6)	20.8 ± 7.1	799 ± 102(x10^6)
Higher plants	500 - 1000(x10^6)	10 - 5	1200 - 2900(x10^6)

POLAR LIPIDS.

Expressed as molecules per plastid, the total amount of polar lipids is particularly low for chloroplasts from *S. hermonthica* and *T. dodoneifolius*. It is higher in *V. album* chloroplasts ranking them within the lowest range of chloroplasts from non-parasitic plants (Table 1). Chloroplasts from *S. hermonthica* and *T. dodoneifolius* (plants from semi-arid countries) present an increased relative content of phospholipids at the expense of glycolipids while the proportions are close to normal for *V. album* (Table 2). Among phospholipids, the proportion of phosphatidylglycerol (PG) is mainly increased, specially in chloroplasts from semi-arid countries. *S. hermonthica* exhibits the largest decreases in glycolipids and displays an MGDG / DGDG ratio lower than 1.

Table 2. Polar lipid composition.

Material	% of total polar lipids		MGDG/DGDG	PG content
	Phospolipids	Glycolipids	(molar ratio)	(molecules plastid^{-1})
S. hermonthica	45	55	0.8	149 ± 67(x10^6)
V. album	20	79	1.14	158 ± 38(x10^6)
T. dodoneifolius	30	70	1.20	119 ± 42(x10^6)
Higher plants	20 - 30	80 - 70	1.1 - 1.4	180 - 300

DGDG, digalactosyldiacylglycerol; MGDG, monogalactosyldiacylglycerol.

FATTY ACID COMPOSITION OF POLAR LIPIDS.

A decrease in the percentage of linolenic acid is observed in all polar lipids of chloroplasts from semi-arid tropics. Phospholipids of all hemiparasitic plants contain oleic acid in large amounts and glycolipids to a lesser degree. The specific *trans*-Δ_3-hexadecenoic acid is present in phosphatidylglycerol of all parasitic plants in normal proportion.

PHOTOSYNTHETIC ACTIVITY.

On a chlorophyll basis, chloroplasts from the hemiparasites display low photosystem II (PS II) and photosystem I (PSI) activities. Mistletoe chloroplasts exhibit large decreases in PSI activity, particularly those from *T. dodoneifolius* .

CHLOROPHYLL-PROTEIN COMPLEXES.

Separation of the main chlorophyll-protein complexes indicate a high proportion of the oligomeric form of the light-harvesting chlorophyll *a/b*-protein complex of photosystem II (LHC II) in *S. hermonthica* chloroplasts and to a lesser degree in *V. album* and *T. dodoneifolius* chloroplasts. The depletion in P700 chlorophyll *a*-protein of photosystem I of *T. dodoneifolius* chloroplasts must account for the extremely low photosystem I activity measured in these plastids.

Conclusion

Chloroplasts isolated from the three hemiparasites display low levels of chlorophyll and polar lipids. Changes occur in polar lipids of *S. hermonthica* chloroplasts and, to a lesser degree, in those of *T. dodoneifolius*, phospholipids increasing at the expense of galactolipids. "Tropical chloroplasts" have higher proportions of PG with *trans*-Δ_3-hexadecenoic acid and present a decrease in the unsaturation of fatty acids, particularly in galactolipids. *S. hermonthica* chloroplasts show low but better photosynthetic activities than mistletoe chloroplasts, which exhibit large decreases in activities of both photosystems. The depletion of *T. dodoneifolius* chloroplasts in P700 chlorophyll *a*-protein of photosystem I (PS I) fits well with the extremely low PS I activity detected in these plastids. The maintenance of a good proportion of PG with C16:1*trans* could allow the stabilization of the oligomeric form of LHCII, optimizing the collection of energy, and thus partly compensate for deficiencies due to parasitism.

References

1 Tuquet C, Farineau N, Sallé G. Biochemical composition and photosynthetic activity of chloroplasts from *Striga hermonthica* and *Striga aspera,* root parasites of field-grown cereals. Physiol Plant 1990;78:574-582.

2. Guillot-Salomon T, Tuquet C, Signol M. Analyse comparative de l'ultrastructure et de la composition lipidique des chloroplastes de plantes d'ombre et de soleil. Cytobiologie 1978; 17:442-452.

3. Harwood JL. Plant acyl lipids: structure, distribution and analysis. In: Stumpf PK, ed. The Biochemistry of plants. A comprehensive treatise. New York: Academic Press, 1980;(4):1-55.

IN VIVO MANIPULATION OF LIPID COMPOSITION IN MUTANTS OF *CHLAMYDOMONAS REINHARDTII*: A TOOL TO STUDY THE ROLES OF LIPIDS IN THE BIOGENESIS OF THE PHOTOSYNTHETIC MEMBRANE

Antoine TREMOLIERES(1), Guy DUBERTRET(1) Harald PAULSEN (2) Stephan HOBE (2) and Chantal DEMANDRE (3)

(1) Laboratoire de Biochimie Fonctionnelle des Membranes Végétales, URA 1128, CNRS, Gif sur Yvette, France.
(2) Botanisches Institüt der Ludwig Maximilian Universität, Menzinger Str., 67, München, Germany.
(3) Laboratoire de Physiologie Cellulaire et Moléculaire , 4 place Jussieu , 75252 Paris cedex 05 , France

Photosynthesis in eucaryotic green organism uses solar energy collected by the light-harvesting chlorophyll a/b protein complex associated with photosystem II (LHCII) which is the most abondant membrane protein in chloroplasts binding half of the pigments involved in plant photosynthesis. Each polypeptide organizes a minimum of 12 chlorophylls, 2 carotenoids (Nuβberger *et al.* 1993)(1) and at least two galactolipids (one mono- and one di-galactosyldiacylglycerol) and one phosphatidylglycerol (Trémolières *et al.* 1994)(2).
Crystallization of the light-harvesting complex in higher plants has allowed to produce an atomic map at a resolution of 3.4 angströms (Kühlbrandt *et al* . 1994)(3). Light-harvesting complex always crystallizes in a trimeric state and lipids play a crucial role in this trimerization process.Trimeric LHC II are stable in mild detergent and thus , can be isolated by electrophoresis in partially denaturing «green « gels at an apparent molecular mass between 70 and 80 kDa. Trémolières *et al.* (1981)(4) and Rémy *et al.* (1982, 1984)(5)(6) showed that phosphatidylglycerol is thighly associated with LHCII trimers. Phosphatidylglycerol in eucaryotic photosynthetic organism contains an unusual fatty acid esterifying specifically the sn-2 position of the glycerol moiety: the trans-Δ_3 hexadecenoic acid (trans-C16:1). The presence of trans-C16:1-containing phosphatidylglycerol could be correlated with the presence of the trimeric LHC II *in vivo* in several physiological situations. . *In vitro* studies on LHC II clearly show a specificic effect of trans-C16:1-containing phosphatidylglycerol on the oligomerization behaviour of the complex : Treatment of thylakoids (Remy *et al* . ,1984)(6) or of isolated LHC II (Krupa *et al* . ,1992)(7) Nuβberger *et al.* 1993)1) with phospholipase A2 results in the dissociation of LHC II into its monomers. Reversely , LHC II monomers can be induced to organize into a trimeric form in the presence of lipids , with liposomes containing trans-C16:1 phosphatidylglycerol exhibiting the strongest effect (Remy *et al.* ,1984)(6). Krupa *et al.* (1992)(7) showed that isolated LHC II monomers , delipidated by the action of lipases could restored to trimer with a yield of about 50% in the presence of trans-C16:1 containing phosphatidylglycerol whereas other molecular species of phosphatidylglycerol or other classes of lipids did not stimulate oligomerization. A stimulating effect of phosphatidylglycerol and particularly trans-C16:1 containing phosphatidylglycerol has also been observed in trimerization experiments using polypeptide from LHC II over-expressed in *Escherichia coli* transformed with Lhcb 1*2 gene from pea and purified pigments (Paulsen *et al* . ,1990)(8)(Cammarata et *al* . ,1990)(9) ; incubation of monomeric LHC II formed by *in vitro* association between

J.-C. Kader and P. Mazliak (eds.), Plant Lipid Metabolism, 179–181.

the pure polypeptide and pigments with trans-C16:1 containing phosphatidylglycerol resulted in more than 50% trimerization whereas no trimerization was observed in the absence of lipids.The interaction bewteen the phosphatidylglycerol and the LHC II likely include hydrophobic interactions to account for the specificity in binding trans-C16:1- containing phosphatidylglycerol. On the other hand , lipid-induced trimerization of LHC II also include ionic interaction since it occurs only at low salt concentration . This interaction appears to be restricted to the N-terminal domain of the apoprotein If this domain is removed by limited proteolytic treatment of LHCII monomers, lipid-stimulated oligomerization is not longer observed (Rémy *et al.*, 1984)(6). Consistently, when LHCII trimers are proteolysed at their N-terminal domain, the trimers disintegrate into monomers and at the same time lose the tighly bound phosphatidylglycerol (Nußberger *et al.*, 1993)(1). Using the genetic expression system of the Lhc b 1*2 gene in *E.Coli* it was possible to manipulate the polypeptide LHC II sequence by mutagenesis. By this way it was shown that two pairs of aminoacids situated exactly in the region where the polypeptide emerges from the thylakoid toward the stromal phase (at around 40 amino acids from the N-terminal) are involved in the interaction with the phosphatidylglycerol (Paulsen*et al.* umpublished data).

In order to definitely prove that the trimeric state is a functional state of organization of the antenna in living photosynthetic organisms, we have developed a method to manipulate *in* vivo the lipid composition in mutants *of Chlamydomonas reinhardtii*. By this method we succeded to demonstrate that trans-C16:1 containing phosphatidylglycerol, catalyzes a crucial step in the trimerization of the light-harvesting antenna(calledCP 2 *in Chlamydomonas*). It was also demonstrated that grana stacks can only be formed in conditions where trimeric antenna can be stabilized in the membrane. These conclusions result from the following set of experiments: mutants *ofChlamydomonas* have been isolated by random mutagenesis with U.V. irradiation followed by screening of the colonies according to their fluorescent properties. Among a set of mutants lacking PSII polypeptides, two have been isolated devoid of trans-C16:1 containing phosphatidylglycerol. These mutants are unable to form trimeric antenna, almost fully devoid of grana stacks and unable to make state I —state II transition according to the reduction state of the plastoquinone pool (Maroc *et* al . 1987)(10). By growing the mutants in the presence of liposomes we were able to re-incorporate significantly trans-C16:1 containing phosphatidylglycerol in the photosynthetic membrane of the mutants. In these conditions, newly synthesized polypeptides of the antenna can be stabilized in a trimeric state, grana stacks reappear and the ability to perform state I —— state II transition was restored.(Garnier *et al* . 1990) (11);(Tremolières *et al* . 1991)(12). We have demonstrated that, in the absence of this specific lipid (PG-C16:1), the polypeptides of the light harvesting complex are largely degraded to give proteins of a lower molecular weight which cannot be assembled to form a trimeric complex.

Our work leads to the conclusion that a precise site in the N-terminal domain of the light-harvesting complex which is oriented toward the stromal phase is recognized specifically by PG-C16:1 trans allowing the organisation of this complex in a trimeric state. This stabilization of the antenna in a trimeric state seems to be crucial for the formation of grana stacks leading to the spatial separation between PSII and PSI. The site of phosphorylation of the mobile antenna is also very close to the site of interaction with the lipid and several experimental arguments support the hypothesis that the PG-C16:1 trans could be involved in the modulation of the mobility of this antenna in the photosynthetic membrane.

Finally experiments using new mutants of *Chlamydomonas* isolated in the lab. have lead to the suggestion that stabilization in a trimeric state is also a perequisite to a correct organization of the light-harvesting complex around the PSII. Starting from the non photosynthetic mutants bringing two mutations as previously described, we have selected , by growing these mutants on a minimum medium , new mutants having recovered PSII activity (likely carrying a mutation suppressing the unability to synthesize PS II). It was a great surprise to observe that some trans-C16:1 was detected in the phosphatidylglycerol of these mutants. A more carefull analysis of the two first non photosynthetic mutants as of the two new photosynthetic mutants have shown that: the mutants are not lacking of the trans-Δ_3-desaturase if weak but significant amount of trans-C16:1 (around 1%) is always found in the monogalactosyldiacylglycerol of all the mutants as of the wild type cells (suggesting that the monogalactosyldiacylglycerol could be the true substrate for the trans-Δ_3-desaturase). On the other hand analysis of molecular species of phosphatidylglycerol have shown that in the wild type cells most of phosphatidylglycerol is constituted by the sn-1-C18:1Δ_9; sn-2-trans C16:1 (around 50% of the molecular species) with smaller amount of molecular species all containing trans-C16:1 or C16:0 at the sn-2 position but with C18:2 or C18:3 at the sn-1 position. The remainding of the phosphatidylglycerol is constituted with a molecular species containing C16:0 at the sn-2 position and C18:1Δ_{11} at the sn-1 position (the C18:1Δ_{11} have been shown to be present in several lipid classes of *Chlamydomonas* - Giroud *et al*. 1988)(13) which represent about 25% of the molecular species of phosphatidylglycerol. In the two non photosynthetic mutants first isolated only this last molecular species remains present and the mutants show a total phosphatidylglycerol content about 25% of the wild type cells while in the two new photosynthetic mutants a low but significant amount of phosphatidylglycerol containing trans-C16:1 at the sn-2 position and C18:1Δ_9 at the sn-1 position reappears . We are then lead to the conclusion that the molecular species of phosphatidylglycerol containing trans-C16:1 at the sn-2 position (and mainly C18:1Δ_9 at the sn-1 position) play a crucial role in the biogenesis of the main photosynthetic antennae and its stabilization under a trimeric state but also that this special molecular species could be protected against a degradative enzymatic action trough its interaction with the PSII which reappears in the «photosynthetic mutants» that we have isolated .

REFERENCES:

1-NUßBERGER S.,DÔRR K.,WANG D.N., KUHLBRANDT W. 1993.
J.Biol.Mol. 234, 347-356
2-TREMOLIERES A., DAINESE P., BASSI R. 1994.
Eur.J.Biochem. 221, 721-730
3- KUHLBRANDT ., WANG D.N., FUJIYOSHI Y. 1994.
NATURE 367, 614-621
4-TREMOLIERES A., DUBACQ J.P., AMBARD-BRETTEVILLE F., REMY R. 1981.
FEBS lett. 130 , 27-31
5-REMY R., TREMOLIERES A.,DUVAL J.C., AMBARD-BRETTEVILLE F.,DUBACQ J.P. 1982.
FEBS Lett., 137, 271-275
6-REMY R.,TREMOLIERES A., AMBARD-BRETTEVILLE F. 1984.
Photobiochem. Photobiophys. 7, 267-276
7-KRUPA Z.,WILLIAMS J.P., KHAN M.U., HUNER N.P.A. 1992.
Plant physiol .100, 931-938
8-PAULSEN H. , RUMLER U., RUDIGER W., .1990.
Planta 181, 204-211
9- CAMMARATA K.V., SCHMIDT G.W. 1991.
Biochemistry 31, 2779-2789
10- MAROC J. , TREMOLIERES A., GARNIER J.. GUYON D.1987.
Biochim. Biophys. Acta 893, 91-99
11- GARNIER J., WU B., MAROC J., GUYON J., TREMOLIERES A. 1990.
Biochim. Biophys. Acta 1020, 153-162
12- TREMOLIERES A. , ROCHE O., DUBERTRET G., GUYON D., GARNIER J. 1991.
Biochim. Biophys. Acta 1059, 286-292
13-GIROUD C., GERBER A., EICHENBERGER W. 1988.
Plant cell physiol. 29, 587-595

TEMPERATURE AND LIGHT CONDITIONS DURING THE GROWTH OF SQUASH SEEDLINGS INFLUENCE GREATLY THE RELATIVE CONTENT OF PHOSPHATIDYLGLYCEROL MOLECULAR SPECIES IN COTYLEDONS

Yinong Xu and Paul-André Siegenthaler
laboratoire de Physiologie végétale, Université de Neuchâtel
Rue Emile-Argand 13, CH-2007 Neuchâtel, Switzerland

Photosynthetic membranes contain 10 molecular species of phosphatidylglycerol (PG) [1,2]. These species have been divided into two groups: the first one contains 16:0 at the sn-2 position (18:3/16:0, 18:2/16:0, 18:1/16:0, 18:0/16:0 and 16:0/16:0); the second one contains 16:1(3t) at the sn-2 position (18:3/16:1(3t), 18:2/16:1(3t), 18:1/16:1(3t), 18:0/16:1(3t) and 16:0/16:1(3t)). The thylakoid transmembrane distribution of total PG and of its molecular species is asymmetric i.e. the outer monolayer is highly enriched (2/3) in PG [2].

PG is a unique phospholipid in the thylakoid membrane because half of its molecular species contains *trans*-Δ^3-hexadecenoic acid (16:1(3t)). In contrast to the other acyl lipid classes, PG contains high levels of saturated fatty acids (16:0, 18:0) and of the particular 16:1(3t). Moreover it has been suggested that PG is involved in the molecular organization of the light collection receptor complex (LHCII) [3] and in the mechanism of chilling sensitivity of plants [4]. Indeed, a distinct difference between chilling-sensitive and chilling-resistant plants is found in the composition of PG molecular species [1]. However, Murata et al. [4] pointed out that the relative content of 16:0 and 16:1(3t) varies greatly in both types of plants.

The aim of this study was to determine the relative content of PG molecular species in cotyledons from squash, a chilling-sensitive plant, which was grown under various temperature and light conditions.

Materials and Methods

Plants. After germination of squash plants (var. Quintal) in darkness at 20, 30, and 35°C, seedlings were grown at the same above temperature and in continuous light (20,000 lux).

Lipid extraction and separation. Immediately after the cotyledons were harvested, lipids were extracted with chloroform/methanol (1/1, v/v) [5]. Total lipids obtained were separated into individual lipid classes by two-dimensional TLC on a Silica gel 60 plate using aceton / toluene / acetic acid / H_2O (100/25/1/9, v/v) then chloroform / methanol / NH_3 25% / H_2O (65/35/3/2, v/v).

HPLC analysis of 3,5-dinitrobenzoyl derivatives. Purified PG were hydrolyzed by phospholipace C (*Bacillus cereus)* and the resulting 1,2-diacylglycerides were esterified with 3,5-dinitrobenzoylchloride. 3,5-dinitrobenzoyl derivatives of PG were separated by HPLC C18 column (5 mm, 4.6 x 250 mm) with acetonitrile / 2-propanol (80/20, v/v) [6].

HPLC peak analysis. Each peak was collected and fatty acids were transesterified with 5% H_2SO_4 in methanol for 1 h at 85°C and then analyzed with GC.

J.-C. Kader and P. Mazliak (eds.), Plant Lipid Metabolism, 182–184.

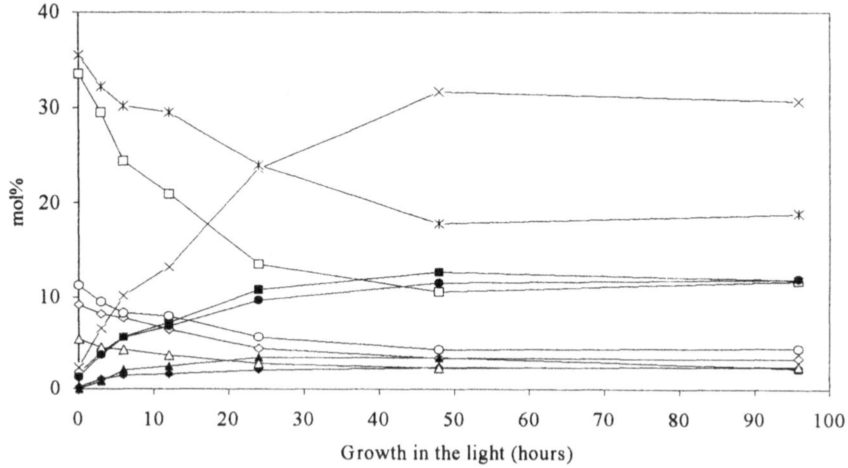

Fig. 1. Changes in the content of PG molecular species from squash cotyledons during the growth of seedlings. After germination in darkness, plant were grown in continuous light (20,000 lux). The temperature was 30°C during germination and growth.
■ 18:3/16:1(3t), □ 18:3/16:0, ♦ 18:2/16:1(3t), ◊ 18:2/16:0, ▲ 18:1/16:1(3t), Δ 18:1/16:0, ● 18:0/16:1(3t), O 18:0/16:0, X 16:0/16:1(3t), ⁎ 16:0/16:0.

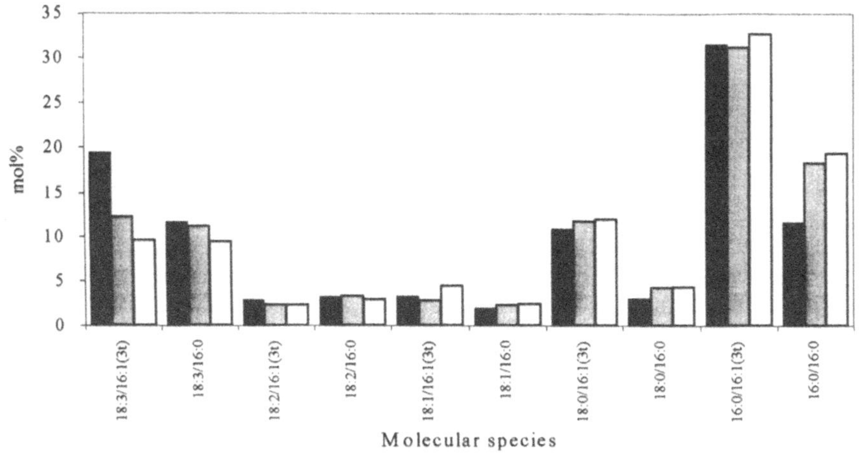

Fig. 2. Molecular species composition of PG from squash cotyledons grown at 20°C ■, 30°C ■ and 35°C □. Analysis of the PG molecular species was carried out after 48 h growth under continuous light (20,000 lux).

Results and Discussion

Fig. 1 shows that when squash seedlings were grown at a constant temperature (30°C), an exposure to continuous light (20,000) lux resulted in a progressive increase in all PG molecular species containing 16:1(3t), at the expense of the 16:0 ones. The PG molecular species undergoing the highest increase were 16:0/16:1(3t), 18:3/16:1(3t) and 18:0/16:1(3t), whereas those showing the greatest decrease were 16:0/16:0, 18:3/16:0 and 18:0/16:0. These changes occurred during the first 48 h in the light, then the content of the above molecular species remained constant. Decreasing the light intensity and photoperiod resulted in a slower rate and extent of these changes (results not shown). This suggests that light is the main factor which determines the content of fatty acid 16:1(3t). Therefore, it is important to control the light intensity and the photoperiod during plant growth when 16:1(3t) is analyzed.

As can be seen in Fig.2, decreasing the growth temperature from 35°C to 20°C resulted in an increase in 18:3/16:1(3t) from about 10 to 19%, in 18:3/16:0 from 9 to 12 %, and a concomitant decrease in 16:0/16:0 from about 19 to 13% in squash cotyledon PG. Diminishing further the growth temperature down to 10°C resulted in a greater and faster increase in 18:3/16:1(3t) and decrease in 16:0/16:0 (data not shown). This suggests that the fatty acid unsaturation induced by low temperature was due to an increase in the molecular species associated with C18:3, especially 18:3/16:1(3t).

It has been reported that growth at low, cold-hardening temperature resulted in a specific 67% (thylakoids) to 74% (whole leaves) decrease in the 16:1(3t) acid level associated with winter rye PG [8]. However, our results show that in squash cotyledons, lowering the growth temperature induced an increase in 16:1(3t), mainly in PG 18:3/16:1(3t) molecular species (Fig. 2).

We conclude that both light intensity and photoperiod are the main factors affecting the content of the 16:1(3t)-containing molecular species of PG. Higher light intensity and longer photoperiod result in an increasing content of these molecular species. On the other hand, temperature affects the degree of unsaturation in PG molecular species. Decreasing the growth temperature resulted in an increase in molecular species containing 18:3 (18:3/16:1(3t) and 18:3/16:0) and a decrease in 16:0/16:0 in squash cotyledon PG. (Supported in part by the Swiss National Science Foundation)

References

[1] Murata, N. Molecular species composition of phosphatidylglycerol from chilling-sensitive and chilling-resistant plants. Plant Cell Physiol. 1983;24(1):81-86.

[2] Siegenthaler PA, Xu YN, Bovet L, Smutny J, Rawyler A. Molecular organization and functions of acyl lipids in spinach chloroplast membranes. 1992 Proceeding of the 10th International Symposium on the Metabolism Structure and Utilization of Plant Lipids, Jerba (Tunisia)

[3] Dubacq JP, Trémolières A. Occurrence and function of phosphatidylglycerol containing Δ^3-trans-hexadecenoic acid in photosynthetic lamellae. Physiol. vég. 1983;21(2):293-312.

[4] Murata N, Sato N, Takahashi N, Hamazaki Y. Compositions and Positional Distributions of Fatty Acids in Phospholipids from Leaves of Chilling-Sensitive and Chilling-Resistant Plants. Plant and Cell Physiol. 1982;23(6):1071-1079.

[5] Bligh, EG, Dyer, WJ. A rapid method of total lipid extraction and purification. Can. J. Biochem. Physiol. 1959;37:911-917.

[6] Kito M, Takamura H, Narita H, Urade R. A Sensitive Method for Quantitative Analysis of Phospholipid Molecular Species by High-Performance Liquid Chromatography. J. Biochem. 1985;98:327-331.

[7] Huner NPA, Krol M, Williams JP, Maissan E, Low PS, Roberts D. Thompson JE. Low Temperature Development Induces a Specific Decrease in trans-Δ^3-Hexadecenoic Acid Content which Influences LHCII Organization. Plant Physiol. 1987;84:12-18.

INVESTIGATIONS OF PYRUVATE DEHYDROGENASE IN PEA ROOT PLASTID PREPARATIONS

Salvatore A. Sparace, Qungang Qi and Kathryn F. Kleppinger-Sparace

Plant Science Department, Macdonald Campus of McGill University, 21111 Lakeshore Road., Ste. Anne de Bellevue, Quebec, Canada H9X 3V9

Synopsis

Pyruvate dehydrogenase and fatty acid biosynthesis from pyruvate in pea root plastid preparations have been been studied. The requirements of pyruvate dehydrogenase from these plastids are similar to those of other plastids and pyruvate is readily utilized for fatty acid biosynthesis. The results presented indicate that pyruvate dehydrogenase plays an important role in promoting carbon flow from glycolytic metabolism to fatty acid biosynthesis in pea root plastids.

Introduction

Nonphotosynthetic plastids have considerable potential for the formation of pyruvic acid. They contain almost all of the enzymes for the complete glycolytic metabolism of hexoses [1]. Further, the glycolytic pathway in these plastids interacts considerably with the cytosolic compartment through the uptake or exchange of various hexoses and trioses [2]. Pyruvate dehydrogenase in nonphotosynthetic plastids may thus play an important role in routing glycolytic carbon towards lipid biosynthesis by providing a supply of acetyl-CoA and reduced nucleotide (NADH) which are required for fatty acid biosynthesis. Our current efforts are aimed at understanding the regulation of plastidic carbon flow from glycolysis to fatty acid and glycerolipid biosynthesis in nonphotosynthetic plastids. We have partly characterized pyruvate dehydrogenase in lysed pea root plastid preparations as well as the utilization of [2-^{14}C]pyruvate in comparison to [1-^{14}C]acetate for fatty acid biosynthesis in intact plastids from pea roots.

Materials and Methods

Plastids isolated from germinating pea roots [3] were lysed [4] and the lysate was concentrated by ultrafiltration. This concentrate was for the measurememt of

J.-C. Kader and P. Mazliak (eds.), Plant Lipid Metabolism, 185–187.

pyruvate dehydrogenase [5]. Intact plastids were used directly for studies of pyruvate- and acetate-dependent fatty acid biosynthesis [6].

Results and Discussion

Pea root plastid pyruvate dehydrogenase was completely dependent on exogenously supplied pyruvate, NAD^+ and CoA (Table 1). Estimated K_m's for these

Table 1. Cofactor Requirements for Pyruvate Dehydrogenase from Pea Root Plastids.

	Optimum Conc., mM	% of control when deleted
Complete control	-	100[a]
Pyruvate	1	4.2
NAD^+	1.5	5.1
CoA	0.1	2.2
Mg^{2+}	5	22.7
Cysteine-HCl	3	38.3
Thiamine-PPi	0.15	67.5

[a]Data are from several experiments where the average rate of control activity was 16 $nmol \cdot min^{-1} \cdot mg^{-1}$.

substrates were 0.23 mM, 0.25 mM and 18 μM, respectively. Mg^{2+} was also required. Cysteine-HCl and thiamine pyrophosphate improved activity 2.6 fold and 1.5 fold, respectively. Enzyme activity was greatest at pH 8 in Tricine buffer. [2-^{14}C]Pyruvate is approximately 1.5 x's more efficiently incorporated into fatty acids than acetate (Table 2). Further, pyruvate-dependent fatty acid biosynthesis

Table 2. Fatty Acid Synthesis From Pyruvate and Acetate by Isolated Pea Root Plastids.

	Pyruvate	Acetate
	($nmol \cdot min^{-1} \cdot mg^{-1}$)	
Control[a]	2.04	1.38
Control, - NAD(P)H	1.61	0.61

[a]Plastids were incubation under optimum conditions for fatty acid biosynthesis from acetate [6].

shows less than half the dependency on exogenously supplied reduced nucleotides (NADH and NADPH) as acetate-dependent fatty acid biosynthesis. The results presented here are in general agreement with related studies [5,7-9]. They indicate that pyruvate dehydrogenase in pea root plastids has sufficient activity to promote the utilization of glycolytic pyruvate for fatty acid biosynthesis in nonphotosynthetic pea root plastids by providing acetyl-CoA and reduced nucleotide.

We are currently performing related studies of acetyl-CoA carboxylase from pea root plastids. Preliminary results indicate that the activity of this enzyme can be coupled to pyruvate dehydrogenase and that these enzymes may be regulated in a coordinated fashion.

References

1. Emes MJ, Tobin AK. Control of metabolism and development in higher plant plastids. Int Rev Cytol 1993; 145: 149-216.
2. Borchert S, Harborth J, Schünemann D, Hoferichter P, Heldt HW. Studies of the enzymic capacities and transport properties of pea root plastids. Plant Physiol 1993; 101: 303-312.
3. Kleppinger-Sparace KF, Stahl RJ, Sparace SA. Energy requirements for fatty acid and glycerolipid biosynthesis from acetate by isolated pea root plastids. Plant Physiol 1992; 98: 723-727.
4. Reid EE, Thompson P, Lyttle CR, Dennis DT. Pyruvate dehydrogenase complex from higher plant mitochondria and proplastids. Plant Physiol 1977; 59: 842-848.
5. Camp PJ, Randall DD. Purification and characterization of the pea chloroplast pyruvate dehydrogenase complex. A source of acetyl-CoA and NADH for fatty acid biosynthesis. Plant Physiol 1985; 77: 571-577.
6. Stahl RJ, Sparace SA. Characterization of fatty acid biosynthesis in isolated pea root plastids. Plant Pysiol 1991; 96: 602-608.
7. Williams M, Randall DD. Pyruvate dehydrogenase complex from chloroplasts of *Pisum sativum* L. Plant Physiol. 1979; 64: 1099-1103.
8. Thompson P, Reid EE, Lyttle CR, Dennis DT. Pyruvate dehydrogenase complex from higher plant mitochondria and proplastids: Kinetics. Plant Physiol 1977; 59: 849-853.
9. Smith RG, Gauthier DA, Dennis, DT, Turpin, DH. Malate- and pyruvate-dependent fatty acid synthesis in leucoplasts from developing castor endosperm. Plant Physiol 1992; 98: 1233-1238.

FORMATION OF AUTOPHAGIC VACUOLES AND ACCUMULATION OF DEACYLATION PRODUCTS OF MEMBRANE POLAR LIPIDS DURING THE COURSE OF SUCROSE STARVATION IN HIGHER PLANT CELLS

Roland DOUCE, Serge AUBERT, Richard BLIGNY, Albert-Jean DORNE, Elizabeth GOUT and Francis MARTY
Laboratoire de physiologie cellulaire végétale, DBMS, CEN-G, 17 rue des Martyrs, 38054, Grenoble cedex 9

Synopsis

Transfer of sycamore (*Acer pseudoplatanus* L.) cells into flasks containing sucrose-free culture medium triggered the following cascade of reactions: a) initially cellular carbohydrate reserves (vacuolar sucrose and starch present in plastids) were consumed; b) when almost all the intracellular carbohydrate pools had disappeared the cell fatty acids deriving from membrane polar lipids such as phosphatidylcholine were utilized as oxidizable substrates for ATP production. Using electron microscopy we demonstrated the presence of a very active autophagic activity during sucrose starvation in sycamore cells. Autophagic vacuoles (double membrane-bounded vesicles) were formed in the peripheral cytoplasm and were expelled into the large central vacuole as single membrane-bounded vesicles. Our results shed new light on the regression of the mitochondrial system previously observed during the course of sucrose deprivation in plant cells [1]: following 5 days of sucrose starvation the total mitochondrial volume within the cell cytoplasm decreased to less than 15% of that of the normal cells. The degradation of intracellular membrane polar lipids by autophagy led to a transient accumulation of phosphodiesters and to a steady accumulation of phosphocholine in the cell cytoplasm which is not further metabolized. The total amount of phosphocholine present within the cell should reflect the extent of intracellular membrane degradation during the course of sucrose starvation. Finally during the course of leaf senescence phosphocholine molecules, derived from phosphatidylcholine degradation, may escape out of the cell and be exported toward the xylem vessel where they accumulate. Under these conditions, phosphocholine would be transported in the xylem fluids to meristematic cells or fast growing tissues where it could be used as a source of choline for membrane synthesis. We have demonstrated that phosphocholine was rapidly hydrolyzed outside the cell and that choline and Pi entered the cytosolic compartment where choline kinase re-forms phosphocholine.

Introduction

Classical autophagy is an important process by which cells sequester and degrade parts of their own cytoplasm, including organelles (except the nucleus). This process has been well documented in animal kingdom [2,3]. Autophagic vacuoles arise when smooth membrane-

J.-C. Kader and P. Mazliak (eds.), Plant Lipid Metabolism, 188–196.

bounded envelopes (smooth endoplasmic reticulum ?) wrap themselves around portions of cytoplasm. Once the sequestered portions of the cytoplasm have been completely enclosed by lateral fusion of the enveloping membranes, various hydrolases (lysosomal enzymes) are probably released from the surrounding exoplasmic space into the sequestered cytoplasm to form an autophagic vacuole (autophagolysosome). Active autophagy also occurs in *Saccharomyces cerevisiae* transferred to a synthetic medium devoid of nutrients [4].

Occurrence of autophagy in plant cells

The occurence of classical autophagy in higher plant cells remains to be established although this phenomenom is reminiscent of the formation of vacuoles in root meristems [5].Vacuoles containing numerous hydrolytic enzymes including proteases are thought to be functionally equivalent to animal lysosomes [6].Interestingly in previous publications Journet *et al.* [1], Roby *et al.* [7], and Dorne *et al.* [8] have shown that transfer of sycamore cells into flasks containing sucrose-free culture medium triggered the following cascade of reactions: a) vacuolar sucrose is first consumed. During this time the starch, Pi, and phosphate ester contents remain constant. The same holds true for the respiration rate; b) when a threshold of intracellular sucrose is attained, the rate of sucrose efflux from the vacuoles becomes a limiting factor for cytoplasmic glycolysis rate and starch breakdown progressively replaces sucrose hydrolysis; c) when almost all the intracellular sucrose pool has disappeared the cell respiration rates (normal and uncoupled) decline progressively (Fig. 1) and fatty acids deriving from membrane polar lipids, such as phospholipids and galactolipids, were utilized for respiratory purposes [during the period of the most rapid

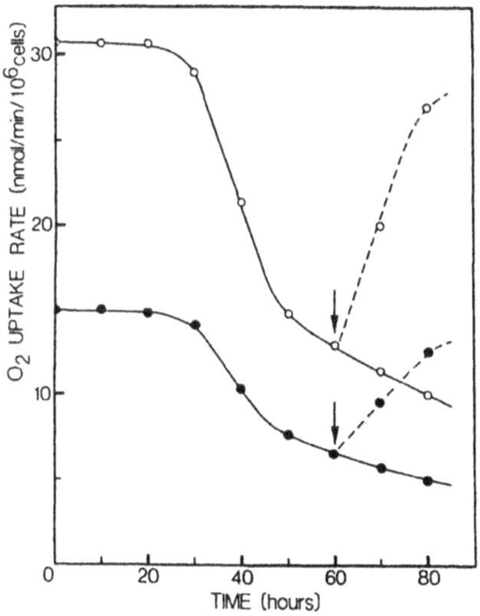

Figure 1: Time course of sycamore cell respiration rate when cells are grown in a medium devoid of sucrose. Cells harvested from the culture medium were rinsed three times by successive resuspensions in fresh culture medium devoid of sucrose and incubated at zero time (20 mg wet weight / ml) into flasks containing sucrose-free culture medium. At each time, cells were harvested and the O_2 consumption rate was measured; ●, normal respiration; ○, uncoupled respiration; the dotted line corresponds to the enhancement of cell respiration that is observed when 50 mM sucrose is added to the culture medium (arrows).

disappearance of polar lipids, the respiratory quotient (RQ) of the cell was as low as 0.7]. This observation is reminiscent of respiration in maize root tips deprived of sucrose during several hours where a connection between respiration of lipids and low RQ has been verified [9]. Interestingly during the course of polar lipid breakdown the total amount of sterols, including sterylglucoside and acylated sterylglucoside, appeared to be constant.

These observations strongly suggest that a large proportion of the endomembrane system disappears in starving cells whilst tonoplast and plasmalemma enriched in sterol compounds [unpublished data] escape the autophagic process. Indeed, the slow decrease in the uncoupled rate of O_2 consumption observed in sycamore cells following the removal of sucrose from the culture medium is attributable to a progressive diminution of the number of mitochondria / cell. The arguments in favour of this autophagic activity are: a) the decrease in the rate of respiration triggered by sucrose starvation is not attributable to the availability of substrate for mitochondrial respiration; b) on a protein basis, the rates of O_2 uptake in state III (i.e. in the presence of ADP) are about the same for normal and sucrose-starved mitochondria. Likewise the cardiolipin (diphosphatidylglycerol) and cytochrome aa_3 (cytochrome oxydase) content of sycamore mitochondria (two typical mitochondrial markers) is invariant to sucrose depletion; c) the uncoupled respiration rates decrease in the same ratio as the decrease in total intracellular cardiolipin or cytochrome aa_3. For example,

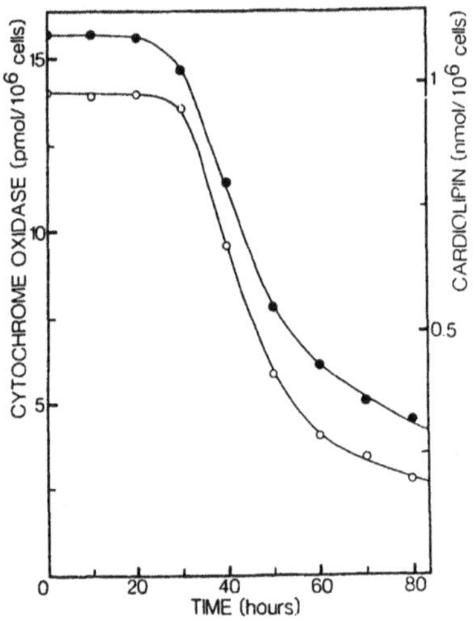

Figure 2: Effect of sucrose starvation of sycamore cells on intracellular cardiolipin (●) and cytochrome aa_3 (O) contents. Cells harvested from the culture medium were rinsed three times by successive resuspensions in fresh culture medium devoid of sucrose and incubated at zero time (20 mg wet weight / ml) into flasks containing sucrose-free culture medium. At each time, cells were harvested and cardiolipin (diphosphatidylglycerol) and cytochrome aa_3 contents were measured.

following 5 days of sucrose starvation, the number of mitochondria per cell decreased to less than 15% of that of the normal cells (Fig. 2). It is possible, however, that the number of mitochondria / unit volume of cytoplasm may be maintained while the total protein of the cell drops sharply during the starvation period. Similar results were also observed during the course of glucose starvation in the meristematic and submeristematic regions of maize roots [10] and in heterotrophic cucumber cells [11]. Likewise Baysdorfer *et al.* [12] have shown in pearl millet seedlings that after 8 days of starvation in the dark the respiratory rate had declined to less than 50% of its initial value and the plant had lost half of their protein content. These authors also indicated the presence of numerous electron-transparent areas in the cell cytoplasm from 8 days-starved plants maintained in total darkness suggesting that intracellular organelles and cytosol are sequestered away from the remaining cytoplasm and then degraded within autophagic vacuoles.

Formation of autophagic vacuoles during the course of sucrose starvation

Using electron microscopy we demonstrate the presence of double membrane-bounded autophagic vacuoles in the cytoplasmic compartment of sycamore cells which were deprived of sucrose for 20 h and 4 days. Abundant membrane remains were seen inside the vacuole and figures indicative of complete sequences of autophagy were readily observed in the cytoplasm of cells deprived of sucrose for only 20 h. These vacuoles were observed in the cells when all the endogenous reserves of carbohydrates (starch and sucrose) had been consumed but they were scarce in cells starved for 6 days, by which time the cytoplasm appeared as a thin film of living matter appressed along the cell wall. These autophagic vacuoles were formed in the peripheral cytoplasm during sucrose starvation and were eventually expelled into the large central vacuole for final deposition (Fig. 3). Membrane rearrangements must have occured when the double membrane-bounded vacuoles are released into the central vacuole as single membrane-bounded vesicles. These observations led to the conclusion that higher plant cells have the ability to conduct controlled autophagy of their cytoplasmic contents during periods of sucrose deprivation.

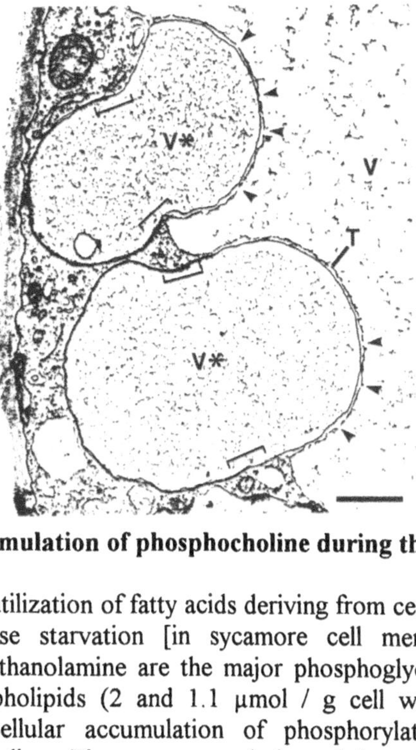

Figure 3: Autophagic activities in sycamore cells grown for 4 days in sucrose-free culture medium. Cells harvested from the sucrose-free culture medium were fixed, embedded in epoxy resins, sectioned, and stained for electron microscopy. Vacuoles with a clear content are protruding into the central vacuole. Note that the tonoplast of the central vacuoles is disrupted in many places, precisely in both areas where the peripheral vacuoles protrude into the main central vacuole (arrow heads).; bar, 250 nm. V, vacuole; T, tonoplast.

Accumulation of phosphocholine during the course of sucrose starvation

The utilization of fatty acids deriving from cell membrane polar lipids after a long period of sucrose starvation [in sycamore cell membranes phosphatidylcholine and phosphatidylethanolamine are the major phosphoglycerides, representing up to 80% of the total phospholipids (2 and 1.1 μmol / g cell wet weight respectively)] should lead to the intracellular accumulation of phosphorylated compounds deriving from polar lipid catabolism. Time-course evolution of the most abundant ^{31}P nuclear magnetic resonance spectroscopy (NMR)-detectable phosphorus compounds in sycamore cells subjected to sucrose starvation indicates that the degradation of intracellular membrane polar lipids by autophagy led to a transient accumulation of phosphodiesters (glycerylphosphorylethanolamine, glycerylphosphorylcholine) and to a steady accumulation of phosphocholine (Fig. 4). Hydrolysis of membrane polar lipids (the whole cell phosphatidylcholine level dropped to less than 35% of its initial value by six days) occurred in several distinct steps: phosphodiesters, which accumulated transiently, appeared before the steady accumulation of phosphocholine and phosphoethanolamine. It is very likely that after sequestration of the membrane systems within the autophagic vacuoles, net lipid breakdown commences: phosphodiesters derive from the attack of membrane polar lipids by nonspecific lipid acyl hydrolase(s) [8] and phosphocholine and phosphoethanolamine derive from phosphodiester hydrolysis to give glycerol for further metabolism. Surprisingly, the final cytoplasmic phosphocholine concentration remained remarkably stable up to cell death which occured approx. 6-7 days after the begining of sucrose starvation. Furthermore, the total amount of phosphocholine that appeared in the cytoplasmic compartment corresponded to the total amount of phosphatidylcholine that disapeared within the same period of time. For example after 4 days of sucrose starvation the total amount of phosphocholine present in intact sycamore cells was approx. 1.1 µmol per g of cell wet weight and the total amount of phosphatidylcholine that had been hydrolyzed was approx. 1 µmol per g of cell wet weight. Phosphocholine exhibits, therefore, a remarkable

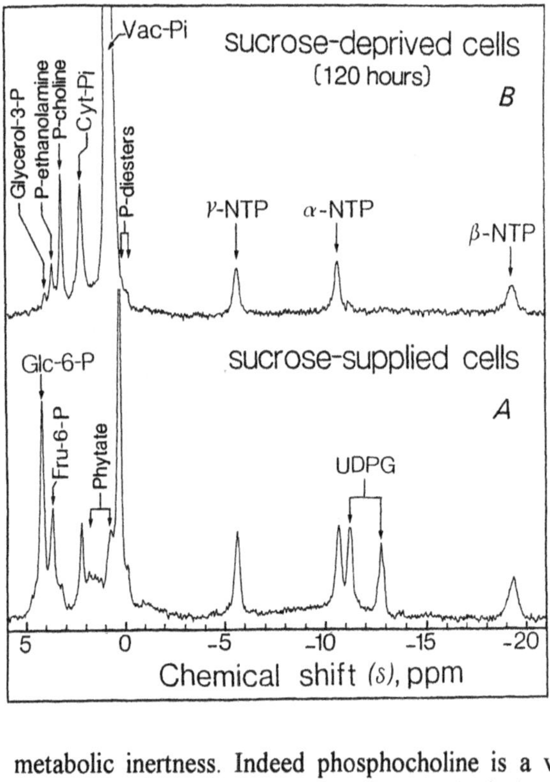

Figure 4: Representative NMR spectra (161.93 MHz) of sucrose-supplied and sucrose-deprived (120 h) sycamore cells. Cells harvested from the culture medium were rinsed three times by successive resuspensions in fresh culture medium devoid of sucrose and incubated at zero time (20 mg wet weight / ml) into flasks containing sucrose-free culture medium. Cells (normal cells and 120-h sucrose starved cells (9 g) were harvested for in vivo examination. The spectra recorded at 20°C with a 50° pulse angle and a 0.6-s repetition time were the result of 6000 transients (1 h). Perfusion medium: A, standard culture medium; B, culture medium devoid of sucrose since 120 h . Peak assigments: Glc-6-P, glucose-6-P; Fru-6-P, fructose-6-P; Cyt-pi, cytoplasmic phosphate (at pH 7.5); vac-Pi, vacuolar phosphate (at pH 5.7); NTP, nucleotides triphosphate; UDPG, uridine-5'-diphosphate-α-D-glucose

metabolic inertness. Indeed phosphocholine is a very stable molecule: it is resistant to strong acid hydrolysis and is hardly hydrolyzed by unspecific acid phosphatases. Consequently, the total amount of phosphocholine present within the cell should reflect the extent of intracellular membrane degradation during the course of sucrose starvation. It has not yet been determined whether particular genes are induced by phosphocholine accumulation.

Effect of glycerol on autophagy

The main point to be addressed from these observations is the nature of the metabolic signal that gives rise to autophagy and therefore to a dramatic changes in the expression of a variety of genes encoding proteolytic enzymes and enzymes from different degradative pathways (fatty acids, amino acids such as branched-chain amino acids, etc..,). For example we have shown that expression of the gene encoding β - methylcrotonyl-CoA carboxylase, a mitochondrial enzyme involved in leucine catabolism, strongly responds to sucrose starvation in sycamore cells [Alban and Aubert, unpublished]. It is possible that induction of autophagy is correlated with a decrease in intracellular sucrose and / or glucose-6-P below threshold concentrations. To test this hypothesis we conducted experiments using glycerol [13]. The effects of glycerol on plant cell metabolism were studied with sycamore cells using ^{31}P NMR. After a long period of sucrose starvation addition of 50 mM glycerol to the medium did not restore the original glucose-6-P pool and led to a rapid accumulation of *sn*-glycerol-3-P in the cytoplasmic compartment. The synthesis of *sn*-glycerol-3-P was rapid and occurred first at the expense of cytoplasmic Pi. Accumulated cytoplasmic *sn*-glycerol-3-P (15-20 mM) competitively inhibited glucose-6-P isomerase activity (Ki = 4 mM) when fructose-6-P was the varied substrate. Such a situation prevented the rapid recycling of triose phosphates back to hexose phosphates and led to an arrest of cytosolic and plastidial pentose phosphate pathways functioning. Under these conditions, glucose-6-P and sucrose concentrations remained very low as was the case with starved-cells, and the flow of carbon to drive cell respiration derived almost exclusively from glycerol. In addition, this polyalcohol was not used as a source of carbon skeletons for biosyntheses (Fig. 5). Surprisingly glycerol added to sucrose-starved cells stopped the accumulation of

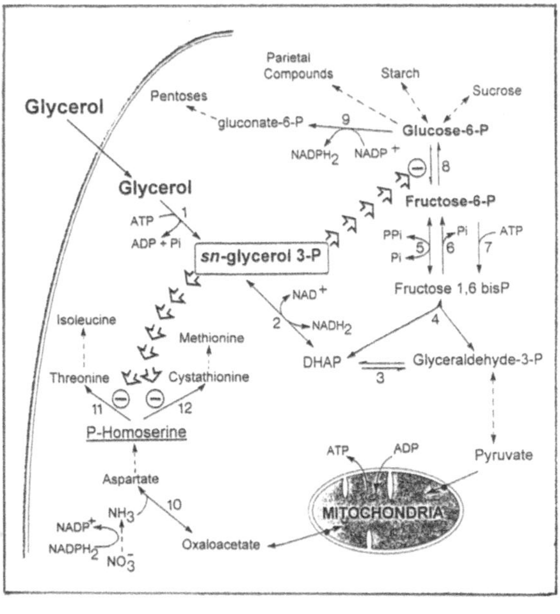

Figure 5 : Scheme summarizing the effect of glycerol on plant cell metabolism: Open arrows and minus signs indicate enzymic reactions inhibited by *sn*-glycerol-3-P. The numbered arrows correspond to the following enzymic reactions: 1, glycerol kinase (EC 2.7.1.30); 2, *sn*-glycerol-3-P:NAD$^+$ oxidoreductase (EC 1.1.1.8); 3, triose-P isomerase (EC 5.3.1.1.); 4, fructose-1,6-bisphosphate aldolase (EC 4.1.2.13); 5, pyrophosphate:fructose-6-P phospho transferase (EC 2.7.1.90); 6, fructose-1,6-bisphosphatase (EC 3.1.3.11); 7, fructose-6-P kinase (EC2.7.1.11); 8, glucose-6-P isomerase (EC 5.3.1.9); glucose-6-P dehydrogenase (EC 1.1.1.49); 10, glutamate-oxaloacetate transaminase (EC 2.6.1.1); 11, threonine synthase (EC 4.2.99.2); 12 cystathionine γ-synthase (EC 4.2.99.9). P-homoserine, *O*-phosphohomoserine;DHAP,dihydroxy-acetone-P.

phosphocholine and prevented a further decline in the uncoupled rate of O_2 consumption by the cells. These last observations strongly suggest that glycerol prevented the triggering of autophagy induced by sucrose starvation in sycamore cells. From these experiments we can conclude that there is no correlation of a fall in sucrose and glucose-6-P concentrations with induction of autophagy. One mechanism by which glycerol might prevent autophagy is by the supply of substrates via the glycolytic flux to the mitochondrial electron transport chain and phosphorylating system. In support of this suggestion we have observed that the removal of extracellular glycerol led to a rapid accumulation of phosphocholine, almost without a lag phase.

Transport of phosphocholine in higher plant cells

Tolbert's group suggested that many plants accumulate large amounts of phosphocholine in their xylem sap during starvation in long periods of storage in the dark or during leaf senescence (during the night or after a long period of drought leading to the closing of stomatal apertures in the leaf epidermis, most of the plant cells are deprived of sucrose)[see 14 and 15]. Thus during the course of sucrose starvation phosphocholine molecules,

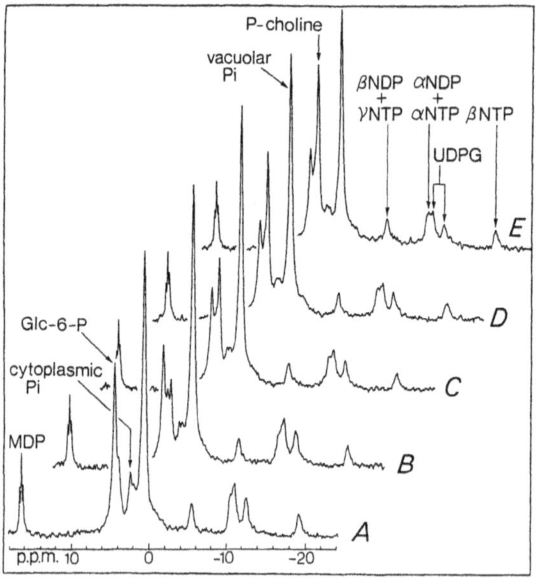

Figure 6: Representative ^{31}P NMR spectra (81 MHz) of sycamore cells after addition of 10 μM choline in the perfusing culture medium. Cells (9 g) are compressed by hand between two circular polymer filters in a 25-mm (diameter) NMR tube and maintained in a continuously aerated solution (Pi-free culture medium). The spectra, recorded at 25°C, with a 45° pulse angle were the result of 12,000 transients with a repetition time of 0.6 s (2 h). *A*, standard spectrum; *B*, *C*, *D* and *E*, spectra obtained after 2, 6, 12, and 18 h, respectively, of perfusion of the compressed cells with a culture medium containing 10 μM choline. A 20-liters reservoir of perfusion medium was utilized in order to maintain a constant choline concentration. Note the steady accumulation of P-choline (phosphocholine). Peak assigment are the same as in Fig. 4. The reference capillary contained 1 μmol of methylene diphosphonate (*MDP*) at pH 8.9.

derived from phosphatidylcholine degradation, may escape out of the sucrose-starved cells and be exported toward the xylem vessels where they accumulate, as has been shown for various amino acids such as asparagine. Phosphocholine would be subsequently transported in the xylem fluids to meristematic cells or fast-growing tissues where it could be used as a source of choline for membrane proliferation. In fact we have demonstrated [14] that when fast-growing cells such as sycamore were suspended in basal medium containing choline, the latter was taken up by the cells very rapidly (Fig. 6). A facilitated diffusion system appertained at low concentration of choline and exhibited Michaelis-Menten kinetics (K_m was 2-4 μM and the V_m was 8 ± 2 nmol choline incorporated per h/10^6 cells at 25°C). Furthermore, perfusion of sycamore cells with phosphate (Pi)-free culture medium containing small amounts of choline triggers a steady accumulation of phosphocholine in the cytoplasmic compartment at the expense of cytoplasmic Pi. When a threshold of cytoplasmic Pi concentration is attained, Pi is drawn from the vacuole to sustain choline phosphorylation although at a much slower rate. Consequently, cytoplasmic choline kinase plays a crucial role in the rapid accumulation of choline deriving from the external medium, in the form of phosphocholine, and prevents its possible uptake into the acidic vacuole. We have also demonstrated that perfusion of cells with phosphocholine instead of free choline also triggers a massive accumulation of phosphocholine in the cytoplasmic compartment [15]. In fact phosphocholine was hydrolyzed outside the cell prior to transport through the cell membrane because a powerful acidic phosphocholine phosphatase activity was found almost exclusively outside the cell, probably associated with the cell wall. Indeed, when sycamore protoplasts devoid of phosphocholine phosphatase are fed with phosphocholine instead of choline, no accumulation of phosphocholine in the cytoplasm compartment was observed. It is clear, therefore, that the low half-saturation of choline transport would make possible a very active system in conjunction with the cytosolic choline kinase and the cell wall phosphocholine phosphatase, to act as a drain for sequestering phosphocholine from its site of formation (i.e.sucrose-starved tissues or senescing leaves).

References

1. Journet E, Bligny R, and Douce R. Biochemical changes during sucrose deprivation in higher plant cells. J. Biol. Chem. 1986; 261: 3193-3199.
2. Holtzman E. Lysosomes. Plenum Publishing Corp, 1989: 439 pp.
3. Marzella L, and Glaumann H. Autophagy; microautophagy, and crinophagy as mechanisms for protein degradation; In: Glaumann H, Ballard FJ, editors. Lysosomes: their role in protein breakdown. New york: Academic Press, 1987: 319-367.
4. Takeshige K, Baba M, Tsuboi S, Noda T, and Ohsumi Y. Autophagy in yeast demonstrated with proteinase-deficient mutants and conditions for its induction. J. Cell Biol. 1993; 119: 301-311.
5. Marty F, Branton D, and Leigh RA. Plant vacuoles. In Tolbert NE, editor. The biochemistry of plants. A comprehensive treatise Vol. 1. New York: Academic Press, 1980: 625-658.
6. Matile P. The lytic compartment of plant cells. Vienna: Springer-Verlag, 1975.
7. Roby C, Martin JB, Bligny R, and Douce R. Biochemical changes during sucrose deprivation in higher plant cells. ^{31}phosphorus nuclear magnetic resonance studies. J Biol Chem. 1987; 262: 5000-5007.
8. Dorne AJ, Bligny R, Rebeillé F, Roby C, and Douce R. Fatty acid disappearance and phosphorylcholine accumulation in higher plant cells after a long period of sucrose deprivation. Plant Physiol Biochem 1987; 25: 589-595.

9. Saglio PH, and Pradet A. Soluble sugars, respiration, and energy charge during aging of excised maize root tips. Plant Physiol 1980; 66: 516-519.
10. Couée I, Jan M, Carde JP Brouquisse R, Raymond P and Pradet A. Effects of glucose starvation on mitochondrial subpopulations in the meristematic and submeristematic regions of maize root. Plant Physiol 1992; 100: 1891-1900.
11. Graham IA, Denby KJ, and Leaver CJ. Carbon catabolite repression regulates glyoxylate cycle gene expression in Cucumber. The Plant Cell 1994; 6: 761-772.
12. Baysdorfer C, Warmbrodt RD, and Van der Woude WJ. Mechanisms of starvation tolerance in pearl Millet. Plant Physiol 1988; 88: 1381-1387.
13. Aubert S, Gout E, Bligny R, and Douce R. Multiple effects of glycerol on plant cell metabolism. J Biol Chem 1994; in press.
14. Bligny R, Foray MF, Roby C, and Douce R. Transport and phosphorylation of choline in higher plant cells. Phosphorus-31 nuclear magnetic resonance studies. J Biol Chem 1989; 264: 4888-4895.
15. Gout E, Bligny R, Roby C, and Douce R. Transport of phosphocholine in plant cells: ^{31}P nuclear magnetic resonance studies. Proc Natl Acad Sci USA 1990; 87: 4280-4283.

LIPID COMPOSITION OF INTACT AND FUNCTIONAL MITOCHONDRIA ISOLATED FROM *ARABIDOPSIS THALIANA* SUSPENSION CULTURES

Thérèse Guillot-Salomon, Jacques Davy de Virville, Ivone Aaron, Marie-Françoise Alin, Catherine Cantrel and François Moreau

Laboratoire de Physiologie Cellulaire et Moléculaire, Université Pierre et Marie Curie, C.N.R.S. URA 1180, 4 place Jussieu 75252 Paris Cedex 05, France.

Introduction

The crucifer *Arabidopsis thaliana* has become the most widely used model for plant molecular biology [1]. Mutants with alterations in chloroplast lipid composition have been isolated and characterized providing considerable informations about the pathways of lipid synthesis in plastid membranes [2]. By contrast, little is known about lipid biogenesis in plant mitochondria. These organelles, like plastids, are semi-autonomous with regard to the formation of their membrane lipids [3]. The isolation of mitochondria could be very useful to investigate the extent to which mitochondrial membranes import lipids from the endoplasmic reticulum. In addition, these mutants should prove very useful to study the consequences of lipid alteration on functional properties of these organelles. Here we report on the lipid analysis of mitochondria isolated from cell suspension cultures of *Arabidopsis thaliana* [4]. This work is a part of a program aimed at developing methods to fractionate *Arabidopsis* tissues.

Materials and methods

The isolation and the characterization of mitochondria from cell suspension culture of *Arabidopsis thaliana* (Heynh), ecotype Columbia, were carried out as previously [4] Electron microscopy and lipid analysis are performed as in [5].

Results and discussion.

The growth characteristics of the cells and parallel changes in lipid composition throughout the cycle have been studied (Fig. 1). During the two first days a decrease in the level of plastidial lipid (MGDG and DGDG) takes place accompanied by an increase of extraplastidial lipids (PC and PE). From 3 days, the evolution of galactolipid to phospholipid ratio is inversed. This evolution can be related to the differenciation of cells with regard to chloroplast content. Mitochondria have been isolated during the exponential growth phase (4 days) when the cell respiration reaches a maximum [4].

J.-C. Kader and P. Mazliak (eds.), Plant Lipid Metabolism, 197–199.

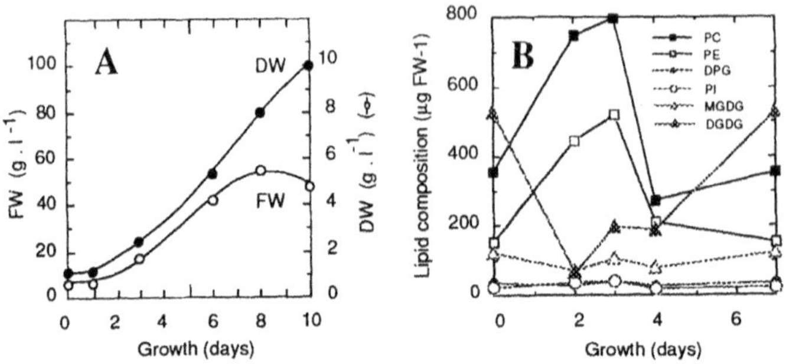

Figure. 1. A : Growth characteristics of *Arabidopsis* cells expressed in fresh (FW) and dry (DW) weigths. B : Changes in lipid composition of cells during the growth cycle.

The purity and integrity of *Arabidopsis* mitochondria was found to be about 90% based on galactolipid analysis and chlorophyll content (data not shown) and electron microscopy (Fig. 2). In addition, purified mitochondria exhibit good oxidative and phosphorylative properties as indicated by the high values of oxidation rates, respiration control and membrane potential (data not shown).

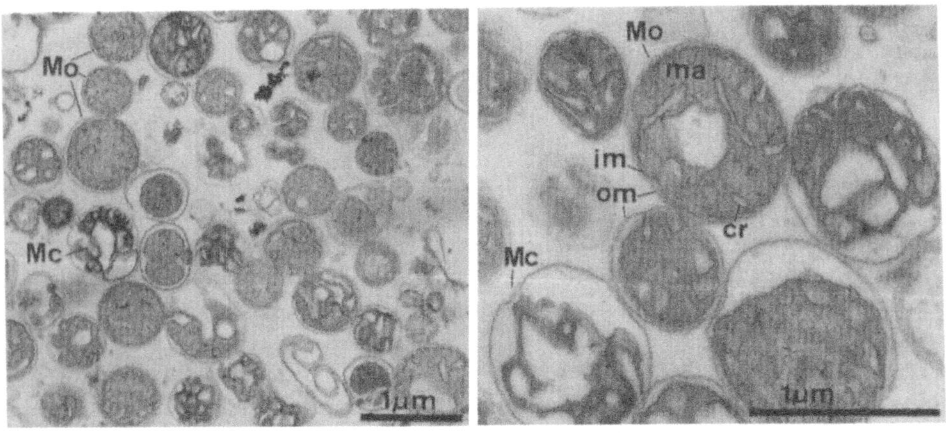

Figure 2. Ultrastructure of isolated *Arabidopsis* mitochondria.

As shown in figures 3 and 4, the glycerolipid and fatty acid compositions of *Arabidopsis* mitochondria are similar to these of mitochondria from other sources [6]. However, when compared to mitochondria isolated from ungreen tissues (potato tuber, maize epicotyl and *Acer* cells) the major difference noticed is the the 18:3 to 18:2 ratio which was found higher in *Arabidopsis* mitochondria.

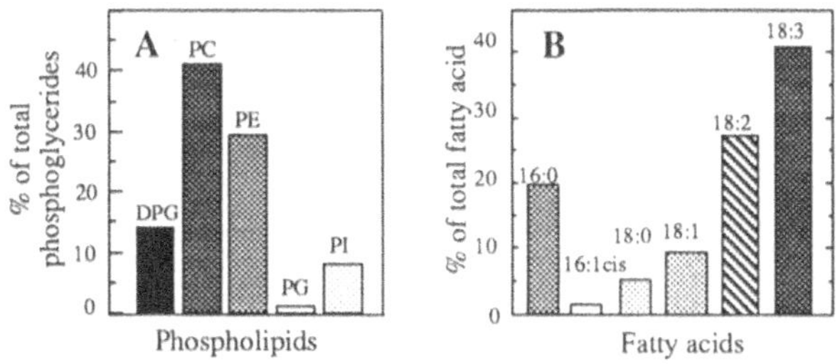

Figure 3. Phospholipids (A) and fatty acid content (B) of *Arabidopsis* mitochondria.

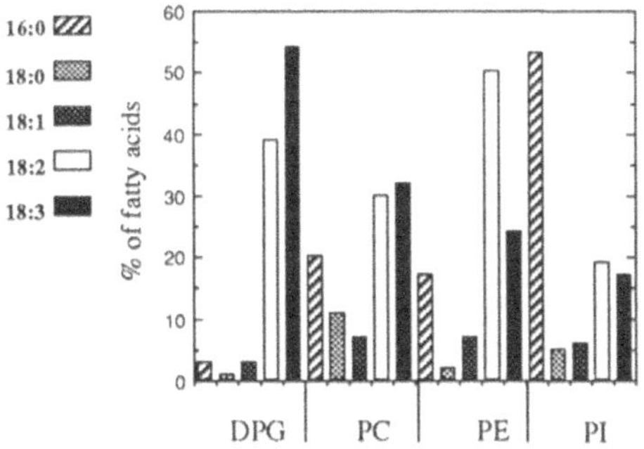

Figure 4. Fatty acid composition of individual phospholipids in *Arabidopsis* mitochondria.

Abbreviations: 16:0, palmitic acid; 16:1 cis, cis - Δ 9-hexadecenoic acid; 18:0, stearic acid; 18:1, oleic acid; 18:2, linoleic acid; 18:3, linolenic acid; DPG, diphosphatidylglycerol ; DGDG, digalactosyldiacylglycerol; FW, fresh weight; MGDG,monogalactosyldiacylglycerol; PC, phosphatidylcholine; PE, phosphatidylethanolamine; PI, phosphatidylinositol; cr, cristae; im and om, inner and outer membranes, respectively; ma, matrix; Mc and Mo, mitochondria in "condensed" or "orthodox" configuration, respectively.

References

1. Meyerovitz EM. *Arabidopsis*, a useful weed. Cell 1989; 56:263-269.
2. Somerville CR, and Browse J. Plant lipids: metabolism, mutants and membranes. Science 1991; 252:80-87.
3. Moore TS. Phospholipid biosynthesis. Ann Rev Plant Physiol 1982; 33:235-259.
4. Davy de Virville J, Aaron I, Alin M-F and Moreau F. Isolation and properties of mitochondria from *Arabidopsis thaliana* cell suspension cultures. Plant Physiol. Biochem 1994; 32:159-166.
5. Guillot-Salomon T, Bahl J, Ben-Raïs L, Alpha M-J, Cantrel C and Dubacq J-P. Heat stress and changes of lipid and carotenoid composition in spinach, a temperate plant, and jojoba, a desert plant. Plant Physiol Biochem 1991; 29:667-679.
6. Harwood JL. Phosphoglycerides of mitochondrial membranes. Methods Enzymol 1987; 148:475-485.

PURIFICATION AND CHARACTERIZATION OF CARNITINE ACYLTRANSFERASE FROM MUNG-BEAN (*Vigna radiata* L.) MITOCHONDRIA

Heidrun Schwabedissen-Gerbling* and Bernt Gerhardt#

*Unité mixte, Rhône-Poulenc Agrochimie, 14-20 rue P. Baizet, 69263 Lyon, France
#Institut für Botanik der Universität Münster, Schloßgarten 3, 48149 Münster, Germany

Synopsis

Carnitine acyltransferase was purified to apparent homogeinity from mung-bean (Vigna radiata L.) mitochondria. A rapid method was employd to overcome the instability, of the enzyme and its activity could be restorted through the presence of phospholipids. The carnitine acyltransferase differs in its properties from the well described mammalian carnitine acyltransferases.

Introduction

Carnitine acyltransferases are differentiated depending on the acyl-CoA favourized to transacylate onto carnitine. Carnitine acetyltransferases (CAT, EC 2.3.1.7), carnitine octanoyltransferases (COT, EC 2.3.1.137), and carnitine palmitoyltransferases (CPT, EC 2.3.1.21) have been purified from mammalian cells and are immunologically distinct [1]. In contrast, knowledge of plant carnitine acyltransferases is rather meager. Plant mitochondria exhibit carnitine acyltransferase activity [2,3,4]. Since the activity was demonstrated with acetyl-CoA and palmitoyl-CoA as substrates, the presence of two enzymes, a CAT and a CPT in pea mitochondria has been claimed [5,6]. On contrary, inhibition studies performed on the plant mitochondrial carnitine acyltransferase activities gave hint for the presence of only one single enzyme, which is active with short-, middle-, and long-chain acyl-CoAs as substrates [4,7]. The present investigation was undertaken in an attempt to purify the mitochondrial inner membrane bound carnitine acyltransferase(s) from mung-bean (*Vigna radiata* L.) hypocotyl, whose function in the plant mitochondria is/are not yet established.

Material and Methods

Mitochondria were isolated and carnitine acyltransferase activity was assayed according to [4]. Mitochondria obtained from the sucrose density gradients were diluted with distilled water and pelleted. The concentrated mitochondria were resuspended in 10 mM phosphate buffer (pH 7.6) containing 50 mM $MgCl_2$ and 0.02 % Triton X-100. After 1h incubation at 4°C the mixture (broken mitochondria) was loaded onto a Reactive Green-A-(crosslinked 5% agarose with covalently coupled dye)-column (Amicon, USA),

J.-C. Kader and P. Mazliak (eds.), Plant Lipid Metabolism, 200–202.

previously equilibrated with 25 mM phosphate buffer, pH 7.6. Unbound proteins were removed by washing with the same buffer. The enzyme was eluted from the column by a gradient of 0 to 1 M KCl. Active fractions were pooled and overnight concentrated in the presence of palmitoylcholine iodide (12 mM at the beginning of the concentration) to a protein concentration of about 6 mg ml^{-1}. The concentrated Green A-eluate was loaded on a Superdex 200 column (1.6 x 60 cm), equilibrated with 25 mM phospate buffer (pH 7.6). Fractions exhibiting carnitine acyltransferase activity and no contaminating proteins were pooled and directly used.

Results and Discussion

Carnitine acyltransferase has been purified for the first time from plant mitochondria. Solubilization of the membrane bound carnitine acyltransferase was the crucial step in the purification procedure. Carnitine acyltransferase could be separated from membrane components without distroying or loosing its activity by Green A-agarose chromatography. Over 90% of the initial mitochondrial activity could be recovered in the Green A-eluate. After concentration of the enriched fraction the mitochondrial enzyme could be separated from the remaining proteins by gelfiltration. The procedure resulted in an overall 100-fold purification. The yield was about 20% (Table 1).

This purification procedure provides carnitine acyltransferase in good yield and of a high level of purity in short working time. Unfortunately, the purified enzyme is not very stable, but can be stored in the presence of polyethylene glycol 6000, palmitoylcholine, phospatidylcholine, or phosphatidylethanolamine without activity loss. Interistingly, activity already lost could also be restorted by phospholipids.

On the basis of SDS/PAGE experiments, mitochondrial carnitine acyltransferase from mung bean hypocotyl is a monomer with a M_r of 45,000. Similar characteristics have been reported for CATs from mammalian mitochondria [1]. The apparent M_r of the native enzyme was estimated by gel filtration chromatography and was found to be 45,000. The values are consistent with a monomeric structure for the mitochondrial enzyme. The results are in accordance with those obtained for mammalian CATs but differ from those ontained for mammalian CPTs which are heteromeric enzymes exhibiting very high molar masses and being composed of several subunits [1].

Isoelectric focusing of the purified carnitine acyltransferase revealed an isoelectrical point of pH 6.0. The isoelectrical point of mitochondrial carnitine acyltransferase from mung-bean hypocotyl lies in the range obtained for commercial CAT (Boehringer, pH 6.0) or reported for CAT and COT from mouse liver peroxisomes (pH 6.8 and pH 5.2 [8]), respectively.

The mitochondrial carnitine acyltransferase exhibits a broad substrate specificity including fatty acids from C2 to C18 chain lenght. The activity ratio determined for the substrates acetyl-CoA and palmitoyl-CoA remained the same throughout purification (Table 1).

The specific activities obtained for carnitine acyltransferase were 159 nmol min^{-1} mg^{-1} with palmitoyl-CoA and 120 nmol min^{-1} mg^{-1} protein with acetyl-CoA as substrate. Apparent K_m values for the enzyme substrates acetyl-CoA, palmitoyl-CoA, and L-carnitine were: 8.5 µM, 2.5 µM, and 5 µM, respectively. Our findings concerning the substrate specificity are completely different from results obtained with mammalian carnitine acyltransferases.

Results obtained from substrate competition studies carried out according to the protocol of [9], revealed that acetyl-CoA and palmitoyl-CoA are apparently metabolized at the

same activity site of the carnitine acyltransferase without antagonistic effects.
With respect to the K_M values obtained from our investigations the plant mitochondrial carnitine acyltransferase should be named a CPT. However, other enzyme properties differ considerably from those of CPTs characterized from mammalian tissues. According to its biochemical properties the enzyme rather appears to be a CAT in spite of its good activity with palmitoyl-CoA. The different metabolic functions of the organells in mammalian and plant cells might be reflected in the properties of the carnitine acyltransferases.

References

1 Bieber LL. Carnitine. Ann Rev Biochem 1988; 57: 261-283
2 Thomas DR, Wood C. The two β-oxidation sites in pea cotyledons. Carnitine palmitoyltransferase: location and function in pea mitochondria. Planta 1986;168: 261-266
3 Burgess N, Thomas DR. Carnitine acetyltransferase in pea mitochondria. Planta 1986; 167: 58-65
4 Gerbling H, Gerhardt B. Carnitine-acyltransferase activity of mitochondria from mung-bean hypocotyls. Planta 1988; 74: 90-93
5 Wood C, Jalil MNR, Ariffin A, Yong BCS, Thomas DR. Carnitine short-chain acyltransferase in pea mitochondria. Planta 1983; 158: 175-178
6 Wood C, Jalil MNR, McLaren I, et al. Carnitine long-chain acyltransferase and oxidation of palmitate, palmitoyl coenzyme A and palmitoylcarnitine by pea mitochondria preparations. Planta 1984; 161: 255-260
7 Gerbling H, Gandour RD, Moore TS, Gerhardt B. Carnitine-acyltransferase activity in plant mitochondria. In: Quinn PJ , Harwood JL, editors. Plant Lipid Biochemistry, Structure, Function and Utilization. London: Portland Press, 1990: 181
8 Farrel SO, Fiol CJ, Reddy JK, Bieber LL. Properties of purified carnitine acyltransferases of mouse liver peroxisomes. J Biol Chem 1984; 259: 13089-13095
9 Chevillard, C, Cardenas ML, Cornish-Bowden A. The competition plot: a simple test whether two reactions occur at the same site. Biochem J 1993; 289: 599-604

Table 1:. Purification of carnitine acyltransferase from mung-bean hypocotyl mitochondria. The activity was determined by the release of CoASH from the acyl-CoA. With purified preparations the cleavage of the thioester bond of acyl-CoA catalyzed by carnitine acyltransferase was determined at 232 nm (ε_{232nm} = 5.0 cm^2 μmol^{-1}).

Step	Total Protein	Total Activity with ACA	Total Activity with PCoA	Specific Activity with ACoA	Specific Activity with PCoA	Ratio	Recovery	Purification
	mg	nmol min^{-1}		nmol min^{-1} mg $protein^{-1}$			%	-fold
Broken mitochondria	60	70.0	90.0	1.17	1.5	1.28	100	1
Green A-agarose chromatography	12	67.2	84.4	5.6	.70	1.26	95	4.7
Gelfiltration	0.12	14.5	19.1	120.8	159.1	1.31	21	105

LIPID CHANGES IN MITOCHONDRIA OF *ARUM ITALICUM* SPADIX DURING THE INFLORESCENCE DEVELOPMENT

Jacqueline Bahl*, Chantal Demandre**, Michèle Chauveau* Marie-José Alpha* and Jean Roussaux*

* Laboratoire de Biologie Cellulaire Végétale, Université P. et M. Curie, 12 rue Cuvier, 75005 - Paris, France.

** Laboratoire de Biologie Cellulaire et Moléculaire, Université P. et M. Curie, URA/CNRS 1180, 4 place Jussieu, 75252 - Paris cedex 05, France.

Introduction

During the inflorescence development six stages have been described: stages α, β, γ, δ, ε, ζ (1). The mitochondrial respiratory activity increases and reaches a maximum (respiratory crisis) at anthesis (stage ε), then decreases at stage ζ (1). Young stages are characterized by mitochondria of high density. From stage γ their density decreases down to standard values. These changes are in relation with differences in the matrix contents, the volume of the intermembrane space and the contrast of the outer membrane of mitochondria (2).

All these modifications suggest a variation in membrane structure and permeability. In order to obtain some additional informations, the lipid composition of mitochondria was analysed.

Materials and methods

Arum italicum L spadices were collected in woods near Paris and mitochondria were isolated at various developmental stages and purified on sucrose gradient, according to (1). Cytochrome oxidase and malate dehydrogenase activities were measured by spectrophotometry in a 300 mOsm medium. The ratio of the activities with or without Triton X-100 (0,2%) give an estimation of outer membrane permeability to cytochrome *c* or of inner membrane permeability to NADH. and thus of mitochondrial integrity. Lipids were extracted and analysed according to (3) and separated by TLC (4). Lipids were separated into molecular species by HPLC on a 5 µm ODS column by methanol/water/acetonitrile (90.5/7/2.5, v/v/v) solvent with 0.28 % of choline chloride (5). Methyl ester of fatty acids from each molecular species were analysed by GLC (6).

Results

MITOCHONDRIAL INTEGRITY

A weak cytochrome *c* permeability was observed in the youngest stages α and β (11%). The permeability increased up to the stage ζ (34%), showing a reduction of the outer membrane integrity.

J.-C. Kader and P. Mazliak (eds.), Plant Lipid Metabolism, 203–205.

On the other hand, the inner membrane integrity, measured as the permeability to NADH (malate dehydrogenase activity) did not change during the development of the inflorescence.

LIPID COMPOSITION

The amount of total lipids decreased from stage γ till the senescence (stage ζ), specially phosphatidylcholine (PC), phosphatidylinositol (PI) and diacylglycerol (DAG). In early stages of spadix development PC was the major lipid. From the stage δ, PE became predominant. At all stages, the amount of diphosphatidylglycerol (DPG) remained constant (table 1).

STAGES	PI	PS	PG	DPG	PC	PE	AP	DAG	TOTAL
β	11.8 *(7.0*	5.1 *(3.2)*	7.0 *(4.6)*	16.6 *(10.5)*	46.8 *(30.0)*	28.3 *(19.0)*	2.7 *(1.3)*	35.8 *(22.0)*	154.1
γ	10.8 *(5.9)*	1.6 *(0.9)*	6.1 *(3.1)*	19.9 *(10.5)*	50.0 *(32.5)*	43.0 *(26.3)*	3.4 *(1.9)*	35.1 *(18.4)*	169.9
δ	10.8 *(6.9)*	2.5 *(1.5)*	6.6 *(4.2)*	17.6 *(11.0)*	32.7 *(20.0)*	51.2 *(33.0)*	4.0 *(2.6)*	31.2 *(19.8)*	156.6
ε	5.4 *(4.5)*	1.9 *(1.5)*	4.5 *(3.6)*	14.0 *(11.0)*	28.7 *(20.0)*	46.9 *(34.7)*	8.1 *(6.3)*	24.9 *(18.0)*	134.4
ζ	5.5 *(4.9)*	4.4 *(4.0)*	4.0 *(3.6)*	16.8 *(15.1)*	22.2 *(20.5)*	40.1 *(36.1)*	5.3 *(4.7)*	12.8 *(11.5)*	111.1

Table 1 - Lipid composition of *Arum* mitochondria at various stages of the inflorescence development. The results are given in $\mu g\ mg^{-1}$ protein. Percentages are given brackets.

FATTY ACIDS OF PC AND PE

In PC molecules, the inflorescence maturation correspond to a loss of C 16:0 and C 18:2 with a concomitant increase of C 18:1. In PE molecules, the amount of C 18:1 and C 18:3 increased during the whole development while the amount of C 18:2 per mg of protein remained more or less constant.

PC AND PE MOLECULAR SPECIES

The main molecular species of PC in young stages (β,γ) was C18:2/18:2. From the maturation (δ) onwards C 18:1/18:2 and C 18:1/18:3 became predominant (fig. 1). In PE molecules during the development, C 18:1/18:2 and C 18:2/18:3 molecular species increased, but C 18:2/18:2 remained always the major molecular species.

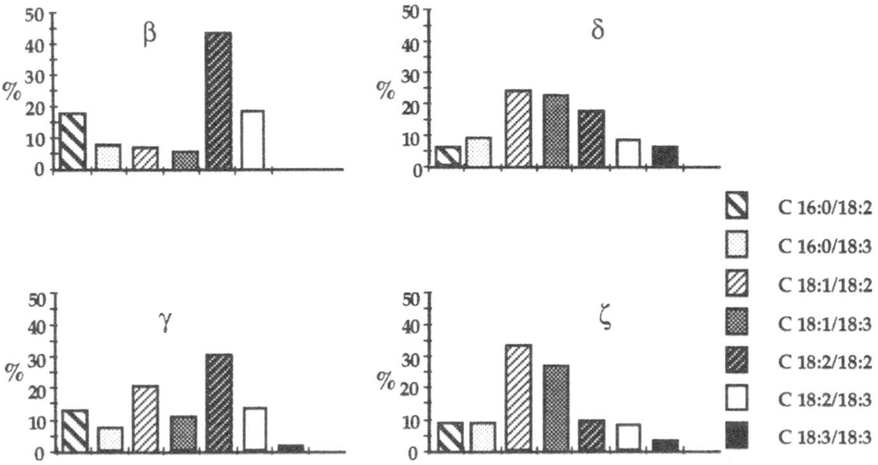

Figure 1 - Molecular species of PC during the development of *A. italicum* inflorescence.

Conclusions

During the development of *A. italicum* inflorescence, the amount of PC, the major mitochondrial lipid decreases, specially the C 18:2/18:2 molecular species. This fact could be in relation with the increase of the outer membrane permeability to cytochrome *c*. On the opposite, the amount of PE, known as being more abundant in the inner membrane, remains constant.

References

1. LANCE C. AND CHAUVEAU M. Evolution des activités oxydatives et phosphorylantes des mitochondries de l'*Arum maculatum* au cours du dévelopement de l'inflorescence.. *Physiol.Vég.*, 1975: **13**, 83-94.
2. CHAUVEAU M. AND TUQUET C. Changes in structure and activities of *Arum maculatum* mitochondria during spadix development. *Biol . Cell.*, 1985: **53**, 51-60.
3. BLIGH E.G. and DYER W.J. A rapid method of total lipid extraction and purification. *Can.J. Biochem. Physiol.,* 1959: **37**, 103-106.
4. LEPAGE M. Identification and composition of turnip root lipids. *Lipids*, 1967: **2,** 244-250.
5. DEMANDRE C., TREMOLIERES A., JUSTIN A.M. AND MAZLIAK P. Analysis of molecular species of plant polar lipids by high-performance and gas liquid chromatography. *Biochim. Biophys. Acta*, 1986: **877**, 380-386.
6. BEN RAIS L., ALPHA M.J., BAHL J., GUILLOT-SALOMON T. AND DUBACQ J.P..Lipid and protein contents of jojoba leaves in relation to salt adaptation *Plant Physiol. Biochem.*,1993: **1**, 547-557.

APPROACH TO *IN VIVO* FUNCTION OF NONSPECIFIC LIPID TRANSFER PROTEINS IN HIGHER PLANTS

M. Yamada[1], S. Tsuboi[2], M. Kosone[1], T. Osafune[3], T. Ehara[4], C. Masuta[5], A. Koiwai[5], S. Muto[6], T. Yuasa[6], H. Ariga[7] and M. Nishimura[8]

[1] Department of Bioscience and Technology, Hokkaido Tokai University, Sapporo, [2] Department of Biology, University of Tokyo, Tokyo, [3] Department of Life Science, Nippon College of Physical Education, Yokohama, [4] Department of Microbiology, Tokyo Medical College, Tokyo, [5] Life Science Research Laboratory, Tobacco Inc., Yokohama, [6] BioScience Center, Hagoya University, Nagoya, [7] Department of Pharmacology, Hokkaido University, Sapporo and [8] National Institute of Basic Biology, Okazaki, Japan.

Early work started by Kader's and Yamada's groups for the isolation of nonspecific lipid transfer protein (nsLTP) from Spinach leaves (Kader et al. 1984), maize seedlings (Douady et al. 1985) and castor bean seedlings (Watanabe & Yamada 1986). nsLTPs were widely distributed in higher plants. Amino acid analysis of purified nsLTPs indicated that nsLTP was a basic 9 kD-simple protein which contains four disulfide bridges (Takishima et al. 1986, 1988). nsLTP enhanced *in vitro* transfer of phosphoglycerides and glycoglycerides from donor to acceptor membranes, but did not that of tri- and di-acylglycerol (Watanabe & Yamada 1986, Tsuboi et al. 1992). However, *in vitro* function of nsLTP by which polar lipids are transferred between biomembranes was questioned from accumulated evidence for the localization and gene expression of nsLTP in plant tissues.

Approach from properties of nsLTP isoproteins.

We found organ-specific occurrence of nsLTP isoforms, A～D, in castor bean seedlings; B in cotyledons, A and C in endosperm and D in hypocotyl (Tsuboi et al. 1991). Among the isoforms, amino acid sequences varied from 30 to 70 %, but specificity for polar lipids was unchanged. Organ-specific distribution of nsLTP isoforms was also confirmed from cloning of cDNAs (Molina & Garcia-Olmendo 1993, Gausing 1994, Pyee et al. 1994). No involvement of nsLTP in the formation of roots or the biosynthesis of root constituents is suggested from missing of nsLTP in roots as common evidence.

Approach from the expression of nsLTP genes in plant orgens.

Table 1 shows cDNAs and genomic DNAs cloned from various plants. Attention must be paid for the expression of nsLTP gene in surface layers of aerial organs in seedlings, flower buds and developing fruits. This suggests that nsLTP concerns the biosynthesis of cuticle substances and/or lignins in epidermal tissues. In this respect, finding of nsLTP as the major protein contained in leaf wax by Kolattukndy's group (1994) is noticeable. Binding capacity of nsLTP with fatty acid or

J.-C. Kader and P. Mazliak (eds.), Plant Lipid Metabolism, 206–209.

Table 1. cDNAs or genomic DNAs for nsLTPs from higher plants

Species	cDNA or gene	Expression in organ	Reference
Spinach	pWS2	Leaf & petiole	Bernhard et al. 1991
Castor bean	**pnsLTP-C**	Cotyledons	Tsuboi et al. 1991
	pnsLTP-D	Axis	Tsuboi et al. 1991
	nsLTPC1, C2	Cotyledons	Weig & Komor 1992
Tobacco	**TobLTP**	Inflorescence	Masuta et al. 1992, 199X
	Ltp1	Leaf epidermis	Fleming et al. 1992
Tomato	**le16**	ABA-treated leaf	Plant et al. 1991
	TSW12	NaCl-stressed stem	Torres-Schumann et al. 1992
Carrot	EP2	somatic embryo	Sterk et al. 1991
<u>Brassica napus</u>	E2	Pollen	Foster et al. 1992
Broccoli	**wax9A wax9D**	Wax & cell wall of leaf epidermis	Pyee et al. 1994
<u>Arabidopsis thaliana</u>	pSLT3	Cell wall	Thoma et al. 1992
	LTP1	Peripheral tissues of shoot & flower	Thoma et al 1994
Maize	9C2	Peripheral layers of shoot and developing seeds	Sossountzov et al. 1991
Barley	**Ltp1**	Seed aleurone layer	Linnestat et al. 1991
	blt4.2, -4.6 and -4.9	Low temperature-, drought- stressed shoot meristem	Hughes et al. 1992 White et al. 1994
	Ltp2 Ltp4	Leaf & shoot, fungal pathogen-induced	Molina & Garcia-Olemendo 1993
	pKG285 pKG2316	Leaf & coleoptile	Gausing 1994
Rice	**rab**	Aleurone tissue	Mundy et al. 1990
		Elicitor chitosan-treated culture	Masuta et al. 1991
	Ltp	Coleoptile, seed	Vignols et al. 1994

Bold indicates genomic gene.

fatty acyl CoA (Tsuboi et al. 1992) suggests that nsLTP functions as a carrier of precursor fatty acids for the biosynthesis of wax and/or cutin. In relation to this, we examined the levels of nsLTP in cucumber leaves after UV irradiation, since leaf wax was shown to increase by UV irradiation (Steinmuller & Tevini 1985). There was little increase in the irradiated leaves under our experimental condition, although further work is needed.

As shown in Table 1, plant nsLTP was found to be a protein which is induced by ABA, drought and pathogen stresses, and directly serves as an antifungal substance (Terras et al. 1992). Plant nsLTP was also induced by low temperature, NaCl and an elicitor. Since drought, relevant to ABA, and microbe infection, relevant to elecitor, and environmental stresses, cause the thickness and lignification of epidermal cell wall, it is suggested that nsLTP is involved in the synthesis of lignin or its related substances in cell walls. In this respect, it is reasonable that the promoter of genomic sequence for nsLTP contains a drought responsive element, ABA responsive element (Plant et al. 1991) or low temperature responsive element (White et al. in press). Of great interest is that PAL regulatory element and chalcone systhase regulatory element were found in *Arabidopsis* nsLTP gene, as found by Sommerville's group (1994).

Approach from intracellular behaviors of nsLTPs.

Most of plant nsLTP possesses a signal peptide and was shown to be translocated to cell walls (Tsuboi et al. 1992, Thoma et al. 1992) or further secreted outside the epidermis (Sterk et al. 1991) by a secretary pathway through endoplasmic reticulum and presumably Golgi body (Madrid & Wettstein-knowls 1991, Barnhard et al. 1991). However, we found another behavior of nsLTP after the translation. In developing caster bean fruits, while nsLTPs in mesocarp and pericarp were transported by the secrete pathway, nsLTPs in seeds were first imported into dense vesicles, as shown previously (Tsuboi et al. 1992), which were in turn assembled in a large vacuole as an aggregate. At a moment, we can not explan this phenomenon, but it would give a clue to *in vitro* function of nsLTP.

Approach from involvement in signal transduction reaction.

Wheat nsLTP was found to be phosphorylated by Ca^{2+}-dependent protein kinase, CDPK, (Newmann et al. 1993), suggesting an involvement of nsLTP in Ca^{2+} regulation and/or signal transduction reaction. We also confirmed that tobacco nsLTP was phosphorylated by *Dunaliella* CDPK. This line is a hopeful way to attack how nsLTP functions *in vivo*.

References

Bernhard W.R., Thoma S., Botella J. and Somerville C.R. (1991).*Plant Physiol,* 95 : 164-170.

Hughes, M. A., Dunn, M. A., Pearce, R. S., White, A. J. and Zhang, L. (1992). *Plant Cell Environment,* 15 : 961-865.

Kader, J. C. (1993). In : *Biochemistry and Molecular Biology of membrane and storage lipids of plants* (Murata,N and Somerville,C.R., Eds), pp. 207-214, The American Society of Plant Physiologists, Rockville.

Kader, J. C., Julienne, M. and Vergnolle (1984). *Eur. J. Biochem.,* 139 : 411-416.

Linnestad C., Lönneborg A., Kalla E. and Alsen O.A. (1991). *Plant Physiol,* 97 : 841-843.

Masuta, C., Van den Bulcke, M., Bauw, G., Van Montagu, M. and Caplan, A. B. (1991) *Plant Physiol.,* 97 : 619-629.

Molina, A., Segura, A. and Garcia-Olmedo, F. (1993) FEBS Lett, 316, 119-122.

Molina, A., Segura, A. and Garcia-Olmedo, F. (1993). *FEBS Lett,* 316 : 119-122.

Mundy, J. and Rogers, J. C. (1986). *Planta,* 169 : 51-63.

Plant A.L., Cohen A., Moses M.S. and Bray E.A.(1991). Nucleotide sequence and spatial expression pattern of a drought- and absissic acid-induced gene of tomato. *Plant Physiol.,* 97: 900-906.

Polya, G. M., Chandra S., Chung, R., Neumann, G. M. and Hoj., P. B. (1992). *Biochim.Biophys. Acta,* 1120 : 273-280.
Pyee,J., Yu, H. and Kolattukudy, P.E. (1994) Arch. Biochem. Biophys., 311, 460-468.
Sossountzov, L., Ruiz-Avial, L., Vignols, F., Jolliot, A., Arondel, V.,Tchang, F., Grosbois, M., Guerbette, F., Miginiac, E., Delseny, M., Puigdomenech, P. and Kader, J. C. (1991). *Plant Cell,* 3 : 923-933.
Sterk, P., Booij, H., Scheleekens, G. A., Van Kammen, A. and de Vries, S. C. (1991). *Plant Cell,* 3 : 907-921.
Thoma, S., Kaneko, Y. and Somerville , C. (1993) Plant J., 3, 427-436.
Torres-Schumann S., Godoy J.A. and Pintor-Toro J.A. (1992). . *Plant Mol. Biol.,* 18 : 749-757.
Tsuboi S., Suga T., Takishima K., Mamiya G., Matsui K., Ozeki Y. and Yamada M. (1991). . *J. Biochem.* (Tokyo), 110 : 823-831.
Vignols,F., Lund G., Pammi, S., Trémousaygue, D., Grellet, F., Kader, J.C., Puigdomenech,P. and Delseny, M. (1994) Gene, 142, 265-270
Weig A. and Komor, E. (1992). *Planta,* 187 : 367-371.
Yamada, M. (1992). . *Plant Cell Physiol.,* 33 : 1-6.

RECONSTITUTED LIPID TRANSFER: COMPARISON OF THE REGULATION OF ACYL LIPID RELEASE FROM ENDOPLASMIC RETICULUM AND CHLOROPLAST ENVELOPE

Anna Stina Sandelius and Mats Räntfors, Department of Plant Physiology, Botanical Institute, Göteborg University, Carl Skottsbergs Gata 22, S-413 19 Göteborg, Sweden

Introduction

Transport of membrane lipids from the site of synthesis to the target membranes has been postulated to be mediated by membrane vesicles, by lipid transport proteins and/or to occur at contact sites between membranes [1-3]. Vesicular-mediated lipid transport between plant membranes has been reconstituted in cell-free systems. The results suggest that transport of lipids from endoplasmic reticulum (ER) to the Golgi apparatus [4,5] or to the chloroplast envelope (Sandelius and Räntfors, unpublished) and from the envelope to the thylakoid [6] is metabolically regulated, as shown by temperature dependence and requirement of and/or stimulation by ATP and cytosol or stroma.

To further identify the regulatory mechanisms involved in vesicle-mediated lipid transfer, the different steps involved need to be characterized separatedly. This work concerns the first step: the release of lipids, presumably as vesicles, from the lipid synthesizing membrane. Cell-free release of radiolabelled lipids from soybean ER was stimulated by palmitoyl-CoA and required cytosol. The release of radiolabelled lipids from chloroplast envelope was stimulated by ATP and GTP hydrolys, palmitoyl-CoA and required stroma. N-ethylmaleimide (NEM) had no effect on the lipid releases. The results point to similarities as well as differences between the regulation of galactolipid transfer from chloroplast envelope to the thylakoid and the regulation of vesicle trafficking among cytosolic membranes.

Materials and Methods

PREPARATION OF RADIOLABELLED DONOR MEMBRANE

Endoplasmic reticulum. Hypocotyls of 4-days-old dark-grown soybean were harvested and incubated overnight with [1-^{14}C]acetate. A fraction enriched in ER was isolated by discontinuous density gradient centrifugation followed by aqueous polymer two-phase partition [4].

Chloroplast envelope. Highly purified chloroplasts were isolated from leaves of wheat or spinach and, after osmotic rupture, subjected to density gradient centrifugation to isolate an envelope fraction [7]. This fraction was incubated with UDP[^{3}H]galactose to label monogalactosyldiacylglycerol [7].

J.-C. Kader and P. Mazliak (eds.), Plant Lipid Metabolism, 210–212.

RECONSTITUTED LIPID TRANSFER FROM DONOR MEMBRANES

Strips of nitrocellulose, 0.7 cm x 0.7 cm, were incubated with suspensions of radiolabelled envelope or ER for 1 h. The strips were rinsed and used for the acyl lipid release assay, except for strips used for determination of protein or total radioactivity. For both envelope and ER, each strip contained 1-3 μg membrane protein (the variations refer to different experiments).

The strips were individually incubated in 200 μl buffer (30 mM KCl, 2.5 mM Mg-oxaloacetate, 30 mM Hepes/KOH, pH 7.0) with various additions (see Table 1). Stroma was obtained from highly purified chloroplasts, which were osmotically ruptured and centrifuged for 2 x 30 min at $100{,}000g_{max}$. The second supernatant was used directly as total stroma or was subjected to fractionation by low speed centrifugation through a 10 kDa cut-off filter (Centricon 10, Amincon). A fraction of soluble proteins, here referred to as cytosol, was obtained correspondingly from a $100{,}000g_{max}$ supernatant of a homogenate of soybean hypocotyls. The incubation was stopped by removing the strip. Released radioactivity as well as radioactivity remaining on the strip were monitored by liquid scintillation counting. Data from representative experiments are shown. All experimental conditions were tested in two (palmitoyl-CoA and envelope) or at lest three independent experiments for each membrane.

Results and Discussion

Cell-free lipid trafficking between animal ER and Golgi membranes has been shown to require ATP, GTP-binding protein(s), palmitoyl-CoA and a NEM-sensitive protein [3]. Of these factors, ATP and the NEM-sensitive protein appears to be involved in vesicle fusion with the accceptor membrane [8].

Lipid release from immobilized soybean ER required cytosol and was stimulated by palmitoyl-CoA (Table 1). Myristoyl-, palmitoleoyl- and stearoyl-CoA were also stimulatory (not shown). The regulation of lipid release thus resembled the regulation of vesicle budding from mammalian ER and the result supports the assumption that lipids were released from soybean ER in the form of vesicles. The lack of effect of added GTP or GTPγS could have been due to that GTP was present already in the cytosol. Alternatively, vesicle release from soybean ER did not require GTP-binding proteins and/or coat formation. As expected, NEM had no effect. Vesicle budding from the ER may have been underestimated as ER is a very efficient acceptor of ER-derived membrane vesicles (Harryson, Morré and Sandelius, unpublished).

Lipid release from immobilized chloroplast envelope required stroma and was stimulated by ATP, GTP and palmitoyl-CoA (Table 1). As with ER, other acyl-CoAs were stimulatory and NEM was without effect. GTPγS abolished both the GTP effect as well as a significant proportion of the stroma effects (Table 1). Thus, the stimulatory effect of GTP probably was connected with GTP hydrolysis interacting with stroma protein(s) rather than with a GTP-binding protein. Since stroma inhibits envelope to envelope vesicular transfer [6], lipid release from the envelope in the presence of stroma probably was not underestimated. The result supports previous findings of a metabolically regulated lipid transport between envelope and thylakoids [6].

The stimulatory effect of acyl-CoAs on lipid release from the envelope and on non-GTP-dependent lipid release from the ER indicates that acyl-CoAs may have a general role in the regulation of membrane vesicle formation, in addition to its known function in stimulating fission of coated vesicles from mammalian ER or Golgi membranes [8].

Table 1. Regulation of the release of radiolabelled lipids from immobilized ER and chloroplast envelope. The incubations were for 60 min at 22°C. 500 μM ATP or ATPγS and 200 μM GTP or GTPγS were used. The results are presented as % of the radioactivity originally present on the strip that was released into the medium during incubation. The different experiments were separately performed for ER and chloroplast envelope.

	endoplasmic reticulum		chloroplast envelope	
	buffer	+ cytosol	buffer	+ stroma
Experiment 1:				
No additions	-	-	0.7	2.5
ATP + GTP	-	-	-	4.0
ATPγS + GTP	-	-	-	2.4
Experiment 2:				
No additions	0.7	2.4	0.5	3.1
ATP + GTP	0.7	2.5	-	3.8
ATP + GTPγS	0.8	2.3	-	1.8
Experiment 3:				
No additions	1.0	3.5	1.4	4.7
ATP + GTP	0.9	3.6	2.2	5.2
ATP + GTP + 16:0-CoA	-	6.9 (25 μM)	-	6.9 (50 μM)

References

1. Kader JC. Intracellular transfer of phospholipids, galactolipids and fatty acids in plant cells. In: Hilderson HJ, editor. Subcellular biochemistry. New York: Plenum Publishing Corporation, 1990: 69-111.
2. Wirtz KVA. Phospholipid transfer proteins. Annu Rev Biochem 1991; 60: 73-99.
3. Allan D, Kallen K-J. Transport of lipids to the plasma membrane in animal cells. Prog Lipid Res 1993; 32: 195-219.
4. Morré DJ, Penel C, Morré DM, Sandelius AS, Moreau P, Andersson B. Cell-free transfer and sorting of membrane lipids in spinach. Protoplasma 1991; 160: 49-64.
5. Sturbois B, Moreau P, Maneta-Peyret L, Morré DJ, Cassagne C. Cell-free transfer of phospholipids between the endoplasmic reticulum and the Golgi apparatus of leek seedlings. Biochim Biophys Acta 1994; 1189: 31-37.
6. Morré DJ, Morré JT, Morré SR, Sundqvist C, Sandelius AS. Chloroplast biogenesis: Cell-free transfer of monogalactosylglycerides to thylakoids. Biochim Biophys Acta 1991; 1070: 437-445.
7. Joyard J, Douce R. Préparation et activités enzymatiques de 'enveloppe des chloroplastes d'épinard. Physiol Vég 1976; 14: 31-48.
8. Ostermann J, Orci L, Tani K, Amherdt M, Ravazzola M, Elazar Z, Rothman JE. Stepwise assembly of functionally active transport vesicles. Cell 1993; 75: 1015-1025.

Supported by the Swedish Natural Science Research Council and the Futura Foundation.

ISOLATION OF TRANSFER VESICLES ISSUED FROM THE ENDOPLASMIC RETICULUM OF LEEK SEEDLINGS

Sturbois Bénédicte[1], Moreau Patrick[1], Loreau Olivier[2] and Cassagne Claude[1]

1 Laboratoire de biogenèse membranaire , URA 1811, Université de Bordeaux II, 1 rue Camille Saint-Saëns 33077 BORDEAUX Cedex
2 C.E.A. Centre de Saclay 91191 Gif sur Yvette

Cell-free systems have been largely used to study the vesicular transport of proteins between the endoplasmic reticulum (ER) and the Golgi apparatus (GA) or within the Golgi apparatus [1-2].
Morré and coworkers have also developed a cell-free system from rat liver with which the vesicular transport of proteins between the ER and the GA was reconstituted in vitro. This approach uses the donor membranes (ER) in solution and the acceptor membranes (GA) immobilized on nitrocellulose strips. This system has also been adapted to reconstitute the vesicular transfer of lipids between the ER and the GA in rat liver [3-4]. The reconstitution of the vesicular (i.e. ATP-dependent) transfer of lipids between different plant membrane fractions has also been obtained with spinach [5].
In our laboratory, studies of lipid transfer in leek seedlings in vivo, have already shown the existence of a vesicular process for the transfer of phospholipids and particularly of very long chain fatty acid-containing lipids [6]. This process follows the vesicular endoplasmic reticulum- Golgi apparatus- plasma membrane pathway. Using the cell-free system developed by Morré and coworkers, we have reconstituted in vitro the vesicular transfer of some phospholipids between the ER and the GA. This transfer is ATP and cytosol-dependent, is N Ethyl Maleimide and temperature sensitive and specific for the ER as donor and the GA as acceptor. The phospholipids transferred via an ATP-dependent manner in vitro between the ER and the GA were phosphatidylcholine (PC; +79%), phosphatidylethanolamine (PE; +67%) and phosphatidylserine (PS ; +123%) [7]. All those results are in favour of a vesicular transport of phospholipids between the ER and the GA of leek seedlings, and brought us to purify these transition vesicles issued from the ER.
The optimal conditions for the formation of the transition vesicles in vitro have first been determined. For these reasons, we have checked whether a sustained lipid synthesis in vitro could increase the ATP-dependent transfer of membrane material i.e. the amount of vesicles formed. For that, the reaction mixture already described [7] was completed with CDP-choline (12,5μM), CDP-ethanolamine (12,5μM), C16:0 (2,5μM), C18:1 (7,5μM) and CoA (500μM). We have measured in vitro the ATP-dependent and ATP-independent lipid transfer between the ER and the Golgi in the presence or absence of the components described above, for three times of incubation : 5, 15 or 30 minutes. The results are presented in the figure 1.
They showed that the addition of the components which allowed a sustained synthesis of lipids (and particularly of PC and PE) could stimulate the ATP-dependent (i.e.

J.-C. Kader and P. Mazliak (eds.), Plant Lipid Metabolism, 213–215.

vesicular) but not the ATP-independent transfer of lipids between the ER and the GA in vitro. Moreover, the stimulation of the ATP-dependent transfer of lipids by CDP-choline, CDP-ethanolamine and the other molecules, was maximal for a time of incubation of 15 minutes (+60%). For the formation of the transition vesicles in vitro, the ER-enriched fraction was therefore incubated in the presence of ATP, cytosol, an ATP-regenerating system and the components which stimulated lipid synthesis and ATP-dependent transfer (Figure 1).

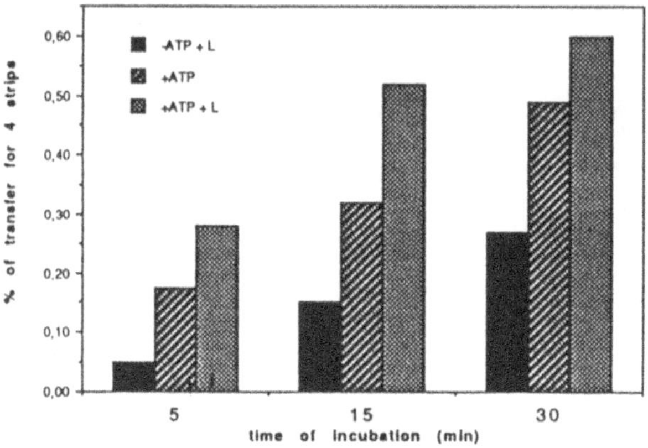

Figure 1 : Transfer of labeled lipids between the ER and the GA.
ATP : reaction mixture described in [7]
L : CDP-choline, CDP-ethanolamine, CoA, C16:0 and C18:1

After incubation, the membranes were lowered onto a discontinuous sucrose gradient (18% ; 37%) or a continuous sucrose gradient (15%-> 45%). After centrifugation, the fractions

interface sample-18% : transition vesicles (TV) — discontinuous gradient
interface 18%-37% : endoplasmic reticulum (ER)

or

fractions 16%- 20% : transition vesicles (TV) — continuous gradient
fractions 24%-30% : endoplasmic reticulum (ER)

were collected, resedimented and resuspended in Hepes buffer, pH 7.8. The lipids, extracted by chloroform/ methanol (1/1) were analysed. The quantity of proteins was determined by the method of Bradford (8). The ratios lipids (μg)/ proteins (μg) for the different fractions were calculated. We found higher ratio for the TV-enriched fraction (0.96±0.33 ; n=5) than for the ER-enriched fraction (0.23±0.1 ; n=6). The fractions enriched with ER-derived transition vesicles showed a similar lipid/protein ratio than the fraction named lipid rich vesicles (LRV,contening both vesicles originating from the ER and from the GA) which was shown to be involved in vivo in the vesicular transfer of lipids to the plasma membrane (6). Therefore, we may suggest that we have isolated, at least to some extent, ER-derived transfer vesicles that were also present in the LRV fraction. After separation of the different phospholipids of TV-enriched fraction and ER-enriched fraction on HPTLC plates (eluted by the method of Heape et al. (9)), the amounts and radioactivity of each phospholipids were determined. The TV-enriched fraction present higher amounts of PS and PE. Similar observations were made for the radioactivities of these phospholipids. The enrichment of the amounts of labeled PS and

PE in the TV is in excellent agreement with the fact that these phospholipids have been shown to be transferred in an ATP-dependent manner between the ER and the GA in vitro (7).
Therefore, it is likely that the transition vesicles isolated correspond, at least to some extent, to those that are operative in the transfer of phospholipids between the ER and the Golgi apparatus.

References :

1. Balch WE, Elliot MM and Keller DS. ATP coupled transport of vesicular stomatosis virus G protein between the endoplasmic reticulum and the Golgi. J. Biol. Chem. 1986;261:14681-14689.
2. Rothman JE. Transport of vesicular stomatosis glycoprotein to trans-Golgi membranes in a cell-free system. J. Biol. Chem. 1987;262:12502-12510.
3. Moreau P and Morré DJ; Cell-free tranfer of membrane lipids. J. Biol. Chem. 1991;266:4329-4333.
4. Moreau P, Rodriguez M, Cassagne C, Morré DM and Morré DJ. Trafficking of lipids from the endoplasmic reticulum to the Golgi apparatus in a cell-free system from rat liver. J.Biol.Chem. 1991;266:4322-4328.
5. Morré DJ, Penel C, Morré DM, Sandelius AS, Moreau P and Andersson B. Cell-free transfer and sorting of membrane lipids in spinach. Protoplasma 1991;160:49-64.
6. Bertho P, Moreau P, Morré DJ and Cassagne C. Monensin blocks the transfer of very long chain fatty acid containing lipids to the plasma membrane of leek seedlings. Evidence for lipid sorting based on fatty acyl chain lenght. Biochim.Biophys. Acta 1991;1070:127-134.
7. Sturbois B, Moreau P, Morré DJ and Cassagne C. Cell-free transfer of phospholipids between the endoplasmic reticulum and the Golgi apparatus of leek seedlings. Biochim. Biophys. Acta 1994;1189:31-37.
8. Bradford MM. A rapid and sensitive method for the quantitation of microgram quantities of protein utilizing the principe of protein-dye binding. Anal Biochem. 1976;72:248-254.
9. Heape AM, Juguelin H, Boiron F and Cassagne C. Improved one dimensional thin-layer chromatographic technique for polar lipid. J. Chromatogr. 1985;322:391-395.

BIOSYNTHESIS OF AN UNUSUAL PHOSPHOLIPID, *N*-ACYL-PHOSPHATIDYL-ETHANOLAMINE IN COTTON COTYLEDONS

Thomas S. Moore, Jr. and Kent D. Chapman

Department of Plant Biology, Louisiana State University, Baton Rouge, LA 70803-1705, USA (TSM) and Department of Biological Sciences, Division of Biochemistry and Molecular Biology, University of North Texas, Denton, TX 76203, USA (KDC)

N-Acyl-phosphatidylethanolamine (NAPE)was first described from plants by Bomstein [1]. Further work on its occurrence and biosynthesis were discontinued when a similarly behaving compound was found to be an artifact of the extraction procedures and identified as *N*-methyl-phosphatidylethanolamine [2]. In 1991, however, Chapman and Trelease [3] noted a spot on thin-layer chromatograms of radiolabeled phospholipids from cotton cotyledons which was tentatively identified as NAPE In light of these results, we chose to examine the occurrence, structure and origin of the putative NAPE in more detail.

Occurrence

NAPE has been reported from a variety of species. The highest concentration described was 70% of the total polar lipids of soybean seeds radiolabeled with acetate or pyruvate [4]. We found that NAPE comprised about 7.2% of the total microsomal phospholipid by weight, and when [^{14}C]ethanolamine was introduced to a variety of tissues, as much as 14% of the label could be found in NAPE, with the majority of the remainder in phosphatidylethanolamine [5].

We detected NAPE in the seeds of a variety of plants, with highest concentrations being in the cotyledons. Both photosynthetically capable and nongreen cotyledons contained NAPE, and a small amount also was found in rice suspension cultures [5]. In summary, a variety of tissues may contain NAPE, and a thorough survey of its occurrence in several plant families and tissues would seem to be called for. Such a survey may provide valuable clues to the role of NAPE in plants.

Structure

A two-dimensional representation of the structure of NAPE is presented in Figure 1 (next page). This phospholipid is quite unique in having three acyl groups, two associated with the glycerol moiety in standard *O*-ester linkages, but the third is attached by an amide linkage to the *N*-terminus of the ethanolamine headgroup.

The most common acyl units found in the amide linkage of cotton microsomal NAPE are 16:0 (57%) and 18:0 (28%), while the most common acyl unit of the *sn*-1 position of this NAPE is 16:0 (82%), while *sn*-2 most often contains 18:0 (58%) or 16:0 (27%) [5].

J.-C. Kader and P. Mazliak (eds.), Plant Lipid Metabolism, 216–218.

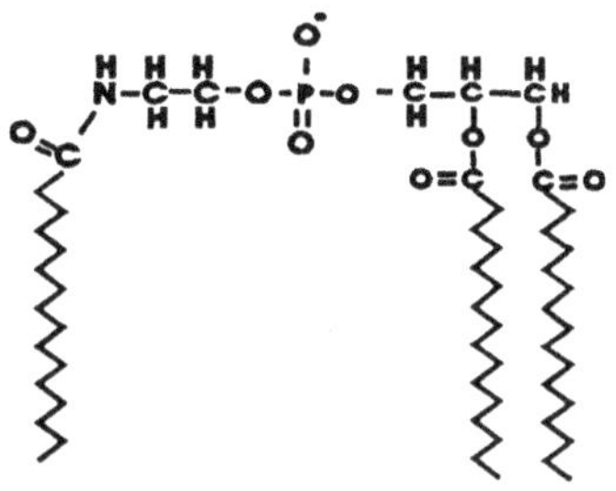

FIGURE 1. Diagram of the structure of *N*-acylphosphatidylethanolamine.

Biosynthesis

PRECURSORS

Biosynthesis of NAPE utilizes phosphatidylethanolamine (PE) as the immediate phospholipid precursor. We have demonstrated this both by direct feeding of radiolabeled phosphatidylethanolamine (unpublished data) and by use of radiolabeled PE formed by feeding microsomes either [^{14}C]ethanolamine or CDP-[^{14}C]ethanolamine under conditions for PE synthesis, followed by removal of the water soluble label and assaying NAPE synthesis under conditions of the latter synthesis. The former procedure produces PE by an exchange reaction, the second by ethanolaminephosphotransferase [6]. PE formed by either reaction could be utilized for NAPE synthesis.

One of the most intriguing aspects of NAPE synthesis is the source of the *N*-linked acyl group. Assays of incorporation of acyl units from dipalmitoylphosphatidylcholine (as in mammals), palmitoylCoA, or free palmitate demonstrated that the free fatty acid serves as the immediate precursor [7]. Indeed, the incorporation from the CoA derivative was only about 0.5% of the free fatty acid, while from PC the incorporation was less that 0.05% of the free acid incorporation. Incorporation from PC or palmitoylCoA generally could be accounted for as having been through degradation to the free fatty acids. Tests of a wide range of potential intermediates and energy sources have been negative, but there is sufficient precedent for such a reaction being possible even without formation of high energy bonds. Indeed, the apolar nature of the precursors may contribute to this process. A precise mechanism remains to be demonstrated, however.

Table 1. Acyl competition with palmitate with the three isoforms of NAPE synthase.

Acyl unit	Isoform A	Isoform B	Isoform C
		% of palmitate	
[^{14}C]palmitate	100	100	100
+ myristate	322	55	74
+ stearate	271	28	71
+ oleate	113	73	91

ENZYMES

The enzymes tested were associated with a microsomal fraction from cotton cotyledon. We were able to solubilize the enzyme through the use of *n*-dodecyl-β-D-maltoside. Following solubilization, three isoforms of the enzyme were separated by preparative isoelectricfocusing [8]. The three forms were found to have different affinities for palmitate, but all were in the μM range. The acyl preferences for each form were distinctly different, however, when compared to palmitate; various fatty acids were added to test for competition with [^{14}C]palmitate, with results similar to those found in Table 1 being obtained. Note that with isoform A, the additional fatty acids stimulated palmitate incorporation, little effect of the added fatty acids was observed with isosoform C, but all three appeared to compete with palmitate incorporation in the isoform B catalyzed reaction.

Summary

Three isoforms of NAPE synthase exist in cotton cotyledons. Although the precise role is unknown, the stability of newly synthesized NAPE coupled with the high palmitate in the acyl units would suggest a role in membrane stabilization. Efforts are underway to clarify this role with respect to temperature tolerances in plants.

Acknowledgements

This research was supported by NSF postdoctoral fellowship DIR9104341 to KDC and grants from NSF (DCB 9003817) and USDA (93-37304-9368) to TSM.

References

1. Bomstein RA. A new class of phosphatides isolated from soft wheat flour. Biochem Biophys Res Commun 1965;21:49-54.

2. Roughan PG, Slack CR, Holland R. Generation of phospholipid artefacts during extraction of developing soybean seeds with methanolic solvents. Lipids 1978;13:497-503.

3. Chapman KD, Trelease RN. Acquisition of membrane lipids by differentiating glyoxysomes - role of lipid bodies. J Cell Biol 1991; 115: 995-1007.

4. Wilson RF, Rinne RW. Phospholipids in the developing soybean seed. Plant Physiol 1974;54:744-747.

5. Chapman KD, Moore TS. *N*-Acylphosphatidylethanolamine synthesis in plants: Occurrence, molecular composition, and phospholipid origin. Arch Biochem Biophys 1993; 301:21-33.

6. Shin S, Moore TS. Phosphatidylethanolamine synthesis by castor bean endosperm. membrane bilayer distribution of phosphatidylethanolamine synthesized by the ethanolaminephosphotransferase and ethanolamine exchange reactions. Plant Physiol 1990;93:154-159.

7. Chapman KD, Moore TS. Catalytic properties of a newly discovered acyltransferase that synthesizes *N*-acylphosphatidylethanolamine in cottonseed (*Gossypium hirsutum* L.) microsomes. Plant Physiol 1993; 102:761-769.

8. Chapman KD, Moore TS. Isozymes of cottonseed microsomal *N*-acylphosphatidylethanolamine synthase: Detergent solubilization and electrophoretic separation of active enzymes with different properties. Biochim Biophys Acta 1994;1211:29-36.

REGULATION OF PHOSPHATIDYLINOSITOL 4-KINASE BY PROTEIN PHOSPHORYLATION: A PLASMA MEMBRANE-CYTOSKELETAL CONNECTION

Wendy F. Boss, Wannian Yang, Zheng Tan, and Myeon Cho

Botany Department, Box 7612, North Carolina State University, Raleigh NC 27695, USA

The polyphosphorylated inositol phospholipids, phosphatidylinositol 4-monophosphate (PIP) and PI-4,5-bisphosphate (PIP_2), are important intermediates in transducing extracellular signals through the production of the second messengers, inositol-1,4,5-trisphosphate (IP_3) and diacylglycerol (1). In addition, they can play a key role in regulating cell growth and development by directly affecting the activity of protein kinases (2), P-type ATPases (3-5), and actin-binding proteins (6-8).

Although the inositol phospholipids are present in plants, the relative amount of PIP_2 is very low compared to that present in many animal cells. This has made it difficult to study the changes in PIP_2 and IP_3, and as a result, the role of PIP_2 in plant signal transduction remains an enigma.

Small changes in inositol phospholipids have been observed in some higher plant systems (9-14). Although no plant stimulus has resulted in the 3 to 5 fold increase in IP_3 observed in animal cells, treatment with mastoparan, the wasp venom tetredecapeptide, causes rapid turnover of PIP_2 and a 3 to 7 fold increase in IP_3 in plants (15,16, Cho et al., unpublished results). These observations are important because they demonstrate that if there had been a large, rapid turnover of PIP_2 with other stimuli, it would have been detected. In addition, the mastoparan experiments suggest that the primary pool of PIP_2 in higher plants is either not sensitive to most plant stimuli or that it normally turns over slowly and only a small amount of IP_3 is released at a time.

Our work has focused on the regulation of PIP biosynthesis via PI 4-kinase. We have characterized two soluble activators of the plasma membrane PI 4-kinase (17,18). One of these, a soluble, 49-kDa protein (PIK-A49) has been purified and sequenced (18). PIK-A49 decreases the K_m of PI 4-kinase for ATP from 0.40 mM to 0.19 mM (19). Most interestingly, PIK-A49 is a multifunctional protein that binds and bundles F-actin and has translational elongation factor-1α (EF-1α) activity (18). Western blot analysis indicates that PIK-A49 is present in the plasma membrane, cytoskeletal and soluble fractions. GTP and GDP, which affect the EF-1α function of the protein, inhibit the activation of PI 4-kinase by PIK-A49 suggesting that when EF-1α functions as an elongation factor it is not functioning as PI 4-kinase activator.

J.-C. Kader and P. Mazliak (eds.), Plant Lipid Metabolism, 219–223.

Phosphorylation of PIK-A49 is essential for PI 4-kinase activation (19). Phosphorylation of the purified PIK-A49 by a calcium-dependent protein kinase (CDPK) enhances activation of PI 4-kinase. Dephosphorylation of PIK-A49 by alkaline phosphatase completely eliminates PI 4-kinase activation, and activation is restored by phosphorylation of PIK-A49 with CDPK. These data suggest one mechanism for feedforward regulation of polyphosphoinositide biosynthesis in response to increased in cytosolic calcium. Such a mechanism could be involved in the recovery of PIP levels after a transient increase in cytosolic calcium. Furthermore, because PIK-A49 is multifunctional, it may be involved in cross-talk between the cytoskeleton, protein synthesis machinery and plasma membrane of the cell. This would be consistent with a role of inositol phospholipids in regulating cytoskeletal structure and would predict that PIP and PIP_2 would change when there were changes in cytoskeleton. Figures 1. A, B, and C depict a working model for the role of PIK-A49 in signal transduction.

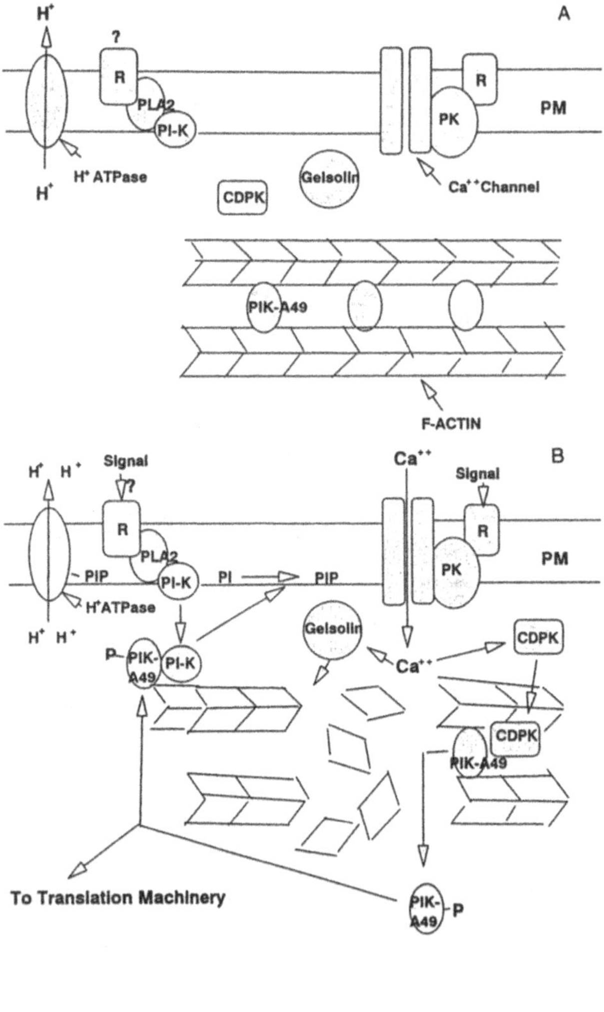

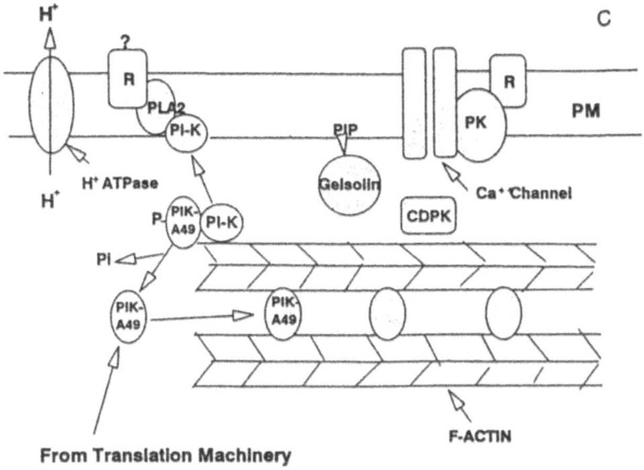

Figure 1. A. The resting stage of the cell. When the cell is in the resting stage, PIK-A49 is primarily associated with F-actin and stored in actin bundles. Because the intracellular Ca^{2+} level is low, the cytosolic CDPK and actin-severing or capping proteins (e.g. gelsolin, profilin) are inactive. **B.** When cytosolic calcium increases, as a result of an influx of extracellular calcium or by the release of intracellular calcium from calcium stores, PLA_2, PLC, actin-severing and capping proteins can be activated. Activated PLA_2 can release PI kinase from the plasma membrane. The released PI kinase will be activated by PIK-A49. Activated PLC can cleave actin-severing protein-associated PIP or PIP_2 and release actin-severing proteins from membranes. Activated actin-severing proteins cut F-actin into actin oligomers. Activated CDPK can phosphorylate actin-associated PIK-A49. The cytosolic free calcium is either pumped out of the cell or into intracellular stores. Thus, the increase of cytosolic calcium ion is transient. Once dissociated from actin, PIK-A49 can move to two targets. One is the ribosomes where it can function as EF-1α and enhance protein synthesis. The other target is the membrane-released PI kinase. The activation of PI kinase will result in an increase in PIP. **C.** The cell returns to the resting stage after a growth phase. As the concentration of intracellular calcium decreases, PLA_2, PLC, actin-severing proteins and CDPK are no longer activated. As the concentration of PIP increases in the membranes, actin-severing proteins and capping proteins will dissociate from actin oligomers and bind PIP. Once actin-severing proteins and capping proteins are dissociated, actin polymerization takes place and new F-actin is formed. At the same time, PIK-A49 is dephosphorylated, binds to F-actin and enhances the formation of new actin-bundles. Enhanced protein synthesis and changes in cytoskeletal structure are necessary for cell elongation, cell division and cell differentiation. The cell returns to its resting stage and is ready to respond to the next signal. **Abbreviations: CDPK,** calcium-dependent protein kinase; **EF-1α,** elongation factor 1 alpha; **PI-K,** phosphatidylinositol kinase; **PIP,** phosphatidylinositol monophosphate; **PK,** protein kinase; **PLA_2,** phospholipase A_2; **PLC,** phospholipase C; **PM,** plasma membrane; **R,** receptor.

In summary, evidence is accumulating to suggest that a major role for inositol phospholipids in higher plants is to regulate cytoskeletal structure (13,20,21). Our thesis is that one of the major differences in plants and animals that has made it so difficult to study PIP_2 is that very little PIP_2 is available for plant stimulus-response pathways. We hypothesize that most of the plasma membrane PIP and PIP_2 pools are associated with cytoskeletal proteins in plants and that turnover is associated primarily with changes in cytoskeleton.

Acknowledgements: This work was supported by a grant from the National Science Foundation, USA and by the North Carolina Agricultural Research Service.

References:

1. Berridge MJ, Irvine RF. Inositol phosphates and cell signalling. Nature 1989;341:197-205.
2. Chauhan VPS, Brockerhoff H. Phosphatidylinositol-4,5-bisphosphate may antecede diacylglycerol as activator of protein kinase C. Biochem Biophys Res Commun 1988;155:18-23.
3. Missiaen L, Wuytack F, Raeymaekers L, De Smedt H, Casteels R. Polyamines and neomycin inhibit the purified plasma-membrane Ca^{2+} pump by interacting with associated polyphosphoinositides. Biochem J 1989;261:1055-1058.
4. Memon AR, Chen Q, Boss WF. Inositol phospholipids activate plasma membrane ATPase in plants. Biochem Biophys Res Commun 1989;162:1295-1301.
5. Chen Q, Boss WF. Neomycin inhibits the PIP and PIP_2 stimulation of plasma membrane ATPase activity. Plant Physiol 1990;96:340-343.
6. Lassing I, Lindberg U. Specific interaction between phosphatidylinositol 4,5-bisphosphate and profilactin 1985;314:472-474.
7. Goldschmidt-Clermont P, Machesky LM, Baldassare JJ, Pollard TD. The actin-binding protein profilin binds to PIP_2 and inhibits its hydrolysis by phospholipase C. Science 1990;247:1575-1578.
8. Aderem A. Signal transduction and the actin cytoskeleton: the roles of MARCKS and profilin. Trends in Biochem Sci 1992;17:438-443.
9. Morse MJ, Crain RC, Satter RL. Light-stimulated inositol phospholipid turnover in *Samanea saman*. Proc Natl Acad Sci USA 1987;84:7075-7078.
10. Cho MH, Shears SB, Boss WF. Changes in phosphatidylinositol metabolism in response to hyperosmotic stress in *Daucus carota* L. cells grown in suspension culture. Plant Physiol 1993;103:637-647.
11. Gross W, Boss WF. Inositol phospholipids and signal transduction. In: Control of Plant Gene Expression, DPS Verma Ed, Boca Raton:CRC Press, 1993:17-32.
12. Einspahr KJ, Thompson GA Jr. Transmembrane signalling via phosphatidylinositol 4,5-bisphosphate hydrolysis in plants. Plant Physiol 1990;93:361-366.
13. Drøbak BK. Plant phosphoinositides and intracellular signaling. Plant Physiol 1993;102:705-709.

14. Coté GG, Crain RC. Biochemistry of phosphoinositides. Annu Rev Plant Physiol Plant Mol Biol 1993;44:333-356.
15. Legendre L, Yueh YG, Crain RC, Haddock N, Heinstein PF, Low PS. Phospholipase C activation during elicitation of the oxidative burst in cultured plant cells. J Biol Chem 1993;268:24559-24563.
16. Quarmby LM, Yueh YG, Cheshire JL, Keller LR, Snell WJ, Crain RC. Inositol phospholipid metabolism may trigger flagellar excision in *Chlamidomonas reinhardtii*. J Cell Biol 1992;116:737-744.
17. Gross W, Yang W, Boss WF Release of carrot plasma membrane-associated phosphatidylinositol kinase by phospholipase A_2 and activation by a 70 kDa protein. Biochim Biophys Acta 1992;1134:73-80
18. Yang W, Burkhart W, Cavallius J, Merrick WC, Boss WF. Purification and characterization of a phosphatidylinositol 4-kinase activator in carrot cells. J Biol Chem 1993;268:392-398.
19. Yang W, Boss WF. Regulation of phosphatidylinositol-4-phosphate biosynthesis by phosphorylation of a protein activator, PIK-A49. J Biol Chem 1994;269: 3852-3857.
20. Tan Z, Boss WF. Association of phosphatidylinositol kinase, phosphatidylinositol monophosphate kinase, and diacylglycerol kinase with the cytoskeleton and F-actin fractions of carrot *Daucus carota* L. cells grown in suspension culture: response to cell wall-degrading enzymes. Plant Physiol 1992;100:2116-2120.
21. Staiger CJ, Goodbody KC, Hussey PJ, Valenta R, Drobak BK, LLoyd CW. 1993 The profilin multigene family of maize: differential expression of three isoforms. The Plant Journal 4:631-641

DO THE MAIZE PHOSPHATIDYLINOSITOLPHOSPHATE-KINASE DISCRIMINATE BETWEEN THE PHOSPHATIDYLINOSITOL MOLECULAR SPECIES ?

Ahmed Chicha, Chantal Demandre, Anne-Marie Justin, and Paul Mazliak
Laboratoire de Physiologie Cellulaire et Moléculaire, URA CNRS 1180, Université Pierre et Marie Curie, Paris 05, France

Introduction

We have recently shown [1] that phosphatidylinositol (PI) was synthesized mainly (or perhaps exclusively) in the endoplasmic reticulum and subsequently transferred to the plasma membrane of cells to allow the phosphoinositide cycle of signal transduction to proceed ; we have also shown that there is nearly the same quantitative distributions of PI molecular species in plasma membranes on one side and microsomes on the other side. In both membrane preparations [2, 3] the molecular species 16:0/18:2 was largely predominant (75-80 mol%). All other molecular species (16:0/18:3, 18:2/18:2, 18:1/18:2, 16:0/18:1, 18:0/18:2) presented percentages lower than 8%. However the question remained : do the PI-kinase and phosphatidylinositolphosphate (PIp)-kinase, which initiate the phosphoinositide cycle in plasma membranes, discriminate between the PI molecular species ?

Material and Methods

Maize (Zea mays L. cv Mona) seeds were furnished by France-Maïs (Toulouse, France). Germination conditions were as following : 30°C, in the dark, in moist vermiculite.
For labelling of lipids, *in vivo*, coleoptiles and first leaves fragments (250 mg ; 5 mm long) were incubated, for 1 to 24 hours, in 0.5 ml of a ^{14}C-inositol solution (1.6 10^6 dpm) or in 0.5 ml of a ^{14}C-acetate solution (220 10^6 dpm), at 30°C, in the dark.
Lipids were extracted in an acid medium : Chloroform/Methanol/HCl 1N (1:1:1, by volume) and separated on potassium-oxalate-impregnated silica-gel. The plate was developed using Chloroform/Methanol/NH4/H2O (45/45/4,8/9,7 by volume) [4]
To analyze the molecular species of PIp, chromatographic spots, scraped from the plate, were incubated with 10 µl of acid phosphatase (Boehringer, Germany, 60 U/µl) in citrate buffer (100 mM, pH 5.8) at 37°C in a final volume of 0.5 ml. The PI molecular species derived from PIp were analyzed by combining radio-HPLC and radio gas-chromatography of labelled fatty acid methyl esters, as previously described [1].

J.-C. Kader and P. Mazliak (eds.), Plant Lipid Metabolism, 224–226.

Results

The inositol phospholipids present in the coleoptiles and first leaves of maize have been labelled *in vivo* from [^{14}C]*myo*-inositol or [^{14}C]-acetate. Figure 1 shows that for incubation periods between 1 and 24 h, more than 90% of total radioactivity was recovered in phosphatidylinositol after only 1 h of incubation time, 5 to 7% of total radioactivity in phosphatidylinositolmonophosphate and about 1% in phosphatidylinositol4,5-biphosphate (PIp$_2$). Similar results have been obtained when the lipids were labelled either from ^{14}C-inositol or from ^{14}C-acetate. The molecular species of PI and PIp have been separated and analyzed by radio-HPLC combined with gas chromatography of fatty acids (table 1). Because of insufficient total labelling, PIp$_2$ molecular species could not be analyzed. The same molecular species (16:0/18:2) was predominant (forming at least 60 mol%) in PI and PIp. Five other molecular species : 16:0/18:1, 16:0/18:3, 18:0/18:2, 18:1/18:2, 18:2/18:2, less abundant in mass and less heavily labelled (containing each from 7 to 13% of total radioactivity), were present also in PI and PIp.

<table>
<tr><th rowspan="3">molecular species</th><th>%moles</th><th colspan="3">% of total radioactivity</th></tr>
<tr><th></th><th>inositol-^{14}C[b]</th><th colspan="2">acétate-1-^{14}C[b]</th></tr>
<tr><th>PI[a]</th><th>PI</th><th>PI</th><th>PIp</th></tr>
<tr><td>16:0/18:3</td><td>4,6</td><td rowspan="2">17,8</td><td>13,3</td><td rowspan="2">27,3</td></tr>
<tr><td>18:2/18:2</td><td>5,0</td><td>8,4</td></tr>
<tr><td>16:0/18:2</td><td>75,4</td><td>79,2</td><td>60,3</td><td>59,0</td></tr>
<tr><td>18:1/18:2</td><td>7,4</td><td>traces</td><td>traces</td><td>traces</td></tr>
<tr><td>18:0/18:2</td><td>3,4</td><td rowspan="2">3,0</td><td>10,8</td><td rowspan="2">13,7</td></tr>
<tr><td>16:0/18:1</td><td>4,2</td><td>7,2</td></tr>
</table>

a : Only PI was enough abundant in all tissues to allow an analysis of molecular species

b : After 24 hours of incubation

Table 1 : Quantitative distribution of PI molecular species and distribution of radioactivity found in PI and PIp molecular species.

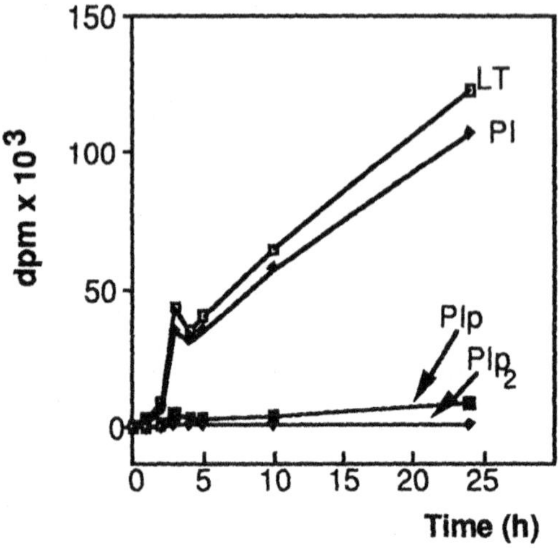

Figure 1 : Labelling kinetics of inositol lipids, from [^{14}C]*myo*-inositol

Conclusion

Those results suggest that, at least after long incubation periods, there is no apparent descrimination by PI-kinase among the molecular species of PI present in maize tissues.

References

1. Chicha A., Justin A.-M., Jolliot A., Demandre C., Mazliak P. Biosynthesis of the molecular species of phosphatidylinositol found in microsomes and plasma membranes from etiolated maize coleoptiles. *Plant Physiol. Biochem.*, 1993; 31 : 507-513.
2. Justin A.-M., Demandre C., Mazliak P. Molecular species synthesized by phosphatdylinositol synthases from potato tuber, pea leaf and soya been. *Biochim. Biophys. Acta*, 1989; 1005, 51-55.
3. Widell S., Lundborg T., Larsson C. Plasma membrane preparations from oat prepared by partitioning in an aqueous polymer two-phase system. Plant Physiol., 1982; 47, 329-334.
4. Sandelius A. S., Sommarin M. Phosphorylation of phosphatidylinositols in isolated plant membranes. *FEBS letters*, 1986; 201, 282-287.

ON THE PRESENCE OF GLYCOSYLPHOSPHATIDYLINOSITOL-ANCHORED PROTEINS IN PLANTS

Guy A. Thompson, Jr., Naoki Morita, Hidetoshi Okuyama, Youngsoo Kim, Young-Gyu Ko, and Chiung-Yu Hung, Department of Botany, University of Texas, Austin, Texas 78713, USA

Introduction

There are now more than 100 different proteins known to be attached to the cell surface of animal or yeast cells by a glycosylphosphatidylinositol(GPI) anchor [1]. The structure of these anchoring units (Fig 1) is relatively highly conserved, although additional

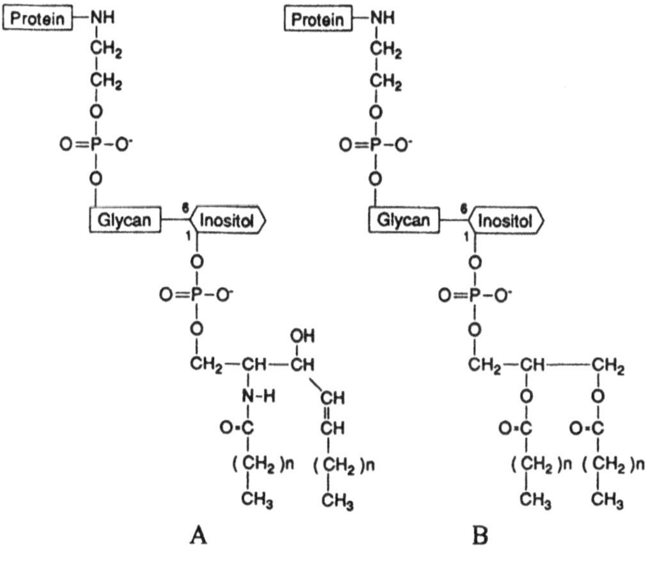

Figure 1. Two diverse lipid moieties that can be associated with GPI-anchored proteins. A, ceramide anchor; B. diacylglycerol anchor.

sugar moieties are sometimes attached to the basic oligoglycan backbone, and the diacylglycerol lipid unit is in certain molecules replaced by an alkyl,acyl-glycerol or a ceramide structure. Among the more familiar proteins now known to be GPI-anchored are alkaline phosphatase, acetylcholinesterase, and the variable surface glycoprotein coating the surface of trypanosomes. Surprisingly, there have been no reports of GPI-anchored proteins in plants. Because my laboratory has an ongoing project to study the GPI-

J.-C. Kader and P. Mazliak (eds.), Plant Lipid Metabolism, 227–229.

anchored proteins of the protozoan *Tetrahymena*, we have been able to apply this technology in a search for these proteins in higher plants.

Experimental Approach

Initially we surveyed several plant species for GPI-anchored proteins, using radiolabeled precursors of the anchor components as indicators of their possible presence. We have obtained tentative evidence for anchored proteins in tobacco, maize, and *Chlamydomonas*. However, for more detailed study we decided to select a plant species known to contain a protein which is consistently GPI-anchored in animals. *Spirodella oligorrhiza*, a species of duckweed, has been observed to produce alkaline phosphatase (AP) when grown in an aquatic environment depleted of inorganic phosphate [2]. We confirmed that AP was induced in *Spirodela* incubated for several days in an inorganic phosphate-free medium, which ultimately curtailed growth, and observed that the plants could induce AP while continuing to grow if placed in a medium containing β-glycerophosphate as the sole source of phosphorus. Experiments were designed to examine the AP of these plants for the possible presence of a GPI anchor.

A widely used method for identifying GPI-anchored proteins involves labeling the tissues with radioactive precursors of the anchor components. Accordingly, we exposed plants to [^{3}H]ethanolamine or to [^{3}H]myristic acid and then analyzed the extracted proteins on SDS-PAGE gels. Both radiotracers labeled a protein band having an apparent molecular weight of 50 kDa. On the denaturing gel it was not possible to correlate the labeling with AP enzymatic activity. However, when proteins from [^{3}H]ethanolamine-labeled plants were electrophoresed on non-denaturing polyacrylamide gels, radioactivity was associated with the AP activity band (which was not stained with Coomassie blue) and not with the major protein bands.

Because of their hydrophobic lipid anchors, GPI-anchored proteins partition into the detergent phase following a Triton X-114 / water phase partition [3]. When this procedure was tested with control samples of mammalian AP, most of the AP activity partitioned into the aqueous phase, in agreement with the known action by an endogenous phospholipase activity capable of cleaving off the anchoring lipid moiety [4]. Similar phase partitioning of *Spirodela* extracts also yielded only 5 % or less of the AP activity entering the TX-114 phase. When the aqueous and the detergent forms of AP from both plant and mammalian sources were run on non-denaturing gels, the traces of TX-114-associated protein having AP activity migrated differently from the aqueous forms, indicating that they retained their lipid anchor.

Surprisingly, the bulk of the AP, recovered in the aqueous phase, was found to still contain some covalently-bound [^{3}H]fatty acid. It appears that this is present in the form of fatty acid linked to a hydroxyl group of the inositol moiety. AP (and other GPI-anchored proteins) of certain animal tissues have been shown to be fatty acylated on inositol.

The limited amounts of the TX-114-extractable proteins recovered from [^{3}H]myristate-labeled plants were hydrolyzed in methanolic HCl, and the lipid products were analyzed by thin-layer chromatography. Two labeled products were obtained; one migrated with authentic fatty acid methyl esters and the other moved just faster than standard sphingosine. The fact that the latter product was extractable from organic solvents only under basic conditions further suggested its identity as a long chain base. Efforts are currently underway to accumulate enough of these products to identify them definitively by mass spectrometry.

As luck would have it, the properties of the *Spirodela* GPI-anchored protein's lipid moiety, as determined by following radioactivity incorporated into the trace amounts of lipid that could be recovered by hydrolysis, were virtually identical to those of the *Tetrahymena* anchor lipid being studied in a parallel project [5,6]. Using the larger amounts of GPI-anchored protein available from the protozoan cells, we have recently recovered the [^{3}H]myristate-labeled hydrolysis products and identified them by GC and mass spectrometry. One was confirmed as fatty acid methyl esters and the other was found to be a unique long chain base derivative, 3-O-methyl-sphinganine (C.-Y.Hung, Y-G. Ko, and G. A. Thompson, Jr., submitted for publication). The initial conclusion that the methylation of the LCB was an artifact of hydrolysis was refuted by several control experiments. At the present time all evidence suggests that the *Spirodela* GPI-anchored protein contains the same type of lipid anchor, even including the unusual methylated LCB. Direct analysis of the *Spirodela* LCB will of course be required to confirm that its structure is identical to that of the *Tetrahymena* LCB.

Although *Spirodela* AP, when induced, becomes localized to a considerable extent on the lower side of the floating fronds, it is most concentrated in the roots. Small amounts of AP can be recovered from the growth medium, but even that which has lost its lipid anchor remains primarily associated with the tissue. In this respect *Spirodela* is quite different from animal cells, which characteristically shed GPI-anchored proteins promptly following the cleavage of their lipid anchor. We speculate that in *Spirodela* the AP that is released by phospholipase action becomes ensnared in the cell wall and lodges there in active form for considerable periods of time. In its natural environment, namely, shallow ponds and marshes, growth of *Spirodela* is often limited by low inorganic phosphate levels. However, these same waters may be relatively rich in organic phosphates, which can only be utilized as a phosphate source if AP or another phosphatase is available to convert the phosphate to an inorganic form.

Conclusions

There is strong but still indirect evidence for the presence in plants of GPI-anchored proteins. However, on the basis of our limited survey, their levels would appear to be considerably lower in plants than in the animal cells which have been used for their study. Further efforts will be required to confirm the identity of the putative GPI-anchored proteins of plants.

References

1. Low, MG. The glycosyl-phosphatidylinositol anchor of membrane proteins. Biochim Biophys Acta 1989; 988: 427-454.
2. Bieleski, RL. Development of an externally-located alkaline phosphatase as a response to phosphorus deficiency. In: Bieleski,RL, Ferguson, AR, Cresswell, MM, editors. Mechanisms of regulation of plant growth. Wellington, Roy Soc N Z, 1974: 165-170.
3. Malik, A-S, Low, MG. Conversion of human placental alkaline phosphatase from a high M_r form to a low M_r during butanol extraction: an investigation of the role of endogenous phosphoinositide-specific phospholipases. Biochem J 1986; 240:519-527.
4. Wong, YW, Low, MG. Phospholipase resistance of the glycosyl-phosphatidylinositol membrane anchor on human alkaline phosphatase. Clin Chem 1992; 38: 2517-2525.
5. Ryals, PE, Pak, Y, Thompson, GA, Jr. Phosphatidylinositol-linked glycans and phosphatidylinositol-anchored proteins of *Tetrahymena mimbres* . J Biol Chem; 266: 15048-15053.
6. Pak, Y, Ryals, PE, Thompson, GA, Jr. Phosphatidylinositol glycan formation and utilization by the ciliate *Tetrahymena mimbres*. J Biol Chem; 266: 15054-15059.

PATHWAY OF SYNTHESIS OF 3- AND 4-PHOSPHORYLATED PHOSPHOINOSITIDES IN STOMATAL GUARD CELLS OF *COMMELINA COMMUNIS* L.

Paroo Parmar and Charles A. Brearley

Department of Plant Sciences, University of Cambridge, Downing Street
Cambridge CB2 3EA, U.K.

Introduction

In the plant kingdom, the pathway of synthesis of 3- and 4-phosphorylated phosphoinositides has been described only in *Spirodela polyrhiza* L. [1]. If animal models of phosphoinositide metabolism are to be applied to plants, it is imperative to determine the route of synthesis of phosphoinositides in other members of the plant kingdom in which signalling functions for phosphoinositides and inositol phosphates have been implicated. Recently, we have rigorously identified 3- and 4-phosphorylated phosphoinositides in stomatal guard cells of *Commelina communis* L. [2]. In the present work we have set out to determine the metabolic route by which PtdIns(3,4)P_2 and PtdIns(4,5)P_2 synthesis occurs and to examine the metabolic relationship between 3- and 4-phosphorylated lipids in stomatal guard cells.

Materials and methods

Extraction and dephosphorylation of GroPInsP and GroPInsP_2 species

Isolated epidermal strips (in which all cells other than guard cells have been killed by low pH treatment) from *Commelina communis* L. were radiolabelled with ^{32}P-orthophosphate ([^{32}P]P_i, carrier-free; 0.2-0.3 mCi/ml) for 3.5-4 h. The tissue was washed, frozen in liquid N_2 and extracted with perchloric acid. Lipids were extracted, deacylated and resolved on Partisphere SAX hplc [2].

The [^{32}P]P_i-labelled GroPIns3P and GroPIns4P fractions were desalted, mixed with their respective ^{3}H-labelled standards, and dephosphorylated with alkaline phosphatase [1] to release phosphates in the 3 and 4 positions respectively. The [^{32}P]P_i-labelled GroPIns(3,4)P_2 and GroPIns(4,5)P_2 fractions were desalted, mixed with their respective ^{3}H-labelled standards, and dephosphorylated with human erythrocyte ghosts [1] under conditions designed to selectively remove either the phosphate in the 5 position or that in the 3 position.

[^{32}P]P_i labelling of the ATP pool and of the gamma-phosphate of ATP

Isolated epidermal strips were radiolabelled with [^{32}P]P_i for 0.5-4 h. extracted with

J.-C. Kader and P. Mazliak (eds.), Plant Lipid Metabolism, 230–232.

perchloric acid and resolved on Partisphere SAX hplc. The ^{32}P content of the ATP peak was determined by Cerenkov counting and the ATP content by A_{254} measurement.

[^{32}P]ATP-containing fractions were desalted, mixed with an excess of unlabelled ATP and presented in an assay with D-glucose as substrate to hexokinase [1]. The products; glucose 6-phosphate, ADP and unreacted substate ATP were resolved on Partisphere SAX hplc. The proportion of total ^{32}P released from ATP and incorporated into glucose 6-phosphate, in concert with the specific radioactivity of desalted [^{32}P]ATP yields the specific activity of the gamma-phosphate of ATP.

Results

Kinetics of [^{32}P]P_i-labelling of the ATP pool and the gamma-phosphate of ATP

There was an increase in the specific activity of the gamma-phosphate of ATP and of ATP as a whole over the labelling period (Figure 1).

This allows us to use the strategy of Stephens and Downes [3] to determine the sequence of addition of phosphates to lipids, that is, over a period during which the specific activity of the ATP pool is increasing any phosphate added at a particular point in a metabolic sequence will be more strongly labelled than its predecessors and less so than its successors. The rank order of specific activity of the individual phosphates reveals the order in which the phosphates were esterified.

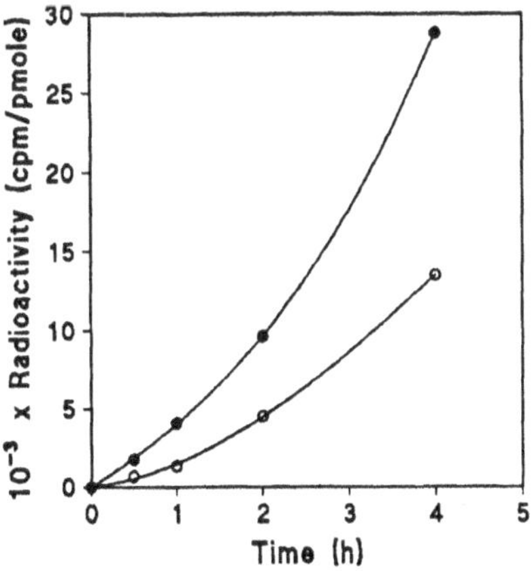

Figure 1 Incorporation of [^{32}P]P_i into ATP pool (●) and gamma-phosphate of ATP (O) in stomatal guard cells.

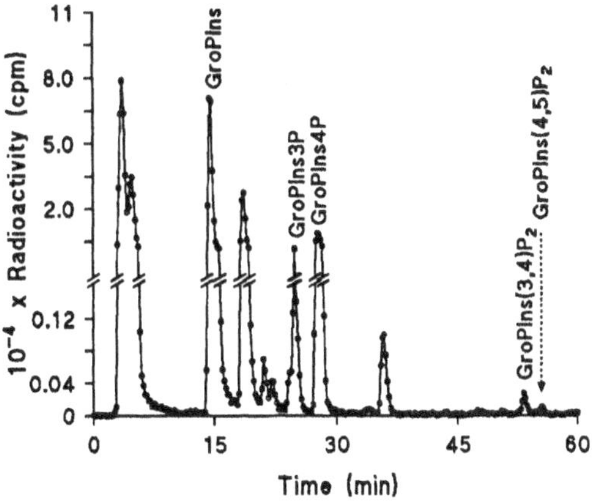

Figure 2 Hplc separation of deacylation products from [^{32}P]P_i-labelled stomatal guard cells.

Distribution of label between phosphates of GroPInsP and $GroPInsP_2$ species

Hplc analysis of deacylates from $[^{32}P]P_i$-labelled isolated guard cells reveals the full complement of 3- and 4-phosphorylated phosphoinositols (Figure 2), commonly found in non-stimulated animal cells. The peak hplc fractions of GroPIns3P, GroPIns4P, GroPIns(3,4)P_2 and GroPIns(4,5)P_2 were desalted and dephosphorylated. The products of dephosphorylation of GroPInsP and $GroPInsP_2$ species were resolved on hplc. The ^{32}P: ^{3}H ratio of the products when compared to that of the substrate yields the proportion of the ^{32}P in each of the phosphate moieties (Table 1).

The monoester phosphates of PtdIns3P and PtdIns4P accounted for 95-97% of the ^{32}P label in PtdInsP species. The 3- and 4-monoester phosphates accounted for 40% and 60% respectively of the label in PtdIns(3,4)P_2 whereas the 4- and 5-monoester phosphates of PtdIns(4,5)P_2 accounted for 39% and 61% respectively.

Table 1 Distribution of ^{32}P label between individual phosphate groups in phosphoinositides.

Phosphoinositides	Identity of phosphate (% of label) 1	3	4	5
PtdIns3P	5	95	--	--
PtdIns4P	3	--	97	--
*PtdIns(3,4)P_2	-	40	60	--
*PtdIns(4,5)P_2	-	--	39	61

* assumes contribution from diester is negligible

Conclusions

Since the 3 phosphate of PtdIns(3,4)P_2 is less strongly labelled than the 4 phosphate, synthesis of PtdIns(3,4)P_2 in *Commelina communis* L. occurs by successive 3- and 4-phosphorylation of PtdIns. In contrast to all animal and plants cell systems described to date, in stomatal guard cells PtdIns(3,4)P_2 predominates, in labelling terms at least, over PtdIns(4,5)P_2. PtdIns(4,5)P_2 is synthesised in *Commelina communis* L. by successive 4- and 5-phosphorylation of PtdIns.

References

1. Brearley CA, Hanke DE. Pathway of synthesis of 3,4- and 4,5-phosphorylated phosphatidylinositols in the duckweed *Spirodela polyrhiza* L. Biochem J 1993;290: 145-150.

2. Parmar PN, Brearley CA. Identification of 3- and 4-phosphorylated phosphoinositides and inositol phosphates in stomatal guard cells of *Commelina communis* L. Plant J 1993; 4:255-263.

3. Stephens LR, Downes CP. Product-precursor relationships amongst inositol polyphosphates. Biochem J 1990; 265:435-452.

SOLUBILIZATION AND SELECTIVITY OF THE PHOSPHATIDYLINOSITOL-SYNTHASE FROM ETIOLATED MAIZE COLEOPTILE MICROSOMES

Anne-Marie Justin[1], Aziz Hmyene[2], Jean-Claude Kader[1] and Paul Mazliak[1]
1 Laboratoire de Physiologie Cellulaire et Moléculaire des Plantes, URA CNRS 1180, Université Pierre et Marie Curie, Paris, France
2 Université Cadi Ayyad, FST-Béni Mellal, Maroc

Introduction

The enzyme responsible for the final step of *de novo* biosynthesis of phosphatidylinositol (PI) is an endoplasmic reticulum enzyme : the cytidine diphosphate-diacylglycerol (CDP-DAG) : *myo*-inositol transferase (EC 2.7.8.11), or briefly PI-synthase. This enzyme catalyses the formation of PI and cytidine monophosphate (CMP) from CDP-DAG and *myo*-inositol [1]. In a previous work [2], we found that the PI-synthase displayed a strong selectivity among the endogenous CDP-DAG molecular species used as substrates, in the microsomes from several plant tissues : 16:0/18:2 and 16:0/18:3-CDP-DAG were markedly more utilized than other molecular species. Here, we are interested in the PI-synthase selectivity among CDP-DAG molecular species, both when the enzyme was left within microsomal membranes or when it was solubilized and pre-purified. As a continuation of our work upon the solubilization of the PI-synthase [3], we used CHAPS to solubilize the microsomal membranes as previously described.

Materials and methods

Microsomes were isolated [2] from 4 day old maize seedlings, grown at 30°C in darkness. Microsomal suspension (250mg of proteins) was shaken at 4°C for 1h in a medium containing 1% CHAPS (detergent/protein ratio : 3/1), 0.5M KCl, 1mM EDTA, 3mM sodium azide, 10mM 2-mercapto-ethanol, 1mM phenylmethyl-sulfonyl-fluoride (PMSF), 10% glycerol, 2mM $MnCl_2$, in 10mM Tris-HCl buffer (pH 8). The solubilized fraction was obtained after centrifugation at 100,000x*g* for 90min. The supernatent was dialyzed overnight and loaded on a DEAE-trisacryl M column (equilibrated with the solubilization medium without CHAPS and KCl). Non-retained proteins were washed and elution was achieved by a continuous gradient of NaCl from 0 to 1M.
Currents assays of PI-synthase activity were performed at 30°C for 30min in a medium containing 300µM dipalmitoylglycerol-CDP, 5mM EDTA, 7.5mM $Mncl_2$, 4µCi *myo*-[2-^{3}H]inositol, 50mM Tris-HCl buffer (pH8) and 250µg of proteins (microsomes and solubilized fraction) or 200µl of DEAE fraction, in a final volume of 0.5ml. Lipids were extracted according to Bligh and Dyer [4]. PI synthase activity was determined by measuring the radioactivity in total lipids. Labelled PI molecular species were separated by HPLC on a 5μ ultrasphere C18 column [5]. The radioactivity of each molecular species was mesured by scintillation in a continous flow counter.

J.-C. Kader and P. Mazliak (eds.), Plant Lipid Metabolism, 233–235.

Results

SOLUBILIZATION AND PRE-PURIFICATION

After solubilization, 50% of the initial PI-synthase activity was recovered in the supernatent. The PI-synthase was bound on a DEAE column ; Fig 1 shows a peak of PI-synthase activity corresponding to the fractions eluted between 0.18 and 0.27M NaCl.

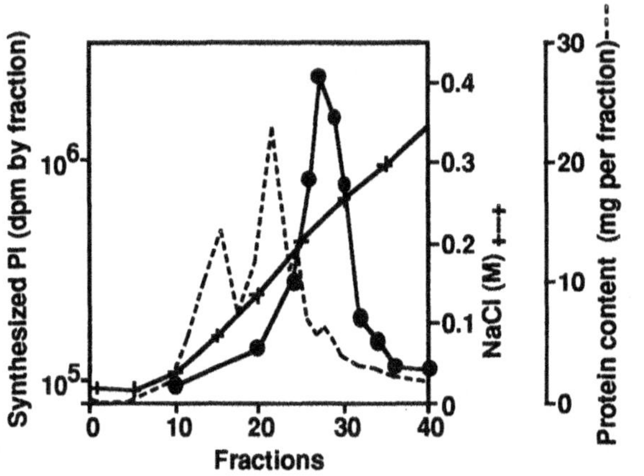

Figure 1. Elution profile of the solubilized microsomal fraction from a DEAE-trisacryl M column (experimental conditions are indicated in Materials and Methods) : A, (—●—) profile of the PI-synthase activity in the fractions collected after elution, (+—+) NaCl concentration (M) determined by conductivity, (-----) proteins contents determined by Lowry method.

SELECTIVITY

When [^{3}H]-inositol was furnished to maize coleoptile microsomes without any exogenous CDP-DAG, 16:0/18:2-PI was the most actively synthesized molecular species (77% of total labelling) ; labelled 16:0/18:3-, 18:2/18:2-, 16:0/18:1- and 18:0/18:2-PI were also found in little amounts. When adding increasing concentrations of 16:0/16:0-CDP-DAG up to 300µM in the incubation medium (Fig 2A), we unexpectedly obtained only little 16:0/16:0-PI ; the most heavily labelled molecular species were the same as above. Thus, when the PI-synthase was located in the microsomal membranes, it utilized endogenous CDP-DAG to synthesize PI *de novo* ; though the furnished 16:0/16:0-CDP-DAG was present in great excess in the medium (about 132 time more than the endogenous), it was very poorly utilized by the enzyme.

In absence of any 16:0/16:0-CDP-DAG, the solubilized PI-synthase synthesized the same PI molecular species as microsomes. In marked contrast, with increasing concentrations of 16:0/16:0-CDP-DAG (Fig 2B), the solubilized fraction synthesized very actively 16:0/16:0-PI, up to 97% for 300µM of 16:0/16:0-CDP-DAG. The same results were obtained with protein fractions having a high PI-synthase activity eluted from the DEAE column.Thus, in marked contrast with the PI-synthase located in microsomal membranes, the solubilized and pre-purified enzyme accepted 16:0/16:0-CDP-DAG to synthesize 16:0/16:0-PI.

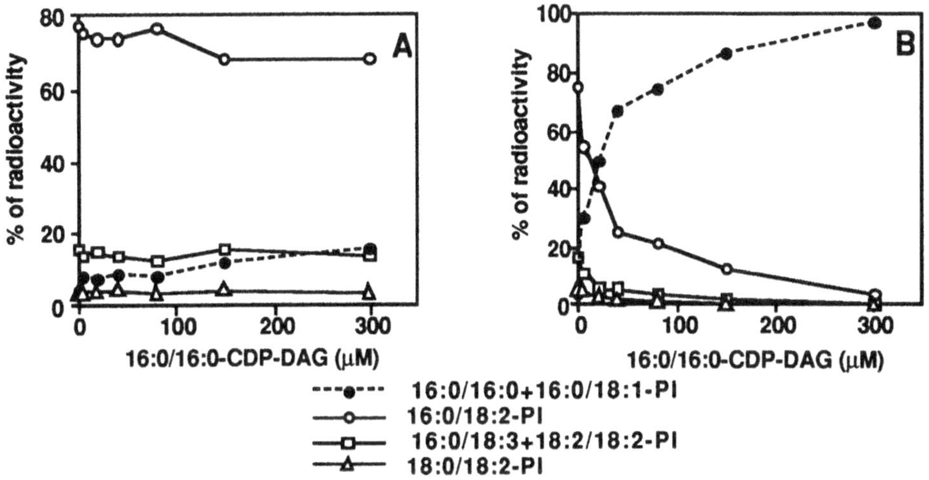

Figure 2. Labelling of the PI molecular species synthesized by the microsomes (A) or the solubilized fraction (B) from etiolated maize coleoptiles, incubated with [^{3}H]inositol, in presence of increasing quantities of 16:0/16:0-CDP-DAG. The same results as in B were obtained with an active fraction eluted from the DEAE column.

Conclusion

When the PI-synthase was left within the membrane and when it was able to choose between endogenous and exogenous substrates, it selected strongly endogenous CDP-DAG substrates. On the other hand, solubilized and partially purified PI-synthase displayed no selectivity among substrates and utilized the furnished CDP-DAG, proportionally to its quantity.

References

1. Paulus H, Kennedy EP. The enzymatic synthesis of inositol monophosphatidate. J Biol Chem 1960;235:1303-1311.
2. Justin AM, Mazliak P. Comparison of the molecular species patterns of phosphatidic acid, CDP-diacylglycerols and phosphatidylinositol in potato tuber, pea leaf and soya-bean microsomes : conseqences for the selectivity of the enzymes catalizing phosphatidylinositol biosynthesis. Biochim Biophys Acta 1992;1165:141-146.
3. Hmyene A, Guerbette F, Justin AM, Mazliak P, Kader JC. Solubilization of phosphatidylinositol-synthase from maize seedlings. In: Quinn PJ, Harwood JL, editors. Plant Lipid Biochemistry, Structure and Utilization. Londres: Portland Press, 1990: 263-265.
4. Bligh EG, Dyer WJ. A rapid method of total lipid extraction and purification. Can J Biochem 1959;37:911-917.
5. Demandre C, Tremolières A, Justin AM, Mazliak P. Analysis of molecular species of plant polar lipids by High-Performance and Gas Liquid Chromatography. Phytochem 1985;24:481-485.

LIPIDS IN BROWN ALGAE: IDENTIFICATION OF A NEW PHOSPHOLIPID AND THE USE OF THE BETAINE LIPID DGTA AS A TAXONOMIC MARKER

Waldemar Eichenberger, Department of Biochemistry, University of Bern, Switzerland
Dieter G. Müller, Faculty of Biology, University of Konstanz, FRG
Christine E. Schmid, present address: Monash University, Clayton, Victoria, Australia

Introduction

The membrane lipid pattern of brown algae differs from that of higher plants and green algae by additional lipid components [1-2]. A hitherto unknown phospholipid PX occurring in all brown algae tested so far, had been characterized [3] but the structure of its polar part remains to be determined. The betaine lipid diacylglycerylhydroxymethyl-ß-alanine (DGTA) is typical of certain orders, genera or species and may therefore be used as a taxonomic marker in cases, where morphological characters are unsatisfactory.

Materials and Methods

PX was isolated from *Fucus serratus* (Brittany coast) as described before [3], then deacylated with 0.1 N NaOH in methanol for 20 min at room temperature and hydrolyzed with 6 N HCl for 5 hours at 100°C. The product was treated with *N*-methyl-*N*-trimethylsilyltrifluoroacetamide (MSTFA) followed by *N*-methyl-bis-trifluoroacetamide (MBTFA). *N*-(2-hydroxyethyl)glycine (HEG) was synthesized by reacting 2-bromoacetic acid ethylester and ethanolamine in a molar ratio of 1:10, respectively, in tetrahydrofurane. The product was treated with MSTFA and MBTFA and the *O,O*-bis-trimethylsilyl-*N*-trifluoroacetyl derivative characterized by GC-MS.
Brown algal cultures used for chemotaxonomic work were unialgal clonal isolates from various coasts of the world which are maintained as stock cultures [4].

J.-C. Kader and P. Mazliak (eds.), Plant Lipid Metabolism, 236–238.

Results

Identification of PX

Positive reactions on TLC plates with both ninhydrin and molybdene blue reagents, as well as IR spectra indicated PX to be an amino acid-containing glycerophospholipid [3]. Data of ^{1}H, ^{13}C, DEPT, C-H and H-H COSY NMR spectra from deacylated PX (Fig. 1) strongly suggest a phosphoglyceryl moiety linked to *N*-(2-hydroxyethyl)glycine (HEG). Acid hydrolysis of deacylated PX gave a product from which after derivatization with MSTFA and MBTFA by GC-MS analysis prominent signals at m/z 359 (M^+), 344 $(M\text{-}CH_3)^+$ and 329 $(M\text{-}2\ CH_3)^+$ were obtained. The value of m/z 359.11963 corresponded to the expected brutto formula $C_{12}H_{24}O_4NF_3Si_2$ (calculated m/z 359.11960). The same product was obtained by synthesis from 2-bromoacetic acid ethylester and ethanolamine in tetrahydrofuran, subsequent hydrolysis of the ester, silylation and trifluoroacetylation.
All the data obtained from lipid PX are in accordance with the structure of phosphatidyl-*O*-[*N*-(2-hydroxyethyl)glycine] (PHEG) in which, according to the fatty acid composition with 80% 20:4 and 20% 20:5 acids [3], the 20:4/20:4 combination predominates.

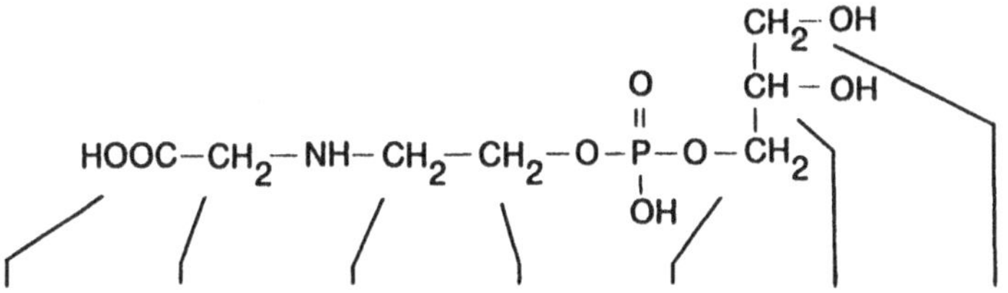

δ ppm							
^{1}H	-	3.7	3.35	4.15	3.9	3.9	3.6
^{13}C	174.00	52.01	50.52 50.41	63.85 63.79	69.41 69.34	73.54 73.43	64.24
J_{CP} (Hz)			7.9	5.0	5.7	7.8	

Fig. 1. NMR data of deacylated PX in D_2O (Bruker AC-300 (300.133 MHz for ^{1}H and 75.469 MHz for ^{13}C)

Betaine lipid DGTA as a taxonomic marker

The betaine-type lipid diacylglycerylhydroxymethyltri-methyl-ß-alanine (DGTA) is, unlike PHEG, synthesized by 8 orders (out of 16 orders) of Phaeophyceae [2]. Its occurrence seems even to be tied to taxonomic entities distiguishing genera and species. The cosmopolitan order of the Ectocarpales comprising organisms of filamentous habitus, was of special interest. Screening of more than 60 isolates from different geographic origin demonstrated a correlation of species and presence of DGTA. Within the genus *Ectocarpus*, *E. fasciculatus* always contains DGTA, while *E. siliculosus* lacks it. Of the genus *Feldmannia*, all species are DGTA-positive. In the genus *Hincksia*, in contrast, *H. conifera* and *H. mitchelliae* produce DGTA, while *H. granulosa* does not. These correlations suggested DGTA to be a useful chemotaxonomic marker. Thus, the lipid pattern was used for the classification of 40 isolates of *Ectocarpus* from different geographic origin which in culture were asexual and consequently could not be classified. Judged by the lack of DGTA, the majority of these strains belong to *E. siliculosus*, while some are DGTA-positive and thus represent the *E. fasciculatus* type. These results demonstrate that, on some coasts (Helgoland, Roscoff, Canarian Islands, Chile and Juan Fernandez Archipelago), both lipid types of *Ectocarpus* co-exist.

Acknowledgements

This work has been supported by the Swiss National Science Foundation (Grant 31-36078-92).

References

1. Dembitsky VM, Rozentsvet OA, Pechenkina PE. Glycolipids, phospholipids and fatty acids of brown algae species. Phytochemistry 1990;29:3417-3421.
2. Eichenberger W, Araki S, Müller DG. Betaine lipids and phospholipids in brown algae. Phytochemistry 1993; 34:1323-1333.
3. Schmid CE, Müller DG, Eichenberger W. Isolation and characterization of a new phospholipid from brown algae. Intracellular localization and site of biosynthesis. J Plant Physiol 1994;143:570-574.
4. Müller DG, Eichenberger W. Betaine lipid content and species delimitation in *Ectocarpus, Feldmannia* and *Hincksia* (Ectocarpales, Phaeophyceae). Eur J Phycol 1994;in press

THE TRANSBILAYER DISTRIBUTION OF GLUCOSYLCERAMIDE IN PLANT PLASMA MEMBRANE

Daniel V. Lynch and Alexi J. Phinney

Department of Biology
Williams College, Williamstown MA 01267, USA.

Glucosylceramides, the predominant sphingolipids in plant tissues, are quantitatively important structural components of the plasma membrane and tonoplast. Here, we describe studies undertaken to determine the transbilayer distribution of glucosylceramides in the plasma membrane of squash. The results indicate that glucosylceramides are present predominantly in the outer monolayer of the plasma membrane.

Introduction

Previous studies of the transmembrane distribution of glycosphingolipids in erythrocytes and myelin membrane preparations suggested that the sphingolipids are located predominantly in the external monolayer of the respective membranes [1,2]. These and related studies in animal tissues have led to the general view that sphingolipids are present exclusively on the cell surface [3]. Glucosylceramides account for 6-30% of the total lipid of highly purified plant plasma membrane preparations [4,5] and changes in glucosylceramide content of the plasma membrane following acclimation to environmental stresses have been reported [5,6]. Knowledge of the lipid composition of plant plasma membranes and changes accompanying acclimation have led to investigations of physical properties and lipid-lipid interactions in model membrane systems of defined composition. Such studies aimed at defining the contributions of different lipid constituents to the properties and function of the plasma membrane would benefit from information on the transbilayer distribution of specific lipids. This is especially true for glucosylceramides, given their distinctive physical properties [4,7]. In addition, a picture of plant sphingolipid biosynthesis is emerging [4,8,9] that demonstrates many similarities to the metabolic pathways present in animals and fungi. Thus, it is of interest to know if the translocation process responsible for establishing the transbilayer distribution of sphingolipids in animal cells is also present in plant cells. For these reasons, we have examined the transbilayer distribution of glucosylceramides in plasma membrane vesicles isolated from squash fruit.

Experimental

Plasma membrane vesicles (oriented predominantly right side-out) were isolated from immature fruit of yellow summer squash (*Cucurbita pepo* L. cv Early prolific straight neck) using an aqueous polymer two-phase system similar to that described previously [10]. More

J.-C. Kader and P. Mazliak (eds.), Plant Lipid Metabolism, 239–241.

heterogeneous mixtures of inside-out and right side-out vesicles were prepared by subjecting plasma membrane preparations to four freeze-thaw excursions as described by Palmgren et al. [11]. The orientation (sidedness) of the vesicles in the preparations was determined by assaying NADH-ferricyanide reductase activity in the absence and presence of Triton X-100 [11,12]. This assay takes advantage of the fact that the reductase is located on the cytoplasmic side of the plasma membrane and is not accessible to substrate in sealed right side-out vesicles. Plasma membrane glucosylceramides were oxidized using 10mM $NaIO_4$ in 100mM potassium phosphate pH 6 at room temperature for 60 min. The reaction was stopped by the addition of glycerol (10% final concentration) and the lipids extracted using chloroform and methanol. The lipid extracts were subjected to alkaline hydrolysis using methanolic NaOH and then quantitatively transferred to thin layer chromatography plates and separated. After visualization of the intact glucosylceramides using orcinol-ferric chloride spray, the plates were photocopied onto clear acetate sheets and the bands were scanned using a laser densitometer to quantitate the remaining glucosylceramides.

Based on assays of NADH-ferricyanide reductase activity, freshly isolated native plasma membrane preparations were comprised predominantly (91%) of sealed right side-out vesicles whereas frozen-thawed samples were more heterogeneous (Table 1). Treatment of native plasma membrane with periodate resulted in the oxidation (degradation) of 74% of the glucosylceramide. If we take into account the proportion of right side-out vesicles, an adjusted value for glucosylceramide degradation of 82% (74/91%) was obtained. The results using frozen-thawed plasma membrane validated the results obtained with native membrane for two reasons. First, although only 32% of the glucosylceramide was degraded, when adjusted to account for the orientation of the membrane vesicles, we obtained a value for the oxidation of glucosylceramide present in right side-out vesicles in the frozen-thawed preparations of 80% (32/40%), comparable to that obtained for native plasma membrane. Second, the diminished ability of periodate to oxidize glucosylceramides in frozen-thawed preparations is evidence that periodate did not penetrate the vesicles. Inclusion of Triton X-100 during periodate oxidation did not, surprisingly, yield complete degradation of glucosylceramides. In both native and frozen-thawed membrane preparations, glucosylceramide oxidation was only 84% complete in the presence of detergent, suggesting the presence of a pool of sphingolipid resistant (inaccessible) to periodate. Taking this into account when considering the extent of glucosylceramide oxidation in native plasma membrane vesicles, the results indicate that approximately 98% (82/84%) of the plasma membrane glucosylceramides are present in the outer (apoplastic) monolayer of the plasma membrane.

Table 1. Periodate Oxidation of Glucosylceramides in the Plasma Membrane of Squash. Values are averages of duplicate experiments performed on separate plasma membrane preparations.

Membrane Preparation	Native Plasma Membrane	Frozen-Thawed Plasma Membrane
Orientation of Membrane		
Right side out	91%	40%
Inside out	9%	60%
Glucosylceramide Oxidation	74%	32%
Glucosylceramide Degradation (Adjusted for orientation)	82%	80%

Glucosylceramide oxidation in the presence of Triton X-100: 84%.

Conclusion

These results suggest that glucosylceramides are present predominantly, if not exclusively, in the outer monolayer of the plasma membrane from squash. Results from experiments using glucocerebrosidase to enzymatically hydrolyze glucosylceramides (not shown) are consistent with the findings above demonstrating a pronounced asymmetric distribution of sphingolipid in the plasma membrane. Such an asymmetric distribution is a potentially important consideration when attempting to relate the glucosylceramide content of the plasma membrane to the membrane's physical properties and behavior. Further, the asymmetric distribution of sphingolipid in plant plasma membrane points out another feature of sphingolipid form, function and metabolism common to plants and other eukaryotic organisms.

Acknowledgements
This work was supported in part by grants from the National Science Foundation (DCB 89-03816) and the Howard Hughes Medical Institute.

References

1. Steck TL, Dawson G. Topographical distribution of complex carbohydrates in the erythrocyte membrane. J Biol Chem 1974; 249:2135-2142.
2. Linington C, Rumsby MG. Accessibility of galactosyl ceramides to probe reagents in central nervous system myelin. J Neurochem 1980; 35:983-992.
3. Karlsson K-A. Glycosphingolipids and surface membranes. In: Chapman D, ed. Biological Membranes. Vol 4. New York: Plenum Press, 1982:1-74.
4. Lynch DV. Sphingolipids. In: Moore TS, ed. Lipid Metabolism in Plants. Boca Raton: CRC Press, 1993:285-308.
5. Uemura M, Steponkus PL. A contrast of the plasma membrane lipid composition of oat and rye leaves in relation to freezing tolerance. Pl Physiol 1994; 104:479-496.
6. Norberg P, Liljenberg C. Lipids of plasma membranes prepared from oat root cells. Pl Physiol 1991; 96:1136-1141.
7. Lynch DV, Caffrey M, Hogan JL, Steponkus PL. Calorimetric and x-ray diffraction studies of rye glucocerebroside mesomorphism. Biophys J 1992; 61:1289-1300.
8. Lynch DV, Fairfield SR. Sphingolipid long-chain base synthesis in plants: characterization of serine palmitoyltransferase activity in squash fruit microsomes. Pl Physiol 1993; 103:1421-1429.
9. Lynch DV, Spence RA, Theiling KM, Thomas KW, Lee MT. Enzymatic reactions involved in ceramide metabolism. In: Somerville CR, Murata N, eds. Biochemical and Molecular-Biological Aspects of Membrane and Storage Lipids of Plants. Rockville: ASPP, 1993:183-190.
10. Lynch DV, Steponkus PL. Plasma membrane lipid alterations associated with cold acclimation of winter rye seedlings (*Secale cereale* L cv Puma). Pl Physiol 1987; 83:761-767.
11. Palmgren MG, Askerlund P, Fredrikson K, Widell S, Sommarin M, Larsson C. Sealed inside-out and right-side-out plasma membrane vesicles. Pl Physiol 1990; 92:871-880.
12. Askerlund P, Larsson C, Widell S. Localization of donor and acceptor sites of NADH dehydrogenase activities using inside-out and right-side-out plasma membrane vesicles from plants. FEBS Lett 1988; 239:23-28.

CHANGES IN THE SYNTHESIS OF UNUSUAL GLYCOLIPIDS DURING SOMATIC EMBRYOGENESIS IN *DAUCUS CAROTA.*

Luis A. Quesada-Allué, Belén Cadenas, Karen Hagelin,
Flavia Guzzo, and Fiorella LoSchiavo.

Instituto Luis F. Leloir, Fundacion Campomar, Antonio Machado 151
1405, Buenos Aires, Argentina

A number of mono- and oligosaccharides bound to plant lipids have been described. In particular, steryl glucosides (SG) and acylated steryl glucosides (ASG) were found in all plant tissues [1-3]. Kleinig & Kopp [4] showed, by radiolabelling of carrot cultures, that SG and ASG represent 0.8 and 1.0 percent, respectively, of carrot membrane lipids. However, relatively little is known on pentose-containing plant glycolipids . A ß-sitosterol xyloside (SX) was chemically characterized in *Bauhinia candicans* [5] and the *in vitro* formation of pentosyl derivatives has been reported in pea [6].

The main objective of the present work was to study the conditions for the synthesis of unusual glycolipids containing xylose, to partially characterize the major ones and to try to understand their physiological role, particularly during the process of somatic embryogenesis. Subcellular fractions from cultured cells and somatic embryos of the edible carrot *Daucus carota* were used as source of enzymes

We recently found that carrot microsomal vesicles were able to catalyze the synthesis of xylose-bound glycolipids [7]. We report here the *in vitro* microsome-catalyzed formation of carrot β-sitosterol-xylose (SX) and the conversion of this substance into a saponifiable derivative. The described enzymatic activities were also found associated to the membrane of lipid bodies.

The changes in the rates of synthesis observed during somatic embryogenesis suggest a possible involvement of these substances in the mobilization of xylose in the vicinity and/or through membranes and this, in turn, may be related to cell-wall elongation.

Methods

Culture conditions for the obtention of somatic embryos were as described [8]. Membranes preparation, reagents and cell-free incubation conditions were reported previously [7]. Membranes partially depleted of sterols were obtained by extraction of the microsomal preparations with acetone. Analytical techniques were as described [7,9].

Results

Experiments of *in vivo* metabolic labelling showed that unusual neutral and anionic glycolipids were synthesized by carrot cells [10]. The synthetic rate of these glycolipids seemed to depend both on the differentiation stage and on the embryogenic potential of the

J.-C. Kader and P. Mazliak (eds.), Plant Lipid Metabolism, 242–246.

cells (not shown). Cell free assays using microsomal membranes as source of enzymes and UDP-[^{14}C]Xylose as donor substrate demonstrated that a main neutral xylolipid as well as minor neutral and negatively charged xylolipids were synthesized. The rate of xylolipids synthesis was 114.7 pmoles/mg protein in globular embryos as compared to 613.6 pmoles in proliferating undifferentiated cells committed to embryogenesis [7]. Xylose incorporation into proteins, instead, was 7.9 pmoles/mg prot. in globular embryos and decreased to 3.1 pmoles in proliferating cells.

Several techniques were used in order to characterize the neutral radiolabelled xylolipids. According to silicic acid and anion exchange column chromatography, acid and alkali treatments and comparative thin layer chromatography in neutral, basic and acidic solvents (Table 1 and [7]) the major neutral xylolipid "B" was found to behave as a sterol derivative, whereas one of the minor neutral xylolipids "D" behaved as an acylated steryl derivative. The main anionic xylolipid "x" behaved as a dolichyl-phosphate derivative (Table 1)[9].

Table 1: Xylolipids mobility in Thin layer chromatography

	Rf of xylolipids and standards in silica gel 60 TLC			
SOLVENT#	NEUTRAL C/M/H_2O		ACID C/I/E/1N AcOOH	ALKALINE C/M/7M NH_4OH
	65/28/4	85/15/1	2/2/3/1	65/25/4
Xylolipid B	0.66	0.36	0.75	0.40
Xylolipid D	0.89	--	0.86	0.77
Xylolipid x	0.20	--	0.52	--
Ceramide-Glc	0.67	0.43	0.71	--
β-sitosteryl-Glc*	0.67	0.35	0.75	0.40
Acyl-β-sitost-Glc *	0.89	--	0.85	0.78
Dol-P-[^{14}C]Glc**	0.19	--	0.54	--

Thin layer chromatography plates were developped at 21±1 ^{0}C in solvent saturated tanks.(#) Abbreviations: M=Methanol; C=Chloroform; E=ethanol; I= isopropanol; Acyl=acylated; Dol= dolichyl (C_{100}). (*) Purified from fresh pea seeds. (**) Dolichyl phosphate[^{14}C]glucose synthesized *in vitro* by carrot microsomes.

IDENTITY OF THE ACCEPTOR LIPID

β-sitosterol was found to stimulate the synthesis of xylolipids, thus appearing as a suitable substrate for the enzymes involved.(Table 2). Preliminary data indicate that ergosterol or cholesterol are not good acceptors for xylose.

Table 2 shows that, after selective removal of the endogenous sterols present in the microsomal membranes no synthesis of xylolipids could be detected. When the corresponding extracted lipids were added back, the xylolipids synthesizing activity was recovered (Table 2). When β-sitosterol was added instead of the endogenous lipids a 415% enhancement of the xylolipids synthesis occured (Table 2). Since, in our conditions, no other sterols were found to be good acceptors of xylose and β-sitosterol usually represent more than 60% of the carrot cell sterols, we concluded that the main synthesized endogenous xylolipid is likely to be β-sitosterol-xylose (SX). This agrees with previous phytochemical data since this compound is the only xylosylated steryl derivative described in plants up to now[5].

Table 2: Effect of β-sitosterol on [^{14}C]Glucose transfer

EXPERIMENT	cpm/μg protein	%
STANDARD	34.1	100
PLUS β-sitosterol	123.8	363
Lipids-depleted enzyme*	0.0	-
L-dep.enz.+ ext. lipids**	1.3	100
L-dep.enz + β-sitosterol	5.4	415

All the assays contained 0.4 % Triton X-100.(*) Microsomal fractions extracted with cold acetone. (**) The lipids corresponding to the extracted microsomes were added back.

Table 3: Synthesis of an acylated steryl derivative of xylose

	radioactivity in fluorographic spots (cpm)		
MOBILITY	Rf=0.75 (SX)	Rf=0.86 (ASG)	%
Control (heated enzyme)	1.800	0	
Standard (no detergent)	2.735	31	1.1
+ 0.04% Triton X-100	2.695	78	2.8
+0.1% Triton X-100	2.676	142*	5.0

After incubation, the radioactive lipidic material was separated in TLC with acid solvent as in Table 1 and the fluorographic spots were scraped and counted in a liquid scintillator. The numbers are the average of two experiments.(*) Similar material from a parallel experiment was alkali-treated and the resulting product behaved as SX [Rf= 0.75].

SYNTHESIS OF AN ACYLATED DERIVATIVE

The *in vitro* synthesized SX was recovered from TLC plates and incubated with variable amounts of Triton X-100. As shown in Table 3 the microsomal preparation was able to catalize the synthesis of a substance that showed the same chromatographic mobility as the acylated SX described above (Table 1). The synthesized substance was found to be saponifiable and the labelled product recovered was SX The synthesis reaction had low efficiency and this might be due to the scarcity of endogenous acyl donors or to a sub-optimal solubilization of the labelled substrate since Triton X-100 was able to increase the amount of labelled product (Table 3).

XYLOLIPIDS SYNTHESIS BY MEMBRANES OF LIPID BODIES

Surprisingly, membranes of lipid bodies were able to synthesize some of the xylolipids described above (Table 4)

Table 4: Xylose transfer to lipids in lipid bodies and microsomes

	cpm/ mg of protein			
	PROLIFERATING CELLS		GLOBULAR EMBRYOS	
	Lipid bodies	Microsomes	Lipid bodies	Microsomes
Steryl-Glc *	510	2,804	220	379
Steryl-Xyl*	70	633	Not detected	31
RATIO Glc/Xyl	(7.28)	(4.42)	--	(12.22)

Lipid bodies were separated in a discontinuous gradient of 0.1M/0.2M/0.4 M sucrose in a 0.1M Hepes Buffer pH=7.4 and washed 3 times with the same buffer. (*) including the acylated derivatives

Discussion

The above results indicate that disorganized proliferating cells and, to a lesser extent, somatic embryos of carrot, catalyze the synthesis of SX and ASX. These enzymatic activities have not been previously described in any organism. Up to now, the scarcity of the labelled xylolipids that might be used as tracers and of unlabeled xylosyl derivatives prevented further chemical analysis. However, we know that acid hydrolysis liberated [^{14}C]Xylose from the above described substances as assessed by paper chromatography, thus excluding any substantial interconversion of the labeled xylose. Since it has been reported [11] that UDP-Xyl might be a weak substrate for UDP-Glc: steryl glucose transferase, we made a number of comparative tests for the transfer of glucose and xylose to lipids. From the results we obtained on thermal inactivation, sensitivity to detergents, isotopic dilution, cross competitive inhibition and others (not shown) it seems that the synthesis of xylose-steryl derivatives is physiological and different from the synthesis of the glucose-steryl ones.

The enzymatic activity of our lipid bodies preparations was unexpected. Lipid bodies membranes seem to be generated by the outer moiety of endoplasmic reticulum [12]; our results are in line with that theory. On the other hand, the possibility exist that hydrophobic proteins from different cellular compartments might become bound to, or inserted into, lipid bodies membranes during subcellular fractionation, thus generating an artifact. Since we extensively washed our lipid bodies and the transferase activities are still present, we presume that the detected xylosyl transferases are tighly bound.

Somatic embryo cells are smaller and have a much denser cytoplasm than proliferating undifferentiated cells, hence the apparent decrease in the specific activity of the synthetases might be due, at lest in part, to a higher protein content of embryo cells. However, the synthesis of xylolipids might be required for xylose to be transported across or within membranes in order to somehow be used in the building of certain xylose-containing glycoproteins like some cell wall components. In such a case, since globular embryos do not elongate the cell walls the requirements for xylolipids synthesis would be lower. We know that heart-shaped embryos that contain a higher proportion of enlarged cells show a slightly higher xylose transferase specific activity than that of globular embryos.

Acknowledgements

L.A.Q-A acknowledges the partial support of RAISA and CNR (Italy) and CONICET, University of Buenos Aires and YPF S.A.(Argentina) to this work.

References

1. Elbein A D. Glycolipids and other glycosides. In: Loewus F, Tanner W, editors. Encyclopedia of Plant Physiology, Vol 13A,Heidelberg: Springer,1982,pp 601-612.
2. Mudd JB. Sterol interconversions. In: Stumpf PK, Conn EE, editors. The Biochemistry of Plants, Vol.4. New York:Academic Press, 1980: 509-534.
3. Axelos M, Péaud-Lenoel C. Steryl-glycosides In: Loewus F, Tanner W, editors. Encyclopedia of Plant Physiology, Vol 13A,Heidelberg: Springer,1982,pp. 613-631.
4. Kleinig H, Kopp C. Lipids,lipid turnover and phospholipase D in plant suspension culture cells (*Daucus carota).* Planta 1978; 139:61-65.
5. Iribarren AM, Pomilio, AB. Sitosterol-3-o-β-D-xylopyranoside from *Bahuinia candicans.* Phytochemistry 1984, 23: 2087-2090.
6. Maclachlan G. Pentosyl lipids in pea. In: Brett CT, Hillman JR, editors. Biochemistry of plant cell walls. Cambridge: University Press, 1985, pp 199-221.
7. Quesada-Allué LA, Hagelin K, LoSchiavo, F. Xylose transfer to lipids by membranes of carrot cell suspensions. Phytochem 1993; 33: 1347-1352.
8. Filippini F, Terzi M, Cozzani F, Vallone D, Lo Schiavo F. Modulation of auxin-binding proteins in cell suspensions. II Isolation and initial characterization of carrot variants impaired in somatic embryogenesis. Theor Appl Genet 1992; 84:430-434.
9. Quesada-Allué LA, Maréchal LR, Belocopitow E. Formation of Dolichyl-phosphate mannose in insects. FEBS Lett 1976; 67: 243-250.
10. Hagelin K. Glycolipids synthesis during the somatic embryogenesis of the carrot.[dissertation]. Buenos Aires: University of Buenos Aires, 1990.
11. Axelos M, Péaud-Lenoel C. Glucose transfer to sterols . Biochimie 1978; 60:35-42
12. Huang AHC. Oil bodies and oleosins in seeds.Annu. Rev. Plant Physiol. Mol. Biol. 1992; 43: 177-189.

ATPase AND H^+ TRANSPORT ACTIVITIES IN MICROSOMES FROM COTYLEDONS OF SUNFLOWER (*Helianthus annuus* L.) GENOTYPES DIFFERING IN OLEIC ACID CONTENT

Antonio Chica, M. Pilar Rodríguez-Rosales, Nuria Ferrol and Juan Pedro Donaire
Departamento de Bioquímica, Biología Celular y Molecular de Plantas, Estación Experimental del Zaidín, CSIC, Profesor Albareda 1, 18008 Granada, Spain.

The plant plasma membrane ATPase has been described as a key regulatory enzyme in cellular events such as adaptation to abiotic stresses and development changes [1,2]; however, the mechanism by which this activity is regulated is still uncertain. The modulation by lipids of ATPases of many sources is a recognized phenomenon which has been discussed [3]. This lipid dependency suggested a possible mode of regulation of plasma membrane H^+-ATPase activity *in vivo* via the modification of its lipid environment [4]. The modulation by lipids may occur via a direct interaction with the enzyme (the annulus hypothesis) or through changes in membrane fluidity, which causes an alteration in the enzyme conformation, and thus in activity.

The aim of the present work was to study whether phopholipids, fatty acids, free sterols and membrane fluidity modulate vanadate-sensitive H^+-ATPase and vanadate-sensitive ATP-dependent H^+ transport activities in microsomes from etiolated sunflower cotyledons of a normal genotype and a mutant with high oleic acid content [5] during seedling establishment.

Seeds were germinated in the dark in moist rolled filter paper for 4 days at 24 °C and then grown hydroponically for 3 days in 0.2 mM $CaCl_2$. ATPase activity, H^+ transport and fluorescence polarization, performed at 25 °C using diphenyl hexatriene (DPH) as probe, were carried out in isolated microsomes [2]. In this membranes, total lipids were extracted and total fatty acids and those from phospholipids [6] determined by gas chromatography. Free sterols were also analized [7].

The level of total phospholipids and free sterols refered to mg of protein did not change in microsomes from both genotypes, which led to similar free sterol to phospholipid ratios (Table 1). Likewise, a lower proportion of 16:0 and 18:0 and a higher proportion of unsaturated 18 C fatty acids, above all 18:1, were found in phospholipids of microsomes from the high oleic acid cotyledons compared to the normal ones. In both genotypes, free sterols had a similar composition with sitosterol being the most abundant compound, followed by stigmasterol and campesterol (Table 2).

To determine whether an increase of unsaturated fatty acids could account for the rise in bulk lipid fluidity of microsomes from the high oleic acid mutant, this physical parameter was determined by fluorescence polarization. The results have shown that the DPH polarization values of microsomes from high oleic acid cotyledons decreased markedly in comparison to those of the normal genotype (Table 1). This drop in DPH polarization for microsomes prepared from the high oleic acid mutant could be attributable to a decrease in saturated/unsaturated fatty acid ratio. In this respect, it has been proved that an increase in bulk lipid fluidity of carnation flower membranes is associated with a selective rise of unsaturated fatty acids [8].

On the other hand, H^+ transport in sealed microsome vesicles from high oleic acid cotyledons was stimulated compared to the normal genotype, while vanadate-sensitive H^+-ATPase activity, responsible for H^+ pumping, was not significantly different in both membranes (Table 3). This table also shows that

J.-C. Kader and P. Mazliak (eds.), Plant Lipid Metabolism, 247–249.

Table 1. Total phospholipids, free-sterols and fluorescence polarization in microsomes from normal and high oleic acid sunflower cotyledons after 5 d of seedling establishment. Values are mean ± SD for n= 4 to 5.

Genotype	Phospholipid nmol/mg prt	Sat:Unsatur Ratio	Free Sterols nmol/mg prot	Sterol:PL Ratio	Fluorescence Polarization
Normal	82.3 ± 5.4	1.10 ± 0.10	50.9 ± 4.7	0.62 ± 0.08	0.268 ± 0.005
High 18:1	76.8 ± 2.1	0.60 ± 0.05	54.6 ± 4.2	0.71 ± 0.02	0.225 ± 0.007

Table 2. Fatty acid composition of the total phospholipids and major free sterols of microsomes from normal and high oleic acid sunflower cotyledons after 5 d of seedling establishment. Values (mol %) are mean ± SD of 4 independent experiments.

	Fatty acid				
Genotype	16:0	18:0	18:1	18:2	18:3
Normal	41.5 ± 3.6	10.9 ± 1.1	16.7 ± 2.0	30.9 ± 2.6	n.d.
High 18:1	29.3 ± 3.0	8.2 ± 0.8	39.7 ± 4.2	18.3 ± 1.4	4.6 ± 0.7

Genotype	Sterol Class		
	Campesterol	Stigmasterol	Sitosterol
Normal	10.4 ± 1.5	22.3 ± 3.5	67.3 ± 5.9
High 18:1	12.7 ± 1.2	25.4 ± 4.1	61.9 ± 3.3

n.d., not detected.

Table 3. Basal ATPase, total vanadate-sensitive ATPase, and total and vanadate-sensitive initial rate of H^+ pumping of microsomes isolated from normal and high oleic acid cotyledons. Passive H^+ conductance (expresed as % of initial rate of fluorescence recovery relative to Q maximum) is also showed. Values are mean ± SD of 4 independent experiments. Q, quenching.

Genotype	ATPase activity (μmol Pi mg^{-1} protein h^{-1})		H^+ pumping activity (% Q mg^{-1} protein min^{-1})		Passive H^+ conductance
	Basal	ΔVa	Total	ΔVa	
Normal	4.19 ± 0.14	3.89 ± 0.11	5.12 ± 0.50	4.82 ± 0.34	31.95 ± 3.02
High 18:1	3.88 ± 0.12	3.42 ± 0.33	18.31 ± 1.06	12.88 ± 1.10	12.31 ± 0.31

microsomes from the high oleic acid mutant had a lower passive H^+ conductance. Our results indicate that, although there is a significant difference in the saturated:unsaturated ratio and in the bulk lipid fluidity between microsomes from both genotypes, the vanadate-sensitive H^+-ATPase activities are very similar. It can be concluded that no clear relationship exist between changes in membrane fluidity and H^+-ATPase activity, which is in agreement with data obtained in *Citrus* root plasma membrane [9].

Taken together, these data suggest that stimulation of the H^+ gradient in microsome vesicles from high oleic acid mutant could be due to decreased H^+ conductance accross the membrane and not to the stimulation of the H^+-ATPase. The results also support the contention that the high oleic character, expresed in the developing sunflower mutant seed during the synthesis of reserve lipids [10], is maintained in phospholipids of cotyledons during seedling establishment.

References

1. Sussman MR, Surowy TK. Physiology and molecular biology of membrane ATPases. Oxford Surveys of Plant Molecular and Cell Biology 1987; 4: 47-70.
2. Ferrol N, Belver A, Roldán M, Rodríguez-Rosales MP, Donaire JP. Effects of boron on proton transport and membrane properties of sunflower (*Helianthus annuus* L.) cell microsomes. Plant Physiol 1993; 103: 763-69.
3. Serrano R. Structure and function of proton translocating ATPase in plasma membranes of plants and fungi. Biochem Biophys Acta 1988; 947: 1-28.
4. Palmgren MG. Regulation of plant plasma membrane H^+-ATPase activity. Physiol Plant 1991; 83: 314-23.
5. Fernández-Martínez J, Muñoz J, Gómez-Arnau J. Performance of near-isogenic high and low oleic acid hybrids of sunflower. Crop Sci 1993; 33: 1158-63.
6. Gadner HW. Preparative isolation of monogalactosyl and digalactosyl diglycerides by thin layer chromatography. J Lipid Res 1968; 9: 139-41.
7. Couture R, Arul J, Gosselin C, Willemot C. The sterols of straberry fruit. Phytochem 1989; 28: 1276-77.
8. Fobel M, Lynch DV, Thompson JE. Membrane deterioration in senescing carnation flowers. Coordinated effects of phospholipid degradation and the action of membranous lypoxigenase. Plant Physiol 1987; 85: 204-11.
9. Douglas TJ, Walker RR. Phospholipids, free sterols and adenosine triphosphatase of plasma membrane-enriched preparations from roots of *Citrus* genotypes differing in chloride exclusion ability. Physiol Plant 1984; 62: 51-8.
10. Garcés R, Mancha M. Oleate desaturation in seeds of two genotypes of sunflower. Phytochem 1989; 28: 2593-95.

Acknowledgements

We gratefully acknowledge to DGICYT (Spain) for the provision of grant PB 91-0100 .

LIPID MODULATION OF THE ACTIVITY OF TONOPLAST H^+-ATPASE PURIFIED FROM MUNG BEAN (*Vigna radiata* L.) HYPOCOTYLS

Kunihiro Kasamo and Hiroyasu Yamanishi
Molecular Function Laboratory, National Food Research Institute, 2-1-2 Kannondai, Tsukuba, Ibaraki 305, JAPAN.

INTRODUCTION

The vacuoles in higher plants are essential for the maintenance and regulation of the homeostatic environment of the cells [1]. The vacuolar membranes of plant cells contains an electrogenic H^+-ATPase which provides the proton-motive force for the active transport of solutes across the vacuolar membranes. The activities of certain membrane-bound enzymes are lost by the removal of constituted lipids, but are restored after the addition of exogenous phospholipids [2,3]. Kasamo [4] showed that the various molecular species of PC activated to different extents the H^+-ATPase solubilized from the plasma membrane of rice cultured cells.

Solubilized an purified tonoplast H^+-ATPase also requires exogenously added phospholipids for expression of maximal activity [5,6]. However, to date, neither the effect of specific molecular species of phospholipids on the activation of purified-tonoplast ATPase nor the membrane components associated with the ATPase have been examined. Highly delipidated enzymes are useful for the studies on the relationship between activation of enzymes and the physical state of membranes. We have been able to purified and delipidate the tonoplast ATPase to such an extent that essentially no activity is detectable without added phospholipid and full activity is restored by the addition of an appropriate phospholipid.

In the present study, we examined the effects of individual phospholipids on the activity of the tonoplast ATPase using a highly delipidated enzyme purified from mung bean hypocotyls. A mechanism for activation of the ATPase by phospholipids is proposed and discussed.

J.-C. Kader and P. Mazliak (eds.), Plant Lipid Metabolism, 250–252.

MATERIALS AND METHODS.

Tonoplast vesicles from mung bean hypocotyls were isolated as previously described [6]. Lipids were extracted by the method of Bligh and Dyer [7]. Phospholipid fraction was partitioned on Sep-Pak cartridges [8]. ATPase activity was measured at 38°C with 3-5 μg of protein per assay [9]. Proteins were quantitated by the method of Peterson [10] using bovine serum albumin as the standard.

RESULTS AND DISCUSSION.

Phospholipid was hardly detectable in ATPase fraction purified by Mono Q column [6]. Exogenously added phospholipids at a concentrations of 50 μg to 100 μg/ml (0.005% to 0.01%) was sufficient to induce the activity. Constituents of phospholipids, such as Polar head groups, fatty acids and glyceride failed completely to activate the ATPase. In case of fatty acids, unsaturated fatty acids were 10-times more inhibitory than saturated fatty acids (Fig.1.). PC with unsaturated and saturated faty acids and dimyristoyl-PC activated ATPase activity. Fig.2 summarizes the relationship between ATPase activity and liquid-crystalline phase transtion temperature of molecular species of PC. Every molecular species of PC with a phase-transition temperature below 50°C seemes to be able to activate the purified ATPase at 38°C. Dipalmitoy-pc having at 41.5°C phase transition temperature was slightly activated the ATPase activity at 38°C, however, when it was assayed at above 40°C, the activity was sharply increased. Thus, the ATPase activity increased sharply at around the pahse-transition temperature of the phospholipids that had been added. It is clear that the phospholipids must be in a liquid-crystalline state to activate the ATPase. These date suggest that fluidity of tonoplast is closely related to the

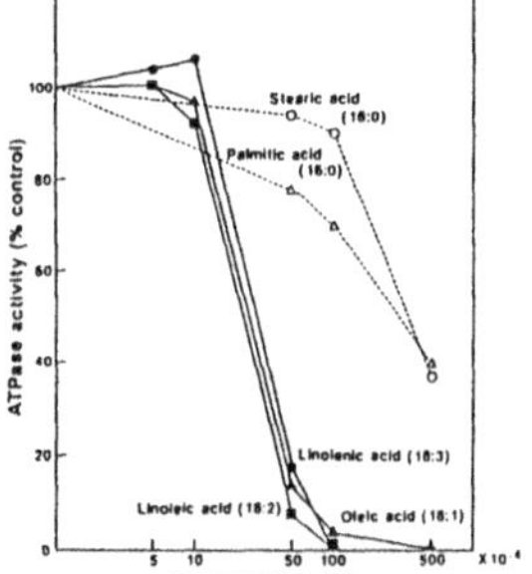

Fig.1 Effects of fatty acids on the ATPase activity.

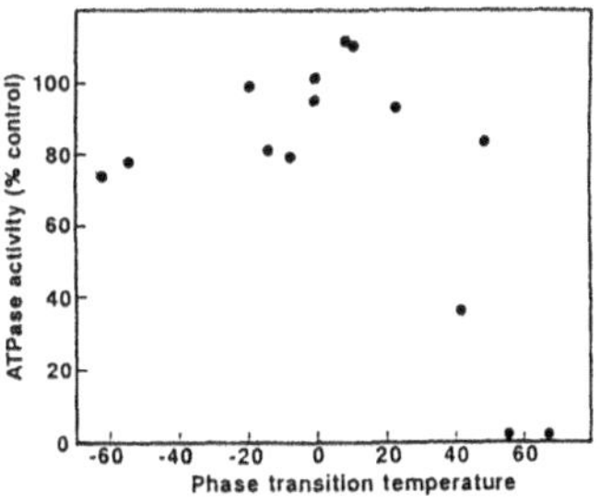

Fig.2 Relationship between phase transition temperature of the molecular species of PC and the ATPase activity.

ATPase activity. The changes in balk lipid composition of the tonoplast alter the its fluidity,which causes a change in the enzyme conformation, and thus the activity. Fluidity is a dynamic concept related to the structural order of membranes-lipids head groups, lipid acyl chain length and saturation/unsaturation and molecular packing.

REFERENCE

1. Boller T, Wiemker A. Dynamics of vacuolar compartmentation. Ann. Rev. Plant Physiol. 1986; 37:137-164.
2. Sandermann H. Regulation of membrane enzymes by lipids. Biochim. Biophys. Acta 1978;515:209-237
3. Kasamo K, Nouchi I. The role of phospholipids in plasma membrane ATPase activity in Vigna radiata L. (mung bean) roots and hypocotyls. Plant Physiol. 1987;83:323-328.
4. Kasamo K. Mechanism for the activation of plasma membrane H^+-ATPase from rice (Oryza sativa L.) culture cells by molecular species of a phospholipid. Plant Physiol. 1990;93:1049-1052.
5. Kasamo K. Inhibition of tonoplast and plasma mambrane H^+-ATPase activity in rice (Oryza sativa L.) culture cells by local anesthetics.Plant Cell Physiol. 1988;29:215-222.
6. Yamanishi H, Kasamo K. Modulation of activity of purified tonoplast H^+-ATPase from mung bean (Vigna radiata L.) hypocotyls by various lipids. Plant Cell Physiol 1993;37;411-419.
7. Bligh EG, Dyer WJ. A rapid method of total lipid extraction and purification. Can. J. Biol. Physiol. 1959;37:911-917.
8. Sandstrom RP, Cleland RE. Selective delipidation of the plasma membrane by sufactants; enrichment of sterols and activation of ATPase. Plant Physiol. 1989;90:1524-1531.
9. Kasamo K. Purification and properties of the plasma membrane H^+-translocating adenosine triphosphatase of Phaseolus mungo L. roots. Plant Physiol. 1986;80:818-824.
10. Peterson GL. A simplification of the protein assay method of Lowry et al. which is more generally applicable. Anal. Biochem. 1977;83:346-356.

Section 3

Lipid Oxidation and Degradation

DEVELOPMENTAL IMPLICATIONS OF MEMBRANE HYPERFLUIDITY INDUCED BY LIPID PEROXIDATION AND SUBSEQUENT JASMONATE FORMATION

Ya'acov Y. Leshem
Department of Life Sciences, Bar-Ilan University, Ramat Gan 52900, Israel

Abstract

In terms of possible relation of JA formation, structure and membrane biophysical architecture, it is proposed that membrane lipid peroxidation following the pathway to JA formation "jolts" bilayer structure-and thus also of embedded enzyme function-by initially, but transiently causing increase in rigidity, immediately followed by a long term stage of hyperfluidity which is non-conducive to growth.

Introduction

Lipoxygenase (LOX) mediated hydroperoxidation of glycerolipids in plant membranes serves as an early event in the usually catabolic sequence of events whose final outcome is pivotal to plant growth regulation, adaptation to stress and to processes leading to senescence. In addition to 9 or 13 hydroperoxidation, in certain cases as in linoleate, *double* peroxidation, i.e. at both a 9 and a 16 site, may also occur [1].

The main steps of one (but not the sole) pathway of PUFA hydroperoxidation in higher plants are as follows:

1. Influx of Ca^{2+} into the cytosol
2. Phospholipase A_2 activation
3. De-esterification of *sn*-2 (usually 18:2 Δ 9,12 or 18:3 Δ 9,12,15) fatty acids (FAs)
4. Hydroperoxidation by LOX
5. Cylization of the de-esterfied PUFA chain (biologically by cyclase, or spontaneously [3]
6. Shedding of a 6C fragment to obtain a 13C "backbone"
7. Formation (at time of cylization) of a double C=C bond (at the 9 and 13 sites of hydroperoxidation) which in the newly constituted molecule are now (physiologically misleadingly) the 3 and 7 sites.
8. Formation of (+)-7-*iso*-jasmonic acid which is sequentially enzymatically converted to (-)-jasmonic acid JA and its methyl ester methyl jasmonate (MJ) [4], and also to other jasmonate compounds such as tuberonic acid, etc.

The above sequence of events and organic chemical terminology pertains to 18:2 Δ 9,12 peroxidation which apparently has its parallel in 18:3 Δ 9,12,15 hydroperoxidation, the latter assumed to be the main source of JA in higher plants. However, in particular stress conditions, 18:2 Δ 9,12 may also contribute to JA formation.

J.-C. Kader and P. Mazliak (eds.), Plant Lipid Metabolism, 255–258.

Experiments and Discussion

Some Physiological Consequences Of Membrane Lipid Hydroperoxidation

The surface tension of a monolayer comprising a mixture of *sn*-1 palmitoyl, *sn*-2 linoleoyl and phosphatidylcholine (PLPC) was determined with a Wilhelmy Tensiometer and a Du Nouy platinum ring. When applied, 10 µl of MJ was added before readings commenced. The buffer system of the subphase in the hour-glass was 0.01 M HEPES [4-2(hydroxyethyl)-1-piperazine ethane-sulphonic acid] at pH 7.5 containing 10^{-5} M $CaCl_2$. In the case of the non-treated control, an identical quantity of hexane solvent lacking MJ was added to the monolayer. Table 1 presents the results which clearly indicate a fluidizing effect of MJ.

Earlier work of our laboratory [2,3] in monolayers and Langmuir tensiometry has indicated that the release of the *sn*-2 PUFA chain increases membrane fluidity since it permits packing of more individual glycerolipid molecules into a given area. Moreover, in keeping with this trend, the individual de-esterified PUFA-OH's also produces significantly greater surface tension-related parameters (such as contact angles) than their corresponding intact PUFA's. In keeping with this, it would be logical to assume that the initial stage of JA formation (steps 1-4 above) involve membrane rigidification. However, as shown presently, the final product of the catabolic sequence, MJ, which *increases* membrane fluidity [2] (see also Table 1), never-the-less has an overall inhibitory effect on carefully monitored pea leaf growth and expansion.

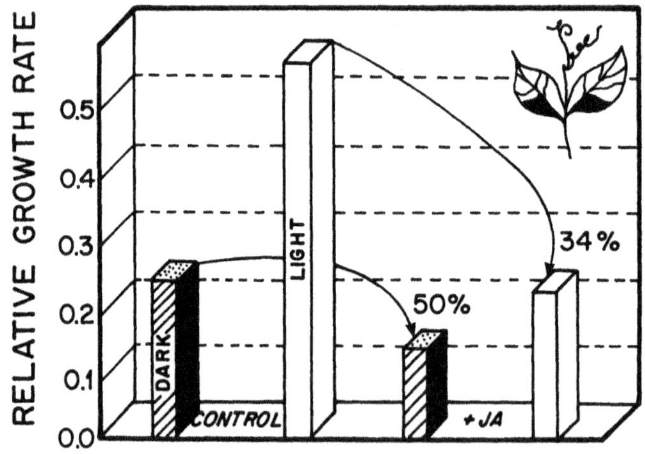

Figure 1: Effect of 10^{-5} M JA on leaf expansion of *Pisum sativum* (*argenteum* mutant) leaf sections. Lower quarters of leaves of 3 week old plantlets were incubated for 20 hr with or without 10^{-5}M JA in 10 mM buffer MES pH 5.5 also containing 10 µM KOH. Light source was from an incandescent source providing 125 µ Einsteins/cm^{-1}. Figures are four replicate means obtained by weighing sections initially and again after 20 hrs.

These observations may well be reconciled with classic membrane biophysical dynamics: increased fluidity does not necessarily imply enhanced enzyme function - since many enzymes manifest a given range of membrane surface tensions for optimal activity. Values exceeding or *below* optimum prevent optimal enzyme substrate recognition. This particularly applies to membrane/cytosol interface located enzymes such as phospholipases.
Figure 1 shows that JA has an overall inhibiting effect on pea leaf expansion and that this is more pronounced in the light than in dark, suggesting some relationship with factors related to the 'acid growth' hypothesis of auxin action [5]. However, when growth kinetics are closely followed by a position transducer, it can be seen (Fig. 2) that the initial effect of JA (the first 30 secs after JA application) is a marked "spike" of growth promotion, later followed by a definite long term trend of growth inhibition as seen in the former figure.

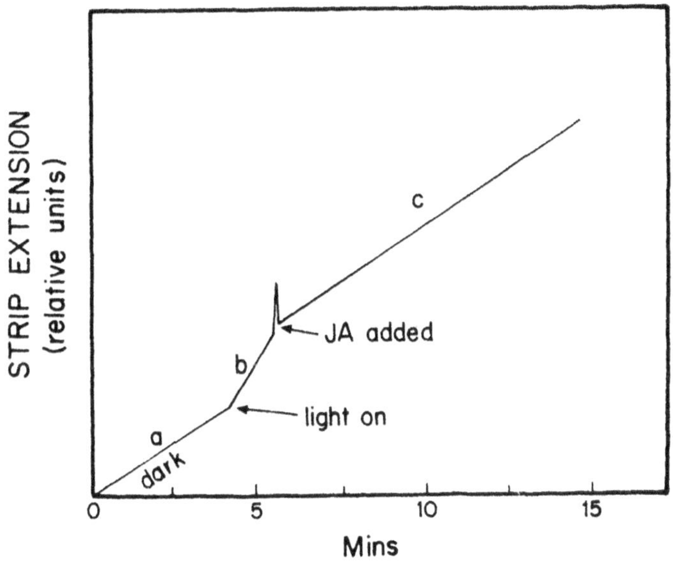

Figure 2: Kinetics of effect of adding JA to pea leaf sections (as above) attached to a position transducer geared to a growth recorder. Experimental details as in Figure 1.

In terms of possible relation of JA formation, structure and membrane biophysical architecture, it is proposed that membrane lipid peroxidation following the pathway to JA formation "jolts" bilayer structure - and thus also of embedded enzyme function - by initially, but transiently, causing increase in rigidity immediately followed by a long term stage of hyperfluidity which is non-conducive to growth. In human erythrocytes, a similar phenomenon of marked reduction in the otherwise high surface pressure of the lipid bilayer upon infection with *Plasmodium*, the malaria-causing parasite, has been found [6].

Acknowledgements: For their aid, advice and encouragement, the author is deeply indebted to Prof. Pieter Kuiper, Dr. Hidde Prins and Dr. Elizabeth van Volkenburgh of the University of Groningen's Biological Centre, The Netherlands, where part of this research was carried out.

Table 1. Surface tension of a PLPC monolayer as affected by methyl jasmonate. Experimental details are given in the text. Numbers in parentheses are standard deviations.

Treatment	**Surface tension ($mN\ m^{-1}$)**
Buffer control	56.6 (0.2)
PLPC	47.1 (1.7)
PLPC + MJ	42.7 (1.5)

References

1. Grechkin AN, Karamshin RA, Safonava EY et al. Double peroxidation of 2-linolenic acid by potato tuber lipoxygenase. Biochem Biophys Acta 1991;1081:7879-9884.
2. Leshem YY, Gottlieb HE, Bach D. Surface tension related mechanical effects of methyl jasmonate on membrane phospholipid. Pl Phys 1994; (in press).
3. Leshem YY, Margel S, Aurbach D, Sofer Y. Biophysical parameters of linoleic acid hydroperoxides as assessed by surface behavior and Fourier Transform Infra Red spectrometry: possible pertinence to senescence. Pl Growth Reg 1993; 12:263-272.
4. Parthier B, Bruckner C, Dathe W et al. Jasmonates: metabolism biological activities and modes of action in senescence and stress responses. In: Karsen CM, van Loon LC, Vreugdenil D, editors. Progress in plant growth regulation. The Netherlands: Kluwer Academic Publishers, 1992: 276-285.
5. van Volkenburgh E. Leaf cell expansion. In: Baker NR, Thomas H, editors. Crop photosynthesis: spatial and temporal determinants. Amsterdam: Elsevier Publishers, 1992: 235-252.
6. van den Bosch H. Biochemistry of lipids. Ann. Rep. Centre for Biomembranes and Lipid Enzymology. Utrecht Univ. 1993: 13.

ESTIMATION OF FREE AND BOUND MDA IN PLANT EXTRACTS: COMPARISON BETWEEN SPECTROPHOTOMETRIC AND HPLC METHODS

Daniel Hagège[1], Sabine Feutry[1], Marijana Krsnik-Rasol[2], Danielle Poder[1] and Jean François Menez[3].

[1] Laboratoire d'Amélioration des Végétaux, Physiologie et Biochimie Végétales, UBO, BP 809, F-29285 Brest cedex.

[2] Laboratory of Molecular Biology, University of Zagreb, Croatia.

[3] Laboratoire de Biochimie-Nutrition, Faculté de Médecine, F-29285 Brest cedex.

Introduction

Lipid peroxidation is a complex process in which polyunsaturated fatty acids (PUFAs) are subjected to attack by oxygen-derived free radicals resulting in the formation of lipid hydroperoxides. In biological tissues, these lipid hydroperoxides are broken down into a variety of products including aldehydes and ketones [1]. The extend of lipid peroxidation is considered to be an important parameter for the identification of the oxidative stress [2]. The improvement of a technique for the measurement of this lipid peroxidation in living tissues is, consequently, of importance. A large variety of methods are available for the detection of lipid peroxidation products. These include determination of diene conjugation, lipid hydroperoxides, chemiluminescence, hydroxy acids, ethane and Thiobarbituric acid (TBA)-reactive materials. The estimation of TBA reactive products is the most widely used. The success of the TBA reaction method depends on the accuracy of the determination of MDA content. MDA is a three carbon dialdehyde which is widely produced in organisms as a end product of polyunsaturated lipid peroxidation. This method is of particular interest because of its simplicity and relatively good sensitivity, but it lacks specificity and often leads to overestimation due to other interfering compounds ([3],[4]). Treatment of biological sample with TBA under appropriate conditions results in the formation of pink-colored products (λmax = 532 nm) and many substances which are not related to lipid peroxidation (sugars, amino acids, biliverdin, chemicals, glyoxal,...) may also form a TBA complex with significant absorbance around 530-535 nm. The absorbance of 532 nm is not indicative uniquely of MDA and it therefore seems more appropriate to express the results in terms of thiobarbituric acid reactive substances (TBARs) rather than MDA [5] One should keep in mind that the purpose of such modified assays is to obtain an index of lipid peroxidation and not to measure MDA.

When plant materials are used, many substances such as pigments, sucrose, glucose and fructose react with TBA to produce a similar chromogen ([6],[7]). We are interested in the determination of lipid peroxidation in plants. Such research requires a sensitive and accurate method of determining MDA (TBARs) to give an index of lipid peroxidation in plant materials. In previous work, we proposed a spectrophotometric method for bound MDA estimation in plant tissues after acetone depigmentation and TBA reaction using the trichloracetic acid precipitated pellet. But MDA may exist both in free and hydrolysable bound forms involving primary amino acid groups of proteins, nucleic acids and other biological nucleophiles. This study was conducted to develop a highly sensitive, specific and simple method for determining free and bound TBARs. Using plant tissues, results

J.-C. Kader and P. Mazliak (eds.), Plant Lipid Metabolism, 259–261.

obtained by two spectrophotometric (difference and triangulation)- and by an HPLC -methods are presented. The role of BHT (used to prevent formation of new peroxides during the assay) has also been investigated.

Materials and Methods.

Plant material. All the experiments were conducted using a fourteen day-old sugarbeet callus (*Beta vulgaris* L., var. *altissima*). The initiation and culture conditions of this callus have been described elsewhere [8]. This callus is chlorophyllous and grows on basic media supplemented with 2,4-dichlorophenoxyacetic acid (2,4-D 0.1 $mg.L^{-1}$) and benzylaminopurine (BAP 0.1 $mg.L^{-1}$).

Samples Preparation. 500 mg callus were homogenized using either 1 mL TCA 10% (P/V) or $HCLO_4$, 586 $g.L^{-1}$. In order to test the dose effect of BHT (butylated hydroxytoluene; Sigma), 50 µL were added (0, 0.001, 0.01, 0.05, 0.1, 1, 10, 20 %).

Free and bound MDA (TBARs) determinations. The mixture was centrifuged (4000 g, 10 min, 4°C) and the supernatant was used for the determination of free MDA. The pellet was washed (twice) with 5 mL acetone, removed after centrifugation (4000 g, 10 min, 4°C). The new pellet obtained was used for the TBA reaction.

TBA reaction. Supernatants (free MDA) or pellets (bound MDA) were incubated with 3 mL H_3PO_4 1% (V/V) and 1 mL TBA (thiobarbituric acid; Sigma) 0.6 % (P/V) at 100°C, 30 min., then cooled on ice. 3 mL of n-butanol was added. After homogeneization and centrifugation (4000 g, 10 min, 4°C), 1 mL of the butanolic supernatant was used directly for spectrophotometric determination. One another mL was evaporated to dryness at 37°C and resuspended with 500 µl of potassium phosphate buffer (50 mmol. L^{-1}, pH 5.8) and used for HPLC determination of MDA.

Spectrophotometric methods. Two methods were compared. According to [6], the difference between the absorbance at 532 and 590 nm (persistence) of the butanolic phase was measured. Justifications of this difference has been extensively explained in a previous paper. This method has been modified by triangulation and three wavelengths have been considered. Some spectrophotometer are able to give the result obtained, using three wavelength, according the following calculation :

$DO_2 - ((WL_1-WL_2).DO_3+(WL_2-WL_3).DO_1)/(WL_1-WL_3)$

For MDA estimation, wavelengh have defined as follow : WL_1= 565 nm, WL_2= 534 nm and WL_3 = 510nm.

HLPC method. An HPLC system Spectra-Physic Analytical (San Jose, CA, USA), composed of Model P2000 pump, an AS 2000 autosampler, equiped with a Rheodyne automatic 20 µL injection valve and an UV 2000 spectrophotometric detector connected to an SP 4400 data jet integrator was used The HPLC column (225x4.6 mm I.D.) was packed with reversed-phase 5µm Hypersil ODS Flow S.F.C.C. The mobile phase was potassium phosphate buffer (10 $mmol.L^{-1}$, pH 5.8)-methanol (70:30, V/V) and the flow rate was 0.8 mL/min. MDA was detected at 532 nm. Peak area measurements were used for quantification and compared with those obtained with standard solutions.

Standard preparation and calibration. MDA standards (0 to 6 $nmol.L^{-1}$) were prepared in deionised water, from a 10 $mmol.L^{-1}$ stock solution of 1,1,3,3-tetraethoxypropane (TEP; Aldrich). 500µl of each standard solution were treated as described for plant tissues. The standard curves obtained using TEP were linear with a correlation coefficient of R_1= 0.9993 , R_2= 0.9992 (for spectrophotometric methods) and R_3= 0.9992 for HPLC.

Results

Effect of TCA or $HClO_4$ on TBARs determination. Whatever the method used, recovery of bound MDA was very low using $HClO_4$: (Difference: 0.61 ± 0.09 nmol.g FW^{-1} v.s.

2.26 ± 0.28 nmol.g FW^{-1}; Triangulation: 0.49 ± 0.04 nmol.g FW^{-1} v.s. 1.89 ± 0.26 nmol.g FW^{-1} ; HPLC 0.71 ± 0.09 nmol.g FW^{-1} v.s. 1.81 ± 0.25 nmol.g FW^{-1}). For free MDA, TCA precipitation gave satisfactory recovery (Triangulation: 4.90 ± 0.35 nmol.g FW^{-1} v.s. 2.04 ± 0.28 nmol.g FW^{-1} ; HPLC: 3.79 ± 0.19 nmol.g FW^{-1} v.s. 0.21 ± 0.01 nmol.g FW^{-1}) except when the difference method was used (6.91 ± 0.50 nmol.g FW^{-1} for TCA precipitation and 14.67 ± 0.36 nmol.g FW^{-1} with $HClO_4$). In this case, the spectrum did not exhibit any clear peak near 532 nm when compared to the other spectra. This could explain the great overestimation of the values : over 700 % when compared to triangulation method and nearly 7000 % when compared to HPLC method. Consequently, precipitation of proteins with TCA was prefered for the following experiments.
Effect of BHT. The dose effect of BHT was investigated. Seven concentrations were tested. For bound MDA, an apparent antioxidant effect of BHT could be noted from 1 %. For free MDA, the results obtained with BHT between 0.001 and 0.01 % suggest a pro-oxidant effect. Higher concentrations (from 0.05 to 20 %) gave values very close to the results obtained without BHT.

Discussion and conclusion

Protein precipitation is an essential step in the preparation of plant tissues for quantification of free or/and bound MDA (TBARs). The results presented in this work clearly indicate that $HCLO_4$ did not allow reproductive results. TCA appeared to be the best reagent to separate free and bound MDA. This could be due to the fact that TCA not only precipitates proteins, but also lipid material (lipoproteins).
Because some lipids may still be present even after protein precipation, the effect of BHT was tested at different concentration. BHT is considered as an antioxidant which prevent formation of new peroxides during the assay [9]. The results presented show a surprisingly pro-oxidant effect of 50 µl BHT 0.001 and 0.01 % when working with free MDA. The conclusion which could be drawn from these results is that BHT become efficient over 0.1%, for both types of MDA. Furthermore, it has been observed that a BHT concentration over 1% has an effect for the complete precipitation of proteins.
HPLC method is usually considered to give lower results than a spectrophotometric methods. In the present report, the triangulation method gives the most accurate values. This could be explained by the fact that the HPLC method relies only on a single measure of the absorbance of the chromogen at 532 nm. The triangulation allows a better calculation of the peak. This method appears to be easy to undertake and to give reproducible results for free and bound MDA (TBARs), using plant tissues.

References

[1] Esterbauer H. In: Brien DCH, Slater TF, editors. Free radicals, lipid peroxidation and cancer. London: Academic Press, 1982: p 101. [2] Petruska JM, Wong SHY, Sanderman W, Mossman BT. Free Rad Biol Med 1990; 9: 51-55. [3] Wade CR, Jackson PG, Van Rij AM. Biochem Med 1985; 33: 291-296. [4] Ichinose T, Miller MG, Shibamoto Lipids 1989; 20: 895-898. [5] Esterbauer H, Schaur RJ, Zollner H. Free Rad Biol Med 1991; 11 : 81-128. [6] Hagège D, Nouvelot A, Boucaud J, Gaspar Th. Phytochem Analysis 1990; 1: 86-89. [7] Du Z, Bramlage WJ. J Agri Food Chem 1992; 40: 1566-1570. [8] Kevers C, Coumans M, De Greef W, Hofinger M, Gaspar Th. Physiol Plant 1981; 51: 281-286. [9] Terao J, Matsushita S. Lipids 1981; 16: 98-101.

Acknolewledgements: This work was supported by a grant from the Conseil Régional de Bretagne. The authors thank Dr Lucas, Mrs Le Bot, Mrs Joannet and Mr Goubin for advices and technical assistance all along this work.

OXYLIPINS IN PLANTS : THE PEROXYGENASE PATHWAY

Elizabeth Blée and Francis Schuber

Départment d'Enzymologie Moléculaire et Cellulaire, Institut de Biologie Moléculaire des Plantes, 28 rue Goethe, 67000 Strasbourg, France

Introduction

Oxylipins are commonly found metabolites in higher plants and most of them originate from polyunsaturated fatty acid hydroperoxides by enzymatic transformations which have been extensively studied (for recent reviews, see 1,2). Two well characterized enzymes : a lyase and an allene oxide synthase were shown to degrade hydroperoxides into compounds of physiological importance since they can ultimately yield fragrances or plant hormones such as jasmonic acid. We have recently reported a new fate for fatty acid hydroperoxides: the peroxygenase pathway. It involves two enzyme activities i.e. a peroxygenase and an epoxide hydrolase which lead to the formation of epoxidized fatty acids and their derived dihydrodiols which are relevant to plant defense mechanisms.

Properties of the peroxygenase

Peroxygenase is a membrane-bound hemoprotein whose activity is independent of any cofactors such as NAD(P)H or flavins, but requires the absolute presence of hydroperoxides. The reactions catalyzed by this enzyme, involve a transfer of an oxygen atom from the hydroperoxide to the reaction product, *via* an intermolecular process (3). The peroxygenase catalyzes the hydroperoxide reduction exclusively through a heterolytic mechanism leading to a ferryl-oxo complex analogous to peroxidase compound I. By this mechanism, peroxygenase was shown to catalyze cooxidation reactions such as hydroxylation, sulfoxidation and epoxidation of compounds such as unsaturated fatty acids. In this latter case, the stereoselectivity of peroxygenase seems absolute since e.g. octadecenoic acids with *trans* configuration were not epoxidized at detectable rates. This enzyme exhibits also a high stereospecificity: e.g. oleic acid was converted mainly (80% e.e.) into 9*R*,10*S*-epoxystearate. Linoleic acid was oxidized with a high enantiofacial selectivity 9*R*,10*S* and 12*R*,13*S* epoxy enantiomers were preferentially formed. In these reactions, fatty acid hydroperoxides act as cosubstrates, but we have demonstrated that these molecules are also as such as substrates for the peroxygenase, according to a

J.-C. Kader and P. Mazliak (eds.), Plant Lipid Metabolism, 262–264.

mechanism of intramolecular oxygen transfer, leading to the formation of epoxy-alcohol derivatives (4).

Characteristics of fatty acid epoxide hydrolase

In soybean seedlings, epoxide hydrolases specific for *cis*-epoxidized fatty acids exist both as membrane-associated and soluble forms. We have purified to apparent homogeneity the soluble enzyme, which is among the rare epoxide hydrolases known to act on endogenous substrates (5). This hydrolase catalyzes *trans*-addition of water across the oxirane ring of *cis*-9,10-epoxystearic acid with inversion of configuration at the attacked carbon, yielding *threo*-9,10-dihydrostearic acid. The water molecule attacks both *cis*-9,10-epoxystearic enantiomers on the oxirane carbon which has the *S*-chirality, thus soybean epoxide hydrolase produces exclusively *threo*-9*R*,10*R*-dihydrostearic acid, a naturally occurring metabolite in plants (6). Moreover, this enzyme is highly enantioselective. It hydrates preferentially 9*R*,10*S*-epoxystearic acid and 9*R*,10*S* and 12*R*,13*S*-epoxyoctadecenoic acid (7), which are precisely the major metabolites formed by epoxidation of oleic and linoleic acids by soybean peroxygenase. For epoxide hydrolase to show high enantioselectivity there must be a free carboxylic acid functionality on the substrate which probably influences its positioning within the active site. This selectivity is much reduced with methyl *cis*-9,10-epoxystearate.

Physiological roles of the peroxygenase pathway

The stereochemical features of the compounds formed from oleic acid by the peroxygenase pathway are summarized in Figure 1.

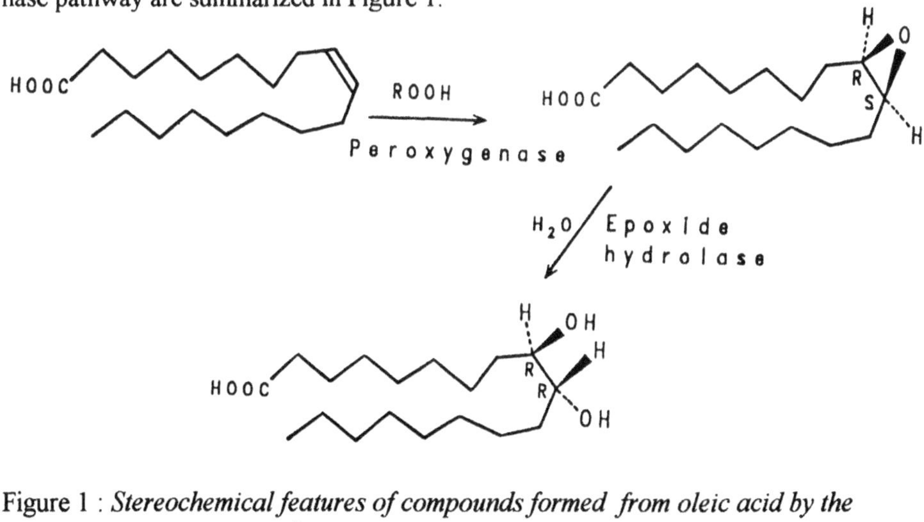

Figure 1 : *Stereochemical features of compounds formed from oleic acid by the peroxygenase pathway*

This pathway is a part of the oleic acid cascade involved in the biosynthesis of cutin monomers. Cutin is the chief component of cuticle which covers all the aerial parts of plants and constitutes the first barrier against invasion by pathogens and also limits the penetration of compounds such as pesticides and herbicides. By studying i) the substrate specificity and the stereoselectivity of the involved enzymes and ii) the transformation of oleic acid into cutin monomers by a cell free system, we recently proposed for the formation of components of cutin a new biosynthetic scheme composed by a lipoxygenase/peroxygenase/epoxide hydrolase cascade (8). When pathogens succeed to cross the cuticle, the plant copes with such aggressions by an array of defense mechanisms and specially by triggering the synthesis of toxic compounds. Oxylipins derived from linoleic acid were shown to act as self-defense compounds. They have been found by Kato and his group in rice plants infected by blast fungus and could also be isolated from rice resistant to this pathogen (9). Moreover, the epoxy derivatives were shown to induce resistance to blast infection in rice plants (10). No enzyme systems had actually been identified for the biosynthesis of such oxygenated fatty acids. But the linoleic acid cascade we have discovered in higher plants could explain the formation of all oxylipins described by Kato as antifungal-substances.

References

1. Gardner HW. Recent investigations into the lipoxygenase pathway of plants. Biochem. Biophys. Acta 1991;1084:221-239.
2. Hamberg M. Pathways in the biosynthesis of oxylipins in plants. J. Lip. Med.1993;6:375-384.
3. Blée E and Schuber F. Efficient epoxidation of unsaturated fatty acids by a hydroperoxide-dependent oxygenase. J. Biol. Chem. 1990;265:12887-12894.
4. Blée E and Schuber F. Mechanism of reaction of fatty acid hydroperoxides with soybean peroxygenase. J. Biol. Chem. 1993;268:1708-1715.
5. Blée E and Schuber F. Occurence of fatty acid epoxide hydrolases in soybean (*Glycine max.*) Purification and characterization of the soluble form. Biochem J. 1992;282:711-714.
6. Blée E and Schuber F. Regio- and enantioselectivity of soybean fatty acid epoxide hydrolase. J. Biol. Chem. 1992;267:11881-11887.
7. Blée E and Schuber F. Enantioselectivity of the hydrolysis of linoleic acid monoepoxides catalyzed by soybean fatty acid epoxide hydrolase. Biochem. Biophys. Res. Commun. 1992;187:171-177.
8. Blée E and Schuber F. Biosynthesis of cutin monomers : involvement of a lipoxygenase/peroxygenase pathway. Plant J. 1993;4:113-123.
9. Kato T, Yamaguchi Y, Namai T and Hirukawa T. Oxygenated fatty acids with anti-rice blast fungus activity in rice plants. Biosci. Biotech. Biochem. 1993; 57:283-287.
10. Namai T, Kato T, Yamaguchi Y and Hirukawa T. Anti-rice activity and resistance induction of C-18 oxygenated fatty acids. Biosci. Biotech. Biochem 1993;57:611-613.

PEROXISOMAL CATABOLISM OF LINOLEIC ACID

Bernt Gerhardt and Anneliese Kleiter
Institut für Botanik, Universität Münster, Federal Republic of Germany

Introduction

Plant peroxisomes metabolize linoleate [1]. However, complete peroxisomal degradation of linoleate to acetyl-CoA has still to be demonstrated. In addition, the pathway surmounting the 12-*cis* double bond of the parent linoleate during linoleate degradation in plant peroxisomes has yet to be established. The 12-*cis* double bond which forms a barrier to linoleate degradation by repetitive passages through the ß-oxidation reaction sequence can be surmounted by involvement of either 2,4-dienoyl-CoA reductase and Δ^2,Δ^3-enoyl-CoA isomerase at the C_{10}-intermediate level or D-3-hydroxyacyl-CoA dehydratase at the C_8-intermediate level (Fig. 1). Activity of the reductase has been demonstrated in plant peroxisomes [2, 3] and isomerase and dehydratase have been purified from cucumber cotyledons [4, 5]. The objective of this study was to examine both the completeness and the pathway of linoleate degradation in glyoxysomes from sunflower cotyledons.

Results and Discussion

COMPLETENESS OF LINOLEATE DEGRADATION

Using [U-^{14}C]linoleate as ß-oxidation substrate, > 7/9 or > 78% of the [^{14}C]acetyl-CoA which can be generated maximally from [U-^{14}C]linoleoyl-CoA must be formed in order to demonstrate complete degradation of at least part of the linoleate employed. Acetyl-CoA was well separated from butyryl-CoA by thin-layer chromatography performed according to [6]. Acetyl-CoA formation was also quantified by estimating [^{14}C]citrate formation from [^{14}C]acetyl-CoA generated at ß-oxidation. In contrast, the amount of [U-^{14}C]linoleoyl-CoA metabolized could not exactly be determined since separation of linoleoyl-CoA from some intermediary, long-chain unsaturated acyl-CoAs was not guaranteed by the thin-layer chromatographic system used. However, the radioactivity detected on the scannograms in the range corresponding to the R_f value of linoleoyl-CoA was extremly low when the acetyl-CoA or citrate formation had ceased in the reaction mixtures (after about 40 min of incubation). Calculation of the percentage of radioactivity associated with acetyl-CoA or citrate was, therefore, based on the total radioactivity of a scannogram less the radioactivity of the linoleate peak. This calculation results in an underestimation rather than an overestimation of the percentage of radioactivity associated with acetyl-CoA and citrate, respectively.
Glyoxysomes isolated from sunflower cotyledons metabolized [U-^{14}C]linoleate (1.31 nmol; 5.35 MBq μmol^{-1}) to [^{14}C]acetyl-CoA in the presence of ATP, CoASH, and NAD (for principal composition of reaction mixtures see [6]). The percentage of radioactivity detected in acetyl-CoA varied from 44 to 67% after the acetyl-CoA formation had ceased in the reaction mixture. Neither this percentage of radioactivity

J.-C. Kader and P. Mazliak (eds.), Plant Lipid Metabolism, 265–267.

nor the rate of acetyl-CoA formation or the kinetics of the peak patterns of the scannograms changed when the reaction mixture containd additionally NADPH (0.1 mM) to allow the 2,4-dienoyl-CoA reductase reaction to proceed (experiments with and without NADPH were allways performed in parallel). In the presence of an acetyl-CoA removing system consisting of added oxalacetate (3.75 mM) and the endogenous citrate synthase of the glyoxysomes, [^{14}C]citrate formation exceeded the [^{14}C]acetyl-CoA formation observed in the absence of oxalacetate. The percentage of radioactivity associated with citrate amounted to 84 to 87%. As in the case of acetyl-CoA formation the reaction characteristics of citrate formation were not influenced by NADPH. Citrate was identified by the R_f value of a reference standard after thin-layer chromatography performed as described [6] or using the solvent system ethylacetate:formic acid:water (5:2:0.5, by vol.). The percentage of radioactivity associated with the citrate spot was identical for both chromatographic systems. When reaction mixtures were treated with citrate lyase after termination of ß-oxidation, scannograms which lacked radiocactivity at the position of citrate were obtained, indicating that the radioactivity was carried by citrate.

Using linoleoyl-CoA (1.31 nmol) as substrate of ß-oxidation and following its degradation by HPLC analysis of the reaction mixture, transient accumulation of acyl-CoA intermediates was observed besides accumulation of acetyl-CoA. After 20 min of incubation the sole acyl-CoA detectable in the reaction mixture was acetyl-CoA. The results presented demonstrate that linoleate can be degraded completely to acetyl-CoA in glyoxysomes from sunflower cotyledons.

PATHWAY OF LINOLEATE DEGRADATION

When the glyoxysomal degradation of linoleyol-CoA (3 to 5 µM) in the absence and presence of NADPH (0.1 mM) was followed by HPLC analysis (according to [7] with slight modifications) of reaction mixtures, the HPLC elution profiles monitored at 260 nm were identical with respect to both the peak patterns and the kinetics of the transient accumulation of acyl-CoA intermediates. Absorbance peaks of the HPLC chromatograms were attributed to specific compounds on the basis of the retention time of reference standards prepared from commercially available acyl-CoAs by enzymatic reactions. Acyl-CoA intermediates were observed only at and below the C_{10}-intermediate level after 1 min of incubation. Occurrence of the intermediate 2-*trans*,4-*cis*-decadienoyl-CoA where linoleate degradation diagrammatically branches off into the reductase and the dehydratase pathway (Fig. 1) was indicated by monitoring the HPLC elution profiles at 260 nm and 296 nm. A single absorbance peak was observed at 296 nm where 2,4-dienoyl-CoA esters absorb, and its retention time was identical to that of one of the absorbance peaks detected at 260 nm. The retention time of this single absorption peak was 1.25 min shorter than that of the reference standard 2-*trans*,4-*trans*-decadienoyl-CoA (a generous gift from W-H Kunau, Bochum, Germany). This difference appears to be due to the replacement of a *cis* double bond by a *trans* double bond. Absorbance peaks which could be attributed to intermediates of the reductase pathway (Fig. 1) were not detected on the HPLC chromatograms. Octanoyl-CoA, 3-hydroxydecanoyl-CoA, and 2-*trans*-decenoyl-CoA which are intermediates of both the reductase pathway in linoleate degradation and the catabolism of palmitoyl-CoA accumulated transiently only at palmitoyl-CoA catabolism which proceeded at an overall rate similar to that of linoleoyl-CoA degradation. In contrast, 3-hydroxy-4-decenoyl-CoA was detected on the HPLC chromatograms when peroxisomes catabolized linoleoyl-CoA, and the transient accumulation of 3-hydroxyoctanoyl-CoA started before that of 2-octenoyl-CoA. At peroxisomal palmitoyl-CoA degradation, 2-*trans*-octenoyl-CoA and L-3-hydroxyoctanoyl-CoA appeared at once in the reaction mixture. The observation that linoleate degradation was not influenced by NADPH and the results of the HPLC analyses of reaction mixtures indicate that the linoleate degradation in glyoxysomes from sunflower cotyledons proceeds prevalently *via* the dehydratase pathway as also suggested for glyoxysomes from cucumber cotyledons [2, 5].

Acknowledgement. The work was supported by the Deutsche Forschungsgemeinschaft.

References

1. Gerhardt B. Fatty acid degradation in plants. Prog Lipid Res 1992; 31:417-446.
2. Behrends W, Thieringer R, Engeland K, Kunau W-H, Kindl H. The glyoxysomal ß-oxidation system in cucumber seedlings: identification of enzymes required for the degradation of unsaturated fatty acids. Arch Biochem Biophys 1988; 263: 170-177.
3. Berger RG, Dettweiler GR, Kollmannsberger H, Drawert F. The peroxisomal dienoyl-CoA reductase pathway in pineapple fruit. Phytochemistry 1990; 29: 2069-2073.
4. Engeland K, Kindl H. Purification and characterization of a plant peroxisomal Δ^2,Δ^3-enoyl-CoA isomerase acting on 3-*cis*-enoyl-CoA and 3-*trans*-enoyl-CoA. Eur J Biochem 1991; 196: 699-705.
5. Engeland K, Kindl H. Evidence for a peroxisomal fatty acid ß-oxidation involving D-3-hydroxyacyl-CoAs. Characterization of two forms of hydro-lyase that convert D-(-)-3-hydroxyacyl-CoA into 2-*trans*-enoyl-CoA. Eur J Biochem 1991; 200: 171-178.
6. Gerhardt B. Peroxisomes and fatty acid degradation. Methods Enzymol 1987; 148: 516-525.
7. Gerbling H, Gerhardt B. Oxidative decarboxylation of branched-chain 2-oxo fatty acids by higher plant peroxisomes. Plant Physiol 1988; 88: 13-15.

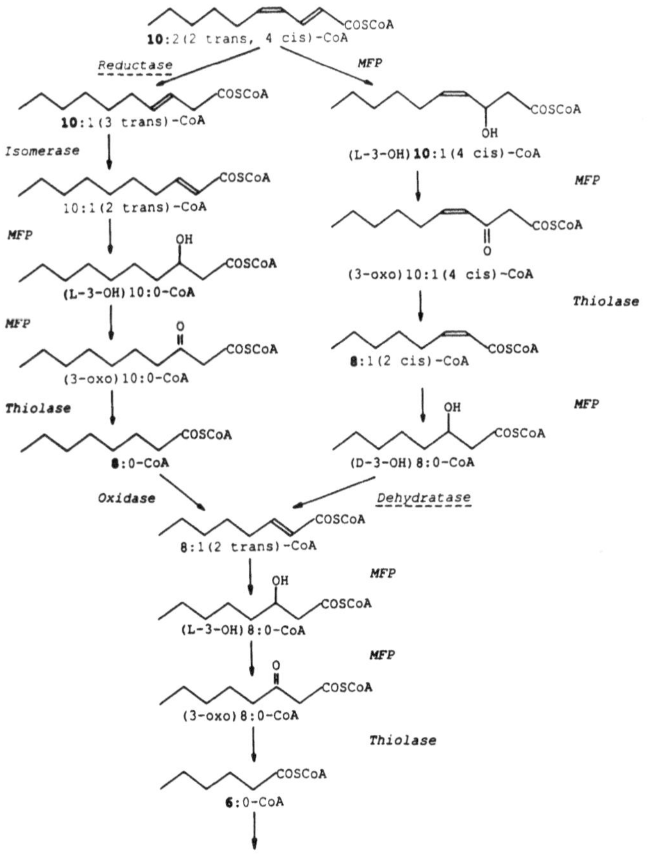

Figure 1. Degradation of 2-*trans*,4-*cis*-decadienoyl-CoA, intermediate in linoleoyl-CoA catabolism, by the alternative routes involving 2,4-dienoyl-CoA reductase and D-3-hydroxyacyl-CoA dehydratase, respectively.

FATTY ACID OXIDATION ENZYMES OF THE YEAST *CANDIDA CLOACAE*

Mark A. West, Judy Hill, Martin Watson, William Simon, Nigel Lindner, John Casey and Antoni R. Slabas.

Lipid Molecular Biology Group, Department of Biological Sciences
University of Durham, South Road, Durham, DH1 3LE, U.K.

Introduction

There is currently a large surplus of low-value plant oils on the world market and any process for converting them to high value products is of great commercial interest. Certain yeasts (e.g. *Candida* sp.) possess ω- and β-oxidation pathways to metabolise fatty acids such as those derived from plant oils [1], and mutant strains have been generated that are capable of accumulating industrially useful compounds[2].

We are interested in fatty acid oxidation by wild-type and mutant strains of the yeast *Candida cloacae* under industrial conditions. The aim is to efficiently generate novel compounds of fatty acid oxidation. To develop procedures for assaying an industrial fermentation, we have grown cultures of parent and mutant organisms under laboratory conditions and compared induction patterns over time of several enzymes of fatty acid oxidation. We have purified alcohol oxidase, an enzyme in the ω-oxidation pathway. The protein is of M_r 70 000 and oxidises long-chain fatty alcohols. Antibodies and amino acid sequence are being obtained to allow us to isolate the gene for this enzyme.

Both strains possess ω-oxidation enzymes. However, the β-oxidation system is uninduceable in the mutant, resulting in low enzyme levels compared to the parent and explaining why the mutant but not the parent accumulates dicarboxylic acids from ω-oxidation. Purification of the alcohol oxidase has revealed that the *C. cloacae* enzyme is similar to the enzyme from *C. tropicalis* [3,4]

Results and Discussion

INITIAL WORK

Methods were determined for the preparation of cell extracts and membranes. Assays were set up for ω-oxidation enzymes (cytochrome P-450 hydroxylase, cytochrome P-450 reductase, alcohol oxidase, aldehyde dehydrogenase) and β-oxidation enzymes (acyl CoA oxidase, hydratase). Conditions were established for cell growth, and induction of fatty acid oxidising enzymes, under laboratory conditions using *C. cloacae* parent and mutant strains.

EXPRESSION OF FATTY ACID-OXIDISING ENZYMES

Shake flask cultures (200 ml) of *C. cloacae* parent were grown on sucrose overnight then induced with oleic acid. Samples were taken at several timepoints for enzyme assays. Results are shown in Fig. 1. A comparison of induction in parent and mutant strains showed that alcohol oxidase was induced to levels

J.-C. Kader and P. Mazliak (eds.), Plant Lipid Metabolism, 268–270.

that were 20 - 25% of the parent, whilst acyl CoA oxidase and hydratase were not induced, remaining at levels less than 5% of the parent. The parent strain is capable of utilising fatty acids as an energy source by β-oxidation in a similar way to other yeast species [5.]. The two β-oxidation enzymes assayed were induced in the parent, but not in the mutant. The mutant strain will not utilise fatty acids as a carbon source, but metabolises them via ω-oxidation to dicarboxylic acids. This is probably due to the absence or very low activity of the β-oxidation pathway.

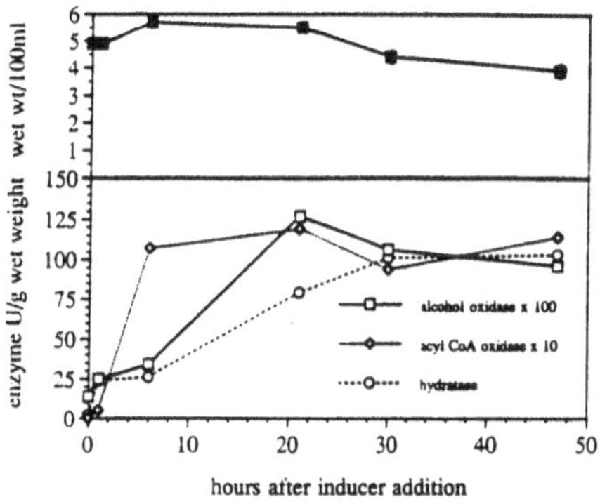

FIGURE 1: ENZYME ACTIVITIES IN OLEATE-INDUCED *C CLOACAE* (PARENT STRAIN) CULTURES

Figure 1. Cultures, 200ml, were grown on sucrose, then induced with 1% oleic acid. Samples were taken at several timepoints, cell-free extracts prepared and wet weight (-■-), alcohol oxidase (-□-), acyl CoA oxidase (-◇-) and hydratase (·○·) activities measured.

ALCOHOL OXIDASE PURIFICATION

The purification procedure is similar to one published for *C. tropicalis* [4]. A 5 l fermenter culture of *C. cloacae* parents strain induced on hexadecane was the source of starting material. Purification involved membrane pelleting by ultracentrifugation, solubilisation with sodium cholate and CHAPS, ammonium sulphate precipitation, ion exchange (QAE-cellulose) and gel permeation chromatography. The final preparation is approximately 90% pure by SDS-PAGE. The purification data are shown in Table 1.

TABLE 1: PURIFICATION OF ALCOHOL OXIDASE

Stage	total units	total protein	fold purified	recovery
cell extract	1500	4250	1.0	100
membrane pellet	1028	1500	2.0	69
solubilisation	947	1500	1.8	63
ammonium sulphate	821	280	8.3	55
ion exchange	269	22	35.0	18
gel permeation	117	-	-	7.8

There is a prominent band at 70 kDa in the most active gel permeation fraction, and activity in the other fractions was correlated with the amount of the 70 kDa band present (Fig. 2). SDS-PAGE of the active band in a native gel stained for alcohol oxidase activity gave a Coumassie-staining band at 70kDa. The K_m for decanol was approximately 15 μM and preliminary data gave a value between 5 and 15 μM for dodecanol. The pH optimum is very broad, but maximal activity occurs between pH 8.5 and 9.5.

Several attempts have been made to gain amino acid sequence of the alcohol oxidase. Material from gel permeation (90% pure) gave multiple signals. An excised 70 kDa band from a partially pure preparation electroblotted onto PVDF gave none. In-gel V8 protease digestion of the 70 kDa band gave several peptide fragments. The three bands applied to the sequencer gave no signal. Partially pure preparation has been transferred to nitrocellulose. The 70 kDa band was excised and used to raise alcohol oxidase antibodies in mice.

The alcohol oxidase purified from *C. cloacae* has many similarities (K_m, pH optimum, molecular weight, detergent solubility, purification behaviour) to the enzyme from *C. tropicalis* [4].

Figure 2A

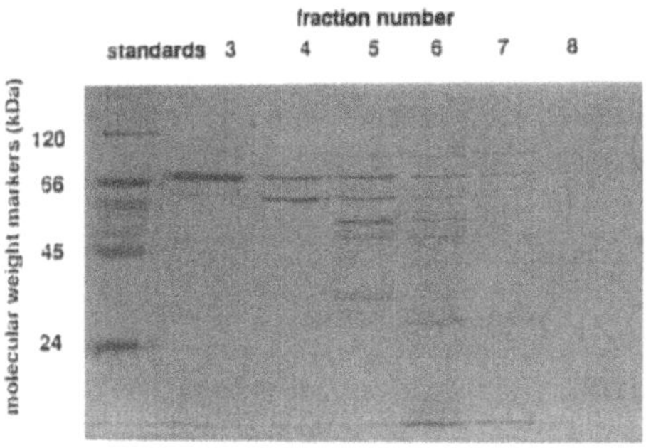

Figure 2B

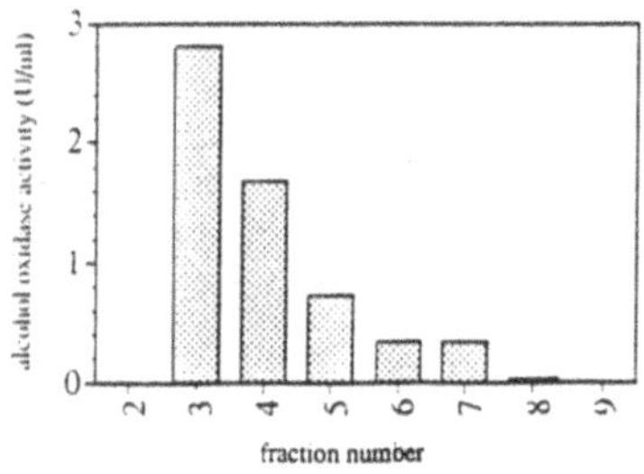

FIGURE 2: GEL PERMEATION OF ALCOHOL OXIDASE. A) SDS-PAGE of Fractions 3 - 9.
B) Alcohol oxidase activity of fractions 2 - 9.

Figure 2. Partially pure alcohol oxidase from QAE-cellulose chromatography was applied to a Superose 12 column and eluted in 200 mM-HEPES pH 8.0. Fractions were collected and either assayed for alcohol oxidase activity or precipitated and run on SDS-PAGE.

References

1. Stephan Mauersberger, Wolf-Hagen Schunck, and Hans-Georg Muller. 1984. The induction of cytochrome P-450 in the alkane-utilizing yeast *Lodderomyces elongisporus:* Alterations in the microsomal membrane fraction. Appl. Microbiol. Biotechnol 19: 29 - 35.

2. F. Meussdoerffer. 1987. Omega-Hydroxylations. Fat Sci. Technol. 89: 578 - 581.

3. Glenwyn D. Kemp, F. Mark Dickinson and Colin Ratledge. 1988. Inducible long chain alcohol oxidase from alkane-grown *Candida tropicalis*. Appl. Microbiol. Biotechnol. 29: 370 - 374

4. Francis M. Dickinson and Catherine Wadforth. 1992. Purification and some properties of alcohol oxidase from alkane-grown *Candida tropicalis*. Biochem. J. 282: 325 - 331.

5. Susumu Kawamoto, Chikateru Nozaki, Atsuo Tanaka and Saburo Fukui. 1978. Fatty acid β-oxidation system in microbodies of *n*-alkane-grown *Candida tropicalis*. Eur. J. Biochem. 83:609 - 613.

α-OXIDATION OF FATTY ACIDS IN CUCUMBER

Grethe Iren Andersen, Pernille Baardseth, Erik Slinde, Jon Bremer[1] and Astrid Nilsson,
MATFORSK - Norwegian Food Research Institute, Oslovn. 1, N-1430 Ås, Norway.
[1]Institute of Medical Biochemistry, University of Oslo, N-0317 Oslo, Norway.

Introduction

A major pathway for fatty acid oxidation in several plants is α-oxidation. In animals α-oxidation is an alternative pathway to β-oxidation and the activity is low. The localization and mechanism of α-oxidation is not fully understood either in plants or in animals. An α-oxidation system in plants has been suggested [1] (Fig. 1), the first step involving decarboxylation of a long-chain fatty acid (C_n) to a long-chain aldehyde (C_{n-1}). This aldehyde is rapidly oxidized by a NAD^+ dependent long-chain aldehyde dehydrogenase to the corresponding free fatty acid (C_{n-1}) which can then be recycled into the sequence for further α-oxidation. Earlier we observed high α-oxidation activity in cucumber, the activity being located in particles sedimenting with a $\bar{s}$-value of 10 500 ± 850 S [2].
The 3-thia fatty acid TTA and the 4-thia fatty acid TTP (Fig. 1) are metabolized by animals [3,4]. In rats β-oxidation and ω-oxidation of 4-thia fatty acids have been observed [3]. The main metabolic pathway for 3-thia fatty acids is ω-oxidation since β-oxidation is blocked [4,5]. Another possibility is α-oxidation, but this pathway was found to be negligible in rats [4].

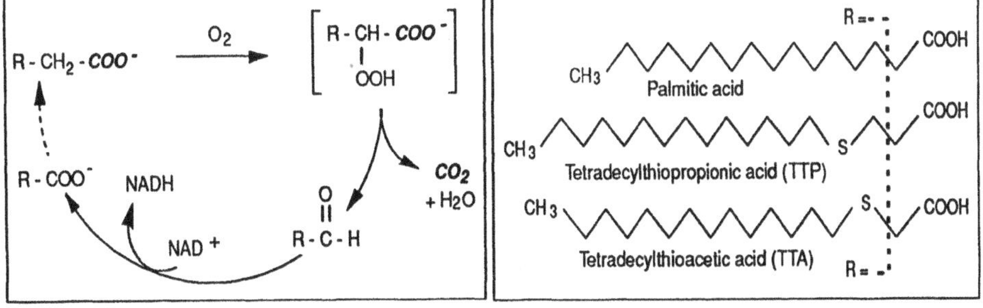

Fig. 1. Proposed mechanism [1] and substrates for α-oxidation in plants.

J.-C. Kader and P. Mazliak (eds.), Plant Lipid Metabolism, 271–273.

Aim

The aim of the present study has been to purify and characterize the α-oxidation enzyme system in cucumber using fatty acid analogues with a sulphur atom at different positions in the carbon chain as substrates.

Experimental

α-Oxidation activity was determined by release of $^{14}CO_2$ from [1-^{14}C]palmitic acid. The incubation mixture (2 ml) contained: Buffer (0.25 M sucrose, 50 mM HEPES, 1 mM EDTA, pH 7.2), protein (0.08 - 0.1 mg) and substrate (100 nmol [1-^{14}C]palmitic acid (0.03 μCi) in buffer with 0.1 % (w/v) Triton X-100). Incubation was performed in closed vessels for 20 min at 25 °C. The reaction was terminated by adding 0.2 ml 2.5 M H_2SO_4 and $^{14}CO_2$ trapped in a hanging well with 0.2 ml β-phenylethylamine in methanol (1:1 v/v). Cucumbers were homogenized in ice-cold buffer (0.25 M sucrose, 50 mM HEPES, 1 mM EDTA, pH 7.2) with a vertical cutter. After filtration (nylon filter), the homogenate was fractionated by differential centrifugation into a nuclear (N, 1000 g_{max}), a mitochondrial (M, 12000 g_{max}), a microsomal (P, 100000 g_{max}) and a soluble (S) fraction. High α-oxidative activity was observed in the M- and P-fractions (Fig. 2).

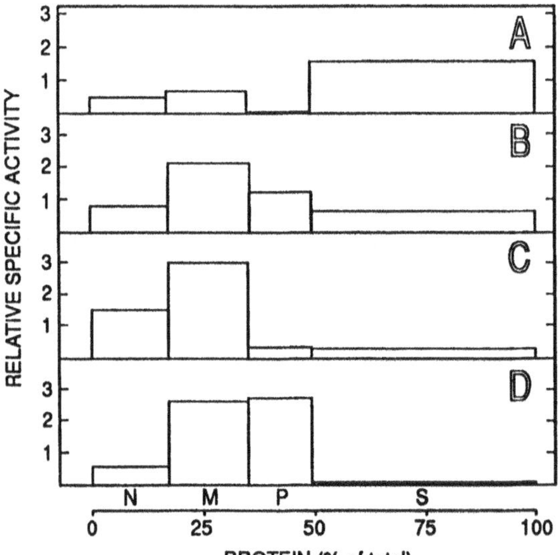

Fig. 2. Subcellular distribution of α-oxidation activity. Fractions were isolated by differential centrifugation of cucumber homogenate. Recoveries were between 71 and 84 % for the marker enzymes (A) malate dehydrogenase, (B) succinate cyt c reductase, (C) catalase, and 94 % for the (D) α-oxidation activity.

The M-fraction was dissolved in buffer (0.25 M sucrose, 50 mM HEPES, 1 mM EDTA, pH 7.2) with 0.1 % (w/v) Triton X-100. After stirring for 30 min on ice, the solution was centrifuged at 120000 g_{max} for 45 min. The proteins in the supernatant were precipitated using $(NH_4)_2SO_4$. The fraction precipitating between 45 and 60 % $(NH_4)_2SO_4$ saturation contained about 9 % of the protein and 23 % of the α-oxidation activity. It was desalted and subjected to FPLC. The anion exchange column (Mono Q HR 5/5, Pharmacia) was equilibrated with buffer (50 mM HEPES, 0.05 % (w/v) Triton X-100, pH 7.5). The proteins were eluted by a linear KCl-gradient from 0-450 mM in 20 min. The flow rate was 1.0

ml/min and 2.0 ml fractions were collected. In the peak fraction, 7 % of the total protein applied and 120 % of α-oxidation activity were eluted. This step gave a 5-fold purification. A continuous decrease in total activity was found during purification. Time, as well as freezing, thawing and desalting, decreased the specific activity more than 50 % (Table 1).

TABLE I
Purification scheme for α-oxidation activity

Purification step	Total protein	Specific activity	Yield	Purification
	mg	*nmol min $^{-1}$ (mg protein)$^{-1}$*	*%*	*-fold*
1. M-fraction	93.2	2.2	100	1
2. Triton X-100 solubilized fraction	52.0	3.1	79	1.4
3. 0-45 % $(NH_4)_2SO_4$ supernatant	19.5	4.0	38	1.8
4. 45-60 % $(NH_4)_2SO_4$ pellet	4.5	8.5	19	3.9
5. Peak fraction from anion exchange column (Mono Q)	0.3	45.1	9.9	20.5

A particulate nature of the α-oxidative activity was observed (Fig. 2), and the enzyme was purified 20 fold (Table 1). The partly purified enzyme α-oxidized both 3-thia and 4-thia fatty acids. TTP and palmitic acid were α-oxidized at equal rates, TTA at a 50 % lower rate. Tetradecylthioacetal and pentadecanal were observed as products from α-oxidation of TTP and palmitic acid, respectively. No release of $^{14}CO_2$ was observed from [2-^{14}C]-TTA which may be a suitable substrate for further studies of the α-oxidation intermediates.

References

1. Shine WE, Stumpf PK. Fat metabolism in higher plants. Arch Biochem Biophys 1974;162:147-157.
2. Baardseth P, Slinde E, Thomassen M. Studies on fatty acid α-oxidation in cucumber. Biochim Biophys Acta 1987;922:170-176.
3. Hvattum E, Skrede S, Bremer J, Solbakken M. The metabolism of tetradecylthio-propionic acid, a 4-thia stearic acid, in the rat. Biochem J 1992;286:879-887.
4. Bergseth S, Bremer J. Alkylthioacetic acids are metabolized and excreted as shortened dicarboxylic acids in vivo. Biochim Biophys Acta 1990;1044:237-242.
5. Lau SM, Brantley RK, Thorpe C. Studies with thiaoctanoyl-CoA and oxaoctanoyl-CoA analogues. Biochemistry 1988;27:5089-5095.

PHYSIOLOGICALLY ACTIVE PRODUCTS OF PLANT LIPOXYGENASE PATHWAY

Alexander Grechkin
Institute of Biology, Russian Academy of Sciences, P.O.Box 30, Kazan, 420503, Russia

Introduction

Unlike studies on animal eicosanoids, involving large forces of biologists and specialists in chemistry and medicine, research of the plant lipoxygenase (LOX) pathway until the recent time was developed only by few small groups. Thus, the progress in field of plant LOX pathway was much slower. Now, in spite of these circumstances, probably, the most important routes of LOX pathway in angiosperms are discovered. At the same time, the physiological importance of octadecanoids seems to be largely incomplete.

The aim of present paper is to review briefly the main routes of the LOX pathway in flowering plants and the existing data on the physiological importance of octadecanoids.

The main routs of lipoxygenase pathway in plants

PATHWAYS OF METABOLISM UNDER THE DIRECT LIPOXYGENASE CONTROL

The single and double dioxygenations of polyenoic fatty acids. The LOX activity is commonly abundant in higher plant species. Usual substrates of LOX are methylene-interrupted polyenoic fatty acids, in plants presumably linoleic and linolenic acids. Most of LOX act regio- and stereospecifically [1,2]. Like soybean LOX, many of plant enzymes convert linoleic and linolenic acids into corresponding 13(*S*)-hydroperoxides. At the same time, many plant tissues, in particular, potato tubers and tomatoes, possess the activity of 9-LOX. Such enzymes are able to form double hydroperoxide of α-linolenic acid via intermediary formation of 9(*S*)-HPOT [3]. The double dioxygenation of linolenate occurs in potato tubers [3], rice and wheat seeds and in other tissues, exhibiting the 9-LOX activity.

Abbreviations: LOX, lipoxygenase; HPL, hydroperoxide lyase; H(P)OD, hydro(pero)-xyoctadecadienoic acid; H(P)OT, hydro(pero)xyoctadecatrienoic acid.

J.-C. Kader and P. Mazliak (eds.), Plant Lipid Metabolism, 274–279.

The oxygenation of the allylic ketones. Direct evidence for such type of lipoxygenase reaction was obtained in 1991 independently by ourselves [4] and by German group [5]. The last one incubated an allylic ketone, prepared by oxidation of ricinoleic acid, with soybean LOX. They have identified an α,β-unsaturated γ-diketone as the predominant product. Studying the oxidation of α-ketols, we have suggested, that the primary products of their LOX oxidation are the corresponding hydroperoxides [4].

Oxygenations are not the sole type of LOX reaction. A number of different free radical rearrangements occur under oxygen deficiency. Such anaerobic LOX reactions afford epoxyalcohols and conjugated oxodienes [2].

THE METABOLISM OF FATTY ACID HYDROPEROXIDES

Hydroperoxide lyase route. Hydroperoxide lyase (HPL) belongs to the most abundant enzymes of fatty acid hydroperoxide metabolism in higher plants. HPLs from different sources convert either 9-HPOD and 9-HPOT into C_9-aldehydes and C_9-aldoacids, or 13-HPOD and 13-HPOT into C_6-aldehydes and C_{12}-aldoacids [1,2].

Allene oxide synthase route. Allene oxide synthase dehydrates hydroperoxides, affording allene oxide fatty acids. Allene oxides are rapidly hydrolyzed into ketols [2,6]. Allene oxide, corresponding to 13-HPOT, along with hydrolysis undergoes cyclization, affording cyclopentenone 12-oxo-10,15(*Z*)-phytodienoic acid. The last upon its reduction and β-oxidation is converted into jasmonic acid:

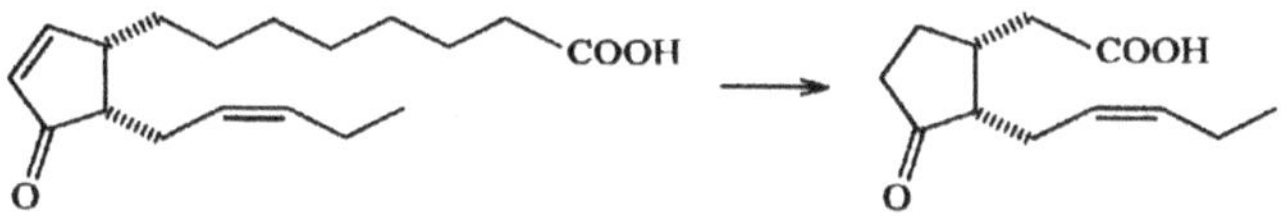

Hydroperoxide-dependent peroxygenase (epoxygenase). This enzyme, detected in soybeans [7] and broad beans [8], controls the fatty acid hydroperoxide-dependent epoxidation of unsaturated fatty acids. Hydroperoxide molecule plays a role of oxygen donor for epoxidation. This pathway was proposed as the source of natural epoxides of linoleate, vernolic and coronaric acids, occurring in different amounts in many seed oils.

The physiological importance of octadecanoids

As seen from the above discussion, the lipoxygenase pathway is the source of a large diversity of oxylipins (octadecanoids). The most important known physiological functions of octadecanoids in flowering plants are briefly described below.

OCTADECANOIDS AS STRESS RESPONSE COMPOUNDS

Octadecanoids and wound reparation in plants. The natural product 2(*E*)-dodecen-1,12-dioic acid, named "traumatic acid", as discovered as the plant wound hormone [1]. As shown later, traumatic acid is the product of non-enzymatic oxidation of 12-oxo-10(*E*)-dodecenoic acid ("traumatin"), which in its turn is formed from 12-oxo-9(*Z*)-dodecenoic acid, a product of HPL enzyme [1]. Like traumatic acid, traumatin possesses the growth stimulating activity. One more related compound, 12-hydroxy-9(*Z*)-dodecenoic acid, the product of NADPH-dependent reduction of 9(*Z*)-aldehyde [10], stimulates the increase of soybean callus culture much stronger, than traumatic acid.

Wounding of plant tissue induces the large increase of LOX activity [2]. Leaves of many plants are rich in 13-LOX and HPL activities. Thus, wounding of leaves leads to rapid formation of traumatin and related compounds (as mentioned above) and C_6-aldehydes. The fates of hydroperoxides in other tissues can be different. For example, in young potato sprouts, having the 9-LOX activity, the wounding causes the formation of 9-hydroperoxides and 9,16-dihydroperoxide of linolenic acid [11]. As shown previously, the wounding of young corn leaves leads to rapid increase of endogenous levels of α-ketol and 12-oxo-10,15(*Z*)-phytodienoic acid [12].

Octadecanoids as self defensive compounds of plants. Recently Kato et al. [13] have found, that the LOX activity in tomato leaves and some other tested plants is rapidly and dramatically increased upon tissue inoculation with pathogens. Testing of linolenic acid and some products of its oxidation revealed exceptional antimicrobial activity of 13(*S*)-HPOT [13], while linolenate itself had no effect. Several isomers of hydroxy and epoxy derivatives of linoleate and linolenate inhibited spore germination and germ tube elongation of rice blast fungus *Pyricularia oryzae* at concentrations 10-45 ppm [14]. The largest effect was displayed by 9-HOT [14]. Previously Kato and his collaborators have identified coronaric and vernolic acids [15], as well as some epoxyalcohols [16] and trihydroxyacids [17] as the self defensive compounds against rice blast disease. Another Japanese group has identified several isomeric hydroxy acids (18:1 and 18:2) as endogenous antifungal components of timothy plants [18]. Taro tubers contain an trihydroxyoctadecenoic acid a defensive substance against the pathogenic fungus *Ceratocytis fimbriata* [19].

The above examples suggest, that many C_{18}-oxylipins are the potent natural fungicides. The existing data show, that aldehydes, forming by HPL enzyme, namely, hexanal and 2(*E*)-hexenal, are involved into the plant antimicrobial defense [1;2,20]. The results of recent work [20] show, that 2(*E*)-hexenal is produced as a defensive substance during the hypersensitive resistance response in *Phaseolus vulgaris* L., infected by pathogenic bacterium *Pseudomonas syringae* [20].

THE ROLE OF SOME LOX PRODUCTS IN PLANT RESISTANCE TO INSECTS.

The antifeedant lactone from alfalfa. Alfalfa species *Medicago rugosa* Desr., resistant to an important pest (the alfalfa weevil), was found to contain the lactone 9(*Z*)-dodecen-12-olide [21]:

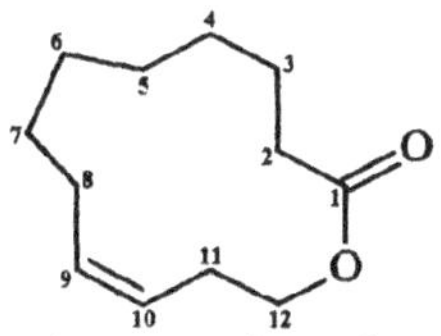

This compound, isolated after the screening of antifeedant activity, is responsible for insect resistance [21]. It is of interest, that this lactone corresponds to the above mentioned 12-hydroxy-9(*Z*)-dodecenoic acid, one of the products of HPL activity.

Pyrethroids, the natural insecticides. The insecticides "pyrethroids", naturally occurring in *Pyrethrum* species, are the esters in which the acidic part is the cyclopentenone fatty acid, structurally similar to jasmonate:

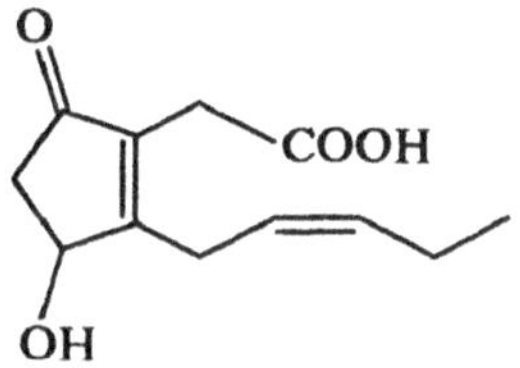

This structural similarity allows one to propose, that the acidic residue of pyrethroids is formed as result of LOX metabolism of linolenic acid. Pyrethroids possess the remarkable toxicity for insects.

OCTADECANOIDS AND PLANT DIFFERENTIATION.

Traumatic acid was proposed in some early works as "florigen", the plant flowering hormone. The new literature data revealed the complex picture of flowering induction. At the same time, if not the sole, traumatic acid is one of the flowering initiation factors [1].

Recently the jasmonic acid derivative, 12-(5'-β-D-glucopyranosyloxy)-jasmonate:

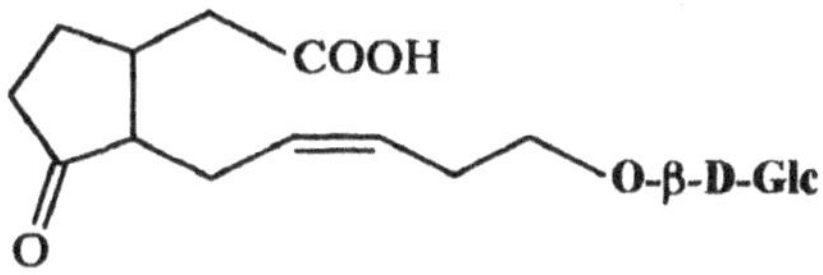

was identified as endogenous component of potato leaves, exhibiting tuber-inducing activity [22]. The aglycon part of this compound was named "tuberonic acid" [22].

THE LOX PRODUCTS AS SIGNALLING MOLECULES IN PLANTS.

Jasmonic acid was suggested to participate in cell signalling in plants. Jasmonate induces the expression of two proteinase inhibitor genes *pin* 1 and *pin* 2 [23]. Abscisic acid has the same effect. Recent data, obtained in our department, show, that this effect is not specific. Succinate and salicylate, as well as jasmonate, induce the same expression. According to the recent data of Farmer et al. [24], metabolic precursors of jasmonate, namely, 13(*S*)-HPOT and 12-oxo-10,15-phytodienoic acid, exhibit the same gene expression, like jasmonate.

LOX PRODUCTS AS ALLELOCHEMICALS.

Some cyclic oxylipins were shown to be involved into the interplant allelopathic relationships between aquatic plants [25]. For example, *Eleocharis microcarpa* produces two compounds, inhibiting the growth of blue-green algae [25].

Recently the allelopathic function was ascribed to methyl jasmonate [23]. Farmer and Ryan [23] have shown, that the vapours of methyl jasmonate, breathing into leaves, induces the formation of proteinase inhibitors.

Studies of last few years have led to important progress in understanding of physiological importance of plant octadecanoids. However, it seems likely, that this important field of research is still very promising and many new regulatory properties of plant octadecanoids are to be discovered.

References

1. Vick BA, Zimmerman DC. Oxydative systems for modification of fatty acids: The lipoxygenase pathway. In: Stumpf PK, Conn EE, editors. The Biochemistry of Plants: A Comprehensive Treatise. Vol. 9. New York: Academic Press, 1987: 53-90.
2. Gardner HW. Recent investigations into the lipoxygenase pathway of plants. Biochim. Biophys. Acta 1991; 1084: 221-39.
3. Grechkin AN, Kuramshin RA, Safonova E et al. Double hydroperoxidation of α-linolenic acid by potato tuber lipoxygenase. Biochim. Biophys. Acta 1991; 1081: 79-84.
4. Grechkin AN, Kuramshin RA, Latypov SK, Safonova EY, Gafarova TE, Ilyasov AV. Hydroperoxides of α-ketols. Novel products of the plant lipoxygenase pathway. Eur. J. Biochem. 1991; 199: 451-7.
5. Kühn H, Eggert L, Zabolotsky OA, Myagkova GI, Schewe T. Keto fatty acids not containing doubly allylic methylenes are lipoxygenase substrates. Biochemistry 1991; 30: 10269-73.
6. Hamberg M, Gardner HW. Oxylipin pathway to jasmonates: biochemistry and biological significance. Biochim. Biophys. Acta 1993; 1165: 1-18.
7. Blée E, Schuber F. Efficient epoxidation of unsaturated fatty acids by a hydroperoxide-dependent oxygenase. J. Biol. Chem. 1990; 265: 12887-94.
8. Hamberg M, Hamberg G. Hydroperoxide-dependent epoxidation of unsaturated fatty acids in the broad bean (*Vicia faba* L.). Arch. Biochem. Biophys. 1990; 283: 409-16.
9. Vick BA. Metabolism of fatty acid hydroperoxides by sunflower plants. Plant Physiol. 1989; 89 (Supplement): 132.
10. Grechkin AN, Korolev OS, Kuramshin RA, Yefremov YJ, Musin RZ, Ilyasov AV, Latypov SK, Tarchevsky IA. Dokl. AN SSSR 1987; 297: 1257-60.

11. Kuramshin RA. The Lipoxygenase Pathway of Polyenoic Fatty Acid Metabolism in Higher Plants (Doctoral Thesis). Kazan, 1991: 23 pp.
12. Vick BA, Zimmerman DC Levels of oxygenated fatty acids in young corn and sunflower plants. Plant Physiol. 1982; 69: 1103-8.
13. Kato T, Maeda Y, Hirukawa T, Namai T, Yoshioka N. Lipoxygenase activity increments in infected tomato leaves and oxidation product of linoleic acid by its *in vitro* enzyme reaction. Biosci. Biotech. Biochem. 1992; 56: 373-5.
14. Namai T, Kato T, Yamaguchi Y, Hirukawa T. Anti-rice blast activity and resistance induction of C-18 oxygenated fatty acids. Biosci. Biotech. Biochem. 1993; 57: 611-3.
15. Kato T, Yamaguchi T, Uyehara T, Yokoyama T, Namai T, Yamanaka S. Self-defensive substances in rice plant against rice blast disease. Tetrahedron Lett. 1983; 24: 4715-8.
16. Kato T, Yamaguchi Y, Ohnuma S et al. Structural elucidation of 11-hydroxy-12,13-epoxyoctadeca-(9Z,15*Z*)-dienoic acids from rice plants suffering from rice blast disease. J. Chem. Soc., Chem. Commun. 1986; 743-4.
17. Kato T, Yamaguchi Y, Abe N et al. Structure and synthesis of unsaturated trihydroxy C_{18} fatty acids in rice plants suffering from rice blast disease. Tetrahedron Lett. 1985; 26: 2357-2360.
18. Koshino H, Togiya S, Yoshihara T, Sakamura S. Four fungitoxic unsaturated fatty acids from stromata of *Epichloe typhina*. Tetrahedron Lett. 1987; 28: 73-6.
19. Masui H, Kondo T, Kojima M. An antifungal compound, 9,12,13-trihydroxy-(*E*)-10-octadecenoic acid, from *Colocasia antiquorum* inoculated with *Ceratocystis fimbriata*. Phytochemistry 1989; 28: 2613-5.
20. Croft KPC, Juttner F, Slusarenko AJ. Volatile products of the lipoxygenase pathway evolved from *Phaseolus vulgaris* (L.) leaves inoculated with *Pseudomonas syringae* pv *phaseolicola*. Plant Physiol. 1993; 101: 13-24.
21. Doss RP, Gould SJ, Johnson KJR, Flath RA, Kohnert RL. (*Z*)-Oxacyclotridec-10-en-2-one, an alfalfa weevil feeding deterrent from *Medicago rugosa*. Phytochemistry 1989; 28: 3311-5.
22. Koda Y, Kikuta Y, Tazaki H, Tsujino Y, Sakamura S, Yoshihara T. Potato tuber-inducing activities of jasmonic acid and related compounds. Phytochemistry 1991; 30: 1435-8.
23. Farmer EE, Ryan CA. Interplant communication: Airborne methyl jasmonate induced synthesis of proteinase inhibitors in plant leaves. Proc. Nat. Acad. Sci. USA 1990; 87: 7713-6.
24. Farmer EE, Ryan CA. Octadecanoid precursors of jasmonic acid activate the synthesis of wound-inducible proteinase inhibitors. Plant Cell 1992; 4: 129-34.
25. Van Aller RT, Pessoney GF, Rogers VA, Watkins EJ, Leggett HG. Ogygenated fatty acids: allelochemicals from aquatic plants. In: Thompson AC, editor. The Chemistry of Allelopathy: Biochemical Interactions among Plants. Washington: Amer. Chem. Soc., 1985: 387-400.

TEMPORAL AND ORGAN-SPECIFIC EXPRESSION OF ENZYMES OF FATTY ACID HYDROPEROXIDE METABOLISM IN DEVELOPING SUNFLOWER SEEDLINGS

Brady A. Vick
Northern Crop Science Laboratory
U.S. Department of Agriculture, Agricultural Research Service
Fargo, ND 58105

Introduction

Lipoxygenase is an enzyme of both plant and animal metabolism that catalyzes the addition of molecular oxygen to certain polyunsaturated fatty acids, resulting in the formation of fatty acid hydroperoxides. In plants these compounds can be further converted to other fatty acid metabolites by several enzymatic pathways, which vary according to species. However, there are two major pathways of fatty acid hydroperoxide metabolism that are present in most plant species [1]. One pathway, the hydroperoxide dehydrase pathway, leads to the biosynthesis of jasmonic acid, which is known to have growth regulatory properties and probably serves as a chemical signal in plant defense reactions. The second pathway, the hydroperoxide lyase pathway, leads to the biosynthesis of short-chain aldehyde products that are known to be toxic to certain fungal, bacterial, or insect pests. This report examines the level of expression of these two enzymes in sunflower leaves, cotyledons, hypocotyl, and roots during the first twenty days after germination. The results suggest that there are distinct organ-specific functional roles for each of the two enzymes. Dynamic alterations were observed in the expression of the two enzymes in the cotyledons during sunflower seedling development. In addition, evidence for the presence of two organ-specific isozymes of hydroperoxide lyase is presented.

Methods

Sunflower (*Helianthus annuus*) seeds (Interstate 7111) were imbibed in moist paper toweling for 3 days, planted in a sand/vermiculite mixture, and then grown in a chamber at 25 C with a 16L-8D photoperiod. They were harvested at various developmental stages and separated into roots, hypocotyl, cotyledons, and leaves. The tissues were homogenized with a Polytron homogenizer in an extraction buffer (1:3, w/v) consisting of 0.1 M K-phosphate buffer pH 6.0, 3 mM dithiothreitol, and 0.5% Triton X-100R, then filtered through cheesecloth, and centrifuged at 12,000*g* for 10 min. The supernatant was used as the enzyme source. Enzymes were partially purified on a DEAE Sepharose CL-6B column (2.6 cm x 30 cm), which was eluted with 0.1 M K-phosphate buffer containing 0.1% Triton X-100R and 3 mM dithiothreitol. Enzyme molecular weights were estimated by gel filtration chromatography on a Sephacryl S-300 column (2.6 cm x 36 cm) with the same elution buffer. Enzyme activity was measured either by hydroperoxide decomposition detected at 234 nm, or by NADH oxidation measured at 340 nm in a coupled enzyme assay [2]. The products of the hydroperoxide lyase reaction were characterized by gas chromatography-mass spectrometry (GC-MS) of their methyl ester/methyloxime derivatives.

J.-C. Kader and P. Mazliak (eds.), Plant Lipid Metabolism, 280–282.

Results

Effect of Detergent on Hydroperoxide Lyase Solubilization. Triton X-100R was added in varying concentrations from 0 to 1.0% to the sunflower leaf extraction buffer. The optimum concentration of Triton X-100R for the extraction of hydroperoxide lyase from sunflower leaf tissue was 0.5%, which resulted in more than a three-fold increase in activity compared to extraction without the detergent. Extraction efficiency declined at concentrations above 0.5%. Hydroperoxide lyase activity in sunflower roots was much lower than in leaves. Furthermore, the extraction efficiency of hydroperoxide lyase from sunflower root tissue was not affected by detergent. The results indicate that sunflower roots and leaves each have a different isozyme of hydroperoxide lyase. The leaf enzyme is highly hydrophobic, while the root enzyme has low hydrophobicity.

Product Characterization. The products of sunflower leaf hydroperoxide lyase are *cis*-3-hexenal and 12-oxo-*cis*-9-dodecenoic acid. Analysis of the methoxylamine derivative of the volatile C_6-aldehyde product by GC-MS showed that it had a mass spectrum and GC retention time identical to that of the methyloxime derivative of commercial *cis*-3-hexenal, with a small molecular ion at *m/z* 127 and a large M-29 peak at *m/z* 98. In contrast, the methyloxime derivative of commercial *trans*-2-hexenal had a large molecular ion at *m/z* 127 and a smaller M-29 peak (*m/z* 98), and it eluted later on the chromatogram. The oxoacid product, 12-oxo-*cis*-9-dodecenoic acid, was characterized as its methyloxime, methyl ester derivative. The compound had a molecular ion at *m/z* 255, with a larger M-31 fragment at *m/z* 224. Its GC retention time was earlier than that of a methoxylamine, methyl ester derivative of 12-oxo-*trans*-10-dodecenoic acid standard, which had an M-31 fragment (*m/z* 224) smaller than the molecular ion at *m/z* 255.

Distribution of Enzyme Activity among Sunflower Organs. At 7 days after germination the roots, hypocotyl, and cotyledons all possessed dehydrase and lyase activities (Fig. 1). The roots had an approximately equal distribution of the two enzymes, but dehydrase accounted for more than 90% of the activity in the hypocotyl and cotyledons. At 7 days leaves had not yet formed. By 13 days after germination, the dehydrase activity decreased sharply in both the hypocotyl and cotyledons, and disappeared in the roots. Hydroperoxide lyase activity increased only slightly in the three organs after 13 days, but assumed a larger proportion of the total hydroperoxide decomposition activity due to the large decrease in dehydrase activity. Sunflower leaves possessed lyase activity, but not dehydrase. Ion exchange chromatography of cotyledon extracts at various stages of development also showed the shift from high hydroperoxide dehydrase in young cotyledons to predominantly hydroperoxide lyase in older cotyledons (Fig. 2).

Substrate Specificity. Hydroperoxide dehydrase from sunflower cotyledons was about four-fold more active with 13-hydroperoxylinoleic acid than 13-hydroperoxylinolenic acid. In contrast, leaf hydroperoxide lyase was more than 10-fold more active with 13-hydroperoxylinolenic acid than with 13-hydroperoxylinoleic acid. However, root hydroperoxide lyase showed no preference between the two substrates, again suggesting that roots have a different form of the enzyme than leaves. Neither 9-hydroperoxylinoleic nor 9-hydroperoxylinolenic acids were suitable substrates for any of the hydroperoxide metabolizing enzymes in sunflower.

Molecular Weight Estimation. The molecular weights of the hydroperoxide lyase enzymes from leaves and roots were different. Root lyase was 122,000 Da, whereas the leaf lyase was larger at 178,000 Da. Hydroperoxide dehydrase from sunflower cotyledons had a molecular weight of 196,000 Da. Molecular weight standards were thyroglobulin (669,000), apoferritin (440,000), ß-amylase (200,000), alcohol dehydrogenase (150,000), bovine serum albumin (66,000), ovalbumin (43,000), and chymotrypsinogen (25,000).

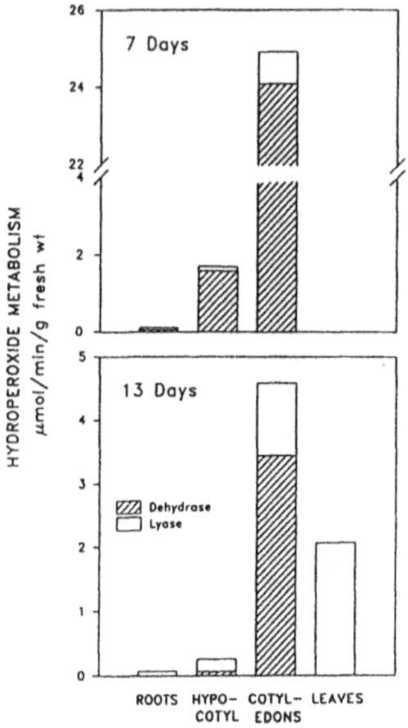

Figure 1. Distribution of hydroperoxide dehydrase and hydroperoxide lyase among organs in sunflower seedlings.

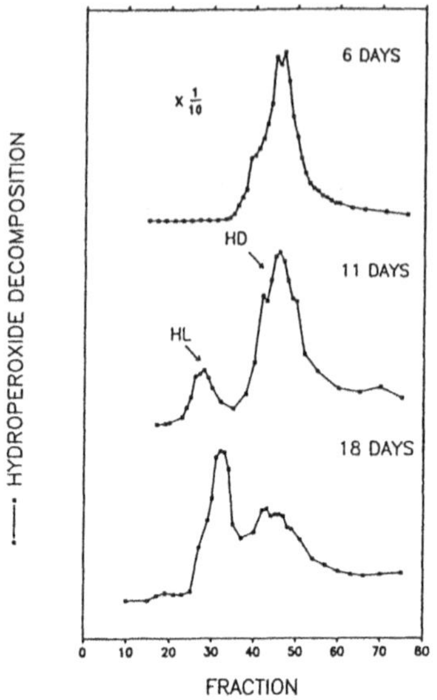

Figure 2. Ion exchange separation of sunflower cotyledon hydroperoxide dehydrase (HD) and hydroperoxide lyase (HL) at various plant ages.

Discussion

This study demonstrated that fatty acid hydroperoxides have multiple functions within the plant. A striking increase and decrease in the activity of hydroperoxide dehydrase, a jasmonic acid pathway enzyme, occurred in the cotyledons during the first three weeks of growth. The high activity of this enzyme suggests a potential role for jasmonic acid during early seedling growth. Later, as the role of the cotyledon changed from a lipid storage organ to a photosynthetic organ, hydroperoxide lyase activity became the predominant enzyme of fatty acid hydroperoxide metabolism. Lyase activity was very high in leaves, but dehydrase was not detectable there. It is possible that the high activity of lyase masked any dehydrase present. Lyase activity was also present in the roots, but its total activity was less than 1% of that in the leaves. Roots and leaves clearly had different isozymes of hydroperoxide lyase. The root enzyme was less hydrophobic, was lower in molecular weight than the leaf enzyme, and had different substrate specificity. Since hydroperoxide lyase is thought to have a defensive role in plant metabolism, the presence of organ-specific isozymes with differing substrate specificities suggests that hexenal may be a principal defensive metabolite in leaves, while both hexenal and hexanal may be important in root defense.

References

1. Vick, B. Oxygenated fatty acids of the lipoxygenase pathway. In: Moore TS, Jr, editor. Lipid metabolism in plants. Boca Raton: CRC Press, Inc., 1993: 167-191.

2. Vick, B. A spectrophotometric assay for hydroperoxide lyase. Lipids 1991;26:315-320.

FORMATION OF KETOLS OF 9 - HYDROXYLINOLENIC ACID IN CORN SEEDS

Tatiana E. Gafarova, Dmitry M. Tyshkivsky
Institute of Biology, Russian Academy of Sciences, P.O.Box 30, Kazan, 420503, Russia

Introduction

The most recent findings relating to lipoxygenase pathway in plants concern allene oxides, the primary products of enzymatic transformations of fatty acid hydroperoxides. Spontaneous hydrolysis of allene oxides leads to the formation of α- and γ-ketols [1]. It has been also found that hydroperoxide dehydrase from flax seeds converts 8-hydroxy-15*S*-hydroperoxy derivatives of arachidonic and eicosapentaenoic acids into α-ketols and prostaglandin A3 analogues [2]. These data allow us to hypothesize formation of ketols of 9-hydroxy derivative of α-linolenic acid.

Recent work has revealed that lipoxygenases from some plant tissues oxidize α-ketols of α-linolenic acid to hydroperoxides of α-ketols [3]. This paper deals with the formation of similar products by the other way: by hydroperoxide dehydration of double dioxygenation products. Incubation of 9-hydroxy-16-hydroperoxy-10(*E*),12(*Z*),14(*E*)-[1-^{14}C]octadecatrienoic acid with enzyme preparation from corn seeds led to the formation of three polar metabolites. After RP- and SP-HPLC purifications two of these polar metabolites were identified by UV spectroscopy and electron-impact mass spectrometry as α- and γ-ketols of 9-hydroxy derivative of α-linolenic acid: 9,16-dihydroxy-15-oxo-10(*E*),12(*Z*)-octadecadienoic acid and 9,12-dihydroxy-15-oxo-10(*E*),13(*E*)-octadecadienoic acid, respectively. Structure of the most polar metabolite is discussed.

Methods

Acetone powder (2 g) of corn seeds (*Zea mays* L.) was prepared and extracted as described previously [3]. The crude extract was immediately incubated with 32 KBq 9-hydroxy-16-hydroperoxy- 10(*E*),12(*Z*),14(*E*)-[1-^{14}C]octadecatrienoic acid (1.3 MBq/mmol) at 22° C for 1 h. Substrate was prepared by incubation of potato (*Solanum tuberosum* L.) tuber homogenate with 74 KBq [1-^{14}C]linolenic acid (1.3 MBq/mmol) as described before [4]. The products were extracted by ethyl acetate and analyzed by RP- and SP- HPLC as described previously [3]. Radioactivity in the eluate was detected by HPLC radioisotope detector Beckman 171 (Beckman, USA). Ultraviolet spectra (190-370 nm) were recorded in line during HPLC analysis using RSD 2140 diode-array detector (LKB, Bromma, Sweden). High-resolution electron-impact mass

J.-C. Kader and P. Mazliak (eds.), Plant Lipid Metabolism, 283–285.

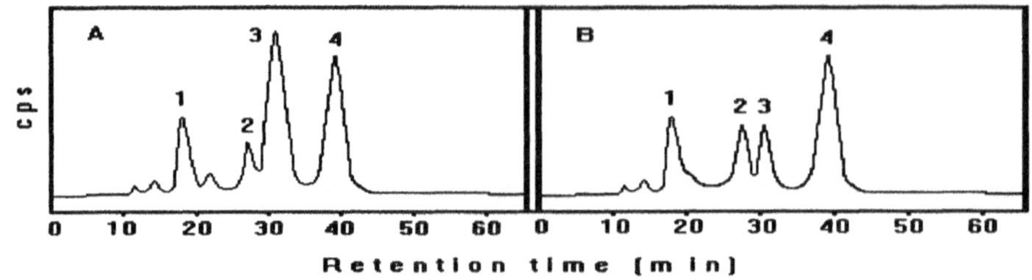

Figure 1. The reversed-phase radio-HPLC analysis of the products from 9-hydroxy-16-hydroperoxy-10(*E*),12(*Z*),14(*E*)-[1-^{14}C]octadecatrienoic acid incubation with the hydroperoxide dehydrase preparation from corn seeds. Incubation was performed at pH 7.5 (**A**) or 6.3 (**B**). Peak **4**, 9-hydroxy-16-hydroperoxy-10(*E*),12(*Z*),14(*E*)-octadecatrienoic acid; peaks **1,2,3**, metabolites 1,2,3, respectively.

spectra (70 eV) of purified metabolites were recorded on MKh-1310 spectrometer and Hitachi M-80 instrument using direct sample introduction.

Results and discussion

In our recent work [4] formation of three polar metabolites from 9-hydroxy-16-hydroperoxy-10(*E*),12(*Z*),14(*E*)-octadecatrienoic acid in corn enzyme preparation has been described. The structure of most prominent metabolite was identified by ultraviolet spectroscopy and electron-impact mass spectrometry as 15,16-α-ketol of 9-hydroxy derivative: 9,16-dihydroxy-15-oxo-10(*E*),12(*Z*)-octadecadienoic acid.

The amount of this metabolite, retention time 31 min (compound 3, fig.1**A**) accounted for approximately 40% of total recovered radioactivity at pH 7.5. When incubation medium pH was 6.3 the amount of compound 2, retention time 27 min. (fig. 1**B**) was the same as the yield of the main product (about 10%). The change of direction of the substrate metabolism corresponds to different formation of α- and γ-ketols under weak acidic and neutral pH of incubation medium. The similar stimulation of γ-ketol formation under weak acidic pH values have been observed in the studying of metabolism of α-linolenic acid hydroperoxides by hydroperoxide dehydrases from different plant species [3]. The UV spectrum of metabolite 2 exhibits a typical oxoene chromophore, λ_{max} at 227 nm. The data obtained allow us to suggest that compound 2 is γ-ketol of 9-hydroxy derivative, 9,12-dihydroxy-15-oxo-10(*E*),13(*E*)-octadecadienoic acid.

Incubation of 8(*S*)-hydroxy-15(*S*)-hydroperoxy derivative of eicosapentaenoic acid with enzyme preparation from flax seeds has also led to formation of ketols. Reaction of the (8*S*)-hydroxy-(15*S*)-hydroperoxy-eicosapentaenoic acid also gave four prostaglandin A3 analogues which differed in the configuration of the side chains [2]. Basing on the wavelength of absorption maximum and the shape of spectrum derivative we conclude that the most polar metabolite with retention time 18 min (Fig. 1, peak **1**) is an conjugated oxoene as well as γ-ketol. The electron impact mass spectrum of metabolite 1 showed the prominent ions $[M-C_2H_5]^+$at m/z 279, $[M-C_2H_5-H_2O]^+$at m/z 261, $[M-C_4H_6O-H]^+$at m/z 237, $[M-C_6H_8O]^+$at m/z 212, $[M-C_9H_{11}O-2H]^+$at m/z 171, $[M-C_9H_{11}O-H_2O]^+$at m/z 155, $[M-C_8H_{15}O_2-H_2O+2H]^+$ at m/z

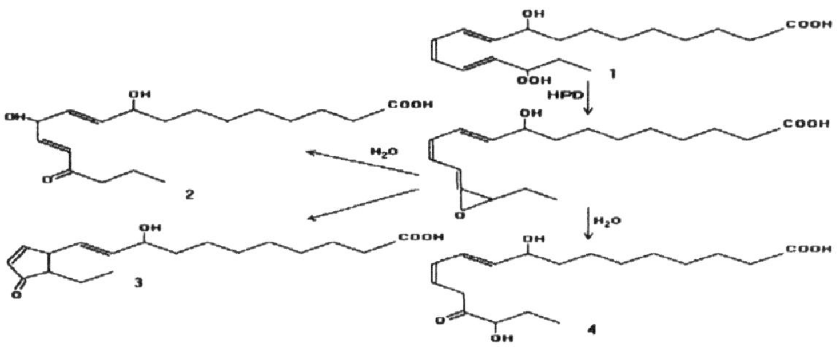

Scheme 1. The pathway of 15,16-α-ketol (**4**) and 12,15-γ-ketol (**2**) formation from 9-hydroxy-16-hydroperoxy-10(*E*),12(*Z*),14(*E*)-octadecatrienoic acid (**1**). **3,** supposed structure of metabolite 1. HPD, hydroperoxide dehydrase.

149, $[M-C_9H_{17}O_3]^+$ at m/z 135, $[M-C_8H_{15}O_2-2H_2O]^+$ at m/z 129, $[M-C_{11}H_{19}O_3]^+$ at m/z 109. The main peaks of the spectrum correspond to the cleavage of the C-C bonds next to C-9, C-11 and C-16. Metabolite 1 is believed to be 9-hydroxy-11-[2-ethyl-3-oxo-cyclopent-4-enyl]-10(*E*)-undecenoic acid.

Recent work has demonstrated that hydroperoxides of α-ketols are formed from the corresponding hydroperioxide precursors by conversion of hydroperoxides into α-ketols and following lipoxygenase oxidation of α-ketols [3]. The present work has revealed that the similar metabolites (α- and γ-ketols of hydroxy derivatives) may be formed from double dioxygenation products by dehydration of hydroperoxide groups and further hydrolysis of hydroxy allene oxides.

The obtained results allow us to conclude that corn seed hydroperoxide dehydrase catalyzes dehydration of hydroperoxide groups at C-16 position with formation of α- and γ-ketols of 9-hydroxylinolenate. We hypothesize the formation of cyclopentenone: 9-hydroxy-11-[2-ethyl-3-oxo-cyclopent-4-enyl]-10(*E*)-undecenoic acid.

References

1.Hamberg M. Mechanism of corn hydroperoxide dehydrase: detection of 12,13(*S*)-oxido-9(*Z*),11-octadecadienoic acid. Biochim Biophys Acta 1987;920:76-84.

2. Brash AR, Baertschi SW, Harris TM. Formation of prostaglandin. A analogues via an allene oxide. J Biol Chem 1990;265:6705-6712.

3. Grechkin AN, Kuramshin RA, Latypov SK, Safonova EYu, Gafarova TE, Iliasov AV. Hydroperoxides of α-ketols. Novel products of the plant lipoxygenase pathway. Eur J Biochem 1991;199:451-457.

4. Gafarova TE. Metabolism of 9-hydroxy-16-hydroperoxy-10(*E*),12(*Z*)-14(*E*)-octadecatrienoic acid in corn seeds. In: Cherif A, Ben Miled-Daoud D, Marzouk B, Smaoui A, Zarrouk M, ditors. Metabolism, structure and utilization of plant lipids. Proceedings of the Tenth International Symposium on Plant Lipids; 1992 Apr 27-May 2; Jerba. Tunisia: Centre National Pedagogique, 1992.

CLONING OF AN ELICITOR-INDUCED LIPOXYGENASE cDNA FROM TOBACCO

Christophe Véronési, Joëlle Fournier, Martina Rickauer, Marie-Laure Pouénat and Marie-Thérèse Esquerré-Tugayé.
Université Paul Sabatier, Centre de Biologie et Physiologie Végétale, URA 1457 CNRS, 118 route de Narbonne, 31062 Toulouse cedex, France.

Introduction

Plant lipoxygenases (linoleate:oxygen oxidoreductase, E.C. 1.13.11.12, **LOX**) are dioxygenases which catalyze the oxidation of polyunsaturated fatty acids (PUFAs) such as linoleic acid or linolenic acid by molecular oxygen. The hydroperoxides formed by LOX action can be converted into a series of compounds, some of them showing biological activity (*e.g.* traumatic acid, jasmonic acid) . We previously reported an increase of LOX activity in tobacco (*Nicotiana tabacum*) cell suspensions treated with an elicitor preparation from the cell walls of *Phytophthora parasitica* var. *nicotianae* (*Ppn*), the causal agent of black shank disease in this plant [1, 2]. The induced LOX is supposed to be involved in a signaling pathway leading to the production of defense proteins *via* membrane lipid metabolisation, and biosynthesis of PUFA-derived compounds such as jasmonic acid. The latter is able to induce defense gene expression in plants [3]. In order to obtain a suitable probe for studying LOX gene expression in tobacco, we prepared and screened a cDNA library of elicitor-treated tobacco cells. Here, we report the isolation and characterization of a LOX cDNA clone. It corresponds to a gene which is not expressed in tobacco cells under standard culture conditions and responds to elicitor treatment.

Results and Discussion

ISOLATION AND CHARACTERIZATION OF THE LOX cDNA CLONE TL-35.

A LOX cDNA clone, pTL-35, was isolated from a cDNA library prepared from elicitor-treated tobacco cells in λZAP-II. Screening was performed with a 555 bp Hind III fragment of a pea LOX cDNA (pCD45, [4]). The tobacco clone (1.4 kb) encodes the C-terminal moiety of the enzyme (figure 1). The amino acid sequence deduced from TL-35 includes two regions which are highly conserved among lipoxygenases, one of them being a histidine-rich stretch of 40 amino acids proposed to contain putative iron-binding residues. The TL-35 sequence shows 80% identity to a partial genomic clone of tobacco (6), 61 to 73% to those of other plant LOX (Table I) and no similarity to other proteins from plants or animals. Southern hybridization of tobacco DNA to the TL-35 probe suggests the occurrence of only 1 gene corresponding to this cDNA (data not shown). We conclude that at least two distinct genes might encode putative isoforms of LOX in tobacco. Indeed, most plants possess several LOX isoenzymes and genes [5].

J.-C. Kader and P. Mazliak (eds.), Plant Lipid Metabolism, 286–288.

A

```
TL35     1 EFPPKSKLDPKIYGNQNSTITREQIEDKLDGLTIDEAIKTNRLFILNHHDILMPYLRRIN  60
        61 TSTDTKTYASRTLLFLQDNGTLKPSAIELSLPHPDGDQFGAVSKVYTPADQGVEGSIWQL 120
       121 AKAYAAVNDSGVHQLISHWLNTHAAIEPFVIATNRQLSALHPIYKLLHPHFRETMNINAL 180
       181 ARQILINGGGLLELTVFPAKISMEMSAVVYKD WVFPEQALPTDLIKRGVAVEDSSSPLG 239
       240 IRLLIQDYPYAVDGLKIWSAIKSWVTEYCNYYYKSDDAVQKDTELQAWWKELREEGHGDK 299
       300 KDEPWWPKMQTVQELIDSCTITIWIASALHAAVNFGQYPYAGYLPNRPTLSRNFMPEPGS 359
       360 PEYEELKTNPDKVFLKTITPQLQTLLGISLIEILSRHSSDTLYLGQRESPEWTKDQEPLS 419
       420 AFARFGKKLSDIEDQIMQMNVDEKWKNRSGPVKVPYTLLFPTSEGGLTGKGI         471
```

B

```
           #    #    #                  #        #                      #
TL35  133 HQLISHWLNTHAAIEPFVIATNRQLSALHPIYKLLHPHFR 172     325 ASALHAAVNFGQYPY 339
T-L   442 ************V*********SE**VI***H******** 481     634 T************** 648
A-L   514 *******MQ***S*************V***VF***E**** 553     707 **************V 721
P-L3  517 ***V********VV*********H**C********Y**Y* 556     709 *************S* 723
S-L3  513 ***V********VV***I*****H**VV**********Y* 552     705 *************** 719
S-L1  494 ***M*********M*******H*H**V********T**Y* 533     687 *************** 700
```

Figure 1 . A) Amino acid sequence deduced from the nucleotide sequence of TL-35. B) Comparison of the TL-35 sequence with those of other plant LOXs : tobacco LOX gene,T-L [6], *Arabidopsis* LOX-1, A-L [7], pea LOX-3, P-L3 [8], soybean LOX-3, S-L3 [9], and soybean LOX-1, S-L1 [10] within two regions which are highly conserved among LOXs. # indicates putative iron- binding sites.

Table 1. Percentage of identity between the TL-35 deduced amino acid sequence and other LOX sequences.

Clone	Origin and Reference	%
genomic	tobacco [6]	80
A-L	*Arabidopsis* [7]	73
P-L3	pea [8]	64
S-L3	soybean [9]	64
S-L1	soybean [10]	62

LOX EXPRESSION IN TOBACCO CELLS AND PLANTS.

LOX transcripts accumulated following *Ppn* elicitor treatment. Using TL-35 as a probe, LOX transcripts (~3kb) were detected in tobacco cells early after addition of the elicitor to the cell suspension, they were absent in the RNA preparations obtained from untreated cells (fig.2). This induction was transient and correlated well with the stimulation of LOX activity that has been measured in elicitor-treated cells (2). Besides elicitors, LOX substrates (C18:2 and C18:3), and methyl jasmonate, have the ability to induce LOX gene expression. This might be an important regulation mechanism, greatly amplifying the initial response. When young tobacco plants were root-inoculated with zoospores from *Ppn*, LOX transcripts appeared as early as 8 hours after inoculation in roots and to a lesser extent in shoots. Induction was transient as in elicitor-treated cells.

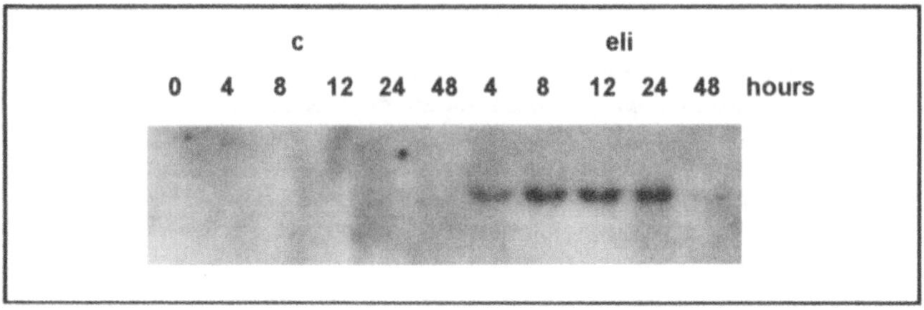

Figure 2. Northern blot analysis of LOX gene expression in tobacco cells treated or not with *Ppn* elicitors. Total RNA (15 µg) was separated on a formaldehyde agarose gel, transferred to a nitrocellulose filter and hybridized with the radiolabeled TL-35 insert.

Conclusion

We have obtained a partial cDNA clone of tobacco LOX and have studied the expression of the corresponding gene in cell cultures upon elicitor treatment. This gene is not constitutively expressed in tobacco but responds to treatment with elicitors. In our biological system, LOX induction always precedes that of defense genes, such as β-1,3-glucanase or chitinase. Our results support the hypothesis that LOX may be part of a transduction cascade leading from perception of a pathogen to defense gene expression *via* the biosynthesis of PUFA-derived compounds such as jasmonic acid.

References

1. Rickauer M, Fournier J, Pouénat ML, Berthalon E, Bottin A, Esquerré-Tugayé MT. Early changes in ethylene synthesis and lipoxygenase activity during defense induction in tobacco cells. Plant Physiol Biochem 1990;28:647-653.
2. Fournier J, Pouénat ML., Rickauer M, Rabinovitch-Chable H, Rigaud M, Esquerré-Tugayé MT. Purification and characterization of elicitor-induced lipoxygenase in tobacco cells. Plant J 1993;3:63-70.
3. Rickauer M, Bottin A, Esquerré-Tugayé MT. Regulation of proteinase inhibitor production in tobacco cells by fungal elicitors, hormonal factors and methyl jasmonate. Plant Physiol Biochem 1992;30:579-584.
4. Casey R, Domoney C, Nielsen N C. Isolation of a cDNA clone for pea (*Pisum sativum*) seed lipoxygenase. Biochem J 1985;232:79-85.
5. Siedow JN. Plant lipoxygenase : structure and fonction. Annu Rev Plant Physiol Plant Mol Biol 1991;42:145-188
6. Bookjans G , Altschuler M, Brockman J et al. Molecular biological studies of plant lipoxygenases. In: Applewhite TH editor. Proceedings of the World Conference on Biotechnology for Fat and Oil Industry. Urbana: Am Oil Chem Soc Press, 1988: 45-71.
7. Melan MA, Dong X, Endara ME et al. An *Arabidopsis thaliana* lipoxygenase gene can be induced by pathogens, abscisic acid and methyl jasmonate. Plant Physiol 1993;101:441-450.
8. Ealing PM, Casey R. The complete amino acid sequence of a pea (*Pisum sativum*) seed lipoxygenase predicted from a near full-length cDNA. Biochem J 1988;253:915-918.
9. Yenofsky RL, Fine M, Liu C. Isolation and characterization of a soybean (*Glycine max*) lipoxygenase-3 gene. Mol Gen Genet 1988;211:215-222.
10. Shibata D, Steczko J, Dixon JE, Hermodson M, Yazdanparast R, Axelrod B. Primary structure of soybean lipoxygenase-1. J Biol Chem 1987;262:10080-10085.

ENZYME ASSISTED FORMATION OF FATTY ACID HYDROPEROXIDE

George J. Piazza and Marjorie B. Medina
Eastern Regional Research Center, ARS, USDA

Introduction

Lipoxygenase (EC 1.13.11.12) catalyzes the dioxygenation of polyunsaturated fatty acids to form their corresponding hydroperoxides. Reduced derivatives of these have the potential to replace ricinoleic acid in industrial products. Currently ricinoleic acid is obtained from castor oil, a commodity that is imported into the U.S. at the level of thirty thousand metric tons per year (1). A method for immobilizing lipoxygenase was developed, and the use of this immobilized preparation to produce the hydroperoxide of linoleic acid (HPOD) in aqueous buffer-organic solvent mixtures was investigated.

Methods

Soybean lipoxygenase (Sigma, lipoxidase, type 1-B) was covalently immobilized on a commercially available carbonyl di-imidazole activated matrix (Reacti-Gel 6X, Pierce) (2). In a typical assay, 40 mg of linoleic acid was dissolved in 15.0 ml of water-saturated octane in a 125 ml glass-stoppered Erlenmeyer flask. After washing the immobilized lipoxygenase (IMM-LOX) with 0.2 M borate buffer (pH 9.0), 0.3 g of IMM-LOX containing 1.5 mg protein and 6 ml of 0.2 M borate buffer, pH 9.0, were added to the solution of linoleic acid. The reaction was allowed to proceed at 15°C while the mixture was agitated at 250 rev./min. After 1 h, the gel was filtered from the reaction mixture. The pH of the water layer was lowered to 3.0, and HPOD was extracted with diethylether. The amount of HPOD generated was determined with Xylenol Orange reagent (3). Reaction products were also examined with thin layer chromatography. Silica gel preadsorbent HPTLC-HL TLC plates (10 X 10 cm, coating thickness: 150 microns) were dipped in 5% boric acid in methanol and allowed to air dry prior to spotting. The TLC plates were developed sequentially in the following solvent systems: diethyl ether, benzene, ethanol, acetic acid (40:50:2:0.2; v/v/v/v); air drying; isooctance, diethyl ether, acetic acid (50:50:2; v/v/v) (4).

Results and Discussion

Reaction medium parameters were methodically varied to find those in which the greatest amount of HPOD was generated. It was found that a milieu containing water-saturated octane and 35% (v/v) 0.2 M borate buffer, pH 9.0, supported the maximal rate of HPOD generation. The rate of oxidation of LA catalyzed by IMM-LOX was determined when

J.-C. Kader and P. Mazliak (eds.), Plant Lipid Metabolism, 289–291.

different organic solvents were added to aqueous buffer (Figure 1). Water saturated solvents were used to ensure that the level of free water was constant

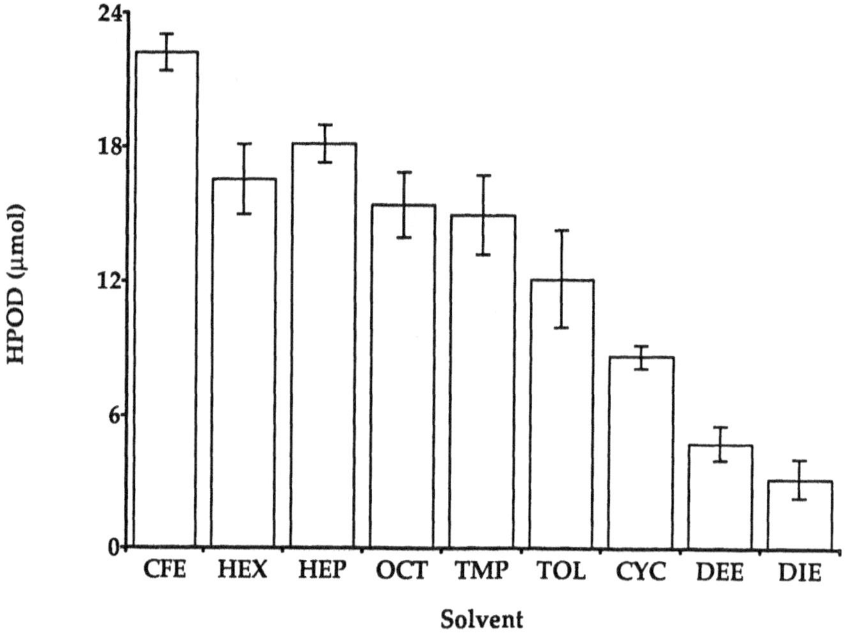

Figure 1. The influence of organic solvent upon HPOD formation catalyzed by IMM-LOX.

as the solvents were varied. The highest amounts of HPOD were obtained with 1,1,2-trichlorotrifluoroethane (CFE). Approximately equal amounts of HPOD were formed when the solvents hexane (HEX), heptane (HEP), octane (OCT) and 2,2,4-trimethylpentane (TMP) were added, but the amount of HPOD was slightly reduced from that formed in 1,1,2-trichlorotrifluoroethane. Use of toluene (TOL) and cyclohexane (CYC) resulted in a further decrease in the amount of HPOD formed. The lowest detectable levels of HPOD were found in reactions that contained diethyl ether (DEE) and diisopropyl ether (DIE). HPOD formation was not detected in reactions conducted with butanone or octanone. That the ethers and ketones inhibit the action of IMM-LOX more than the less polar solvents tested is consistent with other work that shows that generally polar organic solvents are detrimental to enzymatic activity.

The time course of HPOD formation was examined with nonlinear regression analysis over a 6 h period at 15°C using IMM-LOX that contained 3.0 mg protein. Curve fit estimates were obtained using Equation 1,

$$P_{obs} = \frac{P_{max} k_1^n t^n}{1 + k_1^n t^n} \tag{1}$$

where P_{obs} is the amount of product observed at time **t**, P_{max} is the maximum amount of product formed. k_1 is the rate of reaction at one-half P_{max}, and **n** is the cooperativity parameter that is constrained to integer values. The reciprocal of k_1 yields the time needed to generate one-half P_{max} and is therefore the reaction half time (5). That a cooperative model (n = 2) provided the best fit of the data was demonstrated by several observations. First, TLC analysis showed that a portion of LA was converted to byproducts as indicated by materials that had lower R_f values than HPOD. Thus a potential HPOD yield of about 100%, as was given when **n** was set equal to one, is not realistic, while when **n** was equal to two, the yield was 84%. Second, when the value of **n** was set equal to two, the calculated reaction half time, $t_{1/2}$, is 42 min which closely corresponds to that indicated by visual observation of TLC plates. Finally, when **n** was increased from one to two, the errors of the coefficients decreased.

Conclusion

HPOD yields of greater than 80% can be obtained in mixtures of aqueous buffer and organic solvent using IMM-LOX as the catalyst. This yield is as high as that obtained previously using pure oxygen, rather than air (6), although somewhat lower than that obtained in the presence of both organic solvent and surfactant (7). However, the present procedure avoids the difficult problem of separating the surfactant and HPOD.

References

1. Vignolo R, Naughton F. Castor: a new sense of direction. Inform 1991;2:692-99.
2. Parra-Diaz D, Brower DP, Medina MB, and Piazza GJ. A method for immobilization of lipoxygenase. Biotech and Applied Biochem 1993;18:359-67.
3. Jiang Z-Y, Woolard ACS, Wolff SP. Lipid hydroperoxide measurement by oxidation of Fe^{2+} in the presence of xylenol orange. Comparison with the TBA assay and an iodometric method. Lipids 1991;26:853-56.
4. Gravcland A. Enzymatic oxidation of linoleic acid in aqueous wheat flour suspension. Lipids 1973;8:606-11.
5. Piazza GJ, Farrell HM Jr. Generation of ricinoleic acid from castor oil using the lipase from ground oat (Avena satica L) seeds as a catalyst. Biotech Lett 1991;13: 179-84.
6. Iacazio G, Langrand G, Baratti J, Buono G, Trianlaphylidès C. Preparative enzymatic synthesis of linoleic acid (13S)-hydroperoxide using soybean lipoxygenase-1. J Org Chem 1990;55:1690-91.
7. Piazza, GJ. Lipoxygenase catalyzed hydroperoxide formation in microemulsions containing nonionic surfactant. Biotech Lett 1992;14:1153-58.

JASMONATE- AND STRESS-INDUCED LIPOXYGENASE FORMS IN BARLEY LEAF SEGMENTS (HORDEUM VULGARE CV. SALOME)

Ivo Feussner[1], Jörg Ziegler[2], Otto Miersch[2] and Claus Wasternack[2]

[1]Fb. Pharmazie, Abt. Zellphysiologie, Universität Halle, Weinbergweg 16a, D-06120 Halle, Germany; [2]Institut für Pflanzenbiochemie, Weinberg 3, D-06120 Halle, Germany

Introduction

Barley leaf segments respond to the application of (-)-jasmonic acid (JA) or its methylester (JM) as well as to several stress conditions like osmotic stress with the synthesis of so called jasmonate induced proteins (JIPs). Among these proteins is a thionin of 6 kDa, and a protein of 23 kDa with yet unknown function [1]. Beside these results we present evidence on the specific induction of lipoxygenase (LOX) forms in barley leaf segments by jasmonate or by osmotic stress conditions, like incubation with 1 M sorbitol.

Results and discussion

In order to characterize jasmonate- and stress-induced LOX forms in barley leaves three different approaches are presented. The onset of LOX synthesis was compared on mRNA-, and protein level, and for total LOX activity.
Northern-blot analysis with specific oligonucleotide primers against two LOX isoforms from barley (LOX-B - LOX-C, Fig. 1) reveals a different response of these forms for the tested incubation regimes. JIP-23 and thionin were taken as standards for JIP induction by JM. Both forms were induced by JM and sorbitol, and their transcripts could not be detected in untreated or with water incubated leaves. Whereas LOXb showed no differences between the two induction protocols, exhibited LOXc a different response to osmotic stress or incubation with JM.

Similar results were found by Western-blot analysis of total leaf extracts. Induction of specific sets of LOX forms was discovered upon JM treatment or incubation with sorbitol or water, respectively (Fig. 2). Under these conditions the induction of LOX forms on the protein level by JM is about tenfold higher than by osmotic stress. Additionally the temporal appearance of different LOX forms exhibited different maxima.

J.-C. Kader and P. Mazliak (eds.), Plant Lipid Metabolism, 292–294.

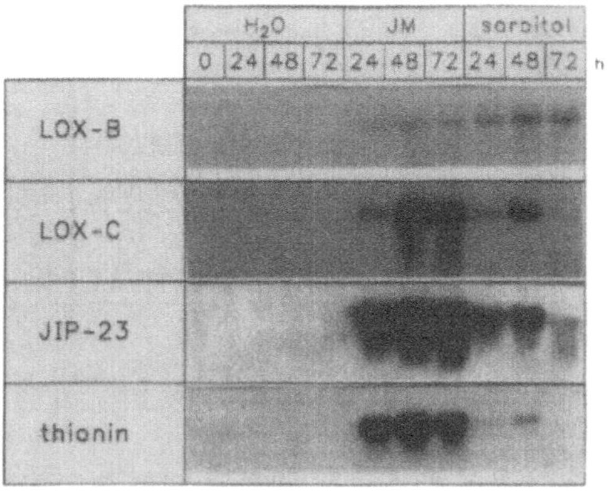

Figure 1. LOX transcripts during treatment of leaf segments with JM or sorbitol. RNA gel blots hybridized with barley cDNAs coding for LOXb and c (LOX derived oligonucleotide primers from arabidopsis [2] were used in PCR reaction with different barley genomic DNAs as templates to generate specific probes. The resulting clones b, and c correspond to LOXb and LOXc from barley; pers. communication J. Rouster), JIP-23, and thionin according to published protocols [1].

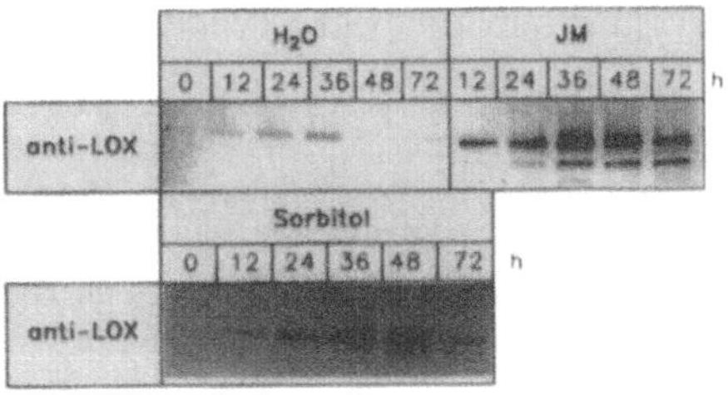

Figure 2. Immunoblot of LOX forms according to different incubation periods with water, JM or sorbitol. Samples (H_2O and sorbitol 3 μg, JM 1 μg) were submitted to SDS-PAGE and transferred to nitrocellulose filters. Blots were immundecorated with LOX antibodies against LOX from cucumber cotyledons as described in [3].

In order to relate mRNA-, and protein level of the LOX forms with the total LOX activity, the extractable enzymatic activity was measured at pH 6.5 during the incubation period in leaf segments treated with water, JM, and sorbitol. In contrast to the induced LOX forms by osmotic stress and JM at the transcriptional- and protein level, the general LOX activity was only induced by JM, whereas the LOX activity of leaves which were stressed for 72 h with sorbitol was even significantly lower than the control. To exclude the possibility of the induction of LOX forms with more acidic or basic pH-optima, the activity was also measured at pH 5.0 and in the range of pH 7.0 to 9.0. Under these conditions no significant activity was detectable.

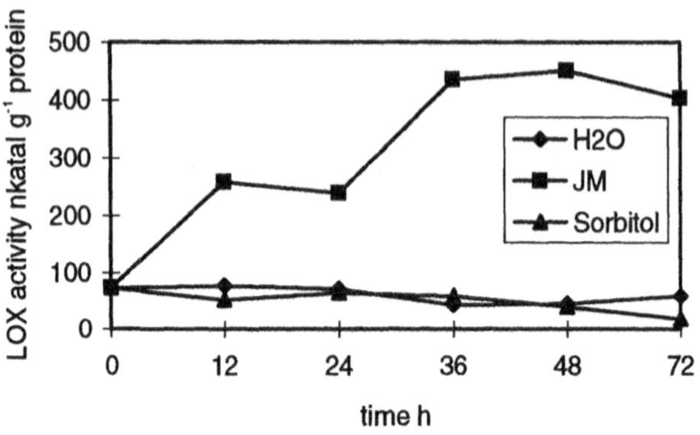

Figure 3. LOX activity in barley leaf segments after different times of incubation with water, JM or sorbitol. Extractable LOX activity was measured at pH 6.5 with linoleic acid as substrate according to [4].

The results presented here show a heterogeneous regulation of LOX forms upon mRNA- and protein level during the treatment of barley leaf segments with water, JM or sorbitol. Whereas the induction of LOX forms with JM showed remarkable effects on enzymatic activity, the induction of new LOX forms by osmotic stress is only visible at the transcriptional- and protein level. From our knowledge so far it has to be assumed, that plants are able to react on distinct environmental conditions with the expression and/or activation of specific sets of LOX forms [5-7].

References

1. Andresen I, Becker W, Schlueter K, Burges J, Parthier B, Apel K. The identification of leaf thionin as one of the main jasmonate-induced proteins of barley (Hordeum vulgare). Plant Mol Biol 1992;19:193-204.

2. Melan MA, Dong XN, Endara ME, Davis KR, Ausubel FM, Peterman TK. An arabidopsis thaliana lipoxygenase gene can be induced by pathogens, abscisic acid, and methyl jasmonate. Plant Physiol 1993;101(2):441-450.

3. Feußner I, Kindl H. Particulate and soluble lipoxygenase isoenzymes. Comparison of molecular and enzymatic properties. Planta 1994;194(1):22-28.

4. Koch E, Meier BM, Eiben HG, Slusarenko A. A lipoxygenase from leaves of tomato (Lycopersicon esculentum Mill) is induced in response to plant pathogenic pseudomonads. Plant Physiol 1992;99(2):571-576.

5. Siedow JN. Plant lipoxygenase - structure and function. Annu Rev Plant Physiol 1991;42:145-188.

6. Bell E, Mullet JE. Lipoxygenase gene expression is modulated in plants by water deficit, wounding and methyl jasmonate. Mol Gen Genet 1991;230:456-462.

7. Grimes HD, Koetje DS, Franceschi VR. Expression, Activity, and Cellular Accumulation of Methyl Jasmonate-Responsive Lipoxygenase in Soybean Seedlings. Plant Physiol 1992;100(1):433-443.

HYDROPEROXIDE DEHYDRASE IN BARLEY GRAINS

Nathalie Schmitt and Anneke C. Douma

Center for Phytotechnology RUL-TNO, department of Molecular Plant Biotechnology, Wassenaarseweg 64, 2333 AL Leiden, The Netherlands

Introduction

In barley grains, lipoxygenase has been well characterized and two isoenzymes have been purified [1-3] but the characterization of hydroperoxide-metabolizing enzymes has received less attention. Hydroperoxide-consuming enzymes can be divided into two types of enzymes. Hydroperoxide lyase cleaves fatty acid hydroperoxides into aldehydes and oxoacids, and hydroperoxide dehydrase (also named hydroperoxide isomerase) catalyzes the formation of α- and γ-ketols.

Until now, no evidence has been presented that hydroperoxide lyase occurs in barley [4], whereas hydroperoxide dehydrase activity was reported [5]. In this paper, the changes of hydroperoxide dehydrase activity were measured in different parts of the grain during germination. The results showed that 90 % of the activity is localized in the germ, principally in the shoot, and increased during germination. Partial purification of the enzyme was performed on a mono Q column. The purified fraction contained two major protein bands of 67 and 55 kDa on SDS-PAGE. The enzyme showed an optimum pH of 6.6 and a low temperature stability. The results were compared with the data previously reported [1, 7].

Materials and Methods

Barley grains (*Hordeum vulgare* L., cv. Pipkin) were germinated in the dark for 4 days on moistened filter-paper at 25°C. Barley greenmalt (cv. Pipkin) was provided by Heineken Technical Services (Zoeterwoude, The Netherlands).

Barley germs or different parts of the grain were extracted with extraction buffer (50 mM phosphate buffer, pH 7.5 containing 0.1 % Triton X-100). The ratio fresh material (g)/ extraction media volume (ml) was 1/20. After grinding with an Ultraturrax, the extracts were incubated on ice for 30 minutes and centrifuged for 30 min. at 10 000g. The supernatant corresponding to the crude enzyme extract was collected.

Hydroperoxide dehydrase activity was determined at 25°C by monitoring the decrease in absorbance at 234 nm [5] with 10 μM of 13-hydroperoxide of linoleic acid as substrate.

Hydroperoxide lyase activity was determined by a specific spectrophotometric assay [7], using 40 μM of 13-hydroperoxide of linoleic acid as substrate. The solution of 13-hydroperoxide of

J.-C. Kader and P. Mazliak (eds.), Plant Lipid Metabolism, 295–297.

linoleic acid was prepared with pure soybean lipoxygenase [8] and extracted on an octadecyl solid phase column [2].
For partial purification of hydroperoxide dehydrase, greenmalt germs (1 g) were extracted in 5 ml of extraction buffer as described above. After centrifugation (30 min. at 10 000g), the supernatant was collected, passed through a 0.22 μm filter and diluted 2 times with water. On an anion exchange column (Mono Q HR 5/5, Pharmacia) equilibrated with buffer A (10 mM phoshate buffer, pH 7.5), 10 ml of extract were loaded. Sequential elution was performed on a FPLC (Pharmacia) at 0.5 ml/min using buffer B, buffer A and Buffer C (buffer B: 1 M NaCl in buffer A, buffer C: 0.1 % Triton X-100 in buffer A).
Dehydrase activity was monitored and the fractions exhibiting a high activity were pooled and concentrated on an Amicon filter (YM 30) under low pressure at 4°C.
Protein concentration in the fractions were determined using the bicinchoninic acid method (Pierce) with BSA as standard.

Results and Discussion

The rate of hydroperoxide consumption, corresponding to lyase and dehydrase activities, was determined in different parts of the grain during germination (Fig. 1). Hydroperoxide lyase activity was measured in different parts of the grain but no activity was detected using the specific spectrophotometric assay. This indicates that hydroperoxide lyase is absent or exhibits low activity in the extracts (competition for the hydroperoxide substrate between hydroperoxide lyase and dehydrase may occur in the incubation media). Consequently, the consumption of hydroperoxides corresponds mainly to the dehydrase activity.

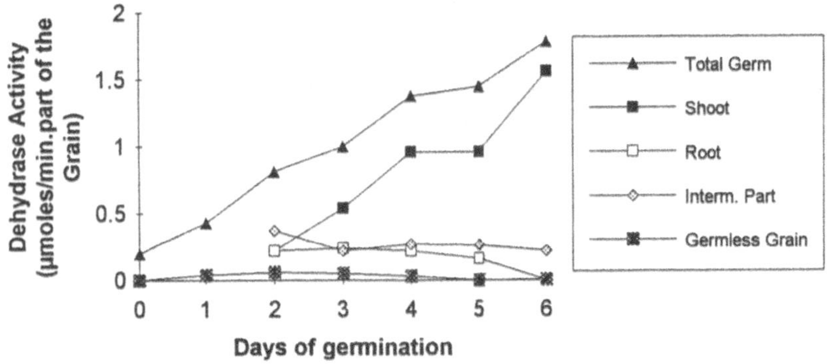

Figure 1 *Changes of hydroperoxide dehydrase activity during germination.*

More than 90 % of the total dehydrase activity of the grain is present in the germ at the first day of germination and this percentage is even higher at the later stages. Activity in the germ is detected in the quiescent grain and is increasing steadily during germination. Most of the activity present in the germ is localized in the shoot and this activity increases during germination, whereas the activity in the roots remains stable and decreases after 6 days of germination. The intermediate part corresponds to a small piece (3 mm) of tissue at the base of the roots and shoots including the scutellum. When the results are expressed on a fresh weight basis, this tissue exhibits the highest dehydrase activity.

Figure 2 shows that most of the dehydrase activity is eluted when Triton X-100 is present in the running buffer. The protocol of purification results in an average of 16-fold purification with a recovery of 50 % of the total activity. The final concentrated fraction has a specific activity of 45 μmoles/min.mg protein. SDS-PAGE characterization shows that the fraction contains two major proteins of 67 and 55 kDa (data not shown).

Previous purification [1] indicated a 118-fold purification, although the final activity of the partially purified preparation was 10 μmoles/min.mg protein.

The optimum pH of the linoleate hydroperoxide dehydrase determined with the purified fraction was 6.6 with a broad activity range (Fig. 3). At pH 4 and pH 10, respectively 30% and 18 % of the maximum activity are still present. Despite the fact that hydroperoxide dehydrase has been shown to be moderately heat stable in crude extracts of commercial malts [6], the enzyme in the purified fraction shows a low temperature stability. A decline of 50 % in activity is observed after 3 hours at 4°C (data not shown).

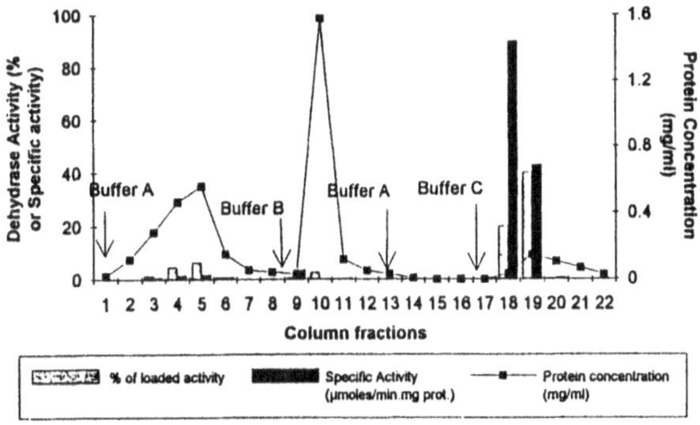

Figure 2 *Mono Q column chromatography profile*

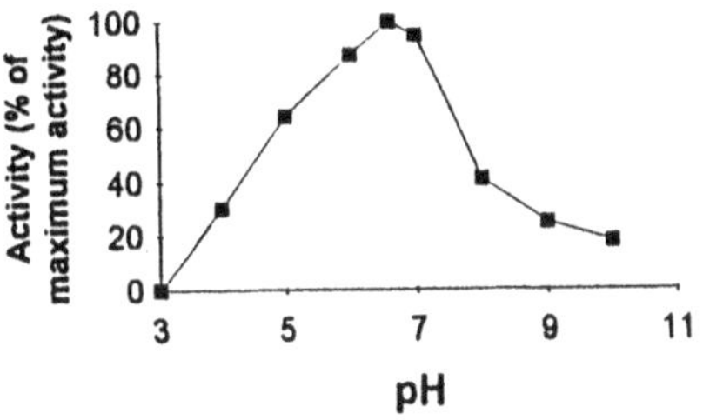

Figure 3 *Optimum pH of hydroperoxide dehydrase*

References

1. Führling D. Lipoxygenase und Linolsaurehydroperoxid-Isomerase in Gerste. Ph.D thesis, Technische Universität Berlin, 1975.
2. Van Aarle PGM, De Barse MMJ, Veldink GA, Vliegenthart JFG. Purification of a lipoxygenase from ungerminated barley. Characterization and product formation. FEBS Lett 1991;280:159-62.
3. Doderer A, Kokkelink I, van der Veen S, Valk BE, Schram AW, Douma AC. Purification and characterization of two lipoxygenase enzymes from germinating barley. BBA 1992;1120:97-104.
4. Vick BA, Zimmerman DC. Lipoxygenase and hydroperoxide lyase in germinating watermelon seedlings. Plant Physiol 1976;57:780-88.
5. Zimmerman DC, Vick BA. Hydroperoxide isomerase. A new enzyme of lipid metabolism. Plant Physiol 1979;53:1-4.
6. Schwarz PB, Pyler RE. Lipoxygenase and hydroperoxide isomerase activity of malting barley. J Am Soc Brew Chem 1984;42:47-53.
7. Vick BA. A spectrophotometric assay for hydroperoxide lyase. Lipids 1991;26:315-20.
8. Hamberg M, Samuelsson B. On the specificity of the oxygenation of unsaturated fatty acids catalyzed by soybean lipoxidase. J Biol Chem 1967;242:5329-35.

This paper is ABIN publication no 142.

ISOLATION AND CHARACTERIZATION OF A LIPASE FROM *BOTRYTIS CINEREA*

Pascal COMMENIL[1], Lionel BELINGHERI[2] and Michel SANCHOLLE[2]

1. Laboratoire de Biologie et Physiologie Végétale, Université de Reims Champagne, BP 347, 51062 Reims cedex France.
2. Laboratoire de Cryptogamie, SN2, Université des Sciences et Technologies de Lille, 59655 Villeneuve d'Ascq cedex France.

SYNOPSIS

A lipase has been purified to homogeneity from *Botrytis cinerea*. Enzyme characteristics are discussed including molecular weight, isoelectric point, substrate specificity, pH optimun and the effect of assay temperature.

INTRODUCTION

Botrytis cinerea is a fungus responsible for severe diseases on a wide range of cultivated plants in temperate regions; it is particularly a very important pathogen for the vine (*Vitis vinifera*). The outer barrier protecting plants is constituted by cuticle (epicuticular waxes and cutin). Cuticle plays an important role in the interaction between plants and fungi and the mode of penetration of this barrier by pathogenic fungi is still controversial.

Enzymatic penetration of cuticle appears involved in pathogenesis and *Botrytis cinerea* has the ability to produce a specific lipase. This enzyme, which is released into the culture medium, hydrolyzes specifically the esters of unsaturated fatty acids. The present study was conducted with the aim of purifying and characterizing the lipase from *Botrytis cinerea*.

MATERIAL AND METHODS

The organism used, *Botrytis cinerea*, has been isolated from grapevine. The cultures of *Botrytis cinerea* were made in 1 l conical flasks containing 200 ml culture medium and incubated on a shaker. *Botrytis cinerea* produces an extracellular lipase only when grown in the presence of an inductor (10 mM methyl oleate). Cultures were harvested after 7 days growth at 23°C in darkness, at which time extracellular lipase activity was maximal.

Proteins were precipitated with ammonium sulfate (51%). After dialysis, ammonium sulfate was added to make the final 1M concentration. The dialyzate was loaded on Octyl-Sepharose CL-4B column (11.5x1 cm). The lipase activity was found in the fraction eluted with Emulphogene BC-720. The passage on Bio-Beads SM-2 allowed to a partial removal of the detergent. The active fraction was applied to Concanavalin A-Sepharose column (4x1 cm). Lipase was eluted with 0.1 M methyl-a-D-mannopyranoside. The final purification step was done with a Sephacryl S-200 column (98x1.5 cm).

J.-C. Kader and P. Mazliak (eds.), Plant Lipid Metabolism, 298–300.

The lipase activity was measured by a sensitive fluorimetric assay using umbelliferyl oleate as a standard substrate (1).

SDS-PAGE was carried out according to the method of Laemmli (2). Isoelectric focusing was performed on 5% polyacrylamide gels. A vertical minigel system with 8x7 cm cell format was utilized (Mini protean, Bio-Rad).

Protein concentrations were determined according to Bradford (3) using the Bio-Rad protein assay kit.

RESULTS AND DISCUSSION

The enzyme was purified to electrophoretic homogeneity using a 4-step procedure involving ammonium sulfate fractionation followed by chromatography on Octyl-Sepharose, Concanavalin A-Sepharose and Sephacryl S-200. The chromatography on an Octyl Sepharose column shows the hydrophobic nature of this lipase. The result with the Concanavalin A-Sepharose column suggests the glycosylation of the lipase. The final preparation resulted in over a 31 fold purification with 21% field (table 1). Analysis by SDS-PAGE revealed a single polypeptide of Mw 60 kDa. The native enzyme has a Mw of 72 kDa when analyzed by gel filtration, suggesting that it is monomeric. This result is similar with most of fungal extracellular lipases which are monomers having molecular weights ranging from 30 to 64 kDa.

TABLE 1: Purification of *Botrytis cinerea* extracellular lipase

Steps	Protein (mg)	Specific activity (μmol UMB/min/mg)	Yield (%)	Purification factor
Culture filtrate	15.5	82	100	1
$(NH4)_2SO4$ precipitation	4.7	170	62.9	2.1
Octyl Sepharose	1.3	498	51	6.1
Con A Sepharose	0.36	1320	37.4	18
Sephacryl S200	0.106	2574	21.5	31.4

The isoelectric point proved to be 6.5 on isoelectric focusing (Fig.1).
The purified enzyme exhibited a pH optimun of 6.0. Maximum activity lipase occured at 38°C.

This lipase showed a high specificity for long-chain unsaturated fatty acid. The best hydrolysis rate was obtained with the oleic acid ester and the presence of double bonds enhances lipase activity.

A cutinase from *Botrytis cinerea* has been purified and characterized (4), showing properties which differ from those of this lipase. The lipase may be considered as the first step in the enzymatic breakdown of the cuticle by hydrolyzing epicuticular waxes or releasing some components of the cutin, then allowing the production of cutinase.

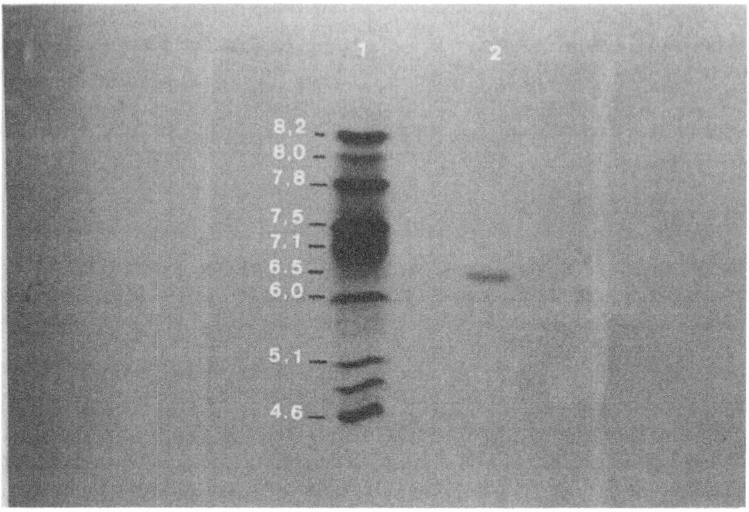

Figure 1. Isoelectric focusing (IEF) of the purified lipase from *Botrytis cinerea.* Lane 1, IEF standard protein preparation; lane 2, purified lipase. The pI values of the standard proteins are indicated in the left margin.

CONCLUSION

A study will be conducted with the aim of checking if the purified lipase is able to hydrolyze epicuticular waxes or cutin.

The purified lipase will be used to prepare antibodies and to obtain protein microsequences. This will allow for the future complete sequence determination following cDNA cloning.

REFERENCES

1. de Laborde de Monpezat T, de Jeso B, Butour JL, Chavant L, Sancholle M. A fluorimetric method for measuring activity based on umbelliferyl esters. Lipids 1990; 25:661-664.
2. Laemmli UK. Cleavage of structural proteins during assembly of the head of bacteriophage T4. Nature 1970; 227: 680-685.
3. Bradford MM. A rapid and sensitive method for the quantitation of microgram quantities of protein utilizing the principle of protein-dye binding. Anal. Biochem. 1976; 72:248-254.
4. Salinas JC. Function of cutinolytic enzymes in the infection of Gerbera flowers by *Botrytis cinerea*, Ph.D. Thesis, University of Utrecht, Netherlands, 1992.

INDUCTION OF LIPASE ACTIVITY IN GERMINATING BARLEY

Jolanda C. Heistek and Anneke C. Douma

Center for Phytotechnology RUL-TNO, department of Molecular Plant Biotechnology, Wassenaarseweg 64, 2333 AL Leiden, The Netherlands

Introduction

Barley grains contain a relatively low amount of lipid (up to 4.8 %), mainly present in the embryo and aleurone. Most of the lipids are stored as triacylglycerols. Lipase (triacylglycerol acylhydrolase [EC 3.1.1.3]) is the first enzyme involved in the degradation of stored triacylglycerols and has been studied extensively in oil-rich seeds [1] and in cereals such as oats, which contain relatively high levels of activity [2]. Less attention has been paid to wheat [3] and barley [4,5], which contain low levels of lipase activity.

In this paper the induction of lipase activity during germination of barley was studied using a titrimetric assay, with which the substrate-dependent acidification is determined. Lipase activity was already present in the quiescent barley kernel and strongly increased (up to 80-fold) during germination at 25 °C. This increase was mainly attributable to an increase in lipase activity in the embryo. Lipase activity in the embryoless grain was low and remained at a constant level.

Materials and Methods

Barley (*Hordeum vulgare* L., cv. Triumph, harvest 1988) grains were germinated between 2 x 3 layers of moistened filter-paper in the dark at 25 °C.

Flour was prepared by grinding maximally 10 g of frozen sample in an Analysenmühle (A10; IKA Labor Technik) for two times 30 sec.

The titrimetric lipase assay was performed as described by Sigma (kit no 800; St. Louis, USA) according to the method of Tietz & Fiereck [6]. Instead of 1 ml serum 0.5 g flour was added. The free fatty acids (FFA) released by lipase activity were determined by back titration against 0.05 N NaOH with an automatic titrator (702 SM Titrino; Metrohm). Controls were used to correct for endogenous acid present in the substrate emulsion and flour extracts.

FFA were quantified by means of the NEFA-C kit (Wako Chemicals Gmbh, Neuss, Germany) after chloroform-methanol extraction according to the method of Lin *et al.* [7].

The enzyme activity is expressed in units (U) per grain or part of the grain or per gram flour. One unit is defined as one micromole of FFA released per minute.

J.-C. Kader and P. Mazliak (eds.), Plant Lipid Metabolism, 301–303.

Results and Discussion

Lipase activity in germinating barley was determined using the titrimetric assay of Tietz & Fiereck [6]. With this assay the substrate-dependent acidification of the incubation medium is determined. Similar activity values were obtained in control experiments, in which the amount of FFA liberated by lipase was measured.

Lipase activity was studied in the course of germination at 25 °C. In the quiescent barley kernels the level of lipase activity was low, namely approximately 8 mU/grain. During the first two days of germination lipase activity remained unchanged, but thereafter strongly increased to a plateau value of about 650 mU/grain, which was reached after four days. A similar pattern was observed when the activity was expressed as U/g flour (Figure 1A).

The maximal lipase activity we observed was much higher than that reported by most other authors (Table 1). This may be due to differences in the assay and germination procedure.

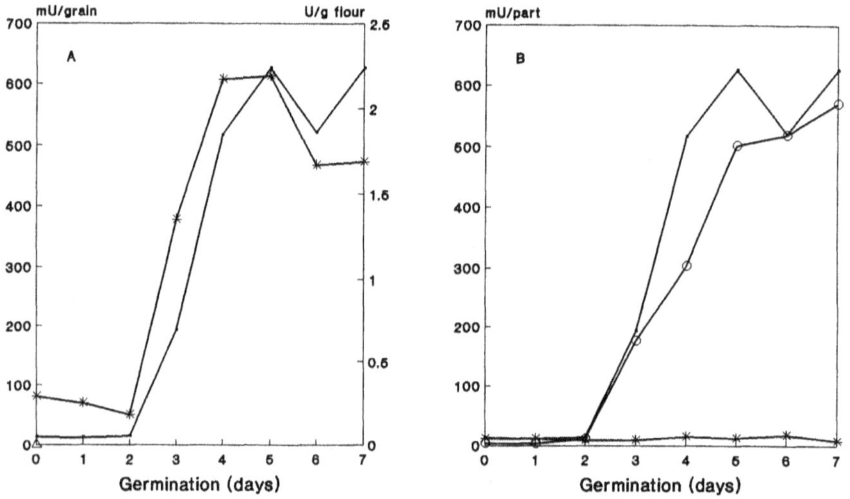

Figure 1 *Induction of lipase activity during germination of cv. Triumph. A; whole grain (——) mU/grain and (*) U/g flour. B; different parts of the grain (——) whole grain, (o) embryo and (*) embryoless grain.*

The induction of lipase activity during germination was completely attributable to the embryo (Figure 1B). In the embryoless grain lipase activity was low and remained at a constant level. The lipase activity in the embryoless grain most likely is derived from the aleurone [4]. According to Baxter [5] two different lipase isoenzymes exist in barley, one of which is present in the embryo and the other in the aleurone.

Table 1 *Literature data on the (maximal) lipase activity in the quiescent and germinated grain*

Authors	Lipase activity U.(g flour)$^{-1}$	
	Dry grain	Germinated grain
our data	0.44	2.18
MacLeod & White [4]		0.014
Narziß & Sekin [8]	0.49	0.71
Matlashewski et al. [2]	1.09	
Anness & Baxter [9]	0.0009	0.004
Baxter [5]	0.004	0.038

Currently, some relevant properties of lipase are studied, e.g. pH-optimum and temperature optimum and stability, to get more insight into the presence of different isoforms of lipase.

References

1. Huang AHC. Plant lipases. In: Borgström B, Brockman HL, editors. Lipases. Amsterdam: Elsevier, 1984: 419-42.
2. Matlashewski GJ, Urquhart AA, Sahasrabudhe MR, Altosaar I. Lipase activity in oat flour suspensions and soluble extracts. Cereal Chem 1982;59:418-22.
3. Tavener RJA, Laidman DL. The induction of lipase activity in the germinating wheat grain. Phytochem 1972;11:989-97.
4. MacLeod AM, White HB. Lipid metabolism in germinating barley II. Barley lipase. J Inst Brew 1962;68:487-95.
5. Baxter ED. Recognition of two lipases from barley and green malt. J Inst Brew 1984;90:277-81.
6. Tietz NW, Fiereck EA. A specific method for serum lipase determination. Clin Chim Acta 1966;13:352-58.
7. Lin Y-H, Wimer LT, Huang AHC. Lipase in the lipid bodies of corn scutella during seedling growth. Plant Physiol 1983;73:460-63.
8. Narziß L, Sekin Y. Über das Verhalten der Lipase während des Mälzungs- und Braupro zesses. Brauwissenschaft 1974;27:311-20.
9. Anness BJ, Baxter ED. The control of lipids and their oxidation products in malting and brewing. Proceedings of the European Brewery Convention Congress 1983;193-200.

This paper is ABIN publication no 141.

PROPERTIES OF A PURIFIED SOLUBLE MGDG-ACYL-HYDROLASE FROM COWPEA (*Vigna unguiculata* L.) LEAVES

Pham Thi Anh Thu *, Sahsah Yamina , Roy-Macauley Harold , d'Arcy-Lameta Agnès and Zuily-Fodil Yasmine .
Laboratoire de Biochimie et Physiologie de l'Adaptation Végétale, Université Paris 7 Denis Diderot, 2 Place Jussieu, F-75241 Paris Cedex 05, France.
* : CNRS, URA 1180.

Galactolipids (monogalactosyl-diacylglycerol, MGDG and digalactosyl-diacylglycerol, DGDG) are the main lipids of leaf membranes. Under environmental stresses and during senescence, degradation of membrane lipids occur, and galactolipids seem to be the preferential targets for lipolytic enzymes [1,2]. Little is known on plant galactolipases and non specific acyl-hydrolases which remove fatty acids from galactolipids, and attempts to purify them led to contradictory and non convincing results [3-8].
With the use of ^{14}C-monogalactosyl-diacylglycerol extracted from leaves as substrate for enzymatic test [9] we have purified to homogeneity a lipolytic acyl-hydrolase (LAH, EC.3.1.1.26) from soluble leaf extract of *Vigna unguiculata* L. cv. EPACE-1 [10]. The procedure involved ammonium sulfate precipitation, ion exchange chromatography (Q-Sepharose Fast Flow) , gel filtration on Sephacryl 300 HR and chromatofocusing on Mono-P, followed by a semi-preparative native electrophoresis on polyacrylamide gel. The purification scheme led to a 263 fold purification with a yield of 2.0 % . In the following text, we present some of the properties of the purified enzyme.

Molecular weight

The estimated molecular mass of the enzyme as determined by gel filtration on Sephacryl 300 HR was found to be about 80 kDa [10].This value was similar to those found by Burns et al [6] on *Phaseolus multiflorus* leaves (70-90kDa) and by Matsuda et al. [7] on leaves of *Phaseolus vulgaris* (90kDa) , but differ noticeably from those found by Helmsing [4] on *Phaseolus multiflorus* leaves (110kDa) and by Anderson et al. [5] on *Phaseolus vulgaris* chloroplasts (55kDa). SDS-polyacrylamide electrophoresis of the purified enzyme, as shown in figure 1, revealed one band of about 40000 molecular weight. These observations demonstrated that the Cowpea leaf MGDG-acylhydrolase is a dimer.

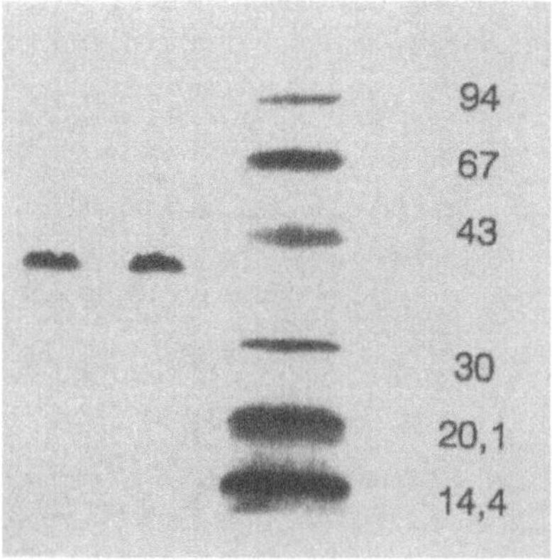

Fig.1: SDS-PAGE of the purified LAH
Lanes 1, 2 : purified enzyme
Lane 3 : Standards (kDa)

J.-C. Kader and P. Mazliak (eds.), Plant Lipid Metabolism, 304–306.

Thermal stability of the enzyme

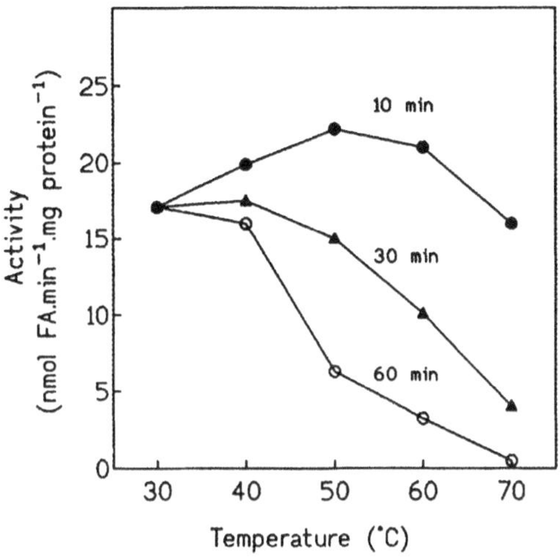

Fig.2. Effect of temperature on LAH activity

When the enzyme was heated for 10, 30, and 60 min in a waterbath at different temperatures, it exhibits a rather high resistance to heat inactivation, being maximally active after 10 minutes incubation at 50°C. (Fig. 2). At more longer incubation times and higher temperatures, however, the activity of the enzyme began to fall. Helmsing [4], Matsuda et al. [7] also found a good resistance to high temperatures of their acyl-hydrolases. Moreover, Anderson et al. [5] suggested the existence in Bean leaves, of a heat-labile inhibitor of the enzyme since they observed an increase in activity when the leaf extract was subjected to 65°C for 2 min.

Effect of substrate concentration

The curve relating the acyl-hydrolase activity and different MGDG concentrations (results not shown) indicates that the enzyme activity tended to reach a plateau beginning at 250mM MGDG. The Km value of the enzyme for this substrate was calculated to be 119μM from a Lineweaver-Burk plot (figure 3)

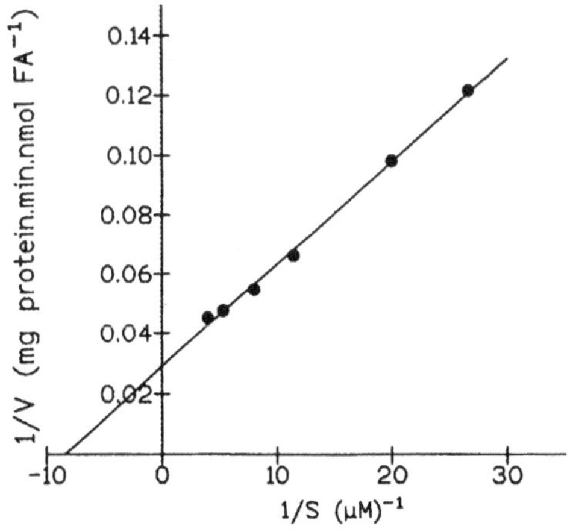

Fig.3: Lineweaver-Burk plots for the hydrolysis of MGDG by the purified LAH.

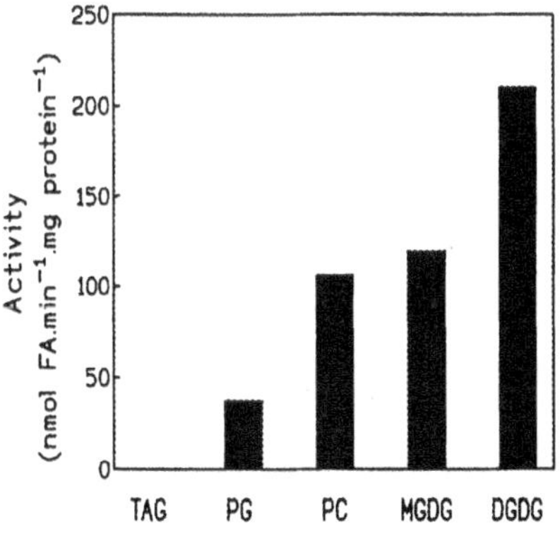

Fig.4 : LAH activity on different lipid substrates

Substrate specificity

The purified enzyme was tested for activity towards the various substrates shown in fig 4. It exhibited no activity towards TAG, on the other hand it deacylated phospholipids as well as galactolipids; it was therefore similar to the LAH first described by Galliard [11] in potato tubers. Similar results were obtained by Matsuda et al. on Bean and Potato leaves [7,8]. These authors suggest that the active site was the same for MGDG and PC .

Conclusion

In the present study, we have presented some properties of a purified MGDG-acyl-hydrolase from the particle-free extract of Cowpea leaves. However, MGDG-acyl hydrolytic activities have been found in the chloroplast membranes [5,9,12,]. It will be of great interest to precise the eventual differences or similarities between them. Further work on this problem is in progress in our laboratory.

On the other hand, the availability of the pure leaf enzyme makes possible more detailed studies on the sub-cellular localization of the enzyme and on the mechanisms of its regulation that may provide a more comprehensive understanding of the process of membrane lipid degradation in plants.

References

1. Kaniuga Z, Gemel J. Galactolipase activity and free acid level in chloroplasts. Novel approach to characteristics of chilling sensitivity of plants. F.E.B.S. Lett 1984;171:55-58.
2. Monteiro de Paula F, Pham Thi AT, Zuily-Fodil Y et al. Effects of water stress on the biosynthesis and degradation of polyunsaturated lipid molecular species in leaves of Vigna unguiculata. Plant Physiol Biochem 1993;31:707-715.
3. Sastry PS, Kates M. Hydrolysis of monogalactosyl diglycerides by specific enzymes in Runner Bean leaves. Biochemistry.1964;3:1280-1287.
4. Helmsing PJ. Purification and properties of galactolipase. Biochim Biophys Acta 1969;178:519-533.
5. Anderson MM, McCarty RE, Zimmer EA. Partial purification of a bean galactolipid lipase and its action on sub-chloroplast particles. Plant Physiol 1974;53:699-704.
6. Burns DD, Galliard T, Harwood JL. Properties of acyl hydrolase enzymes from Phaseolus multiflorus leaves. Phytochemistry 1980;19:2281-2285.
7. Matsuda H, Hirayama O. Purification and properties of a lipolytic acyl-hydrolase from Potato leaves. Biochim Biophys Acta 1979;573:155-165.
8. Matsuda H, Tanaka G, Morita K, Hirayama O. Purification of a lipolytic acyl-hydrolase from Phaseolus vulgaris leaves by affinity chromatography on palmitoylated gause and its properties. Agric Biol Chem 1979;43:563-570.
9. El-Hafid L, Pham Thi AT, Zuily-Fodil Y, Vieira da Silva J. Enzymatic breakdown of polar lipids in cotton leaves under stress; degradation of monogalactosyl-diacylglycerol. Plant Physiol Biochem 1989;27:495-502.
10. Sahsah Y, Pham Thi AT, Roy-Macauley H et al. Purification and characterization of a soluble lipolytic acyl hydrolase from Cowpea leaves. Biochim Biophys Acta 1994 (in press).
11. Galliard T. The enzymatic breakdown of lipids in potato tuber by phospholipid and galactolipid acyl-hyrolase activities and by lipoxygenase. Phytochemistry 1970;9:1725-1734.
12. O'Sullivan JN, Warwick NWM, Dalling MJ. A galactolipase activity associated with the thylakoids of wheat leaves (Triticum aestivum L.). J Plant Physiol 1987;131:393-404.

A PHOSPHOLIPASE ACTIVITY CAUSES CHANGES IN PHOSPHOLIPID COMPOSITION OF TOBACCO CELLS IN RESPONSE TO FUNGAL ELICITORS

Sébastien Roy, Marie-Laure Pouénat, Catherine Caumont, Marie-Claude Prévost and Marie-Thérèse Esquerré-Tugayé. Université Paul Sabatier, Centre de Biologie et PhysiologieVégétale, URA 1457 CNRS, 118 route de Narbonne 31062 Toulouse Cédex. France.

Introduction

The lipoxygenase (LOX) pathway of polyunsaturated fatty acids (PUFAs) metabolism is believed to be involved in the defence of plants against pathogens. It has been shown that LOX is highly induced in cultured tobacco cells treated with elicitors of *Phytophthora parasitica* var. *nicotianae*, a fungal pathogen of this plant [1]. This enzyme leads to the formation of a variety of active compounds. Among them, jasmonic acid is an efficient inducer of defense gene expression [2, 3]. Since prior release of PUFAs might be required to initiate the LOX pathway, we have looked for phospholipid alterations and phospholipase activity in elicitor-treated tobacco cells. The obtained data indicate that a fatty acid releasing-phospholipase activity could be one of the first steps of the LOX pathway in elicitor-treated tobacco cells.

Results

Tobacco suspension-cultured cells were treated with fungal elicitors as previously described [2]. Total lipids were extracted with boiling propanol-2 : chloroform [4]. The phospholipids were separated by two-dimensionnal thin-layer chromatography on silicic acid. The spots obtained were identified by comparison with standards, scrapped off the plates and quantified by assaying phosphorus according to Böttcher et al [5].

The proportion of each phospholipid was determined 24 hours after addition of elicitors to the cell suspension and in control untreated cells (Fig. 1). Phosphatidylcholine (PC) and phosphatidylethanolamine (PE) were the two major phospholipids, representing respectively 43% and 28% of the total lipidic phosphorus content of control cells. In elicitor-treated cells, these yields decreased to 34% and 21%, respectively. In parallel, there was an increase in the amount of lysoPC and lysoPE. The other phospholipids, which showed only minor changes, were not further investigated.

A time course study showed that, following an initial increase probably due to subculturing, there was always less lipidic phosphorus in elicitor-treated cells than in controls, 24 hours after the beginning of the treatment. Over three independent experiments, the observed decrease averaged -15%. The amounts of the two major phospholipids, PC and PE, were determined upon elicitor treatment. In control cells their levels increased for the first 10 hours after beginning of the experiment, *i.e.* after transfer of the cells to the fresh medium. In response to the elicitor, their levels varied moderately, as shown for PC in figure 2. These results are consistant with the net decrease observed at 24 hours. It thus appears that elicitor treatment

J.-C. Kader and P. Mazliak (eds.), Plant Lipid Metabolism, 307–309.

induces changes in the phospholipid pattern of tobacco cells. Since some of them might result from an enzymic release of fatty acids, we have looked for phospholipase activity in tobacco cells.

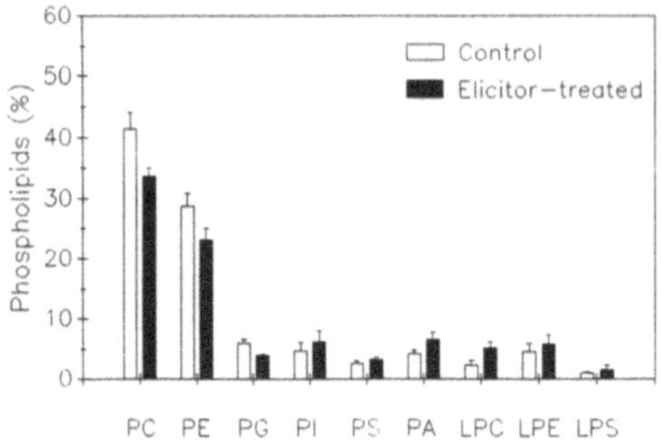

Figure 1 : Changes in phospholipids in tobacco cells in response to elicitor-treatment.

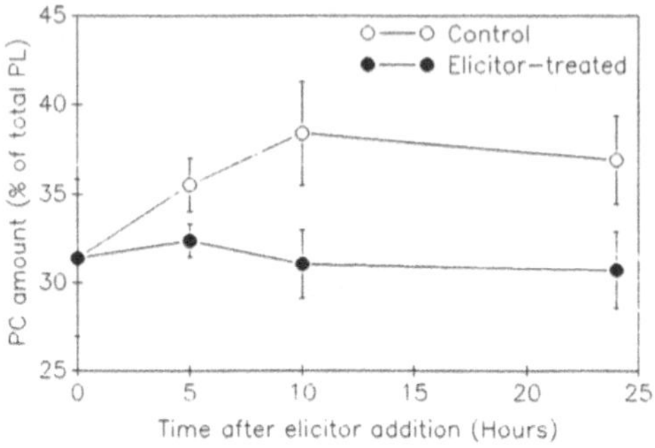

Figure 2 : Time course changes in the amounts of PC following addition of elicitor.

A fluorimetric method, employing the C6-NBD-PC fluorescent analog of PC as a substrate, was used to assay phospholipase activity [6]. The data indicate that a phospholipase activity was induced in tobacco cells in response to elicitors. A time course study performed on control or elicitor-treated tobacco cells showed that phospholipase induction was detectable 1 hour after addition of the elicitor, and quickly increased during the following 6 hours. Maximum activity was reached 24 hours after beginning of the experiment. The induction observed at 24 hours was completely prevented when cycloheximide (10µM) was added to the culture medium (data not shown).

It was found that the enzyme activity is calcium-dependent, inhibited by millimolar concentrations of $MgCl_2$ and has a pH optimum in the range 8.5-9. A partially purified enzyme preparation showed phospholipase-A1, -A2 and lysophospholipase activity *in vitro*.

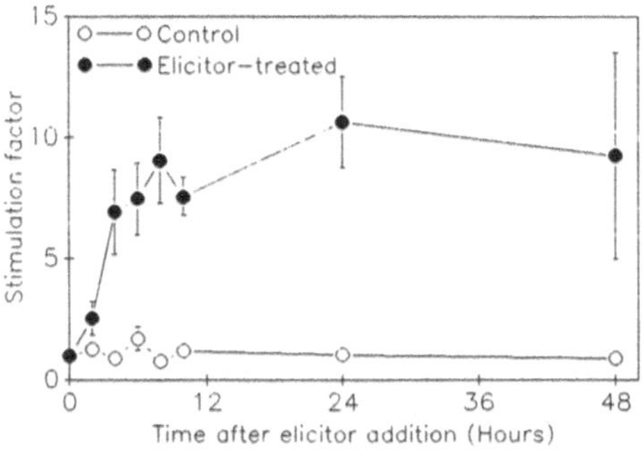

Figure 3 : Time course measurement of phospholipase activity in tobacco cells.

Discussion

This work presents evidence that changes in phospholipid metabolism occur in tobacco cells following treatment with elicitors of *Phytophthora parasitica nicotianae*, a fungal pathogen of tobacco. Over 24 hours, a net decrease of phospholipids was observed. At early stages, this effect was counterbalanced by an apparent increased synthesis of these compounds which might reflect adaptation to subculturing, since the bioassay involve previous washing and re-suspension of tobacco cells into a new medium.

The net decrease in phospholipid amounts correlated well with the elicitation of a phospholipase activity, which started to increase within the very first hours following addition of elicitors, and reached a maximum 10 to 12 hours later. The time course induction of phospholipase was consistent with a role in initiation of the lipoxygenase pathway and defence responses. Indeed, elicitation of lipoxygenase and of defence responses such as proteinase inhibitors are maximum in tobacco cells 24 and 48 hours respectively following addition of elicitors [1, 7]. Decreases of PC and PE and increases in the corresponding lysophospholipids at 24 hours suggested that a phospholipase-A_2 like activity was present in the extract.

Induction of phospholipase or LAH activity has already been shown in a number of plant-pathogen systems, notably during the incompatible interactions of bean leaves with bacteria, and of potato tubers with *Phytophhtora infestans* [8, 9]. Besides initiating the lipoxygenase pathway, these enzymes are supposed to play diverse roles in relation to defence. They might provoke extended membrane damage in the microbial partner on the one hand, and display acyl transferase activity related to reinforcement of the cuticular barrier, on the other hand [10].

The mechanisms by which phospholipases themselves are induced is not known. Activation of potato phospholipase *in vitro* requires Ca^{++} and other effectors [9]. The tobacco enzyme reported in the present study is also activated *in vitro* by calcium. Future work will be aimed at understanding the mechanisms by which this stimulation occurs, and at characterizing the phospholipase activity which is induced in tobacco.

References

1 J. Fournier, M.L. Pouénat, M. Rickauer, H. Rabinovitch-Chable, M. Rigaud and M.T. Esquerré-Tugayé, Purification and characterization of an elicitor-induced lipoxygenase in tobacco cells. Plant J., 3 (1993) 63-70.

2 M. Rickauer, A. Bottin and M.T. Esquerré-Tugayé, Regulation of proteinase inhibitor production in tobacco cells by fungal elicitors, hormonal factors and methyl jasmonate. Plant Physiol. Biochem., 30 (1992) 579-584.

3 E.E. Farmer and C.A. Ryan, Interplant communication : airborne methyl jasmonate induces synthesis of proteinase inhibitors in plant leaves. Proc. Natl. Acad. Sci., USA, 87 (1990) 7713-7716.

4 M. Kates and F.M. Eberhardt, Isolation and fractionation of leaf phosphatides. Can. J. Bot., 35 (1957) 895-905.

5 C.J.F. Böttcher, C.M. Van Gent and C. Pries, A rapid and sensitive submicro-phoshorus determination. Anal. Chim. Acta, 24 (1961) 203-204.

6 L.A. Wittenauer, K. Shirai, R.L. Jackson and J.D. Johnson, Hydrolysis of a fluorescent phospholipid substrate by phospholipase A_2 and lipoprotein lipase. Biochem. Biophys. Res. Commun., 118 (1984) 894-901.

7 M. Rickauer, J. Fournier and M.T. Esquerré-Tugayé, Induction of proteinase inhibitors in tobacco cell suspension culture by elicitors of *Phytophthora parasitica* var. *nicotianae*. Plant Physiol., 90 (1989) 1065-1070.

8 K.P.C. Croft, C.R. Voisey and A.J. Slusarenko, Mechanism of hypersensitive cell collapse : correlation of increased lipoxygenase activity with membrane damage in leaves of *Phaseolus vulgaris* (L.) inoculated with an avirulent race of *Pseudomonas syringae* pv. *phaseolicola*. Physiol. Mol. Plant Pathol., 36 (1990) 49-62.

9 K. Kawakita, K. Senda and N. Doke, Factors, affecting *in vitro* activation of potato phospholipase A_2. Plant Sc., 92 (1993) 183-190.

10 D. Racusen, Lipid acyl hydrolase of patatin. Can. J. Bot., 62 (1984) 1640-1644.

PHOSPHOLIPASE ACTIVITIES ASSOCIATED WITH THE TONOPLAST FROM *ACER PSEUDOPLATANUS* CELLS : IDENTIFICATION OF A PHOSPHOLIPASE A_1 ACTIVITY

Eric Tavernier and Alain Pugin

Laboratoire de Biochimie et Biologie Moléculaire, UFR Sciences et Techniques, 16 route de Gray, 25030 Besançon Cedex, France
UA INRA/Université 692, INRA, B.V. 1540, 21034 Dijon Cedex, France

In higher plants, the lipolytic enzymes and their physiological functions are not well characterized [1]. Most reports demonstrated that phospholipid catabolism in plants is achieved by the concerted actions of membrane-bound enzymes including phospholipase D, phosphatidate phosphatase, lipolytic acyl hydrolases and lipoxygenases [1,2]. With the exception of the phospholipase D, the literature on plant phospholipases is still very limited. We previously reported that tonoplast from *Acer pseudoplatanus* cells contains small amounts of phosphatidic acid and lysophospholipids, which were produced together with free fatty acids, particularly after addition of Ca^{2+}[3]. These data suggested the possible involvement of phospholipase D and phospholipase A in the metabolism of vacuolar membrane lipids. The phospholipase activities were studied by following the hydrolysis of added sn-2-[^{14}C]linoleyl-PC to tonoplast vesicles. Tonoplast was obtained by osmotic lysis of pure preparations of vacuoles isolated from protoplasts derived from *Acer pseudoplatanus* cells [4]. This present work demonstrated clearly the presence of phospholipase D and phospholipase A_1 activities associated with the tonoplast of *Acer*. The phospholipase A_1 showed an optimal activity at pH about 6-6.5, did not necessarily require divalent cations, but was stimulated by Mg^{2+} and particularly by Ca^{2+}. This work presents the first evidence for the presence of phospholipases A_1 in plant cells.

The phospholipase activities in tonoplast were studied by following the hydrolysis of added sn-2-[^{14}C]linoleyl-PC to tonoplast vesicles in the presence of 0.1% Triton X-100. The results presented table I indicated that 90 to 95% of radioactivity lost from PC was recovered in lyso-PC and phosphatidic acid (PA). The production of radioactive PA appeared to be Ca^{2+}-dependent and was stimulated at acidic pH as it was previously described [1,2]. The release of radioactive lyso-PC was, however, the same irrespective of the pH and the addition of EDTA or divalent ions. Almost no radioactivity was recovered in free fatty acids (FFA) in the different assays, but almost all the radioactivity was detected in FFA in the control after addition of bee venom phospholipase A_2 (Table I). The vacuolar sap did not reveal any significant soluble phospholipase activity.

J.-C. Kader and P. Mazliak (eds.), Plant Lipid Metabolism, 310–312.

Lipids	Control (t = 0)	Boiled tonoplast	Tonoplast + exogenous phospholipase A_2	Tonoplast + 10 mM EDTA pH 5	Tonoplast + 2 mM Ca^{2+} pH 5	Tonoplast + 2 mM Mg^{2+} pH 5	Tonoplast + 2 mM Ca^{2+} pH 7
			% of total radioactivity				
lyso-PC	0,6	1,8	0,9	28,4	27	28,1	28,7
PC	98,9	97,3	0,8	67,7	59,3	67,3	61,4
PA	0,1	0,2	0,1	0,8	11,1	0,3	6,8
FFA	0,4	0,5	98,2	2,6	2,1	3,6	2,7
DAG	-	0,2	-	0,5	0,5	0,7	0,4

Table I: Phospholipase activities associated with the tonoplast of Acer

Tonoplast fractions (equivalent to 80 nmol PC) were incubated with 5 nmol. of sn-2-[^{14}C]acyl-PC (9.25 KBq) in the presence of 0.1% (w/v) Triton X-100 for 30 min at 37°C. Lipids were separated by TLC and the radioactivity was measured in a liquid scintillation counter. The values shown are the average of 3 experiments, each performed in triplicate.

These results demonstrated unequivocally the presence of phospholipase A_1 and phospholipase D activities associated with the tonoplast of *Acer*.

To further characterize the phospholipase A_1 activity, we studied the properties of this enzyme *in vitro* by following the hydrolysis of tonoplast endogenous [choline-^{14}C]-PC under various conditions. The phospholipase A_1 showed an optimal activity at pH 6-6.5. The formation of [^{14}C]lyso-PC was remarkably activated by Ca^{2+} and Mg^{2+} but addition of EDTA or EGTA did not totally inhibit the phospholipase A_1 activity. The phospholipase A_1 activity in presence of Ca^{2+} was 3 to 4-fold higher than that measured in presence of EDTA. The higher stimulation was obtained with 1 mM Ca^{2+} but 100 μM Ca^{2+} gave a 55% stimulation. Nevertheless Ca^{2+} effect is not specific since Mg^{2+} yielded 85% of the activity measured in presence of Ca^{2+}. Divalent cations could improve the interactions enzyme/substrate.

The effects of pH, Ca^{2+} and Mg^{2+} were also tested on lyso-PE formation. The results obtained confirmed those previously reported with lyso-PC formation. However, phospholipase A_1 may hydrolyse PC rather than PE.

Preliminary results obtained with whole or sonicated vacuoles suggested that phospholipase A_1 is located on the cytoplasmic face of the vacuolar membrane. Further investigations about acyl chain-specificity of tonoplast phospholipase A_1, its location within the membrane and its mode of regulation will be necessary to clarify its physiological role.

References

1. Huang AHC. Enzymes of lipid degradation. In: Lea PJ, editor. Methods in Plant Biochemistry. London: Academic Press, 1990: 219-227.
2. Paliyath G, Droillard MJ. The mechanisms of membrane deterioration and diassembly during senescence. Plant Physiol Biochem 1992;30: 789-812.
3. Tavernier E, Le Quoc D, Le Quoc K. Lipid composition of the vacuolar membrane of *Acer pseudoplatanus* cultured cells. Biochim Biophys Acta 1993;167: 242-247.
4. Pugin A, Montrichard F, Le Quoc K, Le Quoc D. Incidence of the method for the preparation of vacuoles on the vacuolar ATPase activity of isolated *Acer pseudoplatanus* cells. Plant Sci 1986;47: 165-172.

IN VITRO PHOSPHOLIPASE ACTIVITY OF CAPSICEIN, AN α-ELICITIN SECRETED BY A PHYTOPATHOGENIC FUNGUS, *PHYTOPHTHORA CAPSICI*

Claude Nespoulous, Christine Mella, Jean-Claude Huet, Jean-Claude Pernollet

Laboratoire d'Etudes des Protéines, Départment de Physiologie et de Biochimie végétales, INRA, 78026 Versailles Cedex, France

Most micro-organisms are unable to invade plants; they are defeated in their attempts to establish themselves successfully as parasites by defence mechanisms occurring during the invasion process. As the micro-organism enters the host tissue, the host recognises foreign material, generically termed *elicitors* [1], which results in a signal cascade inducing both short term (host membrane depolarisation, production of active oxygen species, changes in ion membrane permeability and the activation of genes which lead to phytoalexin synthesis) and longer term responses (cell wall alterations, lignin synthesis, expression of pathogenesis related proteins and hydroxyproline-rich glycoproteins). This set of responses may result in a necrotic hypersensitive response in the host, restricting the extent of the invasion [2].

Among plant pathogens, the fungal genus *Phytophthora* is represented by numerous species parasiting many economically important crops. With the exception of *Phytophthora parasitica var. nicotianae*, all the *Phytophthora* species so far studied secrete in culture abundant amounts of one or several elicitin isoforms [3], which are an original family of protein elicitors. Elicitins are 10 kD holoproteins, devoid of side-chain post-translational modifications [4] causing both remote leaf necrosis and induction of systemic resistance of some plants against pathogens, not only *Phytophthora* fungi [5] but also bacteria [6]. The nature of the initial interaction between elicitins and target cells has not been elucidated. One of elicitins, cryptogein, was shown to interact with the plasma membrane of tobacco cells, inducing some of the membrane perturbations mentioned above as hypersensitive short term responses [7]. The presence of specific elicitin-binding sites was suggested.

The physico-chemical properties of the elicitins offer the potential for determination of their 3-dimensional structure by nuclear magnetic resonance (NMR). Resonance assignment, disulphide bridge elucidation and secondary structure determination were conducted on capsicein, an α elicitin, by 3-dimensional homonuclear [8] and heteronuclear [9] NMR. Preliminary data has allowed the assignment of the cysteine pairs and revealed details of the molecular shape of capsicein. It is a protein exhibiting 5 α-helices and 1 antiparallel β-sheet with a β-turn forming a β-wing. The N-terminal α helix edges a cleft that might be involved in the interaction with a substrate. The residue 13, proposed as a key amino acid for toxicity from sequence-function relationships, was found at the rim of this cleft and could be involved in a ligand accessibility. The secondary structure and disulfide bridges shared certain structural analogies with animal phospholipase A2 (PLA2) (Figure 1). The structural analogy with such an enzyme led us to re-examine the interaction of elicitins with plasma membrane through the potential hydrolysis of lipid bilayer.

J.-C. Kader and P. Mazliak (eds.), Plant Lipid Metabolism, 313–315.

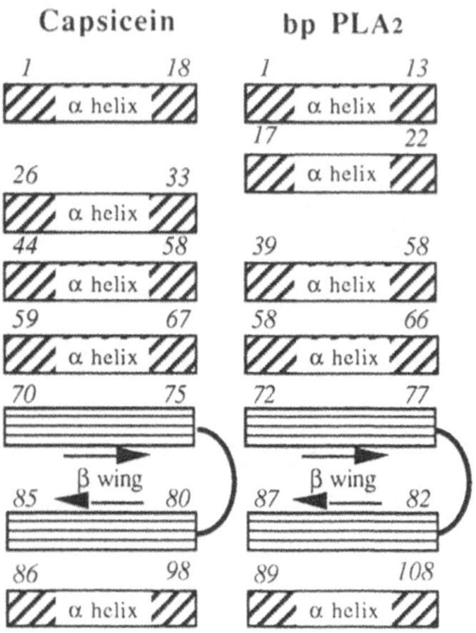

Capsicein S-S bonds	bp PLA2 S-S bonds
3 — 71	**11 — 77**
	27 — 123
27 — 56	**29 — 45**
	44 — 105
51 — 95	**51 — 98**
	61 — 91
	84 — 96

Figure 1: Comparison of the secondary structure and disulphide bridge arrangements of capsicein and bovine pancreas PLA2 (bp PLA2).

With this aim in mind, capsicein was tested for phospholipase activity. When using an agar-plate digestion assay, capsicein showed hydrolytic activity on soybean L-α-lecithin after 7 to 10 days of incubation, 2 to 3 times more slowly than bovine pancreas PLA2 (bpPLA2) which was used as a reference. In those conditions, the optimum pH of capsicein activity was between 7.0 and 7.5. Release of fatty acids from purified phospholipids was measured using the spectrophotometric CoA-coupled assay. A weak but significant phospholipase A activity around 4 nmol. min^{-1}. μmol^{-1} protein was found to characterize capsicein (1/4000 of bpPLA2 activity in the same conditions) (Figure 2). Capsicein hydrolysed at the same rate phosphatidylcholine and phosphatidylglycerol, showing no influence of the charge of the head-group. Calcium was not required for

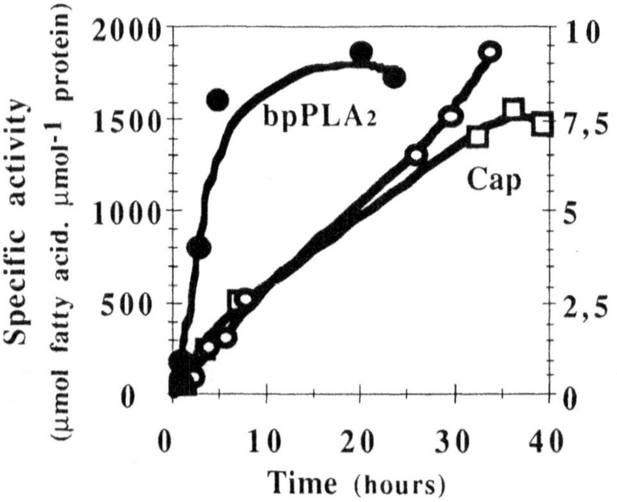

Figure 2: Time-course of hydrolyse of phospholipide by Cap (right scale) compared to bpPLA2 (left scale). Substrates used were: ○ and ●: phosphatidylcholine (DMPC); □: phosphatydylglycerol (DMPG) Experiments were carried out in Tris buffer at pH 7.5 with Ca^{2+} or EDTA 10mM for bpPLA2 and Cap assay respectively.

activity as already reported for some other PLA2's [10]. Furthermore when infiltrated into tobacco leaf lamina, bpPLA2 induced necroses with a pattern akin to that of capsicein. This result support the hypothesis that the plant defence signalling pathway triggered by capsicein involves the jasmonic acid route, urged by the release of linolenic acid [11].

CONCLUSION

The structural analogy found between capsicein and secreted PLA2 was shown to have a functional basis. We are currently characterizing this activity and testing for it in other elicitins. The search for a phospholipid interaction site in capsicein is now under way. A site directed mutagenesis strategy will allow us to determine the residues involved in this site. Detailed informations about this site are of particular interest because elicitins lack the histidyl residue known to be involved in the catalytic site of animal secretory PLA2's. Whether the cleft and location of the key residue 13 play a role remains to be seen. These results will allow the elucidation of the molecular mechanisms involved in the elicitin signalling pathway.

References

1 Dixon RA, Lamb CJ. Molecular communication in interactions between plants and microbial pathogens. Ann Rev Plant Physiol. Plant Molec Biol 1990; 41: 339-367

2 Anderson AJ, Rogers K, Tepper CS, Blee K. Timing of molecular events following elicitor treatment of plant cells. Physiol Mol Plant Pathol 1991; 38: 1-13

3 Pernollet JC, Sallantin, M, Sallé-Tourne M, Huet JC.Elicitin isoforms from seven Phytophthora species: Comparison of their physico-chemical and toxic properties to tobacco and other plant species. Physiol Mol Plant Pathol 1993; 42: 53-67

4 Nespoulous C, Huet JC, Pernollet JC. Structure-function relationships of α and β elicitins, signal proteins involved in the plant-Phytophthora interaction. Planta 1992; 186: 551-557

5 Ricci P, Bonnet P, Huet JC et al. Structure and activity of proteins from pathogenic fungi Phytophthora eliciting necrosis and acquired resistance in tobacco. Eur JBiochem 1989; 183: 555-563

6 Kamoun S, Young M, Glascock CB, Tyler BM. Extracellular protein elicitors from Phtyptophthora: Host-specificity and induction of resistance to bacterial and fungal phytopathogens. Mol Plant-Microbe Interact 1993; 6: 15-25

7 Blein JP, Milat ML, Ricci P. Responses of cultured tobacco cells to cryptogein, a proteinaceous elicitor from Phytophthora cryptogea. Plant Physiol 1991; 95: 486-91

8 Bouaziz S, van Heijenoort C, Guittet E, Huet JC, Pernollet JC. Resonance assignment, cystein-pairing elucidation and secondary-structure determination of capsicein, an α elicitin, by three-dimensional ^{1}H NMR. Eur J Biochem 1994; 220: 427-38

9 Bouaziz S, van Heijenoort C, Guittet E, Huet JC, Pernollet JC. ^{1}H and ^{15}N resonance assignment and secondary structure of capsicein, an α elicitin, determined by three-dimensional heteronuclear NMR. Biochemistry 1994; in press.

10 Vesterqvist O, Sargent CA, Grover GJ et al. Characterization of rabbit myucardial phospholipase A2 activity using endogenous phospholipid substrates. Anal Biochem 1994; 217: 210-19

11 Sembdner G, Parthier B. The biochemistry and the physiological and molecular actions of jasmonates. Annu Rev Plant Physiol Plant Molec Biol 1993; 44: 569-89

POSSIBLE ROLE OF MEMBRANE LIPID DEGRADATION IN TULIP BULB SCALE MICROPROPAGATION

Maarten W.P.C. van Rossum and Linus H.W. van der Plas

Department of Plant Physiology, Wageningen Agricultural University,
Arboretumlaan 4, 6703 BD Wageningen, The Netherlands

Introduction

The traditional methods in tulip propagation have as a main drawback the amount of time (25-30 years) it takes to produce enough bulbs of a new variety before it can be introduced to the market. The time needed could be greatly reduced by applying tissue culture techniques which promise a far higher propagation rate (40-100 instead of 4-5 new bulbs per parent bulb). Unfortunately great problems are encountered in using tissue culture techniques; especially when using bulb scale explants these problems are the most pronounced. The explants suffer from extreme browning and eventually die. This is mainly seen outside a short period (July to half of August) in which regeneration of new shoots can be observed (1). It is also possible to generate shoots (2) on stalk explants but the availability of this material is much more limited.

One hypothesis is that the tulip tissue is extremely sensitive to oxidative stress which leads to excessive membrane damage; this eventually will lead to death of the explant. In order to test this hypothesis we studied the degradation of the membranes of the explant and the activity of a number of stress-related enzymes .

Material

Tulip bulbs cv. "Apeldoorn" and cv. "destiny" were used. Explants were cut from the basal part of the second and third bulb scale and cultured on Murashige and Skoog medium containing supplements according to Nishiuchi (1). As growth regulators 2,4-D (1 mg/l) and BAP (1.5 mg/l) were added.

Results

CHANGES IN TULIP BULB SCALE EXPLANTS DURING INCUBATION

Bulb scale explants turned brown during incubation, probably due to the formation of polyphenols. The degree of browning depended on cultivar and season, but generally the tissue turned deep brown within 4 weeks and lost its viability. Only a few explants developed

J.-C. Kader and P. Mazliak (eds.), Plant Lipid Metabolism, 316–318.

a greenish callus and occasionally some shoot formation occurred.

Although a transient increase in the amount of membrane phospholipids often was observed, the total amount of phospholipids eventually decreased concomittant with this loss of viability. In the same time also the composition of phospolipids changed: generally the linoleic acid content decreased while the oleic and linolenic acid showed a corresponding increase. The percentage of the saturated fatty acids (mainly palmitic acid) remained more or less the same (compare figure 1). Especially the increase in linolenic acid was remarkable: in the beginning it constituted only about 1 % of the total phospholipids while it might increase to values up to 15 %.

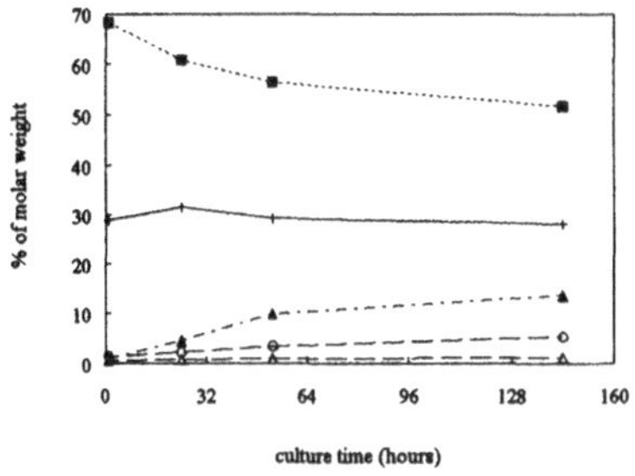

Figure 1: Changes in fatty acid composition of phospholipids during incubation of tulip bulb scale explants (+ palmitic, ▵ stearic, ○ oleic, ■ linoleic, ▲ linoleic acid).

Malondialdehyde (MDA, measured as thiobarbituric acid positive material) increased during incubation, also indicating possible membrane degradation in the explants.

CHANGES IN BULB SCALE EXPLANTS INMEDIATELY AFTER CUTTING

Most of the changes occurring in the explants were already visible in the first days after cutting. The changes in fatty acid composition gradually occurred during the first week (figure 1) and could already be observed after 1 - 2 days. The first accumulation of phenolics started

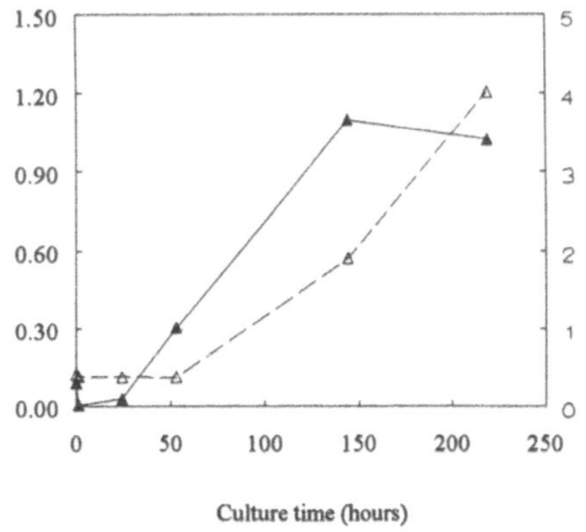

Figure 2: Changes in activity of peroxidase (-▲-)(Δ A470nm) and amount of phenolics(-▵-) during incubation of tulip bulb scale explants.

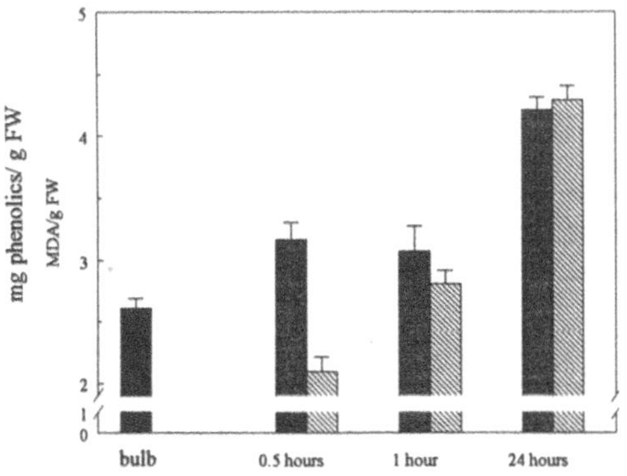

Figure 3 : Changes in MDA content (measured as A_{532}-A_{600} ± SEM) after cutting explants either cut in air (gray bars) or submerged (hatched bars).

after 2 - 3 days; this accumulation was preceded by an increase in peroxidase activity, an enzyme that was almost absent at the time of slicing (figure 2). Katalase activity was present from the beginning and remained more or less the same during the first week of incubation (results not shown).

One of the most rapid changes after slicing was the increase in MDA which seemed to occur within a few hours after slicing (figure 3); this rapid increase was accompanied by a decrease in the amount of membrane lipids.

PROSPECTS FOR FURTHER RESEARCH

The rapid changes in the bulb scale tissue, such as the rapid MDA formation could be prevented by slicing the tissue submerged in water, with or without antioxidants (figure 3), an environment with a low oxygen concentration. However the lagging behind of MDA formation was not permanent: after one day on incubation medium the MDA content was almost the same as in the control tissue sliced in air. Addition of antioxidants to the incubation medium or incubation in a low oxygen atmosphere might lead to a more permanent protection. Preliminary experiments already indicated that tissue 'reactivation', leading to callus formation indeed was improved as a consequence of such treatments.

Conclusions

It appeared that membranes of tulip bulb scale explants are subject to large changes during culturing. Not only breakdown of membrane constituents seems to occur (e.g. formation of MDA), but also there are rapid changes in the phopholipid composition; especially the increase in linolenic acid within two days is remarkable.

Two phases can be observed in these changes after culturing: a rapid one immediately after cutting and a slower one during incubation of the tissue. Membrane damage might negatively affect the regeneration potential of the tissue. Therefore it seems necessary to protect the explants during both phases against oxidative stress, for instance by submerged slicing of the tissue in combination with incubation on medium with anti-oxidants or in a low oxygen atmosphere.

References

1. Nishiuchi, Y. and J. Koster. Difference in organogenic activity in the scale tissue culture between tulip bulbs grown at Asahikawa and bulbs grown in the Netherlands. J. Hokk. Univ. Educ. 1988; 39:51-56.

2. LeNard, M. Obtention d'un bourgeonnement adventif par la culture in vitro de fragments de hampes florales prelévées chez des tulipes (Tulipa Gesneriana L.) en croissance. C. R. Acad. Sci. Paris 1989; 308 :389-394.

Section 4

Isoprenoids

SOME NEW ASPECTS OF ISOPRENOID BIOSYNTHESIS IN PLANTS

Thomas J. BACH
C.N.R.S.-I.B.M.P., Département d'Enzymologie Cellulaire et Moléculaire, Université Louis Pasteur, F-67083 Strasbourg Cédex, France

Introduction

A myriad of natural compounds are derivatives of a common C-5 isoprene unit and are thus classified as isoprenoids. It is impossible to cover all aspects of plant isoprenoid biosynthesis, which would go far beyond the scope of this contribution. I will rather try to focus the reader's attention on some recent aspects, which, at first sight, are not compatible or are even in conflict to each other. Other aspects, partially discussed a long time ago, might be looked at under a new aspect. Of course, with these limitations, not only set by the space available, the selection of literature and subjects must remain quite personal and will also include some aspects of our own studies. The problem of intracellular localization of enzymes, especially for those standing at the entry of the multi-branched isoprenoid pathway, is still a matter of debate and will be discussed in view of new findings. At recent meetings research groups from Japan and Switzerland have presented evidence of alternative pathways of mevalonate and/or isopentenyl pyrophosphate formation in plants, the latter one being reminiscent of the so-called Rohmer pathway occurring in certain bacteria. The possible existence of two independent and strictly compartmentalized biosynthetic pathways for the formation of a single intermediate would explain contradictory results as to the partial or complete independence of organelles from cytoplasmic enzymes, deduced from incorporation and inhibition studies as well as by determination of enzyme activities in subcellular fractions. Moreover, such an operation of an alternative pathway would explain the obviously close linkeage between isoprene production and the Calvin cycle by certain plants as well as unequal labeling of certain isoprenoids from various precursors.

General importance of the multi-branched isoprenoid pathway in plants

Today some 22,000 isoprenoids (= terpenoids = isopentenoids) have been characterized [1], most of them of plant origin. For tobacco alone until 1985 among 2,500 secondary metabolites more than 470 isoprenoid derivatives have been described [2]. It would not be too much of a surprise to find in every plant species in the average some new or unique isoprenoid derivative, an agonizing task for the phytochemists before the devastating activities of the human race will have led to the extinction of most species. It is evident that the multi-branched isoprenoid pathway in plants represents by far the most complicated biosynthetic pathway in nature (Figure 1). There is hardly a single process at the cellular level and/or at the level of plant organs or intact plants that does not require the biochemical and physiological involvement of isoprenoids, be it "pure" isoprenoids being comprised exclusively of isoprene units, or such compounds derived from mixed biosyntheses, with only a moiety of the molecule coming from the

J.-C. Kader and P. Mazliak (eds.), Plant Lipid Metabolism, 321–334.

isoprenoid pathway. Besides their function in photosynthesis (carotenoids, chlorophylls, phylloquinone, plastoquinone), respiration (ubiquinone, cytochrome a), membrane architecture (sterols and triterpenoids), regulation of growth and development (gibberellic acids, abscisic acid, brassinosteroids, certain cytokinins), defense against pathogen attack (numerous isoprenoid phytoalexins), exchange of signals etc., prenylation of proteins might be thought of as one important mechanism for membrane targeting and for cell cycle regulation. However, in view of new findings discussed below, this scheme might need some modifications.

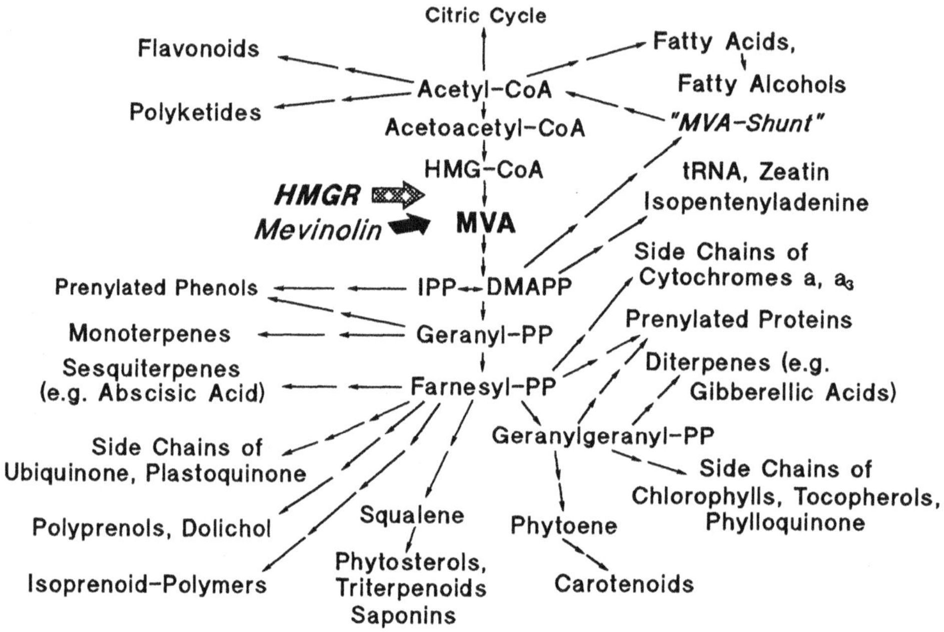

Figure 1. The multi-branched isoprenoid pathway in plants. The central role of mevalonic acid (MVA), and the inhibition by mevinolin of HMG-CoA reductase (HMGR) is indicated.

Mevalonic acid as specific precursor of isoprenoids

Since the discovery of mevalonic acid (MVA)[#] isolated from distillates of brewers yeast as an acetate-replacing growth factor in the culture broth of *Lactobacillus acidophilus* [3], and especially its excellent incorporation into cholesterol in rat liver homogenates [4,5], speculations on the "true" comitted precursor molecule for isoprenoid biosynthesis were practically terminated. By monitoring the incorporation of labeled MVA into

---

[#] *Abbreviations*: AACT, acetoacetyl-CoA thiolase; DMAPP, dimethylallyl pyrophosphate; FPLC, fast performance liquid chromatography; HMG-CoA, 3-hydroxy-3-methylglutaryl-coenzyme A, HMGL, HMG-CoA lyase; HMGR, HMG-CoA reductase; HMGS, HMG-CoA synthase; IPP, isopentenyl pyrophosphate; MVA, mevalonic acid; PDC, pyruvate decarboxylase; PQQ, pyrroloquinoline quinone; SDS, sodium dodecylsulfate, TPP, thiamine diphosphate.

various plant isoprenoids (see [6-8]) the principle that *(R)*-MVA serves as precursor of (all) isoprenoids was generalized. The degree of radiolabel from MVA found in isoprenoids other than sterols or sterol precursors such as squalene, cycloartenol etc. can be rather negligible (*cf.* [7,9] and literature cited therein). Caution must therefore be taken if radiochemical purity of a radiolabeled compound has not been proven. This holds also true for incorporation of acetate into isoprenoids. Since this substrate serves as a building block for many syntheses, especially for the biosynthesis of fatty acids and their derivatives, contamination by such lipophilic compounds may give rise to serious problems. For example, long chain fatty alcohols are known to run on TLC plates with R_f values similar to sterols (see [10]), and complete separation of sterols needs several steps of column chromatography as well as repeated recrystallization. Thus, the radiochemical purity of radiolabeled products of the multi-branched isoprenoid pathway was not really proven, nor in every case had the time of exposure to radiolabeled precursors such as ^{14}C or ^{3}H-labeled MVA excluded *per se* any metabolic degradation and, consequently, re-use of such breakdown products, e.g. of acetate, in the *de novo* biosynthesis of isoprenoids. Further, as already discussed by Banthorpe *et al.* [7], the unavoidable introduction of unphysiologically large quantities of MVA might result in degradation of the additive to products that may also act as enzyme inhibitors or lead to the induction of salvage degradation pathways. On the other hand, a poor incorporation of MVA, due to its limited uptake and transport, does not prove *per se* that it is not an obligate intermediate in the biosynthesis of a certain isoprenoid.

*Properties of plant HMG**R**s*

It is widely accepted that HMG**R** (*R*-mevalonate: $NADP^+$ oxidoreductase, CoA-acylating, EC 1.1.1.34), catalyzing the NADPH-dependent double-reduction of HMG-CoA to MVA, plays a crucial role in the part of the isoprenoid pathway leading to phytosterols (see [11-14]). Treatment of seedlings with mevinolin (= lovastatin), a natural antibiotic and HMG**R** inhibitor, as well by its natural and artificial analogues leads to severe disturbances in plant growth [14-19], due to the specific block of *de novo*-sterol biosynthesis. A further argument in favor of a regulatory role of HMG**R** in phytosterol accumulation was derived from the observation that in triazole-resistant tobacco mutant cells overproducing sterols [20], HMG**R** activity had increased 3 to 4-fold [21]. Overproduced sterols, largely in form of biosynthetic precursors such as cycloartanol, appear as fatty acyl esters localized in lipid droplets of the cytoplasm, but not in membranes [20, 22] Evidently the same phenotype is present in transformed tobacco cells overexpressing the soluble part of hamster HMG**R** [23], or overexpressing rubber tree HMG**R** (H. Schaller, M.-L. Chye, and P. Benveniste, personal communication). Apparently, plants derived from tobacco mutants or from transformed cells overexpressing HMG**R**s exhibit a normal phenotype and growth rate, without significant changes in pigment composition. Furthermore, feeding of exogenous MVA to alfalfa seedlings and cell cultures [24] led to the accumulation of phytosterol intermediates, further underlining the regulatory role of HMG**R** in the coarse control of substrate flux from acetyl-CoA to phytosterols as discussed earlier [11].

Hydrophobicity plots according to Kyte and Doolittle [25] predict that all plant HMG**R**s sequenced so far (see [26-28] for literature) should contain a highly conserved membrane domain consisting of two hydrophobic amino-acid sequences, separated from the soluble and highly conserved domain bearing the catalytic site by a linker region without any significant homology. In plants it appears that HMG**R** isozymes are encoded by a small family of genes, and evidence is emerging that these genes and gene products are placed under developmental or environmental control by factors such as light, xenobiotics, or

microbial elicitors (see [14, 26, 27] for literature), somehow reflecting the greater diversity of mevalonate derivatives in the realm of plants.
In contrast to the sequence areas coding for the sites of catalytic activity, found to be highly conserved in all eukaryotic HMG**R** enzymes (see [26, 29] and literature cited therein), the regions of the membrane anchor and especially the neighboring linker are characterized by evolutionary sequence diversity. Animal and yeast HMG**R**s contain eight such hydrophobic regions in the membrane anchor [30, 31]. When two radish HMG**R** isogenes were expressed in a MVA auxotrophic yeast mutant, we could, however, show that the two membrane-spanning amino-acid sequences are sufficient to direct the targeting of nascent radish HMG**R**s into yeast membranes. Western blot analyses revealed that in the heterologous system proteins of the correct size (about 63 kDa) had been integrated into the membrane(s) [28]. It must be noted, however, that there exists a considerable controversy as to the exact intracellular localization of HMG**R** in plants (see [9, 12-14, 17, 26, 28, 32] and literature cited therein). As known so far, all plant HMG**R** sequences are characterized by the absence of a typical leader peptide indicative of a translocation from the cytoplasm into organelles. Recent *in-vitro* transcription-translation experiments with two rather divergent isogenes encoding *Arabidopsis* HMG**R**s showed a co-translational insertion of plant HMG**R** protein into endoplasmic reticulum-derived microsomes [32], similar to the situation in vertebrate cells. However, for the import of proteins destined for the outer chloroplast envelope apparently no ATP-hydrolysis, no transit sequence and no protease sensitive envelope surface components are needed [33]. Thus, a localization of an isozyme of HMG**R** in the outer plastid envelope as suggested by Wilson and Russell [34] cannot be ruled out.

Conversion of acetyl-CoA to HMG-CoA

Following the pioneering work of Harry Rudney and Feodor Lynen and their associates for yeast and mammalian tissue, it has been well documented (*cf.* [14] and literature cited therein) that the conversion of three units of acetyl-CoA to HMG-CoA is catalyzed by two distinct enzymes: a) AACT (acetyl-CoA acetyl transferase EC 2.1.3.9) catalyzing a Claisen type condensation, and b) HMGS, (*S*-3-hydroxy-3-methylglutaryl-coenzyme A: acetoacetyl-coenzyme A lyase, CoA acylating), EC 4.1.3.5), catalyzing an aldol condensation. Only little was known until recently as to the situation in plants [9, 12-14]. We have recently purified and characterized a membrane-associated enzyme system from 4-day-old etiolated radish seedlings (*Raphanus sativus* L.), capable of converting acetyl-CoA into HMG-CoA [35]. This enzyme system, apparently catalyzing both partial reactions, is further referred to as AACT/HMGS. The quite unusual *in-vitro* properties of this enzyme system lead us to believe that plants might have developed a further mechanism of HMG-CoA synthesis different from other eukaryotes. The most puzzling feature of radish AACT/HMGS is its apparent activation by redox-cycling as mediated by the artificial redox couple Fe(II)/PQQ, or alternatively by Fe(II) and higher concentrations of H_2O_2, suggesting the involvement of hydroxyl radicals formed by a Fenton-Haber-Weiss reaction [$Fe^{2+} + H_2O_2 \rightarrow Fe^{3+} + OH^- + {}^{\bullet}OH$]. This effect is visible with crude solubilized radish AACT/HMGS , but it is more prominent with purified enzyme and it can be further enhanced by addition of PQQ [36]. It is evident that during the purification we get rid of a heat-stable cofactor system (that remains yet to be identified), which cannot only partially replace the PQQ but also to a certain extent the Fe^{2+} . However, at present we cannot exclude that this new system represents a multifunctional enzyme or even an intrinsic side-activity of a plant-specific enzyme, which might be less specific than we thought before. Although on SDS-gels we detected only a prominent band of about 56 kDa [35], in good agreement with values obtained by gel filtration on various materials [14], we would not exclude the possibility of an underlying impurity. This definitely requires further studies. A final proof of the existence of the so-called AACT/HMGS enzyme system needs the isolation of corresponding genes. Unfortunately, until now this goal has not yet been achieved. We have only been able to complement a yeast *erg*-mutant, being deficient in AACT, with a radish cDNA and have thus isolated a gene that encodes a soluble form of AACT (see Vollack and Bach, elsewhere in this volume). This latter observation would rather support the conclusions made by van der Heijden *et al.* [37], although their purification scheme, using cell-free preparations of *Catharanthus roseus* cell cultures, did not (yet) result in a separation of AACT and HMGS activities. A method for the simultaneous determination of all intermediates of the isoprenoid pathway, starting with acetyl-CoA, has recently been developed [38]. While all other intermediates could be found, depending on the substrate fed, in no case free acetoacetyl-CoA was detected in isolated cells capable of synthesizing monoterpenes. In our test system for AACT/HMGS we could not trap free acetoacetyl-CoA as a putative intermediate [14]. Thus the situation is still far from being clear. Furthermore, the exact intracellular localization of HMG-CoA synthesis (*i.e.* cytoplasm versus plastids and/or mitochondria) remains obscure unless corresponding genes have been cloned.

Branched-chain amino acids as precursurs of plant isoprenoids

L-Leucine and L-valine, when fed to germinating pea seeds (*Pisum sativum*) were incorporated into squalene and ß-amyrin [39]. Chemical degradation by ozonolysis of the radioactive squalene revealed an equal distribution of the radioactivity in the IPP-derived (4 portions per molecule) and the DMAPP-derived moieties (2 portions per molecule). However, an unbalanced distribution in favor of the DMAPP-derived moiety of monoterpenoids was found after feeding radioisotopically labeled L-leucine, L-valine, DL-alanine, acetate, and *R,S*-MVA to intact plants of *Cinnamomum camphora* and of *Pelargonium roseum* [40]. The relatively low incorporation of amino acids was explained by several hypotheses: a) the radioactivity of amino acids is scattered into other metabolites rather than monoterpenoids; b) there is a great pool of the amino acids which leads to dilution of radioactivity, and c) the low permeation of the amino acids into the biosynthetic site of monoterpenoids [40]. The DMAPP-derived moiety of monoterpenoids, biosynthesized from [U-^{14}C]leucine and [U-^{14}C]valine was labeled with more than 64% of the incorporated tracers, while this moiety when derived from [2-^{14}C]MVA contained less than 32% of the tracers [40]. This lends support to the interpretation that both amino acids are incorporated not *via* MVA, but by an alternate route, *e.g.* a combination of the well known mammalian leucine degradation pathway and a reversal of the MVA shunt (see below). The distribution pattern in monoterpenoids after administration of [2-^{14}C]alanine was similar to that after incorporation of MVA, which indicates that alanine is first metabolized to acetyl-CoA, which then constructs preferentially the IPP-derived moiety of the monoterpenoids *via* MVA [40].

Contradictory results were obtained by Anastasis *et al.* [41] when they studied the incorporation of [2-^{13}C]- and [3-^{13}C]-leucines into sesquiterpenoid paniculides by tissue cultures of *Andrographis paniculata*. Apparently, HMG-CoA from leucine breakdown was not incorporated being intact but was broken down to acetyl-CoA and subsequently incorporated *via* HMG-CoA and MVA. The same seems to hold true for phytosterols, labeled under the same conditions [42]. However, the incorporation time was 7 days, suggesting degradation of precursors and products, while in the studies summarized by Suga *et al.* [39] and Tange [40] incorporation times ranged from 12 to 72h. From the incorporation pattern in bakuchiol, a meroterpene from *Psoralea corylifolia*, derived from leucine, valine and isovaleric acid, it was concluded that these substrates are metabolized to CO_2, which is then incorporated [43].

Evidence for the occurrence of a MVA shunt pathway in plants

Evidence has been presented for the occurrence of a mevalonate shunt pathway in higher plants [10] with HMG-CoA as an intermediate, as it is most likely true for the degradation of leucine. Given the fact that the membrane fraction used to isolate AACT/HMGS from radish is also a major source for HMGL (EC 4.1.3.4), a key enzyme of the MVA shunt [44], its presence could indicate that in plants AACT/HMGS plays a role in a ketogenic cycle. However, to our knowledge there is no clear information as to the metabolic fate in plants of acetoacetate generated by the HMGL reaction. However, though the activity of 3-methylcrotonyl-CoA carboxylase (EC 6.4.1.4) has been demonstrated in plants ([44-49], see also elsewhere in this volume), and also that of *trans*-3-methylglutaconyl-CoA hydratase (3-hydroxy-3-methylglutaryl-CoA hydrolyase, EC 4.2.1.18) [37], further evidence of an alternative peroxisomal pathway of leucine degradation has been presented by Gerbling 50], including free acids rather than CoA esters. At present the interdependence of those intervening and/or scavenging pathways is not yet known, nor is their regulation.

Isoprene emission by plants

Many plants produce considerable amounts of isoprene (see [51], and literature cited therein) which was reported to cause severe air pollution problems in some areas. Isoprene emission from plants may account for half of the total non-methane hydrocarbon emissions from the biosphere to the troposphere, thereby influencing the oxidative capacity and generating tropospheric ozone and carbon monoxide (see [52] and references cited therein). Environmental stimuli or wounding affect the rate of synthesis [51, 53]. This high amount of isoprene, derived from the IPP isomer DMAPP [54] can hardly be explained by the quite limited substrate flow from acetyl-CoA to MVA, because cellular HMGR activity, at least from what has been found in all plants studied so far, would not be sufficient to saturate such a rate of synthesis. Furthermore, it appears that there is a rather direct coupling to photosynthetic activity, *viz.* intermediates of the Calvin cycle [51, 52, 55-57] Most likely pyruvate derived from 3-phosphoglyceric acid as the primary product of CO_2 fixation is a close precursor (see below for biosynthetic consequences). Isoprene is also a normal constituent of human breath (see [58] and literature cited therein), but in this case a clear coupling to the rate of cholesterol and mevalonate biosynthesis was shown. Blockade of MVA biosynthesis by the hypocholesterolemic drug mevinolin (= lovastatin) leads to a decrease in cholesterol synthesis in liver, and to the same degree to a diminished production of isoprene [58]. Isoprene production might be explained by the enzymatic mechanism of IPP isomerization to DMAPP: Reardon and Abeles [59] and Muehlbacher and Poulter [60] have elegantly shown that IPP isomerase (isopentenyl-diphosphate D-isomerase, EC 5.3.3.2) catalyzes a reaction with an intermediate carbonium ion. This isomerization involves an antarafacial transposition of a proton, suggesting that a conjugate acid-base pair works in tandem to catalyze the reaction [60]. The same carbonium ion intermediate occurs in the acid-catalyzed nonenzymatic conversion of DMAPP to isoprene in liver [61]. Elimination of diphosphate from DMAPP could result in the formation of a delocalized carbocation intermediate followed by the loss of a single proton, which would yield isoprene (*cf.* [54]). It might be speculated that the Mg^{2+}-dependent reaction is in fact identical to that catalyzed by the so-called isoprene synthase partially purified by Silver and Fall [54], and that the plant enzyme cannot closely control the isomerization, ultimately leading to an abortion of enzyme-bound substrate and hence to the production of isoprene (see [58]).

Bacterial isoprenoid synthesis - the Rohmer pathway

Current biosynthetic evidence indicates that the steps from IPP to isoprenoids in Eubacteria are the same as those in eukaryotes [62-69] (see [70-73] for literature). Especially the incorporation of ^{13}C-labeled precursors into prokaryotic hopanoids, sterol surrogates in bacterial membranes (see [73] and literature cited therein), or into ubiquinone-8 [70] has revealed that the classical pathway of IPP formation starting from acetyl-CoA *via* acetoacetyl-CoA, HMG-CoA, MVA, MVA-P, and MVA-PP does not exist in a great variety of bacteria, including *E. coli, Zymomonas mobilis, Methylobacterium organophilum, Rhodopseudomonas palustris, R. acidophila, Acetobacter aceti* ssp. *xylinum,* but also in the thylakoids of the cyanobacterium *Synechocystis* sp. [74].

Label from acetate could be incorporated after being metabolically "scrambled" (see [70]), *i.e. via* the glyoxylate cycle: The condensation of acetate (as acetyl-CoA) with glyoxylate by action of malate synthase (EC 4.1.3.2) forms malate, which could then be converted into pyruvate by malic enzyme (L-malate: $NADP^+$ oxidoreductase [oxalacetate-decarboxylating], EC 1.1.1.40). The incorporation of the C-3 and C-2 unit of the resulting pyruvate into IPP would explain the observed results. The model as proposed by Rohmer *et al.* [73] is as follows: The formation of the structure C-5 of IPP (better: of the structure "isoprene unit") would proceed in a similar way to the biosynthesis of L-valine. A TPP-activated acetaldehyde generated by pyruvate decarboxylation would be added to the C-2 carbonyl group of a dihydroxyacetone derivative (such as dihydroxyacetone, monohydroxyacetone or their mono- or diphosphates, methylglyoxal. . .), followed by a transposition reaction. The following reactions (reduction, isomerisation, elimination of water) remain speculative, but the authors argue that they could again be similar to those found in L-valine biosynthesis. However, at least in the case of the ethanol producing bacteria *Zymomonas mobilis*, they could exclude L-valine or one of its precursors as substrate of IPP formation (see [73] for details).

An alternative IPP synthesis in plants akin to the bacterial Rohmer pathway ?

Hano *et al.* [75] have used callus and cell suspension cultures of *Morus alba* L. to study the biosynthesis of optically active prenylchalcones such as chalcomoracin. [2-^{13}C]MVA was not incorporated into the prenyl moiety of chalcomoracin while it was incorporated into ß-sitosterol. The labeling pattern of chalcomoracin from [2-^{13}C]-L-leucine was the same as from [1-^{13}C]acetate. Feeding D-[U-$^{13}C_6$]glucose, [1,3-$^{13}C_2$]- and [2-^{13}C]glycerol (which formed [1,3,4,6-$^{13}C_4$]- and [2,5-$^{13}C_2$]glucose, respectively) demonstrated that of the three acetate units composing the hemiterpene moiety the starter acetate unit is of glycolytic origin, whereas the second and third acetate units originate from the pentose phosphate cycle. Triose phosphates such as dihydroxyacetone phosphate and gyceraldehyde-3-phosphate are postulated to occupy the junction of this chimeric hemiterpene pathway with glycolysis and the pentose phosphate cycle. The major difference in the isotopic enrichments between the Rohmer pathway [73] and the mechanism by Hano *et al.* [75], is that labeling ^{13}C-label coming from [1-^{13}C]acetate appears in C-4 of IPP, while the hemiterpene moieties of prenylchalcones were not enriched, and second in *M. alba* cell cultures, label from [1,3-$^{13}C_2$]- and [2-^{13}C]glycerol labeled the C-1 and C-5 positions, or the C-2, C-3, and C-4 positions of IPP, respectively. Although the NMR data are quite conclusive, it is still difficult to rationalize in this model of how acetyl-CoA units can be distinguished in the formation

of HMG-CoA as as an intermediate [[75] if not assuming intermediates that are recognizable through tight binding to proteins/enzymes. This work needs to be complemented by a) incorporation studies using specifically ^{13}C-labeled glucose (see below), and in particular b) the application of inhibitors of the mevinolin type, which should be capable of affecting the synthesis of chalcomoracin if the model is true.
In another study, presented at a previous isoprenoid meeting at Strasbourg/Bischenberg, Cartayrade *et al.* [76] compared the biosynthesis of ß-sitosterol and of diterpenoid ginkgolides in embryos of *Ginkgo biloba.* (Ginkgolides act apparently as antagonists of platelet-activating factors and are thus of high pharmaceutical interest for the treatment of thrombosis, inflammatory reactions and cardiovascular disorders (for literature see [77]). In *Ginkgo biloba* the IPP precursor of ß-sitosterol is assembled *via* MVA formed by the classical pathway from acetyl-CoA *via* acetoacetyl-CoA, and HMG-CoA. In contrast, the IPP units used for formation of ginkgolides are apparently assembled through decarboxylative condensation of pyruvate with a triose phosphate, similar or identical to the Rohmer pathway [76]. Further, these two independent biosynthetic machineries are strictly compartmentalized. However, there seems to exists a small "leakage" at the C-15 level, so that a small quantity of farnesyl-PP made in the cytoplasm like for ß-sitosterol synthesis, is then linked by action of geranylgeranyl-PP synthase to an IPP unit made by the alternative pathway, most possibly in the plastid [76]. Soler *et al.* [78] presented evidence of an IPP translocator in the plastid envelope. However, the K_m of ~ 500 M with respect to IPP does not convincingly show that it is really a very specific protein. It might function as a (pyro)phosphate translocator that could perhaps also accept higher IPP homologs such as farnesyl-PP.

Evidence for an enzymatic machinery for the Rohmer pathway ?

A crucial point in this new model of IPP formation in bacteria and eukaryotes as well is the fact that until now individual enzyme reactions have not yet been identified. An exception could be the initial reaction, the formation of TPP-acetaldehyde, catalyzed by pyruvate decarboxylase (PDC, 2-oxoacid carboxylase, EC 4.1.1.1), an enzyme that is not present in animal tissues, but has been isolated from yeast (see [79] and literature cited therein), and also from some plants such as maize [80]. Yeast PDC can also catalyze an aldol-type condensation reaction between two molecules of acetaldehyde to form the a-ketol acetoin in a concerted reaction mechanism [79]. It seems reasonable to assume a similar reaction pathway for the reaction of the TPP-bound "acetaldehyde" (2-(1-hydroxyethyl)TPP) with the dihydroxyacetone derivative as postulated by Rohmer *et al.* [73]. Given the fact that the hyperbolic kinetic properties and apparent molecular weights of the yeast (iso-)enzyme(s) (tetramers, native mass of 242 kDa) and the maize enzymes (kernels: 61 and 61 kDa subunits, roots: 59 and 58 kDa subunits, tetrameric structure for both isozymes are so similar [80], then it would be no surprise if both enzymes could catalyze the same condensation reaction as well. However, PDC is a typical enzyme of anaerobic metabolism (see [81]) and it should therefore be difficult for acetaldehyde to "escape" from the action of alcohol dehydrogenase (ADH, EC 1.1.1.1). Furthermore, the syntheses of PDC and ADH are specifically enhanced under anaerobic conditions. Maize PDC, having a sharp pH optimum of 5.8, is inactive at the pH > 7 of aerobically metabolizing tissues [80]. However, as an alternative it seems reasonable to assume that the entry reaction of the pyruvate dehydrogenase complex, pyruvate decarboxylation which is mechanistically identical, could give rise to the same intermediate 2-(1hydroxyethyl) derivative of the thiazol ring of TPP, which somehow would escape from the subsequent dehydrogenation and the resulting esterification of the acetyl unit to coenzyme A. Plant cells contain two pyruvate dehydrogenase complexes, one in the mitochondria and the other in the plastids (see [82] and literature cited therein). If such a "side reaction" would exist, then incorporation data using pyruvate as a precursor of plastid isoprenoids and fatty acids [83-85] could be re-interpreted, so that pyruvate decarboxylation could lead to a reactive intermediate being exclusively used for IPP synthesis, while full conversion of pyruvate to acetyl-CoA would yield a pool that is only used for fatty acid biosynthesis. In this way the problem of two independent acetyl-CoA pools in the same subcellular compartment with little or no exchange, as postulated by Schulz and coworkers, could easily be circumvented.

Consequences for the so-called segregation model of isoprenoid biosynthesis

The presence of such alternative pathways as outlined above could provide a further explanation for the observations obtained with the HMGR inhibitor mevinolin, *viz.* its great efficiency in blocking cytoplasmic sterol biosynthesis, but its low efficiency to affect ubiquinone biosynthesis, and its practically complete inefficiency to block the accumulation of isoprenoids and prenyl lipids of the chloroplast (summarized and discussed earlier 14, 17]. It could also explain the puzzling observation that, in contrast to phytosterols, prenyl-subsituted coumarins were not labeled from [2-^{14}C]acetate or [2-^{14}C]MVA in elicitor-treated cultures of *Ammi majus*, suggesting that the DMAPP needed for the (plastid ?) umbelliferone: dimethylallyl transferases (EC 2.5.1.3) is formed from IPP synthesized *de novo* within the plastid, rather than from IPP imported into the plastid from the cytosol [86]. Elicitor-induced inhibition of phytosterol biosynthesis should occur by specific inhibition of one of the enzymes on the cytosolic pathway from MVA to DMAPP.

Conclusions

From the observations discussed above, partially in conflict to each other, it should become obvious that we are still quite a long way from completely understanding of how the individual steps of the multi-branched isoprenoid pathway in plants are compartmentalized or interacting. In some way, we have to revise well estabished textbook knowledge (which as far as plant biochemistry is concerned was not always experimentally proven, but rather deduced by analogy from results obtained with animal tissue or with yeast). The co-existence of alternative enzymatic machineries in plants for the synthesis of one and the same precursors, be it MVA or more likely IPP, is an even more intriguing challenge for future studies in biochemistry, molecular biology and physiology. This possibly extends to evolutionary aspects, for instance as to the symbiotic origins of organelles [87] and of (prokaryotic ?) pathways in higher plants. In summarizing, nothing is as simple as it looks at first sight!

References

1. Conolly JD, Hill RA. Dictionary of Terpenoids, Chapman and Hall, New York 1992.
2. Wahlberg I, Enzell, CR. Tobacco isoprenoids [review]. Nat Prod Rep, 1987; 4: 237-276.
3. Wolf DE, Hoffman CH, Aldrich PE, Skeggs HR, Wright LD, Folkers K. ß-Hydroxy-ß-methyl-δ-valerolactone (divalonic acid), a new biological factor. J Am Chem Soc, 1956; 78: 4499.
4. Tavormina PA, Gibbs MH. The metabolism of ß,δ-dihydroxy-ß-methylvaleric acid by liver homogenates. J Am Chem Soc, 1956; 78: 6210.
5. Tavormina PA, Gibbs MH, Huff JW. The utilization of ß-hydroxy-ß-methyl-δ-valerolactone in cholesterol biosynthesis. J Am Chem Soc, 1956; 78: 4498.

6. Goodwin TW (ed.). Natural substances formed biologically from mevalonic acid. Biochem Soc Symp No. 29, Liverpool 1969, Academic Press, New York, 1970.
7. Banthorpe DV, Charlwood BV, Francis MJO. The biosynthesis of monoterpenes. Chemical Reviews, 1972; 72:115-149.
9. Kleinig H. The role of plastids in isoprenoid biosynthesis. Annu Rev Plant Physiol Plant Mol Biol 1989; 40: 39-59.
10. Nes WD, Bach TJ. Evidence for a mevalonate shunt in a tracheophyte. Proc Roy Soc Lond B, 1985; 225: 425-444.
11. Bach TJ. Hydroxymethylglutaryl-CoA reductase, a key enzyme in phytosterol synthesis? Lipids 1986; 21: 82-88.
12. Gray JC. Control of isoprenoid biosynthesis in plants. Adv Bot Res, 1987; 14: 25-91.
13. West C. Terpene biosynthesis and metabolism. In: Dennis DT, Turpin DH, editors Plant physiology, biochemistry and molecular biology. Essex: Longman Scientific & Technical, 1990: 353-369.
14. Bach TJ, Motel A, Weber T. Some properties of enzymes involved in the biosynthesis and metabolism of 3-hydroxy-3-methylglutaryl-CoA in plants. Rec Adv Phytochemistry, 1990; 24: 1-82.
15. Bach TJ, Lichtenthaler HK. Mevinolin: A highly specific inhibitors of microsomal 3-hydroxy-3-methylglutaryl-coenzyme A reductase of radish plants. Z Naturforsch, 1982; 37c: 46-50.
16. Bach TJ, Lichtenthaler HK. Inhibition by mevinolin of plant growth, sterol formation and pigment accumulation. Physiol Plant, 1983; 59: 50-60.
17. Bach TJ, Lichtenthaler HK. Plant growth regulation by mevinolin and otherr sterol biosynthesis inhibitors. In: Fuller G, Nes WD, editors, Ecology and Metabolism of Plant Lipids, Am Chem Soc Symposium Series 325, American Chemical Society, Washington, 1987: 109-139.
18. Hata S, Takagishi H, Egawa Y, Ota Y. Effects of compactin, a 3-hydroxy-3-methylglutaryl coenzyme A reductase inhibitor, on the growth of alfalfa (*Medicago sativa*) seedlings and the rhizogenesis of pepper (*Capsicum annuum*)explants. Plant Growth Regul, 1986; 4: 335-346.
19. Hata S, Takagishi H, Kouchi H. Variation in the content and composition of sterols in alfalfa seedlings treated with compactin (ML-236B) and mevalonic acid. Plant Cell Physiol, 1987; 28: 709-714.
20. Maillot-Vernier P, Gondet L, Schaller H, Benveniste P, Belliard G. Genetic study and further biochemical characterization of a tobacco mutant that overproduces sterols. Mol Gen Genet, 1991; 231: 33-40.
21. Gondet L, Weber T, Maillot-Vernier P, Benveniste P, Bach TJ. Regulatory role of microsomal 3-hydroxy-3-methylglutaryl-coenzyme A reductase in a tobacco mutant that overproduces sterols. Biochem Biophys Res Commun, 1992; 186: 888-893.
22. Gondet L, Bronner R, Benveniste P. Regulation of sterol content in membranes by subcellular compartimentation of steryl-esters accumulating in a sterol-overproducing mutant. Plant Physiol, 1994; in press.
23. Chappell J, Proulx J, Wolf F, Cuellar RE, Saunders C. Is HMG-CoA reductase a rate limiting step for isoprenoid metabolism? Plant Physiol, 1991; 96 (supplement): 127
24. Hata S, Shirata K, Takagishi H, Kouchi H. Accumulation of rare phytosterols in plant cells on treatment with metabolic inhibitors and mevalonic acid. Plant Cell Physiol, 1987; 28: 715-722.
25. Kyte J, Doolittle RF. A simple method of displaying the hydropathic character of a protein. J Mol Biol 1982; 157: 105-132.

26. Bach TJ, Wettstein A, Boronat A, Ferrer A, Enjuto M, Gruissem W, Narita JO. Properties and molecular cloning of plant HMG-CoA reductase. In: Patterson GW, Nes WD, editors, Physiology and Biochemistry of Sterols, American Oils Chemists' Sociandy, Champaign, Illinois, 1991: 29-49:

27. Choi D, Bostock RM. Involvement of de novo protein synthesis, protein kinase, extracellular Ca^{2+}, and lipoxygenase in arachidonic acid induction of 3-hydroxy-3-methylglutaryl coenzyme A reductase genes and isoprenoid accumulation in potato (*Solanum tuberosum* L.). Plant Physiol, 1994; 104: 1237-1244.

28. Vollack K-U, Dittrich B, Ferrer A, Boronat A, Bach TJ. Two radish genes for 3-hydroxy-3-methylglutaryl-CoA reductase isozymes complement mevalonate auxotrophy in a yeast mutant and yield membrane-bound active enzyme. J Plant Physiol, 1994; 143: 479-487.

29. Darney BG, Wang YL, Rodwell VW. Identification of the catalytically important histidine of 3-hydroxy-3-methylglutaryl-coenzyme A reductase. J Biol Chem, 1992; 267: 15064-15070.

30. Olender EH, Simoni RD. The intracellular targeting and membrane topology of 3-hydroxy-3-methylglutaryl-CoA reductase. J Biol Chem, 1992; 267: 4223-4235.

31. Roitelman J, Olender EH, Barnum S, Dunn WA, Simoni RD. Immunological evidence for eight spans in the membrane domain of 3-hydroxy-3-methylglutaryl coenzyme A reductase: Immplications for enzyme degradation in the endoplasmic reticulum. J Cell Biol, 1992; 117: 959-974.

32. Enjuto M, Balcells L, Campos N, Caelles C, Arró M, Boronat A. *Arabidopsis thaliana* contains two differentially expressed HMG-CoA reductase genes which encode microsomal forms of the enzyme. Proc Natl Acad Sci USA, 1994; 91: 927-931.

33. Soll J, Waegeman K. The protein import machinerie(s) of the outer chloroplast envelope membrane. Abstract No. 5.4.3-2, XV International Botanical Congress, Yokohama, Japan, August 28 - September 3, 1993, Abstract Volume p. 132.

34. Wilson TM, DW Russell. The localization, partial purification and regulation of pea plastid HMG-CoA reductase. Biochem Biophys Res Commun, 1992; 184: 530-537.

35. Weber T, Bach TJ. Conversion of acetyl-coenzyme A into 3-hydroxy-3-methylglutaryl-coenzyme A in radish seedlings. Evidence of a single monomeric protein catalyzing a Fe^{II}/quinone-stimulated double condensation reaction. Biochim Biophys Acta, 1994; 1211: 85-96.

36 Bach TJ, Raudot V, Vollack K-U, Weber T, Zeiler S. Further studies on the enzymatic conversion of acetyl-coenzyme A into 3-hydroxy-3-methylglutary-coenzyme A in radish. Plant Physiol Biochem, 1994; in press.

37. Van der Heijden R, de Boer-Hlupá V, Verpoorte R, Duine JA. Enzymes involved in the metabolism of 3-hydroxy-3-methylglutaryl-coenzyme A in *Catharanthus roseus*. Plant Cell Tissue Organ Culture, 1994; in press.

38. McCaskill D, Croteau R. Procedures for the isolation and quantification of the intermediates of the mevalonic acid pathway. Anal Biochem, 1993; 215: 142-149.

39. Suga T, Tange K, Iccho K, Hirata T. Biosynthesis from amino acids in *Pisum sativum*. The distribution of the radioactivity in squalene biosynthesized from radioisotopically labelled L-leucine et L-valine. Phytochemistry, 1980; 19: 67-70.

40. Tange K. The biosynthesis of monoterpenoids in higher plants. The biosynthetic pathway leading to the monoterpenoids from amino acids with a carbon-skeleton similar to mevalonic acid. Bull Chem Soc Jpn, 1981; 54: 2763-2769.

41. Anastasis P, Freer I, Picken D, Overton K, Sadler I, Singh SB. The role of leucine in terpenoid metabolism. Incorporation of [2-^{13}]- and [3-^{13}C]-leucines into

sesquiterpenoids by tissue cultures of *Andrographia paniculata.* J Chem Soc Chem Commun, 1983; 1189-1191.

42. Anastasis P, Freer I, Overton K, Rycroft D, Singh SB. The role of leucine in isoprenoid metabolism. Incorporation of [3-^{13}C]leucine and of [2-^{3}H,4-^{14}C]-ß,ß-dimethylacrylic acid into phytosterols by tissue cultures of *Andrographis paniculata.* J Chem Soc Chem Commun, 1985; 148-149.

43 Banerji A, Chintalwar GJ. Fate of leucine in the biosynthesis of bakuchiol, a meroterpene from Psoralea corylifolia. Ind J Biochem Biophys, 1989; 26: 394-396.

44. Weber T, Bach TJ. Partial purification and characterization of membrane-associated 3-hydroxy-3-methylglutaryl-coenzyme A lyase from radish seedlings. Z Naturforsch, 1993; 48c: 444-450.

45. Wurtele ES, Nicolau BJ. Plants contain multiple biotin enzymes: Discovery of 3-methylcrotonyl-CoA carboxylase, propionyl-CoA carboxylase and pyruvate carboxylase in the plant kingdom. Arch Biochem Biophys, 1990; 278: 179-186.

46. Wurtele ES, Nicolau BJ. Differential accumulatio of biotin enzymes during carrot somatic embryogenesis. Plant Physiol, 1992; 99: 1699-1703.

47 Baldet P, Alban C, Axiotis S, Douce R. Characterization of biotin and 3-methylcrotonyl-coenzyme A carboxylase in plants. Plant Physiol, 1992; 99: 450-455.

48. Alban C, Baldet P, Axiotis S, Douce R. Purification and characterization of 3-methylcrotonyl-CoA carboxylase from higher plant mitochondria. Plant Physiol, 1993; 102: 957-965.

49. Clauss M, Motel A, Lichtenthaler HK. Studies on the plant methylcrotonyl-CoA carboxylase. J Plant Physiol, 1993; 141: 508-511.

50. Gerbling H. Peroxisomal degradation of 2-oxoisocaproate. Evidence for free acid intermediates. Bot Acta, 1993; 106: 380-387.

51. Loreto F, Sharkey TD. On the relationship between isoprene emission and photosynthetic metabolites under different envoronmental conditions. Planta, 1993; 189: 420-424.

52. Fall R, Monson RK. Isoprene emission rate and intercellular isoprene concentration as influenced by stomatal distribution and conductance. Plant Physiol, 1992; 100: 987-992.

53. Loreto F., Sharkey TD. Isoprene emission by plants is affected by transmissible wound signals. Plant Cell Environm, 1993; 16: 563-570.

54. Silver GM, Fall R. Enzymatic synthesis of isoprene from dimethylallyl diphosphate in aspen leaf extracts. Plant Physiol, 1991; 97: 1588-1591.

55. Rasmussen RA, Jones CA. Emission isoprene from leaf discs of *Hammamelis.* Phytochemistry, 1973; 12: 15-19.

56. Tingey DT, Evans R, Gumpertz M. Effects of environmental conditions on isoprene emission from live oak. Planta, 1981; 152: 565-570.

57. Monson RK, Hills AJ, Zimmermann PR, Fall RR. Studies of the relationship between isoprene emission rate and CO_2 or photon-flux density using a real-time isoprene analyser. Plant Cell Environm, 1991; 14: 517-523.

58. Stone BG, Besse TJ, Duane WC, Evans CD, DeMaster EG. Effect of regulating cholesterol biosynthesis on breath isoprene excretion in men. Lipids, 1993, 28; 705-708.

59. Reardon, JE, Abeles RH. Mechanism of action of isopentenyl pyrophosphate isomerase: Evidence for a carbonium ion intermediate. Biochemistry-USA, 1986; 25: 5609-5616.

60. Muehlbacher M, Poulter CD. Isopentenyl-diphosphate isomerase: Inactivation of the enzyme with active-site-directed irreversible inhibitors and transition-state analogues. Biochemistry-USA, 1988; 27: 7315-7328.

61. Deneris ES, Stein RA, Mead JF. Acid-catalyzed formation of isoprene from a mevalonate-derived product using a rat liver cytosolic fraction. J Biol Chem, 1985; 260: 1382-1385.
62. Takahashi I, Ogura K. Farnesyl pyrophosphate synthetase from *Bacillus subtilis.* J Biochem Tokyo, 1981; 89: 1581-1587.
63. Fujii H, Koyama T, Ogura J. Hexaprenyl pyrophosphate synthetase from *Micrococcus luteus* B-P 26. J Biol Chem, 1982; 257, 14610-14612.
65. Fujisaki S, Nishino T, Katsuki H. Isoprenoid synthesis in *Escherichia coli.* Separation and partial purification of four enzymes involved in the synthesis. J Biochem Tokyo, 1986; 99: 1327-1337.
66. Fujii H, Koyama T, Ogura K. Essential protein factors for polyprenyl pyrophosphate synthetases. Separation of heptaprenyl pyrophosphate synthetase into two components. FEBS Lett, 1983; 161: 257-260.
67. Fujisaki S, Nishino T, Katsuki H. Biosynthesis of isoprenoids in intact cells of *Escherichia coli.* J Biochem Tokyo, 1986; 99: 1137-1146.
68. Seckler B, Poralla K. Characterization and partial purification of squalene-hopene cyclase from *Bacillus acidocaldarius.* Biochim Biophys Acta, 1986; 881: 356-363.
69. Siebert M, Bechtold A, Melzer M, May U, Berger U, Schröder G, Schröder J, Severin K, Heide L. Ubiquinone biosynthesis. Cloning of the genes coding for chorismate pyruvate-lyase and 4-hydroxybenzoate octaprenyl transferase from *Escherichia coli.* FEBS Lett, 1992; 307: 347-350.
70. Zhou D, White RH. Early steps of isoprenoid biosynthesis in *Escherichia coli.* Biochem J, 1991; 273: 627-634.
71. Rohmer M, Bisseret P, Sutter B. The hopanoids, bacterial triterpenoids, and the biosynthesis of isoprenic units in prokaryotes. Progr Drug Res, 1991; 37: 271-285.
72 Horbach S, Sahm H, Welle R. Isoprenoid biosynthesis in bacteria: Two different pathways. FEMS Microbiol Lett, 1993; 111: 135-140.
73. Rohmer M, Knani M, Simonin P, Sutter B, Sahm H. Isoprenoid biosynthesis in bacteria: a novel pathway for the early steps leading to isopentenyl diphosphate. Biochem J, 1993; 295: 517-524.
74. Simonin P, Jürgens UW, Rohmer M. 35-O-β-6-Amino-6-deoxyglucopyranosyl bacteriohopanetetrol, a novel triterpenoid of the hopane series from the cyanobacterium *Synechocystis* sp. PCC 6714. Tetrahedron Lett, 1992; 33: 3629-3623.
75. Hano Y, Ayukawa A, Nomura T, Ueda S. Origin of the acetate units composing the hemiterpene moieties of chalcomoracin in *Morus alba* cell cultures. J Am Chem Soc, 1994; 116: 4189-4193.
76. Cartayrade A, Schwarz M, Jaun B, Arigoni D. Detection of two independent mechanistic pathways for early steps in isoprenoid biosynthesis in *Ginkgo biloba.* Abstract N° P1, 2nd Symposium of the European Network on Plant Terpenoids, Strasbourg/Bischenberg, January 23-27, 1994.
77. Flesch V, Jacques M, Cosson L, Teng BP, Petiard V, Balz JP. Relative importance of growth and light level on terpene content of *Ginkgo biloba.* Phytochemistry, 1992; 31: 1941-1945.
78. Soler E., Clastre M, Bantignies B, Marigo G, Ambid C. Uptake of *iso*pentenyl diphosphate by plastids isolated from *Vitis vinifera* L. cell suspensions. Planta, 1993; 191: 324-329.
79. Stivers JT, Washabaugh MW. Catalysis of aceton formation by brewers' yeast pyruvate dacarboxylase isozymes

80. Lee TC, Langston-Unkefer PJ. Pyruvate decarboxylase from *Zea mays* L. I. Purification and partial characterization from mature kernels and anaerobically treated roots. Plant Physiol, 1985; 79: 242-247.
81. Miernyk JA. Glycolysis, the oxidative pentose phosphate pathway and anaerobic respiration. In: Dennis DT, Turpin DH, editors, Plant physiology, biochemistry and molecular biology. Essex: Longman Scientific & Technical, 1990: 77-100.
82. Randall DD, Miernyk JA. The mitochondrial pyruvate dehydrogenase complex. In: Methods in plant biochemistry Vol. 3, Enzymes of primary metabolism, Lea PJ, editor, London, Academic Press, 1990: 175-192.
83. Schulze-Siebert D, Schultz G. Full autonomy in isoprenoid synthesis in spinach chloroplasts. Plant Physiol Biochem, 1987; 25: 145-153.
84. Heintze A, Görlach J, Leuschner C, Hoppe P, Hagelstein P, Schulze-Siebert D, Schultz D. Plastidic isoprenoid synthesis during chloroplast development. Change from metabolic autonomy to a division-of-labor stage. Plant Physiol, 1990; 93: 1121-1127.
85. Hoppe P, Heintze A, Riedel A, Creutzer C, Schultz G. The plastidic 3-phosphoglycerate -> acetyl-CoA pathway in barley leaves and its involvement in the synthesis of amino acids, plastidic isoprenoids and fatty acids during chloroplast development. Planta, 1993; 190, 253-262.
86. Fulton DC, Kroon PA, Matern U, Threlfall DR, Whitehead IM. Inhibition of phytosterol biosynthesis in elicitor-treated cultures of *Ammi majus*. Phytochemistry, 1993; 34: 139-145.
87. Cavalier-Smith T. The number of symbiotic origins of organelles. BioSystems, 1992; 28: 91-106.

MOLECULAR CLONING OF RADISH ACETOACETYL-COENZYME A THIOLASE BY GENETIC COMPLEMENTATION OF A YEAST MUTANT

Kai-Uwe Vollack† and Thomas J. Bach¶
†Botanisches Institut II, Universität Karlsruhe, D-76128 Karlsruhe, Germany
¶C.N.R.S.-I.B.M.P., Département d'Enzymologie Cellulaire et Moléculaire, Université Louis Pasteur, F-67083 Strasbourg Cedex, France

Introduction

For the yeast and mammalian tissue it has been well documented (see [1,2] and literature cited therein) that the conversion of three units of acetyl-CoA to 3-hydroxy-3-methylglutaryl-CoA (HMG-CoA) is catalyzed by two distinct enzymes: a) acetoacetyl-CoA thiolase (AACT, acetyl-CoA acetyltransferase, EC 2.3.1.9), and b) HMG-CoA synthase (HMGS, *S*-HMG-CoA acetoacetyl-CoA lyase, CoA acylating, EC 4.1.3.5). Both reactions are mechanistically distinct, *viz.* Claisen condensation versus aldol condensation. HMG-CoA serves two functions: as the substrate for mevalonate biosynthesis, catalyzed by HMG-CoA reductase (HMGR, EC 1.1.1.34) and as a putative intermediate in the degradation of leucine *via* HMG-CoA lyase (HMGL, EC 4.1.3.4). Only little was known as to the situation in plants [1-3]. Recently we have purified an enzyme system from radish membranes, apparently catalyzing the conversion of acetyl-CoA to HMG-CoA and thus further referred to as AACT/HMGS [4]. Its quite unusual properties *in vitro*, such as apparent activation by the redox couple Fe(II)/pyrroloquinoline quinone (PQQ) suggested that plants might have developed a further mechanism of HMG-CoA synthesis. For the cloning of genes involved in the conversion of acetyl-CoA to HMG-CoA we have chosen the technique of complementation of suitable ergosterol-deficient (*erg*) and thermo-labile yeast mutants, defective either in AACT or HMGS and originally generated in the laboratory of Dr. Francis Karst (Université de Poitiers). Here we are reporting on the isolation of a radish AACT gene, complementing the *erg10* mutation of yeast.

Materials and Methods

Strains and media: *Escherichia coli* strains DH10B from Bethesda Research Laboratories and XL1-Blue MRF' from Stratagene were used as plasmid or phage hosts. Yeast strains F2sp5 [*MATa*, *erg10A*, *ura3-52*, *his3-D200*, *leu2-3,112*, *gal2*$^-$] and W303-1B [*MATa*, *ura3-52*, *his3-D200*, *leu2-3,112*, *trp*, *ade*] were kindly provided by F. Karst (Université de Poitiers, France). F2sp5 (defective in thiolase) is derived from strain FL100, which will not grow on galactose. Therefore, it was crossed with W303-1B in order to get ergosterol auxotrophic yeast strains, in which plant cDNAs placed under the control of the *GAL1* promotor can be expressed, dependent upon the ability to use galactose as a carbon source. Yeast strain KV5 thus obtained [*MATa*, *erg10A*, *ura3-52*, *his3-D200*, *leu2-3,112*, *Gal2*$^+$] was chosen for the complementation experiments with a cDNA library for functional

J.-C. Kader and P. Mazliak (eds.), Plant Lipid Metabolism, 335–337.

expression in yeast, prepared from 4-day-old etiolated radish seedlings (details of its construction must be reported elsewhere). Media for the growth of yeast were as described by Sherman [5]. Complete medium contains 1% yeast extract, 2% peptone and either 2% glucose (YPGlu) or galactose (YPGal); selective media (YNB) contains 0,67% yeast nitrogen base with ammonium sulfate, 2% glucose and appropriate supplements. Ergosterol was dissolved in Tergitol/ethanol (1:1) and was added to liquid and solid media at a concentration of 4 and 80 µg/ml, respectively, for the propagation of ergosterol-auxotrophic yeast strains F2sp5 and KV5.

Complementation of *S. cerevisiae* mutants: About 10 µg from the radish cDNA library were used to transform AACT deficient yeast strain KV5 by the LiAc method [6] but without carrier DNA in the transformation assays. Both ergosterol-auxotrophic strains, F2sp5 and its derivative KV5, exhibit rather low transformation efficiencies with approximately 5 x 10^3 $Ura3^+$-transformants /µg DNA, selected at 28°C on YNB minimal medium, supplemented with ergosterol but lacking uracil. $Ura3^+$-transformants were plated on YPGal complete medium and incubated at 37°C to select for transformants expressing galactose-induced radish AACT and thus complementing the yeast thiolase deficiency, which causes thermolability. From a total of ~3 x 10^5 $Ura3^+$-transformants 53 clones were found to be thermoresistant, but only nine of them showed the wild-type phenotype (growth at 37°C exclusively in the presence of galactose). Plasmid DNA was recovered from these yeast transformants and transformed into *E. coli* DH10B by electroporation. All nine clones carried an identical 1.5 kbp cDNA (referred to as cRS10), inserted in the *Xho*I/*Xba*I cloning site of pYEUra3. The corresponding plasmid referred to as pYRS10 was then retransformed into the yeast strain KV5, thus generating strain KV10 to confirm the galactose-dependent phenotype of thermoresistance at 37°C in 100 % of transformants.

Sequencing analysis: pcRS10 was constructed by ligating the 1.5 kbp *Xho*I/*Xba*I fragment cRS10 into pBluescript (SK^+) for sequence analysis by the dideoxynucleotide method [7]. Double-stranded plasmid DNA was sequenced using Sequenase 2.0 from USB or Taq polymerase from Promega's fmol cycle sequencing kit, both according to the manufacturers' instructions. The nucleic acid sequence of cRS10 and the deduced amino acid sequence were used to search databases for homologies with the DNASIS/PROSIS® software from Hitachi.

AACT/HMGS activity in yeast transformants: Yeast strain KV10, carrying the radish cDNA cRS10 under control of the *GAL1* promoter, the mutant strain KV5 and strain W303-1B (reference), were used to inoculate liquid cultures with 100 ml of complete medium, supplemented with ergosterol and with either glucose or galactose as a carbon source. The cultures were incubated at 26°C until the cells were harvested by centrifugation at 5,000xg for 5 min, when they had reached an OD_{600} of 1.5. Disruption of yeast cells by glass beads, preparation of protein extracts, membrane fractionation as well as protein determination were done as previously described [8]. AACT/HMGS activity was determined by incorporation of [1-^{14}C]acetyl-CoA into ^{14}C-HMG-CoA [4].

Results and Discussion

Sequence analysis of the 1462 kb insert of plasmid pYRS10 (EMBL data base accession number X78116), conferring ergosterol-autotrophy and thermoresistance to the yeast strain KV5 (*erg10*), revealed the existence of an open reading frame containing the entire coding unit for a protein of 404 amino acids (M_r 42,193). There is no evidence for the presence of a leader peptide characteristic of peroxisomal AACT from cucumber (462 amino acids, M_r 48,650), where the precursor form is approximately 4 kDa larger than the 45-kDa subunit of the mature enzyme (see [9]), and where its role is presumably limited to catalyze the last step of fatty acid degradation, rather than its being involved in isoprenoid biosynthesis. At the amino-acid level there exist significant sequence homologies with domains of thiolases from mammals, the yeast *S. uvarum* [11] and from *Zoogloea ramigera* [12], where the homotetrameric (biosynthetic) enzyme has subunits of M_r 40,598, in good agreement with the value determined by SDS-PAGE. Southern blot analysis with the radish cDNA (cRS10) under stringent hybridization conditions suggests the presence of a single gene, but most possibly having several introns. Northern blot analysis of total RNA isolated from 4-day-old radish seedlings indicates the presence of a 1.8 kb transcript. In cell-free extracts prepared from strain KV10, after induction by galactose we measured a 10-fold incorporation of [1-^{14}C]acetyl-CoA into HMG-CoA as compared to the wild-type yeast strain W303-1B. The activity was found in the cytosolic fraction only, but it was not stimulated *in vitro* by Fe(II)/PQQ. Moreover, in contrast to the radish AACT/HMGS system, in these extracts the reaction is inhibited by F-244, a specific HMGS inhibitor. From these observations we conclude a) that the product of the cloned gene is not identical to the membrane-associated AACT/HMGS system described before [4], and b) that at least in the yeast *S. cerevisiae* for the conversion of acetyl-CoA to HMG-CoA (in the presence of endogenous HMGS) AACT appears to catalyze the rate-limiting step. Further attempts are now directed towards cloning of the putative radish HMGS gene(s) by genetic complementation.

References

1. Bach TJ, Motel A, Weber T. Some properties of enzymes involved in the biosynthesis and metabolism of 3-hydroxy-3-methylglutaryl-CoA in plants. Rec Adv Phytochemistry, 1990; 24: 1-82.
2. Bach TJ, Boronat A, Caelles C, Ferrer A, Weber T, Wettstein A. Aspects related to mevalonate biosynthesis in plants. Lipids, 1991; 26: 637-648.
3. Gray, JC. 1987. Control of isoprenoid biosynthesis in plants. Adv Bot Res, 1987; 14: 25-91.
4. Weber T, Bach TJ. Conversion of acetyl-coenzyme A into 3-hydroxy-3-methylglutaryl-coenzyme A in radish seedlings. Evidence of a single monomeric protein catalyzing a Fe^{II}/quinone-stimulated double condensation reaction. Biochim Biophys Acta 1994; 1211: 85-96.
5. Sherman F. Getting started with yeast. Meth Enzymol, 1991; 194: 3-21.
6. Becker, D M, Guarente L. High-efficiency transformation of yeast by electroporation. Meth Enzymol, 1991; 194: 182 - 187.
7. Sanger F, Nicklen S, Coulson AR. DNA sequencing with chain-terminating inhibitors. Proc Natl Acad Sci USA, 1977; 74: 5463-5467.
8. Vollack K-U, Dittrich B, Ferrer A, Boronat A, Bach TJ. Two radish genes for 3-hydroxy-3-methylglutaryl-CoA reductase isozymes complement mevalonate auxotrophy in a yeast mutant and yield membrane-bound active enzyme. J Plant Physiol, 1994; 143: 479-487.
9. Kindl H. Fatty acid degradation in plant peroxisomes: Function and bioynthesis of the enzymes involved. Biochimie, 1993; 75: 225-230.
10. Dequin S, Gloeckler R, Herbert CJ, Boutelet F. Cloning, sequencing and analysis of the yeast *S. uvarum ERG10* gene encoding acetoacetyl CoA thiolase. Curr Genet, 1988; 13: 471-478.
11. Peoples OP, Masamune S, Walsh CT, Sinskey AJ. Biosynthetic thiolase from *Zoogloea ramigera.* III. Isolation and characterization of the structural gene. J Biol Chem, 1987; 262: 97-102.

PLANT STEROL BIOSYNTHESIS. IDENTIFICATION OF THE COMPONENT REACTIONS OF OXIDATIVE STEROL C4-DEMETHYLATION

Alain Rahier, Maryse Taton and Sophie Pascal
CNRS, Institut de Biologie Moléculaire des Plantes ;
Département d'Enzymologie Cellulaire et Moléculaire
28 rue Goethe, 67000 Strasbourg, France

Introduction

Sterols play a crucial role in membrane function [1]. Studies using *in vivo* or model systems for structure-activity correlations have shown that they are specific structural requirements of the sterol molecule to function in the bulk role, including a 3β-hydroxyl group at C3 and a planar tetracyclic moiety. Particularly, lanosterol (**3**) and cycloartenol (**1**), the products of squalene-oxide cyclization in animals and fungi on one hand, and higher plants on the other, become functional only after enzymic removal of the three nuclear methyl groups (fig. 1) [1,2,3]. It has been postulated that the protuding methyl groups at C14 and C4 interfere severely with lipid-lipid interactions in the membrane resulting in the inefficiency of C4 or C14 methylated sterols for membrane function. In addition, recent works suggest that the 3β-hydroxy group is essential for membrane function of sterols and cannot be replaced by a ketone without destabilization of the lipid bilayer structure. Higher plants have elaborated sophisticated enzymic systems which are able to elegantly cleave the non-activated and stable carbon-carbon bonds at C4. Until the work of our group, nothing was known about the enzymology of this process because of the difficulty of obtaining the enzymic complexes catalyzing these demethylations.

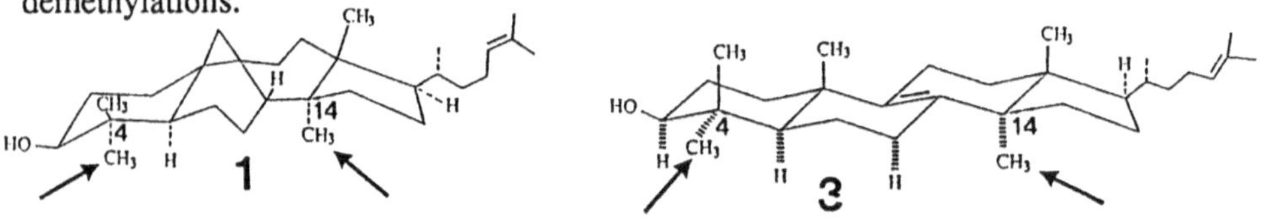

Cycloartenol   Lanosterol

Figure 1. Removal of the 4- and 14-methyl-groups protuding on the α-face of sterols precursors is essential for efficient membrane function of sterols

Identification and characterization of 4,4-dimethyl- and 4α-methylsterol-4α-methyl-oxidases [4,5].

Membrane-bound enzymatic systems obtained from maize embryos catalyze the oxidative C4-monodemethylation of 4,4-dimethyl- and 4α-methylsterols. The

J.-C. Kader and P. Mazliak (eds.), Plant Lipid Metabolism, 338–340.

demethylation process has been interrupted to demonstrate the formation of stable, oxygenated intermediates. The data demonstrate the stepwise oxidation of the C4-methyl group to the C4-carboxylic acid (fig. 2) by a still unidentified cyanide sensitive system capable of activating molecular oxygen for the C4-methyl oxidation. The stereochemistry of the demethylation of 4,4-dimethylsterols has been studied by mean of proton nuclear magnetic resonance experiments and incubation of synthetic substrate analogs. The data indicate that the 4α-methyl group of 4,4-dimethylsterols is oxidized and subsequently removed during its enzymic conversion to the corresponding 4α-methylsterol. The remaining, originally β-oriented methyl group epimerizes to produce the 4α-methylsterol by way of the corresponding 4α-methylsterone (fig. 2).

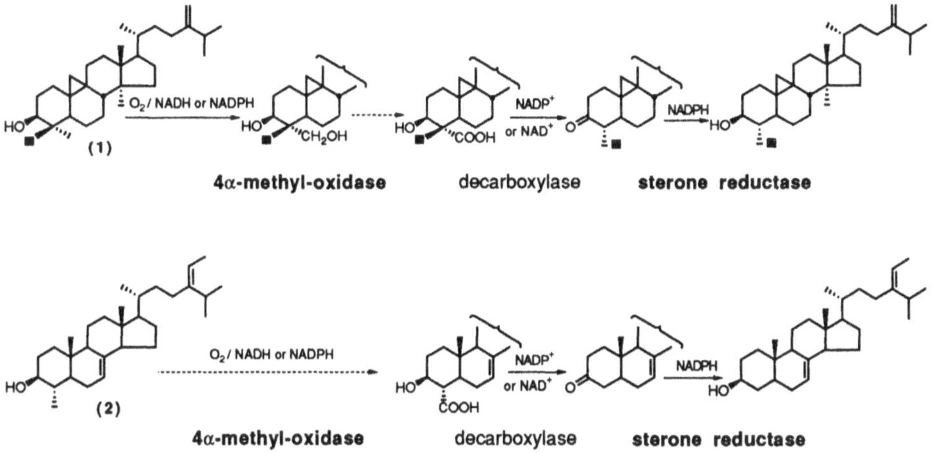

Figure 2. Pathway for the demethylation of 24-methylenecycloartanol (1) and 24-ethylidenelophenol (2) by maize microsomes

Distinct strict structural requirements for the respective oxidation of 4,4-dimethyl- and 4α-methylsterols, differential metabolites patterns observed in response to particular cofactor treatment and distinct sensitivities toward inhibition by cyanide and 3β, 5α, 6α-stigmastatriol, a novel inhibitor of 4α-methylsterol oxidase, are consistent with the conclusion that two distinct oxidative systems are involved in the removal of the first and second C4-methyl group of phytosterols precursors. Inhibition of the 4α-methylsterol oxidase by this cis-diol derivative and by o-phenanthroline, a metal chelating agent, suggest the involvement of a metal ion, presumably iron, in an enzyme bound form in the hydroxylating system. Finally, our results are consistent with the conclusion that the oxidations of (1) and (2) are not catalyzed by a P-450 dependent monooxygenase.

Identification of a NADPH dependent sterone reductase involved in sterol-4-demethylation [6].

We found that a 3-oxosteroid intermediate is involved as a product of the demethylation sequence at C4 (fig. 2). This and other observations suggest the obligatory involvement of an enzymic 3-oxosteroid reduction step. Indeed, maize microsomes were shown to catalyze the reduction of a variety of sterones to produce stereoselectively the corresponding 3β-hydroxy derivatives. The 3-oxosteroid-oxidoreductase identified is reversible, shows exclusive dependence on NADPH and is inactive with NADH. It is not sensitive to the azole inhibitors pyrifenox, ketoconazole and itraconazole nor to phenobarbital nor pyrazole. Based on these coenzyme requirements and inhibitor susceptibility, and according to the common pattern of their classification, the maize microsomal sterone reductase belongs to the family of ketone reductases. Results from a series of incubations with a large number of natural of synthetic sterones support the conclusion that the 3-oxosteroid-oxidoreductase identified is a constitutive component of the microsomal sterol 4-demethylation complex.

Conclusion

Two distinct component reactions of oxidative plant sterol C4-demethylation have been characterized. The 4α-carboxylic acid decarboxylase component will be further investigated.

In addition, the present studies directly establish that remarkably the plant C4-demethylase systems are distinct from those involved in animal and yeast where removal of both C4 methyl-groups are sequentially catalyzed by the same enzymic system. Moreover, they indicate that in photosynthetic organisms, the sterol demethylations occur in the sequence 4,14,4 in contrast to animals and wild type yeast where the sequence is 14, 4, 4. Finally, the enzymes involved in the C4-demethylation complex could allow new opportunities for the design of inhibitors which should show useful herbicidal of fungicidal properties.

References

1. Bloch KE. Sterol structure and membrane function. CRC Crit. Rev. Biochem. 1983;14:47-92.
2. Nes WR, Sekula BC, Nes WD and Adler JH. The functional importance of structural features of ergosterol in yeast. J. Biol. Chem. 1978;253:6218-6225.
3. Benveniste P. Sterol biosynthesis. Annu. Rev. Plant Physiol. 1986;37:275-308.
4. Pascal S, Taton M and Rahier A. Oxidative C4-demethylation of 24-methylene cycloartanol by a cyanide-sensitive enzymatic system from higher plant microsomes. Biochem. Biophys. Res. Commun. 1990;172:98-106.
5. Pascal S, Taton M and Rahier A. Plant sterol biosynthesis : identification of two distinct microsomal oxidative enzymatic systems involved in sterol C4-demethylation. J. Biol. Chem. 1993;268:11639-11654.
6. Pascal S, Taton M and Rahier A. Plant sterol biosynthesis. Identification of a NADPH dependent sterone reductase involved in sterol-4-demethylation. Arch. Biochem. Biophys. 1994;312:000-000.

PLANT STEROL BIOSYNTHESIS. CYTOCHROME P-450 OBTUSIFOLIOL 14α-METHYL DEMETHYLASE A KEY ENZYMATIC STEP

Maryse Taton, Florence Salmon and Alain Rahier
CNRS, Institut de Biologie Moléculaire des Plantes ;
Département d'Enzymologie Cellulaire et Moléculaire
28 rue Goethe, 67000 Strasbourg, France

Introduction

The number of cytochrome P-450-dependent oxygenases which have been identified in higher plants is still limited, and the function and specificity of plant P-450s are much less well known than in animals and fungi, probably because of the low content and supposed lability of P-450s in cell-free plant systems.Plant sterol biosynthesis involves the removal of the 14α-methyl group of obtusifoliol,catalysed by the cytochrome P-450 dependent obtusifoliol 14α-methyl demethylase ($P450_{OBT.14DM}$). The demethylation requires NADPH and molecular oxygen and is photoreversibly inhibited by carbon monoxide [1] (fig :1).

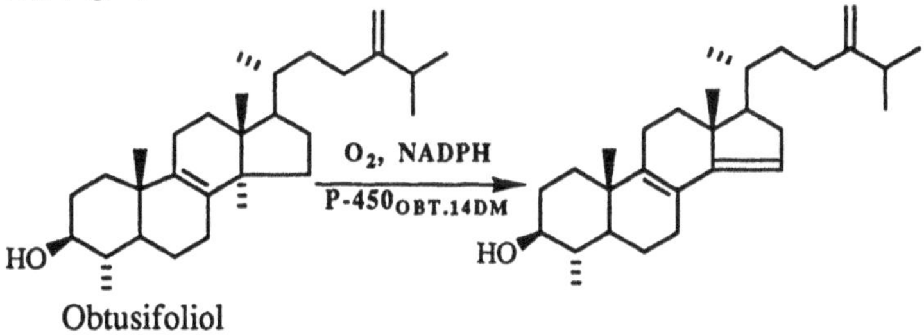

Figure 1. Reaction catalysed by Cyt $P450_{OBT.14DM}$

Sterol 14-demethylase constitutes the only known example of P-450-dependent enzyme carrying out an identical metabolic function in different eukaryotes. Thus studies on sterol 14-demethylases from different organisms should provide interesting informations about evolution of P-450 monooxygenases and differentiation of sterol synthesis pathway. A variety of important agrochemicals such as fungicides and plant growth regulators have been recently shown to inhibit $P450_{OBT.14DM}$ in correlation with plant growth inhibition effects. The phytotoxic effects of such derivatives has been ascribed to the damaging effects of excessive amounts of Δ^8-14α-methyl-sterols and (or) depletion of typical Δ^5-sterols in the plant cell membranes. This property has recently led to the search and development of potent herbicides having $P450_{OBT.14DM}$ as their primary site of action. Therefore, $P450_{OBT.14DM}$ is of interest because of its

J.-C. Kader and P. Mazliak (eds.), Plant Lipid Metabolism, 341–343.

role in plant metabolism and because it is a target for agronomically important compounds.

Critical molecular features for recognition by P-450$_{OBT.14DM}$

Plant P450$_{OBT.14DM}$ displays a high degree of substrate specificity which is clearly distinct from that of the animal and yeast sterol 14-demethylase [2,3].

The strict substrate specificity of P-450$_{OBT.14DM}$ is consistent with the constitutive nature of this new plant P-450. The different substrate specificity of the plant demethylase raises the question of whether it indicates the presence of large differences in amino acid sequence and structure when compared to that of the mammalian or yeast enzymes. Finally, the strict substrate specificity of P-450$_{OBT.14DM}$ constitutes another example of an obligatory sequence of enzymatic steps during plant sterol biosynthesis. These results sustain the hypothesis according to which plant sterol metabolism pathway is linear and does contain only very few parallele routes [3].

Inhibition of P450$_{OBT.14DM}$ by type I steroidal competitive inhibitors.

7-oxo-sterol derivatives have been shown to be potent inhibitors of cholesterol synthesis in various animal cells and yeast both *in vitro* and *in vivo*. Therefore we synthetized and tested the inhibitory potency of a series of 7-oxo-obtusifoliol analogues. 7-oxo-24ζ(24')-dihydroobtusifoliol is a potent competitive inhibitors of P450$_{OBT.14DM}$, binding 125-200 times more tightly than the corresponding substrates. Inhibition of P450$_{OBT.14DM}$ by these analogues shows strict structural requirements. Particularly 7-oxo-24(25)-dihydro-lanosterol is not effective on the plant 14DM while it inhibits strongly the 14DM from yeast. Finally it would be interesting to evaluate the dual action of these derivatives on plant sterol biosynthesis, that is inhibition of P450$_{OBT.14DM}$ and regulatory inhibition of HMG CoA reductase.

Inhibition of P450$_{OBT.14DM}$ by non steroidal type II binding inhibitors

Azoles, pyrimidines and pyridines belong to a large group of fungicides, plant growth regulators and, more recently, herbicides used for plant protection in agriculture.
The heterocyclic sp^2 nitrogen atom is able to ligand the ferric iron atom of the heme of P-450. We have directly demonstrated the potent inhibition of plant P-450$_{OBT.14DM}$ by a series of such derivatives. We have also measured the spectral binding properties of these toxicants. Maize microsomes show characteristic type II binding interactions expected for such azole derivatives. Recently, in the course of a herbicide screening program, a novel class of herbicidally active imidazole carbonic esters has been discovered by the firms Ciba-Geigy and Janssen. We have demonstrated that this series of compounds are extremely potent inhibitors of the plant P450$_{OBT.14DM}$ both *in vitro* in a maize microsomal preparation and *in vivo*. The complete inhibition of the plant 14DM, achieved with a $1x10^{-7}$M concentration of the R-(-) isomer of methyl 1-(2,2-dimethylindan-1-yl) imidazole-5-carboxylate (CGA 214372) would suggest binding interactions in the range of stoïchiometric concentrations of enzyme and this derivative. The molecular features of a variety of analogues of CGA 214372 were related to their potency as inhibitors of P450$_{OBT.14DM}$ *in vitro*, allowing delineation of the key

structural requirements governing inhibition of the demethylase. The binding affinities of the above-mentionned series of analogues of CGA 214372 to the total P450 fraction were also investigated and the corresponding Kd values determined (fig. 2) [4,5].

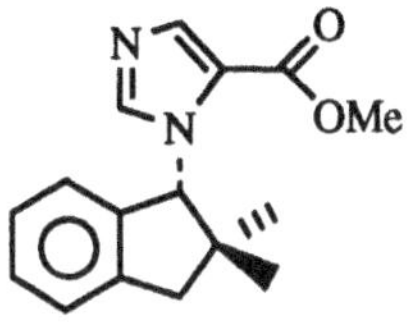

CGA 214372
non-steroidal inhibitor

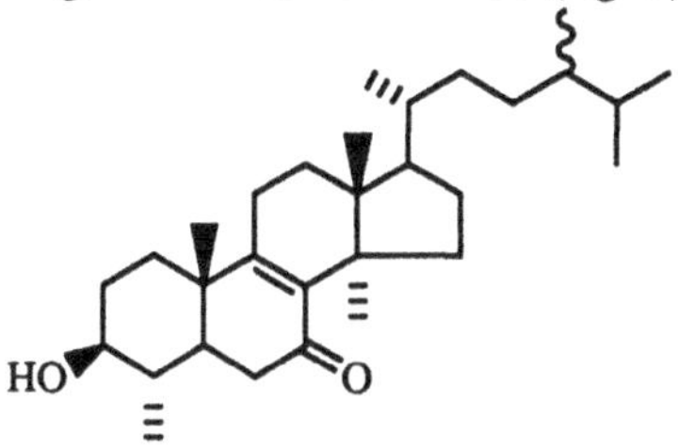

7-oxo-24ξ(24[1])-dihydroobtusifoliol
steroidal-inhibitor

Figure 2. Inhibition of $P450_{OBT.14DM}$ by different type of inhibitors

However the spectral binding data dit not correlate with the corresponding inhibition data indicating that spectroscopic analysis performed with maize microsomes are not suitable for the determination of the specific inhibition of $P450_{OBT.14DM}$ by azoles. A very good correlation between the I_{50} values measured for the *in vitro* inhibition of the plant 14DM and the herbicidal activity *in vivo* is observed, suggesting that the plant 14DM is the primary site of action of these imidazoles. To sustain this hypothesis, we studied the selectivity of CGA 214372 towards $P450_{OBT.14DM}$: among other P450 species present in maize, and 14DM from various organisms. The results gave strong evidence that $P450_{OBT.14DM}$ is a herbicidal target.

Conclusion

The results discussed prove that $P450_{OBT.14DM}$ is a target enzyme for azole herbicide (and antifungal). They show that the mammalian, fungal and plant 14DM show major differences in substrate selectivity,sensitivity towards the 7-oxo-steroïd competitive inhibitors and sensitivity towards imidazole inhibitors, indicating that P450 enzymes having the same catalytic function and metabolic role are not necessarily identical. However strong variation in substrate and inhibitor specificities may result from limited amino-acid replacement provided that they are located in the vicinity of the binding site. Finally, the example discussed here demonstrate that both plant sterol biosynthesis and P-450 systems can be exploited in the search for compounds having potential agrochemical applications.

References

1. Rahier A. and Taton M. Biochem. Biophys. Res. Commun. 1986;140:1064-1072
2. Rahier A. and Taton M. Bioch. Soc. Trans. 1990;18:52-56.
3. Taton M. and Rahier A. Biochem. J. 1991;277:483-492.
4. Salmon F., Taton M., Benveniste P. and Rahier A. Arch. Biochem. Biophys. 1992;297:123-131
5. Rahier A. and Taton M. Biochim. Biophys. Acta. 1992;1125:215-222.

EFFECT OF SALINITY ON THE STEROL CONTENT OF SOYBEAN ROOT MEMBRANES

Agnès SURJUS, Monique DURAND, Yves SAUVAIRE

Laboratoire de Recherche sur les Substances Naturelles Végétales,
Université Montpellier II, CC24, place Eugène Bataillon, 34095 Montpellier
Cedex 05, France

Synopsis

The effect of salinity on the sterol composition of plasma membrane (PM) as well as microsomal fraction (MF) was studied on soybean seedling roots in parallel with the PM ATPase activity.
A salt treatment of 25 mM NaCl in the culture medium induced an increase in the sterol content of the microsomes whereas no modification was observed in the PM. However ATPase PM activity was increased by the NaCl treatment. The relation between sterol, phospholipid compositions and ATPase activity might subsequently be elucidated.

Introduction

In the Soybean, which is a highly Na^+ sensitive plant, retention of Na^+ in the basal parts of roots and stems limits its access to the leaves. Mechanisms of Na^+ exclusion from leaves seemed to be effective only over a very restricted range of external NaCl concentration (less than 50 mM).
Plasma root membranes play an important role in the ability of plants to cope with salinity. Indeed PM exercises a high degree of control on ion fluxes via their H^+ pumps and transport proteins. More precisely, the chemical structure and physical state of the lipid environment (in particular sterol) is important for membrane protein function. Therefore we have chosen to study the sterol composition of root PM to gain further insight into the mechanism of Na^+ exclusion.

Materials and methods

Three day old seedlings of soybean (*Glycine max* L. Merr. cv. Hodgson) were grown for eight days in an aerated nutrient solution with or without 25 mM NaCl. Conditions of culture and of the growth chamber were as previously described by Durand and Lacan [1].

Abbreviations: FS: free sterol, MF: microsomal fraction, PM: plasma membrane, TS: total sterol

J.-C. Kader and P. Mazliak (eds.), Plant Lipid Metabolism, 344–346.

The root plasma membranes (PM) were purified from microsomal fraction (MF) by means of aqueous two-phase partitioning consisting of Dextran T-500 and PEG 3350 according to Sandelius et al. [2].
The purity of the PM was checked with different marker enzymes [3].
The membrane lipids were extracted as outlined by Grandmougin [4] after incorporation of betulinol as an internal standard.
Total sterols (including conjugate sterols plus free sterols) were isolated according to the method of Grunwald [5] slightly modified by Brenac [6].
Free sterols (FS) and total sterols (TS) of the lipid extract were purified by TLC, acetylated and analysed by GC with a DB1 capillary column.

Results

Table 1 shows the distribution of various membrane markers: The PM fraction was enriched in vanadate-sensitive Mg-ATPase by more than 2.6 as compared with the microsomal fraction. Moreover the tonoplastic and mitochondrial Mg-ATPase activities were lower in the PM than in the MF indicating few contaminations by these organelles. Latent IDPase activity was still unchanged in the two fractions indicating little contamination by the Golgi.
Salt treatment did not change the partitioning of the MF between the two phases of aqueous polymer mixtures as the PM from both control and salt-stressed plants were essentially of the same purity.
The only difference between the two PM fractions remained in the specific activity of the PM ATPase which was 1.4-fold more important in the salt-treated PM than in the control.

Table 1: Marker enzyme activities (µmole Pi. mg-1 protein. mn-1) in microsome fractions and PM obtained after aqueous two-phase partitioning using soybean roots from control and salt-stressed plants. *Results show the average of three preparations ± sd.*

Marker enzyme	Control plant		Salt-stressed plant	
	MF	PM	MF	PM
Vanadate-sensitive Mg-ATPase	0.188 ± 0.017	0.496 ± 0.074	0.182 ± 0.025	0.683 ± 0.064
NO_3^--sensitive Mg-ATPase	0.017 ± 0.008	0.011 ± 0.012	0.026 ± 0.008	undetectable
N_3^--sensitive Mg-ATPase	0.038 ± 0.011	0.003 ± 0.003	0.029 ± 0.002	0.013 ± 0.008
Latent IDPase	0.120 ± 0.007	0.128 ± 0.037	0.101 ± 0;045	0.108 ± 0.055

The comparison between the sterol content of PM and microsomal fraction revealed significant differences in their composition (Figure 1). The sterol content of the PM of non salt grown roots was twice the proportion observed for microsomes.
Salt treatment increased microsomal sterol content (TS and FS) but sterol PM content was little affected (in particular FS was not changed).
The sterol identified in the membranes of stressed and unstressed plants were sitosterol + isofucosterol (about 59%), stigmasterol (26%), campesterol (11%) accompanied by small amounts of cholesterol (4%).
Saline treatment induced an increase in the sterol content of each of these compounds in the microsomal fraction but the relative composition of each sterol was not changed.

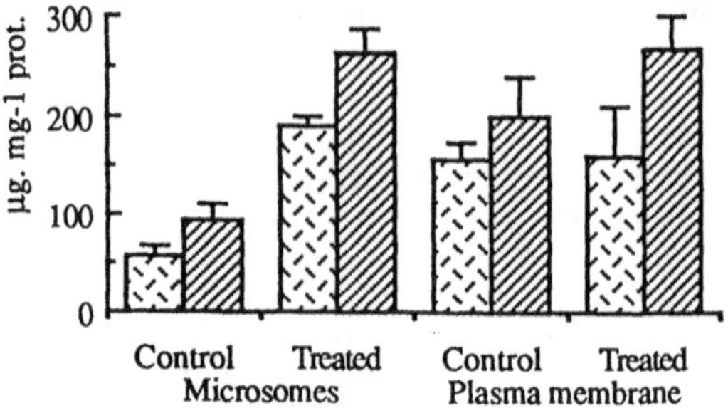

Figure 1: Free (FS:) and total sterol (TS:) content of the microsomal fraction and plasma mebrane of soybean roots from NaCl and control plants. *Results show the averages of three preparations ± sd.*

In the PM fraction, the content of each sterol was generally not changed under salt stress althought the cholesterol content decreased in both total sterol and FS; as a matter of fact the rate of cholesterol decreased significantly by 2-fold in the total sterols and not significantly by 1.6-fold in the FS.

Discussion and conclusion

Under mild conditions of salinity, FS and TS, expressed in terms of protein weight, were increased in the microsomes probably as a result of increasing synthesis. As a matter of fact the modification of the free and conjugate forms in the same proportion does not allow for discussion in terms of conversion. Moreover the lack of major changes in the proportion of the different sterols suggests a homogeneous increase to all biosynthetic fluxes. But these modifications were not recovered in the PM where the FS and TS were maintained. Besides the sterol composition was not greatly affected by salt stress. These results cannot be compared with the permeability and fluidity of the membrane because these properties depend mainly of the free sterol/phospholipid ratio. However the increase in the ATPase PM activity upon NaCl treatment suggests a modification in the fluidity of the membrane as changes in the fluid characteristics of the membranes can strongly influence ATPase mediated active ion transport.

References

1 Durand M, Lacan D. Sodium partitioning within the shoot of soybean. Physiol Plant 1994: 65-71.
2 Sandelius AS, Barr R, Crane FL, Morré DJ. Redox reactions of plasma membranes isolated from soybean hypocotyls by phase partition. Plant Sci. 1986: 1-10.
3 Galtier N, Belver A, Gibrat R, Grouzis JP, Rigaud J, Grignon C. Preparation of corn root plasmalemma with low Mg-ATPase latency and high electrogenic H+ pumping activity after phase partitioning. Plant Physiol 1988: 491-497.
4 Grandmougin A. ATPase-H+ de la membrane plasmique de racines de maïs: rôle des stérols [thesis]. Strasbourg: Univ. Louis Pasteur, 1989.
5 Grunwald C. Sterol distribution in intracellular organelles from tabacco leaves. Plant Physiol 1970: 663-666.
6 Brenac P. Stérols et sapogénines du fenugrec (*Trigonella foenum-graecum* L.). Dynamique de l'accumulation de ces métabolites dans les graines. [thesis]. Montpellier: Univ. Montpellier II, 1993.

REGULATION AND FUNCTIONS OF STEROLS IN YEAST

Leo W. Parks, Department of Microbiology, North Carolina State University, Raleigh NC 27695-7615 USA

Since our earliest work on sterols in fungi, we have obtained evidence that sterols more than simply modulate the physical properties of membranes. Initially, we proposed dual physiological functions [1], but subsequent experiments identified multiple roles for ergosterol [2]. While we have evidence for distinct functions, we are challenged to attach to the features of ergosterol the distinct biochemical activities that are regulated by them or have required sterol involvement. This is the subject of our continuing work in this field. Mutants defective in sterol synthesis have been of enormous benefit in the work on physiological functions of sterols. In this brief article, I shall describe the consequences of two mutations in ergosterol biosynthesis on the cell biology of these organisms.

Structural gene *ERG6* has the information for the sterol-24-methyltransferase. Mutations in that reaction fail to introduce carbon-28 and produce cholesterol derivatives instead of ergosterol [3]. The *erg6* strains are defective in tryptophan transport [4], fail to exclude effectively monovalent cations [5] or Brefeldin A, which disassembles the Golgi structure [6], and have reduced mating frequency [4]. The last defect is consistent with our data which show that sterols other than ergosterol in yeast impair the fusion event following mating-pheromone induced cellular adherence [7].

The *ERG6* structural gene is a weak multi-copy suppressor of a mutant in the secretory pathway. *ERD2* is the structural gene for a receptor responsible for retrieval of luminal endoplasmic reticulum resident proteins that are transported to the Golgi apparatus [8]. Mutants lacking *ERD2* are unable to divide, but have been used to select a variety of genes that can suppress the deletion. It is not clear if *ERG6* functions in that role for its methyltransferase activity or if some other *ERG6*-related function is permitting the replication of the *erd2*Δ cells. Cells that are depleted of the *ERD2*-designated protein accumulate large amounts of membranes that are unlike other components found in other sec-restricted mutants. The *erd2*Δ cells apparently accumulate also large amounts of lipid droplets. We

J.-C. Kader and P. Mazliak (eds.), Plant Lipid Metabolism, 347–349.

have shown [9] that the cellular fraction most enriched with sterol methyltransferase is the microdroplets of lipids that we have called the floating lipid layer.

Although *ERG6* appears to participate in multiple activities, *erg6* mutants are not lethal. The sterol methyltransferase is such a metabolically expensive reaction, it is most likely that the retention of such a reaction has evolutionary selective advantages to the yeast.

ERG3 is the structural gene for the sterol-C5-desaturase. Although we have proposed that the Δ^5 of sterols is essential for the growth of our yeast strains [10], research by Arthington and colleagues failed to see growth inhibition in strains lacking *ERG3* function [11]. The discrepancy appears to be caused by the genetic makeup of the strains used. We found that insertionally inactivated mutants of *ERG3* do not grow without small amounts of Δ^5 sterols in heme-deficient cells. Furthermore, growth of heme-competent cells is precluded on defined medium if the cells are forced to use respiratory substrates, such as glycerol and ethanol [12].

Our group is studying the importance of specific features of sterols on the development or maintenance of respiratory competency. One approach that we have taken is the selection of sterol mutants on respirationally or fermentatively utilized carbon sources. We reasoned that if there are structural features of sterols that are specifically required to permit respiration, we should be able to select mutations in that function on glucose but not on glycerol. We mutagenized three strains of yeast on glycerol or glucose, and selected for resistance on the polyenes, filipin and nystatin. Although we obtained a large variety of sterol mutants in the different strains on glucose with the different polyenes, on glycerol the vast majority of the mutants we selected, regardless of strain or polyene, was those having defects in side-chain modifications. While these experiments are indicative of crucial roles for the several nuclear features of ergosterol in respiration, they must be considered preliminary until biochemical bases for the functions are defined. We have shown previously that the side chain feature of ergosterol are important in regulating sterol biosynthesis [13].

This discussion emphasizes the various functions in which sterols participate in the metabolism of fungi. As more intensive studies of cell biology emerge, more functions will be identified. It is clear, however, that apparently contradictory results obtained in different laboratories generally have a logical biochemical basis. More importantly, understanding these discrepancies contribute productively toward understanding the roles of sterols in the general economy of fungi.

Acknowledgements
Special thanks are given to Dr. J.C. Kader and Dr. P. Mazliak for organizing the International Meeting on Plant Lipids at which this paper was presented. The

original research cited in this paper was conducted with the support of the National Institutes of Health (DK37222) and the North Carolina Agricultural Research Service.

References

1. Adams BG, Parks LW. Evidence for dual physiological roles of ergosterol in *Saccharomyces cerevisiae*. J Cellular Physiol 1967;70:161-168.
2. Rodriguez RJ, Low C, Bottema CDK, Parks, LW. Multiple functions for sterols in *Saccharomyces cerevisiae*. Biochim Biophys Acta 1983;837:336-343.
3. Parks LW. Metabolism of sterols in yeast. CRC Crit Revs Microbiol 1978;6:301-341.
4. Gaber RF, Copple DM, Kennedy BK, Vidal M, Bard M. The yeast ERG6 is required for normal membrane function but is not essential for biosynthesis of the cell-cycle sparking sterol. Mol Cell Biol 1989;9:3447-3456.
5. Welihinda AA, Beavis AD, Trumbly RJ. Mutations in *LIS1* (*ERG6*) gene confer increased sodium and lithium uptake in *Saccharomyces cerevisiae*. Biochim Biophys Acta 1994; In press.
6. Vogel JP, Lee JN, Kirsch DR, Rose MD, Sztul ES. Brefeldin A causes a defect in secretion in *Saccharomyces cerevisiae*. J Biol Chem 1993;268:3040-3043.
7. Tomeo M, Fenner G, Tove S, Parks LW. Effect of sterol alterations on the efficiency of conjugation in sterol auxotrophs of the yeast, *Saccharomyces cerevisiae*. Yeast 1992;8:1015-1024.
8. Semenza JC, Hardwick KG, Dean N, Pelham HRB. ERD2, a yeast gene required for the receptor-mediated retrieval of luminal ER proteins from the secretory pathway. Cell 1990;61:1349-1357.
9. McCammon MT, Hartmann M-A, Bottema CDK, Parks LW. Sterol methylation in the yeast, *Saccharomyces cerevisiae*. J Bacteriol 1984;157:475-483.
10. Rodriguez RJ, Parks LW. Structural and physiological features of sterols necessary to satisfy bulk membrane an sparking requirements in yeast sterol auxotrophs. Arch Biochem Biophys 1983;225:861-871.
11. Arthington BA, Gennetat L, Skatrud PL, Gwynn CJ, Barbuch RA, Ulbright CE, Bard M. Cloning, disruption, and sequence of the gene encoding yeast C-5 sterol desaturase. Gene 1991;102:39-44.
12. Smith SJ, Parks LW. The *ERG3* gene in *Saccharomyces cerevisiae* is required for the utilization of respiratory substrates and in heme-deficient cells. Yeast 1993;9:1177-1187.
13. Casey WM, Burgess JP, Parks LW. Effect of sterol side-chain structure on the feed-back control of sterol biosynthesis in yeast. Biochim Biophys Acta 1991;1081:279-284.

RELATIVE SQUALENE SYNTHASE ACTIVITY AND ITS KINETIC PROPERTIES IN DIFFERENT YEAST STRAINS

Carmen Socaciu, Monique Faye, Marie-Laure Dumas, Ginette Pauly and Michel Gleizes
Laboratoire de Physiologie Cellulaire Végètale,U.R.A.CNRS n°568, Université de Bordeaux I, Avenue des Facultés, 33405 Talence Cedex, France

Squalene synthase (EC 2.5.1.21) known as a membrane-bound enzyme condenses two molecules of farnesyl diphosphate (FPP) in a stable intermediate, the presqualene diphosphate (PS) which is then reduced by NAD(P)H to form squalene (S). In spite of various hypotheses concerning the binding sites and reaction centers, the catalytic machinery for the two reactions and the mechanism(s) responsible for the sterol mediated regulation of squalene synthase (SQS), have not been elucidated. Sterol auxotrophic mutant strains (1) can be used as tools for such studies. Therefore, before going to purify SQS from plant sources (2), we have undertaken the purification and comparative characterization of squalene synthase (SQS) from three yeast strains: the wild-type FL100, the mutant strain *erg10B* and the recombinant strain FK5188 pMF13.
The main objectives of the present work were to investigate the relative SQS activity in the yeast strains using its kinetic properties (PS and S synthesis, affinity for substrate *Km*). For the evaluation of SQS activity, a rapid method consisting in spectrophotometric recording of unlabeled FPP hydrolysis and substrate - dependent NADH consumption, was established and used in addition to the incorporation of [^{3}H]FPP into S. The well correlated results obtained by the both methods (to be published) show a different distribution of the SQS activity between the microsomal (M105) and cytosolic (S105) extracts and also a different PS/S ratio according to the strain.

Material and Methods

STRAINS AND CULTIVATION CONDITIONS

All the strains used derived from *Saccharomyces cerevisiae* FL100 haploïd, a mating type (WS). The mutant strain *erg10B* (MS) was obtained as previously described (1) and the recombinant strain 5188pMF13 (RS) was obtained from FK5188 (erg9, aux 32, ura3) complemented with a 5kb DNA fragment containing the SQS coding region transferred on the bacterial plasmid pMF13 (3).The cells were cultivated and harvested in the conditions previously described (4).

ASSAYS FOR MICROSOMAL AND CYTOSOLIC FRACTIONS

The cells were suspended in 0.1M potassium phosphate buffer used throughout the experiment procedure.After a centrifugation of the cell-free extracts at 105,000 g for 75 min, the pellet corresponding to microsomal fraction (M105) and the supernatant

J.-C. Kader and P. Mazliak (eds.), Plant Lipid Metabolism, 350–352.

corresponding to cytosolic fraction (S105) were separately processed. The SQS activity was assayed in all fractions by two methods: the incorporation of [^{3}H] FPP into squalene and its identification by radio gas-liquid chromatography and the spectrophotometric recording of unlabeled FPP hydrolysis and substrate-dependent NADH consumption. For PS synthesis, the same methods were employed with NADH-free incubation medium.

Results

DISTRIBUTION OF THE SQS

The incorporation rate of [^{3}H]FPP into squalene observed in the microsomal fraction from WS yeasts (Table 1) show that the SQS activity is mainly localized in microsomes (80%). In RS and MS the S105 extracts have an incorporation rate strongly enhanced and especially MS of which the SQS activity is equally distributed between the M105 and S105 The specific activity is positively correlated with the main localization of SQS activity (about 4 nmol S/min/mg inWS and MS where the activity is principally microsomal). The affinity for substrate is higher in the WS than the two other strains.

Table 1 Relative incorporation rates of [^{3}H] FPP into Squalene, specific SQS activities and affinities for substrate (*Km*) in M105 and S105 extracts from the three yeast strains

Yeast strain	Incorporation rate %		SQS Specific activity (nmol S/min/mg)		*K m*	
	M105	S105	M105	S105	M105	S105
WS	80	20	3.83	2.84	10.8	13.3
RS	45	55	2.92	2.75	12.5	14.5
MS	64	36	4.16	2.83	13.2	15.0

PRESQUALENE AND SQUALENE SYNTHESIS

In the MS yeasts, not only the SQS microsomal activity is higher than that of the WS as previously found by M'Baya (4) but also the PS synthesis is higher than that of the RS and WS. The values of PS/S ratios calculated for the microsomal extracts from the three strains are less than 1, slightly higher for the WS and MS than for the RS. The PS/S ratio observed for cytosolic extract from the MS is 2-fold increased regarding to the WS.

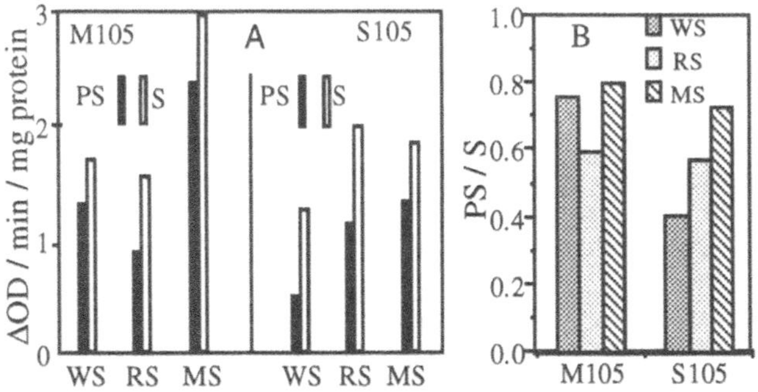

Figure 1 Presqualene and squalene synthesis. A, Relative PS and S activities measured by spectrophotometric assay during the incubation of M105 and S105 extracts from WS, RS and MS yeasts. B, PS /S ratios. The enzyme activity was expressed as Δ OD / min / mg protein.

EFFECTS OF KCl AND LIPOSOME CONCENTRATIONS ON THE SQS ACTIVITY

Figure 2 shows the SQS activity measured in WS microsomal extracts in the presence of different KCl concentrations (A) and the effect of liposome addition to 0.2 and 0.5 M KCl incubation medium. Between 0.1 and 0.5M KCl SQS activity is unchanged but superior to that which is measured in the absence of KCl. The adding of liposomes (PC) from 30 to 75μg results in a increased SQS activity and especially in the presence of 0.2M KCl

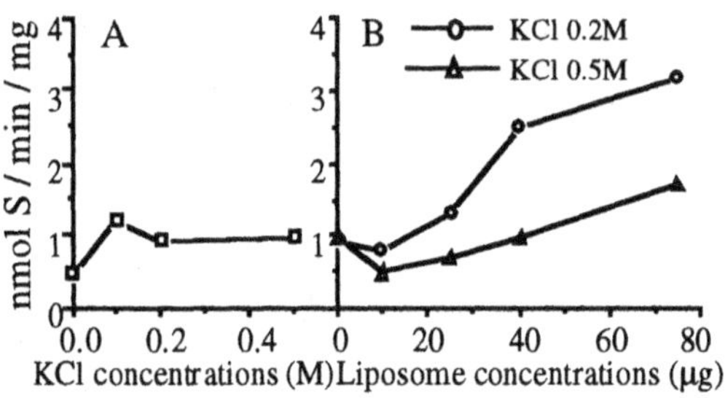

Figure 2 Effects of KCl (A) and liposome (B) concentrations on the SQS activity. The enzyme activity was determined in the absence and in the presence of 0.1, 0.2 and 0.5 M KCl. Different liposome concentrations (PC from 10 to 75μg) were added to the incubation medium containing 0.2 or 0.5M KCl.

Conclusion

Comparisons of SQS activity in the yeast strains show differences in the affinity for FPP, the PS to S ratio, and mainly the distribution of SQS between the microsomal and cytosolic fractions. It could be suggested that in the cytosolic fraction, there is the truncated soluble active form of SQS as already reported by Shechter (5). In order to contribute to a better knowledge of the enzyme structure, the mutant and recombinant strains could be efficiently used. These data together with other experiments made with NAD and NADH (Socaciu et al., to be published), could suggest the existence of a single catalytic site for this enzyme.

Acknoledgements

We thank Prof. Dr. F. Karst for providing the yeasts strains.

References

1 Servouse M, Karst F. Regulation of early enzymes of ergosterol biosynthesis in *Saccharomyces cerevisiae*. Biochem J 1986; 240: 541-547.
2 Belingheri L, Beyer P, Kleinig H, Gleizes M. Solubilization and partial purification of squalene synthase from daffodil microsomal membranes, FEBS Lett, 1991; 292: 34-36.
3 Fegueur M, Richard L, Charles A.D, Karst F. Isolation and primary structure of the ERG9 gene of *Saccharomyces cerevisiae* encoding squalene synthetase, Curr. Genet 1991; 20: 365-372.
4 M'Baya B, Fegueur M, Servouse M, Karst F. Regulation of squalene synthetase and squalene epoxidase activities in *Saccharomyces cerevisiae*. Lipids 1989; 24: 1020-1023.
5 Shechter I,. Klinger E, Rucker M.L et al. Solubilization. purification. and characterization of a truncated form of rat hepatic squalene synthetase. J. Biol. Chem. 1992; 267, 8628-8635.

JASMONIC ACID INDUCED CHANGES IN CAROTENOID LEVELS AND ZEAXANTHIN CYCLE PERFORMANCE

Markus Betz, Christiane Schindler, Jörg Schwender and H. K. Lichtenthaler
Botanisches Institut der Universität, Kaiserstraße 12, D-76128 Karlsruhe, Germany

Introduction

Jasmonates represent new endogenous plant growth regulators which are widely distributed in higher plants. They possess various, often ABA-like physiological potencies when exogenously applied [1]. They inhibit plant growth and induce leaf senescence as indicated by loss of chlorophylls [2, 3]. The exact physiological role of jasmonates, whether hormonal regulators or stress indicators, is still under investigation [4]. Though there are various observations that jasmonates promote chlorophyll degradation during leaf senescence, there is no information whether both chlorophylls a and b and also carotenoids and individual carotenoids are broken down under the influence of jasmonic acid (JA) and methyl jasmonate (JA-Me) at the same rate or in a different way. In order to obtain a detailed insight into the JA-induced senescence processes of the photosynthetic pigment apparatus, we studied in excised leaves of barley the kinetics of chlorophyll and carotenoid degradation using the HPLC technique. Jasmonate was applied in form of its methylester which is easily taken up by intact leaves. Particular emphasis was given to the question whether the pigments of the light harvesting chlorophyll proteins (LHCPs) or those of the reaction center proteins (CPa, CPIa) [5] were broken down at different rates. In recent years the zeaxanthin cycle was established as a stress response and photoprotective system of plants [6-8]. So it was of interest to study the action of jasmonates on the zeaxanthin cycle carotenoids and the zeaxanthin cycle performance i.e. the photoconversion of violaxanthin to zeaxanthin.

Material and Methods

2 cm long primary leaf segments of 6 d old barley seedlings (*Hordeum vulgare* L.) were floated on water and JA-Me (10^{-4}M). The senescence of leaf segments was investigated during 4 d at low to medium light of 100 µmol m^{-2} s^{-1}. Each day samples were taken for pigment determination. Some of the segments were irradiated with 1000 µmol m^{-2} s^{-1} for 10 min to examine the photoconversion of violaxanthin (**v**) via antheraxanthin (**an**) into zeaxanthin (**z**). The photosynthetic pigments were extracted with 100% acetone [5] and separated and quantitatively determined by reversed phase HPLC [9]. The results are the mean of 4 pigment determinations; SD = ± 6% or less.

Results and Discussion

In segments of primary barley leaves floating on JA-Me solution, the breakdown of chlorophylls a and b as well as of total carotenoids x+c was strongly enhanced within 4 days as compared to the water control (Figure 1). The total chlorophyll content decreased about 92.4% under JA-Me treatment as compared to

J.-C. Kader and P. Mazliak (eds.), Plant Lipid Metabolism, 353–355.

only 50% in the H_2O-control. Under the influence of JA-Me the chlorophylls a+b were broken down much faster than the total carotenoids. This is evident from the ratio chlorophylls/carotenoids (a+b)/(x+c), which decreased in the JA-Me treated leaf segments from 6.2 to 1.7, whereas it decreased only very little (to 5.7) in the control segments (Figure 2). Chlorophyll b was broken down much faster than chlorophyll a with JA-Me. Consequently, the chl a/b ratio increased significantly from 3.8 to about 9, but rose only from 3.8 to 4.7 in the water control (Figure 2). Among the carotenoids ß-carotene, a component of the reaction center pigment proteins CPa and CPI [5], was broken down significantly slower than lutein and neoxanthin, which are bound to the LHCPs [5]. Between the 2nd and the 4th day of JA-Me treatment the ß-carotene content decreased about 54%, whereas lutein and neoxanthin decreased by 74% and 92%, respectively.

In contrast, the zeaxanthin cycle carotenoids (**v+an+z**) were retained during the first two days of JA-Me promoted leaf senescence, but on the 3rd and 4th day violaxanthin was broken down quickly, whereas the content of zeaxanthin (**z**) and initially also that of antheraxanthin (**an**) increased (Figure 3). After 4 d of JA-Me treatment the percentage of **v** of the total **v+an+z** decreased from the initial 92% to 32% , whereas **z** had increased from 1% to 56% of total **v+an+z**. The high light-induced photoconversion of **v** into **z** was not affected by JA-Me treatment. In illuminated leaves (1000 μmol m^{-2} s^{-1} for 10 min) **z** amounted to 77% of **v+an+z** after 4 d of JA-Me treatment as compared to 36% in the water control. The dark back-reaction of **z** into **v** was not influenced by JA-Me and proceeded in controls and JA-Me treated leaf segments at similar rates. After 24h darkness or low light (10 μmol m^{-2} s^{-1}) **v** amounted to ca. 90% of the sum of **v+an+z** in both cases.

The treatment of barley leaf segments with JA-Me induced a faster breakdown of all photosynthetic pigments as compared to the water control, except for zeaxanthin and antheraxanthin. This is evidence that jasmonates stimulate and enhance senescence of the photosynthetic pigment apparatus. The faster breakdown of chl b than chl a as well as of lutein and neoxanthin than β-carotene indicates that JA-Me initiates an enhanced breakdown of the LHCPs, whereas the reaction center chlorophyll a-proteins CPa and CPI are not affected by JA-Me treatment or to a much lower degree.

Among carotenoids the zeaxanthin cycle components **v+an+z** showed a temporary increase in the level of **an** and a continuous increase in the level of **z** in JA-Me-treated segments. It appears that these increases are jasmonate induced phenomena. Accumulation of **z** and **an** at low and medium light levels as well as the higher photoconversion of **v** into **z** at high-light conditions represent a photoprotection of the photosynthetic apparatus [6, 7]. This photoprotective mechanism apparently proceeds at higher rates when the total chl content declines as was shown here for the JA-Me treatment. A higher activity of the zeaxanthin cycle had also been demonstrated during greening of tobacco leaves [9].

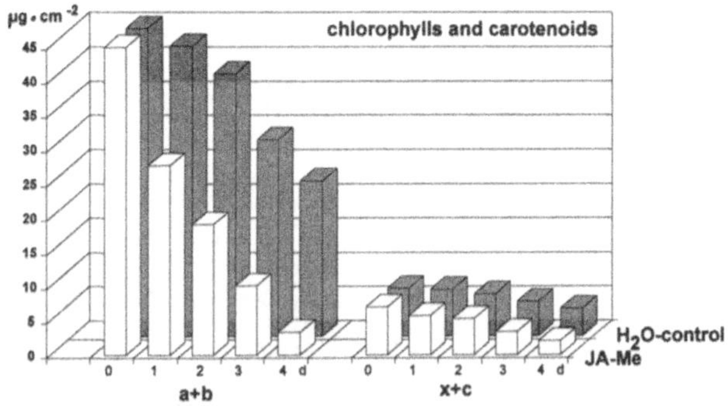

Figure 1: JA-Me enhanced breakdown of total chlorophylls (a+b) and carotenoids (x+c) in barley.

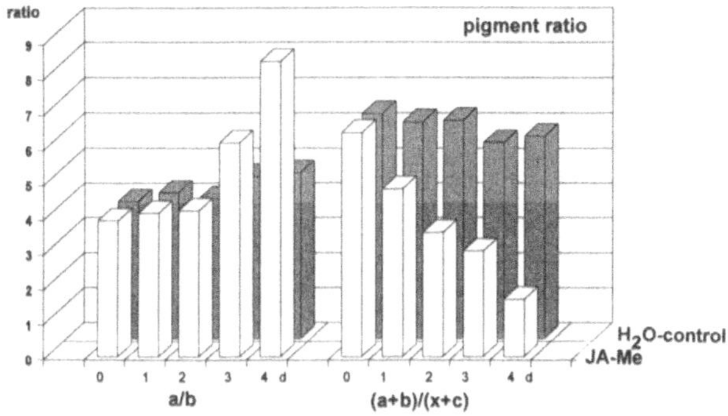

Figure 2: Changes in the pigment ratios of chlorophylls (a/b) and chlorophylls to carotenoids (a+b)/(x+c) in barley upon JA-Me treatment.

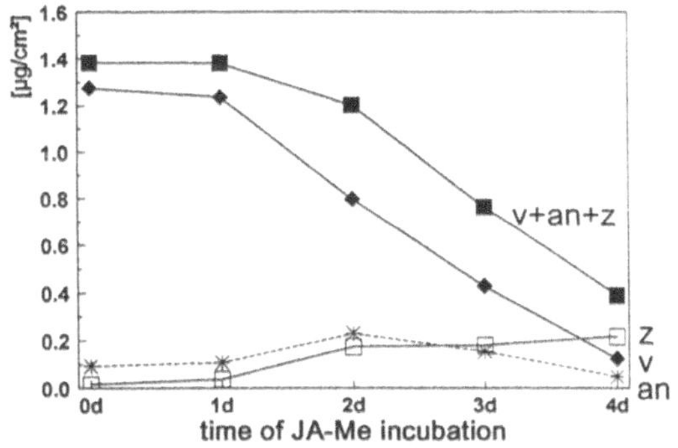

Figure 3: Changes in the level of the zeaxanthin cycle carotenoids violaxanthin (**v**), antheraxanthin (**an**) and zeaxanthin (**z**) during JA-Me enhanced senescence.

References

1. Ueda J, Kato J. Isolation and identification of a senescence-promoting substance from wormwood (*Artemisia absinthium* L.). Plant Physiol 1980; 66: 246-249.
2. Dathe W, Rönsch H, Preiss A, Schade W, Sembdner G, Schreiber K. Endogenous plant hormones of the broad bean, *Vicia faba* L. (-)-jasmonic acid, a plant growth inhibitor in pericarp. Planta 1981; 153: 530-535.
3. Yamane H, Takagi H, Abe H, Yokota T, Takahashi N. Identification of jasmonic acid in three species of higher plants and its biological acivities. Plant Cell Physiol 1981; 22: 689-697.
4. Parthier B. Jasmonates: Hormonal regulators or stress factors in leaf senescence? J. Plant Growth Regul 1990; 9: 57-63.
5. Lichtenthaler HK. Chlorophylls and Carotenoids: Pigments of Photosynthetic Biomembranes. Meth. Enzymol. 1987, 148: 350-382.
6. Demmig-Adams B. Carotenoids and photoprotection in plants: A role for the xanthophyll zeaxanthin. Biochim Biophys Acta 1990, 1020: 1-24.
7. Lichtenthaler HK, Schindler C. Studies on the photoprotective function of zeaxanthin at high-light conditions. In: Murata N, editor. Research in Photosynthesis, Vol. IV. Dordrecht: Kluwer Academic Publishers, 1992: 517-520.
8. Hager A. The reversible, light induced conversion of xanthophyll in the chloroplast. In: Czygan FC, editor. Pigments in Plants. Stuttgart.: Fischer Verlag, 1980: 57-79.
9. Schindler C, Reith P, Lichtenthaler HK. Differential levels of carotenoids and decrease of zeaxanthin cycle performance during development in a green and an aurea variety of tobacco. J Plant Physiol 1994, 143: 500-507.

PURIFICATION AND CHARACTERIZATION OF ABIETADIENE CYCLASE OF MARITIME PINE (*PINUS PINASTER* Ait.)

Jocelyne Walter, Bernard Laprébande, André Laferriere, Haoua and Agnès Saint-Guily
Laboratoire de Physiologie Cellulaire Végétale, U.R.A. CNRS 568
Université Bordeaux I, Avenue des facultés, 33405 Talence Cedex, France

The content of the secretory system of needles of *Pinus pinaster* from corsican origin is remarkable : beside resin acids, high amounts of diterpene hydrocarbons are accumulated, mainly 7,13-abietadiene.
Abietadiene synthase, an enzyme which catalyzes the cyclization of geranylgeranyl pyrophosphate to 7,13-abietadiene (fig. 1) has been partially purified from the 100.000g supernatant of young needle extract by a combination of PEG precipitation, ion exchange, affinity color and gel filtration chromatographies.
The molecular weight of this cyclase was estimated by gel filtration to be 60 ± 5 kDa. It exhibited a pH optimum of 7.5 and preferred Mg^{2+} as the required divalent metal cation. This enzyme possesses a Km of 0.3 µM for GGPP.

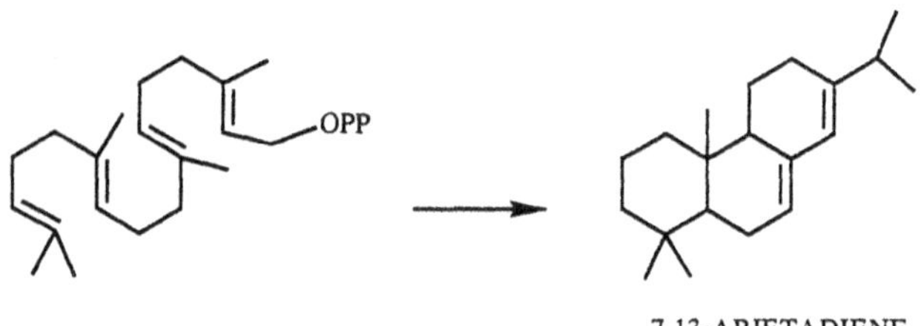

Figure 1 : Cyclization of geranylgeranyl pyrophosphate to 7,13-abietadiene

Material and Methods

ENZYME ISOLATION

In May-June, the growing parts of youg pine needles (10g) are ground in 180 ml of Tris-HCl 100 mM, pH 8.3, 0.15 M saccharose, 2 mM β-mercaptoethanol, soluble PVP 1%. After centrifugation, 100.000g supernatants are frozen at -50°C for subsequent cell-free assays and purification procedure. All the above steps and all subsequent operations were carried out at 4°C.

J.-C. Kader and P. Mazliak (eds.), Plant Lipid Metabolism, 356–358.

ENZYME PURIFICATION

In a first step, the diluted proteins were precipitated with PEG 6000 (10-30%). The precipitated proteins were suspended in a minimum volume of Tris HCl 50 mM, pH 7.5, 1 mM DTT, 10% glycerol, 2 mM $MgCl_2$ (buffer A).
The PEG precipitates have been directly loaded on a DEAE-Sephacel ion exchange column which was previously equilibrated with buffer A and then washed with the same buffer. A linear gradient in the same buffer (total volume of 200 ml) running from 0 M to 0.3 M NaCl was then applied to the column. Fractions containing cyclase activity were pooled and dyalized against buffer A.
The concentrate (centriprep 30-Amicon) was applied to a column of Cibacron Blue 3GA (Sigma) which has been previously equilibrated with buffer A. The column was washed with this buffer to remove unadsorbed proteins and then eluted with increasing NaCl molarity steps, 0.5 M, 1M, 1.5M, 2 M.
Fractions containing cyclase activity were pooled and concentrated. This concentrate was next applied to a Superdex 75 HR 10/30 on a fast protein liquid chromatography system (Pharmacia) previously equilibrated with buffer A.
An Ultrogel ACA 44 column was employed to estimate cyclase molecular weight.

ENZYME ASSAY

Cyclase activity was determined by incubating the preparation (usually 10 µg protein) in 1.0 ml of 50 mM Tris-HCl pH 7.5 containing 1 mM DTT and 10% glycerol with 2 mM $MgCl_2$ and 13 µM labeled geranylgeranyl pyrophosphate at 28°C for 1h. After incubation, 1.5 ml of pentane was added to the reaction mixture. The labeled products contained therein were extracted and olefin were purified on silica gel, then examined by radio GLC following the addition of appropriate carrier standards.

Results

Since diterpene biosynthesis, like monoterpene biosynthesis, is presumed to be compartmentalized within the secretory cells of the resin ducts, only the lower parts of the growing needles where secretory cells are lighly active were ground.
The 100.000g supernatant obtained from the extract showed a diterpene cyclase activity. The resulting olefinic product was identified by radio-GLC and argentation liquid chromatography. This compound has the same retention time on GC, as well as on silice-silver nitrate column, as 7,13-abietadiene (standard marker synthesized by reduction of abietic acid).

PURIFICATION OF ABIETADIENE CYCLASE

Since the needle extraction procedure affords relatively large volumes of dilute protein solution, a PEG precipitate (10 to 30%) was prepared directly from the 100.000g supernatant obtained from the extract.
This procedure not only served to concentrate the desired protein but also to remove competing phosphatases and it provided a roughly 2 fold purification of the abietadiene cyclase activities (table 1). The precipitate obtained was suspended in Tris buffer A and applied to a DEAE Sephacel column that was developped with a gradient of increasing NaCl content. Abietadiene cyclase activity was eluted at about 0,18M NaCl.
This ion exchange chromatography step removed most of the polyphenolic compounds and provided a 5-fold purification of the enzyme.
The fractions containing cyclase activity were dyalized against buffer A, combined and concentrated and further purified by affinity color with increasing NaCl molarity steps. Abietadiene cyclase was eluted principally from 1 to 1,5M NaCl. The affinity extract was dyalized and then subjected to FPLC gel permeation. At this stage of the fractionation, 85-fold

purification of the soluble supernatant of the needle homogenate has been achieved (Table 1) and SDS gel electrophoresis showed the presence of two bands, 62 kDa and 70 kDa respectively.

	Total proteins mg	Total activity pmol/h	Specific activity pmol/mg/h	Purification	yield %
Supernatant	205	594	2.9	1	100
PEG	21	94.5	5	1.7	15.9
DEAE	3	45	15	5.1	7.5
Affinity color	0.1	4.9	49.1	16.7	0.8
Gel permeation	0.01	2.5	250	85	0.4

Table 1 : Protein content, abietadiene cyclase activity of the different chromatographic steps during the purification procedure.

ENZYME CARACTERIZATION

Our studies show a relative stability of the abietadiene cyclase which operates optimally in the pH range 7,2 to 7,5. Abietadiene cyclase prefers Mg^{2+} as cation, Mn^{2+} is not only less effective but also inhibitory at concentration exceeding 2 mM. It is inhibited by NaCl (I50=0,5M). Ionic detergent showed an inhibitory effect whereas non ionic ones (triton X100, Tween 80) enhanced the cyclase activity.
The DEAE extract was subjected to gel permeation. Calibration of Ultrogel ACA 44 indicated the abietadiene cyclase activity to possess a molecular weight of 60 +/- 5 kDa. The Km value for the substrate geranylgeranyl pyrophosphate is 0,3 µM.
The properties of this diterpene olefin cyclase which is the first from conifers to be examined in detail, are quite similar to other monoterpene, sesquiterpene and diterpene cyclases described previously (1,2,3,4)).

Our data suggest the existence of a single soluble abietadiene cyclase catalyzing the cyclisation of GGPP in a single product. The purification of abietadiene synthase from conifer needles is complicated because of the high amounts of polyphenols and resin and the poor development of the secretory tissues.
Through the purification of the enzymes involved in diterpene biosynthesis, our purpose is the understanding of the whole chemical pathway leading to the excretion of resin.Cell cultures from several conifer species are initiated in order to study the regulation of the diterpene biosynthesis and its possible correlation with the differentiation of plastids.

Bibliography

1 Dehal SS, Croteau R. Partial purification and characterization of two sesquiterpene cyclases from sage (*Salvia officinalis*). Arch. Biochem. Biophys. 1988;261(2):346-356.

2 Dudley MW, Green TR, West CA. Biosynthesis of the macrocyclic diterpene casbene in castor bean (*Ricinus communis* L.) seedlings. Plant Physiol. 1986;81:343-348.

3 Gijzen M, Lewinsohn E, Croteau R. Characterization of the constitutive and wound-inducible monoterpene cyclases of grand fire (*Abies grandis*). Arch. Biochem. Biophys. 1991;289(2):267-273.

4 Savage J, Hatch MW, Croteau R. Monoterpene synthases of *Pinus contorta* and related conifers:a new class of terpenoid cyclase. J. Bio. Chem. 1994;269(6):4012-4020.

Section 5

Lipids and Environment

RECENT ENVIRONMENTAL CONCERNS AND LIPID METABOLISM

John L. Harwood

Biochemistry Department, University of Wales, College of Cardiff,
PO Box 903, Cardiff, CF1 1ST, U.K.

Introduction

A wide variety of environmental factors have been shown to produce effects - sometimes lethal - on plant lipid metabolism. These factors range from 'natural' influences (such as light) over which Man has little control to xenobiotics which are used deliberately (e.g. pesticides) or which enter the environment accidently. Some factors such as temperature, sulphur dioxide, oxides of nitrogen and ozone can be classed as 'natural' but their levels (and, therefore, their effects on plants) are exacerbated by human activities.

Plants can react to environmental factors in two ways. The initial exposure may be capable of causing potentially harmful effects on living cells which will be stressed. Plants may respond to the initial exposure and, if stress is involved, then their survival will depend on how well their response counteracts the injurous effects. Secondly, some species may become resistant to stress through genetic adaptation during evolution. Alternatively, some plants are resistant to a particular stress by acclimation gained by a previous exposure to the stress. Furthermore, there are cases where exposure to a given stress of defined intensity and duration may cause increased tolerance to another stress. The breeding of new varieties or creation of transgenic plants which are resistant to, or which can acclimatise better to a given environmental insult, is of clear importance to the agricultural industry.

A number of reviews have been published recently [1-3] and the reader is referred to these for details. In this article I shall concentrate on several areas which have seen advances in the last couple of years. These subjects are temperature effects and chilling tolerance, the combination of temperature and CO_2 changes as predicted in the "Greenhouse Effect" and pesticide action.

J.-C. Kader and P. Mazliak (eds.), Plant Lipid Metabolism, 361–368.

Environmental Factors of Relevance to Lipid Metabolism

Environmental factors which have been shown to alter lipid content and/or lipid metabolism in plants and algae are listed in Table 1. Obviously, light as a prime factor in the life of photosynthetic organisms is bound to have many indirect effects on lipid metabolism. Thus, the development of photosynthetic membranes and their specific arrangement in (chloroplast) thylakoids will necessitate production of those membrane lipids required for that purpose. Some specific requirements seem to hold and the presence of trans- 3-hexadecenoate at the sn-2 position of chloroplast phosphatidylglycerol has intrigued biochemists for over two decades. The synthesis of this particular acid (by desaturation of palmitate on phosphatidylglycerol [4]) has an absolute requirement for light in higher plants [5] and in most algae. Surprisingly, in view of speculations as to its role in membranes [e.g. 6,7], mutants lacking in the acid seem to have virtually unimpaired photosynthesis [8]. The absolute light requirement for trans- 3-hexadecenoate synthesis has qualitative aspects - not all wavelengths being equally active [9]. Thus, the formation of this one lipid is a good example of both the quantitative and the qualitative aspects of light on lipid metabolism.

Table 1 Environmental Factors which can Change Lipid Metabolism

Light	Quantitative, qualitative aspects.
Temperature	Cold or heat stress, average growth temp.
Water stress	Deficit or excess (water-logging).
Soil/water constituents	e.g. cations, fertilizers.
Atmospheric components	e.g. oxides of nitrogen, O_3, CO_2.
Xenobiotics	e.g. herbicides.
Other factors	e.g. physical damage, pest attack.

It is well-known that light stimulates lipid synthesis in photosynthetic tissues generally. This effect can be followed conveniently through the incorporation of radioactivity from ^{14}C-acetate. Some of the stimulation is probably due to increased cofactor supply [see 3]. However, there may be other controls on carbon flux to fatty acids and their incorporation into acyl lipids. Attention has focused particularly on acetyl-CoA carboxylase activity. Since this enzyme catalyses the first committed step in lipid synthesis and is known to be subject to careful regulation in other organisms [10] then such a concept is perfectly reasonable. Experiments with the maize acetyl-CoA carboxylase identified a number of factors which would change in the stroma during active photosynthesis and, when similar alterations were performed in carboxylase assays in vitro increased activity 24-fold [11]. This increase was of the same order as that of light-stimulated fatty acid synthesis [10]. When measuring acyl-ACP pools, Post-Bietenmiller and her associates [12] found direct evidence for an important regulatory function of acetyl-CoA carboxylase in spinach leaves.

In order to quantitate the purported role of acetyl-CoA carboxylase in controlling the light-driven carbon flux to lipids we have measured flux control coefficients in barley and maize leaves. To do this we made use of the specific inhibition of acetyl-CoA carboxylase by two grass-specific herbicides (see later) fluazifop and sethoxydim [13]. Values we obtained varied in the range 0.45-0.61. For the flux control theory the sum of the coefficients of all the enzymes involved in a metabolic pathway is 1. Thus, considering the large number of enzymes involved in the production of lipids by isolated chloroplasts then the values obtained [13] show just how important an enzyme acetyl-CoA carboxylase is in controlling light-driven lipid synthesis.

So far as temperature effects (Table 1) are concerned, then it is usually low temperature stress or freezing injury that is examined [2]. However, phase-separation of membrane lipids is known to cause damage at higher temperatures. Often connected with high temperatures, though not necessarily so, is water stress. A comprehensive review of water deficit stress on plant lipids has been made recently [14]. Water-logging is another well-known stress for many plants producing toxic effects mainly through limiting oxygen availability [3].

Adaptive salt tolerance by plants or salt stress itself (Table 1) has been noted to give rise to changes in the proportions of membrane lipid classes as well as to their acyl composition. Some stresses are due to ionic effects which could be caused by almost any salt while others are due to specific toxicology as a result of particular ions [see 3]. Often agricultural practices will lead to sufficient fertilizers inputs to cause changes in plant lipid metabolism or to interact with other environmental factors (see below for the Greenhouse Effect).

Atmospheric constituents reported to change plant lipids or their metabolism include sulphur dioxide, oxides of nitrogen and ozone [3,6]. Atmospheric CO_2 is changing, because of the "Greenhouse Effect" but, in addition, the partial pressure of CO_2 varies significantly with altitude. Recent experiments (see below) show that CO_2 alone has significant effects on lipid metabolism, perhaps because of its action of leaf development and morphology [15].

The final two categories listed in Table 1 are xenobiotics and other factors. Xenobiotics, which include pesticides (herbicides, insecticides, fungicides), are discussed later but there are many cases of accidental spillages of chemicals (e.g. hydrocarbons) which often have disasterous effects on vegetation. Damage to plants by physical insults or pest attack may alter lipid metabolism. For example, wounding stimulates the plant to repair the cut surface, especially by suberin whose precursors are fatty acids. Insect attack allows microbial entry into the plant tissue and/or may stimulate the wounding responses. Changes in membrane lipids due to wound-response hormones such as jasmonic acid or degradation of lipid stores due to lipase activity are two examples of what may happen [see 3].

Temperature Effects

The effect of environmental temperature on plant lipids has been studied particularly in relation to chilling sensitivity. The way in which (photosynthetic) organisms adapt to lower growth temperatures or the possible genetic manipulation to increase resistance are very important topics in agriculture. The general area has been reviewed [2,16]. Murata and his associates (see chapters in this book) have made particular contributions. In particular, they have been able to manipulate cyanobacteria by increasing unsaturation [17] or to increase unsaturation of the key thylakoid lipid phosphatidylglycerol by introducing an acyltransferase which utilised oleoyl-ACP rather than palmitoyl-ACP into chilling-sensitive plants [18]. In both cases the changes produced in the membranes of transgenic organisms confirmed resistance to chilling temperatures. Very recently the activation of desaturase genes as a result of low growth temperatures has been studied further. By using catalytic hydrogenation methods, Murata and co-workers [19] were able to show that the cyanobacterium senses the fluidity of its plasma membranes and, when this is low, expression of the 12-desaturase gene is induced. The resulting increase in linoleate membrane content restores fluidity to normal levels [19].

The Greenhouse Effect

Emissions of CO_2, primarily as a result of combustion, are predicted to double atmospheric levels to 650-700 p.p.m. by the middle of the 21st century. Associated with this is a rise in ambient temperature thought to be around 4°C. These two environmental increases are often referred to as the "Greenhouse Effect" which is the subject of much current interest. A concerted study involving research into a variety of physiological and biochemical parameters has been mounted in the U.K. under the BAGEC initiative. Wheat has been chosen as the test species because of its importance as a cereal crop. So far as lipid metabolism is concerned, we have been studying two particular aspects. First, does the "Greenhouse Effect" alter membrane lipid synthesis in connection with any changes in the growth and development of wheat plants? Second, does the wheat grain contain altered lipids or lipid amounts which could influence the subsequent use of the seeds for flavour? (In that connection, it may be surprising for readers to learn that wheat lipids play a profound influence on the nutritional and physical properties of flour despite their low amounts [20]. Lipids are also particularly important in the storage properties and wastage of wholewheat flours [21]).

Dealing with wheat grains first, we have studied their composition following growth under ambient temperatures and CO_2 levels, in two successive years. Test plants had a 4°C increase and a doubling of CO_2 from 350 ppm to 700 ppm for their growth conditions. Nitrogen fertilizer application was also varied. Thus, we were able to study, by analysis of variance, the independent or interactive effects of three different environmental factors - temperature, atmospheric CO_2 and nitrogen availability. In addition, because cereal grains contain two distinct locations of accumulated lipids - starch and non-starch - we analysed both independently. The starch lipids, particularly monoacylphosphoglycerides (lysoPtdCho major component in wheat), are tightly associated with the starch grains and require special precautions to extract them fully [22]. The non-starch lipids, dominated by triacylglycerol and, to a lesser extent, glycosylglycerides, are found mainly in the germ and bran components.

We have found that all three environmental factors alter the quantity, the acyl class distribution and the acyl composition of wheat grain lipids. To some extent the decrease in total grain lipid caused by raised temperature is compensated by increases due to raised CO_2. Moreover, manipulation of nitrogen applications can be used to further compensate for the "Greenhouse Effect". These data are discussed in more detail in the article by Williams, Shewry and Harwood (this volume) and in recent publications [23,24].

Unfortunately, insofar as the future is concerned, although it may be possible to change agricultural practices to compensate for changes in total grain lipids caused by the Greenhouse Effect, this action does not restore the lipid class or acyl chain composition to normal levels. Thus, there will still be noticeable changes in the properties of flours [see 20] produced from existing varieties of wheat. It should, however, be possible (knowing the likely changes in grain lipids) to identify other wheat varieties for future use in order to provide the baking industry with good quality flours.

Part of the reason for changes in grain quality and yield is because environmental temperature, CO_2 and N alter significantly the growth and development of the wheat plant. Some general effects of the "Greenhouse Effect" have been reported [25,26]. A very surprising effect of CO_2 is to alter the cellular morphology [15]. The number of mitochondria present goes up about three-fold for comparable sections of normal- and elevated CO_2-grown wheat. Interestingly, when we studied lipid metabolism in leaves from such plants, a highly significant increase in the labelling of the mitochondrial-specific diphosphatidylglycerol was noted. Thus, there is a good correlation between changed membrane lipid formation and organelle biogenesis.

Other effects on leaf lipid metabolism which can be caused by elevated CO_2, temperature or a combination of the two include changed unsaturation and altered partitioning into major lipid classes. Surprisingly, elevated temperature (24°C instead of 20°C) caused an increase in linoleate labelling at the expense of oleate. We interpret at least part of this effect as being due to an increased flux of carbon through PtdCho, which is a 12-desaturase substrate by the extra-chloroplast pathway [27].

Herbicide Action

In previous reviews [e.g. 1,3] I have described the action of three groups of herbicides on fatty acid metabolism. Thiocarbamates inhibit fatty acid elongation, substituted pyridazinones inhibit some desaturases while grass-specific herbicides have acetyl-CoA carboxylase as their target site.

Previous work in our laboratory showed that thiocarbamates altered surface lipid synthesis by inhibiting formation of the precursor very long chain fatty acids [1,3]. Recent work has provided more evidence (P.B. Barrett and J.L. Harwood, this volume) that the thiocarbamates are oxidised to their sulphoxides and it is the latter which are responsible for the specific inhibition of fatty acid elongases. In germinating peas, as in leek, we have found separate chain-length specific elongases. The stearate and arachidate elongases are located in the endoplasmic reticulum and are much more sensitive to sulphoxides than to the parent thiocarbamate.

The aryloxyphenoxypropionates (FOPS) and cyclohexanediones (DIMS) are herbicidal to grasses by virtue of their inhibition of acetyl-CoA carboxylase [3]. This enzyme has often been stated to be chloroplast-located but, because of the need for extra-chloroplast malonyl-CoA predictions have been made that some activity must be present outside the chloroplast [e.g. 29]. Recently, while studying the action of FOPS and DIMS on monocotyledons we have found two acetyl-CoA carboxylases. In maize, both enzymes are multifunctional proteins. The 230KDa isomer is located in mesophyll chloroplasts and is highly sensitive to herbicides. The 220KDa isomer is soluble and probably located in the epidermal cells of maize. Interestingly, the two isomers also differ in their sensitivity to acyl-CoAs but their enzyme kinetics both show similar plots implying a non-ordered Ter-Ter mechanism [30]. Resistant monocotyledonous weeds have been found which have similar molecular weight carboxylases but which are insensitive to herbicides. It is believed that they have mutations to their carboxyltransferase domains.

Conclusions

Research into environmental aspects continues to advance rapidly, especially with the use of molecular biology and the availability of improved assay systems or purified enzymes. However, it is only recently that we have started to consider the interaction of different factors, which is more realistic although much more difficult to interpret. Perhaps in our design of experiments to test environmental factors we should remember Einstein's thoughts that "everything should be made as simple as possible, but not simpler"!

References

1. Harwood, J.L., Abulnaja, A.O., Barrett, P.B., Herbert, D., Rutter, A.J. and Williams, M. In : Cherif, A., Miled-Daoud, D.B., Marzouk, B., Smaoui, A. and Zarrouk, M., editors. Metabolism, Structure and Utilization of Plant Lipids. Tunis : Centre National Pedagogique, 1992 : 319-326.
2. Harwood, J.L., Jones, A.L., Perry, H.J., Rutter, A.R., Smith, K.L. and Williams, M. In : Cossins, A.R., editor. The Temperature Adaptation of Biological Membranes. London : Portland Press, 1994 : in press.
3. Harwood, J.L. Prog. Lipid Res., 1994; 33 : 193-202.
4. Bartels, C., James, A.T. and Nichols, B.W. Eur. J. Biochem., 1967; 3 : 7-10.
5. Harwood, J.L. and James, A.T. Eur. J. Biochem. 1975; 50 : 325-334.
6. Harwood, J.L. In : Siegenthaler, P-A. and Eichenberger, W., editors. Structure, Function and Metabolism of Plant Lipids. Amsterdam, Elsevier, 1984 : 543-550.
7. Garnier, J., Wu, B., Maroc, J., Guyon, D. and Tremolieres, A. Biochim. Biophys. Acta, 1990; 1020 : 153-162.
8. McCourt, P., Browse, J., Watson, J., Arntzen, C.J. and Somerville, C.R. Plant Physiol. 1985; 78 : 853-858.
9. Tremolieres, A., Guillot-Salomon, T.D., Dubacq, J.P., Jaques, R., Mazliak, P. and Signol, M. Physiol. Plant 1979; 45 : 429-436.
10. Harwood, J.L. Lipid metabolism. In : Gunstone, F.D., Harwood, J.L. and Padley, F.B., editors. The Lipid Handbook, second edition. London : Chapman and Hall 1994, in press.
11. Nikolau, B.J. and Hawke, C. Arch. Biochem. Biophys. 1984; 228 : 86-96.
12. Post-Bietenmiller, D., Jaworski, J.G. and Ohlrogge, J.B. J. Biol. Chem. 1991; 266 : 1858-1865.
13. Page, R., Okada, S. and Harwood, J.L. Biochim. Biophys. Acta 1994: 1210 : 369-372.
14. Liljenberg, C. Prog. Lipid Res. 1992; 31 : 332-343.
15. Robertson, E. and Leech, R.M. 1994, personal communication.
16. Harwood, J.L. Trends Biochem. Sci. 1991; 16 : 126-127.
17. Wada, H., Gombos, Z. and Murata, N. Nature 1990; 347 : 200-203.
18. Murata, N., Ishizaki-Nishizawa, O., Higashi, S., Hayashi, H., Tasaka, Y. and Nishida, I. Nature 1992; 356 : 710-712.

19. Vigh, L., Los, D.A., Horvath, I. and Murata, N. Proc. Nat. Acad. Sci., U.S.A.; 90 : 9090-9094.
20. Karpati, E.M., Bekes, F., Lasztity, R., Orsi, F., Smied, I. and Mosonyl, A. Acta Alimen, 1990; 19 : 237-260.
21. O'Connor, J., Perry, H.J. and Harwood, J.L. J. Cereal Sci. 1992; 16 : 153-163.
22. Morrison, W.R. In : Barnes, P.J., editor. Lipids in Cereal Technology New York : Academic Press, 1-32.
23. Williams, M., Shewry, P.R. and Harwood, J.L. Grasas y Aceites, 1993; 44 : 132-134.
24. Williams, M., Shewry, P.R. and Harwood, J.L. J. Exptl. Botany 1994; in press.
25. Mitchell, R.A.C., Mitchell, V.J., Driscoll, S.P., Franklin, J. and Lawlor, D.W. Plant, Cell Environ. 1993; 16 : 521-529.
26. Baxter, R., Ashenden, T.W., Sparks, T.H. and Farrar, J.F. J. Exptl. Botany 1994; 45 : 305-315.
27. Browse, J. and Somerville, C.R. Ann. Rev. Plant Physiol. 1991; 42 : 467-506.
28. Cassagne, C., Lessire, R., Bessoule, J.J. et al. Prog. Lipid Res. 1994; 33 : 55-69.
29. Harwood, J.L. In : Emes, M.J., editor. Compartmentation of Plant Metabolism in Non-Photosynthetic Tissues, Cambridge, C.U.P., 1991; 23-42.
30. Herbert, D., Alban, C., Cole, D.J., Pallett, K.E. and Harwood, J.L. Biochem. Soc. Trans. 1994; 22 : 261.

IS THE MEMBRANE THE PRIMARY TARGET IN THE BIOLOGICAL PERCEPTION OF TEMPERATURE? EFFECT OF MEMBRANE PHYSICAL STATE ON THE EXPRESSION OF STRESS-DEFENCE GENES

László Vigh, Dmitry A. Los[2], Norio Murata[2], Attila Glatz[1], Eszter Kovács[1] and Ibolya Horváth[1].
[1]Institute of Biochemistry, Biological Research Center, 6701 Szeged, POB 521, Hungary and
[2]National Institute for Basic Biology, Myodaiji, Okazaki 444, Japan.

Introduction

Biological organisms are capable of perceiving environmental changes caused by a variety of agents. But whereas sensors of light (rhodopsins in animals or phytochromes in plants) are well-characterized, no knowledge is available concerning the thermal sensors. The only clue hitherto known about the thermal perception is that temperature decrease would result in a reduction of molecular motion in membranes, which is attained mainly via specific desaturation of fatty acids of membrane lipids [1]. It is assumed, that the response of systems to temperature stress via physiological, morphological and biochemical adjustments aimed at offset or compensate for the temperature-induced disturbances. Amongst the many compensatory responses, it seems that the major adjustment is the reorganization of the physical state of the cellular membranes, a phenomenon called "homeoviscous adaptation".

By modulating membrane molecular order of cyanobacterial cells accomplished either by temperature shift or by catalytic hydrogenation of membrane lipids, we have concluded, that the primary signal of the biological thermosensor is the change in membrane fluidity. It is the first demonstration for the involvement of the membrane lipids in the mechanism of biological thermosensing.

Results and discussion

Recently we have examined the temperature-dependent regulation of the expression of the *desA* gene, which encodes Δ12 desaturase of *Synechocystis* PCC6803, this widely used and genetically well-defined strain of the blue-green alga [2]. The level of *desA* transcript increased 10-fold within 1 h upon a decrease in temperature from 36 to 22 °C. This finding suggested, that the rapid cold-induced desaturation of membrane lipid fatty acids is regulated at the level of the expression of the desaturase genes. Both rifampicin and diuron was able to prevent the low-temperature-induced increase of the level of *desA* transcript, whereas no enhancement of desA mRNA was observable when cells were incubated in the dark. Furthermore, it has been established, that the accumulation of the *desA* transcript depended on the extent of temperature change over a certain threshold level, but not on the absolute temperature. It is noted, that at that phase of our studies the mechanism for the perception of the change in temperature, stimulating eventually the expression of desaturase gene, remained unknown. Due to the fact, that temperature-shift is totally nonspecific in its effects on cellular mechanism, it was obvious, that a distinct experimental system is required to enable us to identify the primary factor operative in the regulation.

To accomplish this task, next we employed the technique of homogeneous catalytic

J.-C. Kader and P. Mazliak (eds.), Plant Lipid Metabolism, 369–371.

hydrogenation, which allows the controlled *in situ* saturation of unsaturated membrane lipids with apparently no effects other than the ordering of the membrane matrix [1,3]. The catalyst, palladium-di-(sodium alizarine monosulphonate), $Pd(QS)_2$ is water soluble yet amphipatic enough to hydrogenate very efficiently and still selectively the cis-double bonds of lipid acyl chains in biomembranes either *in vivo* or *in vitro* in most various systems [4]. By using permeaplasts rather than intact cells by partially digesting the cell wall by mild treatment of cells with lysozyme, the membrane lipids of *Synechocystis* could be saturated to an appreciable level under specific hydrogenation conditions [5]. It has also be shown, that *in vivo* hydrogenation of algal lipids up to a certain period of reaction time (less than 5 min) occurred only in the plasma membrane and no statistically significant changes were observed in the thylakoid membranes. It should be added, that during hydrogenation, the physiological activity, measured in terms of the photosynthetic evolution of oxygen, remained always constant. Northern blot hybridization demonstrated that the level of the *desA* transcript was not affected by the presence of either hydrogen or the Pd catalyst. However, following the addition of prehydrogenated B-form of $Pd(QS)_2$ to the cell suspension under hydrogen [4], the level of the *desA* mRNA reached a maximum at 4 min reaction time. The decline detected at longer incubations corresponded closely to the timing of the inactivation of photosynthesis. If permeaplasts derived from cells grown at 36 °C were hydrogenated for 4 min, desA transcript increased to reach a maximum level at 30 min after the hydrogenation process. Clearly, these results indicated close similarities between the effect of low temperature and the effect of hydrogenation in elevating the level of *desA* mRNA. Moreover, the effects of metabolic inhibitors on *desA* gene expression found in hydrogenation experiments appeared to be the same as those found upon low-temperature treatment confirming again, that in both cases a common mechanism operates in the gene regulation.

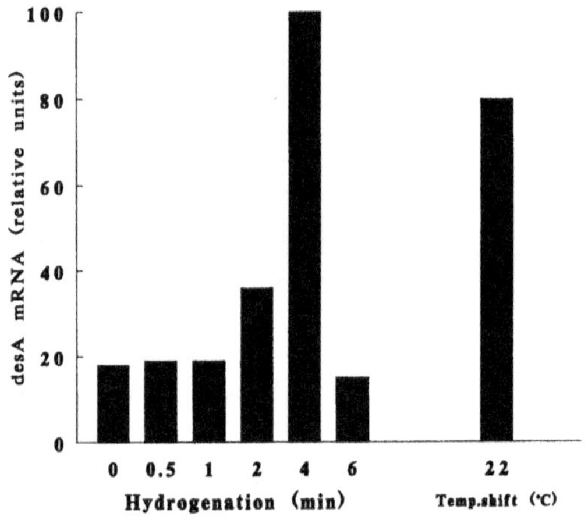

Figure 1. The change in level of mRNA of the *desA* gene 30 min following either hydrogenation for various time intervals under isothermal conditions at 36 °C or temperature shift to 22 °C. For details see [2,5].

Other line of evidences indicate also, that cellular membranes are not only the sites of the thermal damage but they might act as thermal sensors. Maresca, B. and his co-workers suggested, that upon heat shock, perturbation of lipoprotein membrane complexes activates a putative heat shock sensor that monitors temperature changes and stimulates transcription of heat shock genes (personal communication). We predict, that the GroEL and GroES type chaperonins (heat shock proteins) are directly involved in conferring thermotolerance of *Synechocystis* cells and their membrane functions [6,7]. The thermal-acclimation dependent regulation of the groESL operon has been documented recently in our laboratory (Glatz et al, submitted, [8]). By using mutant/transformant cells defective in desaturation, Fad6*desA*::Km^r, we also demonstrated, that it is the modification of the overall level of membrane physical state (fluidity, protein-lipid interactions) rather than the level of lipid saturation *per se* which may control the regulation of the expression of HS genes (unpublished).

While the mechanism of the temperature perception seems to be defined, the understanding of the transcriptional controlling mechanisms of the stress-defending genes is still unknown. It is suggested, that DNA twist may be a transcriptional sensor for environmental changes, ie. alterations in DNA supercoiling may control the rate of transcription [9]. Fully supporting this prediction, a highly enhanced transcription of HS-genes (groEL-analog cpn60, hsp70) has been shown for *Synechocystis* in the absence of heat-treatment, but upon the addition of novobiocin an inhibitor of DNA gyrase [10]. Furthermore, our preliminary studies show, that novobiocin is also able to suspend the formation of *desA* transcript observable following either cold-shock or hydrogenation (unpublished).

The identification of the hypothetical factor(s) which detects the change in membrane physical state and may act eventually through DNA supercoiling or some alternative ways awaits further studies.

References

1. Quinn PJ, Joó F, Vigh L. The role of unsaturated lipids in membrane structure and stability. Prog.Biophys.Molec.Biol. 1989;53:71-103.
2. Los DA, Horváth I, Vigh L, Murata N. The temperature-dependent expression of the desaturase gene *desA* in *Synechocystis* PCC6803. FEBS Lett. 1993;318:57-60.
3. Vigh L, Joó F. Modulation of membrane fluidity by catalytic hydrogenation affects chilling susceptibility of the blue-green alga, *Anacystis nidulans*. FEBS Lett. 1983;162:423-427.
4. Joó F, Balogh N, Horváth LI, Filep G, Horváth I, Vigh L. Complex hydrogenation/oxidation reactions of the water-soluble hydrogenation catalyst palladium di (sodium alizarinmonosulfonate) and details of homogeneous hydrogenation of lipids in isolated biomembranes and livin cells. Anal.Biochem. 1991;194:34-40.
5. Vigh L, Los DA, Horváth I, Murata N. The primary signal in the biological perception of temperature:Pd-catalyzed hydrogenation of membrane lipids stimulated the expression of the *desA* gene in *Synechocystis* PCC6803. Proc.Natl.Acad.Sci. USA 1993;90:9090-9094.
6. Lehel C, Los DA, Wada H, Györgyei J, Horváth I, Kovács E, Murata N, Vigh L. A second groEL-like gene, organized in a groESL operon is present in the genome of *Synechocystis* sp. PCC 6803.J.Biol.Chem. 1993;268:1799-1804.
7. Kovács E, Török Z, Horváth I, Vigh L. Heat stress induces association of the GroEL-analog chaperonin with thylakoid membranes in cyanobacterium, *Synechocystis* PCC 6803. Plant Physiol.Biochem. 1994;32:285-293.
8. Lehel C, Gombos Z, Török Z, Vigh L. Growth temperature modulates thermotolerance and heat shock response of cyanobacterium *Synechocystis* PCC 6803. Plant. Physiol. Biochem. 1993;31:81-88.
9. Wang JY, Syvanen M. DNA twist as a transcriptional sensor for environmental changes. Molec. Microbiol. 1992;6:1861-1866.
10. Chitnis PR, Nelson N. Molecular cloning of the genes encoding two chaperone proteins of the cyanobacterium *Synechocystis* sp. PCC 6803. J.Biol.Chem. 1991;266:58-65.

TEMPERATURE REGULATION OF DESATURATION OF FATTY ACIDS IN CYTOSOLIC AND CHLOROPLASTIC GLYCEROLIPIDS

John P. Williams, Mobashsher U. Khan and Doris Wong
Department of Botany and The Centre for Plant Biotechnology
University of Toronto, Toronto, Ontario, Canada

Introduction

Adaptation or acclimation to low temperatures is a phenomenon found in most poikilothermic organisms which must adjust to changing environmental conditions without an internal mechanism of homeostasis. In plants this acclimation normally takes the form of adjustment to high and low temperature by changes in protein and lipid components of the cell. These changes not only protect the cell from frost or chilling damage but may allow the cell to acclimate to optimal metabolic conditions. Our studies have focussed on one of the most important changes that take place in cell membranes during acclimation: the changing level of unsaturation of the membrane constituent acylglycerols. This, along with changes in the sterol components, is generally believed to affect the degree of fluidity of the membrane at different temperatures.

One of the major problems plant cells face under field conditions is the wide fluctuation of diurnal and seasonal temperatures. Plants grown in temperate climates commonly experience diurnal fluctuations in temperature of as much as 10-20°C and even greater differences over longer periods of time. The plant must adapt and acclimate to long term seasonal changes but cannot respond to rapid and often transient changes without considerable expenditure of metabolic energy. The constant adjustment to rapid diurnal changes appears unnecessary and so the plant must acclimate to a broad optimum level of unsaturation of acylglycerol fatty acids and membrane fluidity. It is also important that plants are able to acclimate to both high and low temperatures, as too high a level of unsaturation could also be detrimental. The system of adaptation may be complicated by the need for different levels of unsaturation (or fluidity) in organelles or membranes with different metabolic functions and components. Similarly, functions performed primarily during the high temperature day period (e.g. photosynthesis) may require different control mechanisms for maintaining optimal membrane environments than other functions that are performed throughout the 24h cycle (e.g. respiration).

Despite the assumption that plants acclimate to different seasonal temperatures, our studies with field-grown plants do not show clearly significant changes in unsaturated fatty acid compositions from Spring to Fall except in the level of trans-16:1 in phosphatidylglycerol (PG) in some species [unpublished data]. In plants grown under controlled conditions we have found significant differences in levels of unsaturation in a wide variety of species[e.g. 1,2]. These plants often show very high levels of unsaturation, particularly in chloroplast lipids, even if grown at

J.-C. Kader and P. Mazliak (eds.), Plant Lipid Metabolism, 372–377.

high temperatures. The question that must be raised is whether plants grown at low temperature acclimate because of a need for higher levels of unsaturated fatty acids or is the plant increasing desaturase activity to maintain the level of unsaturation. Without acclimation low temperatures may well result in significantly lower levels of desaturase activity and consequently lower levels of unsaturated fatty acids.

We, therefore, believe that plants must acclimate slowly to both high and low temperatures. These changes may be organelle or function specific but they eventually allow the plant to metabolize under optimal conditions. Comparisons of plant types and fatty acid compositions of different species under varying physiological conditions suggest that acclimation is a complex problem and may be controlled in many different ways.

Our data show that plants have evolved a low temperature-induced desaturation (LTD) process that results in the very rapid desaturation of fatty acids of newly formed lipids [3]. This is followed by a much slower basal level desaturation (BLD). The two processes result in a final level of desaturation that is determined by a balance between the level of the rapid initial phase and the subsequent slower BLD. We can show that the LTD is induced at low temperature but is thermolabile at high temperatures, thus providing a mechanism for controlling the rate of desaturation of newly formed lipid. The loss and induction of desaturase activity is much more rapid than the changes in level of unsaturation of the bulk membrane lipid. We believe that it is the control of desaturation of newly formed lipid which allows the plant to respond slowly to the environmental temperature over extended periods of time rather than to react immediately to transient fluctuations in diurnal temperature.

CHLOROPLAST LIPID BIOSYNTHESIS

Plant chloroplasts contain four major diacylglycerols (monogalactosyldiacylglycerol (MGDG), digalactosyldiacylglycerol (DGDG), sulphoquinovosyldiacylglycerol (SQDG) and PG). These lipids have distinctive fatty acid compositions and appear to be formed from different pools of diacylglycerol precursors. There are a number of different types of plants which differ significantly in their fatty acid composition: these have been called 16:3- and 18:3-plants based on the presence or absence of hexadecatrienoic acid (16:3) in the MGDG and DGDG of the chloroplast lipids. A third group of plants, which we are calling 18:4-plants, contains octadecatetraenoic acid (18:4) as a major component of the galactolipids [4,5].

Recent studies in our laboratory using *Echium plantagineum*, an 18:4-plant, have suggested that there are different pathways for the synthesis of the bulk of each of the four chloroplast lipids in 16:3-, 18:3- and 18:4-plants. In all three plant types, linolenic acid (α18:3) is synthesised by Δ^{12} desaturation of linoleic acid (18:2); in 18:4-plants, γ18:3 is synthesised by Δ^{6} desaturase of 18:2 in conjunction with cytosolic PC, and 18:4 is synthesised from γ18:3 by Δ^{15} desaturases in the plastids of developing cotyledons [*Borago* 4,5] and in the cytosol and chloroplasts of leaves [*Borago* and *Echium*, our unpublished data]. The pathways suggested for *E. plantagineum* were deduced from fatty acid, molecular species and fatty acid positional analyses of the four chloroplast lipids and the two major cytosolic lipids, phosphatidylcholine (PC) and phosphatidylethanolamine (PE). Of particular importance is the indication of two pathways, a prokaryotic pathway in the inner envelope membrane and a eukaryotic pathway in the outer envelope membrane, for the synthesis of the bulk of MGDG and DGDG, respectively. This has been assumed to be the case in 16:3-plants [6,7,8] but our data indicate that this is also the case in 18:4- and probably

18:3-plants. Figure 1 shows the proposed pathways for 16:3 and 18:3/4-plants. The major difference in our proposal is the assumption that all three types of plants have both prokaryotic and eukaryotic pathways. The difference between the three types of plants is the origins of the precursor diacylglycerols (DAG) for the prokaryotic pathway. We believe that in 16:3-plants the DAG precursor of the prokaryotic pathway may be chloroplastic and/or cytosolic in origin, depending on species and environmental conditions, whereas in 18:3/4-plants the DAG is cytosolic in origin. In the prokaryotic pathway, part of the MGDG is galactosylated to DGDG. Therefore, DGDG is synthesised mainly by the eukaryotic pathway and, to a lesser degree, from the prokaryotic pathway. Our interpretation of these pathways is consistent with the localisation of very active galactolipid:galactolipid galactosyltransferase (GGGT) in the outer envelope membrane [8], which may result in the conversion of most, if not all, of the MGDG to DGDG, and the loss of phosphatidate phosphatase from the inner envelope membrane of 18:3/4-plants [9].

We believe that the biosynthetic pathway of SQDG is similar to the prokaryotic pathway of MGDG, with the origin of the DAG precursor either chloroplastic and cytosolic (16:3-plants) or cytosolic (18:3/4-plants) [10].

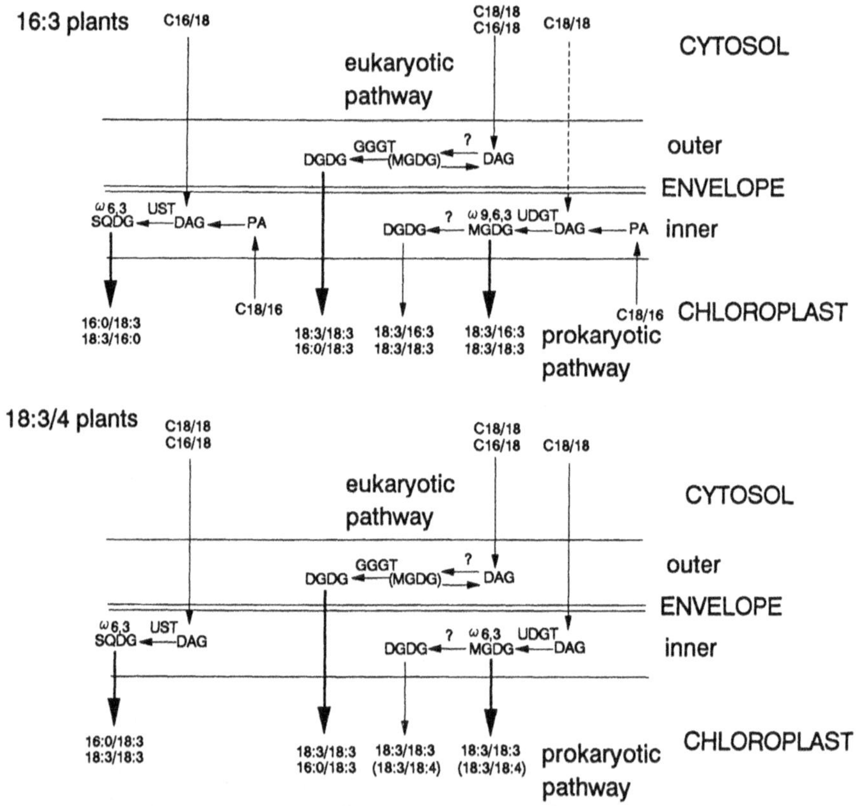

Figure 1. Proposed pathways of biosynthesis of chloroplast glycerolipids in 16:3- and 18:3/4-plants

EFFECT OF TEMPERATURE ON THE PROKARYOTIC AND EUKARYOTIC PATHWAYS

Our data from *Brassica napus* grown at different temperatures are consistent with the view that, at low temperatures, the level of the prokaryotic pathway (synthesising 18:3/16:3) increases at the expense of 18:3/18:3 molecular species in MGDG, while in DGDG the reverse is the case([11] and Figure 2). This indicates that temperature can affect the relative rates of synthesis in these pathways or in the transfer of diacylglycerol from the cytosol.

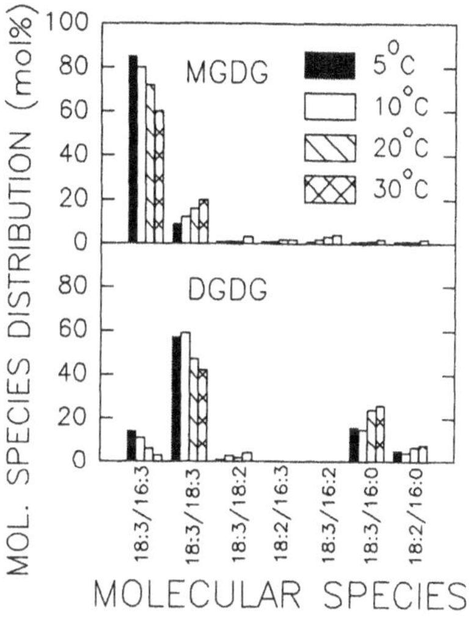

Figure 2. Molecular species of MGDG and DGDG from leaves of *B. napus* plants grown at different temperatures.

EFFECT OF TEMPERATURE ON DESATURATION IN *BRASSICA NAPUS*

We have previously shown [2] that the desaturation in MGDG and PC in leaves is dependent on the temperature at which the plants are grown. The rate of desaturation was determined from the rate of conversion of radioactive saturated to unsaturated fatty acids. We showed that plants grown at 5°C inherently desaturated fatty acids faster than plants grown at 30°C, no matter at what temperature the $^{14}CO_2$-feeding and incubation was carried out. The desaturases, in leaves developed at low temperature, showed normal enzyme kinetics desaturating at lower rates when incubated at lower temperature. This is the reverse of what would be expected if plants are to adapt to lower temperature by raising the rate of desaturating and vice versa. We then showed that the desaturase activity was lost within hours of transferring plants to 30°C, initiating the concept of thermal lability of the activity [3,12]. When plants treated at 30°C were returned to 5°C they once again developed the potential of high rates of desaturase activity which we have called low temperature-induced desaturation. At the same time we were able to show that this high rate of desaturation occurred immediately after the galactosylation of DAG (or choline transfer in PC synthesis). The quantity of fatty acid which desaturated rapidly in newly-formed lipid was dependent on the temperature at which the plants were grown.

To quantify the rates of desaturation more accurately we developed a computer simulation programme that allows us to estimate rates of desaturation of each of the MGDG desaturases after incubation following $^{14}CO_2$-feeding [13]. We have used this technique to compare the relative activities of the desaturases (ω9, ω6 and ω3) involved in the desaturation of 18:1/16:0 molecular species of MGDG in leaves of plants grown at 5, 10, 20 and 30°C. The results suggest that the ω9 and ω6 desaturases are markedly affected by temperature, whereas the ω3 desaturases remain relatively unchanged (Figure 3).

We have subsequently used this technique to follow the initial rate of desaturation in plants grown at 20 and 30°C and the adaptation of these plants to 5°C over a period of two weeks. The data suggest that LTD is due primarily to ω9 and ω6 desaturases of both C16 and C18 fatty acids. ω3 desaturase activity is relatively low (Figure 4).

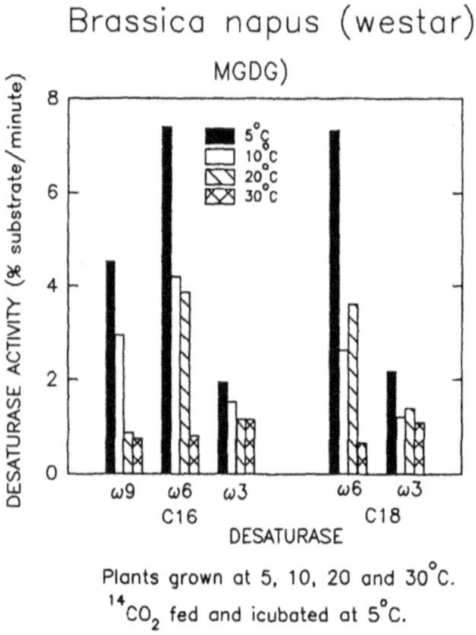

Figure 3. Desaturase activity in MGDG of leaves of *B. napus* grown at different temperatures.

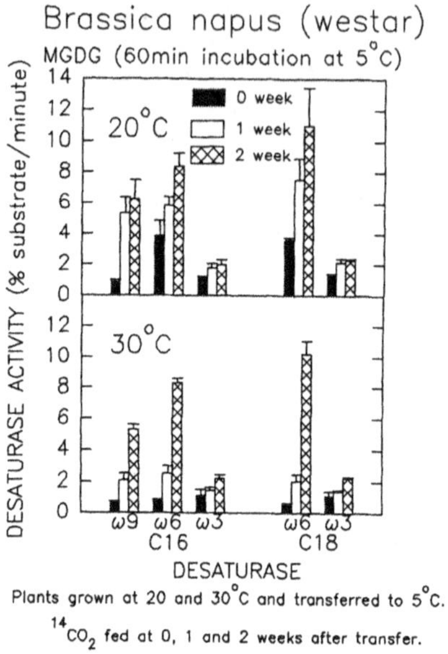

Figure 4. Desaturase activity in MGDG of leaves grown at 20 and 30°C and acclimated at 5°C for 2 wks.

MUTANTS OF *B. NAPUS* DEFICIENT IN LOW TEMPERATURE INDUCTION OF DESATURASE ACTIVITY

We have obtained a double haploid mutant line of *B. napus* that is similar in lipid and fatty acid composition at 20°C but which shows lower levels of chloroplast lipids and trienoic acids (18:3 and 16:3) at low temperature (5°C). With the low trienoic fatty acids there is a concomitant increase in the dienoic fatty acids. This is the reverse of the normal situation where plants grown at constant low temperature increase the level of trienoic fatty acids.

$^{14}CO_2$-feeding experiments and the measurement of desaturase activity using the computer simulation programme show that the lower trienoic acids appears to be the result of a general reduction in desaturase activity at 5°C. Mutant lines grown at 20°C and then transferred to 5°C for 1 and 2 week periods show no increase in desaturase activity, in contrast to non-mutant lines which respond to low temperature in a similar way to *Brassica napus* 'var. Westar'. The transfer of the mutant plants to 5°C results in a gradual reduction in trienoic acids and an increase in dienoic acids in the chloroplast lipids. We believe that the mutant is deficient in its ability to react to low temperature. The resulting lowering of trienoic fatty acid is what would be expected of plants unable to react to lower temperatures with increased desaturase activity. We, therefore, believe that plants have evolved a system of maintaining unsaturated fatty acid levels, which may also result in increased levels of unsaturation. The mutants probably represent the situaton that would occur if this was not possible.

Similar lines have also been isolated from double haploid mutant lines that show a correlation between the level of 18:3 in PC at 5°C and the 18:3 level of the seed oil (2.5-12 wt%, measured at approx. 20°C). These multiple mutants respond in varying degrees to low temperature induction of desaturation in PC in the leaves and appear to effect the level of 18:3 in the seed oil.

Acknowledgements We gratefully acknowledge the financial support of the Natural Sciences and Engineering Research Council of Canada

References

1. Huner NPA, Krol M, Williams JP, et al. A low temperature induced decrease in 3-trans-hexadecenoic acid content and its influence in LHCII organization. Plant Physiol 1987;84:12-18.
2. Williams JP, Khan MU, Mitchell K, Johnson G. The effect of temperature on the level and biosynthesis of unsaturated fatty acids in diacylglycerols of *Brassica napus* leaves. Plant Physiol 1988;87:904-910.
3. Williams JP, Williams K, Khan MU. Low temperature-induced desaturation in *Brassica napus*: thermal lability of the process. Biochim Biophys Acta 1992;1125:62-67.
4. Stymne S, Stobart AK. Biosynthesis of γ-linolenic acid in cotyledons and microsomal preparations of the developing seeds of common borage (*Borago officinalis*). Biochem J 1986;240:385-393.
5. Stymne S, Griffiths G, Stobart K. Desaturation of fatty acids on complex-lipid substrates. *In* PK Stumpf, JB Mudd, WD Nes, eds, The Metabolism, Structure, and Function of Plant Lipids. Plenum Press, New York. 1987:405-412.
6. Roughan PG, Slack CR. Glycerolipid synthesis in leaves. Trends Biochem Sci 1984;9:383-3866.
7. Browse J, Warwick N, Somerville CR, Slack CR. Fluxes through the prokaryotic and eukaryotic pathways of lipid synthesis in the '16:3' plant *Arabidopsis thaliana*. Biochem J 1986;235:25-31.
8. Heemskerk JWM, Schmidt H, Hammer U, Heinz E. Biosynthesis and desaturation of prokaryotic galactolipids in leaves and isolated chloroplasts from spinach. Plant Physiol 1991;96:144-152.
9. Heinz E, Roughan PG. Similarities and differences in lipid metabolism of chloroplasts isolated from 18:3 and 16:3 plants. Plant Physiol 1983;72:273-279.
10. Bishop DG, Sparace SA, Mudd JB. Biosynthesis of sulfoquinovosyl-diacylglycerol in higher plants: the origin of the diacylglycerol moiety. Arch Biochem Biophys 1985;240:851-858.
11. Johnson G, Williams JP. Effect of growth temperature on the biosynthesis of chloroplastic galactosyldiacylglycerol molecular species in *Brassica napus* leaves. 1989;91:924-929.
12. Williams JP, Khan MU, Wong D. Low temperature-induced desaturation in *Brassica napus:* thermal deactivation and reactivation of the process. Biochim Biophys Acta 1992;1128:275-279
13. Williams JP, Khan MU, Wong D. Compouter simulation model for the biosynthesis of galactosyldiacylglycerols and fatty acid desaturation in plants. Plant Physiol 1993;101:977-983.

ACYL TURNOVER IN TRIACYLGLYCEROLS. ITS ROLE IN THE REGULATION BY TEMPERATURE OF THE 18:1/18:2 RATIO IN SUNFLOWER SEEDS

Manuel Mancha, Cristina Sarmiento and Rafael Garcés
Instituto de la Grasa (C.S.I.C.), Apartado 1078
41080 Sevilla, Spain

Introduction

A stimulation of the microsomal oleoyl phosphatidylcholine desaturase (ODS) activity followed by an increase of linoleate (18:2) at the expense of oleate (18:1) was observed in both total and microsomal lipids when developing sunflower seeds were transferred to 10°C even for short times [1]. Under these conditions no "de novo" fatty acid synthesis was detected. Moreover, the ODS activity was successively increased and reduced by alternate shifting to 10°C and 30°C and a concomitant increase and decrease of 18:2 was observed in the microsomal lipids [2]. Also, we have demonstrated that oleate from triacylglycerols (TAG) was desaturated in cold-induced sunflower seeds and the resulting 18:2 was also found in TAG preferentially at position *sn*-2 [3]. This turnover of TAG fatty acids was observed in both microsomal fraction and oil bodies and affected to all TAG molecular species containing 18:1 and 18:2. Now we present evidence of the exchange of 18:1 and 18:2 between TAG and other lipids in response to temperature adaptation.

Results and discussion

Changes in the relative contents of 18:1 and 18:2 of total lipids were observed when developing sunflower seeds were transferred to low (10°C) and high (30°C) temperature (Table 1). After shifting to low temperature the percentage of 18:2 increased in the total lipids at the expense of that of 18:1 but it remained unchanged after shifting to high temperature. It is worthy to note that, while low temperature strongly stimulated ODS activity and increased 18:2 [1], high temperature completely inhibited 18:1 desaturation in spite of producing only a partial inhibition of ODS activity [2]. After shifting to low temperature 18:2 increased in all lipid

J.-C. Kader and P. Mazliak (eds.), Plant Lipid Metabolism, 378–380.

classes of both microsomal fraction and oil bodies. However, after shifting to high temperature 18:2 decreased in total lipids of the microsomal fraction and increased in those of the oil bodies (Table 2). Interestingly, 18:2 decreased in phosphatidylcholine (PC) and diacylglycerols (DAG) (results not shown for DAG) of the two subcellular fractions and in microsomal TAG but increased in TAG of oil bodies. Considering that the mass of oil bodies TAG is much bigger than that of all other lipids, the 1% increase of 18:2 in oil bodies TAG could compensate the decrease in the other lipids. These results indicate that TAG of microsomes and oil bodies form separate pools with different functions. Both TAG can exchange acyl groups with other lipids (see also [3]). Microsomal TAG responded to temperature changes like the other membrane lipids increasing or reducing their 18:2 content. On the contrary, oil bodies TAG acted as a reserve of fatty acids. The composition of these TAG was a result of the exchange of fatty acids with other lipids. Interestingly, under both low and high temperature adaptation conditions oil bodies TAG 18:2 thereby enriching the bulk of seed TAG in 18:2. Whether this acyl turnover is unspecific for the different fatty acids or at least for 18:1 and 18:2 or if, on the contrary, it is specific so TAG only give 18:1 and accept 18:2 needs further research.

Table 1. Fatty acid composition of total lipids from developing (16 DAF) sunflower seeds grown at 24°C constant temperature. Peeled seeds were transferred successively to 10°C and 30°C for the indicated times. For experimental conditions see ref. [3].

	Fatty acid composition (mol%)				
Treatment	16:0	18:0	18:1	18:2	Δ18:2
Control	7.6	7.3	74.0	10.4	---
4h,10°C	7.1	7.0	68.6	16.8	6.4
2h,30°C	7.0	7.1	68.8	16.8	0.0
2h,10°C	6.8	7.1	66.2	19.2	2.4

Results are expressed as mean (two experiments) the SD being less than 5% of mean value.

Table 2. Changes in the percentages of 18:1 and 18:2 of total lipids, TAG and PC of microsomal fraction and oil bodies from developing (15 DAF) sunflower seeds grown at 24°C. Peeled seeds were transferred successively to 10°C and 30°C for the indicated times. Subcellular fractions were prepared as in ref.[3].

Lipid class	Treatment	Fatty acid composition (mol%)					
		Microsomes			Oil bodies		
		18:1	18:2	Δ18:2	18:1	18:2	Δ18:2
Total	Control	64.4	17.1	--	72.0	12.3	--
	4h,10°C	53.4	29.3	12.2	67.0	17.9	5.6
	2h,30°C	58.1	25.3	-4.0	65.9	18.7	0.8
	2h,10°C	52.3	30.4	5.1	65.2	20.1	1.4
TAG	Control	68.2	12.4	--	71.8	12.1	--
	4h,10°C	62.5	21.7	9.3	67.6	17.3	5.2
	2h,30°C	63.8	19.8	-1.9	66.4	18.3	1.0
	2h,10°C	58.8	25.7	5.9	65.9	19.3	1.0
PC	Control	66.3	18.8	--	68.7	16.6	--
	4h,10°C	48.8	36.3	17.5	47.5	41.6	25.0
	2h,30°C	58.8	25.7	-10.6	61.4	25.5	-16.1
	2h,10°C	47.3	36.9	11.2	45.1	43.3	17.8

References

1. Garcés, R., Sarmiento, C. and Mancha, M.(1992) Temperature regulation of oleate desaturase in sunflower (*Helianthus annuus* L.) seeds. Planta 186: 461-465
2. Mancha, M., Garcés, R. and Sarmiento, C. (1992) Control of oleate desaturation in sunflower seeds. In Metabolism Structure and Utilization of Plant Lipids, pp 22-25 (Cherif, A. et al. eds.) Centre National Pédagogique, Tunisia.
3. Garcés, R., Sarmiento, C. and Mancha, M. (1994) Oleate from triacylglycerols is desaturated in cold-induced developing sunflower (*Helianthus annuus* L.) seeds. Planta (In press).

CORRELATION STUDIES BETWEEN CHILLING SENSITIVITY OF STORED BASIL LEAVES AND THEIR COMPOSITION OF MEMBRANE LIPIDS

Shimon Meir, Lilach Twito and Sonia Philosoph-Hadas
Dept. of Postharvest Science of Fresh Produce, ARO, The Volcani Center, Israel.

Sweet basil (*Ocimum basilicum* L.) is a plant of tropical origin which is susceptible to chilling. The objective of the present work was to reveal the relationship between membrane lipid composition and chilling sensitivity of basil leaves. For this purpose, a comparative study was performed with 2 basil cultivars (cv. 79 and cv. 10), which vary in their susceptibility to chilling, analyzing differences in their critical temperatures for appearance of visual chilling injury (CI) symptoms, their phase transition temperatures and their fatty acid (FA) composition of their disaturated membrane phosphatidylglycerol (PG) and phosphatidylcholine (PC). The phase transition temperatures were derived from Arrhenius plots of the respiration rates. The composition of FA in the polar lipids were analyzed by gas-liquid chromatography. Results show that the phase transition temperature of the sensitive cultivar (cv. 79) increased from 12.5 to 17° C during 1-6 days of incubation, whereas that of the less sensitive cultivar (cv. 10) - was much lower and remained between 3-5°C throughout storage. These variations in the transition temperatures of the two cultivars may presumably result from the differences in the FA composition of their disaturated membrane PG and PC. Thus, leaves of the chilling-sensitive cultivar (cv. 79) showed a reduction and an increase in the composition of 18:1 and 16:0 FA, respectively, of PG and PC, as compared with leaves of the less susceptible cultivar (cv 10). The results suggest that these two PG and PC FA, present in the chloroplast membranes of basil leaves, are closely associated with the chilling sensitivity of the two tested basil cultivars. These findings further confirm that the chilling sensitivity of higher plants is affected by the content of saturated PG and PC fatty acids.

Introduction

Tropical and subtropical plants undergo CI when exposed to low temperatures below 10-12°C. It has been proposed that the primary molecular event leading to development of CI, is a low temperature-induced phase transition in the polar lipids of cell membranes [1]. The temperature of this transition correlates with the critical temperature for CI [2]. Hence, the sensitivity of the tissue would be reduced if the membrane polar lipid composition could be altered such that the transition temperature is reduced. Murata [3] has proposed that in chloroplasts of higher plants, PG might play a key role in the initiation of such chilling-associated transition, by virtue of its content of molecular species containing saturated FA. Results obtained with various plants supported his proposal, showing that PG FA became unsaturated during the hardening process [2, 4-7]. Recently, Joh et al. [7] have suggested that unsaturation of FA in not only PG, but also in PC and phosphatidylethanolamine (PE), is involved in chilling sensitivity of *Chlorella*. The

J.-C. Kader and P. Mazliak (eds.), Plant Lipid Metabolism, 381–383.

present work describes experiments designed to test this proposal in basil leaves, which is a tropical and chilling-susceptible plant.

Materials and Methods

Experiments were performed with detached, fresh leaves of two basil cultivars, including a chilling-sensitive one, cv. "79' (Sunset, USA), and a less sensitive one, cv. '10' (Botanical gardens, Swiss cultivar no. 676). The chilling-induced symptoms were evaluated visually according to a CI index scale ranging from 1 (= no damage) to 5 (= severe damage). The phase transition temperatures were derived from Arrhenius plots of the respiration rates [8] of detached basil leaves, exposed for 6 days to temperatures ranging between 0 to 27°C. Determination of the FA composition of PG and PC involved several steps including: extraction of total lipids from the leaves [5], separation of the lipids into 3 classes [9], separation of the phospholipids to PG and PC on TLC, acid hydrolysis and methylation of PG and PC to obtain their FA [6], and analysis of the resultant FA-methyl esters by gas-liquid chromatography [5].

Results and Discussion

The critical temperatures for appearance of visual chilling symptoms in cv. 79 was 12.8°C (Fig.

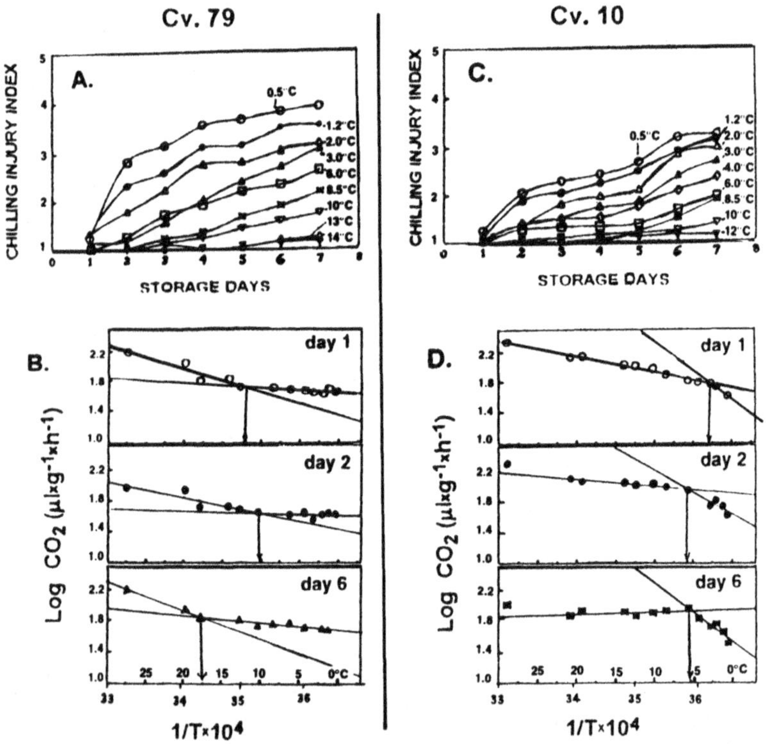

Figure 1: Effect of storage temperatures on CI index (A & C), and Arrhenius plots of respiration rates (B & D) in detached leaves of two basil cultivars (cv. 79 and cv. 10), varying in their chilling sensitivity.

1A) while that for cv. 10 was 10.2°C (Fig. 1C). Also the severity of CI obtained at the same temperature and storage duration was higher in cv. 79 (Fig. 1A) as compared with cv. 10 (Fig. 1C). The phase transition temperature of cv. 79 increased from 12.5 to 17°C during 1-6 days of incubation (Fig. 1B), whereas that of cv. 10, was much lower and remained between 3-5°C throughout incubation (Fig. 1D). Based on the differences in these 2 parameters, it was concluded that the basil cultivar cv. 79 was more susceptible to chilling than the cultivar cv. 10.

Table 1: Fatty acid composition of PG and PC isolated from leaves of the two basil cultivars (cv. 79 and cv. 10), varying in their chilling sensitivity.

FATTY ACIDS	FATTY ACID COMPOSITION			
	PG		PC	
	CV. 79	CV. 10	CV. 79	CV. 10
		(%)		
16:0	38.5 ± 2.5	30.9 ± 1.0	37.5 ± 3.9	27.7 ± 1.1
18:0	12.0 ± 1.0	11.7 ± 2.4	14.1 ± 1.3	13.4 ± 4.7
18:1	6.0 ± 0.9	14.7 ± 2.2	5.0 ± 0.8	15.7 ± 1.0
18:2	25.9 ± 4.7	19.0 ± 2.8	20.2 ± 2.5	21.3 ± 5.3
18:3	17.6 ± 3.3	23.6 ± 0.5	23.3 ± 2.6	21.9 ± 1.5

In order to explain the differences obtained in the chilling sensitivity of the two cultivars on a molecular basis, the FA composition of their membrane lipids was analyzed. The two basil cultivars did not differ significantly in their FA composition of PI and PE (data not shown). However, leaves of the chilling-sensitive cultivar (cv. 79) showed a 40% and 32% reduction in the composition of 18:1-PG and 18:1-PC, respectively, as well as a 26% and 35% increase in the composition of 16:0-PG and 16:0-PC, respectively, as compared with leaves of the less susceptible cultivar (cv. 10) (Table 1). These differences in the FA acid composition of PG and PC, isolated from the two basil cultivars, may be the cause for the variations obtained in their chilling susceptibility (Figs. 1 A,C) and their transition temperatures (Figs. 1B,D).

References

1. Lyons JM. Chilling injury in plants. Annu Rev Plant Physiol 1973;24:455-466.
2. Orr GR, Raison JK. Compositional and thermal properties of thylakoid polar lipids of *Nerium oleander* L. in relation to chilling sensitivity. Plant Physiol 1987;84: 88-92.
3. Murata N. Molecular species composition of phosphatidylglycerols from chilling-sensitive and chilling-resistant plants. Plant Cell Physiol 1983;24: 81-86.
4. Murata N, Nishida I. Lipids in relation to chilling sensitivity of plants. In: Wang CY, editor. Chilling injury of horticulture crops. Florida: CRC Press, 1990: 181-199.
5. Tasaka Y, Nishida I, Higashi S, Beppu T, Murata N (1990). Fatty acid composition of phosphatidylglycerols in relation to chilling sensitivity of woody plants. Plant Cell Physiol 1990;31:545-550.
6. Matsuo T, Ide S, Shitida M. Correlation between chilling sensitivity of plant tissues and fatty acid composition of phosphatidylglycerols. Phytochem 1992;31:2289-2293.
7. Joh T, Yoshida T, Yoshimoto M, Miyamoto T, Hatano S. Composition and positional distribution of fatty acids in polar lipids from *Chlorella ellipsoidea* differing in chilling susceptibility and frost hardiness. Physiol Plant 1993; 89:285-290.
8. Murata T. Investigation on chilling sensitivity of fruits and vegetables using Arrhenius plots. Acta Alimentaria 1990;19:9-17.
9. Hamilton JG, Comai K. Separation of neutral lipids and free fatty acids by high-performance liquid chromatography, using low wavelength ultraviolet detection. J Lipid Res 1984:25:1142-1148.

WHEAT LIPID ACCUMULATION IS ALTERED BY THE "GREENHOUSE EFFECT"

Mark Williams, Peter R. Shewry & John L. Harwood

Biochemistry Department, University of Wales, College of Cardiff, PO Box 903, Cardiff, CF1 1ST, U.K.

Introduction

Computer model systems are predicting that the present carbon dioxide levels will double, rising to 700ppm by the the middle of the twenty-first century with a concommittent 4°C rise in global temperatures (1). Vegetation will change and concern is particularly mounting as to the possible effects on plant growth and development. Since wheat is a very important crop in temperate areas, we have been investigating the possible impact the "Greenhouse effect" may have on it. It is well recognised that these effects will be complex (2).

Elevated temperatures accelerate phenological development of wheat (3) and we have previously reported that this also alters the the lipid composition of the mature grain (4). Cereal crops are characterised by possessing both non-starch and starch lipids. The latter are derived from the starchy endosperm of the grain and consist mainly of monoacyl phospholipids such as lysophosphatidylcholine and lysophosphatidylethanolamine. Non-starch lipids consist primarily of triacylglycerol with some glycolipids and phospholipids. Although the lipid content of wheat grains is relatively low (2.5% - 3.5% dry weight), lipids play an important role in the stability, nutritional quality and baking properties of the flour (5). Herein we report the influence that the "Greenhouse effect" may have on the lipid composition of developing grains.

Materials and Methods

Winter wheat (*Triticum aestivum* L.) cv. Mercia was cultivated in field greenhouses at Rothamsted Experimental Station during the 1991/1992 season. Wheat plants were grown at either ambient temperatures or 4°C above ambient temperatures, and normal carbon dioxide levels (350ppm) or double present levels (700ppm). Developing wheat grains were harvested from each growth condition on the 8th of June (i.e. at the mid-harvest stage). The lipids of developing wheat grains were extracted, separated by TLC and GLC and quantified as previously reported (4).

Results and Discussion

The "Greenhouse effect", as already mentioned, will involve changes in global temperatures and a doubling of atmospheric carbon dioxide levels. Therefore, in this study we have analysed the interactive effects of temperature (ambient and 4°C above ambient temperatures) and carbon dioxide (350ppm and

J.-C. Kader and P. Mazliak (eds.), Plant Lipid Metabolism, 384–386.

700ppm) by carrying out an analysis of variance (ANOVA), a combination of 2^2 factorial set (Table 1). Tests using the standard error of the differences (SED) for the $\log_e$ transformation show that there is evidence of an interaction between carbon dioxide level and temperature.

The effect of temperature on plant growth is well recognised to accelerate phenological development and this was true of the wheat plants in this study (data not shown). In conjunction with this, the more advanced plants also possessed higher amounts of both non-starch and starch lipids, irrespective of the carbon dioxide levels (Table 1). This was reflected in the total lipid levels (0.044 mg fatty acid / wheat grain [ambient temperatures] versus 0.199 mg fatty acid / wheat grain [4°C above ambient], with normal carbon dioxide levels). Carbon dioxide levels, as well as temperature, produced significant changes in grain lipid content. Whereas at ambient growth temperatures, elevated carbon dioxide levels was associated with increased lipid accumulation, at the elevated temperatures there was an unexpected decrease in non-starch lipid content which was reflected in the amounts of the total lipid content. Comparing the above data with the data for the mature harvest (4) was most interesting because higher amounts of grain lipid accumulation were associated with ambient growth temperatures. Therefore, we can propose that with the evident accelerated phenological development, the "Greenhouse effect" will limit the time available for storage lipid to accumulate.

Table 2 shows that the "Greenhouse effect" also influenced the individual lipid classes both quantitatively and qualitatively. Noticeably, non-starch neutral lipids, which consist primarily of triacylglycerol and digalactosyldiacylglycerol (DGDG) were approximately three times greater in wheat grains grown under the Test conditions (i.e. 4°C above ambient temperatures and 700ppm CO_2) as opposed to those grown under the Control conditions (i.e. ambient temperatures and 350ppm CO_2).

Moreover, acyl changes were also observed. For instance, non-starch lipids of wheat grains sampled from Control conditions possessed higher proportions of the saturated fatty acids, palmitate and stearate, than grains cultivated under Test conditions, whereby higher proportions of linoleate were observed. Similar pattern changes were also observed for DGDG. A significant increase in starch lysophosphatidylcholine linoleate was associated with Test conditions. In summary, the "Greenhouse effect" will accelerate phenological development and this is reflected in the lipid composition of the developing wheat grains. Such changes will contribute to changes in the flour quality of mature grains.

Table 1. Effects of temperature and carbon dioxide on lipid accumulation in developing mid-harvest grains

		mg Fatty acid / wheat grain					
Temperature	Carbon dioxide	Non-starch lipid		Starch lipid		Total lipid	
Amb T.	350	0.042	(-3.16)	0.002	(-6.21)	0.044	(-3.12)
Amb T.	700	0.058	(-2.85)	0.003	(-5.81)	0.061	(-2.80)
T.+4°C	350	0.161	(-1.82)	0.038	(-3.27)	0.199	(-1.61)
T.+4°C	700	0.122	(-2.09)	0.041	(-3.20)	0.163	(-1.81)
SED (CO_2xT)	(8df)		(0.0678)		(0.0824)		(0.572)

Data represents means (n=3). Numbers in parenthesis represent $\log_e$ from which ANOVA was performed.

Table 2. Alterations to some major acyl lipids as a result of the "Greenhouse effect"

Growth conditions	Lipid class	mg Fatty acid / wheat grain	Acyl composition (%) 16:0	18:0	18:1	18:2	18:3
Control	NS Neutral	0.0366	23±tr.	16±tr.	21±tr.	34±tr.	7±tr.
Test	NS Neutral	0.1020	18±tr.	N.D.	14±1	62±1	5±tr.
Control	NS DGDG	0.0020	26±tr.	2±tr.	10±tr.	46±1	17±tr.
Test	NS DGDG	0.0060	16±1	1±1	5±tr.	70±tr.	8±tr.
Control	St LysoPtdCho	0.0008	30±1	2±1	4±1	53±tr.	10±tr.
Test	St LysoPtdCho	0.0140	32±tr.	2±1	5±tr.	59±tr.	3±1

Abbreviations: tr.=trace <0.5%, N.D.=none detected, NS=non-starch, St=starch. Data represents mean values±S.D. (n=3).

References

1. Schneider SH. Detecting climatic change signals: are there any "fingerprints"? Science 1994: 263:341-47.

2. Warrick RA, Gifford RM, Parry ML. CO_2, climatic change and agriculture. In: Bolin B, Döös BR, Jager J, Warrick RA, editors. The greenhouse effect, climatic change and ecosystems. Scope 29. New York: John Wiley Publishers, 1986: 393-473

3. Mitchell RAC, Mitchell VJ, Driscoll SP, Franklin J, Lawlor DW. Effects of increased CO_2 concentration and temperature on growth and yield of winter wheat at two levels nitrogen application. Plant, Cell and Environ 1993; 16:521-29.

4. Williams M, Shewry PR, Harwood JL. Influence of the "greenhouse effect" on the lipid composition of wheat (*Triticum aestivum* L.) seeds. Grasas y Aceites 1993;44:132-34.

5. MacRithchie F. The role of lipids in baking. In: Barnes PJ, editor. Lipids in cereal technology.New York: Academic Press,1983: 165-88.

EFFECTS OF A HEAT SHOCK ON THE MOLECULAR SPECIES OF RAPE POLAR LIPIDS

Fatiha AID*, Chantal DEMANDRE**, Ghoziel KESRI-BENHASSAINE* and Paul MAZLIAK**

* Laboratoire de Physiologie Végétale, ISN/USTHB, BP n°39, El Alia, Alger, Algérie
** Laboratoire de Physiologie Cellulaire et Moléculaire, Université P. et M. Curie, URA/CNRS 1180, 4 place Jussieu, 75252 Paris Cedex 05, France

Introduction

Rape (*Brassica napus* L. var Bienvenue) is an oil seed plant of great potential economical interest for Algeria. The main problem for growing rape in this country is the high environmental temperature encountered during summer. Growth at high temperature results in a decrease in the abundance of polyunsaturated fatty acids. Such a decrease is associated with reduced membrane fluidity [1].
Brassica napus is a $C_{16:3}$ plant possessing hexadecatrienoic acid ($C_{16:3}$) in MGDG [2]. It has been shown that in $C_{16:3}$ plants MGDG is synthesized by both chloroplastic and cytoplasmic pathways, the former giving C_{18}/C_{16} molecular species and the latter C_{18}/C_{18} and C_{16}/C_{18} molecular species [3]. We report here the effects of a heat shock on the molecular species composition of the main polar lipids in *Brassica napus* leaves.

Material and methods

Rapeseeds (*Brassica napus* L. var. Bienvenue) were grown in a growth chamber under a 16 hour photoperiod, at 30°C/21°C day/night for a week and then transferred at 12°C. Plants aged from 4 to 5 weeks were transferred from 12 to 36°C for 48 hours to study the effects of heat shock.

LIPID ANALYSIS

Plant lipids were extracted according to [4]. Individual polar lipids were separated by TLC [5]. Methyl esters of fatty acids were obtained using the method of Metcalfe *et al.* [6].

MOLECULAR SPECIES ANALYSIS

Lipids were separated by HPLC on 10mm porasil column according to [7]. Main polar lipids were separated into molecular species on a 5mm ODS column by a methanol/water/acetonitrile, 90.5/7/2.5, v/v/v solvent containing 0.28g of choline chloride. Methyl esters of fatty acids from each molecular species were prepared by transmethylation and analysed by GPC.

J.-C. Kader and P. Mazliak (eds.), Plant Lipid Metabolism, 387–389.

ANALYSIS OF FATTY ACID POSITIONS

Each MGDG molecular species was hydrolyzed by purified lipase A1 from *Rhizopus arrhizus* (Boehringer) according to [8]. The products obtained, 2-acyl lyso MGDG and free fatty acids liberated from position 1, were separated by TLC.

Results and discussion

EFFECTS OF A HEAT SHOCK ON THE POLAR LIPID COMPOSITION OF RAPE LEAVES

The 24°C heat shock (from 12 to 36°C) applied during 48 hours, decreased the lipid content of rape leaves by about 20% (from 3.3 mg/g FW to 2.9 mg/g FW). Fig 1 shows a decrease in MGDG content and an increase in DGDG content : thus a lowering of the MGDG/DGDG ratio was observed, from 2.2 to 1.5. A decrease in PC and PG contents was also observed.

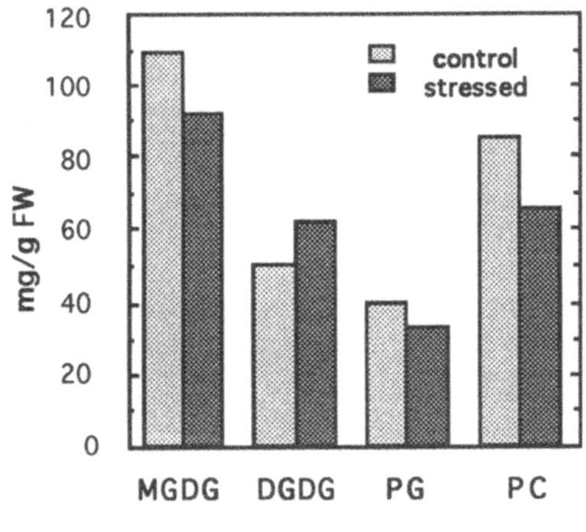

Figure 1 : effect of a heat shock on the main polar lipids of rape leaves

EFFECTS OF HEAT SHOCK ON THE MOLECULAR SPECIES OF PC, MGDG AND DGDG

The effects of a heat shock on the molecular species compositions of the main polar lipids of rape leaves figured in the table 1. MGDG is very rich in molecular species containing six double bonds (18:3/16:3 and 18:3/18:3) which represent more than 90% of total MGDG molecular species. The prokaryotic 18:3/16:3 molecular species is dominant (more than 70 mol%). In marked contrast, the eukaryotic molecular species 18:3/18:3 is the most important molecular species in DGDG (70 mol %).

molecular species	MGDG control	MGDG stressed	DGDG control	DGDG stressed	PC control	PC stressed
18:3/16:3	80.4	72.0	5.3	-	-	-
18:3/18:3	11.4	19.8	71.2	69.0	12.8	15,.0
18:2/18:3	1.8	3.3	1.5	t	27.6	6.2
16:0/18:3	t	4.5	17.8	30.0	23.0	36.5
18:1/18:3	-	-	-	-	-	5.2
18:1/18:2	-	-	-	-	-	2.6
18:0/18:3	-	-	-	-	-	7.0
16:0/18:1						6.2

Table 1 : effects of a heat shock on the main molecular species of MGDG, DGDG, and PC of rape leaf lipids (in mol %)

After the heat shock, we noted first, the apparition of more saturated molecular species like 16:0/18:1, 18:0/18:3, 18:1/18:2 PC and 16:0/18:3 MGDG. Secondly the eukaryotic molecular species 16:0/18:3 increased markely in the three lipid classes. Finally the ratio of total C_{18}/C_{16} to C_{18}/C_{18} molecular species in MGDG was reduced from 6.6 to 2.6 in stressed plants.
These results show that the chloroplastic (prokaryotic) pathway of MGDG biosynthesis was more sensitive to a heat shock than the extrachloroplastic pathway.

References

1. Santarius K.A. and Weis E. Heat stress and membranes. In: J.L. Harwood and T.J. Walton, editors. Plant membranes: structure, assembly and function. 1988; 97-112.
2. Williams J.P. and Khan M.U. Lipid biosynthesis in *brassica napus* leaves Biochem. Biophys. Acta, 1982; 713:177-84.
3. Roughan G. and Slack R. Glycerolipid synthesis in leaves. Trends Biochem. Sci. 1984; 9:383-86.
4. Bligh E.G. and Dyer W.J. A rapid method of total lipid extraction and purification. Can. J. Biochem. Physiol. 1959; 37:103-106.
5. Lepage M. Identification and composition of turnip root lipids. Lipids 1967; 2:244-50.
6. Metcalfe L.D., Schmitz A.A. and Pelka J.R. Rapid preparation of fatty acids esters from lipids for gas chromatographic analysis. Anal. Chem. 1966; 38:514-15.
7. Demandre C., Tremolieres A., Justin A. M. and Mazliak P. Analysis of molecular species of plant polar lipids by high performance and gas chromatography . Phytochemistry 1985; 24:481-85.
8. Fisher W.E., Heinz E. and Zeuz M. The suitability of lipase from *Rhizopus arrhizus* (Delamar) for analysis of fatty acid distribution in dihexosyl deglycerides, phospholipids and plant sulfolipids. Hoppe-Seyler's Z. Physiol. Chem. 1973; 354:1115-23.

THE EFFECT OF TEMPERATURE AND ABSCISIC ACID ON THE LEVEL OF ERUCIC ACID IN THE OIL OF OILSEED RAPE

Jeroen A. Wilmer[1], Johannes P.F.G. Helsper[1] & Linus H.W. v.d. Plas[2]
[1] Department of Biochemistry & Cell Physiology, AB-DLO.
[2] Department of Plant Physiology, Wageningen Agricultural University.

Introduction

In 1965 Canvin reported that the maximum level of erucic acid (22:1) in seed oil of oilseed rape (*Brassica napus* L.) is observed at a growth temperature of 15°C [1]. The aim of the present study is to explore some of the underlying biochemical mechanisms for this observation. The effect of temperature is different in magnitude in the two culivars tested: Reston showed a response similar to that described by Canvin, Gulle did not show any changes at all when comparing oil from plants grown at 15 or 25°C. The influence of growth temperature is most prominent when about 60% of seed development has been completed. This influence could not be observed when assaying microspore-derived embryos under similar conditions, where levels of 22:1 remained at about 20%. Addition of ABA to the medium induced an increase in 22:1, and therefore may be a factor determining the level of 22:1 in the seed oil.

Experimental

Plants of two genotypes of high erucic acid rape, Reston and a breeding selection from Gulle (referred to as Gulle), were grown in a greenhouse at 20°C/17°C day/night with a 16h photo period. In *in vivo* studies plants were transferred to growth chambers at 15°C or 25°C at the onset of flowering. Seeds were harvested at different timepoints during the developmental process. In another experiment plants were transferred from 15°C to 25°C or *vice versa* at different developmental stages during seed development and dry seeds were harvested. For the *in vitro* studies microspore-derived embryos were initiated from flower buds of 3.4 to 3.6 mm, selected from racemes that were at the onset of flowering. The embryos were grown in NLN82-medium [2], with 13% sucrose and without hormones, at 15, 20 or 25°C after an initial period of 14 days at 25°C.

From both embryos and seeds the lipids were extracted using a Bligh & Dyer type extraction and the fatty acid composition of the triacylglycerol (TAG) fraction was determined after transmethylation for three hours in 5% methanolic sulphuric acid using capillary GC. The GC was equipped with a Chrompack CPwax-52CB column (25 m x 0.32 mm) and an FID for detection.

J.-C. Kader and P. Mazliak (eds.), Plant Lipid Metabolism, 390–392.

Results & discussion

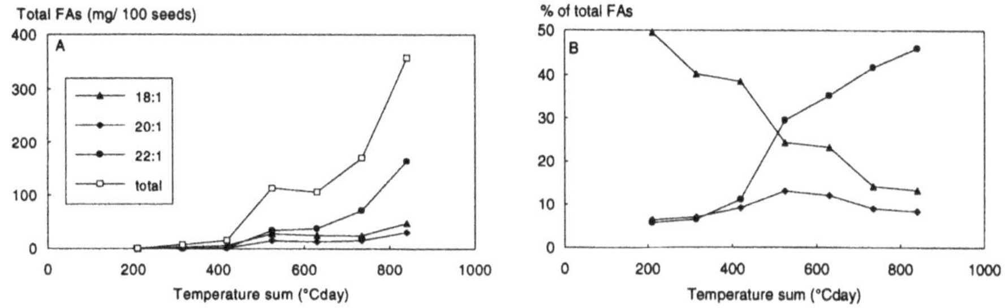

Figure 1: Accumulation of oil in developing seeds of Gulle at 15°C. A: Accumulation of total triglycerides and of the monounsaturated fatty acids 18:1, 20:1 and 22:1. B: Proportion of the monounsaturated fatty acids in the oil (mol%).

During development the level of TAG in the seeds continues to increase until they are almost completely dry (fig 1A). When about 60% of this development has been completed the level of erucic acid starts to increase strongly at 15°C (fig 1B). At 25°C a similar pattern is observed with erucic acid accumulating late compared to total oil synthesis. When plants are transferred from 15°C to 25°C or *vice versa* before this stage the oil composition is similar to that of plants grown at the final temperature only (far left in figs 2A/B). When plants are transferred after this stage the oil composition resembles that of plants grown at the initial temperature (far right in figs 2A/B). The changes in erucic acid are completely compensated for by opposite changes in the level of oleic acid (18:1). The levels of other fatty acids, including the intermediate in 22:1 synthesis eicosenoic acid (20:1), remain stable. The effect of temperature is also dependent upon the cultivar used: while in Reston the level of 22:1 changes from 20% at 25°C to 40% at 15°C, the levels in Gulle remain more or less stable at 40 to 45% (not shown).

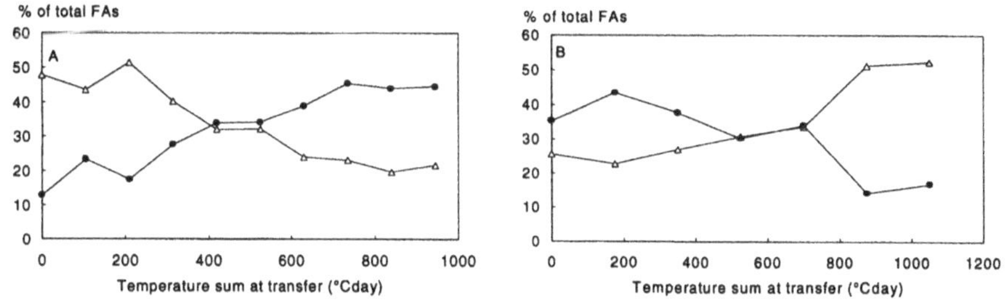

Figure 2: Response to changes in environmental temperature of fatty acid composition in oil from developing seeds of Reston. A: Plants transferred from 15 to 25°C during seed development. B: Plants transferred from 25 to 15°C during development. Open triangles oleic acid, closed circles erucic acid. Harvest at ±1000 °Cday.

In *in vitro* cultured microspore-derived embryos a maximum in erucic acid content at about 15°C could not be observed. Levels of 22:1 never exceeded 30%; average content was about 20% at the temperatures tested. Reston, that showed a clear increase of 22:1 at 15°C in plants, even showed an opposite tendency *in vitro* (fig 3). Obviously some factor is missing in this system which controls the erucic/oleic acids ratio in complete seeds. Abscisic acid (ABA), which is known to be important in seed development and does have a stimulating effect on the level of 22:1 [3], might be the missing factor. Application of ABA (0.5-50μM) leads to a twofold increase in the (low) level of 22:1 (fig 4). These exogenous concentrations have been shown to produce endogenous concentrations in the normal range for *in vivo* material [4].

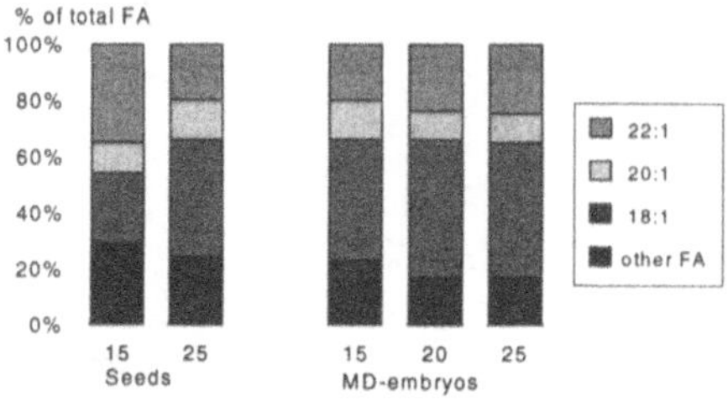

Figure 3: Fatty acid composition of oil from seeds and microspore-derived embryos of Reston grown at 15, 20 and 25°C.

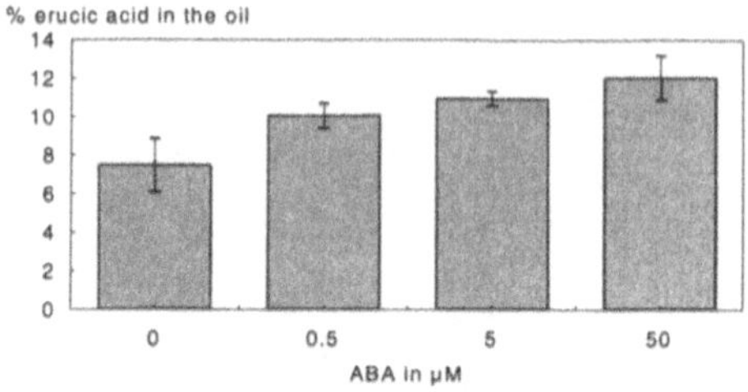

Figure 4: Effect of various ABA concentrations in the medium on the erucic acid level of oil from microspore-derived embryos of Reston. Embryos were treated with ABA during the third part of development and harvested at ±1000 °Cday.

Conclusions & future experiments

The influence of temperature on the fatty acid composition of rapeseed oil *in vivo* is especially effective at the time of maximum lipid accumulation.

At 15°C the level of 22:1 in the seed oil is higher than at 25°C. The magnitude of the effect varies between cultivars.

This effect of growth temperature observed in plants, was not found in microspore-derived embryos

A possible factor that is missing in this *in vitro*-system may be ABA, which is shown to induce an increase in the level of 22:1.

We are currently determining the levels of ABA in seeds developing at 15 and 25°C, and further expanding the tests of the effect of ABA *in vitro*. This will allow us to clarify whether the effect of temperature is an indirect effect due to a change in ABA concentration or to an altered tissue sensitivity for this plant growth regulator.

References

1. Canvin DT. The effect of temperature on the oil content and fatty acid composition of the oils from several oil seed crops. Can J Bot 1965; 43: 63-69.
2. Gland A, Lichter R, Schweiger HG. Genetic and exogenous factors affecting embryogenesis in isolated microspore cultures of Brassica napus L. J Plant Physiol 1988; 132: 613-17.
3. Holbrook LA, Magus JR, Taylor DC. Abscisic acid induction of elongase activity, biosynthesis and accumulation of very long chain monounsaturated fatty acids and oil body proteins in microspore-derived embryos of Brassica napus L cv Reston. Plant Sci 1992; 84: 99-115.
4. Finkelstein RF, Tenbarge KM, Shumway JE, Crouch ML. Role of ABA in maturation of rapeseed embryos. Plant Physiol 1985; 78: 630-36.

EFFECTS OF PESTICIDES ON THE BIOSYNTHESIS AND ACCUMULATION OF LIPOPHILIC COMPONENTS IN PLANT CUTICLES

Hervé Albertin*, Françoise Nurit*, Jean Favre-Bonvin***, Patrick Ravanel** and Michel Tissut*
*Laboratoire de Physiologie Cellulaire Végétale (URA 576), **Laboratoire de Pharmacognosie, Université Joseph Fourier, BP 53, 38041 Grenoble cedex 09 France, ***Service Central d'Analyse, CNRS Solaize, BP 22, 69390 Vernaison France

Introduction

The plant cuticle is an apoplastic production of the aerial parts of plants adapted to terrestrial life. This cuticle stands mainly as a highly lipophilic barrier limiting water loss, the gazeous exchanges being mainly carried out at the level of stomata which are submitted to a physiological control.

Xenobiotics of agronomical interest are very often applied to plants as post-emergence treatments and have therefore to pass through the cuticle lipophilic barrier in order to reach plant cells. Some of these products are highly lipophilic and the cuticle does not impede their penetration [1]. However, several other products are sufficiently hydrophilic to be severely limited in their penetration by the cuticle barrier. It is the case of mineral nutrients, of several insecticides,and endotherapic fungicides and of some herbicides such as glyphosate, paraquat, aminotriazole...[2].
Relatively hydrophilic pesticides are especially interesting insofar as they can penetrate into sieve tubes and be submitted to phloem transport (organophosphorus insecticides, glyphosate...).
It is known for years that cuticle penetration may be improved in this case by adding appropriate surfactants to the active ingredient, and by obtaining unstable lipophilic derivatives of these substances, as is the case of the acidic pesticides (glyphosate, 2,4-D...for instance), which are transformed to lipophilic esters reemitting the free acid inside the cells (Figure 1).

J.-C. Kader and P. Mazliak (eds.), Plant Lipid Metabolism, 393–407.

Figure 1. Leaf penetration and transfer of phloem-mobile xenobiotics.

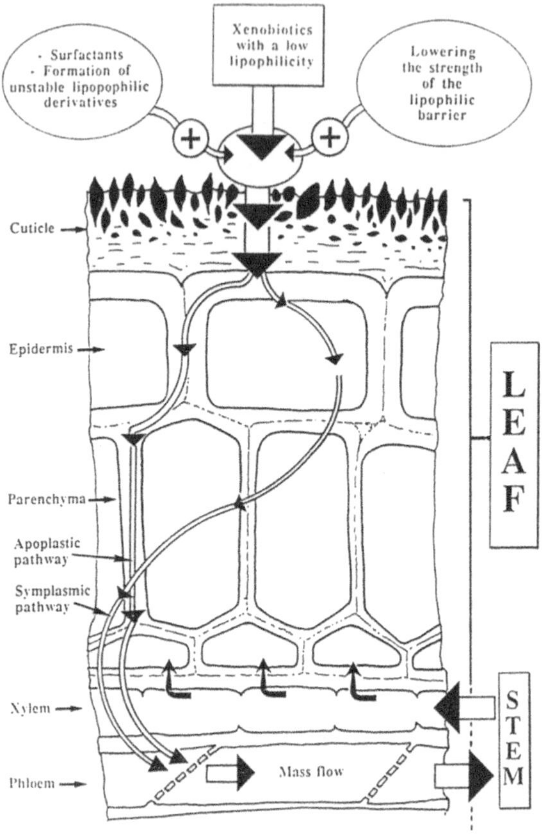

The present report is an attempt to investigate another experimental way and answer the question : is it possible to decrease temporarily the apoplastic accumulation of cuticle components, by acting on enzymes of cuticle biosynthesis, in order to allow a better penetration of hydrophilic substances (Figure 2)?

For this purpose, experimental compounds were selected according to several criteria (Figure 3) :

1. The compounds had to be effective *in vitro* at one or several of the steps of the pathways leading to waxes or cutin biosynthesis, without changing the other chemical activities of the leaf cells. That means that the products had to be selective, in a biochemical point of view
2. The compounds had to penetrate inside plant epidermis cells ; this means that they had to pass through the preexisting cuticle

Figure 2. Scheme illustrating biosynthesis and transfer of waxes and cutin in the leaf epidermis.

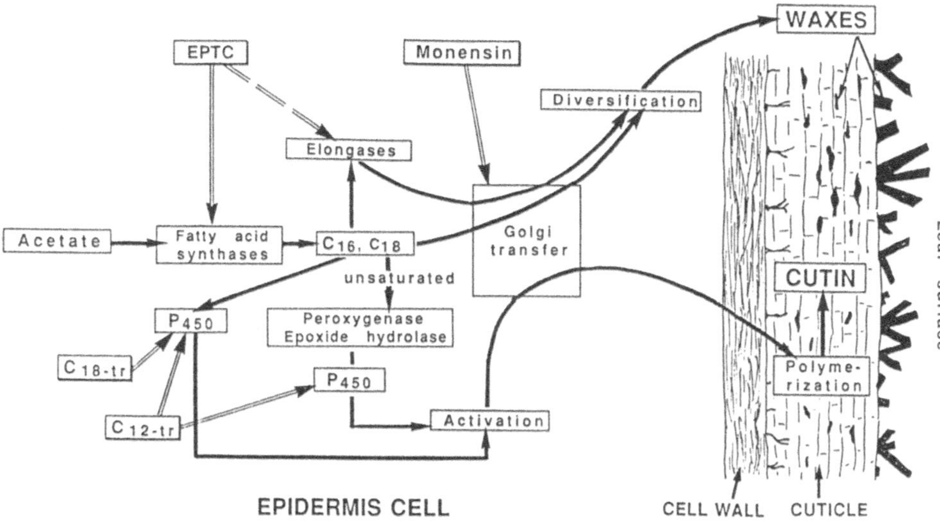

Figure 3. Scheme illustrating the three criteria which are necessary for selective decrease of cuticle formation *in vivo*.

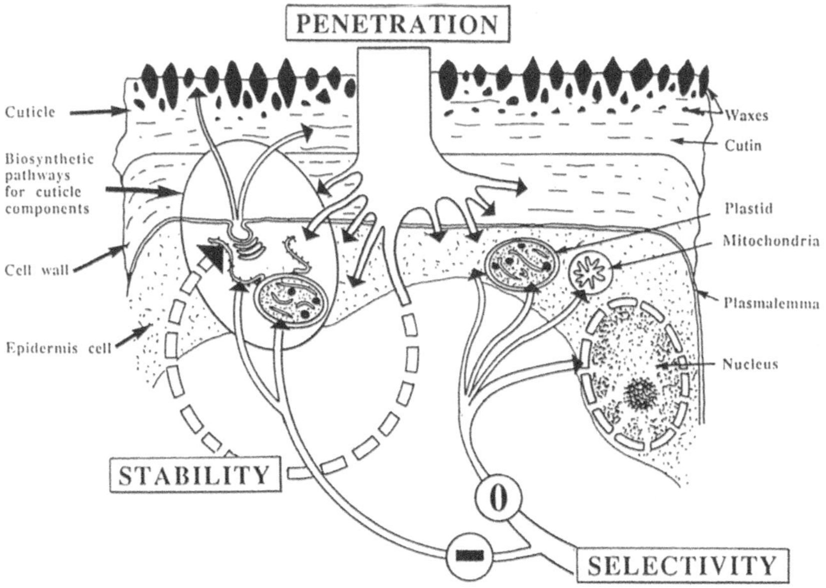

3. Given that the cuticle is the result of a relatively slow accumulation of compounds excreted by the epidermis cells through exocytosis, and that the final steps of their biosynthesis occur in the apoplastic space, it seems likely that decreasing clearly the external accumulation of waxes or cutin requires a long time during which biosynthesis of these compounds has to be severely inhibited. The inhibitors selected for this experimentation had therefore to be stable for days inside the plant.

As a whole, the compounds selected for this experiment had to be biochemically selective, stable *in vivo* and able to penetrate into the leaves (Figure 3).

In this first attempt, four compounds were tested on the three grounds above-mentioned (Figure 4). One was EPTC, an herbicide known to inhibit fatty acid synthases and the elongases responsible of wax formation [3]. Another one was monensin, known to inhibit the Golgi traffic, as well in animal as in plant cells [4-7]. The two last compounds were a C_{18} and a C_{12} compounds (named here C_{12}-tr and C_{18}-tr) representing a whole series of fatty acid analogues having a terminal triple bound, which were demonstrated *in vitro* on isolated microsomal membranes to be suicide inhibitors of the P450 cytochromes which are responsible for ω-hydroxylation leading to cutin polyesterification [8]. The synthesis of this series was carried out by Mioskowski *et al.* (Bioorganic Chemistry Laboratory, L.Pasteur University, Strasbourg, France).

Figure 4. Chemical structure of the four studied compounds.

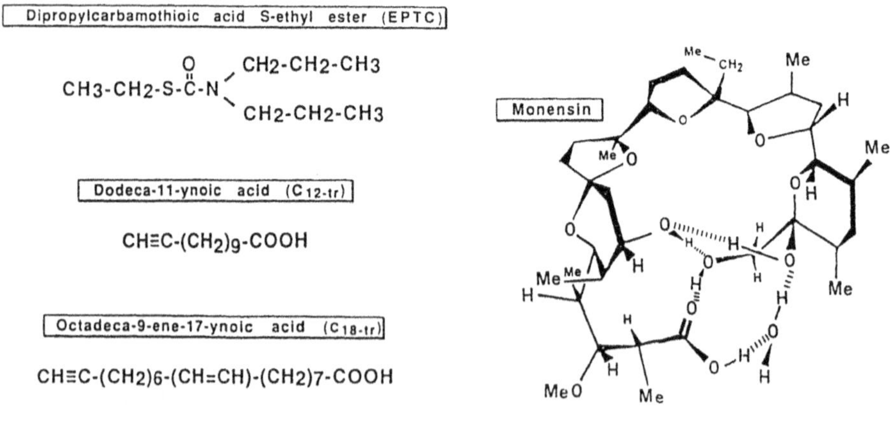

Results

I. THE CHOICE OF AN EXPERIMENTAL PATHWAY

As explained just before, the studied compounds had to be analyzed on three grounds : leaf penetration, stability inside living plant material, selective efficiency on cuticle biosynthesis.

a) Leaf penetration and stability. EPTC (Dipropylcarbamothioic acid S-ethyl ester) is known as a commercially available herbicide acting after a preemergence application. Its agronomical efficiency through such a use requires a good chemical stability in soil and plants. With a maximum water solubility of 370 mg/l at 20°C, a good penetration rate into the leaf under saturating moisture seems likely. Monensin is a very complex structure, with two possible steric configurations, the cyclic one being probably much more lipophilic than the other. It is very stable in neutral and alkaline pH and is sufficiently lipophilic to be highly soluble in organic solvents. It is commonly used as medicine or food additive and is probably stable enough *in vivo* to induce its protective effects. A good leaf penetration seems likely, depending on its lipophilicity, and in spite of a slightly high M.W. (671). As a matter of fact, monensin is responsible for a powerful necrosis effect in *Vicia faba* leaves, at concentrations higher than 10 µM, which shows that this compound can penetrate through the plant cuticle.
The leaf penetration of the fatty acid analogues studied here has been measured with the use of ^{14}C labelled compounds. Under our conditions, 29% ± 1% to 31% ± 1% of the deposited amounts were shown to have penetrated into *Vicia faba* leaves 60 hours after droplet deposition.

b) Selectivity and efficiency of the studied compounds. Our hope was to obtain compounds able to decrease the rate of cuticle accumulation without changing the other biochemical activities of plant cells.
In that view, several points were tested :
- a. effects on the growth of cell suspension cultures
- b. effects on the activities of isolated chloroplasts
- c. effects on the growth of young seedlings under saturating moisture
- d. general effects (photosynthesis rate, cell integrity) on *Vicia faba* leaves, which generally permitted to measure a lethal amount/surface unit, demonstrated through the appearence of necrosis spots. Table 1 summarizes the effects shown during this study.

The solubility in water with 1% DMSO allows a maximum concentration of 100 µM. The studied compounds were without effect on seedlings at this concentration. These results show that specific effects of the products on *Vicia faba* cuticle synthesis may be obtained in the presence of a maximum deposit of product which was 3 nmoles dm^{-2} for monensin, 500 nmoles dm^{-2} for EPTC and 300 nmoles dm^{-2} for the fatty acid analogues.
The changes in waxes and cutin biosynthesis were measured by incorporating labelled precursors into these cuticle components, which were extracted and submitted to extensive purification afterwards.

Table 1. General effects of the studied compounds on plant cell physiology and biochemistry. Chl A : oxygen evolution by class A chloroplasts under light.

	CL_{50} *Acer* cells	I_{50} Chl A	Necrosis effect on *Vicia f.* leaves	Inhibition of seedling growth µmoles dm^{-2}	
				wheat	radish
Monensin	1 µM	0.25 µM	3.6 nmoles dm^{-2}	-	I_{100} at 50 µM
EPTC	100 µM	250 µM	714 nmoles dm^{-2}	I_{100} at 100 µM	I_{100} at 1 mM
C_{18}-tr	CL_{100} at 50 µM	100 µM	375 nmoles dm^{-2}	>> 100 µM	>> 100 µM
C_{12}-tr	CL_{100} at 500 µM	> 200 µM	375 nmoles dm^{-2}	>> 100 µM	>> 100 µM

$(U)^{14}C$-acetate and $(U)^{14}C$-palmitate were used to that purpose. Each of them readily labels the cuticle components with an incorporation ratio which is relatively high : 2.2% for acetate ; 8.5 % for palmitate at 60 h.
The kinetic of wax or cutin labelling is shown in Figure 5.

Figure 5. Incorporation of ^{14}C-acetate and ^{14}C-palmitate into *Vicia faba* leaf waxes and cutin.

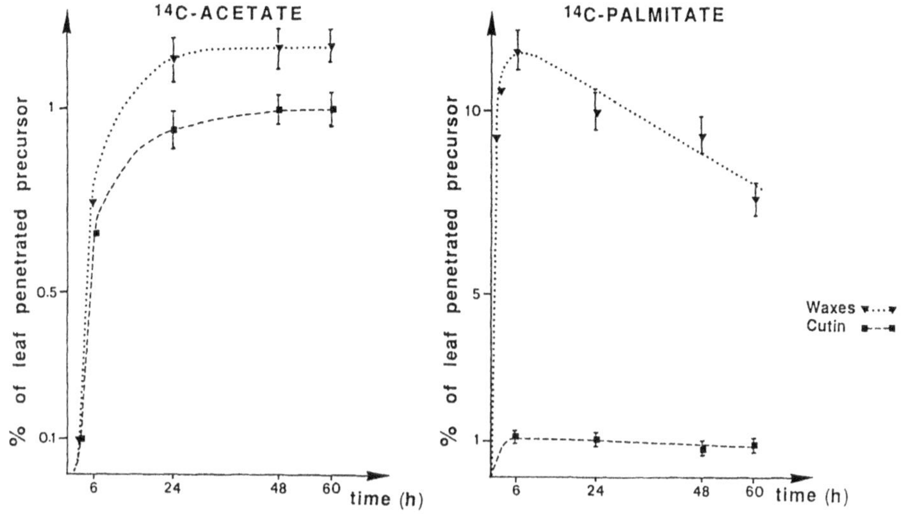

As shown by autoradiography, palmitate incorporates in the leaf cells, near the deposition point, in contrast with ^{14}C-acetate which distributes inside the whole plant through phloem transport. These differences between the two precursors explains partly why the length of the incorporation period differs so much from one case to the other (Figure 6).

Figure 6. Incorporation rate in dpm h^{-1} dm^{-2} of ^{14}C-acetate and ^{14}C-palmitate into *Vicia faba* leaf waxes and cutin.

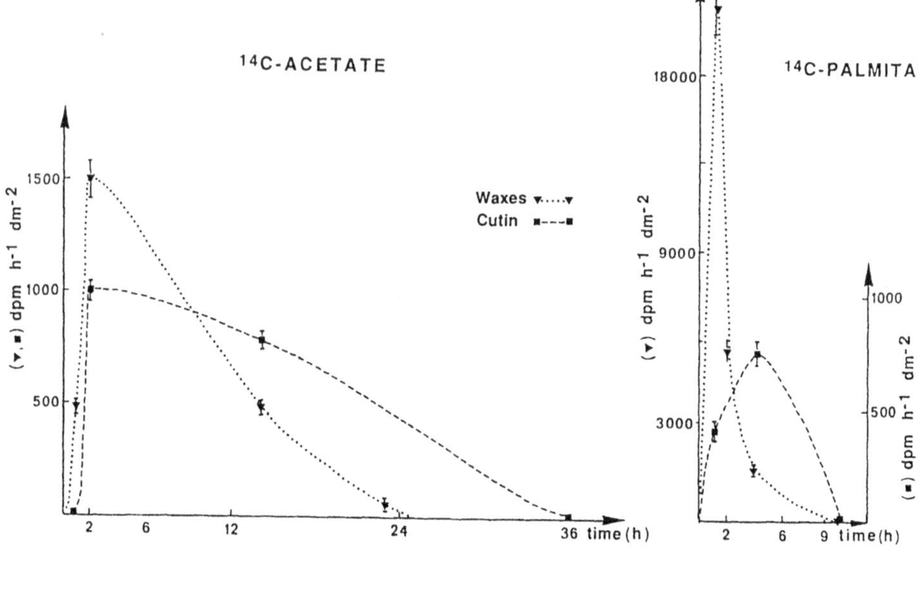

Another striking difference is that we are absolutely sure that free ^{14}C-acetate does not remain either in purified waxes or cutin. For ^{14}C-palmitate, the situation is more ambiguous at the beginning of the experiment in the case of waxes.
Furthermore, for this latter precursor the part of the biochemical pathways tested is comprised between C_{16} and waxes or cutin. In the case of acetate, the test is also concerning the general fatty acid synthesis (acetyl-CoA carboxylase and fatty acid synthases).

II. EFFECTS OF THE STUDIED PRODUCTS ON WAXES AND CUTIN ACCUMULATION IN TREATED *VICIA FABA* LEAVES

In order to show a possible quantitative decrease in waxes or cutin accumulation which might be of agronomical interest, the incorporation of ^{14}C into purified waxes and cutin was compared between the treated and untreated samples, at the end of the labelling period.

a) Effects of EPTC. After several assays, the most appropriate experimental procedure was to treat the broad bean leaves with EPTC 1mM three days before the addition of ^{14}C-acetate. The total incorporation into waxes and cutin was measured 60 hours after labelling, when free ^{14}C-acetate had fully disappeared and before any noticeable effect of waxes or of cutin catabolism under our conditions. The purified waxes and cutin fractions showed a decrease in the amount of ^{14}C dm^{-2} of leaf which was in each case 47% ± 5%. A quite similar result was obtained when considering the specific radioactivity of waxes or cutin which decreased in each case by approximately 50%. These results can be accounted for by the very low ratio between the labelled products and the waxes and cutin previously accumulated in the apoplasm.

b) Effects of monensin. As a powerful cytotoxic effect of monensin appears in each system, a specific inhibition on cuticle biosynthesis could only be studied at concentrations lower than 5 μM. In this case, a comparative incorporation of ^{14}C-acetate and ^{14}C-palmitate was carried out, simultaneously with the monensin leaf treatment on broad bean. The final result of waxes or cutin labelling was observed 60 h later. With palmitate as a precursor, an inhibition concerning equally the radioactivity per leaf surface unit and the specific radioactivity of waxes and cutin was measured. It reached 40% ± 10% for waxes and 38% ± 2% for cutin. The cumulative inhibitory effect observed with acetate was much lower (0%± 10% for waxes; 25% ± 5% for cutin). This discrepancy can be understood when looking at figure 6 which shows that the labelling period is much longer with acetate. Monensin at 5 µM is therefore suspected not to be sufficiently stable in plant material to prevent for more than 10 hours the Golgi transfer necessary for cuticle biosynthesis . For the time intervals chosen in our experiments, the powerful effect of monensin shown on isolated chloroplasts [9] (Figure 7) is probably unable to impede the activity of epidermis non-chlorophyllous cells.

Figure 7. Inhibitory effect of monensin on the light-dependent oxygen evolution by spinach class A chloroplasts.
(A): Inhibitory effect of monensin of the light-dependent oxygen evolution by spinach class A chloroplasts(C = 60 µg chlorophyll ml^{-1}) in the presence of 100 µM $NaHCO_3$, 10 µM KH_2PO_4 (Pi) and light (L). Numbers on traces refer to µmol O_2 produced hr^{-1} mg^{-1} chlorophyll. Insert (B): titration curve showing the relative percent of inhibitory activity of monensin.

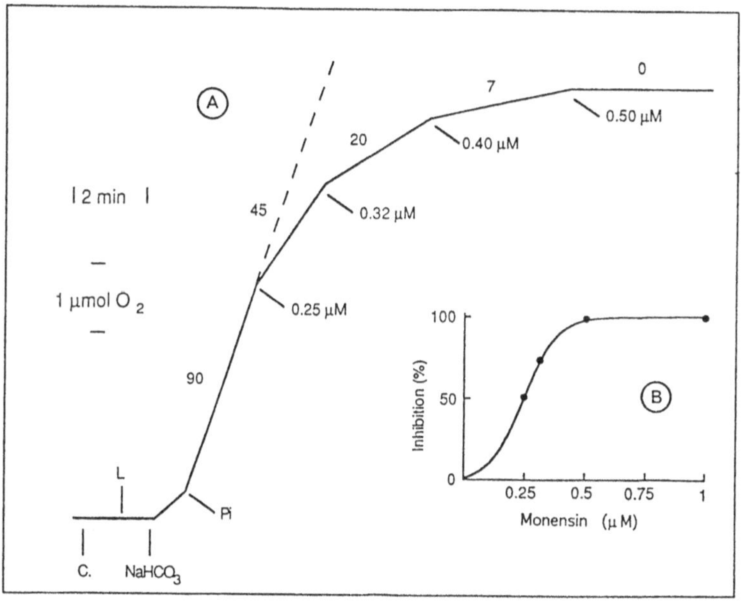

c) Effects of the fatty acid analogues C_{12}-tr and C_{18}-tr. A 500 µM leaf treatment was possible as the general physiological effects of the studied compounds were negligible. The leaf penetration rate of these compounds was high but they were readily metabolized, as shown by the emission of $^{14}CO_2$ from the C_1 labelled compounds (Figure 8).

Figure 8. Leaf penetration and metabolisation rate of the C_{12}-tr and C_{18}-tr inhibitors (labelled on the first carbon).

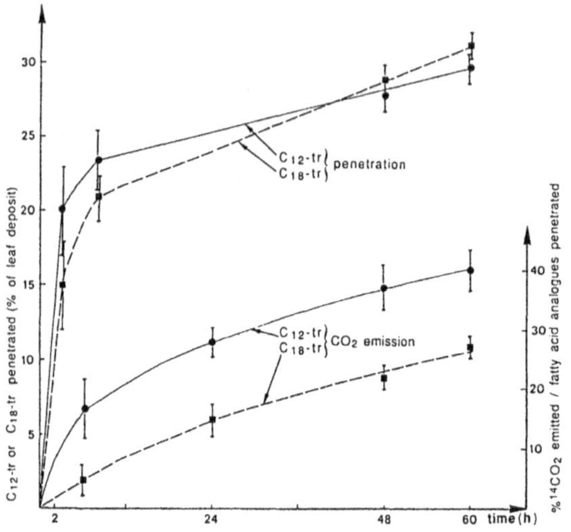

However, a high percentage of initial product was present in the plant material during the first day of experiment. Under these conditions, the two precursors ^{14}C-acetate and ^{14}C-palmitate were used, giving partial inhibitory effects after 60 hours either on waxes or cutin treated with C_{12}-tr or C_{18}-tr (Table 2).

Table 2. Inhibitory effects of C_{12}-tr and C_{18}-tr on waxes and cutin biosynthesis in *Vicia faba* leaves.

Products	Precursor	Time for precursor application	Level of labelling after 60h. % of the untreated sample	
			Waxes	Cutin
C_{12}-tr 500 µM	^{14}C-palmitate	0 h	56% ± 4%	58% ± 4%
C_{18}-tr 500 µM	^{14}C-acetate	6 h	84% ± 6%	66% ± 6%
	^{14}C-palmitate	0 h	50% ± 10%	72% ± 8%

However, neither powerful nor constant inhibitory effects were observed in this case. Furthermore, the global inhibitions concerned cutin as well as waxes, in contrast with what could have been expected from the specific inhibition of P450 hydroxylation shown *in vitro* on purified microsomal fractions [8]. Therefore, the hypothesis was considered of an incorporation of the fatty acid analogues or of their metabolization products, in competition with ^{14}C-acetate or ^{14}C-palmitate. The results shown in figure 9 gave a confirmation of this hypothesis.

Figure 9. Incorporation of the 1-^{14}C C_{18}-tr and of its ^{14}C metabolization products in the waxes and cutin of *Vicia faba* leaves.

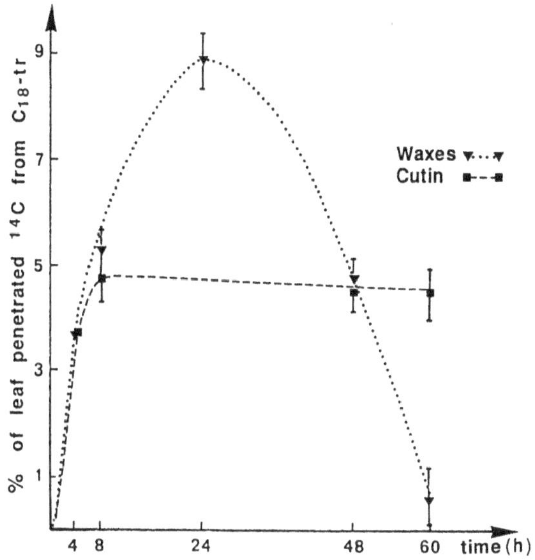

Discussion

Vicia faba leaf cuticle is an amorphous type cuticle [10]. Waxes have been described by Kolattukudy [11]. Cutin studied by Kolattukudy and Walton [12]. is mostly composed of 85% 9,16 or 10,16-dihydroxypalmitic acid, 7% 16-hydroxypalmitic acid and contains minor percentages of palmitic acid, stearic acid and oleic acid (8 %).

The first purpose of this work was fulfilled, which was to elaborate an experimental procedure giving the quantitative measurement of apoplastic waxes and cutin accumulation for broad bean leaves. A direct measurement of the changes in the amounts of accumulated waxes or cutin under the action of inhibitors cannot give an accurate evaluation of their effects. It is therefore necessary for a trustworthy determination to measure the incorporation flux of labelled precursors into purified waxes and cutin. Under our conditions, the incorporation kinetic of acetate corresponds approximatly to a 24 h pulse with a maximum rate of 2 to 4 hours. With palmitate the incorporation period is much shorter and is superimposed on a diffusion of the free acid in the waxes, which requires an extensive separation to evaluate the true incorporation into waxes. None of the studied compounds seems to be a powerful, stable and selective inhibitor of cuticle components biosynthesis. Under our conditions, EPTC seems to give the best results, if we consider that *de novo* synthesis of C_{16} fatty acids in the epidermis is a specific step of cuticle biosynthesis. In the case of monensin, the phytotoxic effect readily obtained on bean leaves might depend on a general alteration of Golgi transfer or of pH decompartimentation affecting the non-chlorophyllous plastids in the epidermis. Even if the cuticle excretion was the main activity of the epidermis cells in non-growing leaves at the studied stage, we are far from being sure that a selective action takes place, although an abundant litterature presents this effect as likely [6,7]. In the case of C_{12}-tr and C_{18}-tr, the biochemical effect shown *in vitro* on fatty acid hydroxylation seems either not to appear *in vivo* in *Vicia faba* leaves or to be without consequence for cutin accumulation rate. Therefore, the question arises of qualitative changes in the position of hydroxylation in the precursors of cutin. From an agronomical point of view, the study of a quantitative decrease in cuticle accumulation now requires experimentation on new groups of xenobiotics, taking into account the same selective factors: selective efficiency, penetration, stability.

Material and Methods

I. PLANT CULTURE

Broad bean (*Vicia faba* L.) plants were grown from seeds on a compost-soil, in a climatic chamber (80% R.H.) in which plants were exposed to a light flux of 70 µmoles photons m^{-2} s^{-1} for 16 h at 25°C and 8 h of night at 20°C. *Vicia faba* plants were used after 14 days of culture, when having 3 levels of leaves.

II. ^{14}C-PRECURSORS

Two labelled precursors were used: ^{14}C (U)-acetate (specific activity: 3.48 GBq mmole^{-1}, used at a 0.1 mM concentration) and ^{14}C (U)-palmitic acid (specific activity: 29.6 GBq mmole^{-1}, used at a 24 µM concentration). In some cases, plants were treated with the labelled precursors and with experimental inhibitors simultaneously. In other cases, the precursors were applied 6 or 72 hours after the inhibitors. Droplets were deposited on the upper surfaces of leaves. Plants were placed in closed chamber with satured moisture under controlled conditions. Carbon dioxide produced by catabolism metabolism was trapped in a KOH solution. The labelling period was of 60 h. Before waxes and cutin extraction, the treated leaves were detached and carefully rinsed with water, to remove any unpenetrated ^{14}C-acetate, or with ethanol-water (7/3: v/v) to remove unpenetrated ^{14}C-palmitate.

III. LABELLED EXPERIMENTAL INHIBITORS

The corresponding ^{14}C-inhibitors of dodeca-11 ynoic acid (C_{12}) and octadeca-9 ene-17-ynoic acid (C_{18}) synthesized by Mioskowski et *al.* were used: specific activity: 1.56 GBq mmole^{-1} for 1-^{14}C C_{12}-tr and 1.36 GBq mmole^{-1} for 1-^{14}C C_{18}-tr.

IV. WAXES AND CUTIN ISOLATION

The leaves were immersed (10 sec) in chloroform at room temperature. Chloroform was evapored under a nitrogen stream and the weight and radioactivity of the solid waxes residue were measured. This residue was separated on TLC successively with petroleum spirit and chloroform-methanol (89/11: v/v) as solvents. The bands were eluted and their radioactivity measured. After this, the leaves were floated in acetone for one night, under agitation at room temperature, allowing inner lipids and pigments to be extracted. Then, on the deshydrated leaves, the residual lipids were finely removed with boiling chloroform under reflux for 4 hours.

After this, the solid residue was dried, then submitted to alkaline hydrolysis with boiling 95% ethanolic 1.8N KOH for 12 h under reflux. The reaction mixture was then diluted with water, acidified and the hydroxy fatty acids

originating from the cutin were extracted with diethylether, then with ethyl acetate. The solvents were evaporated to dryness under a nitrogen stream, and the residue weighed. Its radioactivity was measured and the hydroxylated fatty acids separated on TLC. The radioactivity of these fatty acids was then determined after dilution.
The cutin monomers were transformed into the corresponding trimethylsilyl ethers, using trimethylchlorosilane in pyridine. The separation of the monomers was carried out on a gas chromatograph fitted with a flame ionization detector and an on-column injector. The used capillary column had dimensions 30 m x 0.250 mm, with liquid phase DB-5MS and a film thickness 0.25 µm. Capillary gas chromatography-mass spectrometry was performed under equivalent chromatographic conditions.

References

1. Chamel A. Foliar absorption of herbicides : study of the cuticular penetration using isolated cuticles. Physiol Vég 1986; 24: 491-508.
2. Tissut M, Séverin F. Plantes, herbicides et désherbage. ACTA, 1984: 251.
3. Kolattukudy PE, Brown L. Inhibition of cuticular lipid biosynthesis in *Pisum sativum* by thiocarbamates. Plant Physiol 1974; 53: 903-906.
4. Smilowitz A. Monovalent ionophores inhibits acetylcholine esterase release from cultured chick embryo skeletal muscle cells. Mol Pharmacol 1979; 16: 202-214.
5. Tartakoff A, Moessli D, Vassoli P. Intracellular transport of lymphoïd surface glycoproteins. Role of the Golgi complex. J Mol Biol 1981; 150: 525-536.
6. Mollenhauer HH, Morre DJ, Droleskey RE. Monensin inhibition of growth of rye-grass seedlings. Monensin effects the transhalf of *Euglena gracilis* dictyosomes. Protoplasma 1982; 114: 119-124.
7. Boss WF, Morre DJ and Mollenhauer HH. Monensin-induced swelling of Golgi apparatus cisternae mediated by a proton gradient. Eur J Cell Biol 1992; 34(1): 1-8.
8. Pinot F, Borsch H, Mioskowski C, Durst F, Salaün JP. ω-hydroxylation of oleic acid in *Vicia sativa* microsomes: inhibition by substrate analogues and inactivation by terminal acetylenes. Plant Physiol 1994; 000-000.

9. Albertin H, Nurit F, Ravanel P and Tissut M. Uncoupling activities of monensin in isolated mitochondria, chloroplasts and cells. Phytochemistry 1994; 35 (5): 1105- 1110.
10. Holloway PJ. Structure and histochemistry of plant cuticular membranes: an overview. In : Cutler DF, Alvin KL, Price CE, editors. The Plant Cuticle. London : Academic Press, 1982: 139-165.
11. Kolattukudy PE. Plant waxes. Lipids. 1970; 5 (2): 259-275.
12. Kolattukudy PE, Walton TJ. Structure and biosynthesis of the hydroxy fatty acids of cutin in *Vicia faba* leaves. Biochemistry 1972; 11(10): 1897-1907.

INFLUENCE OF CHLOROACETAMIDE HERBICIDES ON *SCENEDESMUS ACUTUS* FATTY ACIDS. A CENTRAL ROLE OF OLEIC ACID?

Michel Couderchet, Friedhelm Kring and Peter Böger

Lehrstuhl für Physiologie und Biochemie der Pflanzen, Universität,
D-78434 Konstanz, Germany

Although chloroacetamide herbicides have been on the market for almost 40 years, their primary site of action is still unknown. Many physiological and biochemical effects have been described for this class of herbicides and have been reviewed [1] leading to the conclusion that these herbicides might have numerous targets in sensitive plants. However, chloroacetamides were found phytotoxic even at submicromolar concentrations which pleads for a single primary target for the herbicide class. Among the effects of the herbicide, inhibition of lipid metabolism and desaturation has been proposed [2, 3] but is still controversial. In this study we present effects of metazachlor and other chloroacetamides on aspects of lipid metabolism in the unicellular green alga *Scenedesmus acutus* and other systems.

Materials and Methods

The unicellular algae *S. acutus* were cultivated in 200 ml of liquid medium as described before [3]. Metazachlor [2-chloro-*N*-(2,6-dimethylphenyl)-*N*-(1*H*-pyrazol-1-ylmethyl)-acetamide], alachlor [2-chloro-*N*-(2,6-diethylphenyl)-*N*-(methoxymethyl)acetamide], dimethenamid [2-chloro-*N*-(2,4-dimethyl-3-thienyl)-*N*-(2-methoxy-1-methylethyl)acetamide] were added to the culture at inoculation. Fatty acids were isolated after saponification of the algae and methylated before they were separated with gas chromatography [3].

Pea microsomal oleic acid desaturase activity was determined by the conversion of ^{14}C-oleoyl-CoA and labeled 18:1 and 18:2 methyl esters were separated by TLC on $AgNO_3$ impregnated plates [3].

For ^{14}C-oleic acid (0.32 µM; 380,000 dpm) incorporation experiments the algae were incubated in 10-ml-vessels with a chlorophyll content of 50 µg / ml suspension [3], fatty acids were extracted as above, the non-extractable fraction (NEF) precipitated and washed with hexane and methanol. The NEF was also separated by incubating the algae 10 min at 65°C in MeOH as indicated.

J.-C. Kader and P. Mazliak (eds.), Plant Lipid Metabolism, 408–410.

Lipid transfer protein (LTP) was purified from extracts of 72-h-old maize seedlings with gel filtration on a Sephadex G 75 column. Activity was assayed by measuring the ability of a protein fraction to stimulate the transfer of ^{3}H-phosphatidylcholine from liposomes to mitochondria. Purification and assays were done as described by Douady *et al.* [4].

Results and Discussion

In the presence of metazachlor, alachlor or dimethenamid (5 μM) fatty acid desaturation in *S. acutus* was strongly inhibited after 40 h. Under these conditions the relative amounts of polyunsaturated fatty acids decreased while palmitic and oleic acids accumulated (Table 1). In an *in-vitro* assay, desaturation of oleoyl-CoA by microsomes was also decreased by chloroacetamides, suggesting a direct influence of the herbicides on oleic acid desaturation, although 25 μM were necessary for a significant inhibition (Table 2). ^{14}C-oleic acid fed to the algae was rapidly incorporated into the lipids and further desaturated. It was also incorporated into a precipitate obtained after saponification of the cells or lipid extraction. Radioactivity in that fraction was dramatically reduced by metazachlor, a concentration between 10 nM and 100 nM was sufficient for a 50% inhibition [3]. This effect was correlated with

Table 1: Relative fatty acid content of *S. acutus* after the cells were incubated 40 h with 5 μM herbicide.

	Fatty acid species		
	Polyenoic	16:0	18:1
Control	66.67	20.10	7.75
Metazachlor	38.67	25.50	25.45
Alachlor	31.24	29.67	26.95
Dimethenamid	36.59	26.26	28.37

Table 2: Influence of alachlor on the desaturation of ^{14}C-oleoyl-CoA by pea microsomes.

Concentration	% inhibition
0 μM	0.00 ± 0.00
25 μM	19.35 ± 11.15
50 μM	39.50 ± 11.60
100 μM	51.50 ± 6.40
200 μM	42.10 ± 4.80

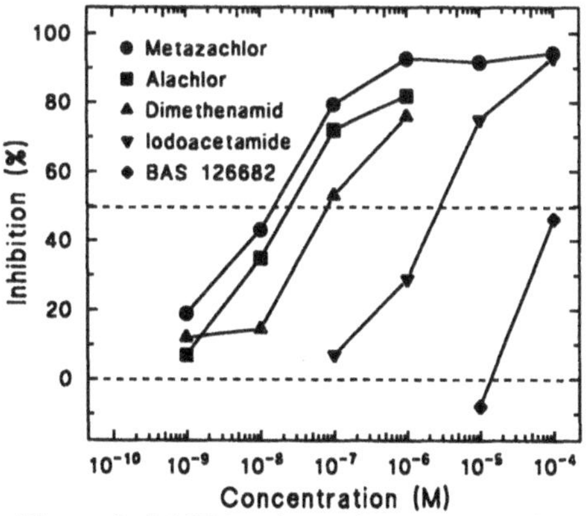

Figure 1: Inhibition by various compounds of ^{14}C-oleic acid incorporation into a non-extractable fraction obtained after lipid or fatty acid extraction of *S. acutus*.

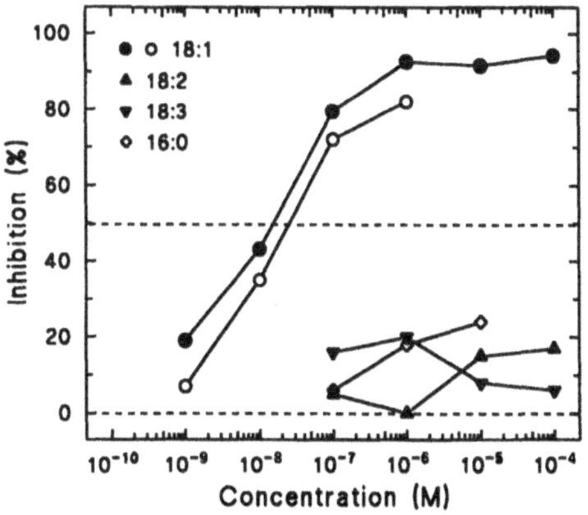

Figure 2: Influence of metazachlor (solid symbols) or alachlor (open symbols) on the incorporation of fatty acids into the NEF of *S. acutus*.

herbicidal activity since BAS 128682 [2-methoxy-*N*-(2,6-dimethylphenyl)-*N*-(1*H*-pyrazol-1-ylmethyl)acetamide], a non-herbicidal derivative of metazachlor and iodo-acetamide were not efficient while other chloroacetamide herbicides (alachlor, dimethenamid) were as efficient as metazachlor (Figure 1). This effect is specific to oleic acid since it could not be reproduced with palmitic, linoleic or linolenic acids (Figure 2). Incorporation of oleic acid in that fraction was not unspecific and the cells had to be metabolically active for chloroacetamides to inhibit incorporation. Indeed, supplying 310 kdpm oleic acid 10 min after the beginning of lipid extraction (MeOH, 65°C) yielded no additional label in the NEF.

The NEF contains rests of the algae's cell-wall and proteins, especially proteins with low molecular weight, possibly involved in lipid metabolism. One of these proteins could be LTP. However, neither the lipid transfer activity nor the fatty acid binding activity of these bifunctional proteins [5] from maize could be inhibited by chloroacetamides (Table 3).

Table 3: Influence of chloroacetamides on LTP activity (in % of the control).

	Concentration		
	1 μM	10 μM	100 μM
Alachlor	97	100	106
Metazachlor	108	108	104

Chloroacetamides induce an accumulation of 18:1 in *S. acutus* and inhibit 18:1 desaturation *in vitro*. Furthermore, they dramatically reduce the incorporation of 18:1 into a non-extractable fraction after lipid or fatty acid extraction. This inhibition is specific to 18:1, requires metabolically active cells, and does not occur in a metazachlor tolerant line [6]. Oleic acid metabolism could thus play a central role in the mode of action of chloroacetamides. The NEF is currently under further investigation and its analysis might lead to the primary target of the herbicides.

References

1. Fuerst EP. Understanding the mode of action of chloroacetamides and thiocarbamate herbicides. Weed Technol 1987; 1: 270
2. Gronwald JW. Lipid biosynthesis inhibitors. Weed Sci 1991; 39: 435-449
3. Couderchet M, Böger P. Chloroacetamide-induced reduction of fatty acid desaturation. Pestic Biochem Physiol 1993; 45: 91-97.
4. Douady D, Guerbette M, Grosbois M, Kader JC. Purification of phospholipid transfer protein from maize seeds using a two-step chromatographic procedure. Physiol Vég 1985; 23: 373-380.
5. Østergaard J, Vergnolle C, Schoentgen F, Kader JC. Acyl-binding/lipid-transfer proteins from rape seedlings, a novel category of proteins interacting with lipids, Biochim Biophys Acta 1993; 1170: 109-117.
6. Couderchet M, Rumbolz J, Böger P. Lipid metabolism in *Scenedesmus acutus* wild type and metazachlor tolerant cell lines. This volume.

LIPID METABOLISM IN *SCENEDESMUS ACUTUS* WILD TYPE AND METAZACHLOR TOLERANT CELL LINES

Michel Couderchet, Joachim Rumbolz and Peter Böger

Lehrstuhl für Physiologie und Biochemie der Pflanzen, Universität,
D-78434 Konstanz, Germany

Availability of resistant plants (mutants) has been very useful in understanding herbicide mechanism of action [1]. Reports of plants resistant to chloroacetamide herbicides are scarce [2,3] and is probably one reason why the mode of action of these herbicides in plants is still unknown after over 40 years of research. The generation time in the unicellular green alga *Scenedesmus acutus* (ca. 8 h) made the isolation of a resistant form for this organism possible. Although the alga is taxonomically far away from the weeds and crops in which the herbicides are used it was preferred to cell suspension cultures of higher plants for which no sensitive form could be found [4]. Characterization of a tolerant line and its comparison with the wild type could shed some light on the mechanism of action of this class of herbicide. Since acyl lipid biosynthesis and desaturation was found to be influenced by the herbicides [5], lipid metabolism was compared in both lines.

Materials and Methods

The unicellular algae *S. acutus* were cultivated in 200 ml of liquid medium as described before [5]. To isolate the resistant line, the algae were subcultured every 2-3 day with increasing concentrations of metazachlor [2-chloro-*N*-(2,6-dimethylphenyl)-*N*-(1*H*-pyrazol-1-ylmethyl)acetamide] as was already done for *Chlamydomonas reinhardtii* [6]. Fatty acids were isolated after saponification of the algae and methylated before they were separated with gas chromatography [5].
For short experiments (up to 5 hours) the algae were incubated in 10 ml-vessels with a chlorophyll content of 50 µg/ml suspension [5], ^{14}C-acetate (2 mM; 430,000 dpm) or ^{14}C-oleic acid (0.32 µM; 380,000 dpm) was fed, fatty acids were extracted as above and the non-extractable fraction ("precipitate") washed with hexane and methanol.
The labeled fatty acid methyl esters were separated by reverse phase HPLC using a C_8 Supelcosil 3 column (150 mm x 4.6 mm, Supelco, Bad Homburg, Germany) and their radioactivity determined on-line with a Ramona radioactivity detector (Raytest,

J.-C. Kader and P. Mazliak (eds.), Plant Lipid Metabolism, 411–413.

Staubenhardt, Germany) equipped with a liquid scintillation flow cell. Elution was achieved using MeOH:H_2O (85:15) at a flow rate of 0.8 ml/min.
The experiments in which ^{14}C-oleic acid was fed to the algae were also performed with *C. reinhardtii* wild type and metazachlor-tolerant (Me1/4) lines [6] (generous gift of Dr. C. Fedkte, Bayer AG, Leverkusen, Germany).

Results and Discussion

Presently the tolerant cells grow in the presence of 4.8 μM metazachlor as fast as the wild line. Microscopically the tolerant cells look like cells treated with herbicide, they are round and single while those of the wild line are lens-shaped and in colonies (coenobia) of 8 cells. This cell line is also resistant against other herbicides from related classes: Dimethenamid (a thiophenamide) and mefenacet (an oxyacetamide).

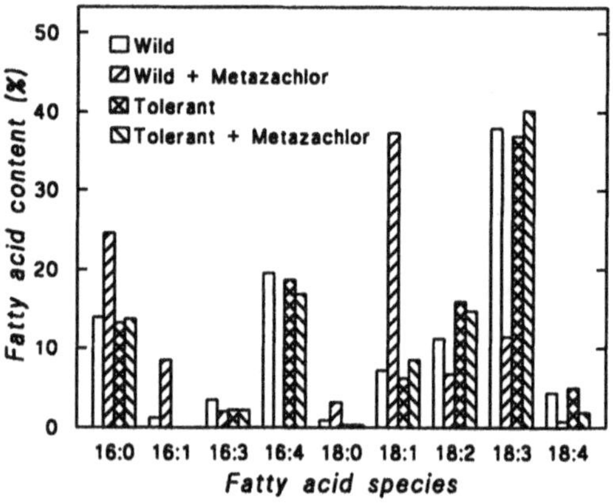

Figure 1: Fatty acid profile of *S. acutus* wild type and tolerant line after a 48-h-incubation with or without 2 μM metazachlor.

In long run experiments (1 to 2 days) metazachlor (2 μM) inhibited fatty acid desaturation in the wild cell lipids but not in those from the tolerant line (Figure 1).
When the algae were incubated for a few hours with ^{14}C-acetate, uptake of radioactivity was faster in the tolerant line than in the wild line and it was not influenced by metazachlor while in the wild line it was sensitive to 10 μM of the herbicide (Table 1). Incorporation of acetate into the fatty acids paralleled uptake and it could not be concluded whether uptake reduction in the wild line was due to reduced incorporation into the fatty acids or vice-versa.

Table 1: Acetate uptake (in % of fed acetate) by 2 cell lines of *S. acutus* incubated with or without 10 μM metazachlor (Mz).

	Time of incubation		
	1:15 h	2:30 h	3:45 h
Wild	46.3	97.8	98.2
Wild + Mz	37.7	81.0	96.6
Tolerant	93.6	97.7	98.4
Tolerant + Mz	95.9	97.7	98.5

Incorporation of ^{14}C-oleic acid into a non-extractable fraction obtained after saponification of the cells is highly sensitive to metazachlor ($I_{50} < 10^{-7}$ M) [5]. This effect could not be demonstrated in the tolerant cells even with 1000 fold higher herbicide concentrations (Figure 2), confirming a causal relation between incorporation of oleic acid into that fraction and herbicidal activity. In the tolerant line less label was recovered in the fatty acid fraction, indicating a lower uptake of oleic acid into the tolerant cells. HPLC analysis of the radioactive fatty acids showed that although

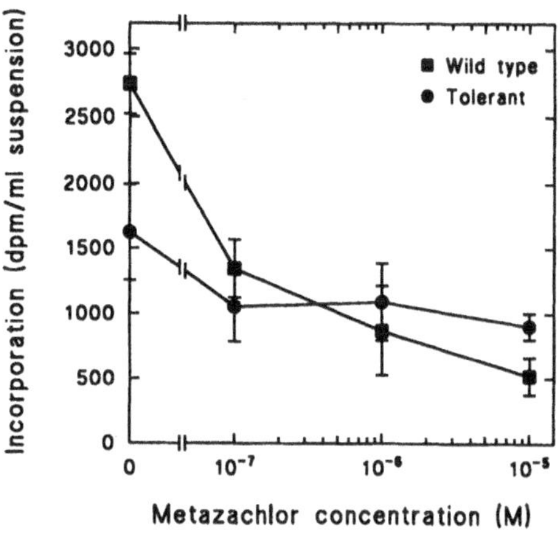

Figure 2: Incorporation of radioactivity into the saponification precipitate after *S. acutus* cells were incubated 3 hours with ^{14}C-oleic acid and metazachlor.

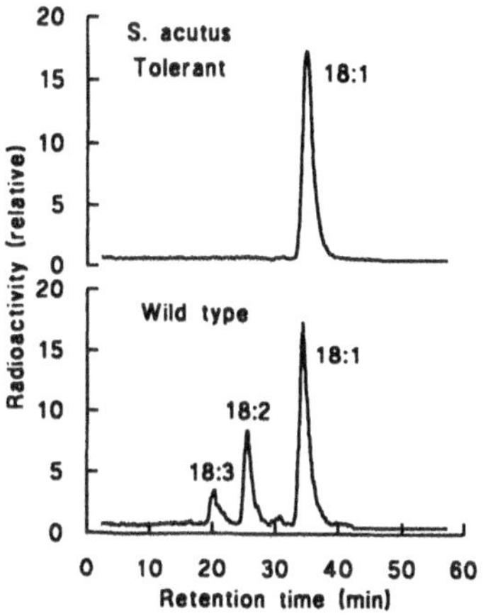

Figure 3: Radio-chromatogram of the fatty acid fraction after 2 lines of *S. acutus* cells were incubated 3 hours with ^{14}C-oleic acid.

desaturation of exogenous oleic acid in the wild cells occurred rapidly, it was strongly reduced in the resistant cells (Figure 3). This effect was confirmed in *C. reinhardtii* (not shown). Reduced uptake of oleic acid in the tolerant cells might be responsible for less desaturation.

These results and herbicide uptake studies show that tolerance might be caused by changes in the uptake properties of the cells, possibly by changes in the membrane and wall of the tolerant cells.

References

1. Mets L, Thiel A. Biochemistry and genetic control of the photosystem II herbicide target site. In Böger P, Sandmann G editors. Target sites of herbicide action. Boca Raton: CRC Press, 1989:1-24.
2. Cottingham CK, Hatzios KK, Meredith SA. Comparative responses of selected corn (*Zea mays* L.) hybrids to EPTC and metolachlor. Weed Res 1993; 33: 161-170.
3. Wilkinson RE, Duncan RR, Meredith SA, Hatzios KK. Growth and physiological responses of sorghum cultivars exposed to excess H^+ and the herbicide metolachlor. Can J Bot; 71: 533-540.
4. Couderchet M, Brozio B, Böger P. Effect and metabolism of the chloroacetamide herbicide metazachlor: comparison of plant cell suspension cultures and seedlings. J Pestic Sci 1994; 19: 127-135.
5. Couderchet M, Böger P. Chloroacetamide-induced reduction of fatty acid desaturation. Pestic Biochem Physiol 1993; 45: 91-97.
6. Fedtke C. Mode of action studies with mefenacet. Pestic Sci 1991; 33: 421-426

EFFECTS OF POWDERY MILDEW ON THE LEAF SURFACE WAXES OF WHEAT

Abdelrahman Almihanna[1] and Dorothy Lösel
Department of Animal and Plant Sciences, University of Sheffield, Sheffield S10 2UQ, U.K.

Various previous studies have indicated a relationship between infection and the ability of fungal pathogens to penetrate wax and cutin layers covering the aerial surfaces of plants but few attempts have been made to correlate the amount and composition of such lipid layers with susceptibility or resistance and to quantify the effects of pathogenic attack on the amounts produced. These aspects have been investigated in a study of wheat (*Triticum aestivum* L.), during infection by the powdery mildew pathogen *Erysiphe graminis* f. sp. *tritici,* which is confined to epidermal surfaces of stems and leaves, penetrating the epidermal cells by means of specialized haustoria. Transpiration losses from the surface mycelium and the abundant spore production (Ayres, 1976; 1978), as well as effects on water supply via the roots (Berryman *et al.*, 1991) result in severe water stress.

To determine whether surface lipid layers may be a significant component in the resistance of certain wheat cultivars to powdery mildew infection and to investigate changes in the production and composition of surface waxes which may accompany infection, the following study was carried out on a range of wheat cultivars which had been previously been shown to differ in their resistance to powdery mildew disease (Almihanna, 1993).

METHODS

Plants of 7 selected cultivars, grown from seed in pots of washed sand, were watered with nutrient solution and maintained under growth room conditions (80 mmol m^{-2}, 12 h photoperiod, 20 $\pm$ 2°). After 3 weeks, the seedlings in half of the pots of each variety were inoculated by spraying with spores of *E. graminis*, suspended in Fluorinert electronic liquid, to give a uniform inoculum of 50 - 60 conidia cm^{-2} of leaf surface. After a further 30 d of growth, 5 g samples of leaves were harvested from 4 replicate pots of healthy or infected plants of each variety, Fresh weight and leaf area were recorded, disease severity estimated using the assessment keys of James (1971) and surface lipids were extracted by immersing the leaf samples in 2 ml of chloroform for 45 sec. Individual classes of wax were separated by TLC on silica gel, using benzene: chloroform (7 : 3, v/v) as solvent and visualized with iodine vapour or by spraying with 3% vanillin in ethanol, mixed with an equal volume of concentrated H_2SO_4 and heating at 100°C. Total lipid content of the extract and the amounts of the separated wax fractions were estimated by the procedure of Marsh and Weinstein (1966).

RESULTS AND DISCUSSION

The amounts of wax recorded per unit area of leaf surface in uninoculated plants of the 7 wheat varieties examined (Table 1) corresponded with their relative susceptabilities to powdery mildew disease, being significantly higher in the resistant cultivars Apollo and Beaver than in the partially resistant Ycora Rojo which, in turn had significantly more wax per cm^{-2} than the susceptible cultivars Sebo, Sham-1, Probred and Westbred. It is of interest that no clear correlation was

[1]Present address: Department of Plant Pathology, College of Agriculture, Bureida, Saudi Arabia

J.-C. Kader and P. Mazliak (eds.), Plant Lipid Metabolism, 414–416.

observed when the amounts of surface wax were compared on a leaf dry weight basis. This suggests that the thickness of the wax layer on the epidermis is an important factor in the relative susceptibilities of these varieties to powdery mildew infection. The function of the surface wax as a barrier to the initiation of powdery miildew infection has been demonstrated by observations of normal development of infection structures and germlings of *Erysiphe graminis* on lower surfaces of leaves of *Lolium* spp., after leaves were washed with chloroform to obtain wax-free surfaces (Carver *et al.*, 1990) and of greater germination rates and appressorial germ-tube lengths on wax-free than on intact leaves of cereals (Carver and Thomas, 1990).

Table 1. Total surface wax of healthy (H) and infected (D) wheat leaves, 30 d after inoculation with *Erysiphe graminis*

Cultivar	Total wax ($mg\ g^{-1}$)		Total wax ($mg\ cm^{-2}$)		Disease level
	H	D	H	D	
Apollo	23.3^{a}	24.1^{ns}	40.7^{a}	43.8^{ns}	Nil
Beaver	19.7^{c}	18.5^{ns}	36.5^{b}	38.2^{ns}	Nil
Ycora Rojo	15.8^{e}	24.3^{*}	31.5^{c}	44.2^{*}	Moderate
Sebo	22.9^{a}	35.4^{*}	24.3^{d}	54.1^{*}	High
Sham-1	21.8^{b}	39.2^{*}	16.2^{f}	70.2^{*}	High
Probred	17.4^{d}	27.1^{*}	19.5^{e}	41.6^{*}	High
Westbred	14.1^{f}	25.6^{*}	16.1^{f}	32.5^{*}	High

Means of leaf extracts from 4 replicate pots of plants. Superscript letters indicate significance of differences in analysis of variance (Waller-Duncan K-test, $P \leq 0.05$). * = significant differences between infected and uninfected plants, ^{ns} = not significantly different from controls.

In all susceptible varieties but not in resistant cultivars, powdery mildew infection resulted in a significant increase in the amounts of wax present on leaf surfaces (Table 1). This is likely to reflect the response of host plants to the development of water stress following infection but the relationship has not been specifically investigated in the present study. Preliminary evidence for an increase in hydrocarbons in leaves of durum wheat during the development of water stress was obtained in this laboratory by Kameli (1990). Bengston *et al.* previously reported increases in alkanes, fatty acids and hexacosanol in oat plants subjected to water stress.

The proportions of individual wax classes separated by TLC from surface lipid extracts from healthy and infected leaves of the different wheat cultivars (Table 2) showed no clear correlations with their susceptibility to powdery mildew and uninfected plants of each cultivar appeared to differ in their profile of surface lipids. Hydrocarbons formed the largest component of the surface wax from leaves of the disease resistant varieties Apollo and Beaver as well as of the susceptible cultivars Sebo and Sham-1, whereas the moderately resistant Ycora Rojo and the susceptable Probred and Westbred produced lower proportions of hydrocarbons with higher proportions of primary alcohols. No consistent differences in wax composition following infection seemed to occur, although some changes in individual components were noted, these varied with the cultivars.

Table 2. Cuticular wax composition of healthy (H) and infected (D) leaves of wheat, 30 d after inoculation with *Erysiphe graminis* (% of total waxes)
(means from 4 pots of plants)

	Apollo		Beaver		Ycora Rojo	
	H	D	H	D	H	D
Carboxylic acids	5.9	6.1	3.6	2.5	8.1	8.8
Hydroxy ß-ketone	6.8	4.9	3.7	5.0	8.0	9.7
Primary alcohols	4.2	5.2	4.4	5.8	36.8	28.4
Secondary alcohols	2.7	4.6	7.7	5.4	1.8	3.8
ß-diketone	5.9	5.1	10.8	11.2	15.9	9.7
Esters	1.6	1.9	2.2	2.3	1.6	6.9
Hydrocarbons	72.8	72.2	67.6	67.7	27.8	32.7

	Sebo		Sham-1		Probred		Westbred	
	H	D	H	D	H	D	H	D
Carboxylic acids	2.1	4.5	5.6	5.6	7.9	12.8	6.7	12.2
Hydroxy ß-ketone	1.3	5.2	5.0	5.0	7.0	6.0	12.9	12.4
Primary alcohols	6.0	4.4	3.1	6.5	19.9	13.4	25.2	21.4
Secondary alcohols	2.7	6.1	2.4	3.5	3.8	7.6	6.3	6.9
ß-diketone	19.2	15.1	15.4	21.3	11.9	7.9	13.9	7.9
Esters	2.8	2.7	2.2	2.6	2.3	7.4	3.7	4.1
Hydrocarbons	66.0	62.1	66.4	55.6	47.2	44.9	31.3	35.1

CONCLUSION

Powdery mildew infection of wheat leaves was accompanied by increased production of cuticular waxes but did not result in consistent changes in the proportions of the component fractions. The amount of wax present per unit area of leaf surface of healthy plants was greater in resistant than in susceptible cultivars, while the partially resistant cultivar investigated was intermediate in this respect. Although exudates from *Erysiphe graminis* f. sp. *hordei* have been shown to have esterase and cutinase activity which is likely to be of importance in the initiation of infection (Kunoh *et al.*,1990; Nicholson *et al.*, 1993), the present results support other evidence for the importance of the thickness of the cuticular wax layer in resistance to powdery mildew.

REFERENCES

Ayres, P.G. (1976) Trans. Brit. Mycol. Soc. **68**, 97 - 100

Bengston, C., Larrson, S. and Liljenberg, C. (1978) Physiol. plant. **44**, 319 - 324

Berryman, C.A., Eamus, D. and Farrer, J.F. (1991) Physiol. Mol. Plant Pathol. **38**, 407 - 415

Carver, T.L.and Thomas, B.J. (1990) Plant Pathol. **39**, 367 - 375

Carver, T.L., Thomas B.J., Ingerson-Morris, S.M. and Roderick, H.W. (1990) Plant Pathol. **39**, 573 - 583.

James, W.C. (1971) Canada Dept. Agric. Publ. 1458

Kunoh, H., Nicholson, R.L., Yosioka, H., Yamaoka N. and Kobayashi, I. (1990) Physiol. Mol. Plant Pathol. **36**, 307 - 407

Marsh, J.B. and Weinstein, D.B. (1966) J. Lipid Res. 7, 574 - 576

Nicholson, R.L., Kunoh, H., Shiraishi, T. and Yamada, T. (1993) Physiol. Mol. Plant Pathol. **43**, 307 - 318

EFFECTS OF HEXACONAZOLE ON THE PLASMA MEMBRANES OF *ALTERNARIA BRASSICICOLA*

Mounia IHABI, Lionel BELINGHERI and Michel SANCHOLLE

Laboratoire de Cryptogamie SN2 , Université des Sciences et Technologies de Lille, 59655 Villeneuve d'Ascq, cedex, France.

SYNOPSIS

Plasma membranes were obtained by differential centrifugation and aqueous polymer two-phase partitionning from mycelium of *Alternaria brassicicola.*; they showed a high affinity for the polyethylene glycol-rich upper phase whereas other microsomal membranes partitioned into the dextran-rich lower phase. The plasma-membrane fraction was identified by the presence or the absence of biochemical markers. Lipids were extracted from control and hexaconazole treated cultures. There were no significant changes in the fatty acid profile except a strong increase of C16 :1 and C 18 :1 in the DPG/PA fraction only. Ergosterol was the predominant sterol. Obtusifoliol and 24-methylene dihydrolanosterol accumulated in the treated cultures.

INTRODUCTION

The introduction of partition in aqueous two-phase polymer systems by Albertsson (1971) contributed, in the recent past, to the isolation of plant membranes and organelles. This method, based upon the membrane surface properties in particular, resulted in the separation of plasma membranes from E. R. , Golgi membrane or plastid envelopes. In the present study, we adapted this procedure in order to isolate and characterize a plasma membrane fraction from fungal material.

Lipids were extracted from these membrane fractions, then fractionnated on T.L.C. with the aim of performing analysis of the fatty acids from the main lipid fractions ; sterols were also analyzed.

In a comparative attempt, plasmalemma was prepared from *Alternaria brassicicola* grown in the presence of Hexaconazole (0,6 ppm) that is the most active growth inhibitor for *A. brassicicola* in vivo. The aim of this study was to see if any changes are induced in the lipid content of this fungus, like it was observed before with some other triazoles and fungi (Weete et al. 1985, 1991).

MATERIAL AND METHODS

Alternaria brassicicola was grown in 1 l Roux flasks containing 100 ml of culture medium (Morquer, 1931) with 20 g/l dextrose and 3.43 g/l $NaNO_3$; pH = 5.5. Cultures were harvested after 4.5 days growth at 23 °C. Mycelia were ground in a homogenization medium containing 0.5 M sorbitol, 1 mM dithiothreitol, 1 mM EDTA, 50 mM Tris-H_2SO_4 ; (pH =7.4). The homogenate was fractionated by differential

J.-C. Kader and P. Mazliak (eds.), Plant Lipid Metabolism, 417–419.

centrifugation (10 min at 6,000g, 15 min at 18,000g and 60 min at 110,000g). The crude microsomal pellet (110,000g) was resuspended in an aqueous polymer two-phase system with a final composition of 5.6% (w/v) dextran T 500, 5.6% (w/v) polyethylene glycol (PEG) 4000, 10 mM K^+-phosphate (pH = 7.6), 30 mM NaCl and 0.25 M sucrose. After centrifugation at 1,500g for 15 min, both phases were diluted 1/10 (v/v) with 0.25 M sucrose, 10 mM K^+-phosphate buffer (pH = 7.6) then pelleted down by centrifugation at 100,000g for 40 min.

Lipids were extracted and analyzed as described by Weete et al. (1985).

RESULTS AND DISCUSSION

The microsomal pellet contained the highest plasma membrane marker activity . It was used as starting material for further purification in the aqueous polymer two-phase system.

The distribution of enzyme activities in the two fractions showed a high activity of glucan synthetase II, marker of plasmalemma, in the upper phase (63%). The contamination of E. R. and Golgi membranes in this fraction was low. Antimycin A-insensitive NADH cytochrome c reductase (6%), marker of E.R., and Glucan synthetase I (2%), marker of golgi membranes, were very low in the PEG-rich upper phase. No activity of succinate cytochrome c reductase, could be detected indicating thus that this fraction was free of mitochondrial membranes.This procedure resulted in a fraction containing highly purified plasma membranes.

Individual percentages of fatty acids and sterols are reported in tables and figure.

Table 1 : Total Fatty Acids

Fatty acids % of total F.A.	16 : 0	16 : 1	18 : 0	18 : 1	18 : 2	18 : 3	*U
Control	19,2	3,1	8,1	24,7	40,9	4,0	1,22
Hexaconzole treated	18,5	3,9	8,4	24,8	40,3	4,0	1,21

*U = unsaturation degree=1 x (% monoène)/100 + 2 x (% diène)/100+3 x (% triène)/100

Figure 1 : Fatty acids from phospholipid classes
(Percent of total fatty acids)

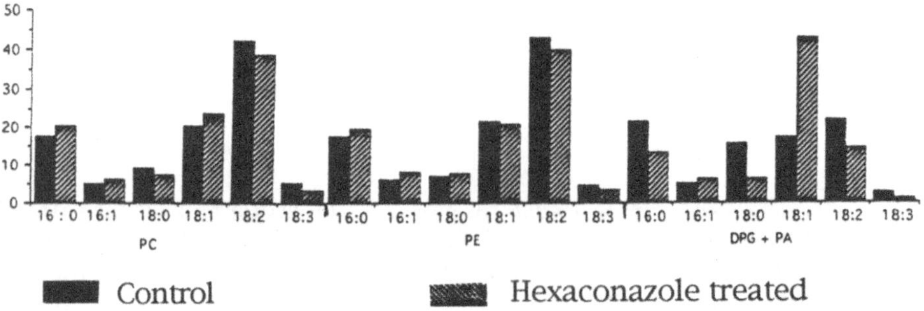

No significant changes can be detected in the total lipid from *A. brassicicola* plasmalemma treated with hexaconazole. Oleic and linoleic acids represent respectively 25 and 41 percent of the total fatty acids. The only phospholipid fraction exhibiting any changes in the presence of hexaconazole is the DPG/PA one : there is an **increase** in monounsaturated fatty acids ; as well **16:1** (light) as **18:1** (very high), accompanied by a strong **decrease** of the corresponding saturated ones (**palmitic and stearic**). These changes suggest a stimulatory effect of haxaconazole on palmitoyl and stearoyl desaturases.

Table 2 : Sterols from plasma membranes
(Percent of total sterols).

	Control	Hexaconazole	*R. R. T.
Ch.	9,4	6,6	1,73
Erg.	83,7	66,4	1,82
Obt.	-	10,1	1,93
24-M.	2,5	10,3	2,60

*R. R.T : relative retention time to 5 α - cholestane.
Ch : cholesterol, Erg : Ergosterol, Obt : Obtusifoliol
24-M : 24 - methylene dihydro lanosterol.

Ergosterol is the major sterol. It decreases in treated cultures while obtusifoliol and 24-methylene-dihydrolanosterol accumulates. This shows that hexaconazole works as a sterol synthesis inhibitor.

CONCLUSION

The phase partition between polymers, in its version adapted to fungal material, is a very convenient and safe method for the preparation of purc plasma membrane fractions.The lipids of *Alternaria brassicicola,* despite the lack of any form of linolenic acid, can be characterized by their high degree of unsaturation, even in the total lipid. This unsaturation can be increased, inside the DPG/PA fraction by the use of hexaconazole, which acts as an inhibitor of sterols synthesis.

The main sterol of *Alternaria brassicicola* is ergosterol, the synthesis of which is perturbated by the use of hexaconazole, which induces the accumulation of obtusifoliol and 24-methylene dihydrolanosterol in the treated cells.

It seems a correlation between the inhibition of sterol synthesis and the increase of fatty acids unsaturation in the DPG/PA fraction.

REFERENCES

1. Albertsson PA. Partition of cell particles and macromolecules. 2nd ed Wiley and Sons. New York, 1971.
2. Morquer R. Recherches morphogénétiques sur le *Dactylium macrosporum* - Thèse Doct. ès Sciences . Toulouse 1931.
3. Weete JD, Sancholle M, Touzé-Soulet JM, Bradley J., Dargent R. Effects of triazoles on fungi. III Composition of a plasma membrane-enriched fraction of *Taphrina deformans*.. Biochim Biophys Acta 1985:812:633-642.
4. Weete JD, Sancholle M. Patterson KA, Miller KS, Huang MQ, Campbell F. Van den Reek M. Lipids 1991:26 (8):669-674.

EFFECTS OF PROLINE AND GLYCINE BETAINE ON LIPID PEROXIDATION IN RAPE (*Brassica napus* L.) LEAVES SUBMITTED TO WATER STRESS

Aziz Aziz, Dominique Rolin and François Larher
Laboratoire de Biologie Végétale, Université de Rennes I, URA- CNRS D1853, Campus de Beaulieu, 35042 Rennes cedex, France

The deleterious effect of water stress on shoot tissues of higher plants might partly be due to the stress induced production of reactive oxygen species which lead to peroxidative damages on polyunsaturated fatty acids (PUFA) [1]. However higher plants exhibit a number of traits which allow some protection against oxygen toxicity [2]. Thus the osmotic solutes which are accumulated in response to water stress have been suggested to play, in addition to osmoregulatory function, a role in scavenging free radicals [3,4]. We have assessed such an assumption using excised rape leaves which accumulate proline (PRO) in response to osmotic stress [5]. We have demonstrated in a preliminary experiment that PRO accumulated is associated with both a decrease in PUFA content (data not shown) and an increase production of malondialdehyde (MDA), a cytotoxic product of PUFA peroxidation. The experiments described in this short paper were designed to compare the putative function of PRO and that of glycine betaine (GB) in preventing stress-induced MDA production in rape leaves.

Water stress induces both proline accumulation and MDA production

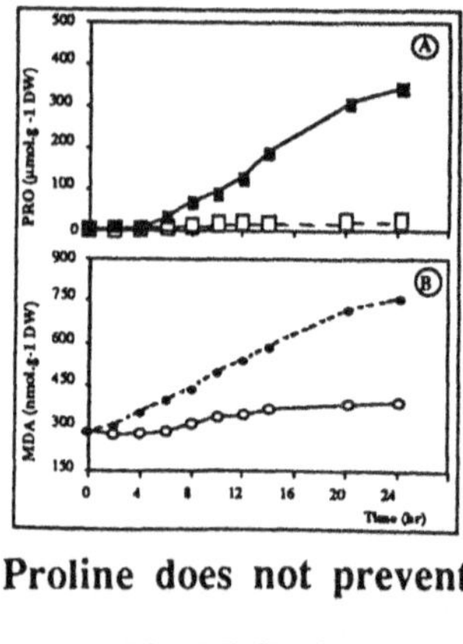

When excised rape leaves were submitted to water stress by lowering the external osmotic potential (-1.5 MPa), PRO was found to be accumulated at a very high level (Fig. 1A) after a 4 hr-lag phase. The MDA content was also enhanced under stress conditions, the stress effect being noticeable before the onset of PRO accumulation (Fig. 1B). In addition, the rate of MDA production was not modified when the PRO level began to increase.

Figure 1: Changes in PRO (■) (A) and MDA (●) (B) contents in rape leaves subjected for 24hr to a stress treatment with PEG 6000. The control leaves(□ ,○) were maintained on the reference medium [5]. For PRO and MDA determination see [6], [7] and [8].

Proline does not prevent fatty acid peroxidation in water stressed leaves.

The PRO effect was investigated *via* experiments where this amino acid was provided to excised leaves from 4 week old plants at 40mM concentration *via* the transpiration stream. This treatment led to an accumulation level for PRO close to that reached in shoots of droughted plants (up to 300 µmoles.g^{-1} DW) (Fig.2A), provoking

J.-C. Kader and P. Mazliak (eds.), Plant Lipid Metabolism, 420–422.

both a slight decrease of π (-1,2MPa) (Fig.2B) and stimulation of MDA production compared to the samples not supplemented with exogenous PRO (Fig.2C). Following this pretreatment, excised leaves were transferred for 18h to a stress medium containing PEG 6000 (-1.5 MPa). In response to stress conditions, PRO was further accumulated (Fig.2A) whereas MDA production was enhanced (Fig.2C). Excised leaves loaded with exogenous PRO behave exactly as those accumulating PRO in response to water stress indicating no inhibitory effect of this solute on MDA production.

Our results are in contrast with those of Smirnoff and Cumbes [3] and Alia *et al* [4] with *B. juncea*. Indeed these last authors showed that PRO accumulation can decrease photoinhibition in isolated thylakoid membranes allowing a decrease in thylakoid lipid peroxidation as measured by MDA formation.

Glycine betaine has an inhibitory effect on fatty acid peroxidation.

When similar experiments were carried out with exogenous GB, a typical compatible solute, excised rape leaves accumulated GB at a high rate (Fig.3A); whereas PRO accumulation (Fig.3B) and MDA production (Fig.3C) remained unchanged compared to the control. During the stress period, lowered water activity induced the typical osmoinduced PRO response (Fig.3B) in non GB pretreated rape leaves. In contrast in GB pretreated rape leaves, the osmoinduced PRO response was strongly inhibited and simultaneously the MDA production was slowed down. These results suggest that the GB which has been supplied to *B. napus* leaves exerts a stabilizing effect on rape metabolism reducing PRO accumulation level as well as that of lipid peroxidation.

Water stress induced lipid peroxidation, mesured by MDA formation, is slightly increased by exogenous PRO but inhibited by GB. If lipid peroxidation is caused by hydroxyl radicals, this suggests that PRO is not an effective OH° scavenger *in vivo* or that lipid peroxidation is triggered by other forms of active oxygen.

Acknowledgements: We thank Dr. N. Smirnoff (Dept of Biological Sciences, University of Exeter, Exeter EX4 4PS, UK) for critical discussion and correcting the English text.

References

1. Ferrari-Iliou R, Pham Thi AT, Mazliak P, Vieira da Silva J. Mécanismes naturels de protection des végétaux supérieurs contre les espèces oxygénées réactives. Ann Biol 1992;31:1-23.
2. Smirnoff N. The role of active oxygen in the response of plants to water deficit and desiccation. New Phytol 1993;125:27-58.
3. Smirnoff N, Cumbes QJ. Hydroxyl radical scavenging activity of compatible solutes. Phytochem 1989;28:1057-1060.
4. Alia, Pardha Saradhi P, Mohanty P. Proline enhances primary photochemical activities in isolated thylakoid membranes of *Brassica juncea* by arresting photoinhibitory damage. Biochem Biophys Res Commun 1991;181: 1238-1244.
5. Larher F, Leport L, Petrivalsky M, Chappart M. Effectors for the osmoinduced proline response in higher plants. Plant Physiol Biochem 1993;31:911-922.
6. Magné C, Larher F. High sugar content of extracts interferes with colorimetric determination of amino acids and free proline. Anal Biochem 1992;200:115-118.
7. Heath RL, Packer L. Photoperoxidation in isolated chloroplasts. I. Kinetics and stoichiometry of fatty acid peroxidation. Arch Biochem Biophys 1968;125:189-198.
8. Du Z, Bramlage WJ. Modified thiobarbituric acid assay for mesuring lipid oxidation in sugar-rich plant tissue extracts. J Agric Food Chem 1992;40:1566-1570.

9. Grieve CM, Grattan SR. Rapid assay determination of water soluble quaternary ammonium compounds. Plant Soil 1983;70:303-307.

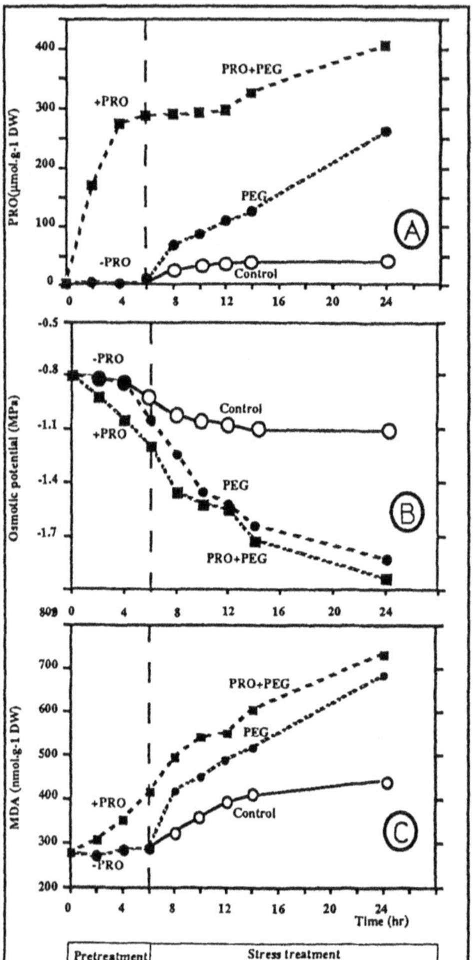

Figure 2: Change in PRO (A) and MDA (C) contents and of internal osmotic potential (B) in *B. napus* leaves during a PRO preincubation period and a stress treatment. Fully expanded leaves were detached and left in the light for 6h with their petioles in a reference medium supplemented with 40 mM proline (■, +PRO) or without (●, -PRO). They were then transferred to a basal medium (○, Control) added with PEG 6000 (-1,5 MPa) for 18h (■, ●). Osmotic potential was determined on sap extracted from 200 to 300 mg FW tissues by measuring freezing point depression.

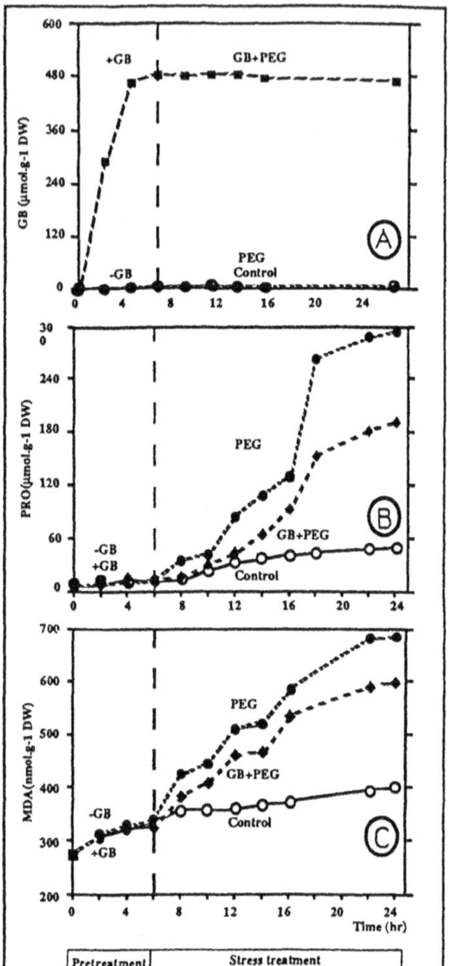

Figure 3: Change in GB (A), PRO (B) and MDA (C) contents in *B. napus* leaves during a GB preincubation period and a stress treatment. Fully expanded leaves were detached and left in the light for 6h with their petioles in a reference medium supplemented with 40 mM glycine betaine (■, +GB) or without (●, -GB). They were then transferred to a basal medium (○, Control) added with PEG 6000 (-1,5 MPa) for 18h (■, ●). GB was measured by the periodide method as described by Grieve and Grattan [9].

SALT EFFECT ON LIPID METABOLISM OF RAPE SEEDS DURING GERMINATION

Douja Ben Miled-Daoud [1] and Abdelkader Chérif [2]

1 UBVRG, INRST. BP 95, 2050 Hammam-Lif, Tunisia.
2 ESIA, 58 Rue Alain Savary, 1003 Tunis, Tunisia.

Introduction

Brackis water salinity is an important problem for plants growing in arid and semi-arid areas of Tunisia.In this meaning,the study of seed behaviour (a culture recently introduced) during germination constitutes a perspective research of plants able to be adapted to higher salt conditions.

Materiel and Methods

Brassica napus seeds Drakkar variety were placed on two sheets of filter paper in Petri-dishes with either distilled water (control) or with 3, 6, 9, 12, 15 and 18 g/l NaCl solutions in darkness at 25±1°C. Cotyledons were separated from embryonic axes for lipid analysis at two day intervals up to the 10th day of germination;lipid extraction and fatty acid analysis have been described in previous paper [1]. Kinetic studies of (1-^{14}C)-acetate incorporation were carried out in the control and treated whole seedlings according to Zarrouk *et al.* [2] methods. The results represent the mean of three experiments.

Results

- NaCl effect on total fatty acids
The total fatty acid (TFA) amounts were determined at different stages of seedling growth in darkness. The fatty acids present in seed were palmitic (C16:0), stearic (C18:0), oleic (C18:1), linoleic (C18:2) and linolenic (C18:3) acids. The oleic acid represented by a high percentage (70%) of TFA decreased during germination procedure and reached 58% of TFA on the 10th day and thus essentialy for the benefit of linolenic acid (Table I)

J.-C. Kader and P. Mazliak (eds.), Plant Lipid Metabolism, 423–425.

Table I - Percentage evolution of oleic and linoleic acids of lipids in control and treated (3 and 18 g/l) cotyledons

day	1 st		3 rd		5 th		7 th		10 th	
	18:1	18:3	18:1	18:3	18:1	18:3	18:1	18:3	18:1	18:3
control	70	8	69	9,5	67,5	10,5	64	12,5	58	14
3 g/l	69,5	8,5	70	9	67	11	63	13	53	15,5
18g/l	70	8	69,5	8,5	69,5	8,4	68	9,5	68	9,5

The figure 1 showed that decrease in TFA amounts control cotyledons was linear, therefore at the 10th day the cotyledons didn't contain more than 1,4 mg of TFA wich represented 10% of initial amount. In cotyledons treated with 3 g/l of NaCl, we observed an enhanced degradation of TFA which represent only 4,6% of initial quantity contained in seeds. On the other hand, at 12 g/l of NaCl, decease in TFA was delayed only the first days of germination. At the higher salt concentration, the degradation was strongly delayed and mainly at the extreme salinity (18 g/l) for wich the fatty acid quantity used, represented 14% of the initial one.

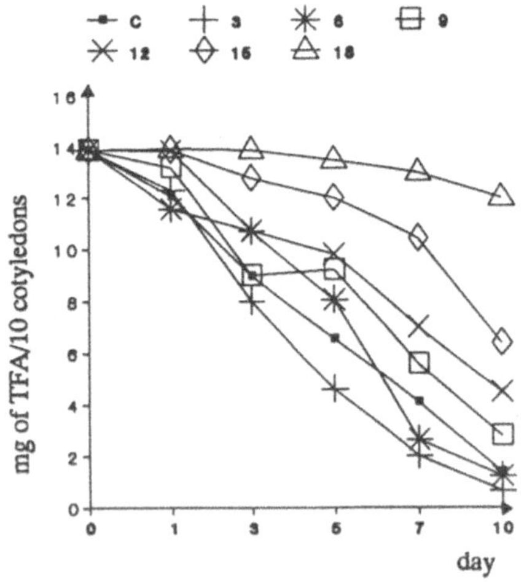

Fig. 1 - NaCl effect on TFA evolution in rape cotyledons during germination in darkness

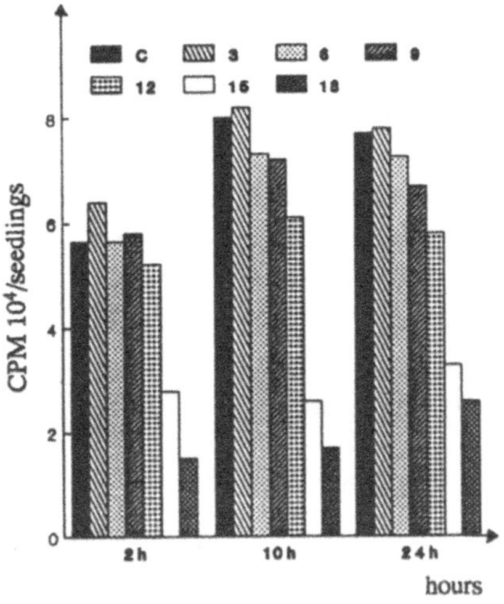

Fig. 2 - NaCl effect on (1-^{14}C) -acetate kinetic incorporation into the total fatty acid fraction of lipid seedlings at the 5th day of germination (measured in cpm 10^4/ seedlings)

- NaCl effect on molecular categories

The triacylglycerols constitute the bulk of reserve lipids(Table II). During germination in control medium,we attended a gradual decrease in the various lipid class amounts (TAG, DAG, MAG, FFA and PL). In the meantime, the TAG percentage changed from 91 at 65% of total lipids whereas PL percentage increased from 5,5 to 27,5%. At 3g/l of NaCl this evolution was more pronounced since at the 10th day the TAG and the PL reached

65% of total lipids whereas PL percentage increased from 5,5 to 27,5%. At 3g/l of NaCl this evolution was more pronounced since at the 10th day the TAG and the PL reached respectively 52 and 42,5% of total lipids. At 18 g/l of NaCl, the metabolic processes were very slowing down and consequently both TAG degradation and PL biosynthesis were inhibited (Table II).

Table II - Lipid composition of rape cotyledons from control and treated (3 and 18 g/l) cotyledons at the 1st, 5th and 10th day of germination

	CONTROL			3 g/l			18 g/l		
time	1 st	5 th	10 th	1 st	5 th	10 th	1 st	5 th	10 th
TAG	91	74	64	92	84	52	94	92	87
DAG	2	5	6,5	2	2	2,4	1,6	2,2	2,8
FFA	0,1	0,2	0,2	0,1	0,3	0,5	0,2	0,2	0,2
MAG	1,5	1,1	2	1	1,1	2,8	1	1	2,7
PL	5,5	20	27,5	5	12,5	42,5	5	5	7,2

- NaCl effect on (1-^{14}C)-acetate incorporation in Rape seedlings

We have choosen the 5th day of germination for supplying (1-^{14}C)-acetate in the medium because at this stage the enzymes reached their maximum activities [1] The kinetic of incorporation in the total lipid presented practically the same evolution until 12 g/l (Fig.2). As in the case of previous TFA degradation we noticed an inhibition of lipid biosynthesis at higher NaCl concentrations.

Conclusion

Brassica napus seed seemed to tolerate high NaCl concentration concerning its germination. It is only at higher concentration more than 12 g/l that rape seed germination and lipid use were affected. It was even noticed, a slight stimulation of the germination at 3 g/l as well as of TAG degradation and PL biosynthesis. In the same meaning, Kuiper [3] underlined an activation of membrane lipid synthesis in salt resistant species under low concentrations. This could be related with a better cellular compartmentation.

References

1. Ben Miled-Daoud D. et Chérif A. Effet du NaCl sur l'utilisation des lipides et activités enzymatiques glyoxysomales au cours de la germination de deux espèces de Medicago. Can.J.Bot., .1992; 70 : 876-883.
2. Zarrouk M., Marzouk B. et Chérif A. Biosynthèse des acides gras polyinsaturés des glycérolipides de la feuille d'olivier. Physiologia Plantarum, 1990; 79 : 45-50.
3. Kuiper P. J. C. Lipid metabolism as factor in enviromental adaptation. In Mazliak P., Benveniste P., Costes C. and Douce R. editors : Biogenesis and function of plant lipids, Biomedical Press, Elsevier North Holland, 1980 : 169-176.

EFFECTS OF DROUGHT STRESS ON ENZYMATIC BREAKDOWN OF GALACTOLIPIDS IN COWPEA (*Vigna unguiculata* L.) LEAVES

Campos Paula[1] and Pham Thi Anh Thu [2]
1: Dep.Fisiologia Vegetal, Estaçao Agronomica Nacional, 2780 Oeiras, Portugal; 2: Laboratoire de Biochimie et Physiologie de l'Adaptation Végétale, Université Paris 7 Denis Diderot, 2 Place Jussieu, F-75241 Paris Cedex 05, France, CNRS, URA 1183.

Introduction

Lipidic composition seems to be one of the main factors related to membrane resistance under water stress. Drought provokes a decrease in polar lipid content of leaves, the galactolipids (GL) being more affected than the phospholipids (PL) [1,2]. The decrease in GL content could be due to an increase in galactolipase activities and/or to a decrease in galactolipid biosynthesis. Studying two cultivars of *Vigna unguiculata*, Monteiro de Paula *et al.* [3] demonstrated that water stress stimulated the catabolism of galactolipids. These authors also found that changes in biosynthesis and degradation rates of leaf polar lipids were less pronounced in the drought-tolerant cultivar, possibly indicating a more stable cell membrane. Lipolytic acyl-hydrolases (LAH) and phospholipases play a major role in plant polar lipid degradation. Fatty acids (FA) resulting from LAH activity are the starting point for lipid peroxidation phenomena, which can in turn cause oxidation of membrane proteins and have damaging effects on the structure and function of the membrane. El-Hafid *et al.* [4] demonstrated that, in cotton leaves, LAH activities for ^{14}C-MGDG increased under water stress.

In this study we investigated changes in MGDG- and DGDG- LAH activities, in soluble and membrane fractions of leaf extracts prepared from three cultivars of *V. unguiculata* submitted to drought and differing in their degree of tolerance to water stress.

Materials and methods.

Three cultivars of *V. unguiculata* (cv.1183, drought-sensitive, cv. EPACE and IT-83D, drought-resistant), were grown in previously described conditions [1]. Drought was imposed by withholding irrigation for a period of 6 to 10 days. ^{14}C galactolipids used as LAH substrates were obtained supplying leaves with ^{14}C acetic acid, sodium salt, for 36h. They were subsequently extracted and purified according to [1]. Water potential (Ψ) was followed with a pressure chamber [5]. 4-5g fresh weight leaves were fixed in liquid nitrogen and homogenized in 0.1M MES buffer, pH6.5, containing 1mM EDTA, 1mM cystein, 5mM sodium metabisulfite, 5mM DTT and 0.5mM PMSF. After centrifuging , the supernatant was used as soluble enzymatic fraction. The pellet was resuspended in the same buffer containing 0.1M Triton X-100, and left for 15min for solubilizing the membrane proteins, recovered after centrifuging. ^{14}C-MGDG- and ^{14}C-DGDG-LAH activities were measured according to [4].

J.-C. Kader and P. Mazliak (eds.), Plant Lipid Metabolism, 426–428.

Results

Figures 1 and 2 present the LAH activity in the soluble (A) and membrane (B) fractions of leaf extracts . Activity levels for membrane enzyme are generally higher than those found for the soluble enzyme, in the 3 cv. under study. For both DGDG and MGDG-LAH activities, the 2 resistant cv. (EPACE and IT-83D) showed significant lower values than the sensitive cv. (1183), at all water potential values. Apart from the soluble MGDG-LAH activity (fig.1A) , in all other cases, LAH activities drastically increased

Fig.1 : Effect of water stress on the activity of MGDG-LAH in soluble (A) and membrane (B) fractions from *Vigna unguiculata* leaf extracts.

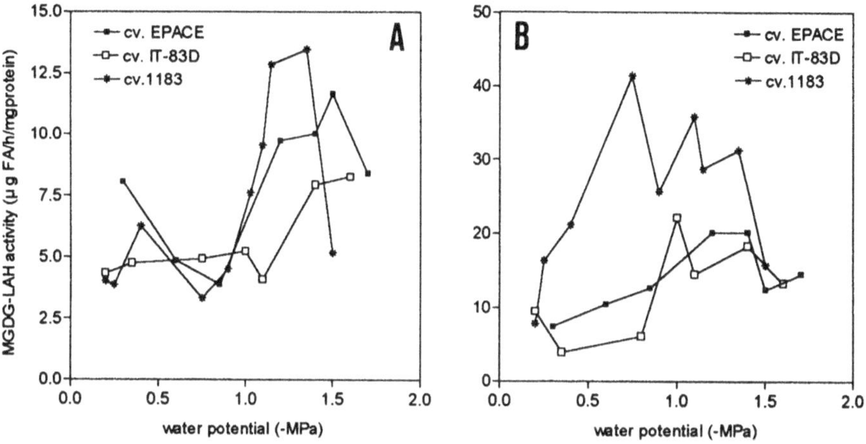

Fig.2 : Effect of water stress on the activity of DGDG-LAH in soluble (A) and membrane (B) fractions from *Vigna unguiculata* leaf extracts

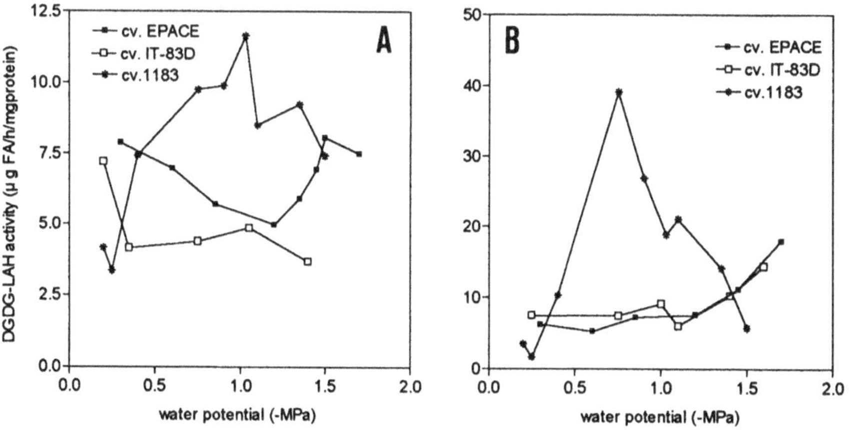

from the earliest stages of drought treatment in the sensitive cv. At the most severe water deficits, LAH activities decreased sharply in this cv., contrarily to the 2 more resistant cultivars. The sensitive cv. 1183 also presented the highest stimulation of the LAH activity, being 5 to 10-fold higher than the controls. Furthermore, those maximum values were reached at rather moderate water deficits (-0.7MPa). In the 2 resistant cv. EPACE and IT-83D, LAH activities did not vary to a great extent, the most significant changes being observed only for soluble MGDG-LAH activity.

Conclusion.

It has been referred that in *Vigna unguiculata*, degradative phenomena in leaves submitted to dehydration are of the greatest importance, especially in species susceptible to drought [3]. An increase of LAH activities as well as a higher susceptibility of leaf lipid extracts to peroxidation may be at the basis of membrane disfunctions occurring under water stress [4,6]. In this study we observe that the activation of this enzyme, under water stress conditions, is different for the membrane or the soluble form of the enzyme. This is particularly evident in the sensitive cv. 1183 .

In previous studies, we have demonstrated that drought susceptible *Vigna* species present higher 18:3-MGDG content of their leaf membranes [7]. On the other hand, LAH preferentially degrades polyunsaturated MGDG molecular species [3,4]. It could therefore be expected that LAH activity causes more damage to membranes of sensitive cv., which seems to be the case from the present results.

References

1 Pham Thi AT, Borrel-Flood C, Vieira da Silva J, Justin AM and Mazliak P. Effects of water stress on lipid metabolism in cotton leaves. Phytochem 1985,24:723-727.
2 Monteiro de Paula F, Pham Thi AT, Vieira da Silva J, Justin AM, Demandre C, Mazlliak P. Effects of water stress on the molecular species composition of polar lipids from Vigna unguiculata L.leaves. Plant Sci 1990,66:185-193.
3 Monteiro de Paula F, Pham Thi AT, Zuily-Fodil Y, Ferrari-Iliou R, Vieira da Silva J, Mazliak P. Effects of water stress on the biosynthesis and degradation of polyunsaturated lipid molecular species in leaves of Vigna unguiculata. Plant Physiol Biochem 1993;31:707-715.
4 El-Hafid L, Pham Thi AT, Zuily-Fodil Y, Vieira da Silva J. Enzymatic breakdown of polar lipids in cotton leaves under stress; degradation of monogalactosyl-diacylglycerol. Plant Physiol Biochem 1989;27:495-502.
5 Scholander PF, Hammel HT, Bradstreet ED, Hemmingsen EA. Sap pressure in vascular plants. Science 1965,148:339-346.
6 Ferrari-Iliou R., d'Arcy-Lameta A, Iliou JP et al. In vitro photodynamic lipid peroxidation of total lipophilic extracts from leaves of bean plants. Biochim Biophys Acta 1993,1166:48-54.
7 Campos P, Pham Thi AT. Drought resistance and membrane lipids in Vigna sp. Physiol Plant 1990, 79:A113.

SALT STRESS EFFECT ON POLAR LIPID METABOLISM OF OLIVE LEAVES

Moktar Zarrouk, Wafaâ Seqqat-Dakhma and Abdelkader Chérif

Unité de Biotechnologie Végétale et de Ressources Génétiques, INRST
BP 95, 2050, Hammam-Lif, Tunisia.

Introduction

Lipids are important for regulation of the selective cell membrane permeability and may play a role in salt resistance of plants. Several studies showed that salt stress induced modifications in lipid and fatty acid composition of higher plants [1, 2]. Most attention was given to polar lipid and sterol changes since they may reflect the cell membrane status under stress conditions. The nature and amplitude of these lipid changes depend on plant species and the intensity of salt stress. In a previous work, we have reported that salt stress decreased polar lipid contents (galacto- and phospho- lipids). A decrease in linolenic acid percentage was also observed leading to a reduction in the degree of unsaturation of membrane lipids.
The aim of the present work is then to investigate the effect of NaCl on the polar lipid and fatty acid biosynthesis of olive leaves by using (1-^{14}C) acetate as precursor.

Material and Methods

Experiments were carried out on young olives (one year old) which were grown in greenhouse under controlled conditions : 14h day at 27°C and 10h night at 22°C, the relative humidity was 70%, under natural light. Plants were cultivated in cases filled with sand and irrigated with nutrient solution [3] added with NaCl at the following concentrations : 0 (control), 3, 6 and 9 g.l^{-1}. For labeling experiments microdroplets of sodium (1-^{14}C)- acetate (specific activity 1,94 TBq.mol^{-1}) (CEA, France) were deposited on the surface of leaves. After 48 hours of incubation, leaves were harvested and lipids were extracted following a modified procedure of [4]. Polar lipid categories were separated by thin layer chromatography on silicagel plates (Merck G60) according to the method of [5]. Fatty acid methyl esters were analysed by radio GLC.

J.-C. Kader and P. Mazliak (eds.), Plant Lipid Metabolism, 429–431.

Results

effect of salt stress on (1-^{14}C) acetate incorporation into polar lipids of olive leaves

Figure 1 showed that the rate of (1-^{14}C) acetate incorporation into individual polar lipids of olive leaves (expressed in cpm/g of fresh weight) decreased regulary with increasing NaCl concentrations. This decrease in labeling was observed mainly in galactolipids (MGDG, DGDG), PC and PE. Although, the decrease in labeling was more pronounced for galactolipids. However the rate of (1-^{14}C) acetate incorporation into PG and PA remained constant.

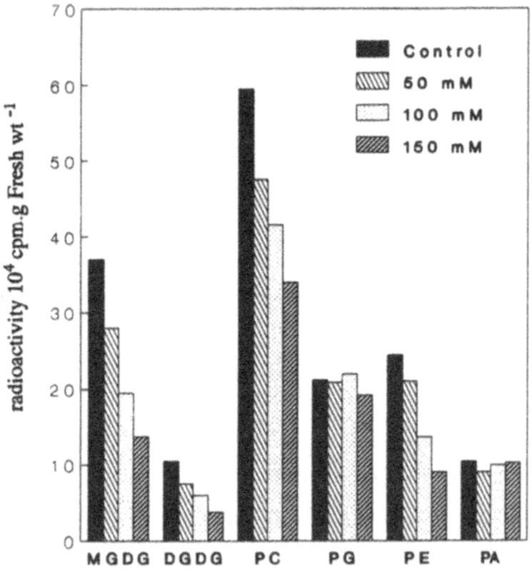

Fig. 1 - Effect of salt stress on the incorporation of (1-^{14}C) acetate into polar lipid classes from olive leaves, after 48h of labeling

Effect of salt stress on the incorporation of (1-^{14}C) acetate into total fatty acids of olive leaves

After 48 hours of incubation, the relative labeling of palmitic and oleic acids increased when NaCl concentrations increased in the culture medium. Whereas the rate of (1-^{14}C) acetate incorporation into linolenic acid decreased. Therefore the level of radiactivity in this component varied from 26% in control to 10% in severe stressed leaves, at 150 mM of NaCl (Table 1)

Table 1 : Effect of salt stress on the relative radioactivity (in %) of total fatty acids from olive leaves, after 48h incorporation of (1-^{14}C) -acetate

	0 mM	50 mM	100 mM	150 mM
16:0	20,4	21,8	26,3	26,6
18:1	32,4	35,2	39,3	44,0
18:2	21,7	22,0	18,6	19,3
18:3	25,4	21,0	15,8	10,1

Effect of NaCl on the incorporation of (1-^{14}C) acetate into fatty acids of individual polar lipids

As shown in table 2, high NaCl concentration (150 mM) provoked a decrease in linolenic acid labeling for all polar lipid classes, but this decrease was more pronounced in galactolipids. In PE and PG there was no linolenic acid labeling detectable after 48h of incubation. In parallel, the level of precursor incorporation into oleic and palmitic acids increased.

Table 2 : Effect of salt stress on the relative radioactivity (in %) of fatty acids from individual polar lipids of olive leaves, after 48h of labelling

Lipids	PC		PE		PG		MGDG		DGDG	
F.A.	0	150	0	150	0	150	0	150	0	150
16:0	17,4	18,5	16,1	19,2	52,4	56,0	9,4	17,0	17,0	33,2
18:1	48,8	57,6	42,3	50,7	25,4	30,1	7,5	18,7	21,0	30,3
18:2	23,5	22,8	31,3	30,1	13,7	13,9	5,8	12,7	11,0	9,5
18:3	10,2	1,1	10,3	--	8,4	--	77,2	51,6	51,0	27,0

Conclusion

These results showed that NaCl treatment induced significant changes in polar lipid metabolism of olive leaves. Thus the biosynthesis of galactolipids, PC, PE and linolenic acid was inhibited under salt stress conditions, which confirmed our previous findings on the amount variations of these lipids [6]. The inhibition of polar lipids biosynthesis should be due to the ion accumulation withing the leaf cells (data not shown). The activity of two enzymes responsible for the acylation of galactolipids were inhibited in leaves of salt - treated barley seedlings [1].

References

1. Müller M, Santarius KA. Changes in chloroplast lipids during adaptation of barley to extreme salinity. Plant Physiol 1978; 62 : 326-329
2. Kuiper PJC. Environmental changes and lipid metabolism of higher plants. Physiol Plant 1985; 64 : 118-122
3. Keldey LS, Bowling DJF. The effect of cycloheximide on uptake and transport of ions by sunflower roots. Ann Bot 1975; 39 : 1023-1027
4. Bligh EG, Dyer WS. A rapid method of total lipid extraction and purification. Can J Biochem Biophys 1959; 37 : 911-917
5. Lepage M. Identification and composition of turnip root lipids. Lipids 1967; 2 : 244-250
6. Zarrouk M, Chérif A. Action du chlorure de sodium sur la teneur en lipides de plants d'oliviers (*olea europea* L.). Z Planzenphysiol 1981; 105 : 85-92

ADAPTATION OF POTATO CELLS TO LOW WATER POTENTIAL AND CHANGES IN MEMBRANE FATTY ACID COMPOSITION AND FLUIDITY

Antonella Leone, Marina Tucci, Stefania Grillo, Antonello Costa, Laszlo Vigh*
Research Centre for Vegetable Breeding, CNR, Via Università 133, Portici, Italy; *Biological Centre, Hungarian Academy of Sciences, P.O. Box 521, H-6701, Szeged, Hungary

Introduction

Disruption of cellular membranes has been observed in plant tissues in response to intense water stress, with loss of membrane functionality. A stepwise adaptation allows biochemical and structural changes of cellular membranes and maintenance of membrane fluidity, which is a prerequisite of acquired tolerance (1). While an extensive literature on changes in membrane fatty acid composition induced by varying external temperatures is available (2), much less is known about the changes induced in plants under drought conditions. It has been reported that, besides variation of the major classes of lipids, the level of unsaturation of fatty acids (FA) decreases (2). The data reported deal with the changes observed in potato cells which were gradually adapted to increasing low water potentials induced by PEG 8000, which reduces free water concentration extracellularly (3). Adapted cells were able to grow actively in the presence of 20% PEG and maintained a normal ultra-structure of the main cellular and sub-cellular constituents. FA of the main extra-chloroplast phospholipids were found to be more saturated in PEG-adapted cells, compared to those in unadapted cells. These changes were associated to an increase of microviscosity of isolated protoplasts from adapted cells, as determined by steady-state fluorescence anisotropy by using diphenil-hexatriene (DPH) as a membrane probe (4).

Results and Discussion

CELLULAR GROWTH

A drastic reduction of growth (50%), which is one of the primary visible effects of water shortage in potato plants (5), was observed in our potato cells when water stress was imposed abruptly, while cells which were gradually adapted to a medium containing increasing amount of PEG 8000 were able to sustain active growth, with a reduction of only 14% in dry weight at the stationary growth phase (6).

J.-C. Kader and P. Mazliak (eds.), Plant Lipid Metabolism, 432–434.

ULTRASTRUCTURE MODIFICATIONS

Cellular organelles, mitochondria and pro-plastids, kept their normal structure in adapted cells at values of low water potential (approx. -2.0 MPa) which caused membrane disruption and appearance of lipid droplets when this intensity of stress was applied to unadapted cells (Fig. 2). Therefore, a gradual adaptation to low water potential allowed the conservation of a particular state of membrane structure, which is critical for a wide spectrum of membrane-associated protein functions.

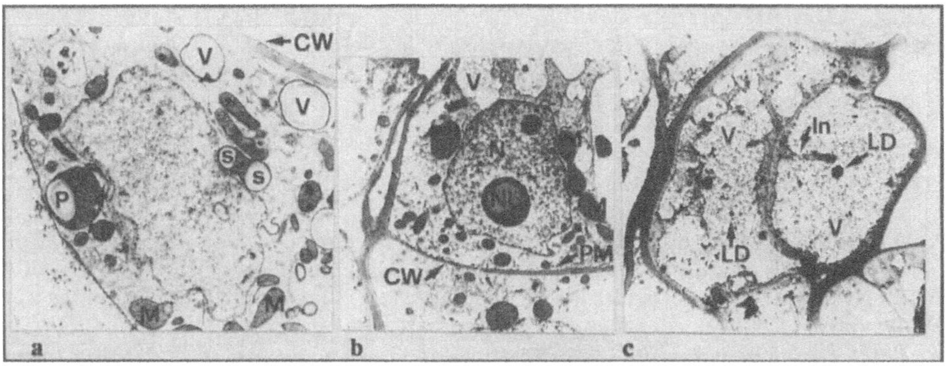

Figure 1. EM ultrastructure of (**a**) unadapted, (**b**) PEG-adapted and (**c**) PEG-shocked potato cells. **CW**, cell wall; **In**, infolding of cell wall; **LD**, lipid droplet; M, mitochondria; **NL**, nucleus; **P**, pro-plastid; **S**, starch; **V**, vacuole.

CHANGES IN MEMBRANE FA COMPOSITION AND MICROVISCOSITY

Compared to unadapted cells, FAs of the main extra-chloroplast membrane phospholipids, phosphatidylcholine (PC) and phosphatidylethanolamine (PE) were more saturated in PEG-adapted cells (Table 1). Measurements of the microviscosity of isolated protoplasts, labeled with DPH, revealed that membranes were more rigid in PEG-adapted cells in good accordance with the variation in FA saturation found (Fig. 2).

Table 1- Fatty acid composition of phosphatidylcholine (PC) and phosphatidylethanolamine (PE) from unadapted and PEG-adapted cells

Acyl Group	PC		PE	
	Unadapted	*PEG-adapted*	*Unadapted*	*PEG-adapted*
	%		%	
16:0	23.2	20.2	19.9	27.0
16:1	1.0	1.0	nd	nd
18:0	2.7	4.6	2.0	5.8
18:1	2.8	8.3	9.4	6.3
18:2	66.2	61.8	74.4	58.0
18:3	4.1	4.3	3.2	2.8

nd not detected (< 0.5%)

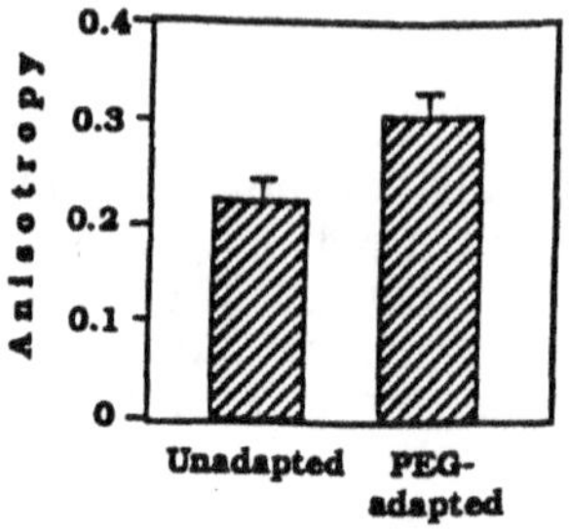

Figure 2. Membrane fluidity measurements of isolated protoplasts from unadapted and PEG-adapted cells, by using DPH probe.

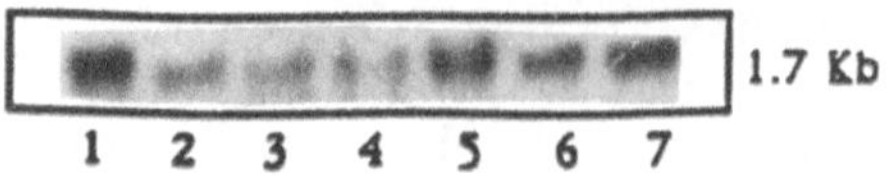

Figure 3. Level of transcript relative to gene encoding a Δ-9-stearoyl-ACP-desaturase in potato cells. **1)** unadapted cells; **2)** 5% PEG-adapted cells; **3)** 10% PEG-adapted cells; **4)** 20% PEG-adapted cells; **5)** unadapted cells abruptly transferred and kept for 24 h in a medium supplemented with 20% PEG; **6)** ABA 100 μM for 24 h; **7)** unadapted cells treated at 37 °C for 3h.

REGULATION OF GENE EXPRESSION OF A Δ-9-STEAROYL-ACP DESATURASE

Total RNA was extracted from unadapted and adapted cells as well as from unadapted cells abruptly transferred to a medium supplemented with 20% PEG. Northern blot hybridization demonstrated that the level of transcript relative to a cDNA clone encoding a Δ-9-stearoyl-ACP-desaturase (6) decreased in cells gradually adapted to PEG-mediated low water potentials. This result shows that the reduction in the level of fatty acid unsaturation was at least partially due to down-regulation of the expression of this gene. The steady state level of the desaturase gene was not affected either in PEG-shocked cells or in unadapted cells which were heat-shocked at 37°C for 3 h, demonstrating that modulation in gene expression of this gene occurs only after a long term adaptation (Fig. 3).

References

1. Stewart GR. Desiccation injury, anhydrobiosis and survival. In: Jones HG, Flowers TJ, Jones M, editors. Plant under stress. Cambridge: Cambridge University Press, 1989: 115-130.
2. Quinn P. Regulation of membrane fluidity in unicellular organisms. In: Aloia RC, Curtain CC, Gordon LM, editors. Physiological regulation of membrane fluidity. New York: Alan R Liss, Inc, 1988: 293-324.
3. Rains DW. Plant tissue and protoplast culture: applications to stress physiology and biochemistry. In: Jones HG, Flowers TJ, Jones M, editors. Plant under stress. Cambridge: Cambridge University Press, 1989: 181-186.
4. Gombos Z, Kiss M, Pali T, Vigh L. Nitrate starvation induces homeoviscous regulation of lipids in the cell envelope of the blue-green alga, *Anacystis nidulans*. Eur J Biochem 1977: 165:461-465.
5. Gregory PJ, Simmonds LP. Water relations and growth of potatoes; In: Harris P, editor. The potato crop. London: Chapman and Hall, 1992: 214-246.
6. Leone A, Costa A, Tucci M, Grillo S. Adaptation *versus* shock response to PEG-induced low water potential in cultured potato cells. Phys Plant 1994 (in press).
7. Tucci M, Grillo S, Costa A, Vigh L, Horvarth I, Leone A. Cloning, Characterization and Expression of a cDNA encoding a Δ-9-stearoyl-Acyl-Carrier-Protein-Desaturase from cold tolerant *Solanum commersonii*. Proceedings of the 4th International Congress of Plant Molecular Biology; 1994 June 19-24, Amsterdam.

SODIUM CHLORIDE EFFECT ON THE EVOLUTION OF FATTY ACID COMPOSITION IN DEVELOPING RAPE SEEDLINGS

Fouzia Najine , Brahim Marzouk and Abdelkader Cherif

Unité de Biotechnologie Végétale et de Ressources Génétiques, INRST
BP 95, 2050, Hammam-Lif, Tunisia.

Introduction :

Different environmental factors may cause many alterations in lipid metabolism : light, temperature, drought, salt, xenobiotics, atmospheric polluants, etc... .These factors may produce qualitative and quantitative changes in plant lipid compositions (1,2). Studies have shown that salt stress induced modifications in lipids and fatty acids composition of plant cell membranes (3,4,5).
Because we have no indication about rape plant changes under saline conditions, we have investigated the sodium chloride effect on polar glycerolipid composition in this 16:3-plant.

Materials and methods :

Experiments were carried out on rape seedlings ; transferred in one litre pots filled with the nutrient medium described by (6). This hydroponic cultivation was made under culture room conditions : 16 h day (460W/m^2) at 25°C and 8 h night at 18°C, the relative humidity was 50%.
When plants were 2 weeks old, they were submitted during 4 weeks to salt stress by adding sodium chloride in the nutrient medium at the following concentrations : 0, 68 and 170 mM.
At the end of each week, lipids of plant leaves were extracted according to a modified method of (7) and polar glycerolipid classes were separated by thin layer chromatography according to the method of (8). Fatty acids were analyzed by gas chromatography (9).

Results :

Table 1 : Effect of sodium chloride on the fatty acid contents in developing rape plants :

Week	1 st	2 nd	3 rd	4 th	5 th
NaCl (mM)		mg	g-1 DM		
0	22,65	42,01	24,19	13,22	6,13
68	-	60,34	23,85	8,61	4,70
170	-	12,09	23,62	8,62	1,91

The content of total fatty acids in rape plants - expressed in mg/g of dry matter (DM) -

J.-C. Kader and P. Mazliak (eds.), Plant Lipid Metabolism, 435–437.

decreased with development stage for the control and stressed plants. But at the 3 rd week, total fatty acid contents were practically constant at all the NaCl concentrations.

Table 2 : Sodium chloride effect on fatty acid compositions in developing rape plants :

		% Total Fatty Acids						
NaCl (mM)	Week	16:0	16:1	16:3	18:0	18:1	18:2	18:3
	2nd	21,8	5,0	7,8	6,8	15,1	14,2	28,0
0	3rd	17,1	3,5	12,4	1,9	4,8	11,1	46,6
	4th	13,9	7,2	13,0	2,4	3,4	10,5	47,7
	5th	13,2	6,2	16,4	1,0	4,1	9,3	47,3
	2nd	19,1	3,2	9,8	3,4	5,0	12,7	44,9
68	3rd	17,9	3,3	9,7	2,2	3,9	12,5	48,8
	4th	15,1	5,0	11,5	2,6	4,8	11,7	47,5
	5th	13,9	5,2	13,0	1,8	3,1	11,5	48,7
	2nd	24,9	6,8	8,1	7,5	5,0	11,1	35,0
170	3rd	21,5	1,5	7,7	3,9	9,2	16,2	39,2
	4th	15,5	5,3	10,5	2,9	4,0	12,2	47,4
	5th	15,8	5,1	9,4	2,4	4,4	14,2	47,8

Independently of stress degree, the percentages of saturated fatty acids, such as palmitate, stearate, oleate and linoleate decrease, while those of triunsaturated acids increased. Under salt stress and at all plant ages, linolenic acid rate was reduced weakly and that of hexadecatrienoic acid decreased significantly , in parallel levels of saturated fatty acids increased.
In presence of high NaCl concentration (170 mM), the polar lipid amounts decreased strongly. Therefore MGDG was reduced at the low level it represented only 20% of the control. Similar changes were observed for PL and DGDG (Fig. 1). These alterations in polar lipids indicated probably a disorganization of the cell membrane structures.

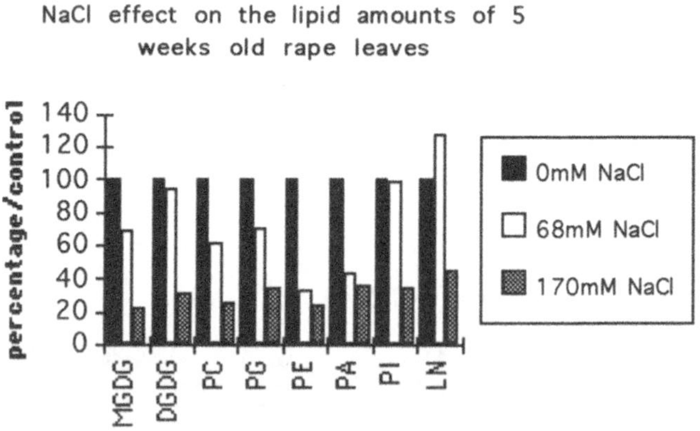

Conclusion :

The rape plant may resist to low salt concentrations (68 mM). Fresh and dry matter decrease when NaCl concentration increased in the culture medium.
Concerning lipids, we noticed a stimulation of GL synthesis to the benefit of PL under the effect of age whereas salt stress inhibited the polar lipid biosynthesis and changed the lipid membrane composition. The altered lipid composition of rape leaves could involve membrane permeability to ions which were essential for plant growth and development.
In the other hand the age induced a decrease in the percentages of 16:0 and 18:0 to the benefit of that of 18:3 and 16:3 . This the phenomenon explained by the fact that as the plant grows there is formation of new membrane structures of cell organelles and particularly those of chloroplast. Consequently, lipid and fatty acid components and mainly galactolipids and triunsaturated fatty acids (16:3 and 18:3) increased considerably. Changes in fatty acid composition of rape leaves under salt-stress were characterized by a decreased levels of triunsaturated acids (16:3 and 18:3) whereas those of saturated, mono-and di-unsaturated acids (16:0 , 18:0, 18:1 and 18:2) increased. This change could be the result of an inhibition or slowing down of desaturation enzymes.

References :

(1) Kuiper P.J.C. Environmental changes and lipid metabolism in higher plants. Physiol. Plant., 1985. 64,118-122.

(2) Harwood J.L., Abulnaja K.O, Barrett P.B., Herbert D., Rutter A.J. and Williams M. Changes in lipid Metabolism caused by Environmental Factors.In "Metabolism, structure and Utilization of plant lipids" .Ed. A. Cherif, D. Ben Miled, B. Marzouk, A. Smaoui and M. Zarrouk -CNP. Tunisia, 1992. 319-321.

(3) Zarrouk M. et Cherif A. Action du chlorure de sodium sur la teneur en lipides de plants d'olivier (Olea europea L.) Z. Pflanzenphysiol., 1981. 105,85-92.

(4) Kuiper P.J.C. Functioning of plant cell membranes under saline conditions : membrane lipid composition and ATPases. In "Salinity tolerance in plants" Richard C. Staples, G.H. Toennisen, J. Wiley and sows Ed., New york, 1984, 77-91.

(5) Marzouk B., Zarrouk M. Cherif A. and Mazliak P. Calcium Chloride effect on glycerolipid metabolism in olive tree leaf. In "The Metabolism, Structure and Function of Plant Lipids", Stumpf P.K., Mudd J.B. and Nes W.D. Ed., USA, 1986, 317-319.

(6) Kelday L.S. and Bowling, D.J.F. The effect of cycloheximide on uptake and transport of ions by sunflower roots. Ann. Bot., 1975. 39, 1023-1027.

(7) Bligh E.G. and Dyer W.J., A rapid method of total lipid extraction and purification. 1959. 37- 911-917.

(8) Lepage M. - Identification and composition of turnip root lipids.Lipids, 1967, 2,244-250.

(9) Marzouk B. et Cherif A. La lipogenèse dans l'olive : Formation des lipides Polaires. Oléagineux, 1981, 36, 387-391.

EFFECT OF SALT ON [1-^{14}C] ACETATE INCORPORATION IN DEVELOPING COTTON SEEDS

Abderrazak Smaoui[1] and Abdelkader Cherif[2],
1. I.N.R.S.T.. BP 95 2050 .Hammam Lif. Tunisia.
2. E. S. I. A. , 58 Rue Alain Savary, 1003 Tunis; Tunisia

Introduction

Previous studies showed that salt affected lipid biosynthesis in cotton seeds [1] essentially by decreasing the linoleic acid amount . In this study we investigated kinetics of [1-^{14}C] acetate in cotton seeds lipids ,*in vivo* at different stages of developing salt stressed seeds.

Material and methods

Two sets of *Gossypium hirsutum* cv S4 were grown in pots in greenhouse, at controlled conditions, and irrigated with tap water for control set and with water added of 150 mM NaCl for stressed one. Developing bolls of cotton were incised to display seeds, 3, and 5 weeks after anthesis (WAA). The [1-^{14}C] acetate(55mCi/mmol, supplyed by CEA,France) was injected into seeds, *in vivo.* Seeds were harvested 1, 6,16 ,and 24 hours after labelling. Lipids were extracted according to Bligh and Dyer[2], and analysed by thin layer chromatography using Mangold's method [3]. The radioactive fatty acid methyl esters were separated by argentation thin layer chromatography [4]. Radioactivity of lipid classes and fatty acids were counted by Radiochromatoplate (Chromelec 101,France)

Results

SALT EFFECT ON COMPOSITION OF LABELLED TOTAL FATTY ACIDS

At 3 WAA after anthesis (WAA),the fatty acid composition of total fatty acids(TFA) showed that radioactive linoleic acid percentage did not changed in stressed seeds during short

J. C. Kader and P. Mazliak (eds.), Plant Lipid Metabolism, 438–440.

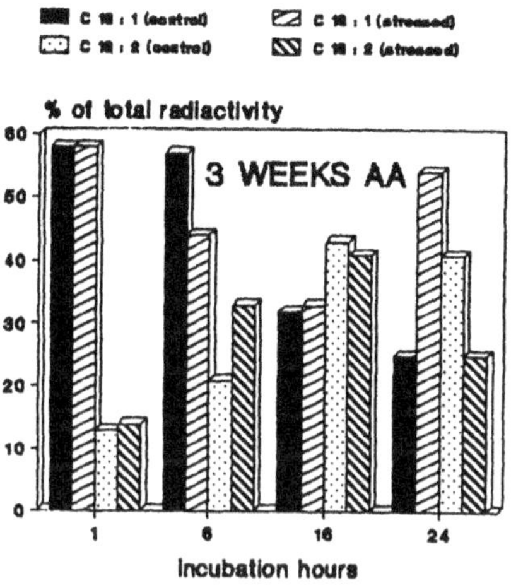

Figure 1
Salt effect on labelled TFA composition

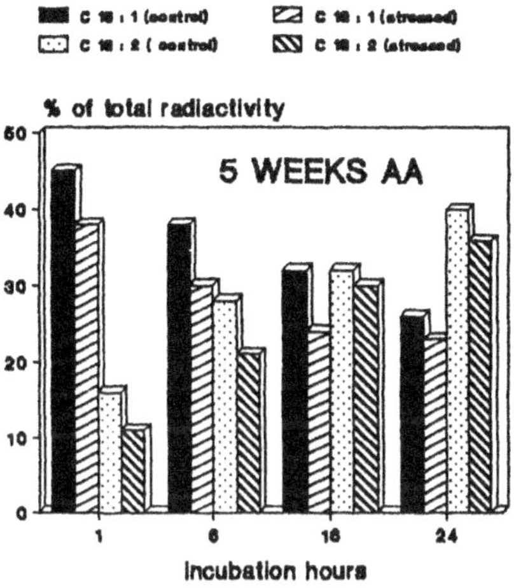

Figure 2
Salt effect on labelled TFA composition

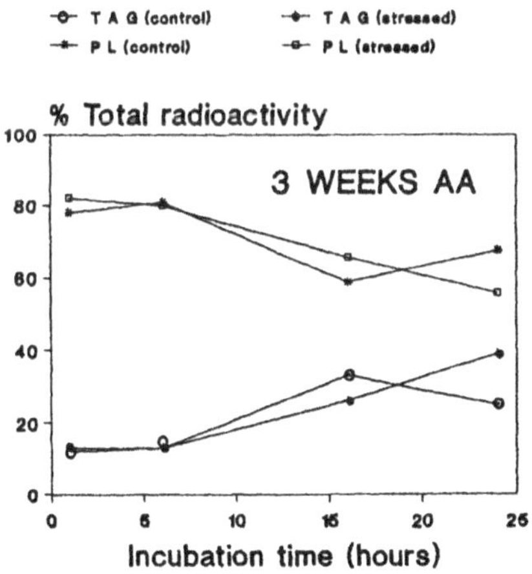

Figure 3
Salt effect on labelled lipid classes

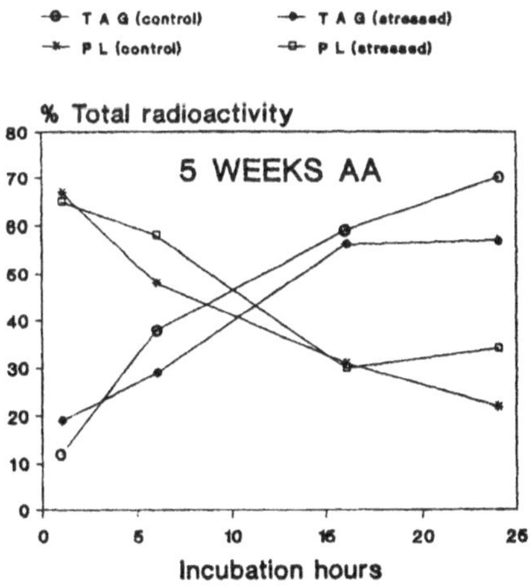

Figure 4
Salt effect on labelled lipid classes

incorporation time (one hour). In contrast 24 hours ago, linoleic acid amount was 41 % in control and 25 % in stressed seeds. Labelled oleic acid represented respectivly 25 % and 54 % of TFA [fig. 1]. However on fith WAA , salt affected slightly TFA composition. In NaCl treated seeds, linoleic acid level is depressed for only 10 % [fig. 2].

SALT EFFECT ON LABELLED LIPID CLASSES

Polar lipids (PL) were the important radioactive component in lipid seeds on 3 WAA ;it represent 82% of total labelled fatty acids after one hour acetate incubation. In these case triacylglycerols (TAG) biosynthesis was slow and reached 13 % [fig. 3].
We noticed a slight salt effect on lipid classes. On fith week this stress effect was more obvious on TAG of treated seeds [fig. 4].
Diacylglycerols (maximum 9 % of total radioactivity) decreased weekly during all developing seeds kitetics and practically did not be affected by salt treatment.

Conclusions

- At early stage of biosynthesis (3 WAA), PL (polar lipids) was the important lipid class labelled (68 % of incorporated radioactivity). Sodium chloride disturbed slightly kinetic [1-^{14}C] acetate incorporation. Beyond 3 WAA, TAG biosynthesis was faster and reached 70 % of total labelled lipids in control durig 24 incorporation hours.
- Salt effect was noticeable essentially 24 hours after [1-^{14}C] acetate supplying during fith WAA .This developing stage agreed with active lipogenesis. Similar results were observed about salt stress depressing on TAG during olive lipid synthesis [2]. Present study on radiolabel kinetics showed that linoleic acid, major fatty acid in cotton seeds, was drastically decreased by salt. This finding confirm previous investigation on non labelled lipids [5].
- Sodium chloride reduced principal fatty acid amount : linoleic acid for cotton seeds lipids,and oleic acid for olive lipids [4]. In both cases, decreasing concern always major fatty acids.

References

1. Smaoui A, and Cherif :.Glycerolipidsin salt stressed cottonseeds In:Cherif A, et al., editors.Metabolism structure and utilization of plant lipids.Tunis: Centre National Pedagogique, 1992:341-343.

2.Mangold H K . Thin layer chromatography of lipids. . Am. Oil Chemist. Soc. 1960;47:762-773.

3.Morris L J, Warry P M, et Hammon W E, Chromatographic behaviour of isomeric log-chain aliphatic compounds.Argentation thin-layer chromatography of isomeric compounds.J Chromatog 1967;31;69-76.

4. Marzouk B, Zarrouk M. , and Chérif A :Sodium effect on glycerolipids changes during ripening of olive. In: Biacs PA, Gruiz K, and Krammer T, editors.Biological role of plant lipids. New York and London: Akademiai Kiado,Budapest and Plenum Publishing Corporation, 1989:537-540

COMPARISON OF FATTY ACID COMPOSITION OF PHOSPHOLIPIDS IN NORMAL AND HABITUATED SUGAR BEET CELL LINES

Sabine Feutry[1], Danielle Poder[1], Marijana Krsnik-Rasol[2], Jean François Menez[3] and Daniel Hagège[1].

[1] Laboratoire d'Amélioration des Végétaux, Physiologie et Biochimie Végétales, UBO, BP 809, F-29285 Brest cedex.

[2] Laboratory of Molecular Biology, University of Zagreb, Croatia.

[3] Laboratoire de Biochimie-Nutrition, Faculté de Médecine, F-29285 Brest cedex.

Introduction

The composition of the lipid bilayer of biological membranes is of importance for the properties of the membrane and consequently for the function of the cell. The biomembranes are dynamic entities. Both the protein and lipid components of the membrane are constantly being turned over [1], [2] so as to maintain its functionnal state. Various features of the membrane lipids such as the composition, degree of unsaturation and surface charge properties affect and regulate the activities of membrane proteins. Hormones have also been implicated in the turn over of membrane lipids and protein components [3], [4]. Thus, the membrane properties are subject to very precise regulation for the maintenance of homeostasis. Habituated cells are considered to be the result of neoplasic progression of plant cells due to in vitro conditions and they exhibit many cytological and biochemical characteristics of animal tumors [5]. The rolc of membrane lipid alterations in the biological behavior of animal tumor cells have been investigated by several laboratories and some relations between altered membrane lipids and neoplasia have been underlined. Important membrane alterations have been reported in fatty acid saturation and phospholipid composition during hepatocarcinogenesis [6]. As membrane phospholipids are supposed to function as precursors of bioactive substances [7], enzyme activators [8] and modulators of plasma membrane-bound ATPases [9], we have investigated phospholipid content and their fatty acid composition in normal (auxin and cytokinin requiring) and habituated (auxin and cytokinin independent) sugarbeet calli.

Materials and methods

Calli cultures : Experimental conditions for obtaining normal (N) and habituated organogenous (HO) or non organogenous (HNO) calli of sugarbeet (*Beta vulgaris L. altissima*) and for maintaining these tissues in solid stock cultures under light (16 h photoperiod of Sylvania Grolux fluorescent light providing 17 W.m $^{-2}$, 25 °C) have been reported by 10].

Tissue labelling : For the determination of the relative percentage of PI, PC, PE, PG, PA, and NL, lipids were labelled with ^{32}P-phosphoric acid (C.E.A. Saclay). 1 g of tissue was incubated in a 10^{-4} mol.L^{-1} ^{32}P-phosphoric acid K_2HPO_4 (1.85. 10^8 Bq.L^{-1}) for two hours, at 22 °C, pH 6.

Lipid extraction and analysis. Lipids were extracted and analysed according to Bligh and Dyer [11]. 1 g of callus was boiled in 20 mL deionized water. Total lipids were extracted with 22,5 ml $CHCl_3$-MeOH (3:2, v/v) and 9 mL of the boiled water containing

J.-C. Kader and P. Mazliak (eds.), Plant Lipid Metabolism, 441–443.

NaCl 1 %. 200 µL butylated hydroxy-toluene (BHT) 1 % were added to prevent the lipoperoxidation. Calli were homogenised with a turax for 30 s. The extract was centrifuged 10 min at 1000 g, 5 °C.

Phospholipids assay : Polar lipids were determined using colorimetric method according to [12]. After acidic hydrolysis, phosphorus was complexed with ammonium molybdate, leading to the formation of a chromogen which absorbed at 820 nm. Standard curve was done using phosphatidyl choline (Sigma) 100 $\mu L.mL^{-1}$. 1 mL of chloroformic phase was evaporated to dryness at 110 °C. Dried samples were dissolved with 1 mL of A reagent countaining 10N H_2SO_4, 0.1 mol. L^{-1} periodic acid and ethanol 95 % (3-1-6, v/v). After 40 min in a boiling water bath, 5 mL of B reagent (ammonium molybdate 2.5 %-sodium bisulfite 6 %-deionized water : 3-5-40, v/v/v) were added. Tubes were replaced in a boiling water bath for 10 min and then cooled on ice. Absorbance was measured at 820 nm (LKB Ultrospec II E 4057).

TLC of phospholipids : Phospholipids were separated by 1-dimensional TLC on silica gel G60 plates (Merck) without fluorescence indicator. Plates were activated at 110 °C for 1h30. Chromatographic solvent was prepared according to [13] : Chloroform-acetone-methanol-acetic acid-deionized water : (50:20:10:10:5, v/v/v/v/v). 13 mL of chloroformic phase was evaporated to dryness and dissolved in 150 µL of Folch solvent: chloroform-methanol (2:1). Phospholipids were identified by co-chromatography with authentic standards : (PI, PC, PE, PS, PG, PA from Sigma) and staining by 2,4 dichlorofluorescein in anhydrous methanol under U.V.

The radioactive compounds were revealed by autoradiography after exposure to Kodak Regulix films, and analysed by scanning (Scanjet II C-Hewlett Packard - Logiciel Desk scan II, Scan analysis).

Fatty acid esterification and gas chromatography (GC) : Phospholipids spots obtained on TLC were scraped to analyse the endogenous (unlabelled) fatty acids of the phospholipids fraction from the different calli. Fatty acid (FA) methyl esters were prepared by transesterification with 0.2 M of sodium methoxyde for 15 min in darkness. Then, 4 mL of chloroform and 1.5 mL of deionized water were added. After homogenisation, the mixture were centrifuged (2000 g, 4 °C, 10 min) to remove remaining silica gel particles. FA were estimated on 3 mL of the chloroformic phase and separated by gas liquid chromatography (Packard 427, fitted with a flame ionisation detector and a Ross injector). FA were analysed with a fused silica capillary column of 25 m x 0.32 mm internal diameter coated with a CP wax 52 CB phase immobilised in situ in a film thickness of 0.21 µm (Chrompack, the Netherlands). The carrier gas was hydrogen at a pressure of 0.4 bars.

Results and discussion

An increase in total phospholipid content was found during the culture cycle for the two calli with a maximum at day 21 for N (12 mg.g DW $^{-1}$) and at day 10 (15,7 mg.g DW $^{-1}$) for HNO. The total phospholipid content in HNO cell line was higher than for N one from day 3 to day 28. This difference was particularly important at day 10 (HNO/N ratio : 249 %). The sum of the values of all the spots revealed by autoradiography, for each days of culture, represented 100 %. On this basis, relative percentage for each lipids (PI, PC, PE, PA, NL) were estimated, and PI and PC were the main lipids found. Lipid classes during the culture cycle was studied. 14 days old N callus spots respectively obtained for PC, PE, PA, NL were used as references (100 %). For N callus, PC and PE exhibit the same evolution during the culture. PC and PE levels increase until day 21 and decreased at day 28. For PI, NL, and PA, a maximum was observed between days 10 and 21 and a minimum was noted at day 3. On the other hand, HNO callus exhibits a similar evolution for each lipid classes. For the three calli, the main fatty acids were C 16:0 > C 18:2 > C

18:3. Taking in account some punctual variations, HNO line has higher content in C 16:0, C 18:0 and C18:2 than N one. This last line exhibits higher values in C 18:3 and C 18:1. The evolution of the bond index indicated a similar evolution for both cell lines all along the culture. Except at day 3, HNO callus has an lower bond index than N line.

Conclusion

The lipid composition of the habituated cell line differs from the normal one. It can be characterized by a high level in total phospholipid, a relative increase in C 16:0, C 18:0 and C 18:2, a decrease in C 18:3. Previous work have stated that habituated sugar beet callus exhibits a higher protection against oxygen-free radicals and hydroperoxide-dependent oxydation [14], [15]. This property could be explained by this different lipid composition. Polyunsaturated fatty acid are the main target for lipid peroxidation and a decrease in C18:3 may provide a best protection to this callus. Interestingly, a low lipid peroxidation capacity, based upon low amounts of polyunsaturated fatty acids as substrate, has been shown in cancerous liver nodules [6]. It has been postulated that lipid peroxidation could lead to a genetic perturbation involved in the carcinogenic process in animal. This resistance could be considered as an adaptative mechanism. Plant cells, submitted to various stress (as in vitro cultures) could be able to onset an adaptative process. The cells react to an hostile environment with adaptative responses including acute transitory reactions affecting most cells, with clonal development of new cell populations resistant to this environment. The processes of adaptation involve alterations at molecular and subcellular levels, able to reestablish steady-state conditions compatible with survival and continuous life. Lipid alterations seem to be consistent in neoplasic cells. Their impact on membrane stability is limited because alterations in one lipid class are balanced by alterations in other lipids that also affect membrane stability. A reduction in polyunsaturated lipid classes increases the resistance to oxidative stress and lipid peroxidation. Similar results have been obtained in periwinkle habituated cultures [16]. The results presented here allow to drive some parallel with animal tumors and to postulate the existence of an ubiquitous genetic program for cellular adaptation to escape a noxious environment. A better knowledge of what is turning the program on and off would be of biological importance.

References

[1] Mazliak P. Prog Phytochem, 1980 ; 6: 49-102. [2] Kinnunen PK. Chem Phys Lip, 1991 ; 57: 375-399.[3] Price-Jones MJ, Harwood JL. Biochem J, 1983 ; 216: 627-631.[4] Hager A, Debus G, Edel HG, Stransky H, Serrano R. Planta, 1991; 185: 527-537.[5] Hagège, 1990. Thèse d'état. Université de Caen. [6] Eriksson LC, Andersson GN. Crit Rev Biochem Mol Biol, 1992; 27: 1-55. [7] Chilton FH, O'Flaherty JT, Ellis JM, Swendser CL, Wykle RL. J Biol Chem, 1983; 258: 7268-7271. [8] Nelson N. Biochem Biophys Acta, 1976; 456: 314-338. [9] Cooke DT, Burden RS. Physiol Plant, 1990; 78: 153: 159. [10] Kevers C, Coumans M, De Greef W, Hofinger M, Gaspar Th. Physiol Plant 1981; 51: 281-286. [11] Bligh EG, Dyer WJ. Can J Biochem Physiol, 1959; 37: 911-917. [12] Shibuya I, Honda H, Maruo B. Agri Biol Chem, 1967; 31: 111-114. [13] Lepage M. Lipids, 1967; 2: 244-250. [14] Hagège D, Werk-Reichhart D, Schmitt P, Gaspar Th. Plant Physiol Biochem, 1990; 30: 649-654. [15] Hagège D, Kevers C, Salabert P, Gaspar Th. In Vitro Cell Dev, 1992; 28P:143-147. [16] Mérillon JM, Filali M, Dupéron P, Chénieux JC, Rideau M. Physiol Plant, submitted.

Acknolewledgements: This work was supported by a grant from the Conseil Régional de Bretagne. The authors thank Dr Lucas, Mrs Le Bot, Mrs Joannet and Mr Goubin for advices and technical assistance all along this work.

INFECTION BY ARBUSCULAR MYCORRHIZAL FUNGUS *GLOMUS MOSSEAE* OF LEEK PLANTS (*ALLIUM PORUM* L.) : EFFECT ON LIPIDS

A. Grandmougin-Ferjani[1], Y. Dalpé[2], E. Veignie[1], M. A. Hartmann[3], C. Rafin[1] and M. Sancholle[4]
1-Laboratoire de cryptogamie et de phytopathologie, Université du littoral, BP 699, F-62228 Calais. 2-CRTB Agriculture Canada, Ottawa, Ontario, Canada, KIA 0C6. 3-Département d'enzymologie cellulaire et moléculaire de l'IBMP, Institut de Botanique, 28 rue Goethe, F-67083 Strasbourg. 4-Laboratoire de Cryptogamie-Phytopathologie, SN2, USTL, F-59655 Villeneuve d'Ascq.

Introduction

It is well recognized that mycorrhizal symbiosis is highly beneficial to plant growth and protection from disease (Strullu, 1991). In arbuscular mycorrhizae, a highly intimate relationship is created between symbionts by the intracellular differentiation of arbuscules that represent a critical site for plant-fungi interactions. Several experiments have demonstrated that metabolites, such as carbon, minerals, and phosphate can be exchanged between mycorrhizal symbionts. In order to investigate whether lipids may be exchanged between plant and fungal symbionts, analysis of fatty acids and sterols were performed from leek plants (roots, bulbs and leaves), colonized and non-colonized with *Glomus mosseae*, as on isolated spores of this species.

Materials and methods

Fungal and plant materials

Spores of *Glomus mosseae* (Nicolson & Gerdemann) were first extracted from a cultivated soil by a sieving and decanting technique. Fifty of the extracted spores were added to 10 cm (diam.) pots filled with vermiculite and sowed with leek. Plants were grown for 12 weeks (22-25° (± 1°C) night and day temperatures, 16/8 h photoperiod, 85-90 % humidity). Twelve week old leek plants were collected from inoculated and non-inoculated substrates. Root colonization of leek plants inoculated with *G. mosseae* was evaluated at 30 % with the grid line technique and no colonization was observed in the control plants. Spores of *G. mosseae* were extracted from the pot cultures.

Lipid analysis

Prior to lipid extraction, plants and spores were lyophilized. Sterols were extracted and analysed by the method described by Costet-Corio and Benveniste (1988), except that the saponifiable matter, containing total fatty acids, was adjusted to a pH of one by adding HCl before being extracted three times with hexane. These extracts were pooled and dried. Fatty acids and sterols were quantified by gas chromatography (GC). Fatty acid and sterols identifications were made by comparing retention times respectively to methyl heptadecanoate (17:0) and cholesterol as internal standards. All sterols isolated were identified by GC / mass spectrometry.

J.-C. Kader and P. Mazliak (eds.), Plant Lipid Metabolism, 444–446.

Results

In leek plants

Δ 11 hexadecenoic acid (16:1, Δ 11), a fatty acid regulary found in *G. mosseae* spores (Sancholle and Dalpé, 1994), was also found abundantly in colonized roots; its presence in bulb tissues may indicate its possible transfer from roots to bulb tissues. In colonized roots, the decrease of amount of linoleic acid (18:2) resulted in an important decrease of unsaturation index (UI) of fatty acid chains. We observed in colonized roots a decrease in the linoleic acid (18:2) content whereas the Δ 11 hexadecenoic acid (16:1, Δ 11) increased. Palmitic acid (16:0) and linoleic acid (18:2) were the most abondant fatty acids in leaves. Mycorrhizal colonization induced an increase of the UI in leaves, due essentially to the higher levels of linoleic acid (18 :2) and linolenic acid (18:3).

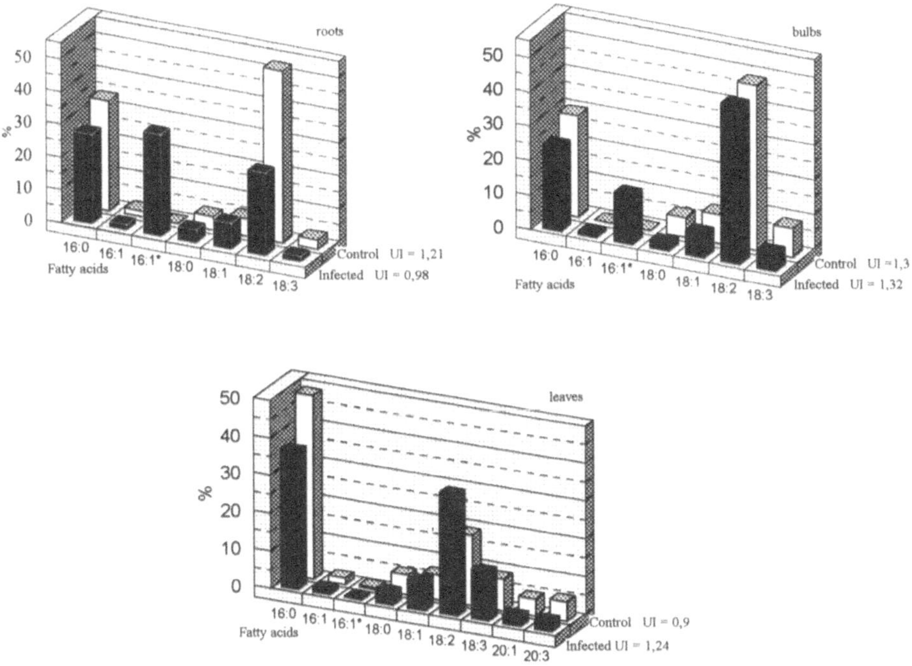

Figure 1 : Fatty acids composition of leek plants colonized and non-colonized with *Glomus mosseae.*
UI : (unsaturation index = 1 * (% monoenes) + 2 * (% dienes) + 3 * (% trienes)) / 100
16:1* = Δ 11 hexadecenoic acid (16:1, Δ 11)

Qualitative sterol composition was modified in bulb and root tissues (Table 1). Sitosterol decreased in bulbs and roots and a new compound, the 24-methylene cholesterol appeared only in roots. Surprisingly cholesterol concentration remained the same, despite the presence of intraradical vesicles and hyphae of *Glomus mosseae* in which cholesterol is the major sterol. The quantitative sterol composition was drastically modified : the *Glomus mosseae* root infection induced an important modification of sterol content such as a decrease in sitosterol content of 58 %.

Table 1 : Qualitative and quantitative sterols compositions of leek plants colonized and non-colonized with *Glomus mosseae*

	Leaves				Bulbs				Roots			
	Control		Infected		Control		Infected		Control		Infected	
sterols	%	µg/g dw[1]	%	µg/g dw	%	µg/g dw	%	µg/g dw	%	µg/g dw	%	µg/g dw
cholesterol	18	69	21	68	12	33	14	36	8	56	10	32
24- methylene cholesterol	nd[2]	nd	nd	nd	nd	nd	nd	nd	nd	nd	21	71
24-methyl cholesterol	6	25	6	21	9	24	12	32	13	84		
stigmasterol	4	15	2	7	5	14	9	24	3	18	2	8
sitosterol	66	253	67	218	74	199	60	158	70	473	58	197
isofucosterol	6	23	4	11	tr[3]	tr	5	12	6	43	9	29
total		385		325		270		262		674		337

1. µg/g dw : µg sterols per g dry weight
2. nd : non detected
3. tr : traces

In *G. mosseae* spores

The major sterol found in *G. mosseae* spores was cholesterol while palmitic (16:0) and Δ 11 hexadecenoic acid (16:1, Δ 11) acids were the major fatty acids (Table 2).

Table 2 : Qualitative sterol and fatty acids composition of *Glomus mosseae* spores

Total sterols			cholesterol		24-methyl cholesterol		sitosterol
%			60		4		36
Fatty acids	14:0	16:0	16:1 Δ 11	18:0	18:1	20:5	UI[1]
%	4.8	20.3	37.6	11.4	17.0	5.3	0.78

1. UI : unsaturation index = (1 * (% monoenes) + 2 * (% dienes) + 3 * (% trienes)) / 100

Conclusion

In roots, the primary site of interaction between symbionts, a decrease in sterol content was observed while the unsaturation index (UI) decreased. From these preliminary results, it appears that very drastical modifications in the lipid content occurred in response to the mycorrhization, especially at the interface site. The increase of Δ 11 hexadecenoic acid (16:1, Δ 11) content in the colonized leek plants can be directly correlated with the fatty acid metabolism of the fungus which is essentially a C16 pathway. Important modifications of the UI of total fatty acids in leaves can be likely associated to the effect of mycorrhizae on plant lipid metabolism. Investigations about lipid modifications may bring valuable biological data regarding plant reaction against detrimental growth conditions, related specifically to edaphic environnemental factors such as chilling and drought stresses where lipids could be deeply involved.

Acknowledgements : Thanks to Dr M. Corlett from Agriculture Canada for revision of the english manuscript and to Dr Couturier for identification of fatty acids by mass spectrometry.

References

Strullu DG. Les mycorhizes des arbres et des plantes cultivées. Technique et Documentation - Lavoisier. Paris. 1991. 250 pages.

Costet-Corio MF, Benveniste P. Sterol metabolism in wheat treated by N-substituted morpholines. Pestic Sci 1988; 22:343-57.

Sancholle M, Dalpé Y. Taxonomic relevance of fatty acids of arbuscular mycorrhizal fungi and related species. Mycotaxon 1994; 49:187-93.

INCREASING THE CONTENT OF POLYUNSATURATED FATTY ACIDS IN SPIRULINA PLATENSIS AND PORPHYRIDIUM CRUENTUM BY VARIATION OF LIGHT AND NUTRIENTS

Monika Springer[1], Nadeshda Rogova[2], Horst Franke[1], Otto Pulz[1]. [1]Institut für Getreideverarbeitung GmbH, Arthur-Scheunert-Allee 40-41, D-14558 Bergholz-Rehbrücke. [2]Institute of Plant Physiology, Botanicheskaya 35, 127 276 Moscow.

Introduction

The investigation of the influence of light on PUFA synthesis in microalgae has brought about a partly controversial pattern. Based on the GLA distribution in lipid fractions (mainly glycolipids) and on investigations of Sukenik et al. [2] who detected an increased MGDG and DGDG content at reduced light intensity, the following items were investigated:

* Influencing the GLA content of Spirulina platensis by
  - additions of acetate,
  - variation of light intensity and exposure to light,
  - combination of acetate addition and light variation.
* Influencing the AA and EPA content of Porphyridium cruentum by variation of light intensity or exposure to light, with special attention to PUFA changes immediately after changed exposure to light.

Algal cultivation was performed at a 2 l scale: Porphyridium cruentum, NÄG. Vischer 1932/107, Greifswald, Spirulina platensis, 85.79, Göttingen. Fatty acid analysis was performed analogeously [1].

The increase of the GLA content at a light intensity which was reduced from 150 to 50 $\mu E \cdot m^{-2} \cdot s^{-1}$ was a significant finding of Spirulina analysis. The highest GLA content is achieved not at continuous low-light exposure, but by a change from high- to low-light. The GLA content also may be increased by the addition of acetate. GLA production further is increased by the combination of light variation and acetate addition. However, the increase rates do not have an additive function. Also in Porphyridium cruentum, the highest AA and EPA contents were achieved at varying exposure to light, namely already 9 h after changed exposure to light. While EPA synthesis is increased to the same level at high-low light and low-high light changes, the highest AA content is achieved at low-high light change.

J.-C. Kader and P. Mazliak (eds.), Plant Lipid Metabolism, 447–449.

Results

FATTY ACID SPECTRUM AND GLA DISTRIBUTION IN SPIRULINA PLATENSIS

Influence of acetate additions:
The influence of increasing acetate concentration was investigated at a light intensity of 150 $\mu E \cdot m^{-2} \cdot s^{-1}$. As may be seen from fig. 1, the addition of acetate generally stimulates fatty acid anabolism thus increasing the content of the individual fatty acids. At optimum acetate concentration the GLA content is doubled.
Influence of light intensity and its change, respectively:
The investigations were based on the following light regimes: continuous exposure to high light (150 $\mu E \cdot m^{-2} \cdot s^{-1}$), to low light (50 $\mu E \cdot m^{-2} \cdot s^{-1}$), change of light intensity (150 to 50 $\mu E \cdot m^{-2} \cdot s^{-1}$).
Fig. 2 is illustrating that a reduced light intensity is increasing the GLA content. The increase is enhanced quantitatively at a high to low light change, after the first half of cultivation time.

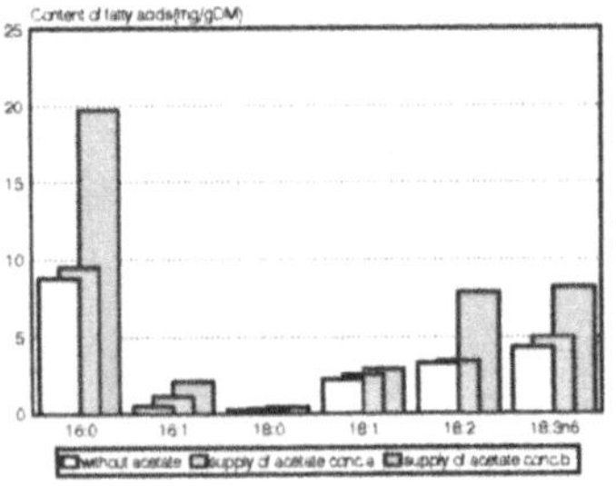

Fig.1: Spirulina platensis Spectrum of fatty acids

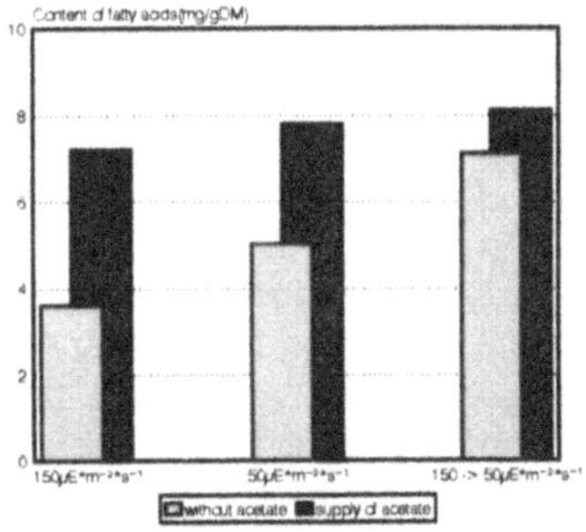

Fig.2: Content of gamma-linolenic acid in Spirulina platensis at variated light conditions and acetate supply

Combining acetate addition and light variation:
The highest GLA content is resulting from an optimum acetate addition and optimum light regime, i.e. change from high to low light (fig. 2). Acetate and light have no additive effect. The influence of the light regime is less distinct in acetate cultivation than in an autotrophic cultivation.

FATTY ACID SPECTRUM OF PORPHYRIDIUM CRUENTUM

Influence of light intensity and its change, respectively:
The light regimes for the cultivation of Porphyridium cruentum were varied similarly to those of Spirulina platensis, as follows: continuous exposure to high light (200 $\mu E \cdot m^{-2} \cdot s^{-1}$), to low light (50 $\mu E \cdot m^{-2} \cdot s^{-1}$), change of light intensity from

high to low and low to high.
The fatty acid spectrum was measured immediately after the change of light at intervals of 3 hours.
Fig. 3 is demonstrating that Porphyridium cruentum is responding to the high-low and low-high light stress with an increased AA synthesis. The highest AA synthesis (27 mg/g d.m.) was achieved 9 h after the increase of light intensity and decreased after 24 h.
At low↔high light stress the AA content is increasing to approx. 11 mg/g d.m. This value is maintained up to 24 h after the stress release.
The analysis of EPA kinetics (see fig. 4) reveals that the highest contents are achieved 9 h after the light change, as in the case of AA. In the case of EPA the same quantitative increase is achieved at high-low and low-high light stress. After 24 h, the EPA content is decreasing, however maintaining a higher level than in the case of continuous exposure to light.

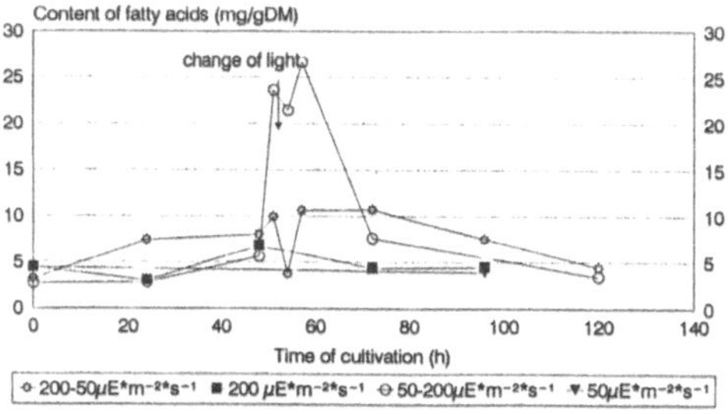

Fig.3:Content of arachidonic acid in Porphyridium cruentum at different light conditions

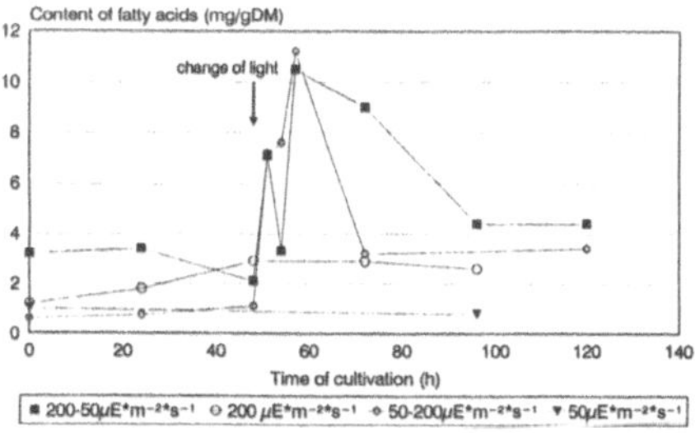

Fig.4:Content of eicosapentaenoic acid in Porphyridium cruentum at variated light conditions

Acknowledgements

The research project was supported by financial funds of the Federal Ministry of Research and Technology.
Thanks for technical assistance are extended to Mrs. Döbel, Mrs. Träger and Mrs. Sasse.

References

1. Springer M, Franke H, Pulz O. Increase of the content of polyunsaturated fatty acids in Porphyridium cruentum by low-temperature stress and acetate supply. J Plant Phys 1994; 4/5:534-537.

2. Sukenik A, Yamaguchi Y, Livne A. Alterations in lipid molecular species of the marine eustigmatophyte Nannochloropsis sp. J Phycol 1993; 29:620-626.

INFLUENCE OF SUMMER EPISODES OF ATMOSPHERIC O_3 ON Δ^5 AND Δ^9 FATTY ACIDS IN LIPIDS OF NORWAY SPRUCE DURING WINTER HARDENING

Alan R. Wellburn[1], Deborah C. Robinson[1], Alan Thomson[1] and Ian D. Leith[2]
[1]Division of Biological Sciences, Institute of Environmental and Biological Sciences, Lancaster University, Lancaster, LA1 4YQ, U.K. and [2]Institute of Terrestrial Ecology, Bush Estate, Penicuik, Midlothian, EH26 0QB, Scotland U.K.

The fatty acid composition of membrane lipids is important during the development of cold tolerance of many plants [1] often because increases in the proportion of polyunsaturated lipids retain the fluidity of membranes. For example, there is a correlation between the different degrees of winter hardiness and membrane lipid composition in Norway spruce [2]. There is also an inverse seasonal relationship in Norway spruce between short chain (C_{12} & C_{14}) saturated and unsaturated C_{18} fatty acids in certain lipid classes [3].

In an earlier seasonal study of the effect of episodes of O_3 throughout the last year of 3 consecutive summer exposure seasons, Wolfenden and Wellburn [4] showed that the molar percentage of octadecatetraenoate (18:4) in monogalactosyl diglyceride (MGDG) increased and the molar ratios of 2 forms of linoleate, $\Delta^{5,9}$18:2 and $\Delta^{9,12}$18:2 in MGDG decreased during frost hardening in filtered-air grown, 5 year-old Norway spruce trees but not in similar trees exposed to episodes of summer O_3. Wellburn *et* al., [5] then repeated that previous study using Norway spuce seedlings of a similar age and provenance having had the same number of summer O_3 exposures but using a different exposure facility, a much larger number of replicates, and analysis of other lipid classes (digactosyldiglyceride, DGDG; phosphatidyl glycerol, PG; and phosphatidyl choline, PC) as well as MGDG. The analyses were also carried out at the time of maximum winter hardiness in order to define the importance of fatty acid composition of membrane during winter hardening following episodes of summer O_3.

Current year needles from 5 year-old Norway spruce trees, which had been exposed to either episodes of atmospheric O_3 throughout 3 summer periods, were analysed for changes in molar percentages and ratios of fatty acids isolated from different lipids at the time of maximum winter hardening. Significant decreases in the degree of unsaturation of both C_{16} and C_{18} fatty acids, the molar percentage of $\Delta^{5,9,12,15}$18:4, and the molar ratio of $\Delta^{5,9}$18:2 to $\Delta^{9,12}$18:2 in MGDG due to summer O_3 exposures were found [5]. Molar percentages and ratios of fatty acids did not change much in other lipids but these changes in plastidic MGDG could be traced to a significant effect of summer O_3 on the Δ^5- and Δ^{12}-desaturases acting upon PC in the endoplasmic reticulum [5]. The replacement of the Δ^6-subset of C_{18} fatty acids by an equivalent Δ^5-series throughout was

J.-C. Kader and P. Mazliak (eds.), Plant Lipid Metabolism, 450–452.

also confirmed by gas chromatography and mass spectrometry [5]. Furthermore, molecular modelling also showed that the Δ^5-forms, which resembled the Δ^9-isomers, are very different in shape to the Δ^6-series [5] and this may account, in part, for the extremely low winter temperatures from which Norway spruce needles may recover.

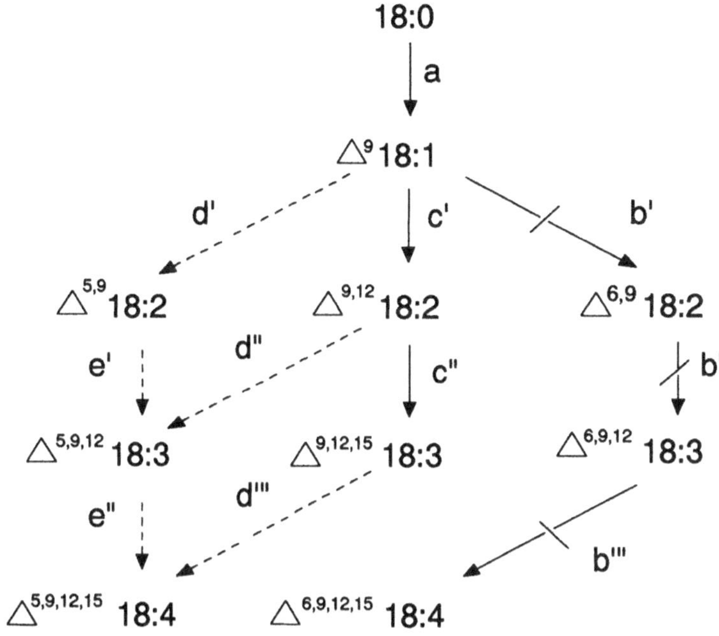

The figure above shows a revised pathway for possible fatty acid biosynthesis in Norway spruce from that shown earlier [4] to account for the results of Wellburn *et al.* [5]. Conversion of fatty acids attached to Route **a-b** is the normal angiosperm pathway and sections marked as **b** (with deleted arrows) do not appear to occur in Norway spruce [4,5]. Assuming that similar compartmental features still apply for final synthesis of MGDG by the so-called 'eukaryotic' route then 18:1 fatty acids arrive from the plastids as CoA complexes after the first Δ^9 desaturation (**a**) and are attached to form PC in the endoplasmic reticulum at the *sn*-1 position. Meanwhile, the *sn*-2 position is occupied in spruce PC mainly by 16:0 which amounts to nearly 50% in molar terms. The next desaturation on PC is then the one that is sensitive to summer O_3 fumigation. This is usually reaction **c'**, but perhaps **d'**, catalysed by oleoyl-phosphatidylcholine Δ^{12} and Δ^5-desaturases respectively. The latter being the most sensitive because the ratio of $\Delta^{5,12}$18:2 to $\Delta^{9,12}$18:2 in MGDG was found to be the most affected [5]. The significant depression by summer O_3 of $\Delta^{5,9}$18:2 in PC [5] strongly supports this contention and the significant fall in $\Delta^9$18:1 in PC and rise in MGDG must mean a premature return of some lipid to plastids due to O_3. Some 18:0-CoA must even return before attaching to endoplasmic reticulum lipid due to O_3 because these levels of 18:0 also rise in MGDG [5].

For the $\Delta^{9,12}$18:2 that is formed on PC, conventional understanding of the 'eukaryotic' higher plant lipid biosynthetic pathway suggests that this residue is transferred to MGDG inside plastids where the final Δ^{15} desaturation takes place [6]. In terms of the above scheme, this means that reaction **d'"** is highly unlikely which still leaves routes **a-c'-d"-e"** and **a-d'-e'-e"** as possibilities for the Δ^5-containing minor fatty acids and **a-c'-c"** for the major product, $\Delta^{9,12,15}$18:3 in MGDG. However, as $\Delta^9$18:1 is the preferred substrate for the Δ^{12}-desaturase [7], the minor route is probably **a-c'-d"-e"**.

The final outcome is a dramatic fall in levels of $\Delta^{5,9,12,15}$18:2 [5] which possibly accounts for the failure of summer O_3 exposed Norway spruce to harden sufficiently to low enough temperatures. Indeed, Cape *et al.* [8] have already shown that 20% of shoot death occurs at -23.5°C in those needles exposed to episode of summer O_3 by October as compared to a similar mortality at -28.5°C for needles from trees growing in filtered air.

The significance of the Δ^5-containing lipids in Norway spruce is of especial interest especially as needles of this species can survive extremely low winter temperatures (-35°C or lower). Less hardy conifers appear to contain lesser amounts of Δ^5-containing fatty acids and those that contain hardy any (e.g. *Pinus halepensis*) show winter injury at temperatures around -10°C. After summer O_3 exposures, amounts of Δ^5-containing fatty acids are much lower in Norway spruce.

1. Levitt J. Responses of Plants to Environmental Stresses, In Kozlowski TJ editor Chilling, Freezing and High Temperatures. 2nd edn, vol. 1., Academic Press, New York 1980:274-289.
2. Senser M, Beck E. Correlation of chloroplast ultrastructure and membrane lipid composition to the different degrees of frost resistance acheived in leaves of spinach, ivy, and spruce. J Plant Physiol 1984;117:41-55.
3. Puchinger L, Stachelberger H. Standortabhängigheit der Lipiddynamik von Fichtennadeln. Phyton 1989;29:233-253.
4. Wolfenden J, Wellburn AR. Effects of summer ozone on membrane lipid composition during subsequent frost hardening in Norway spruce [*Picea abies* (L.) Karst]. New Phytol 1991;118:323-329.
5. Wellbuirn AR, Robinson DC, Thomson A, Leith ID. Influence of episodes of summer O_3 on Δ^5 and Δ^9 fatty acids in autumnal lipids of Norway spruce [*Picea abies* (L.) Karst]. New Phyt 1994; in the press.
6. Ohlrogge JB, Browse J, Somerville CR. The genetics of plant lipids. Biochim Biophys Acta 1991;1082, 1-26.
7. Wada H, Murata N. *Synechocystis* PCC6803 mutants defective in desaturation of fatty acids. Plant & Cell Physiol 1989;30:971-978.
8. Cape JN, Fowler D, Eamus D, Murray MB, Sheppard LJ, Leith ID. (1990). Effects of acid mist and ozone on frost hardiness of Norway spruce seedlings. In: Payer HD, Pfirrmann T, Mathy P. editors. Environmental Research with Plants in Closed Chambers), Air Pollution Research Report No. 26, Commisson of the European Communities, DG XII, Brussels 1990:331-335.

GIBBERELLIC ACID ANALOGS MODIFICATION OF EPICUTICULAR WAX

Robert E. Wilkinson* and John J. Roberts+

*Professor, Department of Crops and Soil Sciences and

+Research Plant Pathologist, USDA-ARS, Georgia Station, Griffin, GA 30223-1797 USA

Synopsis

Gibberellic acid (GA_3) and analogs without the lactone ring induced altered quantity and composition of wheat (Triticum aestivum L. cv Stacy) leaf epicuticular waxes.

INTRODUCTION

Wheat (Triticum aestivum L.) leaf rust (Puccinia recondita Rob. ex. Desm.) urediospore germ tubes utilize leaf epicuticular wax as an energy source (Wilkinson, et al., 1990). Leaf epicuticular wax quantity and composition are genetically influenced (Bianchi, et al., 1980; Tulloch, et al., 1980), but change with age, (Tulloch & Hoffman, 1971; Wilkinson and Kasperbauer, 1972; Wilkinson and Mayeux, 1990) photoperiod, temperature, and light intensity (Wilkinson and Kasperbauer, 1972, 1980). Exogenously applied ^{14}C-fatty acids were recovered from intracellular complex lipids (Roughan, et al., 1987). Plant growth and development are controlled by plant-growth-regulators and GA influences plant size cell enlargement, α-amylase and IL biosyntheses (Krishnamoorthy, 1975). Fatty acids are synthesized in plastids (Mazliak, et al., 1980), transferred to microsomes where they are desaturated, transferred to plastids or other intracellular biosynthetic (i.e., phospholipid) systems, and then incorporated into cellular components. An increased quantity or activity of glyoxysomal IL could indicate an increased β-oxidation rate of fatty acids resulting in an increased supply of sucrose, which would be transferred to seed and stored as starch. This biosynthetic conversion of lipids to carbohydrate requires the activity of IL whose synthesis is controlled by GA (Doig, et al., 1975). GA biosynthesis is modified by photoperiod, temperature, etc. (Gilmore, et al., 1986). Thus, potentially, epicuticular wax quantity and composition could be influenced by GA.

METHODS AND MATERIALS

Hydroponics: Ten-day-old sand-grown wheat plants were transferred to Hoagland and Arnon complete mineral nutrient solution (H&A) containing GA_3 (0, 1, 10 or 100 μM) that was solubilized in 0.5 N KOH. Thirty plants were utilized/treatment. At initiation of treatment, equivalent plants were harvested for time = 0 FW and height. Total waxes were extracted and weighed, and β-diketone contents were determined spectrophotometrically (Horn and Lambertson, 1962). After 28-days treatment, the treated plants were harvested for the above listed parameters. Five replicates were utilized. Data were subjected to analyses of variance on a randomized complete block design.

GA_3 Spray: Twenty-five wheat seed/pot were planted in sand. At 10 DAP, the plants were sprayed with GA_3 (0, 0.1, or 1 mM) and plants were harvested as above for the time = 0 baseline of parameters listed above. At 14 DAS, plants were harvested for parameters listed above and data were analyzed as described.

G-ME Spray: G-ME (0, 0.1, 1.25, 5, or 100 μM) was applied as listed in the previous experiment. Data were collected and analyzed as described in the GA_3 experiment.

Isocitric Lyase: Ten DAP plants were sprayed with G-ME (5 μM), GA_3-ME (10 μM), and G-ME + GA_3-ME (5, 10 μM, IL activity was assayed according to Doig et al. (1975). Protein was assayed by the Lowry method (1951). Data were analyzed as discussed above. Plants were harvested at 7 days after spraying. Five replications/treatment.

ABBREVIATIONS: IL = isocitric lyase; GA = gibberellic acid; G = gibberic acid; GA_3-ME = methyl ester of GA_3; G-ME = methyl ester of G; DAP = days after planting; DAS = days after spraying; FW = fresh weight;

Mention of products or trade names does not imply endorsement of them by the University of Georgia or the U.S. Department of Agriculture. Cooperative investigations of the University of Georgia Agricultural Experiment Stations State Hatch Project 1451 and United States Department of Agriculture, Agricultural Research Service.

J.-C. Kader and P. Mazliak (eds.), Plant Lipid Metabolism, 453–455.

RESULTS AND DISCUSSION

In hydroponics, length and FW were not increased by putative GA_3 (Table 1). Presence of a lactone ring in GA_3 is requisite for the induction of increased height (Krishnamoorthy 1975). Rupture of the lactone ring occurs quantitatively in mild alkali (Cross, et al., 1961). Therefore, the compound applied was not GA_3. But, the increase in total epicuticular wax Δmg/plant) and β-diketone content in the epicuticular wax (Δmg/plant) were decreased (Figs. 1, 2).

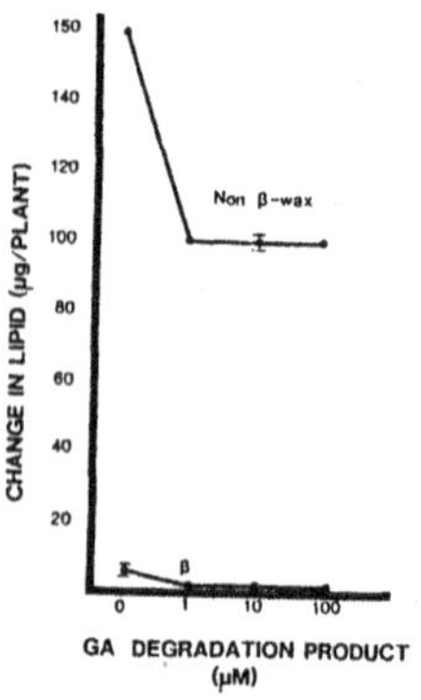

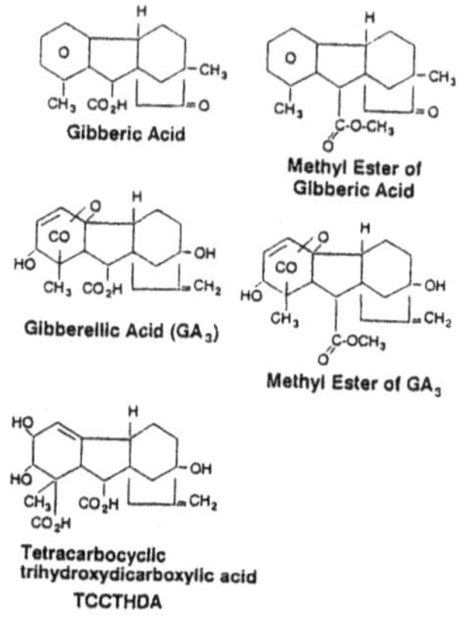

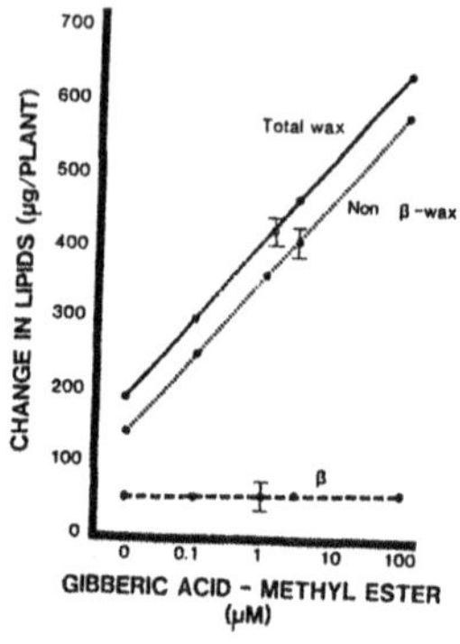

GA_3 topical sprays induced increased height and FW (Δmg/plant) (Table 2) but, also, induced decreased total epicuticular wax (Δmg/plant) and β-diketone (% total wax) contents.

G-ME did not induce an increased height (Δcm) or FW (Δmg/plant) (Table 3). Total wax and total non-β-epicuticular wax Δµg/plant) were increased by G-ME treatment (Fig. 3), but β-diketone content (Δµg/plant) was decreased (%-total wax) (Table 3). Thus, β-diketone synthesis was greatly decreased.

When G-ME and GA_3-ME were evaluated concomitantly, G-ME induced a decreased leaf FW (Table 4), while GA_3-ME induced an increased leaf FW and the combination of G-ME and GA_3-ME was intermediate but not significantly different from the untreated material. β-Diketone content (Δµg/g FW) was decreased by G-ME and GA_3-ME, but where the compounds were applied concomitantly, there was not any significantly different from the untreated plants. Total wax (Δmg/g FW) was increased by G-ME, decreased by GA_3-ME, and the combination was not significantly different from the untreated plants. IL activity (µM hydrazone product/min/mg protein) was increased by G-ME and GA_3-ME. These data show: 1) the lactone ring of GA was requisite for increased height and FW, 2) the rates of total epicuticular wax and β-diketone accumulation was influenced by GA-type molecules without the lactone ring, and 3) IL activity was influenced by GA-type molecules without the lactone ring.

FIGURE HEADINGS

Fig. 1. (Left) β-Diketone and non-β-diketone wax of wheat exposed to GA_3 solubilized in 0.5 N KOH in complete mineral nutrient solution for 28 days.

Fig. 2. (Middle) Structures of compounds used.

Fig. 3. Right) Changes (µg/plant) in total epicuticular wax, non-β-diketone wax, and β-diketone contents of wheat sprayed with G-ME at 14 DAP.

REFERENCES

Bianchi, G., et al. (1980) Planta 148:328.
Cross, B.E., et al. (1961) Adv. Chem. Series 28:3.
Doig, R.I., et al. (1975) J. Exp. Botany 26:387.
Gilmore, S.J., et al. (1986) Plant Physiol. 82:190.
Horn, D.H.S. and J.A. Lambertson (1962) Chem. and Indust. pp 2036.
Krishnamoorthy, H.N. (1975) Gibberellins and Plant Growth. John Wiley & Sons, New York.
Lowry, O.H., et al. (1951). J. Biol. Chem. 193:265.
Mazliak, P., et al. (1980). Biogenesis and Function of Plant Lipids, Elsevier, Amsterdam.
Roughan, P.G., et al. (1987) Arch. Biochem. Biophys. 259.

REFERENCES (continued)

Tulloch, A.P. and L.L. Hoffman (1971) Phytochem. 12:2225.
Tulloch, A.P., et al. (1980) Can. J. Bot. 58:2602.
Wilkinson, R.E. and H.S. Mayeux, Jr. (1990) Bot. Gaz. 151:342.
Wilkinson, R.E. and M.J. Kasperbauer (1972) Phytochem. 11:2439.
Wilkinson, R.E. and M.J. Kasperbauer (1980) Phytochem. 19:1379.
Wilkinson, R.E., et al. (1990) in Plant Lipid Biochemistry, Structure and Utilization, P.J. Quinn & J.L. Harwood (eds) Portland Press, Ltd., London, pp 370-372.

Table 1. Influence on height (Δcm)[1] and fresh weight (Δmg/plant) of GA_3 solubilized in 0.5 N KOH applied in nutrient solutions for 28 days in growth and wax content of wheat.

	KOH (0.5 N) Solubilized GA_3 (μM)			
Parameter	0	1	10	100
Height	8.4a[2]	6.2a	7.0a	7.0a
Fresh Weight	483a	373b	340b	319b

[1]Δ = change from that parameter taken on the day treatment was initiated.
[2]Values on a line followed by the same letter are not significantly different at the 5% level.

Table 2. Influence of GA_3 on growth and epicuticular wax of wheat at 14 DAS.

GA_3 (μM)	Height (Δcm)	Fresh Weight (Δmg/plant)	Total Wax (Δμg/plant)	β-Diketones (% Total Wax)
0[1]	+3.0b[2]	769c	+855a	0.99a
0.1	+6.3a	878b	+725b	0.73b
1.0	+7.3a	934a	+734b	0.73b

[1]Δ = changes in values from the untreated tissue before spraying.
[2]Values in a column followed by the same letter are not significantly different at the 5% level.

Table 3. Influence of G-ME on height (Δcm)[1], fresh weight (Δmg/plant), and wax (percent β-diketone) content of wheat at 14 DAS (24 DAP).

	G-ME (μM)				
Parameter	0	0.1	1.2	5.0	100.0
Height	+4a[2]	+5a	+5a	+5a	+3a
Fresh Weight	+62.8c	+180.5a	+170.8ab	+192.6a	+152.2b
β-Diketones	25a	17b	16b	11c	9c

[1]Δ = change from measurements taken at spraying (time = 0).
[2]Values in a line followed by the same letter are not significantly different at the 5% level.

Table 4. Influence of G-ME and GA_3-ME on growth wax characteristics and IL activity of wheat at 7 DAS.

Treatment G-ME (μM)	Treatment GA_3-ME (μM)	Leaf Fresh Weight (Δ[1]mg/plant)	β-Diketone (Δμg/g FW)	Total Wax (Δmg/g FW)	Hydrazone mg Product/min / mg Protein
0	0	+573b[2]	+3.9a	-0.48b	14.22c
5	-	+327c	+2.0b	+1.99a	52.32a
-	10	+719a	+1.8b	-2.77c	29.62b
5	10	+481b	+3.9a	-0.65b	38.38ab

[1]Δ = change from measurement taken at spraying (time = 0).
[2]Values in a column followed by the same letter are not significantly different at the 5% level.

FORMATION OF ω-FERULOYLOXYPALMITIC ACID AND TETRADECYL FERULATE IN ENZYMIC EXTRACTS OF WOUND-HEALING POTATO TUBER DISCS (*Solanum tuberosum* L.)

Samia Lotfy, Francine Javelle and Jonathan Negrel

Laboratoire de Phytoparasitologie INRA-CNRS, Sation de Génétique et d'Amélioration des Plantes, INRA BV 1540 21034 Dijon Cedex , France

Suberin is a complex heteropolymer comprizing lipidic components attached to a phenolic matrix [1]. It naturally covers the underground parts of plants but it is also synthesized at the surface of wound-healing plant tissues. While the composition and the biosynthesis of the fatty acids and alcohols components of suberin have been extensively studied, very little is known about the composition and the enzymology of the phenolic fraction [1,2]. Soluble ferulic acid esters of saturated fatty alcohols, ranging in carbon chain length from C_{16} to C_{28}, have been isolated from suberized potato periderm and it has been postulated that these alkyl ferulates could be suberin monomers [2,3]. We recently reported [4] the occurence in suberizing potato periderm (*Solanum tuberosum* L. cv Bintje) of a wound-induced acyl-CoA transferase catalysing the formation of ferulic acid esters of several medium-chain 1-alkanols and also, very efficiently, of ω-hydroxypalmitic acid (Fig. 1). It is the first time that an enzyme (HHT= hydroxycinnamoyl-CoA:ω-hydroxypalmitic acid O-hydroxycinnamoyltransferase) conjugating lipids to hydroxycinnamic acids is characterized.

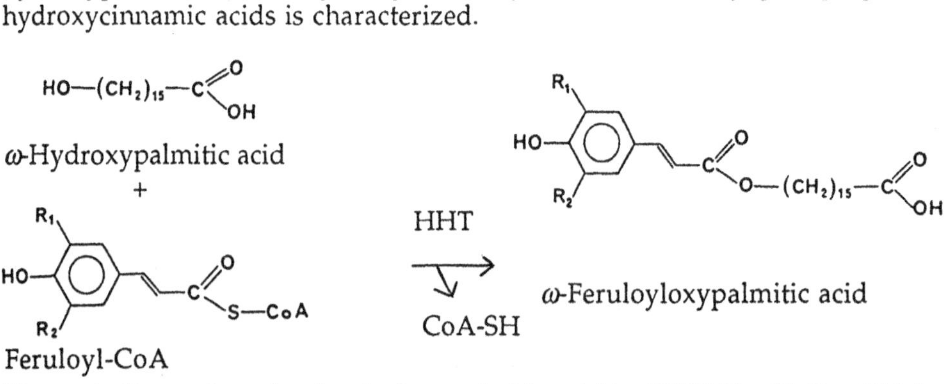

Figure 1: Formation of ω-feruloyloxypalmitic acid by HHT (R_1=H, R_2=OMe)

J.-C. Kader and P. Mazliak (eds.), Plant Lipid Metabolism, 456–458.

Characterization of HHT Activity in Extracts of Wound-Healing Tuber Discs

Using a spectrophotometric assay to detect feruloyl-CoA transferase activity in enzymic extracts of 7-day suberized potato tuber discs, we tested a series of fatty alcohols or acids as potential acceptors of the feruloyl moiety (Table 1). Two kinds of substrates were found to be active: ω–hydroxyfatty acids and 1-alkanols, the best substrates being ω–hydroxypalmitic acid and 1-tetradecanol, respectively. *p*-Coumaroyl-CoA (R_1=R_2=H, Fig.1) and sinapoyl-CoA (R_1= R_2=OMe) also acted as phenolic substrates, but were less active than feruloyl-CoA. The products formed in the incubation medium were separated by HPLC and identified by cochromatography with synthetic standards. Different carbon-chain lengths were required for optimal activity of 1-alkanols and ω-hydroxyfatty acids, suggesting that the two families of fatty compounds could be conjugated to ferulic acid by distinct enzymes. Surprizingly long-chain 1-alkanols (C_{20}-C_{28}) were not conjugated *in vitro* (Table 1). Since the substrate specificity of the transferases present in the crude enzymic extracts did not significantly vary during ageing of the tuber discs, these long-chain alkyl ferulate esters are therefore unlikely to be synthesized by direct conjugation of 1-alkanols to ferulic acid.

Table 1. Relative activity of fatty alcohols or ω-hydroxyfatty acids in the feruloyl-CoA transferase assay

Substrate	Relative activity (%)	Km (µM)
12-hydroxydodecanoic acid	<6	nd
15 hydroxypentadecanoic acid	68	16.8
16-hydroxyhexadecanoic acid	100	1.0
1-Decanol	6	nd
1-Dodecanol	49	5.0
1-Tetradecanol	82	4.0
1-Hexadecanol	42	10.0
1-Octadecanol	<6	nd
1-Eicosanol --> 1-Octacosanol	0	–

Wound Induction of HHT in Potato Tubers

HHT activity was not detectable in intact potato tuber tissue but was induced upon wounding after a lag period of three days (Fig.2). The time course of induction of HHT was therefore comparable with the kinetics of accumulation of alkyl ferulates [3] and of deposition of the aliphatic components of suberin [1]. HHT activity being restricted to the outermost cell layers of the wound-healing tuber discs, the development of HHT activity was both temporally and spatially coincident with the suberization process. The role played by ω–feruloyloxypalmitic acid in the biosynthesis of alkyl ferulates remains to be determined. One cannot rule out the possibility that long-chain alkyl ferulates could be formed from ω-feruloyloxypalmitic acid after elongation by C_2 units originating from malonyl-CoA, followed by decarboxylation [1]. Such a mechanism could explain the occurence of esters of odd chain length 1-alkanols (C_{19} and C_{21}) [3]. In potato, fatty acids ω-hydroxylases have previously been shown to be involved in the biosynthesis of ω-hydroxyfatty acids and of dicarboxylic acids [1]. The finding that ω-hydroxypalmitic acid is involved in the formation of ferulic acid esters therefore confirms the importance of these hydroxylases in the formation of the different components of potato suberin.

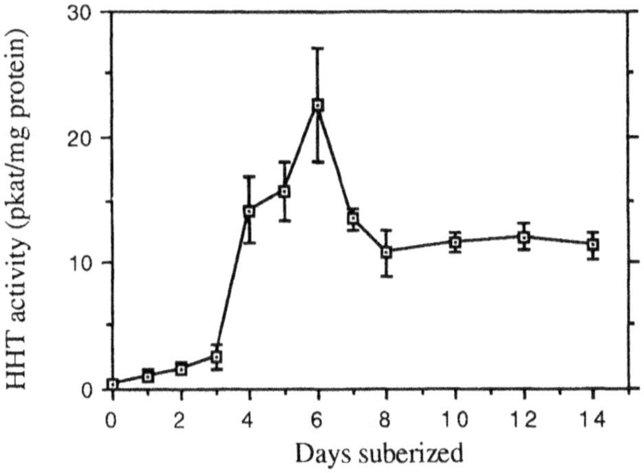

Figure 2 : Time course of induction of HHT activity in wound-healing potato tuber discs. Each point represents the average of three replicates (mean±s.e.)

References

1. Kolattukudy PE. Structure, biosynthesis, and biodegradation of cutin and suberin. Ann. Rev. Plant Physiol. 1981; 32:539-67.
2. Adamovics JA, Johnson G, Stermitz FR. Ferulates from cork layers of *Solanum tuberosum* and *Pseudotsuga menziesii*. Phytochemistry 1977; 16, 1089-1090.
3. Bernards MA. and Lewis NG. Alkyl ferulates in wound healing potato tubers. Phytochemistry 1992; 31, 3409-3412.
4. Lotfy S, Negrel J, Javelle F. Formation of ω-feruloyloxypalmitic acid by an enzyme from wound-healing potato tuber discs. Phytochemistry 1994; 35,1419-1424.

THE LEAF ACYL LIPID COMPOSITION OF PLANTS EXPOSED TO MODERATELY ENHANCED LEVELS OF OZONE: SPECIES, AGE AND DOSE DEPENDENCE

Anna Stina Sandelius, Anders S Carlsson, Håkan Pleijel, Lars I Hellgren, Göran Wallin and Gun Selldén, Department of Plant Physiology, Göteborg University (ASS, ASC, LIH, GW, GS), and Swedish Environmental Research Institute (HP), Göteborg, Sweden.

Introduction

Moderately enhanced levels of ozone have adverse effects on the yield of many crop plants as well as on the growth of conifers [1,2]. Although knowledge about yield and growth are necessary to assess the overall impact of ozone in the environment, these parameters are not suitable when trying to establish casual relationships. For this purpose the direct effects of ozone on plant metabolism and on/or interactions between plants and their environment need to be identified. Several reports have shown that in many plant species, short exposures (< 24 h) to high concentrations of ozone, 150-600 ppb, cause large decreases in the content of galactolipids and in the level of unsaturation of galactolipid fatty acids [3-6]. However, reports concerning the effects of more moderate levels of ozone on plant lipids are scarce.

This report summarizes the results from our investigations of the effects of up to 65 ppb ozone on the leaf lipids of plants grown and exposed in a climate chamber and on wheat grown and exposed in open-top chambers in a field. The levels of ozone used were higher than the daily average level of ozone normally occurring in Northern Europe during the summer, but lower than the levels encountered during ozone episodes. The results show that ≤ 65 ppb ozone had different effects on different plant species and that in the sensitive species, the ratio of chloroplast to non-chloroplast lipids decreased. Furthermore, the effects differed from those reported for much higher concentrations of ozone [3-6].

Materials and Methods

OZONE FUMIGATION

Laboratory experiments. Garden pea (12-days-old), wheat (9-days old) or spinach (2-months-old) plants were transferred to a fumigation system in a climate chamber and exposed to < 5 or 65 ppb ozone, 12 h daily, for 6 days. To study the effects of ozone on lipid degradation, the plants were exposed to $^{14}CO_2$ for 2 h 3 days prior to being transferred to the fumigation system. With pea, lipid synthesis was studied by adding ^{14}C-acetate to leaves of plants that had been exposed to <5 or 65 ppb ozone for 3 days. For a description of the fumigation system, see [7].

Field experiments. Wheat was grown in open-top chambers in a wheat field and, at the time of anthesis, exposed to charcoal-filtered air (6 ppb), non-filtered air (22 ppb) or non-filtered air with added ozone (44 or 56 ppb). The concentrations of ozone refer to a

J.-C. Kader and P. Mazliak (eds.), Plant Lipid Metabolism, 459–461.

7 h daily average. For a description of the field station, see [1].

For both laboratory and field experiments, lipids were extracted from defined leaf areas and further analyzed as described elsewhere [7].

CALCULATION OF ACCUMULATED OZONE DOSE ABOVE THRESHOLD CONCENTRATION

Accumulated ozone doses were calculated for the different open-top chamber treatments in the field experiment as numbers of ppb*h above different thresholds (0, 10, 20, 30, 40, and 50 ppb, respectively) after 27 days of treatment. The proportion of 18:3 (α-linolenic acid) in the total lipids and in monogalactosyldiacylglycerol (MGDG) was tested for best linear fit for the 6 different doses calculated (the calculation method is described in [8]).

Results and Discussion

In the laboratory experiments, 6 days of exposure to 65 ppb ozone had clear effects on the lipid composition of leaves of pea and wheat while spinach was unaffected. In both pea and wheat, the proportion of 18:3 in the total lipid fraction decreased (Tab. 1) which primarily reflected a marked decrease in the contents of MGDG and digalactosyldiacylglycerol (DGDG) (not shown). The radiolabelling experiments showed that the decrease in galactolipids was due to ozone-induced breakdown of galactolipids while the synthesis was unaffected (not shown). It should be noted that, unlike the reports on high ozone levels, we did not observe any significant changes in the MGDG/DGDG ratio (Tab. 2) nor an increase in 18:3 of DGDG (Tab. 1), which indicate that in contrast to high ozone levels [4,6], a moderate ozone level does not induce MGDG:MGDG galactosyltransferase. In pea, but not in wheat, ozone caused a decreased proportion of 18:3 in MGDG, while in wheat only, ozone caused an increased proportion of 18:3 in phosphatidylcholine (PC) (Tab. 1). In senescing barley leaves, MGDG breakdown caused a transient accumulation of PC with increased 18:3 [9] and our results may be taken as support of that in wheat, ozone-stimulated lipid degradation resembles senescence. In both pea and wheat, phosphatidylethanolamine (PE) increased, and with pea, radiolabelling experiments demonstrated that the increase mainly was due to increased rate of synthesis (not shown). The calculated ratio MGDG/PE clearly demonstrates that in both wheat and pea, 65 ppb ozone caused a shift in carbon flux from chloroplast to cytosolic membrane lipids (Tab. 2).

In the field experiment with mature wheat leaf tissue, the effects of elevated ozone

Table 1. The effects of ozone fumigation on the % of trienoic fatty acids (18:3 in pea and wheat, 18:3+16:3 in spinach) in the total fatty acids and major acyl lipids. Mean values of 3-5 independent experiments are shown. Pairs of data (same plant species and lipid class, different ozone treatments) in ***bold italics*** indicate statistically significant difference ($p < 0.05$; Student's t-test).

	Total lipids		MGDG		DGDG		PC		PE	
LABORATORY (6 days exposure)	< 5 ppb O_3	65 ppb O_3	< 5 ppb O_3	65 ppb O_3	< 5 ppb O_3	65 ppb O_3	< 5 ppb O_3	65 ppb O_3	< 5 ppb O_3	65 ppb O_3
Pea, 2nd leaf	***68***	***57***	***92***	***88***	78	79	31	30	25	18
Wheat, 1st leaf	***78***	***71***	94	94	86	89	***41***	***55***	27	31
Spinach	71	68	95	95	80	80	24	26	18	20
FIELD (27 days exposure)	22 ppb O_3	56 ppb O_3	22 ppb O_3	56 ppb O_3	22 ppb O_3	56 ppb O_3	22 ppb O_3	56 ppb O_3	22 ppb O_3	56 ppb O_3
Wheat flag leaf	***61***	***44***	***93***	***83***	83	83	47	50	58	58

Table 2. The effects of ozone fumigation on the molar ratios of MGDG/DGDG and MGDG/PE. Otherwise as in Table 1.

	MGDG / DGDG		MGDG / PE	
LABORATORY (6 days exposure)	< 5 ppb O_3	65 ppb O_3	< 5 ppb O_3	65 ppb O_3
Pea, 2nd leaf	1.5	1.3	***8.9***	***3.9***
Wheat, 1st leaf	1.8	1.4	***12.9***	***2.8***
Spinach	1.9	1.8	8.9	7.7
FIELD (27 days exposure)	22 ppb O_3	56 ppb O_3	22 ppb O_3	56 ppb O_3
Wheat flag leaf	1.2	1.0	***3.8***	***1.1***

resembled the ozone-effects on young pea more than the effects on young wheat in that the decrease in total fatty acid 18:3 reflected not only MGDG breakdown (not shown) but also a decrease in MGDG 18:3 (Tab. 1). As with young leaves, the MGDG/DGDG ratio was unaltered by ozone fumigation, while the MGDG/PE ratio decreased (Table 2).

When best fit between lipid 18:3 and accumulated ozone doses at the different thresholds were calculated, we found that after 27 days of field experiments, the threshold for a linear response was 40 ppb for total lipid 18:3 and 50 ppb for MGDG 18:3. The results indicate that a lipid or lipid-metabolizing process in the cytosol was more sensitive to ozone than MGDG or MGDG-metabolizing reactions.

Our results may suggest different responses to ozone in 18:3 versus 16:3 plants, in monocotyledons versus dicotyledons and in young versus mature leaves. Clearly, a larger selection of plant species and ages as well as ozone levels and doses needs to be analyzed before any such conclusion can be drawn. Presently, the data from our and others' investigations demonstrate that the effects of ozone on plant lipids are varying.

References

1. Pleijel H, Skärby L, Wallin G, Selldén G. Yield and grain quality of spring wheat (*Triticum aestivul* L., cv. Drabant) exposed to different concentrations of ozone in open-top chambers. Environ Pollut 1991; 69:151-168.
2. Pye JM. Impact of ozone on the growth and yield of trees. A review. J Environ Quality 1988; 17:347-360.
3. Fong F, Heath RL. Lipid content in the primary leaf of bean (*Phaseolus vulgaris*) after ozone fumigation. Z Pflanzenphysiol 1981; 104:109-115.
4. Sakaki T, Ohnishi J, Kondo N, Yamada M. Polar and neutral lipid changes in spinach leaves with ozone fumigation: Triacylglycerol synthesis from polar lipids. Plant Cell Physiol 1985; 26:253-262.
5. Nouchi I, Toyama S. Effects of ozone and peroxyacetyl nitrate on polar lipids and fatty acids in leaves of morning glory and kidney bean. Plant Physiol 1988; 87:638-646.
6. Sakaki T, Tanaka K, Yamada M. General metabolic changes in leaf lipids in response to ozone. Plant Cell Physiol 1994; 35:53-62.
7. Carlsson AS, Hellgren LI, Selldén G, Sandelius AS. Effects of moderately enhanced levels of ozone on the acyl lipid composition of leaes of garden pea. Physiol Plant 1994; In Press.
8. Pleijel H. Ozone impact on Swedish cereals [dissertation]. Göteborg (Sweden): Göteborg University, 1994.
9. Wanner L, Keller F, Matile PH. Metabolism of radiolabelled galactolipids in senescent barley leaves. Plant Science 1991; 78:199-206.

Supported by the Swedish Natural Science Research Council and the Swedish Council for Forestry and Agricultural Research.

IN VIVO AND *IN VITRO* INHIBITION OF FATTY ACID DESATURATION BY THE ANTIBIOTIC CERULENIN

Zvi Cohen*, Inna Khozina*, Devorah Shiran*, Helen A. Norman** and Parthasarathy Pillai**

Microalgal Biotechnology Unit, Jacob Blaustein Institute for Desert Research, Ben -Gurion University of the Negev, Sede-Boker Campus, 84990 Israel

Introduction

Polyunsaturated ω3 and ω6 C_{20} fatty acids are the final products of fatty acid biosynthesis in certain microalgae including *Porphyridium cruentum* [1]. There is current interest in developing biological sources particularly of eicosapentaenoic acid (EPA, 20:5ω3) for its use as a pharmaceutical. This requires a comprehensive understanding of biosynthetic pathways. The desaturation of 18:2ω6 appears to be a key step involved, and 18:2ω6 is proposed to be a common precusor in two distinct pathways leading to 20:4ω6 and 20:5ω3 [2]. This has been studied using SAN 9785 (BASF 13-338, 4-chloro-5(dimethylamino)-2-phenyl-3(2H) pyridazinone) [2], a direct inhibitor of 18:2 desaturation [3]. To continue this approach, we have searched for further inhibitors of fatty acid desaturation that can be used to identify steps in EPA biosynthesis. Here we show that cerulenin, an inhibitor of *de novo* fatty acid synthesis [4], is also a desaturase inhibitor.

Results and Discussion

Treatment of *P. cruentum* with 5μM cerulenin decreased total fatty acid content (TFA) to about 3% of dry wt (Table 1). Surprisingly, the relative amount of 20:5ω3 (EPA) was reduced from 44.8% to 22.8% of TFA, while at 7.5μM, 20:5 was decimated to 8.1%. At 5μM cerulenin, 20:4ω6 was increased to about 20% of TFA, while at 7.5μM 20:4ω6 was reduced. Analysis of individual polar lipids revealed an effect of cerulenin on fatty acid desaturation in each case (Table 2). In particular, 20:5 was reduced in monogalactosyldiacylglycerol (MGDG) with increasing cerulenin concentration; yet 20:4ω6 was initially increased before its decline together with an increase in 18:2ω6 and 16:0.

Table 1. Effect of cerulenin on the fatty acid composition of *P. cruentum*

Cerulenin conc.(μM)	Fatty acid composition (% of total fatty acids)									FA content (% of dry wt.)	
	16:0	16:1	18:0	18:1	18:2 ω6	18:3 ω6	20:3 ω6	20:4 ω6	EPA	TFA[a]	EPA
0	31.1	2.7	0.3	0.7	4.9	0.6	0.6	13.7	44.8	4.61	1.80
2	33.1	3.2	0.3	0.9	6.4	0.8	1.3	21.7	31.7		
5	40.0	4.8	1.1	0.6	5.8	1.3	2.0	20.0	22.8	2.97	0.68
7.5	59.8	5.6	5.4	0.7	4.8	0.6	1.6	13.0	8.1	1.49	0.12

a - TFA: total fatty acid. Cultures were treated with cerulenin for 3 days.

J.-C. Kader and P. Mazliak (eds.), Plant Lipid Metabolism, 462–464.

Table 2. Effect of cerulenin on the fatty acid composition of major lipids of *P. cruentum*.

Cerulenin conc.(μM)	Lipid	Fatty acid composition (% of total fatty acids)							
		16:0	18:1 ω9	18:2ω 6	18:3 ω6	20:2 ω6	20:3 ω6	20:4 ω6	EPA
0	MGDG	28.6	0.9	2.1	-	-	-	5.8	62.3
2	MGDG	19.8	0.5	2.7	-	-	0.3	26.3	50.0
5	MGDG	40.2	2.0	8.1	0.6	0.6	0.2	13.9	31.2
0	DGDG	45.9	0.8	6.4	-	-	-	2.0	43.0
2	DGDG	45.3	0.8	5.6	0.2	-	-	9.2	38.6
5	DGDG	45.3	3.6	7.7	0.6	1.1	0.2	7.3	31.7
0	SQ	49.9	1.1	2.1	-	2.1	-	4.4	33.5
2	SQ	58.6	1.4	2.7	-	3.1	0.3	11.7	16.6
0	PC	23.5	1.7	3.8	5.8	2.8	2.7	53.0	5.9
2	PC	21.1	1.4	3.8	5.0	2.8	7.2	53.8	3.4

Cultures were treated with cerulenin for 3 days. Lipids were extracted and fractionated on silica gel cartridges and further separated by TLC (galactolipids). PC was recovered by normal phase HPLC (LiChrospher 125-4, SI-60, 5μm). Fatty acids methyl esters were analyzed by GC. DGDG:digalactosyldiacylglycerol; SQ:sulfoquinovosyldiacylglycerol; PC:phosphatidylcholine.

Separation of MGDG into molecular species by HPLC identified 20:5/20:5, 20:5/20:4ω6, 20:4ω6/20:4ω6, 20:5/16:0, 20:4ω6/16:0, and 18:2/16:0 MGDG. Cerulenin (2μM) caused a reduction in both the eukaryotic 20:5/20:5 MGDG and the prokaryotic 20:5/16:0 MGDG, while 20:4ω6/16:0 was increased (Table 3).

Table 3. Effect of cerulenin on the molecular species composition of MGDG in *P. cruentum*.

Cerulenin concentration (μM)	Molecular species composition					
	20:5/20:5	20:5/20:4	20:4/20:4	20:5/16:0	20:4/16:0	18:2/16:0
0	39.0	2.1	t	57.8	1.1	t
2	19.6	7.8	3.1	48.6	20.4	t

t-trace levels. Separation of molecular species of MGDG was performed by HPLC on Lichrocart 250-4 (RP-18, 5μm) cartridges eluted with 95% MeOH-H_2O. An evaporative light scattering detector (Varex, Model ELSD 2, Burtonsville MD, USA) was used for quantitation of chromatograms. Peaks were detected at 205 nm, isolated, and fatty acid composition was elucidated by GC analysis of methylesters.

Figure 1 details a scheme for EPA biosynthesis in *P. cruentum* which is an extension of that proposed for prokaryotic and eukaryotic pathways of 18:3ω3 in plant cells [5], and is based on the combined data we have obtained using the inhibitors SAN 9785 [2] and cerulenin. The basic assumption is that EPA is derived from 18:3ω3 by two desaturation steps and one elongation step. The proposed pathways for 20:4ω6 involve desaturation of 18:2ω6 to 18:3ω6, a step which may be less sensitive to cerulenin than desaturation of 18:2ω6 to 18:3ω3. Therefore at low cerulenin concentrations there may be increased utilization of 18:2 for 20:4ω6 synthesis. This would explain the initial increase seen in 20:4 in MGDG, while at higher cerulenin concentration 20:4ω6 declined and 18:2 increased (Table 2). Unlike the action of SAN 9785, we propose that cytoplasmic desaturation of PC is also inhibited by cerulenin.

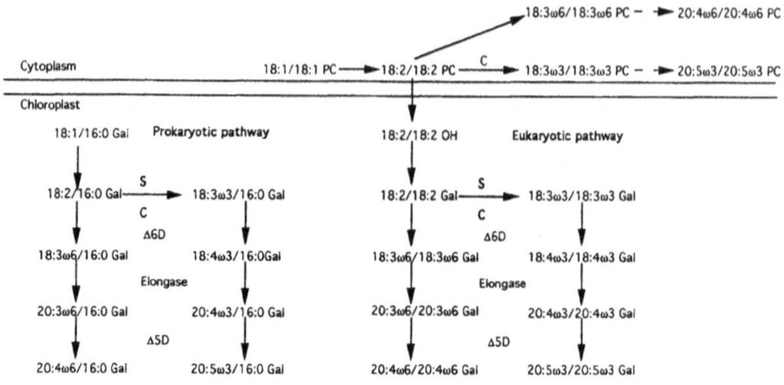

Figure 1. Possible pathways of EPA (20:5ω3) and 20:4ω6 biosynthesis in microalgae. S and C denote suggested primary sites of inhibition of SAN 9785 and cerulenin, respectively. D - desaturase.

In attempt to determine whether or not cerulenin directly affects desaturation reactions we have utilized procedures developed to stabilize chloroplast desaturase activity *in vitro* [3]. We have studied the effect of cerulenin on desaturation of various MGDG and PC molecular species by following desaturase activity during incubation of membrane preparations from chloroplasts of *Arabidopsis thaliana.* Cerulenin differentially inhibited the desaturation of 18:2 or 16:2 in all the substrates tested (Table 4). These results suggest a direct inhibition of enzyme activity, although desaturation could not be entirely eliminated even with higher concentrations of cerulenin. Possibly there is more than one n-3 desaturase present or regions affected are not at the active site of the enzyme.

Table 4. Effect of cerulenin on desaturation of ω–6 fatty acids on MGDG or PC substrates.

	Lipid substrate				
	18:3/16:2 MGDG	18:2/18:3 MGDG	18:2/16:0 MGDG	18:2/18:3 PC	16:0/18:2 PC
Desaturase activity (% of control)	65.8	58.4	76.3	68.2	80.4

Chloroplast membranes solubilized with CHAPS were incubated with ^{14}C-lipid substrates (25-50 nmol) for 45 min with cerulenin (10^{-6}M). Reaction rates were determined by HPLC [3].

References

1. Cohen Z, Vonshak A, Richmond A. Effect of environmental conditions on fatty acid composition of the red alga *Porphyridium cruentum:* correlation to growth rate. J.Phycol 1988;24:328-32.
2. Cohen Z, Norman HA, Heimer YM. Evaluating the potential of substituted pyridazinones for inducing polyunsaturated fatty acid overproduction in algae. Phytochemistry 1993;32:259-63.
3. Norman HA, Pillai P, St John JB. In vitro desaturation of monogalactosyldiacylglycerol and phosphatidylcholine molecular species by chloroplasts homogenates. Phytochemistry 1991;30:2217-22.
4. Omura S. Cerulenin. In: Lowenstein JM, editor. Methods in enzymology, Vol. 72. New York: Academic Press, 1981: 520-532.
5. Browse J, Somerville C. Glycerolipid synthesis: biochemistry and regulation. Ann Rev Plant Mol Biol 1991;42:467-506.

Section 6

Oil Seed and Gene Technology

STRUCTURE OF PLANT ENOYL-ACP REDUCTASE GENES

Gert-Jan de Boer, Martin M. Kater, Alwin R. Wagenaar, Tony Fawcett, Toni R. Slabas, H. John J. Nijkamp and Antoine R. Stuitje

Department of Genetics, Institute for Molecular Biological Sciences, BioCentrum Amsterdam, Vrije Universiteit,De Boelelan 1087,1081 HV Amsterdam, The Netherlands

Lipid Molecular Biology Group, University of Durham, South Road, Durham, DH1 3LE, U.K.

Introduction

In general fatty acids in plants serve two purposes, i.e. as a fuel reserve (storage lipids) or as membrane lipids. The formation of storage lipids occurs only for a short period during midphase of seed development. At this point an abrupt increase in steady state mRNA levels for the different fatty acid biosynthesis genes can be seen, followed by a peak in enzymatic activity [2]. One of these enzymes is enoyl-ACP reductase. It catalyses the reaction in which a *trans*-double bond is removed to form a saturated acyl-ACP (Fig. 1). For the enoyl-ACP reductase genes, a high expression level was also observed in young leafs, flower buds and anthers [2,4]. To be able to determine which elements are involved in regulating this tissue and developmentally dependent expression pattern, the genes from Tobacco and Arabidopsis encoding enoyl-ACP reductase were cloned and characterized, or partially characterized in the case of Petunia, to clarify their genomic structure.

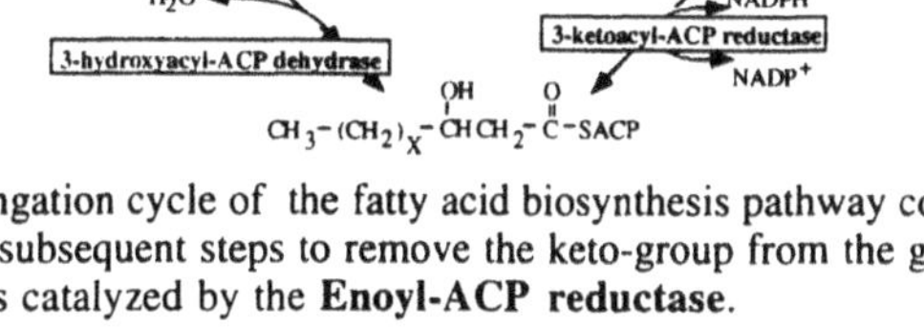

Figure 1. The elongation cycle of the fatty acid biosynthesis pathway consists of a condensation reaction followed by three subsequent steps to remove the keto-group from the growing carbon chain. The last of these three steps is catalyzed by the **Enoyl-ACP reductase**.

Results

PROMOTER ANALYSIS OF ENOYL-ACP REDUCTASE GENES

Although the two enoyl-ACP reductase genes from Tobacco are virtually identical, the homology does not extend any further than to about 200bp upstream of their transcription start site. This suggests that all necessary promoter elements are located within this region. Analysis of a promoter deletion series showed that a 400bp fragment of the Arabidopsis enoyl-ACP reductase promoter, fused to the gus reporter gene, was sufficient to give

J.-C. Kader and P. Mazliak (eds.), Plant Lipid Metabolism, 467–469.

expression (G-J de Boer, unpublished results). Alignment of the promoter sequences from both the Tobacco genes and the Arabidopsis enoyl-ACP reductase promoter revealed a small number of short identical sequences present in all three promoters (Fig. 2). Analysis of the Petunia enoyl-ACP reductase promoter will give more insight in the importance of these sequences. Finally, promoter deletion constructs have been made to be able to analyze the functional importance of these sequences in regulating the expression of the enoyl-ACP reductase genes.

Box I	-670	A A C A G A A	-664	Arabidopsis
	-1632	A A C A G A C	-1626	Tobacco T1
	-1815	A A C A G A C	-1809	Tobacco T2
BOX II	-700	T T T A A A G	-694	Arabidopsis
	-1643	T T T A A A G	-1637	Tobacco T1
	-1828	T T T A A A G	-1822	Tobacco T2
BOX III	-765	A T C T A A	-760	Arabidopsis
	-1697	A T C T A A	-1692	Tobacco T1
	-1890	G T C T A A	-1885	Tobacco T2
BOX IV	-789	T A G C A T G A	-783	Arabidopsis
	-1943	T A G C A T A A	-1936	Tobacco T1
	-2169	C A G C A T A A	-2162	Tobacco T2

Figure 2. Sequences present in both the Tobacco and the Arabidopsis enoyl-ACP reductase promoter. Numbers indicate their position relative to the ATG.

A CONSERVED ABNORMAL 5' SPLICE SITE IS PRESENT IN MOST ENOYL-ACP REDUCTASE GENES

The genes encoding enoyl-ACP reductase all contain a similar high number of introns (Fig. 3). Although in general introns can vary considerably in size, their basic structure is conserved. In the enoyl-ACP reductase genes from Brassica (4), Arabidopsis(1) and Tobacco (2), an abnormal 5' splicing site (GC instead of GT) was found to be present in one specific intron (intron 7 in Arabidopsis or intron 8 in the other genes). Only one previous report about a similar 5' splicing site in higher plants has been published, in this case the Nodulin 24 gene of soybean [3]. In vitro studies on the effects of mutations in the conserved 5' and 3' splicing sites in human HeLa cell extracts have shown that changing the 5' GT splice site into GC still permits accurate splicing although this may happen more slowly [1]. Although the enoyl-ACP reductase gene from Petunia contains a normal 5' splice site in this intron the fact remains that the presence of a 5'GC splice site appears to be strongly conserved over a wide range of other plant species. Whether its presence might serve as a possible site of regulation remains to be investigated. One piece of evidence is that differential splicing of one of the two Tobacco genes would create an insertion of 42 aminoacids. This could explain the two isoforms found in a Tobacco leaf extract (data not shown).

Conclusion

Enoyl-ACP reductase genes in plants are highly conserved both in structure and in DNA and aminoacid sequence. Unlike for example the acyl carrier protein, no seed specific genes are present for enoyl-ACP reductase. Since the genes encoding enoyl-ACP reductase were shown to be developmentally regulated. They should contain sequence elements that control the differential expression of this housekeeping gene. A number of short conserved sequences found in all enoyl-ACP reductase promoters studied so far, could therefore be involved in the transcriptional regulation of these genes. In addition, the abnormal 5' splice junction found in 7 enoyl-ACP reductase genes from four plant species, could also indicate to post-transcriptional regulation of gene expression at the level of intron splicing.

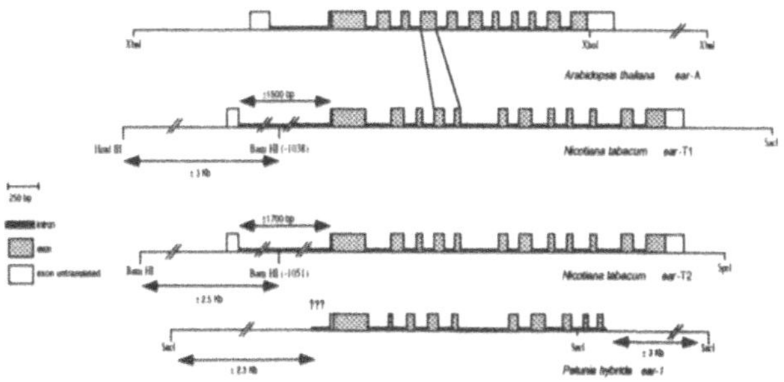

Figure 3. Schematically represented structures of the different plant enoyl-ACP reductase genes. The structure of the gene from Petunia is yet only partially elucidated.

The absence of this abnormal splice junction in the single petunia enoyl-ACP reductase gene, however, questions the importance of the conservation of this sequence in regulating gene expression.

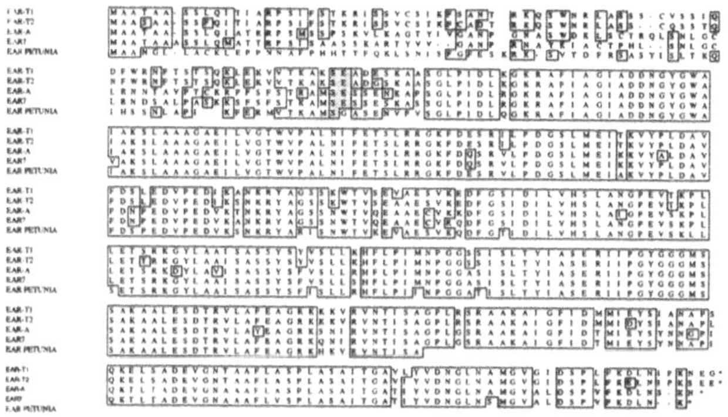

Figure 4. Alignment of the aminoacid sequences of the different enoyl-ACP reductase protein sequences. Ear-T1, -T2, -A, -7, Petunia denote the Tobacco T1, T2, the Arabidopsis, rapeseed and Petunia sequences respectively.

REFERENCES

1 Aebi, M., Hornig, H. and Weissman, C. 5' cleavage site in eukaryotic pre mRNA splicing is determined by the overall 5' splice region, not by the conserved 5' GU. Cell 1987; 50: 237-246

2 Kater, M.M., Koningstein, G.M., Nijkamp, H.J.J. and Stuitje, A.R. cDNA cloning and expression of *Brassica napus* enoyl-acyl carrier protein reductase in *Escherichia coli*. Plant Mol. Biol.1991;17: 895-909

3 Katinakis, P. and Verma, D.P.S. Nodulin-24 gene of soybean codes for a peptide of the peribacteroid membrane and was generated by tandem duplication of a sequence resembling an insertion element. Proc. Natl. Acad. Sci.1985; 82: 4157-4161

4 Slocombe, S.P., Piffanelli, P., Fairbairn, D., Bowra, S., Hatzopoulos, P., Tsiantis, M. and Murphy, D.J. Temporal and tissue specific regulation of a *Brassica napus* stearoyl-acyl carrier protein desaturase gene. Plant Phys. 1994;104: 1167-1176

DECANOYL-ACYL CARRIER PROTEIN INHIBITS THE β-KETOACYL-ACYL CARRIER PROTEIN SYNTHASE III IN *CUPHEA LANCEOLATA* SEED EXTRACT

Ricardo Schuch, Fritzi M. Brück, Monika Brummel, Friedrich Spener
Institute of Chemical and Biochemical Sensor Research Münster and Institute of Biochemistry
University Münster, Mendelstr.11 , D-48149 , Münster, Germany

Introduction

In order to identify a possible mechanism for the regulation of the fatty acid biosynthesis the influence of various acyl-ACPs on the activity of β-ketoacyl-ACP synthase (KAS) III was investigated in cell free seed extracts of *Cuphea lanceolata*, a crop accumulating up to 90 % decanoic acid in triacylglycerols. The activity of KAS III was assayed by determining the incorporation of radioactive acetyl-CoA into acetoacetyl-ACP in the presence of cerulenin. The influence of acyl-ACP on KAS III-activity was investigated by supplying non-radioactive acyl-ACPs of different chain-length (C_4 to C_{10}) to this assay. The products of these reactions were analyzed by monitoring the incorporation of radioactive malonyl-CoA into acyl-ACPs, in a modified assay. All acyl-ACPs induce a reduction in the activity of KAS III towards acetyl-CoA. Acyl-ACPs (C_4, C_6 and C_8) are in part elongated in the assay, but this cannot explain the over-all decrease in enzyme activity. The highest reduction in KAS III-activity was observed in the presence of C_{10}-ACP, which is not elongated by the enzyme. We conclude that C_{10}-ACP inhibits KAS III (50% inhibition at about 0.45 µM). Our data indicate the presence of a new condensing enzyme elongating short-chain acyl-ACPs (KAS IV). Moreover, these *in vitro* figures for the first time give some evidence that product acyl-ACPs may be involved in feed-back modulation of fatty acid biosynthesis in oil seeds.

Experimental procedures

A cell free extract of *Cuphea lanceolata* seeds was prepared from a 0-65 % ammonium sulfate precipitate as described [1] and assayed for KAS III-activity by determining the incorporation of [1-^{14}C]acetate from [1-^{14}C]acetyl-CoA into acetoacetyl-ACP in the presence of 100 µM cerulenin. The enzyme extract is preincubated with cerulenin to ensure complete inhibition of KAS I. Non radioactive acyl-ACPs, synthesized chemically from *E. coli* ACP and fatty acids, were supplied to this assay in different concentrations. For analysis of the reaction products, the assay was modified and the incorporation of [2-^{14}C]malonate from [2-^{14}C]malonyl-ACP (synthesized *in situ* from malonyl-CoA) into acyl-ACPs was monitored, after reduction of ß-ketoacyl-ACPs in the presence of NADH and NADPH using a *C. lanceolata* seed extract treated with N-ethylmaleimide (NEM-extract containing all reducing enzymes of the FAS but no condensing activity). Product acyl-ACPs were separated by urea-PAGE, transferred to Immobilon P, visualized by autoradiography and quantified by densitometry.

J.-C. Kader and P. Mazliak (eds.), Plant Lipid Metabolism, 470–472.

Results and discussion

The cerulenin insensitive KAS III catalyzes the first condensing step of the fatty acid biosynthesis between acetyl-CoA and malonyl-ACP producing acetoacetyl-ACP, which is further reduced to butyryl-ACP in the course of the FAS-reaction using NADH and NADPH as reducing equivalents. The current assay for KAS III-activity is based on the incorporation, in presence of 100 µM cerulenin, of radioactively labeled acetyl-CoA into acetoacetyl-ACP, which is precipitated by acid. NADH and NADPH are not present in this "normal" assay. However, the supplementation of NADH and NADPH, with the consequent reduction of acetoacetyl-ACP to butyryl-ACP, leads to a decrease in the KAS III-activity, as shown in Table 1. This decrease in KAS III-activity can be directly attributed to the presence of butyryl-ACP (Table 1). Similarly as described for spinach leaf

Table 1. Variation of KAS III-activity in *Cuphea lanceolata* seed extract in the presence of reducing equivalents and butyryl-ACP

KAS III-assay	Activity [%]
"normal"	100
+ NADH, NADPH	82
+ butyryl-ACP	
0.5 µM	78
1.0 µM	74
2.0 µM	56

extracts [2], even in presence of cerulenin *C. lanceolata* seed extracts elongate short-chain acyl-ACPs, although at minor extents [3]. Two questions have then to be addressed: i. is the elongation of short-chain acyl-ACPs a consequence of the action of KAS III or another today unknown condensing enzyme? and ii. if KAS III is the responsible enzyme, is the decrease in enzyme activity observed a consequence of a concurrence or an inhibition?

As shown in Fig. 1A acyl-ACPs reduce the activity of KAS III. The intensity of this effect is apparently related to the chain-length, decanoyl-ACP exerting the highest reduction in KAS III-activity. The analysis of the products of these reactions show that C_4-, C_6- and C_8-ACPs are only hardly elongated in the assays (Fig. 1B). The extent of this elongation is not sufficient to explain the levels of reduction in KAS III-activity obtained. These data indicate that acyl-ACPs inhibit KAS III and the elongation of short-chain acyl-ACPs is probably due to the action of another enzyme (KAS IV). Decanoyl-ACP is not a substrate for KAS III - the elongation shown in Fig. 1B (lane 5) is due to KAS II action [3], which is only partially inhibited at the assay conditions. We conclude that decanoyl-ACP strongly inhibits KAS III (50 % inhibition at about 0.45 µM).

These *in vitro* figures give some evidence that product acyl-ACPs may be involved in the modulation of fatty acid biosynthesis in a feed-back mechanism. Previously we demonstrated that condensing enzymes in *C. lanceolata*

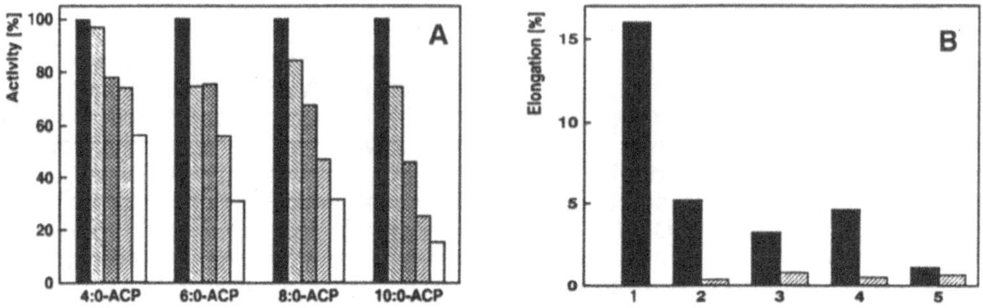

Figure 1. Influence of acyl-ACPs on the activity of KAS III in *Cuphea lanceolata* seed extract. A, Activity of KAS III in the presence 0 µM (■), 0.2 µM (▧), 0.5 µM (▩), 1.0 µM (▨) and 2.0 µM (□) acyl-ACPs; B, Elongation products of KAS III reaction using 10 µM acetyl-CoA (1), supplied with 10 µM butyryl-ACP (2), hexanoyl-ACP (3), octanoyl-ACP (4) and decanoyl-ACP (5). Bars represent elongation (%) of acetyl-CoA (■) and acyl-ACPs (▨).

seeds have a low activity for elongation of decanoyl-ACP [3], which would lead to an accumulation of this substrate in the plastids. We suggest that the synthesis of decanoic acid in these seeds is regulated at one side by a feed-back control of KAS III, at the other side by the activity of an earlier described thioesterase that specifically hydrolyzes decanoyl-ACP [4].

Acknowledgements

This work was supported by the Federal Ministry of Research and Technology and in part by the Fond der chemischen Industrie and the Ministry of Science and Technology of the state Northrhine-Westfalia.

References

1. Schuch R, Brummel M, Spener F. ß-Ketoacyl-acyl carrier protein (ACP) synthase III in *Cuphea lanceolata* seeds: Identification and analysis of reaction products. J Plant Physiol 1994;143:556-560.

2. Clough RC, Matthis AL, Barnum SR, Jaworski JG. Purification and characterization of 3-ketoacyl-acyl carrier protein synthase III from spinach - A condensing enzyme utilizing acetyl-Coenzyme A to initiate fatty acid biosynthesis. J Biol Chem 1992;267:20992-20998.

3. Schuch R, Brummel M, Spener F. Medium-chain acyl-ACP thioesterase is not the exclusive enzyme responsible for early chain-length termination in medium-chain fatty acid biosynthesis. Grassas y Aceites 1993;44:126-128.

4. Dörmann P, Spener F, Ohlrogge JB. Characterization of two acyl-acy carrier protein thioesterases from developing *Cuphea* seeds specific for medium-chain- and oleoyl-acyl carrier protein. Planta 1993;189:425-432.

STUDIES IN THE OPTIMISATION OF THE OVEREXPRESSION OF β-KETOACYL-ACP SYNTHASE I FROM BARLEY AND *ESCHERICHIA COLI*

Jo-Anne Chuck[1], Margit Wissenbach[1], Mads Siggaard-Andersen[1], Johan Gotthardt Olsen[1,2] and Penny von Wettstein-Knowles[1,2]

[1]Department of Physiology, Carlsberg Laboratory, Gamle Carlsberg Vej 10, DK-2500 Copenhagen, Denmark and [2]Department of Genetics, Institute of Molecular Biology, University of Copenhagen, Denmark.

Introduction

The condensation of malonyl-ACP and acyl-ACP to form β-ketoacyl-ACP is the elongation step in fatty acid biosynthesis. This reaction is carried out by a series of enzymes known as β–ketoacyl-ACP synthases or condensing enzymes. In higher plants and *E. coli*, several condensing enzyme activities have been identified (KAS I, KAS II, KAS III and KAS IV). These proteins differ in their ability to elongate acyl-ACP substrates of different chain lengths and in their sensitivities to the inhibitor, cerulenin. Our laboratory has previously isolated clones encoding KAS I from *E. coli* (*fabB*)[1] and from barley (*Kas12*) [2]. Both enzymes are highly sensitive to cerulenin and carry out the elongation of C_4-C_{14} acyl-ACPs. As we are interested in the function and structure of this class of enzyme, we would like a protein expression system so that the products could be used in site directed mutagenesis studies, for the construction of hybrid proteins and in structural studies involving x-ray crystallography.

Construction of expression vectors

Three constructs containing the *Kas 12* gene and one of the *fabB* gene were made in the T7 expression vector pET 5a (Novagen). The mature coding region of *Kas 12* was expressed with eleven extra residues derived from the phage capsid protein 10 (pBE*Kas12*M). A second construct was made which resulted in two extra residues at the amino-terminal (pNB*Kas12*M) while the final construct was designed to included these two extra amino acids as well as a deletion at the amino-terminal of the mature KAS 12 (pNB*Kas12*D). Amino acid sequence alignments lead to the suggestion that the barley enzyme has an extended amino-terminal compared to the FAB B protein (see [2]). It was hoped that the deleted plant enzyme would be considered less foreign to the expression host. The *fabB* gene was cloned into pET5a such that no extra amino acids were produced (pNB*fabB*). All plasmids were transformed into *E.coli* strain BL21(DE3) for expression studies.

The expression

Western blots of total cellular proteins from expression strains indicated that all four constructs made proteins that were recognised by their respective antibodies (Figure 1, ~43kDa). When the cells were lysed, the majority of the protein was found as insoluble inclusion bodies. These were easily detected in the cell by light microscopy.

J.-C. Kader and P. Mazliak (eds.), Plant Lipid Metabolism, 473–475.

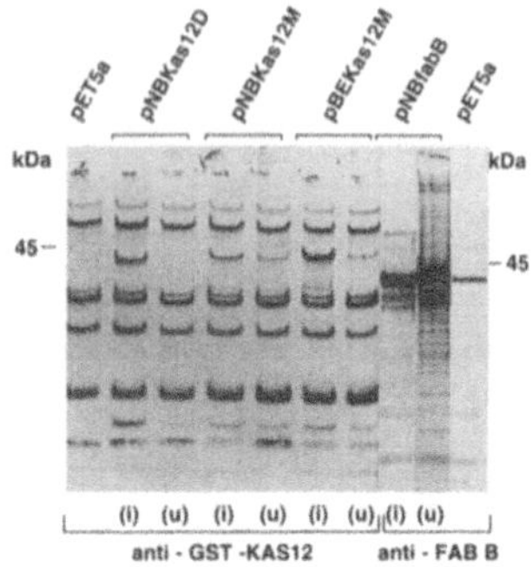

Figure 1 Western blots of total cellular proteins of BL21(DE3) containing KAS I expression constructs. Cells were grown in LB at 37°C to an OD_{600} of 0.6, and then induced for 3 hrs by addition of 0.4 mM IPTG. Aliquots were taken before **(u)** and after induction **(i)** (50μl and 30μl of culture respectively). Cells were harvested and proteins separated on a 12% SDS page gel. After transfer to a nitro-cellulose membrane, western blots were prepared using primary antibodies raised in rabbits. The antigens were a recombinant glutathione-S-transferase fusion with KAS 12 and purified FAB B.

To ascertain whether the recombinant proteins were active, soluble extracts from cells containing the constructs were fractionated by ion exchange chromatography. The fractions were analysed in a complementation assay involving a crude extract of *E. coli* that had been treated with 200 μM cerulenin to inhibit KASes I, II and IV. That is, fractions were added back to the treated extract, and monitored for restoration of fatty acid biosynthesis by measuring the incorporation of radioactively labelled malonyl-CoA into fatty acids.

Two peaks of activity (Figure 2) were found and attributed to FAB J (KAS IV) and FAB B (KAS I) on the basis of (i) the chain lengths of acyl-ACPs produced by the enzyme and (ii) comparison with a crude extract derived from cells with a mutation in *fabB* [1]. Quantification of activity indicated that only extracts derived from cells containing the *fabB* expression construct had elevated KAS I activity (6.2± 0.5 versus 3.5 ± 0.2 nmol/min/mg crude protein). No significant difference was found in the level of KAS IV activity in these two extracts (38.1 ± 4.1 versus 43.2 ± 5.5 nmol/min/mg crude protein). As this level of expression was insufficient for the studies envisaged with the recombinant proteins several avenues were investigated to optimise the recovery of active protein.

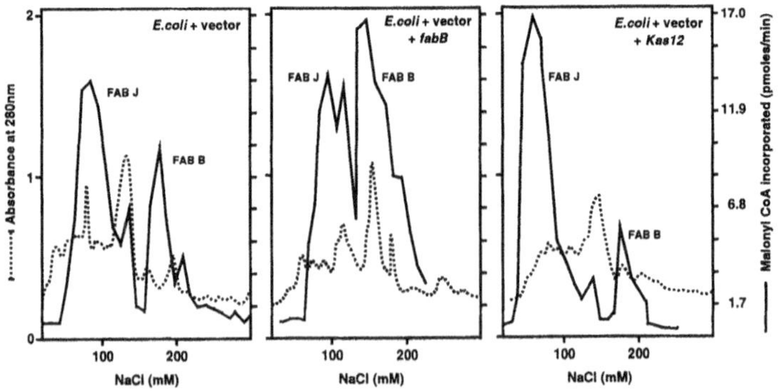

Figure 2 Ion exchange chromatography of crude soluble extracts of BL21(DE3) containing expression plasmids. *E. coli* cultures (100 ml) were grown and induced as described above. Cells were passed through a French pressure cell at 2 000 psi and debris removed by centrifugation (100 000 g for 1.5 hrs at 5°C). The soluble extracts were loaded onto a 5 x 50 mm Mono Q column (Pharmacia) which had been pre-equilibrated with 20 mM TrisHCl, pH 8.0 and 10 mM β-mercaptoethanol. Proteins were eluted by a linear NaCl gradient and assayed as described above.

Optimising expression

Tables 1 and 2 list some of the parameters explored with the objective of improving the yield of active recombinant protein.

Alteration	Conditions tested	Ref.
temperature	15°C - 37°C	[4]
medium	LB, TBYE, tryptone phosphate	[5]
medium additions	DMSO, thioglycerol	-
co-expression of *cis*-acting chaperonins	thioredoxin fusion partner (Thiofusion, Invitrogene)	[6]
co-expression of *trans*-acting chaperonins	GroES and GroEL	[7]

Table 1. Alterations in induction environment of expression strains containing pNB*Kas12*M and pNB*FabB*. Experiments were evaluated by light microscopy and cellular fractionation to ascertain extent of inclusion body formation.

Refolding protocol	Ref.
rapid dilution of denaturant	[8]
gradual dialysis of denaturant from protein	-
surfaces of mixed bed ion exchange resins	[9]
addition of PEG	[10]
addition of GroES and GroEL	[11]

Table 2. Protocols used in attempts to correctly refold KAS I. Inclusion bodies were isolated from induced *E. coli* cells containing expression constructs. Pellets were repeatedly washed in buffer containing 0.2% (w/v) Triton X-100 then in H_20. Protein was denatured in guanidine hydrochloride (2 M) except for the method of [9] in which the washed inclusion bodies were used directly. Results were assessed by enzyme assays or the appearance of soluble protein after removal of denaturant.

Conclusions

T7 expression systems can be used to generate large quantities of recombinant KAS I proteins. They can be easily purified for use in antibody production and for use in refolding studies. Only a small quantity of the recombinant FAB B protein was soluble and active. No elevation in activity was found in cells containing the *Kas12* gene. Manipulation of the growth conditions of expression strains, including co-expression with chaperonins did not increase soluble recombinant protein. Refolding of proteins from inclusion bodies was not possible even under conditions where native FAB B protein has been successfully refolded [8].

References

[1] Kauppinen S, Siggaard-Andersen M and von Wettstein-Knowles P. β-ketoacyl-ACP synthase I of *Escherichia coli*: Nucleotide sequence of the *fabB* gene and identification of the cerulenin binding residue. Carlsberg. Res. Commun. 1988;53:357-370.

[2] Siggaard-Andersen M, Kauppinen S and von Wettstein-Knowles P. Primary structure of a cerulenin binding β-ketoacyl-[acyl carrier protein] synthase from barley chloroplasts. Proc. Natl. Acad. Sci. USA 1991;88:4114-4118.

[3] Siggaard-Andersen M, Wissenbach M, Chuck J, Svendsen I, Olsen JG and von Wettstein-Knowles P. The *fabJ* encoded β-ketoacyl-[acyl carrier protein] synthase IV from *Escherichia coli* is sensitive to cerulenin and specific for short chain substrates. Proc. Natl. Acad. Sci. USA 1994; submitted.

[4] Schein CH and Noteborn MHM. Formation of soluble recombinant proteins in *Escherichia coli* is favored by lower growth temperature. Bio/Technology 1988;6:291-295.

[5] Moore JT, Uppal A, Maley F and Maley GF. Overcoming inclusion body formation in a high-level expression system. Protein Expression Purif. 1993;4:160-163.

[6] LaVallie ER, DiBasio EA, Kovacic S, Grant KL, Schendel, PF and McCoy JM. A thioredoxin gene fusion expression system that circumvents inclusion body formation in the *E. coli* cytoplasm. Bio/Technology 1993;11:187-194.

[7] Goloubinoff P, Gatenby AA and Lorimer GH. GroE heat-shock proteins promote assembly of foreign prokaryotic ribulose bisphosphate carboxylase oligomers in *Escherichia coli*. Nature 1989;337:44-47.

[8] Prescott DJ and Vagelos PR. Acyl carrier protein XIV. Further studies on β-ketoacyl acyl carrier protein synthetase from *Escherichia coli*. J Biol Chem 1970;245:5484-5490.

[9] Hoess A, Arthur AK, Wanner G and Fanning E. Recovery of soluble, biologically active recombinant proteins from total bacterial lysates using ion exchange resin. Bio/Technology 1988;6:1214-1217.

[10] Cleland JL, Builder SE, Swartz JR, Winkler M, Chang JY and Wang DIC. Polyethylene glycol enhanced protein refolding. Bio/Technology 1992;10:1013-1019.

[11] Peralta D, Hartman DJ, Hoogenraad NJ and Høj PB. Generation of a stable folding intermediate which can be rescued by the chaperonins GroEL and GroES. FEBS Lett. 1994;339:45-49.

TRANSGENIC EXPRESSION OF BACTERIAL FAS COMPONENTS IN RAPESEED

Ira I. G. S. Verwoert, Elizabeth C. Verbree, Karin H. van der Linden, H. John J. Nijkamp and Antoine R. Stuitje

Department of Genetics, Institute for Molecular Biological Sciences, BioCentrum Amsterdam, Vrije Universiteit,De Boelelan 1087,1081 HV Amsterdam, The Netherlands

Introduction

Fatty acid biosynthesis in plants is catalysed by the so-called FAS II fatty acid synthetase complex, which very much resembles the bacterial FAS system. The close evolutionary relationship between the plant and bacterial FAS system, as exemplified by structural and functional similarities, was unambiguously demonstrated by Kater *et al.* [1]. Gene-replacement studies showed that a single component of the plant FAS system (enoyl-ACP reductase) can functionally replace its bacterial counterpart. Recently, it was shown that the *E. coli fabD* gene, encoding malonylCoA-ACP transacylase (MCAT), under the control of a napin promoter, can be expressed at a high level and in a seed- and developmental specific manner, in transgenic plants [2]. Plastid targeting of the bacterial FAS component was successfully accomplished by the use of the *B. napus* enoyl-ACP reductase leader sequence [3]. In this study, data are presented on the transgenic expression of the 1) bacterial *fabH*, encoding 3-ketoacyl-ACP synthase III (KAS III), and 2) the bacterial *fabA*, encoding 3-hydroxy decanoyl thioester dehydrase (HDD), in rapeseed. Overexpression of HDD did not result in the production of bacterial-type fatty acids (e.g. *cis*-vaccenic acid) in transgenic plants. Overexpression of bacterial KAS III, however, resulted in increased relative amounts of C18:2 and C18:3, and a decrease in the C18:1 content of the storage lipid fraction. From these data it seems evident that bacterial FAS proteins can interact with the plant FAS system, and that the expression of heterologous FAS components in transgenic plants can contribute to elucidate the regulatory steps in plant lipid biosynthesis.

Results

CHIMERIC CONSTRUCTS.

For seed specific expression in plants, the bacterial KAS III and HDD genes were linked to the *B. napus* enoyl-ACP reductase sequence, encoding the plastid transit peptide, and a napin promoter, resulting in constructs pAR93 (KAS III) and pAR83 (HDD) (Fig. 1). The promoter-terminator cassette and transit peptide were successfully used in an earlier study to express the *E. coli fabD* gene, encoding MCAT, in transgenic plants, and to direct its gene product to the plastids [2].

J.-C. Kader and P. Mazliak (eds.), Plant Lipid Metabolism, 476–478.

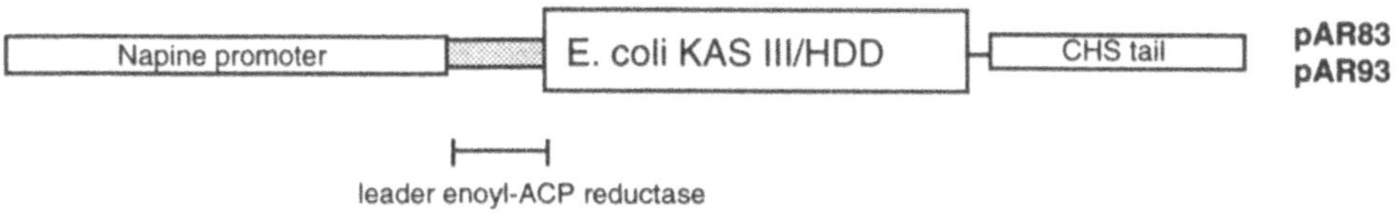

Figure 1. Chimeric constructs used in the transformation experiments

EXPRESSION OF THE CHIMERIC GENES DURING SEED DEVELOPMENT

The presence of the chimeric KAS III and HDD genes in transgenic rape (BN93 and BN83, respectively) was confirmed by PCR and Southernblot analysis (not shown). HDD accumulated, like the bacterial MCAT [2], to a high level in seed tissue (to approximately 4 % of the total soluble protein), and was correctly processed post-translationally (Fig. 2A). HDD enzymic activity in transgenic rape seeds, peaks around 5 weeks after anthesis (Fig. 2C).

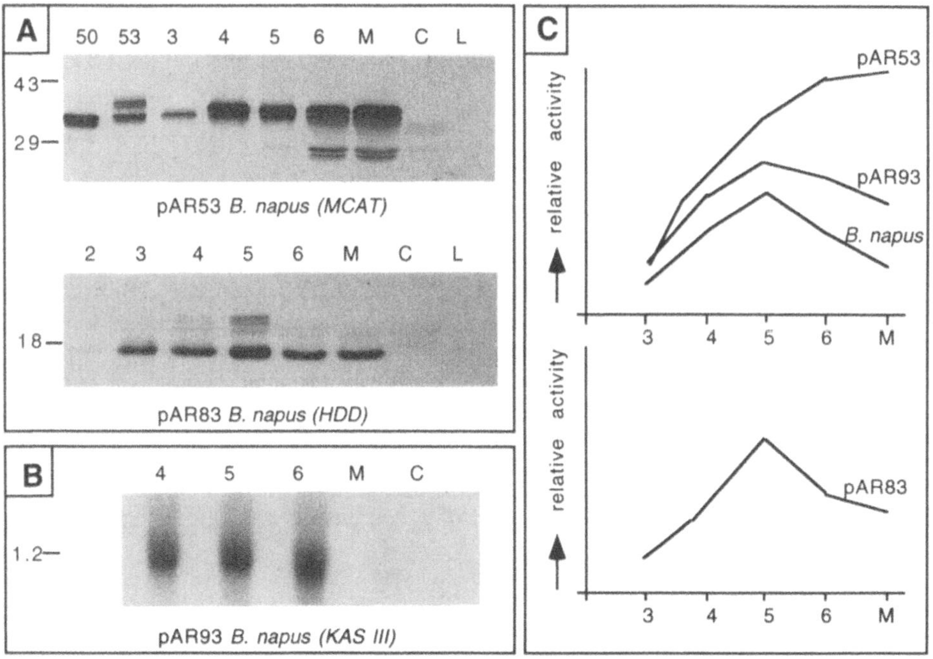

Figure 2. Analyses of transgenic seeds during development. A: Immunoblots with anti-E. coli MCAT and HDD. B: Northernblot analysis of steady state mRNA levels in a BN93 transformant, hybridized with an E. coli fabH probe. C: MCAT, KAS III and HDD relative temporal enzyme patterns, compared to control plants (Not on scale). Numbers indicate the developmental stage of the seeds (weeks after anthesis). M, mature seeds. C, control plant. L, leaf tissue. All these data are quantified per seed. 53 and 50 indicate constructs of the E. coli fabD gene, with and without the transitpeptide, expressed in E. coli.

This expression pattern is somewhat different from that of the *E. coli* MCAT, which increases during seed development and remains high in mature seeds (Fig. 2C). The bacterial KAS III protein was hardly detectable on Westernblots, although mRNA was clearly present (Fig. 2B), and a significantly enhanced enzyme activity was found (Fig. 2C). The highest expressor, BN93.15, showed KAS III levels 3.7 times the endogenous KAS III activity in mature seeds and 1.4 times that at the peak of fatty acid biosynthesis (5-6 weeks after anthesis).

FATTY ACID ANALYSIS OF TRANSGENIC SEEDS

Overexpression of bacterial MCAT [2] or HDD did not have a significant effect on the fatty acid composition of plant storage lipids. MCAT, apparently does not catalyse a controlling step in plant and bacterial fatty acid biosynthesis [2], and the plant FAS system is possibly

Table 1. Fatty acid composition of E. coli KAS III transformed rape seeds

	16:0	18:0	18:1	18:2	18:3
controls[a]	4.9±0.4	1.7±0.4	64.1±4.9	18.0±4.7	6.0±1.8
BN90 (-transit peptide)[a]	4.6±0.3	1.9±0.5	66.1±4.5	17.0±4.2	6.4±2.0
BN93 (+transit peptide)[b]					
BN93.14	5.3±0.3	1.8±0.3	53.5±2.4	26.5±2.5	6.6±1.2
BN93.15	5.1±0.4	1.3±0.3	54.1±2.3	26.5±2.4	8.0±1.5
BN93.50	4.6±0.4	1.2±0.4	53.8±1.8	26.0±3.1	10.8±1.6
BN93.60	5.0±0.2	1.2±0.3	52.5±2.1	27.5±2.3	10.0±1.7

[a]*The values are mol % ± SD (n = 15 for BN90 plants; n = 10 for controls)*
[b]*The values are the average (in mol % ± SD) of three independent analyses per plant*

unable to elongate omega-8 unsaturated acyl-ACP derivatives. Transgenic plants expressing a plastid-targeted bacterial KAS III gene, showed significantly increased relative amounts of C18:2 and C18:3, and decreased amounts of C18:1 in their storage lipid fraction. This could be the indirect result of a decrease in the rate of core fatty acid biosynthesis.

References

1. Kater MM, Koningstein GM, Nijkamp HJJ, Stuitje AR. The use of a hybrid genetic system to study the functional relationship between prokaryotic and plant multi-enzyme fatty acid synthetase complexes. Plant Mol Biol 1994; *in press.*
2. Verwoert IIGS, van der Linden KH, Nijkamp HJJ, Stuitje AR. Developmental specific expression and organelle targetting of the *Escherichia coli fabD* gene, encoding malonyl Coenzyme A-acyl carrier protein transacylase in transgenic rape and tobacco seeds. Plant Mol Biol 1994; *in press.*
3. Kater MM, Koningstein GM, Nijkamp HJJ, Stuitje AR. cDNA cloning and expression of *Brassica napus* enoyl-acyl carrier protein reductase in *Escherichia coli*. Plant Mol Biol 1991; 17: 895-909.

SYNTHESIS OF STARCH AND FATTY ACIDS BY PLASTIDS ISOLATED FROM DEVELOPING EMBRYOS OF OILSEED RAPE

Stephen Rawsthorne and Fan Kang
Department of Brassica and Oilseeds Research, John Innes Centre, Colney, Norwich NR4 7UH, U.K.

We have investigated which metabolites are utilized for starch and fatty acid synthesis by plastids isolated from oilseed rape embryos that are accumulating both oil and starch. A method was developed that was rapid and yielded intact, washed plastids with a very low contamination by the cytosol. Measurements of the recovery and latency of enzyme activities revealed that the plastids contained a glycolytic pathway, pyruvate dehydrogenase (PDC), NADP-malic enzyme, fructose-1,6-bisphosphatase (FBPase), and phosphoglucomutase. The presence of this range of enzymes suggested that several cytosolic metabolites could potentially be used for the synthesis of starch and/or fatty acids within the plastid. Glucose-6-phosphate (G-6-P), fructose, glucose and dihydroxyacetone phosphate (DHAP) were all incorporated into starch by the washed plastids, the former giving the highest rates. Pyruvate and G-6-P gave the highest rates of fatty acid synthesis. The rates of incorporation into fatty acids from DHAP, malate and acetate were much less than that from pyruvate.

Introduction

During the development of oilseed rape embryos starch accumulates transiently during the early stages of oil accumulation (P. da Silva and A.M. Smith, pers. comm.). Both starch and fatty acids are synthesized in the plastids and these processes are dependent upon the import of cytosolic metabolites in non-photosynthetic cells. Plastids have been isolated from several non-photosynthetic tissues and characterized with respect to their enzyme capacities and utilization of exogenous substrates to support starch and fatty acid synthesis *in vivo*. These studies have revealed that the glycolytic pathway is present although the activities of some of the enzymes in the lower half of the pathway can be low. Pyruvate dehydrogenase is also present in these plastids.

Experiments to determine which exogenous metabolites are imported by plastids and utilized for starch synthesis have shown that either G-6-P and glucose-1-phosphate (G-1-P) are the principal substrates depending upon the tissue from which the organelles are extracted [1,2]. Similar experiments have revealed that a much greater range of metabolites are utilized for fatty acid synthesis [3,4]. In this study we have investigated the enzymic capacities of plastids isolated from developing oilseed rape embryos and have then examined which metabolites support starch and fatty acid synthesis by them *in vitro*.

Materials and Methods

Oilseed rape embryos (average fresh weight = 2.5 mg) were harvested, removed from their

J.-C. Kader and P. Mazliak (eds.), Plant Lipid Metabolism, 479–481.

testas and homogenized (500 embryos in 25 ml) in plastid isolation medium (PIM: [medium A; 5]) using a Polytron. The homogenate was filtered with Miracloth. The plastids were pelleted by centrifugation at 750 g for 5 min, resuspended to wash them, and repelleted before resuspension in 1.5 ml PIM to give a washed plastid fraction.

Enzyme activities were determined using established methods from the literature and were optimized to give the maximum rate. The recovery of enzyme activity in the washed plastids was determined as a percentage of that in the initial homogenate. The latency of enzyme activity was determined by assaying the enzyme activities using intact plastids under iso-osmotic conditions, and using plastids that had been osmotically ruptured.

Isolated plastids were incubated with ^{14}C-labelled metabolites in PIM. The incorporation of ^{14}C into methanol/KCl insoluble material was used to measure starch synthesis. For fatty acids the incubations were extracted with methanol/chloroform after saponification and the ^{14}C present in the organic fraction was determined.

Results and Discussion

The yield of plastids from the embryos was on average 10% as judged by the recovery of the plastidial marker enzymes NADP-glyceraldehyde-3-phosphate dehydrogenase (NADP-GAPdH) and ADPglucose pyrophosphorylase (ADPGppase). The contamination of the plastids by the cytosol was low as judged by the low recovery (0.2%) of the cytosolic marker enzymes. The plastids were on average 65% intact as judged by the latency of activity of ADPGppase and NADP-GAPdH. The recovery and latency of enzyme activity measurements revealed that all of the glycolytic enzymes were present in the plastids, as were NADP-malic enzyme, PDC and FBPase. The presence of plastidial FBPase was confirmed using a monospecific antibody to probe western blots of proteins from the washed plastids. Using plastid preparations with increasing levels of contamination by mitochondria it was possible to determine that approximately 65% of the total cellular PDC activity was plastidial. Clearly, the presence of this enzyme complement in the plastids suggested that a large range of metabolites could potentially be utilized for synthesis of starch and or fatty acids. However, which of this range could cross the membrane of these oilseed rape plastids required further study with radiolabelled metabolites.

Figure 1 summarizes the results of experiments in which intact, washed plastids were supplied with ^{14}C-labelled metabolites and the incorporation of ^{14}C into starch or fatty acids determined. In all these experiments the rates of incorporation were linear during the 60 min incubation and incorporation was dependent upon the intactness of the organelles. A major finding was that G-6-P was not only the most effective substrate for starch synthesis but was also capable of supporting high rates of fatty acid synthesis. The utilization of G-6-P for starch synthesis was also reported for plastids from developing pea embryos [1]. Glucose-1-phosphate was not incorporated into starch by the intact rapeseed plastids. Hexoses and DHAP were also incorporated into starch. The later observation is consistent with the presence of FBPase in the rapeseed plastids. Plastidial FBPase is generally absent from non-photosynthetic plastids isolated from various plant tissues [6]. The incorporation of G-6-P into fatty acids reveals that the glycolytic pathway is functional and this is the first report of a hexose phosphate supporting high rates of fatty acid synthesis. Pyruvate was also incorporated into fatty acids at a high rate while the rates from DHAP, malate and acetate were somewhat lower.

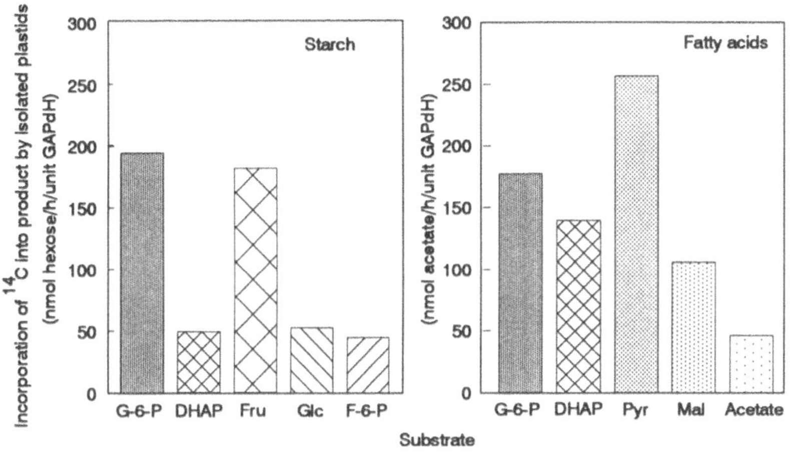

Figure 1. Net incorporation of ^{14}C-labelled metabolites into starch and fatty acids by plastids isolated from developing embryos of oilseed rape. Metabolites were supplied at 1 mM for fatty acid synthesis and 2 mM for starch synthesis. Incubations included 2 mM ATP for starch synthesis and 4 mM ATP, 7 mM $MgCl_2$, 10 mM $NaHCO_3$, 0.5 mM NADPH and NAD, 0.2 mM CoA, and 0.2 mM thiamine pyrophosphate.

Our data show that several cytosolic metabolites could be utilized by the plastids for the synthesis of starch and/or fatty acids. The presence of specific translocators for some of these metabolites is also implied by the intactness-dependence of their incorporation. Which of the metabolites are used *in vivo* during embryo development is likely to be determined by the concentrations of the metabolites, those of other metabolites which compete for the same translocators, and the fine and coarse control of the enzymes in the plastids which convert them into the immediate precursors for starch and fatty acid synthesis.

References

1. Hill LM, Smith AM. Evidence that glucose 6-phosphate is imported as the substrate for starch synthesis by the plastids of developing pea embryos. Planta 1991; 185:91-96.
2. Tyson RH, ap Rees T. Starch synthesis by isolated amyloplasts from wheat endosperm. Planta 1988; 175:33-38.
3. Smith RG, Gauthier DA, Dennis DT, Turpin DH. Malate- and pyruvate-dependent fatty acid synthesis in leucoplasts from developing castor endosperm. Plant Physiol 1992; 98:1233-1238.
4. Masterson C, Wood C, Thomas DR. L-acetylcarnitine, a substrate for chloroplast fatty acid synthesis. Plant Cell and Environ 1990; 13:755-765.
5. Kang F, Ridout CJ, Morgan CL, Rawsthorne S. The activity of acetyl-CoA carboxylase is not correlated with the rate of lipid synthesis during development of oilseed rape (*Brassica napus* L.) embryos. Planta 1994; 193:320-325.
6. Entwistle G, ap Rees T. Lack of fructose-1,6-bisphosphatase in a range of higher plants that store starch. Biochem J 1990; 271:467-472.

IDENTIFICATION OF A LOW DENSITY MEMBRANE FRACTION INVOLVED IN STORAGE TRIACYLGLYCEROL SYNTHESIS

Dominic J. Lacey and Matthew J. Hills

Department of Brassica and Oilseeds Research, John Innes Centre, Colney, Norwich NR4 7UH, U.K.

Introduction

It has recently been reported that fatty acid elongase in developing seeds of *Lunaria annua* and *Brassica napus* was located in two different fractions of organelles separated by sucrose gradient centrifugation [1].The denser of these fractions was found to be endoplasmic reticulum (E.R.) membrane whereas the nature of the less dense fraction was unclear. Storage triacylglycerol (TAG) biosynthesis in developing seeds of *Brassica napus* is now being characterised using the same system. The enzymes involved in TAG synthesis were found in two distinct fractions with equilibrium densities of 1.05 and 1.10. The more dense membranes contained the E.R. marker enzymes CDP-choline: DAG cholinephosphotransferase and *lyso*-phosphatidylcholine acyltransferase (LPC-AT). The less dense membranes did not contain choline phosphotransferase activity and contained only trace levels of LPC-AT. The triacylglycerol/phospholipid (TAG/PL) ratio of the vesicles in the light fraction was 2.9 compared to 0.18 in the E.R.. The fatty acid composition of these two lipid classes were similar in both fractions.

In conclusion, although the main site of storage and membrane lipid synthesis in *Brassica napus* is the E.R., TAG is also synthesised by light vesicles, the nature of which have yet to be fully elucidated.

Materials and Methods

Brassica napus L. cv Topas seeds were harvested at approximately 30 DAF, the embryos were excised and homogenised in grinding buffer [1]. The homogenate was centrifuged at 1000g to remove large cell debris and the resulting supernatant was filtered through two layers of miracloth, before layering onto 10-48% sucrose gradients and centrifuging at 100,000g for 3 hours. The gradients were then fractionated for further analysis. Activities of enzymes of the Kennedy pathway were assayed by measuring the incorporation of [^{14}C]oleoyl-CoA (100μM) or [^{14}C]glycerol-3-phosphate (G-3-P), (500μM) into lipid by incubating samples for 30 minutes at 30^{0}C, extracting the products and separating them by thin layer chromatography [2]. CDP-choline DAG: cholinephosphotransferase was assayed by the method of reference [3]. Methyl ester derivatives of lipids were prepared after purification of lipid classes by preparative TLC and separated on a Perkin Elmer 8310 gas chromatograph using a 2m packed column with 10% CP

J.-C. Kader and P. Mazliak (eds.), Plant Lipid Metabolism, 482–484.

SIL-58 on chromosorb WHP 100-120 mesh.

Results

Incorporation of radiolabelled oleoyl CoA and G-3-P into TAG occurred in two distinct regions with peak equilibrium densities of 1.05 and 1.10. The more dense peak corresponded to the LPC-AT activity, an E.R. marker when [^{14}C] oleoyl CoA was the labelled substrate (figure 1). When radiolabelled G-3-P was used, the same pattern of incorporation into TAG was observed (figure 2), showing the light fraction contains all the enzymes of the Kennedy pathway.

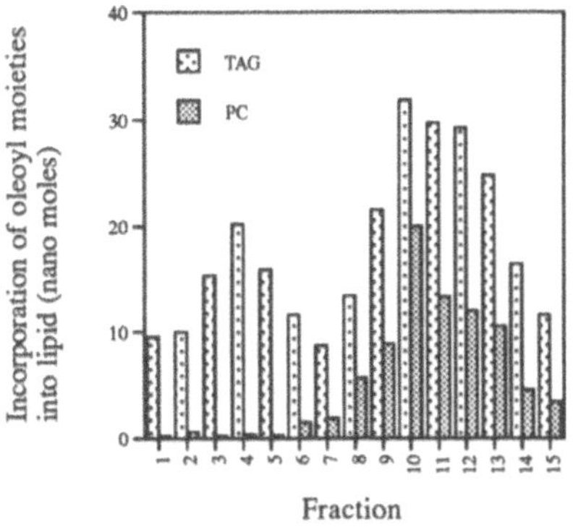

Figure 1, Incorporation of oleoyl moieties into TAG and PC in the top 15 fractions (nano moles).

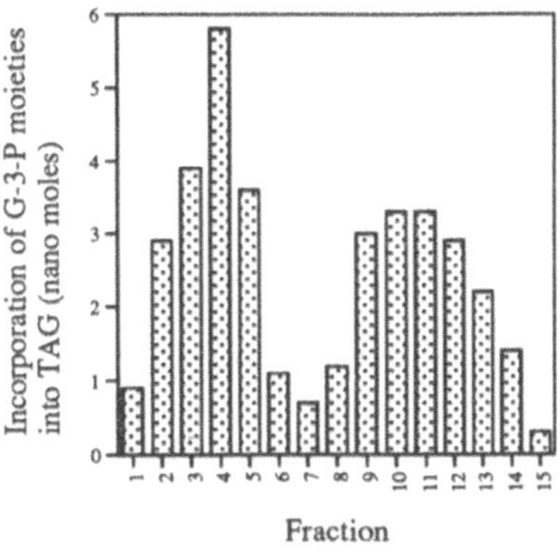

Figure 2, Incorporation of G-3-P into TAG in the top 15 fractions (nano moles).

The activity of the E.R. marker enzyme, CDP-choline: DAG transferase for the top 15 fractions of a gradient is shown in figure 3. CDP-choline: DAG transferase activity like LPC-AT (figure 2) copurifies with the enzymes of TAG synthesis in the dense fraction (9-13),but is almost undetectable in the light vesicle fraction (3-5). Marker enzymes for the plastids, cytosol, mitochondria and chlorophyll were also assayed (data not shown), however these organelles were found not to be associated with TAG synthesis.

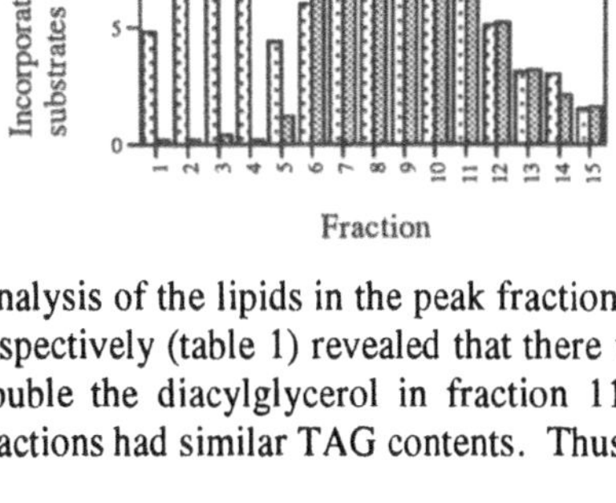

Figure 3, Incorporation of radiolabelled oleoyl-CoA and CDP-choline into TAG and PC respectively in the top 15 fractions (%).

Analysis of the lipids in the peak fractions, 4 and 11 with equilibrium densities of 1.05 and 1.10 respectively (table 1) revealed that there was approximately ten times as much phospholipid and double the diacylglycerol in fraction 11 compared to fraction 4, however within error the fractions had similar TAG contents. Thus the TAG/PL ratio was 16 fold higher in fraction 4 than in fraction 11, the TAG/PL ratio in the light fraction is intermediate between the E.R. and that of oil bodies (oil body TAG/PL ratio is approximately 40). Fatty acid compositions of the lipids in the two fractions were similar (data not shown).

Table 1, Content of TAG, DAG and PL in fractions of density 1.05 and 1.10 (μg +/-SE).

Lipid classes	Lipid in fraction 4 (µg +/- SE)	Lipid in fraction 11 (µg +/- SE)
TAG	17.0 +/- 4.0	13.2 +/- 4.3
DAG	4.8 +/- 2.2	11.0 +/- 1.3
Total PL	5.8 +/- 2.7	73.1 +/- 4.3
TAG/PL	2.93	0.18

Discussion

Sucrose density gradient separation of homogenates made from developing seeds of rape reveal the main site of storage and membrane lipid synthesis to be the E.R., however fractions near to the top of the gradient were also capable of TAG synthesis.This light fraction however, did not contain detectable levels of LPC-AT or CDP-choline: DAG cholinephosphotransferase. The TAG:PL ratio of the light membranes was approximately sixteen times higher than the microsomal fraction, which accounted for their lower equilibrium density. The smaller amount of PL also suggested the less dense fraction contained fewer vesicles than the microsomal fraction.

It is not yet clear what the light vesicles represent or what role they play in storage oil synthesis and deposition. It is possible that the microsomal membranes are compartmentalised for oil and membrane lipid synthesis, containing regions where TAG has accumulated within the phospholipid bilayer [4]. Upon homogenisation of the embryos these areas become detached from E.R. and, due to the decreased buoyant density due to accumulated TAG in the membrane, separate from the E.R. when subjected to sucrose gradient centrifugation. Another possibility is that the light particles are transport vesicles, which are involved in the transfer of TAG from the E.R. to the developing oil bodies. The vesicles may be derived from the E.R. in a process reminiscent of transport within the E.R./ Golgi system. Further work aims to elucidate the nature of the light vesicles.

References

1. Whitfield HV, Murphy DJ, Hills MJ. Sub-cellular localisation of fatty acid elongase in developing seeds of *Lunaria annua* and *Brassica napus*. Phytochemistry, 1993;32:255-258.
2. Stobart AK, Stymne S. TAG biosynthesis. In: Harwood JL, Bowyer JR, editors. Lipids, membranes and aspects of photobiology. London: Academic press, 1990;4:19-46.
3. Lord JM, Kagawa T, Moore TS, Beevers H. E.R. as the site of lecithin formation in castor bean endosperm. J Cell Biology, 1973;57:659-667.
4. Wanner G, Theimer RR. Membrane appendices of Spherosomes (Oleosomes). Planta, 1978;140:163-169.

MOLECULAR CHARACTERIZATION OF ACONITASE IN ETIOLATED PUMPKIN COTYLEDONS.

Makoto Hayashi[1], Luigi De Bellis[2] and Mikio Nishimura[1]

[1]Department of Cell Biology, National Institute for Basic Biology, Okazaki 444, Japan; [2]Dipartimento di Biologia delle Piante Agrarie, Sez. Fisiologia Vegetale, via Mariscoglio 34, I-56124 Pisa, Italy

Aconitase (EC 4.2.1.3) is an enzyme that catalyzes the reaction of reversible conversion between citrate and isocitrate. In cotyledons of oil-storage plants such as pumpkin, aconitase was thought to exist not only in mitochondria as a member of the tricarboxylic acid cycle but also in glyoxysomes as a member of glyoxylate cycle. The glyoxylate cycle is an important pathway for producing sucrose from reserved lipid at early stage of seedling growth under the cooperation of other metabolic pathways such as β-oxidation.

We reported that extremely low aconitase activity was detected in pumpkin glyoxysomal fraction after sucrose gradient centrifugation in spite that aconitase for glyoxylate cycle has been believed to localize in glyoxysomes [1]. Recent immunoblot analysis revealed that isolated pumpkin glyoxysomes as well as caster bean glyoxysomes do not contain aconitase [2, 3]. However, it is still possible that aconitase is released from glyoxysomes during organelle preparation.

In the present study, we isolated and characterized pumpkin cytosolic aconitase cDNA. Fluorescent microscopic observation of the etiolated pumpkin cotyledons using an antibody raised against the aconitase expressed in *E. coli* clearly showed that no glyoxysomal aconitase exists in etiolated pumpkin cotyledons.

Isolation and Characterization of Pumpkin Aconitase cDNA

A λgt11 cDNA library prepared from RNA of 4 day-old etiolated pumpkin cotyledons was screened using an affinity purified antibody against pumpkin aconitase [2]. The aconitase cDNA of 3154 bp was obtained by the combination of immunoscreening and plaque hybridization technique. The aconitase cDNA contains an open reading frame of 2694 bp capable of encoding a 98 kD protein.

Because the molecular mass of the protein encoded by the cDNA well coincides with that of purified aconitase [2], the aconitase is seemed to be translated without N-terminal extrapeptide that is equivalent to the

J.-C. Kader and P. Mazliak (eds.), Plant Lipid Metabolism, 485–487.

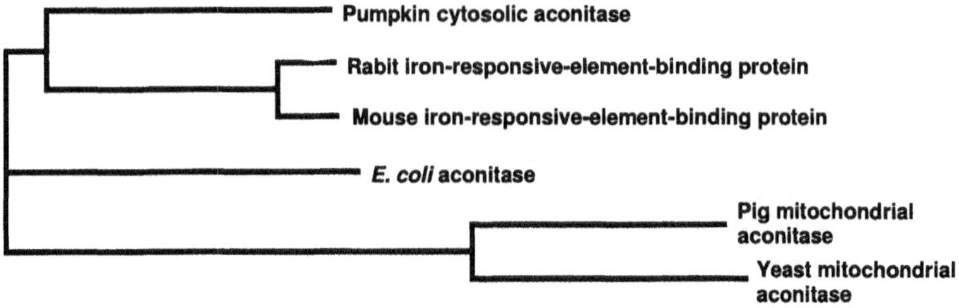

Figure 1. Phylogenic tree of pumpkin cytosolic aconitase and its related proteins.

leader sequence found in mitochondrial proteins and some glyoxysomal proteins. It has been proposed that some glyoxysomal enzymes such as malate synthase and isocitrate lyase contain unique C-terminal tripeptide motifs that are important for targeting into glyoxysomes. However, we could not found any of these C-terminal tripeptide motifs within the aconitase.

The result of homology search for protein databases revealed that the pumpkin aconitase is similar to mammalian iron-responsive-element-binding proteins (IRE-BPs), prokaryotic aconitases and mammalian mitochondrial aconitases. Figure 1 shows the molecular evolutional relationship among these proteins by a phylogenic tree. The result indicated that strongest relationship between the pumpkin aconitase and mammalian IRE-BPs. IRE-BP is reported to be a cytosolic protein that regulates the translation of some iron-related protein such as ferritin. Recent extensive studies revealed that IRE-BP is identical with cytosolic aconitase. High similarity of the pumpkin aconitase with mammalian cytosolic aconitase (IRE-BP) and the lack of transport signal within the deduced amino acid sequence strongly suggests that the pumpkin aconitase cDNA we obtained encodes cytosolic aconitase.

Intracellular Localization of Aconitase

We prepared a specific antibody against 168-330 amino acid region of the pumpkin aconitase using the fusion protein expressed in *E. coli*. Immunoblot analysis was performed using the antibody after subcellular fractionation by sucrose density gradient centrifugation. Almost 80% of aconitase was detected in supernatant fraction. Aconitase is also detected in mitochondrial fractions, although no aconitase is detected in glyoxysomal fractions.

In order to analyze the localization of the aconitase without organelle preparation, we used immunofluorescent microscopic observation. When the cells isolated from etiolated pumpkin cotyledons were treated with the antibody, weak and small fluorescent signals were observed on the weak fluorescent background throughout whole cells as shown in figure 2. This indicates that aconitase localized in an organelle that represent the small

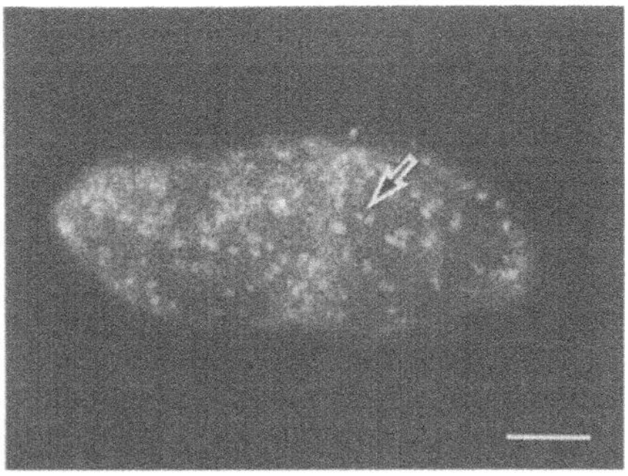

Figure 2. Immunofluorescent photomicrographs of pumpkin cells.
Cell isolated from pumpkin cotyledons was stained with anti-aconitase antibody and visualized with FITC-conjugated secondary antibody. Arrow shows one of the small fluorescent spots. Bar indicates 10μm.

fluorescent signal as well as in cytosol.

Thin sections of etiolated pumpkin cotyledons were also analyzed using immunofluorescent microscopic observation. Similar small fluorescent signals approximately 0.5 mm in diameter were observed when the thin sections were stained with anti-aconitase antibody and with anti-mitochondrial succinate dehydrogenase antibody. Moreover, double immunofluorescent staining using anti-aconitase antibody and anti-glyoxysomal isocitrate lyase antibody clearly demonstrated that both enzymes localize in different organelles, namely the former in mitochondria and the latter in glyoxysomes, and no aconitase is present in glyoxysomes.

Developmental Change of Aconitase after Germination

The enzymes of the glyoxylate cycle such as malate synthase and isocitrate lyase are specifically expressed during a short period after germination. This is because that sucrose produced from stored oil by the glyoxylate cycle is necessary for initial seedling growth. We compared the developmental change in amounts of aconitase and isocitrate lyase after germination. The result indicated that the expression pattern of the aconitase is similar to that of the isocitrate lyase. Our recent analysis for aconitase isoforms revealed that one of the cytosolic aconitase isozymes shows the specific expression pattern [4]. Overall data indicated that aconitase is absent in glyoxysomes and the cytosolic aconitase is involved in the glyoxylate cycle.

References

1. De Bellis L, Tsugeki R, Alpi A. Nishimura M. Purification and characterization of aconitase isoforms from etiolated pumpkin cotyledons. Physiol Plant 1993;88:485-492.
2. De Bellis L, Hayashi M, Biagi PP, Hara-Nishimura I, Alpi A. Nishimura M. Immunological analysis of aconitase in pumpkin cotyledons: the absence of aconitase in glyoxysomes. Physiol Plant 1994;90:757-762.
3. Courtois-Verniquet F. Douce R. Lack of aconitase in glyoxysomes and peroxisomes. Biochem J 1993; 294:103-107.
4. De Bellis L, Hayashi M, Nishimura M. Alpi A. Subcellular and developmental changes in distribution of aconitase isoforms in pumpkin cotyledons. Planta in press.

Effects of the constitutive expression of a medium chain acyl-ACP thioesterase from California Bay on leaf lipid composition of transformed *Brassica* plants

Victoria Eccleston and John Ohlrogge
Michigan State University, Department of Botany and Plant Pathology,
East Lansing, Michigan 48824-1312, U.S.A.

SYNOPSIS

Constitutive expression of California bay medium chain acyl-ACP thioesterase in *Brassica napus* resulted in medium chain acyl-ACP thioesterase activity in transformed leaf tissue without alteration of leaf fatty acid composition. [^{14}C] Labelling studies of intact leaf discs, chloroplasts and protoplasts were used to evaluate if medium chain acyl moieties are produced in transformed leaves. The possible metabolism of medium chain fatty acids in leaf via β-oxidation is discussed.

INTRODUCTION

Medium chain fatty acids (C_8-C_{14}) are unusual fatty acids found in the seed oils of several plant species such as *Cuphea* and California bay [1]. Medium chain fatty acids are produced by premature chain termination of fatty acid synthesis, mediated by a medium chain acyl-ACP thioesterase (MCTE) [2,3]. Since medium chain fatty acids are a valuable renewable resource, one goal of researchers has been to develop a medium chain-producing annual crop plant using genetic transformation [4]. Workers at Calgene transformed *Brassica napus* with California bay medium chain acyl-ACP thioesterase cDNA, testing the ability of a seed specific napin promoter or the constitutive CaMV 35S promoter to alter seed oil fatty acid composition. We have used these transformants to evaluate possible roles for acyl-ACP chain length termination in the control of fatty acid synthesis.

RESULTS AND DISCUSSION

The fatty acid composition of mature seed (with seed coat removed) from either napin-MCTE or 35S-MCTE transformed *Brassica* was determined using GC analysis. Wild type *Brassica* seed contains approximately 60 mol% oleate (18:1) and 30 mol% linoleate (18:2) whereas seed from plants transformed with the napin-MCTE construct produced approximately 50 mol% laurate (12:0), 20 mol% 18:1 and 10 mol% 18:2. A less striking difference was seen in 35S-MCTE transformants which produced a maximum of 7 mol% 12:0 in seeds. The 35S-MCTE transformation event had no effect on leaf fatty acid profile nor did it significantly alter root fatty acid composition.

J.-C. Kader and P. Mazliak (eds.), Plant Lipid Metabolism, 488–490.

Therefore, constitutive expression of the California bay acyl-ACP medium chain thioesterase resulted in altered phenotype of seeds only. To determine whether the Bay thioesterase was functional, leaf homogenates were assayed for MCTE activity. Little MCTE activity was detected in wild type leaves but 35S-MCTE transformant leaves exhibited specific activities of 60-100pmol/min/mg protein. Could this amount of MCTE activity account for an alteration in fatty acid composition? In 35S-MCTE transformed seed, MCTE activity, although significantly higher than wild type activity, was measured at a maximum rate of only 20pmol/min/mg protein. In the napin-MCTE seed, MCTE activity was determined at 100-250pmol/min/mg protein. Therefore, MCTE specific activity at an equivalent rate to that measured in leaf tissue had increased the seed 12:0 concentration by at least 40mol%. This suggests that there is a significant difference in the way that leaves and seeds process any excess 12:0 produced by the introduction of MCTE activity, and this difference may be involved in the regulation of fatty acid synthesis.

The rate of MCTE activity was also compared to that of oleoyl-ACP thioesterase (OTE) activity. Seed OTE activity was fairly constant in all transformants and the wild type, with a specific activity of approximately 70pmol/min/mg protein. However, leaf OTE activity in the 35S-MCTE transformants was slightly increased, at approximately 100pmol/min/mg protein, compared to leaf wild type OTE activity at approximately 65pmol/min/mg protein. This may indicate that the total rate of fatty acid synthesis in 35S-MCTE transformed leaves is increased by the introduction of MCTE activity. Since a similar increase in the rate of OTE activity was not observed in the 35S-MCTE transformed seeds, any proposed increase in leaf fatty acid synthesis does not appear to have any effect on the total amount of seed oil deposition.

Chloroplasts were incubated with [^{14}C]acetate to isolate MCTE activity from β-oxidation. Radiolabelled fatty acids were extracted and separated using successive argentation thin layer chromatography (TLC) and C_{18}-reverse phase TLC. The *Brassica* chloroplasts did not synthesise fatty acids well; whereas control spinach chloroplasts incorporated [^{14}C]acetate into fatty acids at a rate of 100-300nmol/hour/mg chlorophyll, the *Brassica* chloroplasts had a specific activity of only approximately 20nmol/hour/mg chlorophyll.

Since *Brassica* chloroplasts did not uptake [^{14}C]acetate at a sufficient rate to enable the analysis of radiolabelled fatty acids, we repeated our experiments using isolated protoplasts. [^{14}C]Acetate was incorporated into 16:0 initially and then into 18:1 with increasing incubation time, but no radiolabelled medium chain fatty acids were detected. When leaf discs were incubated with [^{14}C]acetate there was no incorporation into medium chains. The majority of radiolabel (>60% of total incorporation) was found in 16:0 and 18:1, although 35S-MCTE leaf discs incorporated 1.5- to 2-fold more radioactivity into 16:0 with a concomitant decrease in radiolabelled 18:1 compared to wild type leaf discs.

To investigate a possible route of degradation of excess 12:0 in the 35S-MCTE transformants *in vivo*, leaf discs were incubated with [^{14}C]12:0 and then subjected to Bligh and Dyer extraction. The distribution of radioactivity between the lipid phase, aqueous phase and a carbon dioxide trap which was included in the incubation vessel was determined. Radioactive incorporation into the leaf tissue was extremely low, the majority of which was present in the lipid phase as 18:2. No difference was seen between wild type and 35S-MCTE leaf discs and the 18:2 radiolabelling was probably a result of acyl-ACP track fatty acid elongation and desaturation. Little radioactivity was found in the aqueous phase or carbon dioxide trap.

We have not been able to detect any accumulation of medium chain fatty acids in 35S-MCTE transformed *Brassica* leaves, even though leaf MCTE activity *in vitro* was comparable to OTE activity. There are two possible explanations for this; either the medium chain fatty acids do not accumulate because they are metabolised immediately, most likely via β-oxidation. MCTE transformation of *E.coli* only results in accumulation of medium chain fatty acids in mutants of *E.coli* that cannot β-oxidise. Alternatively, *in vitro* measurement of MCTE activity may not be a true reflection of leaf MCTE activity *in vivo*. In the chloroplast acyl-ACP substrates may not be accessible to MCTE, a protein that is usually associated with plastid fatty acid synthesis.

Measurement of carbon flux through pools of free CoA-SH and medium to long chain acyl-CoAs in leaf should provide valuable information about the potential differences in chloroplast and plastid fatty acid metabolism. Fatty acid breakdown in leaf is not required to any great extent until senescence, when the glyoxylate cycle uses β-oxidised fatty acid to produce 4-carbon acids which are converted to carbohydrates[5]. Currently, we are using enzyme assays, Western blotting and RNA blot analysis to determine if the glyoxylate pathway is functional in 35S-MCTE transformed *Brassica.*

ACKNOWLEDGEMENTS

We thank Toni Voelker, Maelor Davies and Vic Knauf of Calgene for providing seed material transformed with California bay acyl-ACP medium chain thioesterase.

REFERENCES

1. Davies HM. Medum chain acyl-ACP hydrolysis activities of developing oilseeds. Phytochemistry 1993; 33: 1353-1356.
2. Pollard MR, Anderson L, Fan C, Hawkins DJ, Davies HM. A specific acyl-ACP thioesterase implicated in medium-chain fatty acid production in immature cotyledons of *Umbellularia californica.* Arch. Biochem. Biophys. 1991; 284: 306-312.
3. Davies HM, Anderson L, Fan C, Hawkins DJ. Developmental induction, purification, and further characterisation of 12:0-ACP thioesterase from immature cotyledons of *Umbellularia californica.* Arch. Biochem. Biophys. 1991; 290: 37-45.
4. Voelker TA *et al.* Fatty acid biosynthesis redirected to medium chains in transgenic oilseed plants. Science. 1992; 257: 72-74.
5. Graham IA, Leaver CJ and Smith SM. Induction of malate synthase gene expression in senescent and detached organs of cucumber. Plant Cell. 1992; 4: 349-357.

CLONING AND EXPRESSION IN *ESCHERICHIA COLI* OF A NOVEL THIOESTERASE FROM *ARABIDOPSIS THALIANA* SPECIFIC FOR LONG CHAIN ACYL-ACYL CARRIER PROTEINS

Peter Dörmann[1], Toni A. Voelker[2], and John B. Ohlrogge[1]
[1]Department of Botany and Plant Pathology, Michigan State University, East Lansing, MI 48824-1312; [2]Calgene Inc., 1920 Fifth St., Davis, CA 95616

INTRODUCTION

In most plants the major product of fatty acid biosynthesis in plastids is oleic acid and a highly active oleoyl-ACP thioesterase has been described [1, 2]. In the seeds of California Bay (*Umbellularia californica*) however, a lauroyl-ACP thioesterase has been found that is involved in the production of 12:0 [3]. Recently, a partial cDNA clone (YAP140) from *Arabidopsis thaliana* siliques was isolated [4]. This clone showed a high sequence identity with the California Bay thioesterase cDNA but lower similarity with the safflower oleoyl-ACP thioesterase cDNA [2]. *A. thaliana* seeds show a "normal" fatty acid pattern (6.9 mol % 16:0; 3.1 % 18:0; 14.4 % 18:1; 26.9 % 18:2, 21.2 % 18:3, 2.2 % 20:0, 20.6 % 20:1) and they contain only trace amounts of 12:0 [3]. In addition, the thioesterase from extracts of *A. thaliana* seeds is specific for oleoyl-ACP [3]. In this study we address the question of whether the novel *A. thaliana* cDNA encodes a functional thioesterase and we examine its substrate specificity and expression in different *A. thaliana* tissues and in developing *Brassica napus* seeds.

RESULTS

Using the probe YAP140 [3], a cDNA (TE 3-2) from an *A. thaliana* library was purified. The deduced amino acid was 51 % identical to the lauroyl-ACP thioesterase from California Bay [3] and 39 % identical to the safflower oleoyl-ACP thioesterase [2]. The 1814 bp TE 3-2 cDNA contained an open reading frame from bp 226 to 1551 with the first in frame ATG at bp 316 (Figure 1). This allowed the translation of a 412 amino acid polypeptide. By screening the same library with an oleoyl-ACP thioesterase cDNA probe from coriander [5] a cDNA was found (TE 1-7) that was 66 % identical to the safflower oleoyl-ACP thioesterase and therefore most likely encodes the corresponding enzyme from *A. thaliana*.

In order to determine the enzyme activity of the TE 3-2 gene product, the cDNA was subcloned into an expression vector and transformed into *Escherichia coli*. The recombinant protein was highly toxic to the bacteria. The soluble protein was isolated from the cells and used for thioesterase assays with different acyl-ACPs [1]. The recombinant TE 3-2 protein showed thioesterase activity with 14:0-ACP, 16:0-ACP, 18:0-ACP and $18{:}1^{\Delta 9cis}$-ACP, the best substrate was 16:0-ACP (Figure 2).

J.-C. Kader and P. Mazliak (eds.), Plant Lipid Metabolism, 491–494.

```
3-2        MVATSATSSFFPVPSSSLDPNGKGNKIGSTNLAGLNSTPNSGRMKVKPNAQAPPK

1-7 MLKLSCNVTDSKLQRSLLFFSHSYRSDPVNFIRRRIVSCSQTKKTGLVPLRAVVSADQGSV

3-2 INGKKVGLPG:VDIVRTDTETS:HPAPRTFINQLPDW:MLLAAI:TIFLAAEKQWMMLDWK

1-7 VQGLATLADQLRLGSLTEDGLSYKEKFVVRSYEVGSNKTATVETIANLLQEVGCNHAQSVG
3-2 PRRSDM:V:PFGI:RIVQ:::VFRQN:SI::::I:ADRS:SI::VM:H:::TAL::VKTA:

1-7 FSTDGFATTTTMRKLHLIWVTARMHIEIYKYPAWGDVVEIETWCQSEGRIGTRRDWILKDS
3-2 LLG:::GS:PE:F:KN::::VT::QVVVD:::T::::::VD::VSQS:KN:M::::LVR:C

1-7 VTGEVTGRATSKWVMMNQDTRRLQKVSDDVRDEYLVFCPQEPRLAFPEENNRSLKKIPKLE
3-2 N::TLT::S:V:::::KL::::S::IPEE::G:IEPYFVNSDPV LA: DS:K:T::DD

1-7 DPAQYSMIGLKPRRADLDMNQHVNNVTYIGWVLESIPQEIVDTHELQVITLDYRRECQQDD
3-2 KT:D:VRS::T::WS:::V:::::::K::::I:::A:VG:MERQK:KSM::E:::::GR:S

1-7 VVDSLTTTTSEIGGTNGSATSGTQGHNDSQFLHLLRLSGDGQEINRGTTLWRKKPSS
3-2 :LQ:::AV:GCDI:NLATAGDVECQ:LLRLQDGAEVVR:RTEWSSKTP:TTWGTAP
```

Figure 1. Amino acid sequences of the thioesterase cDNA clones TE 1-7 and TE 3-2 from *Arabidopsis thaliana*

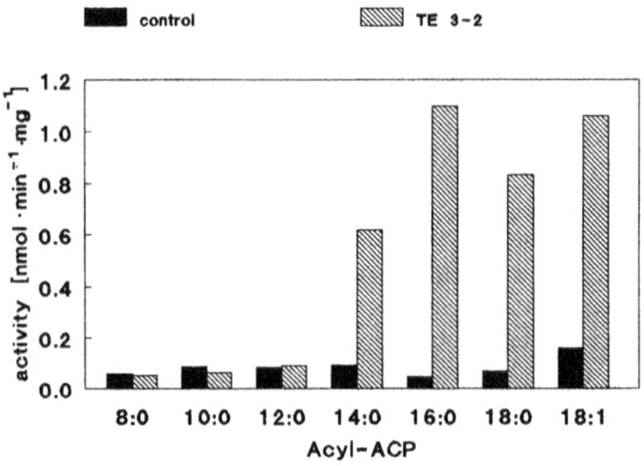

Figure 2. Acyl-ACP thioesterase activity of the TE 3-2 cDNA expressed in *E. coli*.

The thioesterase specifiticity which we observed in *E. coli* cells transformed with the TE 3-2 clone was 10 to 20 fold higher than control cultures but was similar in acyl-chain specificity to the *E. coli* thioesterases I and II [6]. Although these *E. coli* thioesterases markedly prefer CoA esters over ACP esters for hydrolysis, we considered the possibility that by expressing the TE 3-2 polypeptide in *E. coli*, induction of an endogenous *E. coli* thioesterase occurred . In this scenario, the observed increased acyl-ACP thioesterase activity might not be directly related with the introduced TE 3-2 protein but instead reflect an *E. coli* response reaction. However, unlike our observations of strongly impaired growth and substantial free fatty acid production (Table I), it was shown that overexpressing the *E. coli* thioesterase I or II in *E. coli* resulted in no change in lipid composition or growth rate in the transformed bacteria [7,8]. Therefore we believe that the most likely origin of the large increase in acyl-ACP hydrolytic activity and free fatty acid production in the transformed cells is the action of the TE 3-2 gene product. Most plant tissues contain low levels of extractable 16:0-ACP hydrolytic activity which seems inadequate to account for the 16:0 content found in these plants. The 16:0-ACP thioesterase activity of crude plant extracts or purified oleoyl-ACP thioesterase is about 10 % of the 18:1-ACP thioesterase activity [3]. However, palm mesocarp (*Elaeis guineensis*) contains up to 44 % 16:0 and Sambanthamurthi and Oo [9] could detect a high 16:0-ACP thioesterase activity in this tissue.

Transformation of *E. coli* by the TE 3-2 cDNA provides the opportunity to study its *in vivo* activity by analyzing the fatty acid composition of *E. coli* lipids. *E. coli* cells carrying the TE 3-2 cDNA or the vector without insert (control) were grown to OD_{600} 0.9 and total lipids were extracted. Free fatty acids and phospholipids were separated by thin layer chromatography. The fatty acids of different lipid classes were methylated and analyzed by gas chromatography. The transformed *E. coli* contained significant amounts of free fatty acids that were absent in the control cells. These fatty acids mostly consisted of 16:0 (Table 1). Due to the additional free fatty acids, the amount of total lipids was also higher (about 50 %), but the quantity and fatty acid composition of polar lipids was not changed.

Table 1. Fatty acid composition of different lipid classes of *E. coli* transformed with the TE 3-2 cDNA.

Fatty Acid	Total Lipids [μg]		Phospholipids [μg]		Free Fatty Acids [μg]	
	Control	TE 3-2	Control	TE 3-2	Control	TE 3-2
14:0	3.3	15.0	2.3	3.1	**0.1**	**10.0**
16:0	31.3	45.5	26.5	18.4	**0.7**	**27.5**
16:1Δ9*cis*	23.3	36.1	18.4	20.6	**0.0**	**7.8**
cyclopropane fatty acid	3.8	4.8	3.7	6.4	0.0	0.0
18:0	0.0	0.0	0.5	0.4	0.0	0.6
18:1Δ11*cis*	12.1	17.1	12.9	9.9	**0.0**	**5.7**
total	73.8	119.2	64.1	58.9	0.8	52.0

In order to analyze the expression of the TE 3-2 gene, antibodies against the recombinant TE 3-2 protein were raised in rabbits and used to probe Western blots with protein from different *A. thaliana* tissues. The TE 3-2 protein was detected in siliques, leaves and roots. As the TE 3-2 gene product presumably was involved in fatty acid synthesis, we also analyzed the expression of the thioesterase in developing *Brassica napus* seeds. The protein was expressed throughout seed development and peeked between 18 and 31 days after flowering. This expression pattern was similar to other enzymes involved in fatty acid biosynthesis.

Based on the results this study and on experiments by Browse and Somerville [10], it seems possible that in addition to the oleoyl-ACP thioesterase, plants contain a second thioesterase with more broad specificity for C_{14}-C_{18}-ACPs.

LITERATURE CITED

1. Ohlrogge JB, Shine WE, Stumpf PK Fat metabolism in higher plants, characterization of plant acyl-ACP and acyl-CoA hydrolases. Arch Biochem Biophys 1978;189:382-391
2. Knutzon DS, Bleibaum JL, Nelson J, Kridl JC, Thompson GA. Isolation and characterization of two safflower oleoyl-acyl carrier protein thioesterase cDNA clones. Plant Physiol 1992;100:1751-1758

3. Grellet F, Cooke, R, Raynal M, Laudié M, Delseny M. *Arabidopsis thaliana* systematic cDNA sequencing reveals a gene with homology with *Umbellularia californica* C12:0-ACP thioesterase. Plant Physiol Biochem 1993;31:599-602
4. Voelker TA, Worrell AC, Anderson L, Bleibaum J, Fan C, Hawkins DJ, Radke SE, Davies HM. Fatty acid biosynthesis redirected to medium-chains in transgenic oilseed plants. Science 1992;257:72-74
5. Dörmann P, Kridl JC, Ohlrogge JB. Cloning and expression in *Escherichia coli* of a cDNA coding for the oleoyl-acyl carrier protein thioesterase from coriander. Biochim Biophys Acta 1994;1212:134-136
6 Spencer AK, Greenspan AD, Cronan JE Jr. Thioesterase I and II of *Escherichia coli*. Hydrolysis of native acyl-acyl carrier protein thioesters. J Biol Chem 1978;253:5922-5926
7 Naggert J, Narasimhan ML, DeVeaux L Cho H, Randhawa ZI, Cronan JE Jr., Green BN, Smith S. Cloning, sequencing and Characterization of *Escherichia coli* thioesterase II. J Biol Chem 1991;266:11044-11050
8 Cho H, Cronan JE Jr. *Escherichia coli* thioesterase I, molecular cloning and sequencing of the structural gene and identification as a periplasmic enzyme. J Biol Chem 1993;268:9238-9245
9 Sambanthamurthi R, Oo K-C. Thioesterase activity in the oil palm (*Elaeis guineensis*) mesocarp. In: Quinn PJ, Harwood JL, editors. Plant Lipid Biochemistry, Structure and Utilization. London: Portland Press 1990:166-168.
10 Browse J, Somerville C. Glycerolipid synthesis, biochemistry and regulation. Annu Rev Plant Physiol Plant Mol Biol 1991;42:467-506

EXPRESSION OF ACYL-[ACP] THIOESTERASE IN *CUPHEA LANCEOLATA* AND IN TRANSGENIC RAPESEED

Norbert Martini, Jeff Schell, and Reinhard Töpfer
Max-Planck-Institut für Züchtungsforschung, Carl-von-Linné Weg 10, D-50829 Köln, Germany

Summary

A comparison of deduced amino acid sequences from 10 plant acyl-[Acyl Carrier Protein] thioesterases (TE) and their relationship to six *Cuphea* TEs is discussed. Expression studies of one TE from *Cuphea lanceolata* (*Cl*TE13) in roots, flowers, leaves, maturing embryos and seeds showed a preferred expression in flowers. Heterologous expression in transgenic *Brassica napus* revealed 3 % capric acid in immature seeds of the T2 progeny.

Introduction

Acyl-[ACP] thioesterase is thought to play a crucial role in the chain termination reaction during fatty acid biosynthesis in plants while also other chain termination mechanisms were discussed [1]. The enzyme catalyzes the hydrolysis of the thioester bond between the growing acyl chain and its carrier ACP releasing ACP and free fatty acids. It was demonstrated that a C12:0 specific TE from *Umbellularia californica* (*Uc*TE) is responsible for the accumulation of lauric acid by prematurely hydrolyzing the growing acyl-[ACP] thioesters in transgenic plants [2]. Our particular focus is to engineer *B. napus* such that seeds will contain fatty acids of a predetermined chain length (e. g. C10:0). Therefore we are interested in the isolation of genes encoding medium-chain specific acyl-[ACP] thioesterases from *C. lanceolata*, an annual plant, which predominantly accumulates capric acid (C10:0; 83%) in its storage triacylglycerols. Recently TE activity specific for medium-chain acyl-[ACP] substrates in this species was reported [3]. Moreover, a condensing enzyme appears to contribute to early chain termination in *C. lanceolata* [4]. In order to test whether or not the thioesterase from *C. lanceolata* is sufficient to control fatty acid chain length termination we have expressed a cDNA, *Cl*TE13 [5] from *C. lanceolata* in transgenic *B. napus*.

Results and Discussion

Cloning of Acyl-[ACP] Thioesterase genes from *C. lanceolata*

PCR amplification of cDNA 3′ends of embryo-specific poly(A)$^+$ RNA from *C. lanceolata* using the degenerate oligonucleotide TGG AAC/T GAC/T C/TTI GAC/T GTI AAC/T GA derived from the peptide WNDLDVNQ of the *U. californica* thioesterase (*Uc*TE) resulted in the isolation of cDNAs (*Cl*TE5 and *Cl*TE13) and of genomic clones (*Cl*TEg100 and *Cl*TEg200) encoding acyl-[ACP] thioesterases with substantial similarity (>75%) to the *Uc*TE sequence.

J.-C. Kader and P. Mazliak (eds.), Plant Lipid Metabolism, 495–498.

Comparison of Plant Acyl-[ACP] Thioesterases

The alignment of deduced amino acid sequences of TE cDNAs from various plants with *Cl*TE13 depicted in figure 1 reveals two distinct types of TEs with different substrate specificities: the long-chain (C16-C18)specific type I (represented by TEs from safflower, *Carthamus tinctorius*; [6,9]) and the medium-chain (C8-C14)specific type II (represented by *Uc*TE [2]). From this amino acid sequence comparison it is obvious, that *Cl*TE13 from *C. lanceolata* is a type II thioesterase. Three striking features are different in both types of TEs. (i) While in general the chloroplast targeting signal sequences vary, a characteristic hydrophobic domain in the C-terminal part of the presumptive transit peptides previously described for *Uc*TE [2] is present in all type II TEs between positions 88 to 117 and absent in all type I TEs. This domain is thought to play a role in thylakoid targeting of these enzymes within chloroplasts [7]. This is surprising, since the substrate, acyl-[ACP], is found in the stroma [8]. Voelker and Davies argued, this domain in the *Uc*TE preprotein possibly is a membrane anchor for the mature protein, which upon protein purification is cleaved off leading to the identification of an artificial N-terminus (position 112 in Figure 1). The real *in-vivo* N-terminus of the mature protein would then lie at a position further upstream [9]. Furthermore, the type II TEs contain deletions of five (ii) and 8 to 20 (iii) amino acid residues, when compared to type I TEs. Interestingly each of these divergent regions is surrounded by separate blocks of nearly identical residues, one of which contains the presumptive active site cysteine [5,6]. It is tempting to speculate, that these differences in the primary structure of plant thioesterases might account for their distinct substrate specificities. The dendrogram (Figure 1, insert) reflects the close relationship of acyl-[ACP] thioesterases from two species of the genus *Cuphea* and illustrates the classification of long-chain specific TEs and medium-chain specific TEs in type I and type II, respectively.

*Cl*TE13 Transcript Accumulates Predominantly in Flowers *of C. lanceolata*

Northern blot analysis of *Cl*TE13 expression in *C. lanceolata* plants revealed accumulation of TE transcripts in a decreasing order of intensity in flowers, immature embryos, leaves and immature seeds. Moreover, a larger number of molecular species of transcripts were detected in flowers and leaves compared with immature embryos. At present it is unclear, whether this transcript accumulation in flowers is the result of TE expression in different flower organs e. g. young ovaries prior to the development of embryos, or whether it is associated with an as yet unknown function.

Expression of *Cl*TE13 in *B. napus*

We obtained transgenic rapeseed plants containing chimaeric *Cl*TE13 gene constructs driven either by a seed-expressed ACP promoter from rapeseed (A) [10,11] or by the Cauliflower Mosaic Virus 35 S promoter (B) [12]. The presence of the corresponding transgene was confirmed by Southern analyses of transformants and/or by testing for kanamycin resistance of the T2 progeny. 52 out of 102 individuals with construct A and 43

Figure 1: Comparison of Plant Acyl-[ACP] Thioesterases (next page)

Alignment of deduced amino acid sequences of cDNAs encoding acyl-[ACP] thioesterases (TE) from various plants. Amino acid residues are denoted according to the single-letter code. Identical amino acids relative to *Cl*TE13 (lane 6) are omitted. Lane 1 - 5 depict long-chain specific type I and lane 6 - 10 medium-chain specific type II TEs. Among conserved regions through all sequences depicted, one of which comprises the presumptive active site cysteine (highlighted in black [5, 6]), there are three striking regions of divergence (underlined) between the two TE types. The putative transit peptides extend up to position 93 for *Cf*TE5-2 [9] and up to position 111 for *Uc*TE [2] (arrow heads) including up to 80% identical amino acids of type II TEs between positions 88 to 117.The dendrogram of plant acyl-[ACP] thioesterases (insert) shows the clustered relationships used to create the multiple amino acid sequence alignment [16] here with the addition of five further acyl-[ACP] thioesterases from *Cuphea lanceolata* and *Cuphea wrightii* (*Cw*; [17]).

```
    1                                                                                                    ▽ 100
 1  ..............................................M  LSCNATDKLQTL..FSHSHQPDPAHRRTV SVSCSHLRKPVL PLRAIVSADQ
 2  ..............................................M  LSCNVTNHLHTF FFSDS LFIPVNRRTL.AVSSSQPRKPAL PLRAVISADQ
 3  ..............................M  RPL TTA A TTTTNNCNGVNSRGALPH.SRSV FASIRKRSTGSLCN..SPPRTVAPV AVRTGEQ
 4  ..............................M  KGA ..A P VAAMYNASAKDTTFALTH.SRSI SVSIRRRYNVFLCNSSSSSRKVSPLLAV TGEQ
 5  ...............................................MVYPHFKT IQCRFLT.SDSI...SIRRRTAV.......SRWRSPTFSANYNGV
 6  MVATAASSAFFPLPSPDTSSRPGKLGNGSSSLSPLKPKFVANAGLKVKASASAPPKINGSSVGLKSGSLKTQEDTPSVPPPRTFINQLPDWSMLLAAITT
 7  .......................................................................................................
 8  ......................MATTSLA AFCSM AVML RD RGM PRS DLQLRA NA...QTSLKMINGTKF YTESLK...K        F V
 9  ......................MATTSLA AFCSM AVML RD RGM PRS DLQLRA NA...PTSLKMINGTKF YTESLK...R        F V
10  ......................MVTTSLA AFFSM AVML PD SGI PRS GLQVRA KE...QNSCKMINGTKVKDTEGLKGRST HG    P EL
                                                     (i)

    101                                                                                               200
 1  GSVIRA...........EQGLGS A QLR   LTED  SYKEK IV     V SNK  TV    A L    VGC  AQ V FST    AT  T R LH    T
 2  GSISPV...........NSCTP..A R RA RLMED YSYKEK IV     V INK  TV   IA L    V C    QKC FST    AT LT R LH    T
 3  PTGV VGLKEAEAEVE .....S A RLRM  LTED  SYKER I  C   V INK  TV   IA L    VGG  AQ V FST    AT TT R LH    TS
 4  PSGV .SLREADK..E .....S GNRLR   LTED  SYKEK V  C   V INK  T    IA L    VGG  AQGV FST    AT TT R LH    T
 5  NAQVLGVLKQEQKEIEEEKRSSS AEKLR   LTED  SYKEK IV C   V INK  TV   IA L    VGG  AQ V FST    AT  T R LH    T
 6  VFLAAEKQWMMLDWKPKR..PDMLVDPFGLGSIVQGGLVFRQNFSIRSYEIGADRTASIETVMNHLQETALNHVKSAGLLNDGFGRTPEMFKRDLIWVVA
 7  ..............................   D             S                     T   G    S      KN       T
 8  I S      TN E    PNP QL D H  P.....H       RT A       V P   ST  VA          A    A  V I G    T L  S         K
 9  I S      TN E    PKL QL D H  L.....H       RT A       V P   ST  LA      M   AT    A  V I G    T L  S     M   R
10  I S      TN VS   .... QL D HL L.....H       RT A   CS V P   ST  VA      Y   A C   AE L   G    E L  S         R
             △

    201                                                                                               300
 1  R HIEIYK  A   V  IE  CQSE RI T     ILK VA   VTG   T K       D  R Q VS D  D YLVFCPKELRLAFPEENNRSLK IP  E
 2  R HIEIYK  A S V  IE  CQSE RI T     ILR SA N VIG   T K       D  R QRVT    D YLVFCPREPRLAFPEENNSSLK IP  E
 3  R HIEIY   A S V  IE  CQSE RI T     IMK HAS  VIG   T K      ED  R Q VN D  D YLVFCPKTPRLAFPEKNTSSLK IA  E
 4  R HIEIY   A S VI IE   QGE  V T     ILK YAN  VIG   T K      ED  R Q VS D  E YLVFCPRTLRLAFPEENNNSMK IP  E
 5  R HIEIY   A S V  IE  CQSE RI T     I K FA D VIG   T K       D  R Q VS D  D YLVFCPKTPRLSFPEENNKSLK IS  E
 6  KMQVMVNRYPTWGDTVEVNTWVAKSGKNGMRRDWLISDCNTGEILTRASSVWVMMNQKTRKLSKIPDEVRHEIEPHFIDCAPV.....IEDDDRKLRKLD
 7  R   V DK        V   D  SQ            VR              S          KL  R      E   G     Y.................
 8  RTH A E    A         EC  GA   N  R H F VR   K          CT LS       TR  R      E   G  G A    NVA .....KDEEIK PQ  N
 9  RTH A E              EC IGA   N        F VR   K          CT LS L     TR  R  T       G  G A    NVA .....KD EIK  Q  N
10  RTH V GT   A    T    EA IGAA NI      HF VR   K  H  A CT  S       MR  R       Q   G  D L  EKFA .....K GEIK  Q FN
                                                                                          (ii)

    301                                                                                               400
 1  DP. QYSMI  K RRA   M        T    V    I    IVD H  QVI  D        QQDDVVD   T........TTSEIGGTNGSASSGTQGQNDS
 2  DP. QYSMLE K RRA   M        T    V    I    IID H  QVI  D        QQDDIVD   TSE..IPDDPISKLTGTNGSATSSIQGHNES
 3  DP. EYSTL  V RRA   M K      T    V    I     ID H  QTI  D        QHDDIVD   SSE LLDDAAISKLEGTNGSSVPKKDETDLS.
 4  DP. EYS L  V RRS   M K      T    A    I P  IID H  QAI  D        Q DDIVD   SREP.LGNAAGVKFKEINGSVSPKKDEQDLS.
 5  DP. QHS L  S RRA   M        A    V    I K   Y H  ETI  D        QHDD VD   SPE.VDEDTAVTKIIGTNGHAAAATEARDDSL
 6  EKTADSIRKGLTPKWNDLDVNQHVNNVKYIGWILESTPQEVLETQELSSLTLEYRRECGRESVLESLTAVDS....................SGKGFGSQ
 7  ........................................................................................................
 8  DS    Y QG     R       I       I  VD     TV DSIF SHHI   F          TMD   Q    T SG....................GSSEA LV
 9  DS    Y QG     R               L  VA VF TV DSIF SHHI   F          T D   R    T SG....................GSSEA LV
10  DS    Y QG W   R               I  V   FK V DSIY NHH    I          T GRA Q    T CG....................GSSEA II
                                                                                                   (iii)

    401                         444
 1  . L       SGD Q   NR T L   K PSNL...............
 2  . L M     S N Q   NR    Q  K SSR................
 3  R L       SSGD L  LNR      K P KK...............
 4  R M       SAGS L  NRC      K P KR...............
 5  K L F  VSGQ L   NR       K SEKR...............
 6  .FQHLLRL.EDGGEIVKGRTEWRPKTAGVNGAIASGETSHGDS.
 7  ............................................
 8  .CE      Q . G S VLRAK        LTDSFRG SVIPAESSV..
 9  .CD      Q . G S VLRA         LTDSFRG SVIPAEPRV..
10  .CE      Q .   S VLR     D    RTDSFEG SERFPQQEPHN
```

1 *Brassica campestris* TE [9]; 2 *Brassica napus* TE2 [13];
3, 4 *Carthamus tinctorius* TE2-1, TE5-2 [6, 9]
5 *Coriandrum sativum* TE [14]; 7 *Arabidopsis thaliana* TE [15];
6, 11, 12, 14 *Cuphea lanceolata* TE13 [5], TE5, TEg200, TEg100
8 *Cinnamomum camphora* TE [9]
9, 10 *Umbellularia californica* TE, TEd [2, 9]
13, 15 *Cuphea wrightii* TE11, TE48 [17]

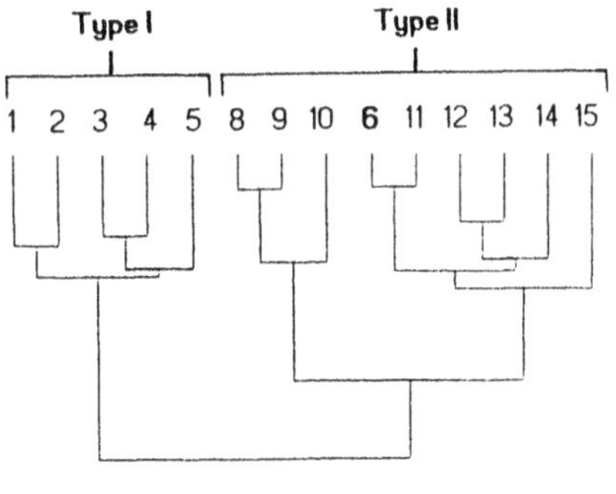

out of 75 individuals with construct B were positive. Mature seeds of these plants were analyzed by gas chromatography for capric acid content along with seeds of non-transformed plants as controls. The results obtained with this material indicate that capric acid does not accumulate to detectable levels in these seeds. However, it might well be that the timing of expression of *Cl*TE13 in these plants is suboptimal. This is supported by our observation that up to 3% capric acid was detected in imature T2 seeds from a transgenic plant containing construct A. The time window of expression in developing seeds might be too short to result in higher levels of capric acid versus the other fatty acid species in storage triacylglycerols in mature seeds. In order to optimize the timing of expression of the transgene we are currently introducing two genomic clones of *C. lanceolata* with their own promoters into rapeseed.

References

1. Harwood JL. Fatty acid metabolism. Ann Rev Plant Physiol Plant Mol Biol 1988; 39: 101-138.
2. Voelker TA, Worrell AC, Anderson L et al. Fatty acid biosynthesis redirected to medium chains in transgenic oilseed plants. Science 1992; 257: 72-74.
3. Dörmann P, Spener F, Ohlrogge JB. Characterization of two acyl-acyl carrier protein thioesterases from developing Cuphea seeds specific for medium-chain- and oleoyl-acyl carrier protein. Planta 1993; 189: 425-432.
4. Schuch R, Brummel M, Spener F. Medium-chain acyl-ACP thioesterase is not the exclusive enzyme responsible for early chain-length termination in medium-chain fatty acid synthesis. Grasas y Aceites 1993; 44: 126-128.
5. Töpfer R, Martini N. Molecular cloning of cDNAs or genes encoding proteins involved in de novo fatty acid biosynthesis in plants. J Plant Physiol 1994; 143: 416 -425.
6. Knutzon DS, Bleibaum JL, Nelsen J, Kridl JC, Thompson GA. Isolation and characterization of two safflower oleoyl-acyl carrier protein thioesterase cDNA clones. Plant Physiol 1992; 100: 1751-1758.
7. Smeekens S, Weisbeek P, Robinson C. Protein transport into and within chloroplasts. Trends Biochem Sci 1990; 15: 73-76.
8. Ohlrogge JB, Kuhn DN, Stumpf PK. Subcellular localization of acyl carrier protein in leaf protoplasts of Spinacia oleracea. Proc Nat Acad Sci USA 1979; 76: 1194-1198.
9. Voelker TA, Davies HM. Plant medium-chain thioesterases. International Patent Application 1992; WO 92/20236.
10. de Silva J, Robinson SJ, Safford R. The isolation and functional characterization of a B. napus acyl carrier protein 5´ flanking region involved in the regulation of seed storage lipid synthesis. Plant Mol Biol 1992; 18: 1163-1172.
11. Voetz M, Töpfer R. Köln, 1992; unpublished data.
12. Töpfer R, Maas C, Höricke-Grandpierre C, Schell J, Steinbiss HH. Expression vectors for high-level gene expression in dicotyledonous and monocotyledonous plants. In: Wu R, editor. Recombinant DNA. Methods in Enzymology. San Diego: Academic Press, Inc. 1993; 217: 66-78.
13. Loader NM, Woolner EM, Hellyer A, Slabas AR, Safford R. Isolation and characterization of two Brassica napus embryo acyl-ACP thioesterase cDNA clones. Plant Mol Biol 1993; 23: 769-778.
14. Dörmann P, Kridl JC, Ohlrogge JB. Cloning and expression in Escherichia coli of a cDNA coding for the oleoyl-acyl carrier protein thioesterase from coriander seed endosperm (Plant Gene Register). Genbank accession number: L20978.
15. Grellet F, Cooke R, Raynal M, Laudie M, Delseny M. Arabidopsis thaliana systematic cDNA sequencing reveals a gene with homology with Umbellularia californica C12:0-ACP thioesterase. Plant Physiol Biochem. 1993; 31: 599-602.
16. Devereux J, Haeberli P, Smithies O. A comprehensive set of sequence analysis programs for the VAX. Nucl Acids Res 1984; 12: 387-395.
17. Leonard J, Slabaugh M, Knapp SJ. Corvallis, 1993; unpublished data.

Acknowledgements

We gratefully acknowledge J. Bautor, C. Bonnesen and E. Filsak for excellent technical assistance as well as J. Leonard, M. Slabaugh and S. Knapp for providing us with unpublished sequence data. Our research was supported by the BMFT (grant number 0319412D) including financial support from Bayer AG, Deutsche Saatveredelung, Hoechst AG, Kleinwanzlebener Saatzucht AG and Norddeutsche Pflanzenzucht GmbH.

GENETIC AND BIOCHEMICAL STUDIES OF MEDIUM CHAIN FATTY ACID SYNTHESIS IN *CUPHEA*

Mary Slabaugh, Jeff Leonard, Gordon Huestis, Jim Crane & Steve Knapp

Department of Crop and Soil Science, Oregon State University, Corvallis, Oregon, USA, 97331

Summary

Cuphea mutants displaying altered seed oil fatty acid profiles have been isolated following EMS mutagenesis. The phenotypes of single and double homozygous mutants are discussed. cDNA clones encoding putative acyl-ACP thioesterases and β-ketoacyl synthases have been isolated from a *C. wrightii* embryo library.

Introduction

Our goal is to understand the metabolic pathway(s) by which *Cuphea* spp. channel more than 90% of their seed storage lipids into medium chain fatty acids (MCFA). Although other genera include species that store predominantly MCFA-containing triacylglycerol (e.g. California bay, elm), *Cuphea* is unique in the variety and extent of specialization exhibited within the genus. We have developed a collection of EMS-induced FAS mutants from the predominantly C10:0-producing species, *C. viscosissima*. These mutants exhibit a variety of deviations from the characteristic wild type fatty acid profile, and represent a valuable base for producing specialized agronomic lines by conventional plant breeding [1], as well as providing clues to the underlying biochemistry of MCFA biosynthesis in *Cuphea*. In order to further investigate the genetics of MCFA synthesis in *Cuphea*, double mutants have been constructed and their oil phenotypes determined.

Enzyme activities that may control MCFA synthesis include thioesterases and β-ketoacyl synthases. A *Cuphea* species that produces predominantly C8:0 might be endowed with a C8:0-ACP-specific thioesterase, analogous to the C12:0-ACP-specific thioesterase isolated from bay [2], and subsequently shown to promote MCFA biosynthesis in transgenic plants [3]. The synthase enzymes may play a role in controlling chain length by virtue of their ability to channel the feedstocks of fatty acid synthesis, acetyl-CoA and malonyl-ACP, toward chain initiation (KASIII) or chain elongation (KASI and II).

J.-C. Kader and P. Mazliak (eds.), Plant Lipid Metabolism, 499–502.

This idea has not been tested in transgenic plants. In order to investigate the role of these enzymes in *Cuphea*, we have isolated a number of cDNA clones with homology to plant thioesterase and synthase coding sequences.

Results and Discussion

EMS-induced fatty acid mutants of *C. viscosissima*

Seed of the inbred line VS-6 were exposed to 0.04 M ethyl methanesulfonate for six hr before germination, and the fatty acid percentages of bulk seed samples of 862 M_2 lines were assayed by GC. M_3 sublines homozygous for mutations affecting FAS were selected from M_2 lines with FA phenotypes different from wildtype. Double homozygous mutants were generated by crossing pairs of homozygotes and selfing the F1. Homozygous *C. viscosissima* mutants were also crossed to *C. lanceolata* f. *silenoides* (LN43). LN43 bears the *cei* allele, which behaves as a single locus decreasing the percentage of C8:0 in interspecific crosses. Table 1 summarizes the MCFA seed oil phenotypes of wild type *C. viscosissima* (VS), *C. lanceolata* (CEI), four VS mutant lines, and four double mutants. Mutant lines produce the same amount of oil as wild type VS (30% of dry seed weight), except for CPR-4, which is deficient in oil production (17%).

Table 1. Medium Chain Fatty Acids as Percentage of Total Seed Fatty Acids

Line	C8:0	C10:0	C12:0	C14:0	C16:0
VS	17.5	71.1	2.2	0.9	1.8
CEI	1.1	87.4	1.8	1.7	2.5
CPY-1	3.9	70.9	14.3	7.4	1.9
CPR-1	24.5	38.3	6.3	14.2	8.2
CPR-4	22.4	65.9	2.8	1.2	2.0
MYR-1	20.5	70.4	2.3	0.3	0.8
CEI + CPR-1	1.1	42.1	4.5	27.2	13.3
CPR-1 + CPR-4	17.2	34.5	6.1	18.6	14.1
CPR-1 + MYR-1	35.8	39.2	5.6	2.4	4.4
CPY-1 + MYR-1	3.7	69.7	18.4	5.2	1.5

The potency of the *cpr-1* allele in decreasing C10:0 by nearly one-half is seen in combination with *cei, cpr-4,* and *myr-1*. However, simultaneous homozygosity of the *cei* allele prevents the *cpr-1*-promoted accumulation of C8:0; instead, the positive effect of both alleles on C14:0 and C16:0 results in a more than additive increase in these fatty acids in the double homozygote. *cpr-1* and *cpr-4* both depress the C10:0 percentage; together, they cause an additive decrease in C10:0. Singly, the *cpr-1* and *myr-1* loci have opposite effects on C14:0 and C16:0, with *cpr-1* increasing and *myr-1* decreasing these fatty acids. When combined, a large increase in C8:0 results. However, the *myr-1* locus combined with *cpy-1*, which has a negative effect on C8:0, results in increased C12:0. In sum, the negative effect(s) of each mutant in decreasing specific fatty acids appears also in double mutant constructions. One interpretation of these results is that several pathways lead to MCFA production in *Cuphea*.

cDNA clones encoding thioesterases and synthases from *C. wrightii*

A cDNA library was constructed using mRNA from immature embryos of *C. wrightii*, a C10:0/C12:0 species. Two full-length cDNAs with homology to California bay thioesterase, two clones with homology to algal and spinach KASIII sequences, and two clones with homology to castor KASII have been characterized. cwTE11 (1295 nt cDNA) and cwTE48 (1449 nt cDNA) contain ORFs encoding proteins of 398 and 408 amino acids, respectively. cwKASIIIa (1696 nt cDNA) and cwKASIIIb (1827 nt cDNA) contain ORFs encoding proteins of 400 and 402 amino acids, respectively. cwKASX34 (2112 nt cDNA) and cwKASX48 (2190 nt cDNA) contain ORFs encoding proteins of 540 and 517 amino acids, respectively. Southern analysis of DNA from several *Cuphea* species suggests that these enzyme classes are encoded by small gene families containing three to four members. Single strand conformational polymorphism (SSCP) analysis, in which oligonucleotide primers selected from regions of uninterrupted identity between homologous clones were used to amplify PCR products from genomic DNA, yielded the same conclusion: three TE-specific products, three KASIII-specific products, and two KASX-specific products were produced. By scoring the SSCP banding patterns produced from DNA isolated from 100 individual plants from an interspecific *C. viscosissima X C. lanceolata* segregating F2 population, these genes were assigned map positions using the program GMendel [4]. Three TE loci, one KASIII locus, the *cei* locus, and several other FAS loci (ketoreductase, enoyl reductase, and KASI) are clustered on one linkage group.

Western analyses of extracts from various plant tissues, using polyclonal antisera prepared in rabbits from recombinant protein antigens, indicate that all three classes of enzyme are specifically expressed in developing *Cuphea* embryos.

References

1. Knapp SJ. Modifying the seed oils of *Cuphea*--a new commercial source of caprylic, capric, lauric, and myristic acid. In: MacKenzie S, Taylor D, editors.

Seed oils for the future, Amer. Oil Chem. Soc. Mono. Champaign, Illinois, 1993:142-154.

2. Pollard MR, Anderson L, Fan C, Hawkins D, Davies M. A specific acyl-ACP thioesterase implicated in medium-chain fatty acid production in immature cotyledons of *Umbellularia californica*. Arch Biochem and Biophys 1991;284: 306-312.
3. Voelker TA, Worrell AC, Anderson L, Bleibaum J, Fan C, Hawkins D, Radke S, Davies HM. Fatty acid biosynthesis redirected to medium chains in transgenic oilseed plants. Science 1992;257:72-74.
4. Holloway J, Knapp S. Unpublished.

Acknowledgements

We gratefully acknowledge Jan Jaworski and Heeyoung Tai for providing a spinach KASIII PCR clone, and R. Topfer and N. Martini for providing unpublished sequence data.

UTILIZATION OF LAURATE BY THE KENNEDY PATHWAY IN DEVELOPING SEEDS OF *BRASSICA NAPUS* EXPRESSING A 12:0-ACP THIOESTERASE GENE

H. Maelor Davies, Cheryl A. Eriqat, and Thomas R. Hayes: Calgene Inc., 1920 Fifth St., Davis, CA 95616, USA.

The seeds of rapeseed plants transformed with 12:0-ACP thioesterase (from *Umbellularia californica*; 1) accumulate up to 50 mol % 12:0 groups in the reserve triglycerides. Published studies on the specificity of the Kennedy pathway enzymes have indicated that the acyltransferases of rapeseed can utilize medium-chain, saturated acyl groups to different extents [2]. In particular the enzyme responsible for the *sn*-2 acylation, acyl-CoA:lysophosphatidate acyltransferase (LPAAT), exhibits very low activity with 12:0-containing substrates *in vitro*. Accordingly, we were interested to examine the distribution of 12:0 within the triglyceride molecules of the transgenic plants.

The most direct method to determine the stereochemical composition of triglycerides is to analyze the products of partial hydrolysis, either from specific lipase action or general Grignard deacylation. There are established procedures for this, but unfortunately many questions have been raised concerning their suitability for the analysis of medium-chain triglycerides (for discussion see; 3). There are concerns regarding the specificity of the lipases, and the possibility of spontaneous re-arrangements of monoglycerides and diglycerides containing medium or short chain acyl groups at *sn*-2 [4]. We therefore took a number of precautions to detect such artifacts and to reduce the possibility of their occurrence. The use of pancreatic lipase rather than the Grignard reagent increased our chances of detecting acyl migration from *sn*-2 to the primary positions; this lipase does not attack at *sn*-2 so any formation of any *sn*-1,3 diglyceride products would suggest migration. The possibility of rearrangement was reduced by conducting the digestion for a very short time (3 min) at neutral pH, and by resolving the products on borate-impregnated TLC plates (W. Christie - personal commun.). We monitored lipase selectivity by arranging for only 50% of the triglyceride to be attacked and then analyzing the acyl composition of the undigested portion.

We conducted this analysis with a TLC-purified rapeseed triglyceride fraction containing 37 mol % 12:0. (All triglycerides and oils mentioned in this account were from T3 transgenic plants whose seeds varied in 12:0 content as a result of genetic segregation. The acyl and triglyceride compositions therefore represent averages.) The undigested triglyceride showed the same acyl composition as the starting material, suggesting minimal chain-length preference of the lipase. No *sn*-1,3 diglycerides were detected, suggesting that acyl migration was minimal. The composition of the monoglyceride fraction was therefore considered to represent the *sn*-2 composition of the oil, and this contained only 10 mol% 12:0, supporting the predictions of discrimination against medium chains and saturated acyl groups by the rapeseed LPAAT. The monoglyceride

J.-C. Kader and P. Mazliak (eds.), Plant Lipid Metabolism, 503–505.

composition and either the original triglyceride or the diglyceride fraction compositions could be used to compute *sn*-1+3. The average 12:0 enrichments in these primary positions was thereby estimated at 51 mol%. (The measured monoglyceride and diglyceride fractions could also be used to back-calculate the original triglyceride value, and this calculation satisfactorily yielded 38 mol%.)

While these results are interesting, and aligned with the predictions from numerous enzymology studies, we regard them as preliminary. A serious limitation with such analytical methods has been the lack of stereochemically defined triglyceride standards, particularly containing medium chains, with which to validate them. However, Dr. Phil Sonnet (USDA Eastern Regional Research Center, Philadelphia, PA) has developed procedures for the chiral synthesis of triglycerides [5], and has provided us with appropriate examples (*e.g.* with 12:0 at *sn*-2 and 18:1 at the primary positions). These will be used to test the above lipase procedure and then to establish methods for distinguishing the *sn*-1 and *sn*-3 compositions of the 12:0-containing triglycerides.

We have also examined the 12:0-containing rapeseed triglycerides by HPLC fractionation of the individual species. An evaporative light-scattering detector was used to detect the triglyceride fractions and the flow was split so that individual peaks could be collected for fatty acid analysis, providing both compositional and quantitation data. Reverse-phase HPLC on C18-silica columns was used to separate triglycerides according to equivalent carbon number. The transgenic (42 mol % 12:0) and control oils yielded quite different chromatograms. In the transgenic oil, 12:0 was detectable in almost every triglyceride peak and the most abundant triglyceride was monolaurin-diolein. In a likely overestimate, the abundance of trilaurin was only 1.2 mol% of all the triglycerides. If 12:0 were selected at random for all three *sn*-acylations the abundance of trilaurin would be 7.4 mol%. This low trilaurin content again evidences discrimination against 12:0 during triglyceride biosynthesis (although the method cannot distinguish the particular acylation(s) which are discriminatory).

Chromatography on silver-phase HPLC fractionates the triglycerides according to degree of unsaturation. Again the chromatograms for control and transgenic (42 mol% 12:0) oils were very different and again 12:0 was detected in every triglyceride class. The transgenic oil contained a small quantity of tri-saturated triglycerides, which contained 72 mol % 12:0. Since this number is less than 100% this fraction was not entirely made up of tri-12:0, and some of the 12:0 must have been combined in triglycerides with 14:0, 16:0, and 18:0. However, assuming all the 12:0 to represent trilaurin, and correcting for the recoveries of authentic trilaurin in the procedure, we estimated a maximum trilaurin content for the oil of 0.9 mol%. This is again much less than the 7.4 mol% expected from random 12:0 incorporation.

In summary, limited analysis of the 12:0-containing rapeseed oil suggests that 12:0 is incorporated into most triglyceride size and saturation classes but that the incorporation is not uniform amongst the *sn*-acyl groups. Considerable discrimination probably occurs at *sn*-2. To enhance the laurate incorporation at *sn*-2 in these transgenic plants we have commenced the isolation of a suitable LPAAT.

References

1. Voelker TA, Worrell AC, Anderson A *et al.* Fatty acid biosynthesis redirected to medium chains in transgenic oilseed plants. Science 1992; 257:72-74.

2. Oo K-C, Huang AHC. Lysophosphatidate acyltransferase activities in the microsomes from palm endosperm, maize scutellum, and rapeseed cotyledon of maturing seeds. Plant Physiol. 1989; 91: 1288-1295.

3. Mattson FH, Volpenhein RA. The use of pancreatic lipase for determining the distribution of fatty acids in partial and complete glycerides. J. Lipids Res. 1961; 2: 58-62.

4. Kodali DR, Tercyak A, Fahey DA, Small, DM. Acyl migration in 1,2-dipalmitoyl-*sn*-glycerol. Chem. Phys. Lipids 1990; 52: 163-170.

5. Sonnet, PE. A short highly regio- and stereoselective synthesis of triacylglycerols. Chem. Phys. Lipids 1991; 58: 35-39.

REDUCING POLYUNSATURATION IN OILS OF TRANSGENIC CANOLA AND SOYBEAN

William D. Hitz, Narendra S. Yadav, Robert S. Reiter, Charles J. Mauvais, Anthony J. Kinney

Agricultural Products, E402-4249 DuPont Experimental Station, Wilmington DE 19880-0402, U.S.A.

Introduction

A key goal in improving the quality of vegetable oil is reducing its level of polyunsaturated fatty acids, linoleic (18:2) and linolenic (18:3), to result in an oxidatively-stable oil not requiring hydrogenation. This will result in reduced processing cost and the absense of *trans* fatty acids, by-products of hydrogenation believed to be unhealthy. Since canola and soybean germplasms with the desired low level of polyunsaturation are not currently available, at least, without an agronomic penalty, a transgenic approach was used.

Oleate and linoleate desaturases catalyze the biosynthesis of 18:2 and 18:3, respectively. In plants there are two pathways for polyunsaturated fatty acid biosynthesis, one localized in the ER and the other in the plastid. While both pathways operate in the leaf tissue with significant exchange between them, the ER one predominates during seed oil biosynthesis. However, the relative contribution of the plastid pathway might become significant in a low polyunsaturate oil. Browse and Somerville's study (1) of *Arabidopsis* desaturation mutants laid the ground work for the cloning of ER linoleate and oleate desaturases genes by genetic approaches (2-4). The ER linoleate desaturase cDNA was used to clone its plastid homolog (3, 5) and oligomers made to a region conserved among known desaturases were used to clone plant cDNAs encoding the plastid oleate desaturase (6). We used*Arabidopsis* cDNAs for all four desaturases described above as heterologous probes to clone their homologs from soybean and canola seed cDNA libraries. We report on the first round of transgenic experiments using ER oleate and linoleate desaturases in reducing polyunsaturation in canola and soybean

Limited Success of Mutagenesis For Low Polyunsaturate Canola or Soybean

Since membrane and oil biosynthesis share a common pathway, modifications in seed oil composition could be limited, especially by mutations in genes

J.-C. Kader and P. Mazliak (eds.), Plant Lipid Metabolism, 506–508.

expressed in both seed and non-seed tissues. This is illustrated by the altered physiology in some high oleate mutants of *Arabidopsis* (7) and canola. Mutation in an oleate desaturase gene of Westar canola resulted in a 78% seed oleate line, IMC129, with normal root and leaf lipid profiles as well as good agronomy. Its remutagenesis produced a double mutant with mutations in two, separate oleate desaturase genes resulting in higher seed oleate (87%) but also with high oleate in root and non-plastid leaf lipids. The agronomic properties of the double mutant are very poor, which might be attributable to the lipid changes in non-seed tissues (Chen, DeBonte, Hitz, and Miao). In soybean, the highest oleate level reported is less than 50% and the trait is temperature-sensitive. The lowest linolenate (2.2%) reported involves at least two loci (8). Thus, breeding of these multigenic or environmentally-sensitive traits is expected to be difficult. Thus, notwithstanding the good agronomy of a sunflower mutant with 90% oleate, mutagenesis has resulted, thus far, in limited reduction in polyunsaturation in canola and soybean.

Reducing Polyunsaturates In Transgenic Canola and Soybean

Seeds of transgenic canola lines homozygous for transgenes comprising antisense canola oleate desaturase cDNA under the control of a seed-specific promoter (napin or phaseolin) had 18:1 contents above 80% (Hitz, Reiter, Mauvais, and Ripp, unpublished). Line # 8 (Table 1), which contains a single copy of a chimeric napin:antisense oleate desaturase cDNA and whose high oleate trait segregates as a single, almost dominant locus, is currently in field trials. Further increase in seed 18:1 content was achieved by crossing it with IMC129. F2 progeny homozygous for both the mutant oleate desaturase gene of IMC129 and the transgene showed 18:1 levels around 88% (Table 1). In these seeds there was also a 2% drop in saturate levels (16:0+18:0). Seeds homozygous for a chimeric napin: antisense linoleate desaturase cDNA gene had a decreased seed 18:3 level, 4.2% compared to 6.5% in control. Further reductions are expected by combining this transgene with ultra-high oleate lines and/or low 18:3 mutations.

Fertile soybean plants were regenerated from somatic embryos transformed with chimeric genes consisting of an antisense soybean oleate or linoleate desaturase cDNA under the control of a seed-specific conglycinin promoter (Kinney, Knowlton, Campbell, Schweiger, and Stecca, unpublished). A single copy of an antisense linoleate desaturase cDNA reduced the seed 18:3 level from 9% to less than 2% (Table 1). An antisense oleate desaturase cDNA increased the 18:1 level from 20% to ca 80% and decreased the saturate (16:0+18:0) level from 15% to less than 11% in seeds (Table 1). A transgene with soybean oleate desaturase cDNA in sense orientation was, at least, as effective in reducing 18:1 desaturation in somatic embryos by cosuppression.

Table 1. Fatty Acid Composition in Transgenic Canola and Soybean Seeds

	Percent Fatty Acids				
	16:0	18:0	18:1	18:2	18:3
Canola					
Control	4.2	2.2	63.0	19.9	10.6
High 18:1 mutant line, IMC129	4.0	2.0	77.8	10.4	5.9
Transgene line #8, napin:OD antisense	3.7	2.2	83.3	5.0	5.8
Transgene line#8 X IMC129	3.0	1.4	88.4	2.9	4.0
Soybean					
Control line	11.3	3.8	21.5	55.0	8.9
Antisense 18:2 desaturase, line #1	10.1	3.4	20.1	65.0	1.4
Antisense 18:1 desaturase, line #2	7.9	3.0	78.9	3.0	5.8

The data for canola line #8 is of T4 seeds homozygous for the transgene, and for line#8 X IMC129 is an average of six F2 half-seeds homozygous for the transgene and the mutant gene. The data for transgenic soybean lines is an average of ten T2 seeds homozygous for the transgene.

References

1. Browse J, Somerville C. Glycerolipid Synthesis: Biochemistry and Regulation. Ann. Rev. Plant Physiol. Plant Mol. Biol. 1991;42:467-506.
2. Arondel V, Lemiewx B, Hwang I, Gibson S, Goodman HM, Somerville CR. Map-based cloning of a gene controlling omega-3 fatty acid desaturation In *Arabidopsis*. Science 1992;258:1353-1355.
3. Yadav NS, Wierzbicki A, Aegerter M, et al. Cloning of higher plant ω-3 fatty acid desaturases. Plant Physiol. 1993;103:467-476.
4. Okuley J, Lightner J, Feldmann K, Yadav N, Lark E, Browse J. *Arabidopsis* FAD2 gene encodes the enzyme that is essential for polyunsaturated lipid synthesis. Plant Cell 1994;6:147-158.
5. Iba K, Gibson S, Nishiuchi T, et al. A gene encoding a chloroplast omega-3 fatty acid desaturase complements alterations in fatty acid desaturation and chloroplast copy number of the *fad7* mutant of *Arabidopsis thaliana*. J. Biol. Chem. 1993;268:24099-24105.
6. Hitz WD, Carlson TJ, Booth, JR, Kinney AJ, Stecca KL, Yadav NS. Cloning of a higher plant plastid ω-6 fatty acid desaturase cDNA and its expression in a cyanobacterium. Plant Physiol. 1994;105:635-641.
7. Miquel MF, Browse JA. High oleate oilseeds fail to develop at low temperature. Plant Physiol. 1994;in press.
8. Fehr WR, Welke GA, Hammond EG, Duvick DN, Cianzo SR. Inheritance of reduced linolenic acid content in soybean genotypes A16 and A17. Crop Sci. 1992;32:903-906.

SOLUBILIZATION OF Δ12- and Δ6-DESATURASES FROM SEEDS OF BORAGE MICROSOMES

Anne-Marie GALLE*, Annette OURSEL°, Madeleine JOSEPH° and Jean-Claude KADER°
*I.N.R.A., Station de génétique et d'amélioration des plantes, route de St Cyr, 78026 Versailles Cedex, France
°Laboratoire de Physiologie Cellulaire et Moléculaire, 4 place Jussieu, 75252 Paris cedex 05, France

Introduction

In plants, the polyunsaturated fatty acids linoleate (octadeca-Δ9,12-dienoic acid) and α-linolenate (octadeca-Δ6,9,15-trienoic acid) are synthetized by Δ12- and Δ15-desaturases, respectively. They are essential fatty acid components of human nutrition as precursors of prostaglandins. Plant Δ12-desaturase (oleic acid → linoleic acid) is a well known membrane-bound enzyme present in endoplasmic reticulum of developing seeds. Δ12-desaturase employs NADH and molecular oxygen and requires cytochrome b5 and NADH-cytochrome b5 reductase as intermediate electron carrier (Demandre *et al.*, 1986). In contrast, *in vitro* Δ15-desaturase activity (linoleic acid → α-linolenic acid) is actually not observed in plant microsomes.

In sharp contrast to animals, few plants only contain γ-linolenic acid (octadeca-Δ6,9,12-trienoic acid) as triunsaturated fatty acid; the major triunsaturated fatty acid in plants is usually α-linolenic acid . The oil of the common borage (*Borago officinalis* L.) seed is one of the most abundant plant source of γ-linolenic acid (Stymne and Stobart, 1986). After several experiments to improve the *in vitro* Δ6- and Δ12-desaturation assays with respectively (1-^{14}C)linoleoyl-CoA and (1-^{14}C)oleoyl-CoA, we optimized the conditions for *in vitro* incubation (Galle *et al.*, 1993). From studies on Δ6-desaturase of rat liver microsomes (Okayasu *et al.*, 1981), we supposed that in plant, the Δ6-desaturase belongs to the same enzymatic complex as Δ12-desaturase.

This report describes the solubilization assays for Δ6- and Δ12-desaturations in developing seeds of borage microsomes with (1-^{14}C)oleoyl-CoA as precursor. Three detergents were selected : sodium deoxycholate, Triton X100 and Chaps.

Material - Methods

Borago officinalis L. seeds were obtained from the Institut National de Recherche Agronomique (I.N.R.A., Versailles, France). Microsomal membranes were obtained after two centrifugations at 15 000x*g* and 100 000x*g*, respectively. The resulting microsomal pellet was resuspended in 70mM phosphate buffer (pH 7.2). To the microsomal preparation, detergent was added by increasing concentrations from 0 to 1% (w/v). The mixture was stirred at 4°C for 30 min, and then centrifuged at 100 000x*g* for

J.-C. Kader and P. Mazliak (eds.), Plant Lipid Metabolism, 509–511.

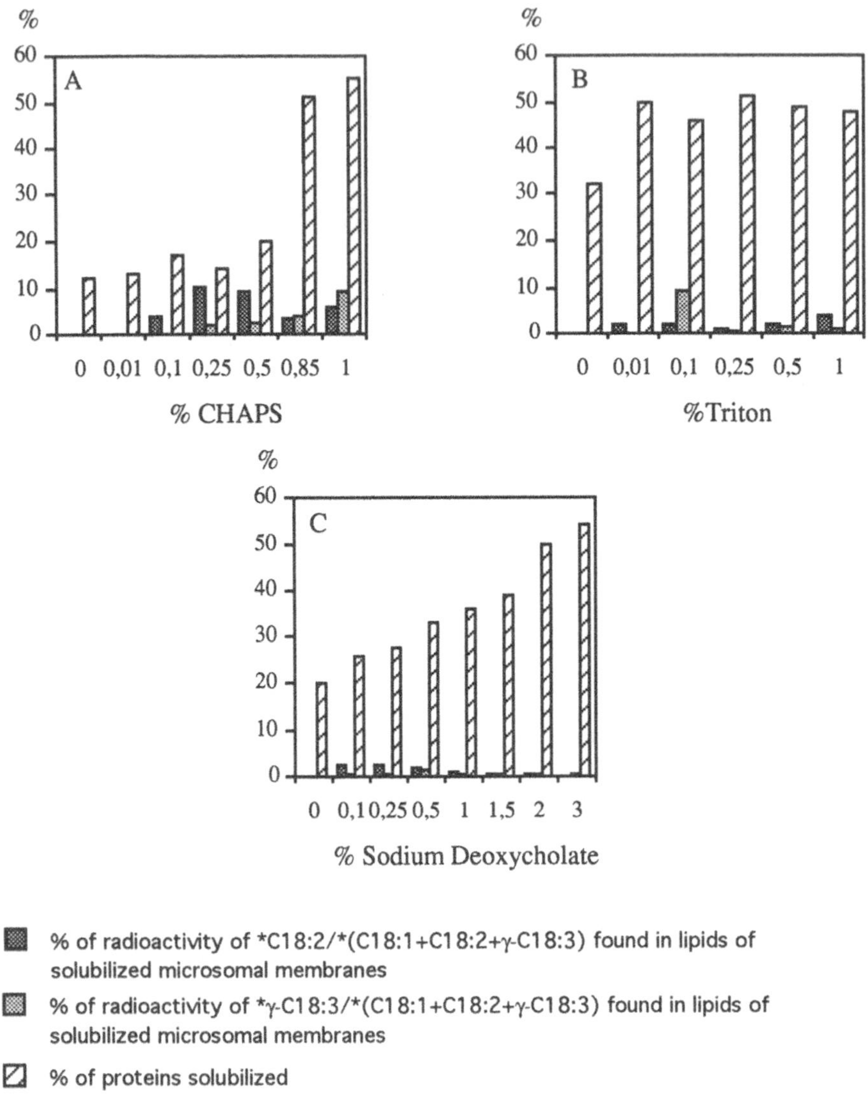

Fig.1. The effect of increasing concentrations of chaps (A), Triton X100 (B), Sodium deoxycholate (C) on the activities of Δ12- and Δ6-desaturase.

90 min. The activities were measured on the supernatant. The acivities of Δ12- and Δ6-desaturases were determined as described by Galle *et al.* (1993) by using 14C-oleoyl-CoA as precursor.

Results - Discussion

The effect of increasing concentration of detergents on Δ12- and Δ6-desaturase activities are shown in fig.1. We succeeded in preparing a solubilized fraction from detergent treated microsomes of borago seeds that was active in Δ12- and Δ6-desaturases. In our study, solubilization of proteins was optimal for 1% of Chaps (around 60% of total proteins was solubilized) in contrast to sodium deoxycholate where 3% of detergent was necessary to obtain the same rate of solubilization. Since early concentration of Triton X100, the maximum of solubilized proteins was recovered (around 50%). The Δ12-desaturase activity in the solubilized mixture of proteins increased from low concentrations of Chaps, to reach an optimum for 0.5% of detergent. Above 0.5%, the presence of detergent inhibited the Δ12-desaturase activity. In contrast, Δ6-desaturase activity increased with detergent concentration, reaching an optimum for 1% Chaps.With Triton X100, the Δ12- and Δ6-desaturase activities were low. 0.1% and 1% of Triton X100 were the most efficient concentrations to solubilize Δ6- and Δ12-desaturase activities, respectively; in contrast to Chaps, the two desaturase activities were lower. With sodium deoxycholate, the Δ12 and Δ6-desaturase activities were low and decreased rapidly when the concentration of detergent increased. The ratio of detergent to protein (results not shown) was also determined with those three detergents. The best result was obtain with Chaps, with the optimum rate of 1mg of Chaps per mg of proteins.

In conclusion, we selected **1% CHAPS** for futher experiments.

References

A.M. Galle, M. Joseph, C. Demandre, P. Guerche, J.P. Dubacq, A. Oursel, P.Mazliak, G. Pelletier and J.C. Kader (1994) Biosynthesis of γ-linolenic acid in developing seeds of borage (*Borago officinalis* L.). Biochim. Biophys. Acta, 1158,52-58.

Demandre, C., Trémolières, A., Justin, A.M. and Mazliak, P. (1986) Oleate desaturation in six phosphatidylcholine molecular species from potato tuber microsomes. Biochim. Biophys. Acta 877, 380-386.

Stymne, S. and Stobart, A.K. (1986) Biosynthesis of γ-linolenic acid in cotyledons and microsomal preparations of the developing seeds of common borage (*Borago officinalis*). Biochem. J. 240, 385-393.

Okayasu, T., Nagao, M., Ishibashi, T. and Imai, Y. (1981) Purification and partial characterization of linoleoyl-CoA desaturase from rat liver microsomes. Arch. Biochem. Biophys. 206, 21-28.

SUNFLOWER MUTANTS WITH ALTERED FATTY ACID COMPOSITION IN THE SEED OIL

Rafael Garcés[1], Jorge Osorio[1], Manuel Mancha[1] and José M. Fernández-Martínez[2].
[1]Instituto de la Grasa (CSIC). Apartado 1078. 41080 Sevilla, Spain
[2]Instituto de Agricultura Sostenible (CSIC). Apartado 4084. 14080 Crdoba, Spain

Introduction

Several sunflower (*Helianthus annuus* L.) mutants with altered seed fatty acid composition have been obtained. These mutants were selected after mutagenesis of sunflower seeds with X-rays, ethyl methane sulfonate (EMS) and sodium azide. The screening was performed at the level of the M_2 and M_3 generations, where individual seeds were analysed for fatty acid composition by GLC [1]. All mutants selected had higher saturated fatty acid content than the normal sunflower. Four had high stearic acid content and two high palmitic acid content. The high palmitic mutants are similar to that reported by Ivanov [2] and were obtained in both normal and high oleic acid backgrounds.

Results and Discussion

Several mutagenic treatments were performed on more than 20.000 seeds. The mutagenized seeds were grown in the field and the M_2 seeds collected. Three seeds from each capitulum were grown to obtain 3 sister M_3 seed populations. Both M_2 and M_3 generations were used to analyse the fatty acid content of the seeds. Any capitulum containing seeds with a particular fatty acid content out of the average in the mutagenized population ± 3 SD was selected. Near 100 putative mutants were found. At least 24 seeds from each were analysed and those having the more extreme fenotypes were planted to obtain the next generations. This procedure was followed until the mutant was fixed or lost. Most of the mutants were discarded after two or three generations because the characters were associated to lethality or were not genetically inheritable. Only 6 with high saturated fatty acid content were completely fixed (Table 1). Four of them had high stearic acid content (Table 1A). In the

J.-C. Kader and P. Mazliak (eds.), Plant Lipid Metabolism, 512–514.

original capitulum it was possible to detect two kinds of seeds, one with high stearic acid content and another with a low (normal) content. It is interesting to note that in some cases like CAS-4 and CAS-8 mutants the difference between the high and low classes was very small. The two high palmitic acid mutants (Table 1B) had two very different palmitic contents in the high and low classes. These mutants were already fixed in some seeds of the original capitulum, suggesting that the character is controlled by a single recessive gen. The genetic control of the high stearic acid mutants is probably more complex. After a few generations the mutant characters of these six lines were fixed (Fig. 1). The high stearic mutants have from 5 to 2 times more more stearic acid content than the normal lines and the high palmitic mutants also around fivefold more palmitic acid than the normal lines. In all these mutants the increase of stearic or palmitic acids was at the expense of oleic and linoleic acids. In the two high palmitic mutants palmitoleic acid was detected at the level of 4-7%.

Table 1. High saturated fatty acid fixed mutants. The content of the abnormal fatty acid of the original capitulum is shown for high stearic mutants (A) and high palmitic mutants (B), compared to that of normal lines.

	Mutant line	Fatty acid phenotype	Fatty acid (%) High	Fatty acid (%) Low	Mutagenic agent
(A)					
	CAS-3	Very high 18:0	12.2	4.5	EMS
	CAS-4	High 18:0	8.5	5.4	Azide
	CAS-8	High 18:0	7.3	3.5	EMS
	CAS-11	High 18:0	10.0	6.1	Azide
	RHA-274	Normal		4.8	-
(B)					
	CAS-5	Very high 16:0	31.0	8.2	X-rays
	CAS-12*	Very high 16:0	27.7	5.3	X-rays
	RHA-274	Normal	-	6.1	-
	G-8**	Very high 18:1	-	2.4	-

* CAS-12 mutant was obtained in a high oleic acid background. It had less than 5% of 18:2.
** High oleic mutant from Soldatov [3].

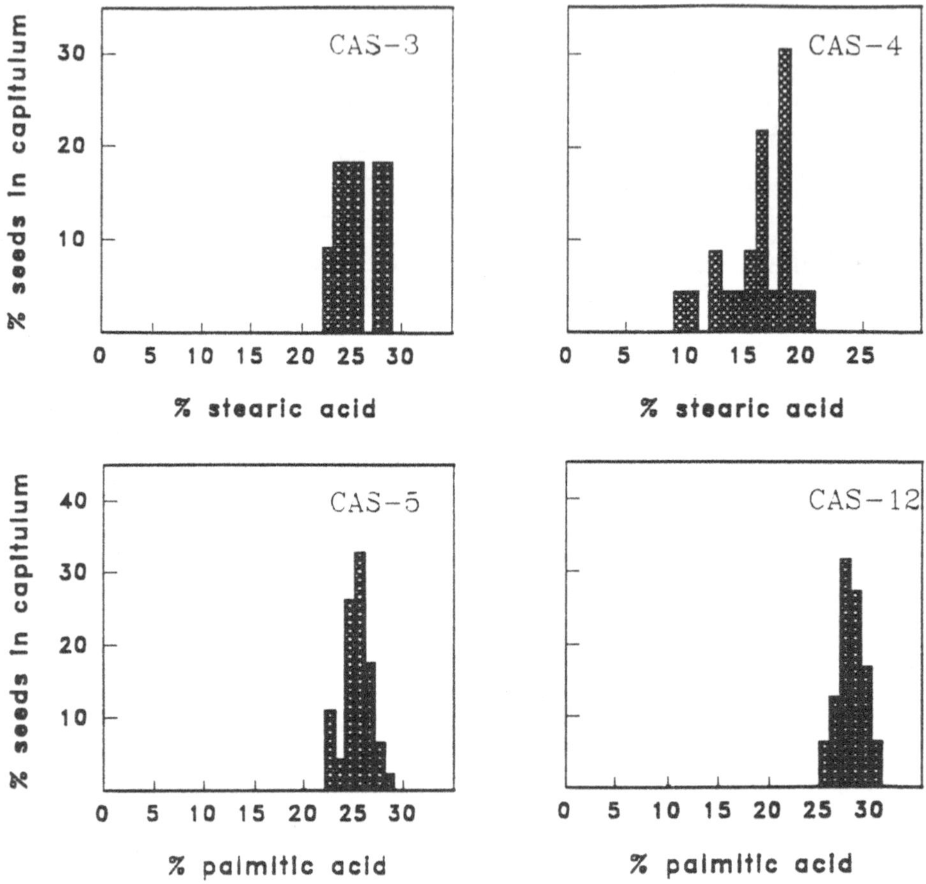

Figure 1. Distribution of stearic and palmitic acids content in four fixed mutants.

References

1. Garcés, R. and Mancha, M. One-step lipid extraction and fatty acid methyl esters preparation from fresh plant tissues. Anal. Biochem. 1993;**211**:139-143.
2. Ivanov, P. Petakov, D. Nikolova, V. and Petchev, E. Sunflower breeding for high palmitic acid content in the oil. Proc. 12th Int. Sun. Conf. 1988; Novi-Sad. Int. Sun. Assoc., Toowoomba, Australia. pp. 463-465.
3. Soldatov, K.I. Chemical mutagenesis in sunflower breeding. Proc. 7th Int. Sunf. Conf.; 1976; Krasnodar. Int. Sunf. Assoc., Vlardingen, Holland. pp. 352-357.

DIRECT GENETIC SELECTION OF A MAIZE cDNA CLONE THAT ENCODES A PROTEIN WHICH HAS SIMILARITIES TO FATTY-ACYLTRANSFERASES

Adrian P. Brown, Jack Coleman, Andrew M. Tommey, Martin D. Watson and Antoni R. Slabas

Lipid Molecular Biology Group, Department of Biological Sciences
University of Durham, South Road, Durham, DH1 3LE, U.K.

Introduction.

Membrane bound acyltransferase enzymes, located in the microsomal fraction of the cell, play a critical role in determining the composition of seed oils and the transfer of such enzymes from one plant source to another is a potential method of engineering plant oils to acquire desired characteristics. Purification of these enzymes to homogeneity has proved highly problematic however and to date no protein sequence information has been reported for any microsomal acyltransferase. Complementation cloning is a technique which circumvents the requirement for protein purification by directly selecting cDNAs which can correct a metabolic defect in a yeast or *E.coli* mutant and the method may prove to be useful for the isolation of acyltransferase genes. This report describes results of experiments designed to select maize cDNAs encoding a 1-acyl-glycerol-3-phosphate acyltransferase (1-acyl-G3P-AT) protein.

A plasmid, designated pMAT1, was isolated which corrected the temperature sensitive phenotype of JC201, an *E. coli* strain with no detectable 1-acyl-G3P-AT activity [1]. Biochemical studies showed that addition of this plasmid to JC201 restored 1-acyl-G3P-AT activity to the bacteria and the sequence of the largest open reading frame in the cDNA was most similar to the yeast protein SLC1, which is thought to be the first eukaryotic 1-acyl-G3P-AT isolated [2]. On the basis of this data we believe we have isolated a maize cDNA encoding a 1-acyl-G3P-AT enzyme.

Results.

Isolation of pMAT1. A pUC13-based cDNA library [3] made from maize endosperm cell suspension cultures was used to transform JC201, which was then plated on ampicillin at 44°C. After three rounds of selection, restriction-mapping revealed that twelve plasmids which supported growth at this temperature were the same and one was designated pMAT1. ^{32}P-labelling of

J.-C. Kader and P. Mazliak (eds.), Plant Lipid Metabolism, 515–517.

total phospholipid showed that JC201 accumulated lysophosphatidic acid to a level of 21% and addition of pMAT allowed further metabolism of it to form mainly phosphatidylethanolamine. Acyltransferase assays demonstrated that 1-acyl-G3P-AT activity was present in JC201 (pMAT1), but not in JC201.

Sequence Analysis. pMAT1 contained a cDNA insert of 1511bp. which encodes only one long open reading frame (ORF) of 374 amino acids - see below.

Figure 1.

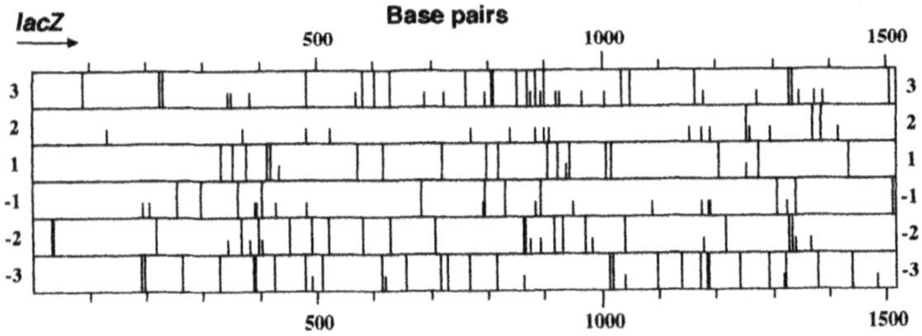

Figure 1: Map of the open reading frames present in the cDNA insert of pMAT1, derived from the cDNA sequence using DNA Strider. Full vertical lines represent stop codons and half vertical lines ATG sequences. The arrow shows the direction of transcription from the lacZ promoter of pUC13.

This was most likely expressed as part of a fusion protein including an extra 10 amino acids of LacZ and 43 corresponding to the proposed 5' untranslated region of the cDNA. The 374 amino acid ORF encodes a protein of relative molecular mass 42,543 which has a hydrophobic N-terminus. The yeast protein SLC1 was the most similar protein to the ORF of pMAT1 although the homology was not extensive (22.7% identity over a 229 residue overlap). Prior to the release of SLC1, the top-matched protein was the glycerol-3-phosphate acyltransferase protein - PlsB [4] of *E. coli.* The result of aligning the maize ORF, SLC1 and the *E. coli* 1-acyl-G3P-AT - PlsC [5] with the program MACAW is shown in figure 2. There are three blocks of conserved residues present, with the order and spacing almost identical in the three proteins.

Acknowledgements.

This work was supported by AFRC grant number PG 12/532. We thank Dr I. Rubenstein, Department of Plant Biology, University of Minnesota for allowing use of the library.

Figure 2.

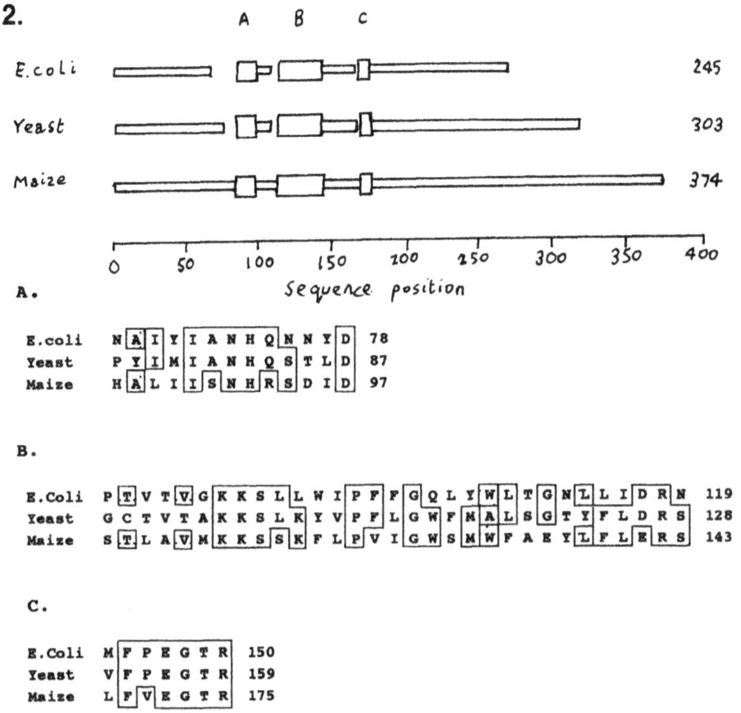

Figure 2: Spacing of conserved blocks of amino acids in the largest ORF in pMAT1 and the *E.coli* and yeast 1-acyl-G-3-P acyltransferases. The schematic diagram shows the distribution of conserved regions along the three sequences, represented by thicker bars. The conserved blocks A, B and C are shown underneath, with identical residues boxed.

References.

1. Coleman J. Characterization of *Escherichia coli* cells deficient in 1-acyl-*sn*-glycerol-3-phosphate acyltransferase activity. J Biol Chem 1990; 265:17215-17221.
2. Marek Nagiec M, Wells GB, Lester RL, Dickson RC. A supressor gene that enables *Saccharomyces cerevisiae* to grow without making sphingolipids encodes a protein that resembles an *Escherichia coli* fatty acyltransferase. J Biol Chem 1993;268:22156-22163.
3. Snustad DP, Hunsperger JP, Chereskin BM, Messing J. Maize glutamine synthetase c DNAs: Isolation by direct genetic selection in *Escherichia coli*. Genetics 1988;120:1111-1124.
4. Lightner VA, Bell RM, Modrich P. The DNA sequences encoding *plsB* and *dgk* of *Escherichia coli*. J Biol Chem 1983;258:10856-10861.
5. Coleman J. Characterization of the *Escherichia coli* gene for 1-acyl-*sn*-glycerol-3-phosphate acyltransferase (*plsC*). Mol Gen Genet 1992;232: 295-303.

INTERACTION OF PHOTOREACTIVE SUBSTRATE ANALOGS WITH DIACYLGLYCEROL ACYLTRANSFERASE FROM MICROSPORE-DERIVED EMBRYOS OF OILSEED RAPE

Randall Weselake[1], Keith Pomeroy[2], Tara Furukawa-Stoffer[1], Dawn Little[1] and Ram Rajasekharan[3]
[1] Department of Chemistry, University of Lethbridge, Alberta, Canada, T1K 3M4; [2] Plant Research Centre, Agriculture Canada, Ottawa, Ontario, Canada, K1A 0C6; [3] Plant Genetic Engineering Laboratory, New Mexico State University, Las Cruces, New Mexico, 88003, USA.

Synopsis

Photoreactive acyl-CoA analogs were used in the characterization of diacylglycerol acyltransferase (DGAT, EC 2.3.1.20) from microspore-derived (MD) embryos of oilseed rape. 12-[(4-Azidosalicyl)amino]dodecanoyl-CoA (ASD-CoA) was more effective than 12-azidooleoyl-CoA in the inhibition of DGAT activity. Photoaffinity labeling of particulate and solubilized fractions with ^{125}I-labeled ASD-CoA resulted in the incorporation of label into a number of polypeptides but fewer polypeptides were labeled in the solubilized fraction. Thus, photoaffinity labeling may be useful in the purification and characterization of DGAT.

Introduction

The properties of oilseed rape DGAT have been examined in particulate fractions from MD cultures [1-3]. Recently, photoreactive acyl-CoA analogs have been used to label microsomal acyl-CoA binding proteins from developing soybean cotyledons [4]. The current study examines the potential of using photoreactive analogs of acyl-CoA in the characterization of DGAT from MD embryos of oilseed rape.

Materials and Methods

Photoreactive substrate analogs, [^{125}I] ASD-CoA (0.12 Ci/mole) and azidooleoyl-[^{32}P]CoA (10 Ci/mmole) were synthesized according to Rajasekharan et al. [5]. [1-^{14}C]oleoyl-CoA (56 Ci/mole) was synthesized according to Taylor et al. [6]. The 1,500-100,000 *g* particulate fraction from MD embryos (*Brassica napus* L. cv Reston) [2] was shaken vigorously for 30 min at 4°C in 10 mM Tris-HCl buffer (pH 8.0) containing 2 M NaCl, 20% (w/v) glycerol and 1% (w/v) octanoyl-N-methylglucamide (MEGA-8) at a detergent to protein ratio of about 2:1. The solubilized enzyme was recovered in the supernatant following centrifugation at 105,000 *g* for 1 h. DGAT was concentrated by precipitation with polyethylene glycol 8,000 in the range of 2-6% (w/v). DGAT activity was assayed as described by Weselake et al. [2] and the protein content of extracts was determined using the Bio-Rad (Richmond, CA, USA) protein

J.-C. Kader and P. Mazliak (eds.), Plant Lipid Metabolism, 518–520.

microassay based on the Bradford [7] method. Photoaffinity labeling of particulate and solubilized fractions (50 μl aliquots containing about 200 μg protein) was conducted using [^{125}I]ASD-CoA (4 μM) as described previously for soybean microsomes [4]. Proteins were precipitated with trichloroacetic acid and washed with acetone. Polypeptides were analyzed by SDS-PAGE (7-15% gel) coupled with autoradiography.

Results and Discussion

Inhibition studies were conducted to determine if photoreactive acyl-CoA derivatives would be potentially useful in photoaffinity labeling of DGAT. Enzyme activity in the particulate fraction (1,500-100,000 *g*) was inhibited by 50% when using 50 μM ASD-CoA and 2 μM [1-^{14}C]oleoyl-CoA in the dark (Fig. 1A). The inhibitory effect of ASD-CoA was overcome to a large extent by increasing the acyl-donor substrate concentration to 14 μM which suggested that ASD-CoA interacted with the active site of DGAT. The ASD moiety was not incorporated into TG. 2-Azidooleoyl-CoA also inhibited DGAT activity (Fig. 1B) but this analog was not as effective as ASD-CoA. DGAT assays with azidooleoyl-[^{32}P]CoA resulted in the appearance of [^{32}P]CoA in the reaction mixture suggesting that the azidooleoyl moiety was incorporated into TG (data not shown).

Results of photoaffinity labeling of particulate and solubilized fractions with [^{125}I]ASD-CoA are shown in Fig. 2. A number of polypeptides were labeled in the particulate fraction (Fig. 2A, lane 5). Photoaffinity labeling of the solubilized fraction resulted in fewer labeled polypeptides (Fig. 2A, lane 2). Among the labeled polypeptides, only a 39 kDa species usually showed an increase in the intensity of labeling in the solubilized fraction. Exogenous diolein (300 μM) had no effect on the labeling patterns observed for the solubilized and particulate fractions, respectively (Fig. 2A., lanes 1 and 6). Protein-staining of separated polypeptides from the solubilized and particulate fractions is shown in Fig. 2B. A preliminary experiment using oleoyl-CoA in an attempt to protect the active site of DGAT resulted in decreased labeling of polypeptides (data not shown). We also plan to conduct photoaffinity labeling experiments with DGAT which has been further purified.

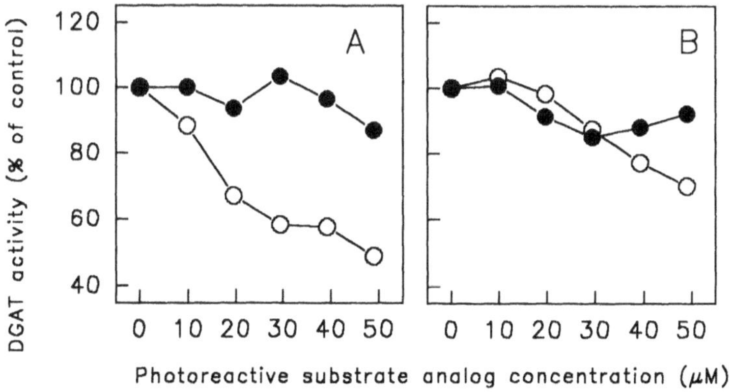

Figure 1. Effect of ASD-CoA (panel A) and 12-azidooleoyl-CoA (panel B) on DGAT activity in the particulate fraction. Membranes containing 68 μg protein were assayed for 10 min in the dark in reaction mixtures containing 2 μM (○) or 14 μM [1-^{14}C]oleoyl-CoA (●) and 330 μM exogenous *sn*-1,2-diolein as substrates. Data points are the means of duplicate assays.

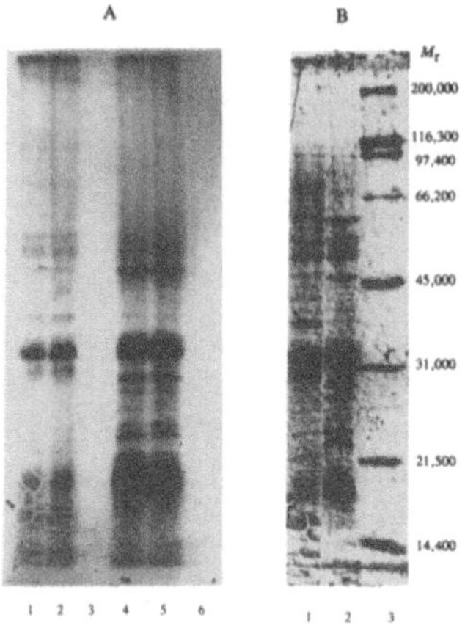

Figure 2. SDS-PAGE and autoradiography of particulate and solubilized fractions photolabeled with [^{125}I]ASD-CoA (4 μM). Photoaffinity labeling reactions were conducted using a hand-held UV lamp (5000μW/cm^2, Model UVG-54, Ultraviolet Products Inc., San Gabriel, CA, USA) with filter removed held 7 cm from the 50 μl sample (containing about 200 μg protein) which was kept on ice in the cap of a microcentrifuge tube. (A) Visualization of labeled polypeptides following autoradiography. Lanes 1 and 4: solubilized and particulate fraction, respectively, in the presence of exogenous diolein (300 μM). Lanes 2 and 5: same as lanes 1 and 4 but without exogenous diolein. Lanes 3 and 6: the analog was photolyzed before addition to the solubilized and particulate fraction, respectively.
(B) Coomassie Brilliant Blue R-250-staining of separated polypeptides in the solubilized (lane 1) and particulate (lane 2) fractions. M_r markers were from Bio-Rad.

Acknowledgements

This work was supported by grants to R.W. from the Natural Sciences and Engineering Research Council of Canada, Imperial Oil Ltd. and the Alberta Agricultural Research Institute, and by a USDA/NRI competitive grant to R.R. The authors thank Nancy Long and Ken Stanley for technical assistance and Doug Bray for photography.

References

1. Weselake RJ, Taylor DC, Pomeroy MK, Lawson SL, Underhill EW. Properties of diacylglycerol acyltransferase from microspore-derived embryos of *Brassica napus*. Phytochemistry 1991;30:3533-3538.
2. Weselake RJ, Pomeroy MK, Furukawa TL, Oishi RL. Partial purification and characterization of diacylglycerol acyltransferase from microspore-derived embryos of oilseed rape. In: MacKenzie SL, Taylor DC, editors. Seed oils for the future. Champaign, Illinois, AOCS Press, 1993:103-115.
3. Weselake RJ, Pomeroy MK, Furukawa TL, Golden JL, Little DB, Laroche A. Developmental profile of diacylglycerol acyltransferase in maturing seeds of oilseed rape and safflower and microspore-derived cultures of oilseed rape. Plant Physiol 1993;102:565-571.
4. Rajasekharan R, Nachiappan V, Roychowdhury H. Photolabeling of soybean microsomal membrane proteins with photoreactive acyl-CoA analogs. Eur J Biochem 1994;220:1013-1018.
5. Rajasekharan R, Marians RC, Shockey JM, Kemp JD. Photoaffinity labeling of acyl-CoA oxidase with 12-azidooleoyl-CoA and 12-[(4-azidosalicyl)amino]-dodecanoyl-CoA. Biochemistry 1993;32:12386-12391.
6. Taylor DC, Weber N, Hogge LR, Underhill EW. A simple enzymatic method for the preparation of radiolabeled erucoyl-CoA and other long-chain fatty acyl-CoAs and their characterization by mass spectrometry. Anal Biochem 1990;184:311-316.
7. Bradford MM. A rapid and sensitive method for the quantification of microgram quantities of protein utilizing the principle of protein-dye binding. Anal Biochem 1976;72:248-254.

ACYLTRANSFERASE ACTIVITIES SPECIFIC TO SN-2 IN LOW LINOLENIC ACID RAPESEED LINES

Christelle JOURDREN and Michel RENARD
INRA, Centre de Recherche de Rennes,
Station d'Amélioration des plantes BP 29, 35650 Le Rheu, France

SYNOPSIS

To promote uses of rapeseed (*Brassica. napus*) oil, some new oil fatty acid profiles are needed. In this paper, we report the development of low linolenic acid (C18:3) lines. The triacylglycerol (TAG) lipid structure of high and low C18:3 rapeseed oils has been determined to test Sn-2 specific acyltransferase activities. No differences were detected among the lines for the TAG lipid structure even with a 0.9% C18:3 content.

INTRODUCTION

Today it is more and more important that the agricultural productions match the request of industrial and food industries. Oilseed rape will have to provide special oil quality (*e.g.* one fatty acid composition, for one industrial application). The rapeseed alimentary oil contains mainly C18 fatty acids, and a few C16 and C20 fatty acids. The α-linolenic acid (C18:3) content of this oil is about 8 to 12% which is not appropriate for cooking nor for margarine production. Because of the poor genetic variability of this character, physical and chemical mutagenesis were used to modify the linolenic acid content of seeds (1). A Canadian variety, Stellar, obtained by chemical mutagenesis treatment on seed, has a α-linolenic acid content lower than 3%. In addition we have obtained in our group, a line "SL" which has only 0.9% of C18:3, by pedigree breeding.

The fatty acid profile of seed oil,(*e.g.* triacylglycerols or TAGs), is very different from membrane lipids. This difference can be assigned to a small number of enzymes (2); some of which being acyltransferases that are specific to the Sn-2 position of glycerol backbone(3). One objective of this study was fo verify if deficiency of Sn-2 specific acyltransferase activities correlates with low linolenic acid content. The TAG lipid structure of 3 low linolenic acid lines, variety Stellar (SL, St T and St P1) and 2 high linolenic acid lines, variety Drakkar, (Drk C and Drk P2) are reported.

J.-C. Kader and P. Mazliak (eds.), Plant Lipid Metabolism, 521–524.

MATERIALS AND METHODS

Plant materials :

The two spring rapeseed varieties Stellar (St P1) and Drakkar (Drk P2) were crossed. The F2 progeny of 25 F1 plants were sown in field conditions. The F2 plants were self pollinated, and harvested separately. Analyses of fatty acid composition on F3 seed bulk (50 seeds / plant) permitted to select 15 F2 plants. Later on, screening in the 15 F3 progeny with the same method, allowed to select 5 plants of one F2 original plant. One of them had a C18:3 content lower than 1.2 %, and we determined the fatty acid composition on 50 F3 half-seeds; the plants obtained, from the half-seeds remained, were grown in field and self-pollinated. Matures seeds (F4 progeny) of each plant were harvested separately and seed bulks (10 seeds / plant) were analysed for fatty acid composition. One of these plants, having 0.9% of linolenic acid, and enough seeds for oil extraction, was kept for this study; This plant was named SL line.

St T was a double haploid line of Stellar. Certified seeds of the French variety Drakkar (Drk C) were used as high linolenic acid control.

Enzymatic digestion :

Oils were extracted according to Bligh (4). Forty mg of oil from each lines were digested by a swine pancreatic lipase during 1mn in Tris buffer, Sodium Cholate and CaCl2 (AFNOR convention). The reaction was stopped by an addition of HCl 11N. One volume of Diethyl ether was added to the suspension which has been homogenised during 2mn then centrifuged. The upper phase was recovered and stored under N_2 at 4°C.

Lipid class separation and FAMES (Fatty Acid Methyl Esters) preparation

The lipid mixtures (40µl) were deposed on thin layer, then migrated in Hexane/Diethyl ether/Formic acid (70:30:1) in two times. The spots were scraped and the FAMES obtained by acidic methanolyse with BF3 (Boron Trifluorure 14% in methanol). The fatty acid composition of the 4 classes, TAG, fatty acids (FAs), Diacylglycerols (DAGs) and Monoacylglycerols (MAGs) was determined by CPG as well as the oil composition. The results are given for the unsaturated C18 fatty acids, and we have considered only the fatty acid composition of oils and MAGs. In fact, even with a very short time of enzymatic digestion, isomerisation between Sn-2 and Sn-1 or 3 position is possible. We therefore preferred not to use the data from DAGs and Fas analyses. The fatty acid composition of [Sn-1 + Sn-3] position and the acylation rate in Sn-2 position (Table 1) were determined for each fatty acid as follows :

- [% Sn-1 + Sn-3] = (3 x [% oil] - [% Sn-2]) / 2

- Acylation rate in Sn-2 position : [% Sn-2] x 100 / 3 x [% oil]

Table 1 : Unsaturated C18 Fatty acid composition of MAGs (I) , oils(II), Sn-1 + Sn-3 positions (III), and acylation rates on Sn-2 position (IV).

Table 1

	Lines	I [%Sn-1+%Sn-3]	II [%Sn-2]	III [%oil]	IV [%Sn-2]/3x[%oil]
C18:1	SL	62.3	43.4	56.0	25.8
	St T	66.7	52.1	61.8	28.1
	St P1	67.2	49.0	61.1	26.7
	Drk P2	70.8	53.5	65.0	27.4
	Drk C	73.8	51.3	66.3	25.8
C18:2	SL	25.1	54.0	34.7	51.9
	St T	20.3	40.7	27.1	50.1
	St P1	19.2	45.3	27.9	54.1
	Drk P2	11.7	29.4	17.6	55.7
	Drk C	10.6	29.5	16.9	58.2
C18:3	SL	0.7	1.3	0.9	47.0
	St T	1.4	3.8	2.2	57.6
	St P1	1.0	4.4	2.1	69.8
	Drk P2	5.9	15.6	9.1	57.1
	Drk C	5.0	15.2	8.4	60.3

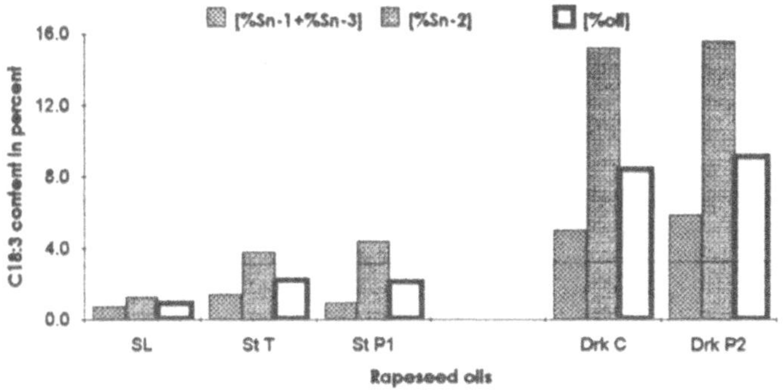

Figure 1 : Linolenic acid content in [Sn-1 + Sn-3] positions, [Sn-2] position and oil .

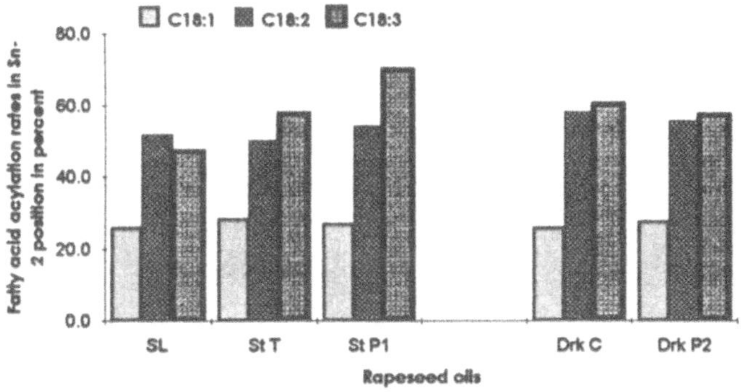

Figure 2 : Unsaturated C18 fatty acid acylation rates on the Sn-2 position.

RESULTS AND DISCUSSION

A stable low linolenic line was obtained. Such a result was confirmed by F5 analysis (data not shown). The oil content of the line was 45% (data not shown). So, it has been shown that it is possible, using traditional plant breeding and mutagenesis, to reduce C18:3 content to a very low level (0.9%). This is lower than the level obtained using genetic engineering techniques (4%)(5).

As expected, the linolenic acid was mostly located on the Sn-2 position. It was the same for linoleic acid (Table 1).

In the low linolenic acid oils, the decrease in C18:3 was observed on all the positions (Sn-2 and Sn-1+Sn-3)(Fig1).

The acylation rate of C18:1 on the Sn-2 position was constant (Fig2). On the opposite, the acylation rate of C18:3 on this position, compared to the Drakkar lines, was :(i) lower in the SL line, (ii) equal in the St T line, and (iii) superior in the St P1 line. The average of the two groups (low C18:3 lines and high C18:3 lines) were nearly the same (Fig3).

It was unexpected that the acylation rate of C18:2, on this same position, was lower in the low linolenic lines. These lines were also characterised by a strong increase in C18:2 oil content, an increase in the order of 50 - 100 percent (Table 1).

The oil fatty acid composition of low linolenic acid lines were widely modified. This transformation affected only the C18 unsaturated fatty acids (C18:1, C18:2 and C18:3). But no major consequence was observed on TAG lipid structure, with this change in the ratios among these 3 fatty acids .We did not find a decrease acyltransferase activities specific to Sn-2, in this experiment. It doesn't mean that there is no one, but even there is one, it does not clearly modify the TAG lipid structure at the end of the storage lipid biosynthesis. It will be interesting to determine the lipid structure of some key components like Phosphatidylcholine (PC) or LysoPC, involved in previous steps of lipid biosynthesis.

References:

1. Robbëlen G.,Nitsch A., **1975.** Genetical and physiological investigations on mutants for polyenoic fatty acids in rapeseed. Z. Pflanzenzüchtg. 75 : 93-105
2. Somerville CR., Browse J.,**1991**.Plant lipids: Metabolism, Mutants, and Membranes. Science. 252: 80-87
3. Ohlrogge JB., Browse J., Somerville CR.,**1991**. The genetics of plant lipids. Biochimica and Biophysica Acta. 1082: 1-26
4. Bligh E.G., Dyer W.S.,**1959** A rapid method of total lipid extraction and purification. Can J.Biochem. Physiol. 37 : 911-917
5. Davies M., **1994.** Proceedings of the 11th plant lipid meeting. Jun 27-02 july. Paris.

PHOTOAFFINITY LABELING OF LYSOPHOSPHATIDYLCHOLINE ACYLTRANSFERASE FROM DEVELOPING CASTOR BEAN ENDOSPERM

Ram Rajasekharan and Hiranya Roychowdhury
Plant Genetic Engineering Laboratory, New Mexico State University. Las Cruces, New Mexico 88003, USA.

Introduction

Acyltransferases play key roles in the biosynthesis of phospholipids and triacylglycerol. The membrane-bound acyltransferases have proven to be extremely recalcitrant to biochemical purification. We have used photoaffinity ligands bearing azido groups for site-directed covalent labeling of proteins [1]. The method described here opens up a new and potentially rewarding approach for the identification of lysophosphatidylcholine (LPC) acyltransferase (EC 2.3.1.23) from developing castor bean (*Ricinus communis* L.) endosperm using [(3-iodo-4-azidosalicyl)-12-amino]dodecanoyl-CoA (ASD-CoA) and [(3-iodo-4-azidosalicyl)-12-amino]dodecanoyl-*sn*-glycerol-3-phosphocholine (N_3-LPC). This enzyme catalyzes the acylation of LPC to form phosphatidylcholine (PC). It has been suggested that the levels of polyunsaturated fatty acids in oilseeds may largely be controlled by the action of this acyltransferase [2].

Materials and Methods

A photoreactive PC analog (N_3-PC) was synthesized by acylating glycerophosphocholine with 12-aminododecanoic acid followed by alkylation of the primary amino group with 4-azidosalicylic acid. N_3-PC was digested with phospholipase A_2 to form N_3-LPC. ASD-CoA was synthesized, iodinated and purified as described earlier [1,3]. Developing castor bean endosperm (35-45 days after flowering) were collected and stored at -70 °C until use. LPC acyltransferase was assayed as described by Rajasekharan, et al [1]. Sucrose step density gradient was used to isolate microsomal membranes (20-30% sucrose interface) [4]. Photolysis was carried out in 50 μl volume containing 60 μg protein in the presence of 2-mercaptoethanol (1 mM) for 2 min with a hand-held UV-lamp with the filter removed at a distance of 7 cm.

J.-C. Kader and P. Mazliak (eds.), Plant Lipid Metabolism, 525–527.

Results and Discussion

The structures of the synthesized photoprobes are presented in Figure 1. The synthesized analogs were used to study the interaction with LPC acyltransferase from developing castor bean endosperm. ASD-CoA acts as a non-competitive inhibitor with an apparent Ki of 8.8 μM in the dark. On the contrary, the LPC analog acts as a substrate (Km 23.3 μM) in the dark. Sucrose density gradient purified microsomal membranes were photolyzed with 25 μM acyl-CoA or LPC analog. The unbound photoprobe was removed by dilution followed by centrifugation to pellet the membranes. The enzyme was assayed using [^{14}C]oleoyl-CoA, which resulted in 78 or 62% inhibition of enzyme activity for ASD-CoA or N_3-LPC, respectively, indicating that both the photoprobes act as irreversible inhibitors. The enzyme activity was unaffected by 3 min of exposure with UV-light. ASD-CoA or N_3-LPC inhibition was protected by oleoyl-CoA or LPC (oleoyl), respectively. These results suggest that both the photoprobes interact with the active site or the substrate binding site of the enzyme.

Figure 1. Structure of ASD-CoA and N_3-LPC

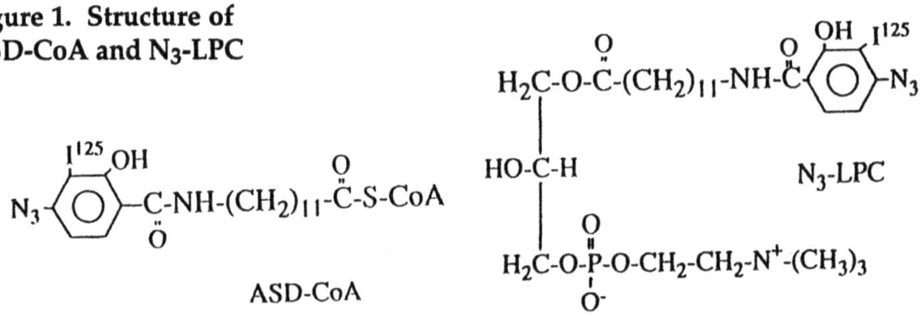

Covalent incorporation of ^{125}I-labeled photoprobes into sucrose step gradient purified microsomal membranes was determined after complete denaturation of proteins followed by 10% SDS-PAGE and autoradiography. Five proteins with the molecular weights of 63, 55, 52, 33 and 29 kDa were labeled with ASD-CoA (Fig. 2). Among the labeled proteins, the 29 kDa polypeptide was most intensely labeled. N_3-LPC labeled 63, 52, 45, 36, 35, 33 and 28 kDa polypeptides. No labeling was observed in the absence of UV-light. Photolysis of the photoprobes prior to the addition of microsomal membranes led to the labeling of the 63 kDa polypeptide (Fig. 2, lane 1,4), suggesting the possibility of the existence of long-lived chemical intermediates. Coomassie-Blue staining of total proteins revealed that the labeled polypeptides are not abundant in microsomal membranes (Fig. 2, lane 7). Among the labeled proteins, 52 and 33 kDa polypeptides were commonly labeled with both photoprobes. However, the 52 kDa polypeptide was most intensely labeled with acyl acceptor analog suggesting that this polypeptide could be LPC acyltransferase or its subunit.

Fig. 2. Photolabeling of microsomal membranes with N_3-LPC and ASD-CoA.

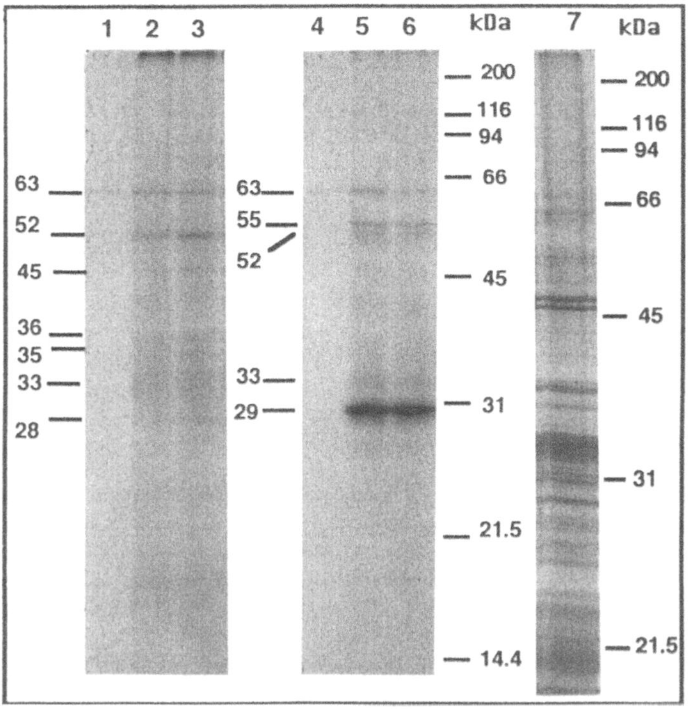

Membranes were photolyzed with N_3-LPC (lane 1-3) and ASD-CoA (lane 4-6). N_3-LPC (lane 1) and ASD-CoA (lane 4) were photolyzed first then added to the membranes. Membranes were photolyzed in the absence of oleoyl-CoA (lane 2) or in the absence of LPC (lane 5). Photolysis was carried out in the presence of oleoyl-CoA (lane 3) or LPC (lane 6). Lane 7, Coomassie-Blue stained protein profile (60 μg).

Acknowledgments

This work was supported by USDA/NRI competitive grants to R.R. We thank Dr. John Kemp for his constant encouragement.

References

1. Rajasekharan R, Nachiappan V, Roychowdhury HS. Photolabeling of soybean microsomal membrane proteins with photoreactive acyl-CoA analogs. Eur J Biochem 1994; 220: 1013-1018.
2. Stymne S, Stobart AK. Triacylglycerol biosynthesis. In: Stumpf PK, editor. The biochemistry of plants. New York: Academic Press. 1987: 175-214.
3. Rajasekharan R, Marians RC, Shockey JM, Kemp JD. Photoaffinity labeling of acyl-CoA oxidase with 12-azidooleoyl-CoA and 12-[(4-azidosalicyl)-amino]dodecanoyl-CoA. Biochemistry 1993; 32: 12386-12391.
4. Lord JM. Isolation of endoplasmic reticulum. Methods Enzymol 1987; 148: 576-584.

ROLE OF CHOLINEPHOSPHOTRANSFERASE AND DIACYLGLYCEROL ACYLTRANSFERASE IN CHANNELING UNUSUAL FATTY ACIDS INTO THE TRIACYLGLYCEROL POOL DURING OILSEED DEVELOPMENT

Guido Vogel and John Browse, *Institute of Biological Chemistry, Washington State University, Pullman, WA 99164, USA*

Introduction

Diacylglycerols (**DAGs**) synthesized through the Kennedy pathway are precursors of both membrane (phosphatidylcholine: **PC**) and storage glycerolipids (triacylglycerol: **TAG**) **[Fig. 1]**. In plants that store unusual fatty acids in their TAGs the question arises of how these components are excluded from the membrane lipids. Several different strategies have been proposed [1]: a) selectivity of CDP-choline:diacylglycerol cholinephosphotransferase (**CPT**) for DAGs containing C16 and C18 fatty acids, b) selectivity of acyl-CoA:diacylglyerol acyltransferase (**DAGAT**) for DAGs with membrane incompatible fatty acids, c) separate localization of storage and membrane glycerolipid biosynthesis, d) fast turnover of PC species with wrong fatty acids.
These hypotheses were examined by characterizing the selectivity of **CPT** and **DAGAT** in microsomal preparations from developing cotyledons of a series of oilseed plants towards different DAG species. Four different plants were used as model systems: safflower (*Carthamus tinctorius*), a plant producing no unusual fatty acids; rapeseed (*Brassica napus)*, which stores erucic acid (22:1) in its TAG species; *Cuphea*, a plant accumulating medium chain fatty acids (C8 - C12) in its seed triacylglycerols; and castor bean (*Ricinus communis*) which stores up to 90% ricinoleic acid (18:1-OH) in its seed oils.

Materials and Methods

Microsomes were isolated from developing cotyledons (or endosperm in the case of castor bean) as described by Slack et al. [2].
The substrates (di-10:0-, di-18:1-, di-18:2- and di-22:1-DAG) were synthesized starting from L-[U-^{14}C]-glycerol-3-phosphate and the fatty acid anhydride. DAGs containing ricinoleic (di-ric- and 1-ric-DAG) acid were obtained after incubation of developing castor bean endosperm with [U-^{14}C]-glycerol.
CPT and DAGAT activity was assayed using a modified method of Slack et al. [2].
Lipids were separated by TLC and identified by staining with iodine vapour or primulin spray reagent. The incorporation of labeled DAG into PC and TAG was measured by scintillation counting.

J.-C. Kader and P. Mazliak (eds.), Plant Lipid Metabolism, 528–530.

Table 1: Initial rates of CPT activity (nmol PC x (mg protein x min)$^{-1}$) in microsomes from developing cotyledons of safflower, rapeseed and *Cuphea* and developing endosperm of castor bean using a) 150 or b) 1000 nmol DAG x mg protein^{-1}

substrate	safflower(a,b	rapeseed(a	*Cuphea*(a	castor bean(b
di-18:1-DAG	5.09±0.98	0.53±0.01	1.48±0.51	7.98±0.37
di-18:2-DAG	5.26±1.06	-	-	8.72±0.47
di-10:0-DAG	4.46±0.22	0.55±0.13	1.31±0.08	8.38±1.61
di-22:1-DAG	3.84±0.52	0.20±0.03	-	6.65±0.10
di-ric-DAG	-	-	-	7.49±2.00
1-ric-DAG	-	-	-	7.49±0.84

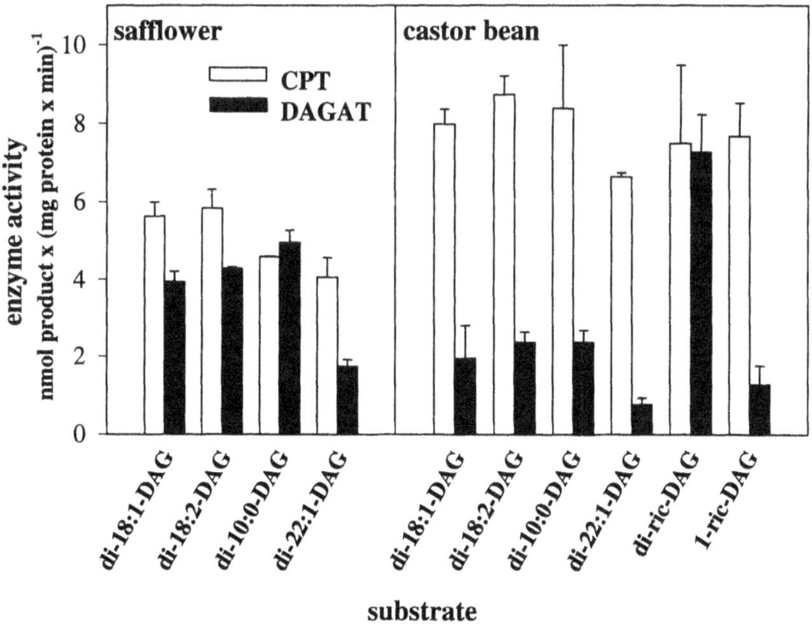

Figure 1: Activity of CPT and DAGAT in microsomal preparations from developing cotyledons of safflower and developing endosperm of castor bean towards different DAG molecular species

Results

Microsomes from developing cotyledons of safflower and developing endosperm of castor bean were able to use DAG for the *in vitro* synthesis of PC and TAG.
Using safflower microsomes and di-18:1-DAG as substrate we were able to achieve a saturation of PC synthesis at around 700 nmol DAG / mg protein and of TAG synthesis at 500 nmol DAG / mg protein for an assay period of 15 minutes at 30°C (data not shown).
The rates of PC and TAG synthesis in assays with safflower microsomes were linear for at least 20 minutes using 1000 nmol di-18:1-DAG / mg protein (data not shown).
CPT from safflower and castor bean showed no selectivity towards any of the tested substrates **[Table 1]**.
Rapeseed microsomes exhibited only 12% of the CPT activity / mg protein found in safflower **[Table 1]**. The cholinephosphotransferase from this organism synthesized PC at ~ 50% the rate with di-22:1-DAG compared to di-10:0 and di-18:1-DAG. **[Table 1]**.
In *Cuphea* microsomes CPT showed similar reaction rates using either di-10:0- or di-18:1-DAG as substrate **[Table 1]**. PC synthesized with di-10:0-DAG however seemed to be hydrolized to lyso-PC in the course of the experiments (data not shown).
DAGAT from safflower and castor bean exhibited the lowest activity towards di-22:1-DAG **[Fig. 1]**.
In both plants DAGAT could not distinguish between di-10:0-DAG and di-18:1-DAG or di-18:2-DAG **[Fig. 1]**.
DAGAT from castor bean, a plant storing up to 90% ricinoleic acid in its TAGs, used di-ric-DAG at three times the rate of any other tested substrate **[Fig. 1]**.

Conclusions

CPT from developing cotyledons of safflower, rapeseed, *Cuphea* and developing endosperm from castor bean were unable to discriminate between DAG species containing C18 and those containing 10:0, 22:1 and/or 18:1-OH fatty acids. It is therefore unlikely that this enzyme plays a central role in excluding unusual fatty acids from the membrane pool of PC.
In contrast **DAGAT** from castor bean endosperm showed a strong preference for DAG species containing two ricinoleic acid residues. This indicates, that this enzyme is involved in channeling ricinoleic acid, a membrane incompatible fatty acid, into the storage TAGs. The fact that CPT in the same plant tissue cannot exclude this DAG molecular species from the production of PC suggests that the biosynthesis of membrane glycerolipids is spatially separated from that of TAGs.

References

1. Stymne S. Biosynthesis of uncommon fatty acids and their incorporation into triacylglycerols. In: Murata N, Somerville C, editors. Biochemistry and Morlecular Biology of Membrane and Storage Lipids of Plants. Current Topics in Plant Physiol 9, Rockville: ASPP, 1993: 150-158

2. Slack CR, Roughan PG, Browse JA and Gardiner SE. Some properties of cholinephosphotransferase from developing safflower cotyledons. Biochim Biophys Acta 1985; 883: 438-448

cDNA CLONES FROM *LIMNANTHES DOUGLASII* ENCODING AN ERUCOYL-CoA SPECIFIC 1-ACYLGLYCEROL-3-PHOSPHATE ACYLTRANSFERASE

Christiane Hanke, Gabriele Peterek, Frank P. Wolter and Margrit Frentzen
Universität Hamburg, Institut für Allgemeine Botanik, Ohnhorststr. 18,
D-22609 Hamburg, Germany

Introduction

Rapeseed oil represents a renewable resource suitable for different technical purposes. Its industrial applicability can be improved by a homogenous fatty acid composition consisting of 90 % of one technically attractive fatty acid, such as erucic acid [1]. The content of erucic acid in the oil of current rapeseed varieties is limited to about 60 % since erucic acid is esterified in the *sn*-1 and *sn*-3 position of the glycerol backbone only, whereas the *sn*-2 position carries unsaturated C_{18} acyl groups [2]. This fatty acid pattern of rapeseed oil is established by the microsomal acyltransferases involved in triacylglycerol biosynthesis [3]. Due to the pronounced specificity and selectivity of the 1-acylglycerol-3-phosphate acyltransferase for unsaturated C_{18} acyl groups, erucic acid is excluded from the *sn*-2 position of the glycerol backbone and, thus, the formation of trierucin is prevented [3,4]. In contrast to the acyltransferase of rapeseed, the respective seed-specific enzyme from *Limnanthes douglasii* effectively utilizes erucoyl-CoA as substrate [5]. To achieve trierucin biosynthesis in rapeseed by genetic engineering, we have isolated cDNAs from *Limnanthes douglasii* encoding an erucoyl-CoA specific 1-acylglycerol-3-phosphate acyltransferase.

Results and Discussion

ISOLATION OF cDNA CLONES FROM *LIMNANTHES DOUGLASII*

To isolate cDNAs encoding erucoyl-CoA specific 1-acylglycerol-3-phosphate acyltransferases, a cDNA expression library from about ten days old embryos from *Limnanthes douglasii* was prepared in λZAP (Stratagene). The primary library consists of 3.5×10^8 independent cDNA clones with an average insert size of 1000 base pairs (bp). This library was converted into the respective plasmid library and screened via heterologous complementation of the thermosensitve *E. coli* mutant JC201 deficient in 1-acylglycerol-3-phosphate acyltransferase activity [6]. Nine cDNA clones of *Limnanthes douglasii* were

J.-C. Kader and P. Mazliak (eds.), Plant Lipid Metabolism, 531–533.

isolated which enabled the *E. coli* mutant to grow at nonpermissive temperatures. According to their restriction patterns and partial sequence analyses, six of the cDNA clones represent the same mRNA species differing in length of their 5' ends only.

The insert of pCH21 was sequenced. It has a length of 1030 bp plus a stretch of adenosyl residues and contains one long open reading frame. Starting with the first ATG the open reading frame covers 852 nucleotides. The stop codon detected within the same reading frame upstream to this ATG is taken as evidence that the complete coding region is covered by pCH21. From the open reading frame a polypeptide of a molecular mass of about 31 kD can be deduced which shows substantial amino acid sequence similarity to 1-acylglycerol-3-phosphate acyltransferase of *E. coli* encoded by *plsC* [7].

FUNCTIONAL AND EMBRYO SPECIFIC EXPRESSION

Functional expression studies in *E. coli* were carried out in order to demonstrate directly that the isolated cDNA clone of *Limnanthes douglasii* encodes a 1-acylglycerol-3-phosphate acyltransferase. For this purpose the insert of pCH21 was cloned in the expression vector pQE60 (Qiagen) giving the construction pQEL21. Membrane fractions were isolated from JC201 cells harbouring either pQEL21 or pQE60 and used for enzyme assays. In agreement with the results of Coleman [6], no acyltransferase activity was detectable in membrane fractions of JC201 cells only harbouring the vector (Fig. 1). On the other hand, membrane fractions of JC201 cells harbouring the construct pQEL21 displayed acyltransferase activity (Fig. 1). This activity was inducible in the bacterial cells by isopropyl-ß-D-thiogalactopyranoside (data not shown). The acylation rates of 1-acyl-

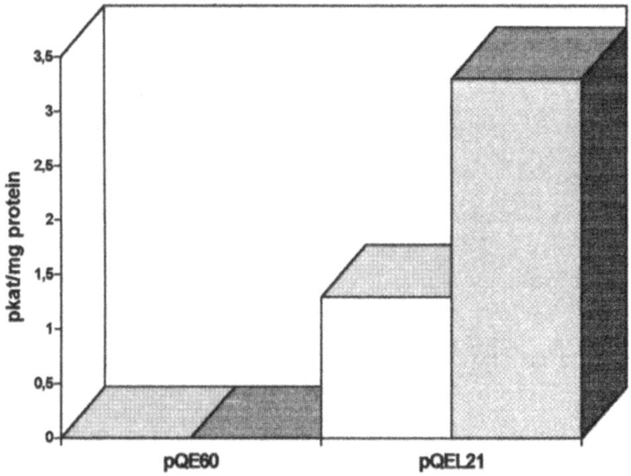

Fig. 1. Functional expression of the cDNA from *Limnanthes douglasii* in *E. coli* JC201. Activities of 1-acylglycerol-3-phosphate acyltransferase were determined with either oleoyl-CoA (light bars) or erucoyl-CoA (dark bars) in membrane fractions of JC201 cells harbouring the vector pQE60 or the construct pQEL21, respectively, as described before [5].

glycerol-3-phosphate catalysed by the membrane fractions were found to be threefold higher with erucoyl-CoA than with oleoyl-CoA (Fig. 1). Hence, the results univocally show that the cDNA clone from *Limnanthes douglasii* encodes an erucoyl-CoA specific 1-acylglycerol-3-phosphate acyltransferase. Furthermore, Northern blot analysis revealed that the corresponding transcript of the isolated cDNA clone was detectable in developing embryos, but not in leaves of *Limnanthes douglasii* (data not shown). These data correlate very well with the results of enzymatic studies [5,8].

In summary, we succeeded in the isolation of a cDNA from *Limnanthes douglasii* which encodes an erucoyl-CoA specific 1-acylglycerol-3-phosphate acyltransferase. The deduced amino acid sequence shows substantial sequence similarity to the respective acyltransferase of *E. coli*. Consequently, the isolated cDNA is suitable for transforming rapeseed with the aim to increase the content of erucic acid in the fatty acid composition of rapeseed oil and, thus, to improve its industrial applicability.

Acknowledgements

We are indebted to Jack Coleman for providing us with the *E. coli* mutant JC201. We thank Monika Rade and Anja Sievers for their skillful technical assistance. These investigations were supported by the Bundesminister für Forschung und Technologie (Förderkennzeichen 1310528F).

References

1. Ohlrogge J B. Design of new plant products: engineering of fatty acid metabolism. Plant Physiol 1994; 104: 821-826.
2. Norton G, Harris JF. Triacylglycerols in oilseed rape during seed development. Phytochemistry 1983; 12: 2703-2707.
3. Bernerth R, Frentzen M. Utilization of erucoyl-CoA by acyltransferases from developing seeds of *Brassica napus* L. involved in triacylglycerol biosynthesis. Plant Science 1990; 67: 21-28.
4. Cao Y, Oo KC, Huang AHC. Lysophosphatidate acyltransferase in the microsomes from maturing seeds of meadowfoam (*Limnanthes alba)*. Plant Physiol 1990; 94: 1199-1206.
5. Löhden I, Frentzen M. Triacylglycerol biosynthesis in developing seeds of *Tropaeolum majus* L. and *Limnanthes douglasii* R. Br. Planta 1992; 188: 215-224.
6. Coleman J. Characterization of *Escherichia coli* cells deficient in 1-acyl-*sn*-glycerol-3-phosphate acyltransferase activity. J. Biol Chem 1990; 265: 17215-17221.
7. Coleman J. Characterization of the *Escherichia coli* gene for 1-acyl-*sn*-glycerol-3-phosphate acyltransferase (*plsC*). Mol Gen Genet 1992; 232: 295-303.
8. Laurent P, Huang AHC. Organ- and development-specific acyl coenzym A lysophosphatidate acyltransferases in palm and meadowfoam. Plant Physiol 1992; 99: 1711-1715.

CLONING OF A cDNA CODING FOR A GLYCEROL-3-PHOSPHATE DEHYDROGENASE FROM *CUPHEA LANCEOLATA*

Ludger Hausmann, Jeff Schell and Reinhard Töpfer
Max-Planck-Institut für Züchtungsforschung
Carl-von Linné-Weg 10, D-50829 Köln, Germany

Introduction

The formation of glycerol-3-phosphate (G3P) in plants as a substrate for glycerolipid biosynthesis can be accomplished by either a glycerol-3-phosphate dehydrogenase (GPDH, EC 1.1.1.8) or a glycerol kinase (GK). However, it is likely that GPDH rather than GK is responsible for the major part of G3P formation, since GPDH the activity of GK is particularly low in oil-synthesizing plant tissues [1]. Two highly regulated isoforms of GPDH, one plastidic and the other cytosolic, have been described [2]. In order to investigate the role of GPDH in glycerolipid biosynthesis a cDNA encoding GPDH was isolated from an expression library of immature embryos from *Cuphea lanceolata* [3] by functional complementation of an *Escherichia coli* (*E. coli*) G3P auxotroph. The cDNA was completely sequenced and the longest open reading frame (ORF) consisting of 372 codons was found to share about 50% amino acid identity with animal GPDH sequences.

Results and Discussion

A functional complementation experiment was performed using the *E. coli* G3P auxotrophic mutant BB26-36 carrying the *pls*B26 mutation [4] and a cDNA expression library of *Cuphea lanceolata*. Six complementing clones were obtained which also restored the mutant phenotype after retransformation. Sequencing the ends of these cDNAs revealed that four of them were identical and show considerable homology to GPDHs (EC 1.1.1.8) of mouse, *Drosophila* and fission yeast. The complete sequence of one of these four clones, *Cl*GPDH20, is shown in figure 1. The 1331 bp insert contains a 1116 bp ORF, coding for a 372 residue polypeptide (40.8 kDa). Interestingly, the ORF of the complementing cDNA clone was found to be fused to the *lacZ* gene of the cloning vector resulting in a fusion protein of 393 residues. The deduced amino acid sequence of *Cl*GPDH20 shows the highest similarity with the GPDH sequence of mouse as illustrated in figure 2. Database searches suggest that this is the first plant GPDH so far cloned.

J.-C. Kader and P. Mazliak (eds.), Plant Lipid Metabolism, 534–536.

Based on sequence comparisons and the capacity to complement the *E. coli* mutation *pls*B26 we propose that the isolated cDNA codes for a GPDH. The possible role of GPDH in providing the G3P backbone for triacylglycerol biosynthesis is currently under investigation.

```
CAATGGCTCCCTCTGAGCTCAACTGCACCCACCAGAACCAGCATTCAAGCGGTTACGACGGACCCAGATCGAGGGTCACC   80
 M  A  P  S  E  L  N  C  T  H  Q  N  Q  H  S  S  G  Y  D  G  P  R  S  R  V  T

GTTGTCGGTAGTGGAAACTGGGGTAGTGTTGCTGCCAAGCTCATTGCTACCAATACCCTCAAGCTTCCATCTTTTCATGA  160
V  V  G  S  G  N  W  G  S  V  A  A  K  L  I  A  T  N  T  L  K  L  P  S  F  H  D

TGAAGTGAGAATGTGGGTATTTGAGGAGACGCTACCGAGCGGCGAGAAGCTTACTGATGTCATCAACCAGACCAATGAAA  240
 E  V  R  M  W  V  F  E  E  T  L  P  S  G  E  K  L  T  D  V  I  N  Q  T  N  E  N

ATGTTAAGTATCTCCCCGGAATTAAGCTCGGTAGGAATGTTGTTGCAGATCCAGACCTCGAAAACGCAGTTAAGGATGCA  320
 V  K  Y  L  P  G  I  K  L  G  R  N  V  V  A  D  P  D  L  E  N  A  V  K  D  A

AATATGCTCGTGTTTGTGACACCGCATCAGTTCATGGAGGGCATCTGCAAAAGACTCGAAGGGAAAATACAAGAAGGAGC  400
N  M  L  V  F  V  T  P  H  Q  F  M  E  G  I  C  K  R  L  E  G  K  I  Q  E  G  A

ACAGGCTCTCTCCCTTATAAAGGGCATGGAGGTCAAAATGGAGGGGCCTTGCATGATCTCGAGCTTAATCTCTGATCTTC  480
 Q  A  L  S  L  I  K  G  M  E  V  K  M  E  G  P  C  M  I  S  S  L  I  S  D  L  L

TCGGGATTAACTGCTGTGTCCTAATGGGGGCAAACATCGCTAATGAGATTGCTGTTGAGAAATTCAGTGAAGCGACAGTC  560
 G  I  N  C  C  V  L  M  G  A  N  I  A  N  E  I  A  V  E  K  F  S  E  A  T  V

GGGTTCAGAGAAAATAGAGATATTGCAGAGAAATGGGTTCAGCTCTTTAGCACTCCGTACTTCATGGTCTCAGCTGTTGA  640
G  F  R  E  N  R  D  I  A  E  K  W  V  Q  L  F  S  T  P  Y  F  M  V  S  A  V  E

AGATGTTGAAGGAGTAGAACTTTGTGGAACACTGAAGAATATCGTGGCCATAGCAGCCGGTTTTGTGGATGGATTGGAGA  720
 D  V  E  G  V  E  L  C  G  T  L  K  N  I  V  A  I  A  A  G  F  V  D  G  L  E  M

TGGGAAACAACACAAAAGCAGCAATTATGAGGATCGGGTTACGGGAGATGAAGGCATTCTCCAAGCTTTTGTTTCCATCT  800
 G  N  N  T  K  A  A  I  M  R  I  G  L  R  E  M  K  A  F  S  K  L  L  F  P  S

GTTAAGGACACTACTTTCTTCGAGAGCTGTGGAGTCGCTGACCTCATCACAACTTGTTTGGGCGGGAGAAACAGAAAAGT  880
V  K  D  T  T  F  F  E  S  C  G  V  A  D  L  I  T  T  C  L  G  G  R  N  R  K  V

TGCTGAGGCTTTTGCAAAGAATGGCGGGAAAAGGTCATTCGATGATCTCGAAGCAGAGATGCTCCGGGGGCAAAAATTAC  960
 A  E  A  F  A  K  N  G  G  K  R  S  F  D  D  L  E  A  E  M  L  R  G  Q  K  L  Q

AGGGTGTCTCAACAGCAAAGGAGGTCTATGAAGTCTTGGGGCACCGAGGCTGGCTCGAGCTGTTCCCGCTCTTCTCAACC 1040
 G  V  S  T  A  K  E  V  Y  E  V  L  G  H  R  G  W  L  E  L  F  P  L  F  S  T

GTGCACGAGATATCCACTGGCCGTCTGCCTCCTTCAGCCATCGTCGAATACAGCGAACAAAAAACCATCTTCTCTTGGTA 1120
V  H  E  I  S  T  G  R  L  P  P  S  A  I  V  E  Y  S  E  Q  K  T  I  F  S  W  *

GAGCAAGAGGCTGCCCTTGAAAGACTAAGAGCCACCCTGCCCTGTTTAAAGGGCTAAAAGTTTAATATTTCTCTGCAGCC 1200

TAAACAGTCGGAAACATTGAAAATCTAGGATGTATAAGAAAAAAAAAAGAAGGTTTGAAGGAAGTATGGATGGGCATGAA 1280

TGTATTTATTTTCGGTATACTCTTTTTCTGCAAAAATAATTTCTTCAGAAA                              1331
```

Figure 1: Nucleotide sequence of cDNA clone *Cl*GPDH20 and the predicted amino acid sequence. The underlined sequence represents the consensus of a polyA signal. The invariably conserved amino acid residues of the NAD binding domain are circled [5].

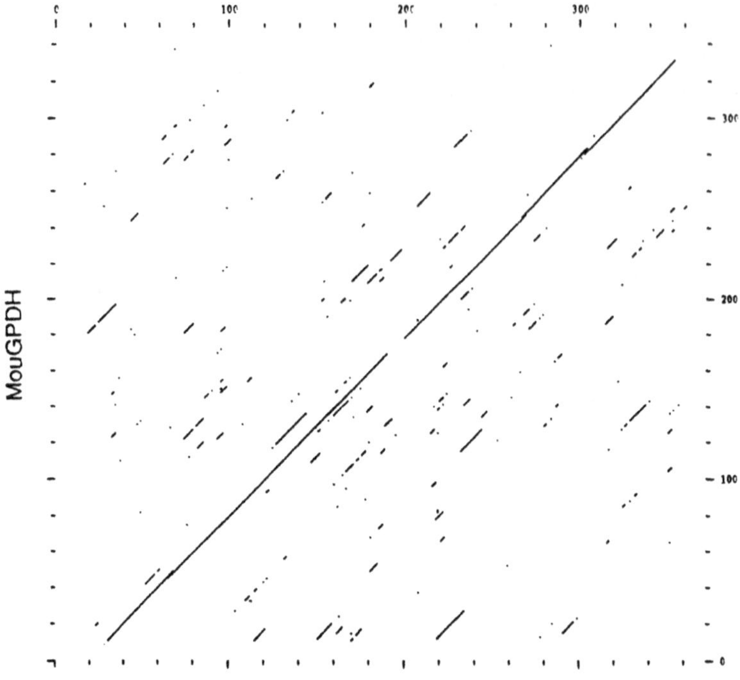

Figure 2: Dot-matrix plot of *Cuphea lanceolata* (*Cl*GPDH20) versus mouse (MouGPDH) amino acid sequence of GPDH.

References

1. Stymne S. and Stobart A. K. (1987) In *The Biochemistry of plants Vol. 8* ed. Stumpf P. K. and Conn E. E. , pp188-189, Orlando: Academic Press
2. Gee R. W., Byerrum R. U., Gerber D. W. and Tolbert N. E. (1988) Plant Physiol. 86: 98-103
3. Klein B., Pawlowski K., Höricke-Grandpierre C., Schell J. and Töpfer R. (1992) Mol. Gen. Genet. 233: 122-128
4. Bell R. M. (1974) J. Bacteriol. 117: 1065-1076
5. Otto J., Argos P. and Rossmann M. G. (1980) Eur. J. Biochem. 109: 325-330

We gratefully acknowledge the BMFT (grant number 0319412D) as well as Bayer AG, Deutsche Saatveredelung, Hoechst AG, Kleinwanzlebener Saatzucht AG and Norddeutsche Pflanzenzucht for financial support of this work.

The nucleotide sequence data reported in this paper will appear in the EMBL, GenBank and DDBJ Nucleotide Sequence Databases under the accession number X79677.

CHARACTERIZATION OF THE DIACYLGLYCEROL ACYL-TRANSFERASE IN THE LIPID BODY FRACTION FROM AN OLEAGINOUS FUNGUS

Yasushi Kamisaka
National Institute of Bioscience and Human-Technology, Tsukuba-shi, Ibaraki-ken 305, Japan

Mortierella fungi have been found to accumulate lipids, especially triacylglycerol (TG) which contains polyunsaturated fatty acids, such as γ-linolenic acid and arachidonic acid [1,2]. To elucidate mechanism responsible for TG accumulation and fatty acid composition of TG in the fungi, diacylglycerol acyltransferase (DGAT) in *Mortierella ramanniana* var. *angulispora* has been investigated. In this report, subcellular fractionation of the fungus was performed to examine distribution of DGAT in lipid bodies and microsomes. Although it has been reported that the lipid body fractions in fungi and plants have DGAT activity [for review see 3], only few quantitatively compared DGAT activity in the lipid body fraction and the membrane fraction. Around 77% of DGAT activity was recovered in the lipid body fraction, which was much higher than that in the crude membrane fraction in terms of both total and specific activity. The recovered DGAT activity in the lipid body fraction had similar characteristics to that in the crude membrane fraction except for its sensitivity to SH-reagents. Several lines of evidence suggest that DGAT is localized in lipid bodies in the fungus and DGAT activity in the lipid body fraction is not due to artificial contamination of DGAT activity in the membrane fraction during the fractionation.

Materials and Methods

Fungal cells suspended in phosphate buffer (pH 7.0) containing 0.5 M sucrose were usually homogenized with a Braun Homogenizer. The Cryo-Clean Blaster (CCB) was also used for homogenization in some experiments. The homogenates were centrifuged at 1,500 x *g* for 5 min to remove cell debris and nuclei. The supernatant, upon which phosphate buffer (pH 7.0) containing 0.3 M sucrose was overlayered, was centrifuged at 58,000 x *g* for 3 hr. A floating fraction served as the lipid rich fraction. The layer between the supernatant and 0.3 M sucrose solution, and the pellet served as the interface fraction and the membrane fraction, respectively. The remaining supernatant served as the soluble fraction.

The assay mixture for DGAT activity contained 10 mM phosphate buffer (pH 7.0), 150 mM KCl, 3.4 mM (0.2 mCi/mL) [1-^{14}C]oleoyl-CoA, 1.5 mM 1,2-diolein, 0.1% Triton X-100 and enzyme sources in a final volume of 100 mL. The reaction was carried out at 30°C for 5 min, and DGAT activity was calculated from radioactivity into TG.

J.-C. Kader and P. Mazliak (eds.), Plant Lipid Metabolism, 537–539.

Results and Discussion

Subcellular fractionation of the fungus revealed that the lipid body fraction occupied 77% of DGAT activity recovered after the ultracentrifugation and that specific activity of DGAT increased about 6-fold in the lipid body fraction compared with in the 1,500 g supernatant (Table 1). The total and specific activity of DGAT in the lipid body fraction was 3.3- and 5.2-fold, respectively, higher than those in the crude membrane fraction. The result that DGAT activity was highly enriched in the lipid body fraction was also obtained when the fungal cells were homogenized with the CCB which was claimed to be a milder homogenization [4]. This suggests that homogenization does not cause adsorption of DGAT to the lipid body fraction. When the lipid body fraction was washed twice by the homogenization buffer, total activity of DGAT in the lipid body fraction was still larger than that in the crude membrane fraction, indicating that DGAT activity detected in the lipid body fraction is topologically distinguished from DGAT in the crude membrane fraction.

The lipid body fraction in the fungus had lipid profiles which were distinct from those in the crude membrane fraction (Table 2). The lipid body fraction contained around 10-fold higher amount of lipids per unit mass of protein, in which TG and diacylglycerol (DG) were major components. Assuming mean molecular weight of TG, DG, and phospholipids as 885, 621, and 770 respectively, weight percentage of TG, DG, phospholipids, and protein in the lipid body fraction was calculated from Table 2 as 72, 16, 1.1, 11 % respectively. The composition was comparable to that of lipid bodies in fungi and plants [3], presumably reflecting lipid bodies in the fungus.

Optimal pH of DGAT in the lipid body fraction was 7.0 and metal ions were not essential for the enzyme activity, which was basically the same as DGAT activity in the crude membrane fraction [5]. Dithiothreitol and *N*-ethylmaleimide slightly increased DGAT activity in the lipid body fraction, whereas they inhibited DGAT activity in the crude membrane. It was confirmed that the ^{14}C incorporation into TG was catalyzed by DGAT but not by the reverse reaction of lipase, because 1) the ^{14}C incorporation into TG was decreased by unlabeled oleoyl-CoA but not by unlabeled oleic acid, and 2) the ^{14}C incorporation into TG was not observed when using [1-^{14}C]oleic acid as a substrate under the same condition as using [1-^{14}C]oleoyl-CoA.

TABLE 1. Subcellular distribution of DGAT activity in *Mortierella ramanniana* var. *angulispora*. Data are presented as means of triplicates ± S.D.

Fractions	Protein (mg)	DGAT activity		
		Specific Activity (nmol/min/mg protein)	Total Activity (nmol/min)	% Recovery
1,500 x g supernatant	10.4 ± 1.4	1.35 ± 0.24	14.0 ± 2.5	114
Lipid body fraction	1.17 ± 0.18	8.02 ± 0.35	9.38 ± 0.41	77
Interface fraction	0.37 ± 0.05	0.14 ± 0.02	0.05 ± 0.01	0.41
Soluble fraction	4.77 ± 0.56	Not Detected	0	0
Membrane fraction	1.85 ± 0.13	1.53 ± 0.14	2.83 ± 0.26	23

TABLE 2. Lipid profiles of subcellular fractions of *Mortierella ramanniana* var. *angulispora*. Data are presented as means of triplicates ± S.D.

	Lipid body fraction	Membrane fraction	1,500 g sup.
Lipids/Protein (mg/mg)	9.40 ± 0.87	0.86 ± 0.07	2.68 ± 0.74
TG (µmol/mg protein)	7.24 ± 1.22	0.96 ± 0.07	1.17 ± 0.10
DG (µmol/mg protein)	2.12 ± 0.05	0.28 ± 0.02	0.62 ± 0.02
PC (nmol/mg protein)	89 ± 3.7	86 ± 2.7	7.8 ± 2.2
PE (nmol/mg protein)	34 ± 5.5	54 ± 6.6	5.0 ± 1.5

Specific activity of DGAT in the lipid body fraction rapidly increased after 2 day cultivation when lipids in the fungus increased most rapidly. This suggests a close relationship between the lipid accumulation and the DGAT activity in the lipid body fraction. When a carbon to nitrogen (C/N) ratio in media was changed, it is known that lipid content in oleaginous fungi increased with increase in the C/N ratio [6]. Specific activity of DGAT in the lipid body fraction, however, decreased with increase in the C/N ratio, which indicates that DGAT activity and lipid accumulation is not correlated under the conditions. Other factors would be strongly rate-limiting under the conditions.

In summary, the present study suggests localization of DGAT activity in lipid bodies in *Mortierella* fungi. Physiological importance of the distribution of DGAT in TG assembly and lipid body formation would be a future question.

References

1. Nakahara T, Yokochi T, Kamisaka Y, Suzuki O. Gamma-linolenic acid from genus *Mortierella*. In: Kyle DJ, Ratledge C, editors. Industrial applications of single cell oils. Champaign: American Oil Chemists' Society, 1992: 61-97.
2. Yamada H, Shimizu S, Shinmen Y, Akimoto K, Kawashima H, Jareonkitomongkol S. Production of dihomo-γ-linolenic acid, arachidonic acid and eicosapentaenoic acid by filamentous fungi. In: Kyle DJ, Ratledge C, editors. Industrial applications of single cell oils. Champaign: American Oil Chemists' Society, 1992: 118-138.
3. Murphy DJ. Storage lipid bodies in plants and other organisms. Prog Lipid Res 1990; 29: 299-324.
4. Omori Y, Ichida T, Ukita R, Osumi M, Ueda M, Tanaka A. A novel method -a "freeze-blast" method- to disrupt microbial cells. J Ferment Bioeng 1989; 67: 52-56.
5. Kamisaka Y, Yokochi T, Nakahara T, Suzuki O. Characterization of the diacylglycerol acyltransferase activity in the membrane fraction from a fungus. Lipids 1993; 28: 583-87.
6. Ratledge C. Biotechnology of oils and fats. In: Ratledge C, Wilkinson SG, editors. Microbial lipids. London: Academic Press, 1989: 567-668.

IDENTIFICATION OF JOJOBA SEED ACYL-COA:FATTY ALCOHOL ACYLTRANSFERASE BY PHOTOLABELING WITH ACYL-COA ANALOG

Jay M. Shockey, John D. Kemp, and Ram Rajasekharan
Plant Genetic Engineering Laboratory, New Mexico State University.
Las Cruces, NM, 88003, USA

Introduction

The desert shrub jojoba (*Simmondsia chinensis*, Link) appears to be the only plant system that synthesizes large quantities of liquid wax. Two enzymes are specific to the wax biosynthetic pathway: fatty acyl-CoA reductase, which catalyzes the NADPH-specific reduction of the fatty acyl moiety of acyl-CoA to fatty alcohol, and acyl-CoA:fatty alcohol acyltransferase, which catalyzes the coupling of fatty acid to fatty alcohol.

In our efforts to purify the acyltransferase, we have used a non-traditional approach, tagging the enzyme using 12-[*N*-(4-azidosalicyl)]aminododecanoyl-CoA (ASD-CoA). In the present study, we have identified and characterized the acyl-CoA:fatty alcohol acyltransferase from developing jojoba cotyledons using this photoreactive acyl-CoA analog.

Materials and Methods

Jojoba seed were frozen, homogenized, then mixed with extraction buffer that contained 3 M NaCl and 0.3 M sucrose. This mixture was spun at 20,000 x g and the pellet discarded. The floating layer, called fat pad, was washed and saved. The 20,000 x g supernatant was centrifuged at 80,000 x g for 1 h. The supernatant was dialyzed then centrifuged at 200,000 x g for 1 h to pellet the microsomal membranes.

The fatty alcohol acyltransferase was assayed in the presence of 50 mM Tris-HCl, pH 8.0, 15 μM ^{14}C-labeled oleoyl-CoA, 500 μM dodecanol, and 1 mg/ml bovine serum albumin. The product was extracted with hexane, dryed in a Speed-vac, dissolved in hexane and spotted on a silica gel G TLC plate. The plates were developed, the wax spots were scraped, and radioactivity was quantified by liquid scintillation counting.

12-[*N*-(azidosalicyl)]aminododecanoic acid was synthesized as previously described [1]. The fatty acid analog was coupled to CoA to form the CoA thioester [2]. Iodination of ASD-CoA was performed using chloramine-T [3].

All photolysis reactions were carried out in the caps of eppendorf tubes on ice. The mixture was incubated in the dark before exposure to UV light. The reaction mixture contained 25 mM Tris-HCl, pH 8.0, 7.2 μM (0.25 μCi) ^{125}I-labeled ASD-CoA, and 70 μg of protein, in a final volume of 50 μl. All photolysis experiments were
accompanied by a control in which the photoprobe was photolyzed for 5 min before

J.-C. Kader and P. Mazliak (eds.), Plant Lipid Metabolism, 540–542.

adding it to the microsomal membranes. After photolysis, the samples were quenched with an equal volume of 4X SDS-PAGE sample buffer. The samples were boiled and analyzed by 11% SDS-PAGE. The gels were dried and labeled bands detected by autoradiography.

Results and Discussion

Previous studies on subcellular fractionation of jojoba seed found that wax biosynthetic activity was found in the fat pad, a floating layer that separates from other fractions after centrifugation at about 12,000 x g. Consistent with these findings, we have also observed most of the activity in the fat pad and negligible activity in the 200,000 x g pellet. It has been suggested that microsomes may become strongly associated with the fat pad during seed homogenization, resulting in the lack of wax synthesizing activity in the high speed pellet [4]. We attempted to eliminate the association between microsomes and fat pad while maintaining enzymatic activity, by using a tissue homogenization buffer containing 3 M NaCl. This treatment yields a 200,000 x g pellet which contains fatty alcohol acyltransferase activity.

We used the microsomal membranes, in the presence of 15 μM oleoyl-CoA, to determine the apparent kinetic constants for dodecanol. The V_{max} and K_m were 71.1 pmol/min/mg and 99.5 μM, respectively. Using 500 μM dodecanol, an apparent K_m of 26 μM was determined with respect to oleoyl-CoA. The acyl-CoA analog was not accepted as substrate by the enzyme but did, however, act as an inhibitor of enzyme activity. An apparent K_i for ASD-CoA of 24.7 μM was determined. Plots of the kinetic data suggests that ASD-CoA acts as an effective mixed-type inhibitor of acyltransferase in the dark. The acyl-CoA analog also acted as an irreversible inhibitor of acyltransferase activity upon exposure to UV light. Photolysis of microsomal membrane with ASD-CoA followed by dialysis resulted in 40-50% inhibition of the enzyme activity compared to control samples incubated with ASD-CoA then dialyzed in the absence of UV light.

The photoreactive acyl-CoA analog was used to identify the membrane-bound fatty alcohol acyltransferase. Photolabeling of microsomal membranes with [^{125}I]ASD-CoA resulted in major labeling of 57 and 52 kDa polypeptides. Of the two, the 57 kDa polypeptide was present in much lower abundance, while labeling nearly as strongly as the other polypeptide, suggesting that this protein could be the fatty alcohol acyltransferase. This conclusion is supported by the results of dodecanol-dependant labeling experiments, in which the labeling intensity of the 57 kDa polypeptide increased significantly in the presence of dodecanol. The 52 kDa polypeptide showed only a slight increase in labeling under these conditions (Figure 1).

In addition to the dodecanol effect, addition of increasing concentrations of oleoyl-CoA to the photolysis mixtures led to substantial differences in the levels of bound radioactivity between the two polypeptides, as quantified by densitometric scanning. Low concentrations of oleoyl-CoA (10 μM) caused a large reduction in the labeling of the 57 kDa polypeptide but only a slight reduction in the labeling of the 52 kDa polypeptide, compared to the dodecanol-containing lane. At 100 μM oleoyl-CoA, nearly all of the radioactivity in the 57 kDa polypeptide had been competed out, as compared to an approximately 50% reduction in radioactivity in the 52 kDa polypeptide. These data suggest that the photoprobe is interacting specifically with the acyltransferase.

We have shown that a 57 kDa polypeptide from the microsomal membrane of developing jojoba seed is indeed the fatty alcohol acyltransferase. The results presented in this paper show that ASD-CoA can be particularly useful for identification of membrane-bound acyltransferases in plants.

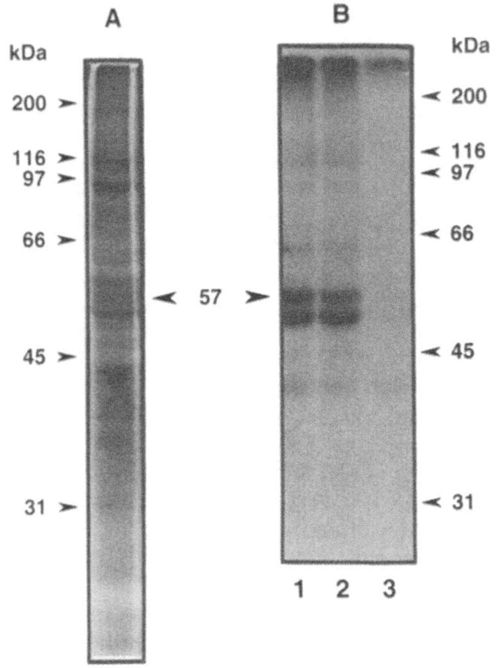

Figure 1. The effect of dodecanol on photolabeling of jojoba seed microsomal membrane proteins

Panel A: Coomassie Blue R-250 stain of jojoba microsomal membrane SDS-PAGE profile

Panel B: Autoradiogram of ASD-CoA labeled microsomal membrane proteins
Lane 1: ASD-CoA + 250 μM dodecanol
Lane 2: ASD-CoA alone
Lane 3: pre-photolyzed ASD-CoA

References

1. Rajasekharan, R, Marians, RC, Shockey, JM, Kemp, JD. Photoaffinity labeling of acyl-CoA oxidase with 12-azidooleoyl-CoA and 12-[(4-azidosalicyl)amino]dodecanoyl-CoA. Biochemistry 1993; 32: 12386-12391
2. Bishop, JE, and Hajra, AK. A method for the chemical synthesis of ^{14}C-labeled fatty acyl coenzyme A's of high specific activity. Analytical Biochemistry 1980; **106**: 344-350
3. Ji, I, Shin, J, and Ji, TH. Radioiodination of a photoactivatable heterobifunctional reagent Analytical Biochemistry 1985; **151**: 348-349
4. Pollard, MR, McKeon, T, Gupta, LM, and Stumpf, PK. Studies on biosynthesis of waxes by developing jojoba seed. II. The demonstration of wax biosynthesis by cell-free homogenates Lipids 1979; **14**: 651-662

ACYL CoA BINDING PROTEINS IN *BRASSICA NAPUS* L.: AMINO ACID SEQUENCE, GENES AND EXPRESSION

Matthew J. Hills, Rachel Dann, Derek Lydiate and Andrew Sharpe

Department of Brassica and Oilseeds Research, John Innes Centre, Colney, Norwich NR4 7UH, U.K.

Introduction

Acyl-CoAs lie at the centre of lipid metabolism in developing seeds since they are the products of fatty acid synthesis by the plastid which are transferred to the ER where they act as substrates for triacylglycerol and phospholipid synthesis [2]. At present it is not clear how the acyl-CoAs are transported within the plant cell and whether the pool of acyl-CoAs resides free in the cytosol or is bound to membranes or proteins. Acyl CoA binding proteins (ACBP) were first reported by [1] as a factor which stimulated fatty acid synthesis in mammary gland extracts. They have been found subsequently in several animals, *Drosophila* and yeast. ACBPs are cytoplasmic proteins with an Mr of about 10 kDa and they have a very high affinity for binding long chain acyl-CoAs, but they do not bind fatty acids or CoA individually. The concentration of ACBP in some animal systems has been calculated to be higher than that of the acyl-CoA suggesting that the entire cellular pool of acyl-CoA is bound to ACBP. Over expression of the ACBP gene in yeast causes a corresponding increase in acyl-CoA concentration suggesting that the acyl-CoA pool size is influenced by the expression of ACBP genes. Some recent *in vitro* studies showed that phospholipid and TAG synthesis by microsomes isolated from rat liver are strongly influenced by the ratio of [ACBP] to [acyl-CoA]. When ACBP was in excess, TAG synthesis was strongly reduced but phospholipid synthesis remained unaffected. It is therefore important to determine whether ACBPs are present in plants.

In studies using radiolabelled acyl-CoAs as photo-affinity probes we found a low molecular weight protein in the cytoplasmic fraction of homogenates of developing seeds

J.-C. Kader and P. Mazliak (eds.), Plant Lipid Metabolism, 543–545.

of rape (approx. 10 kDa) which became very strongly labelled by the probes when illuminated under ultraviolet light. Since the size of this protein was very similar to the ACBPs found in animals we investigated the possibility that plants also express acyl-CoA binding protein genes which are similar to those in animals. Attempts at purification of the protein from these extracts were not successful so a PCR strategy was taken.

A full length cDNA was cloned from a λt 11 library and sequenced. The amino acid sequence showed 40-45% identity with those from human, yeast and fruit fly and 61-65% with conservative substitutions. The rape protein was 92 amino acids in length with a calculated Mr of about 10.2 kDa which is slightly bigger than the animal proteins. Southern blots of genomic DNA from rape showed the presence of 6 ACBP genes. It was found that the ACBP cDNA hybridized strongly with messenger RNA with a molecular weight of 550 bp which had been extracted from developing seeds and the cotyledons of 4 day old seedlings. It therefore appears that plants express genes encoding acyl-CoA binding proteins which are very similar to those found in animals and yeast at least in tissues which are maintaining high rates of lipid metabolism.

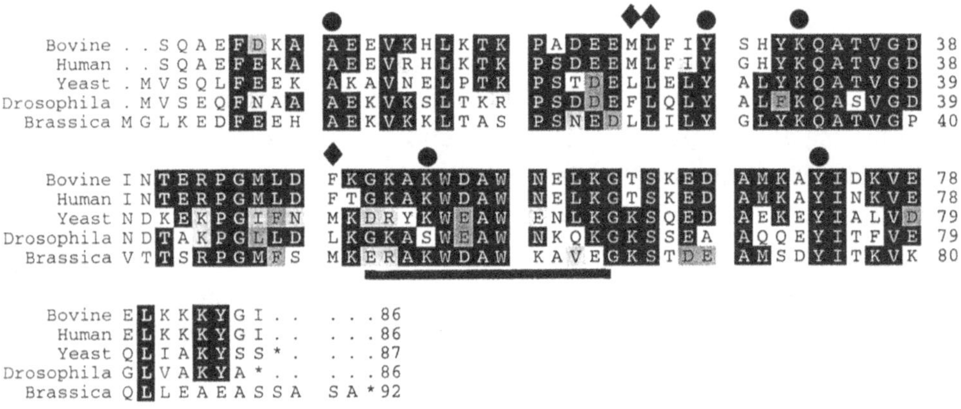

Fig. 1. Amino acid sequences of ACBPs from bovine, human, *Drosophila*, yeast and oilseed rape

The sequence of the oilseed rape ACBP and comparison with amino acid sequences from other organisms is given in Fig. 1. The regions of sequence identity are indicated by a black background. Structural studies of the bovine protein by 2D NMR spectroscopy in the presence and absence of bound acyl-CoA have shown in great detail which amino acid residues are involved in the binding of the different parts of the acyl-CoA molecule and give indications why fatty acid and CoA alone cannot bind [3]. All of the amino acid residues which are known to be involved in the binding of the adenine-3-phosphate of the CoA moiety are indicated by the filled circles. The region of the ACBP which forms a large part of the fatty acid binding site is underlined and other residues which form the rest of the hydrophobic cleft protecting the acyl chain from solvent are indicated by filled diamonds.

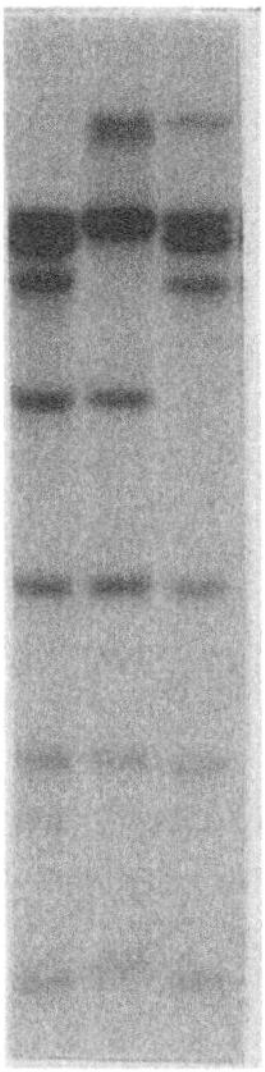

Fig. 2. ACBP genes in oilseed rape

Genomic DNA from *Brassica napus* was digested with *EcoRI*, size fractionated on an agarose gel, blotted onto Nylon membranes and probed with the ACBP cDNA. Three lines from a segregating population are shown here.

A Southern blot of rape DNA digested with *EcoRI* (Fig 2.) reveals 7 fragments which hybridize the ACBP cDNA. Digestion with *XbaI* (not shown) showed that two of those fragments represent one sequence and thus there are probably 6 related ACBP genes in *Brassica napus* L.

References

1. Mogensen IB, Schulenberg H, Hansen HO, Spener F, Knudsen J. A novel scyl-CoA binding protein from bovine liver. Biochem J 1987;241:189-192

2. Murphy DJ, Rawsthorne S, Hills MJ. Storage lipid formation in seeds. Seed Sci Res 1993;3:79-95

3. Kragelund BB, Andersen KV, Madsen JC, Knudsen J and Poulsen FM. Three-dimensional structure of the complex between ACBP and palmitoyl-CoA. J Mol Biol 1993;230:1260-1277

AN UNCOMMON PATHWAY IN THE BIOSYNTHESIS OF ACETYLENIC FATTY ACIDS IN MOSSES

Peter Beutelmann, Institute of General Botany, University of Mainz, D-55099 Mainz, Germany

Sten Stymne, Department of Plant Physiology, Swedish University of Agricultural Sciences, P.O.Box 7047, S-47450 Uppsala, Sweden

Introduction

Although more than 600 acetylenic compounds of various structures have been found in plants [1], hardly anything is known about the formation of the triple bonds. The acetylenes so far known from Bryophytes, have been identified as long chain fatty acids with a characteristic arrangement of one triple bond and one to three double bonds.

The moss *Ceratodon purpureus* (Hedw.) Brid. contains large amounts of the acetylenic fatty acids 9,12-octadecadien-6-ynoic acid **[18:2A]** and 6,9,12-octadecatrien-6-ynoic acid **[18:3A]**, which are in close structural relationship to the common fatty acids linoleic acid [18:2(n-6)] and α-linolenic acid [18:3(n-3)], and which are obviously restricted to the triacylglycerol fraction [2].

Protonema cells of *Ceratodon purpureus* which originate from spores can be grown in simple liquid media as a permanent cell culture, consisting of filaments of uniform green cells, multiplying at a high rate and producing large amounts of triacylglycerols, particularly rich in acetylenes. Since labelled compounds like [^{14}C]-acetate and [^{14}C]-fatty acids are taken up by the protonema cells at high rates, the cells are an excellent tool for the study of lipid biosynthesis, and particularly for the study of the biosynthesis of acetylenic fatty acids.

In pulse/chase experiments [^{14}C]-labelled acetate, linoleic acid 18:2(n-6)], γ-linolenic acid [18:3(n-6)], α-linolenic acid [18:3(n-3)], stearidonic acid [18:4(n-3)] and 18:2A were applied to protonema cells. Per sample 50 mg (fr. wt.) of protonema cells were suspended in 4 ml of inorganic growth medium supplied with the respective radioactive precursor. Lipids were extracted and analyzed by TLC and GC according to established methods.[3]

Results

18:3(n-6) and 18:4(n-3) both were rapidly converted into their corresponding acetylenic acids, 18:2A and 18:3A, demonstrating that the formation of the triple bond in both these acids occurs by a second desaturation of a Δ6 double bond (fig.1). From 18:3(n-6)

J.-C. Kader and P. Mazliak (eds.), Plant Lipid Metabolism, 546–548.

not only 18:2A but also 18:3A is formed, at rates which are very similar to those observed with the formation of 18:3A from 18:3(n-3). This indicates an active Δ15 desaturase, using fatty acids with a Δ6 double bond as substrate.

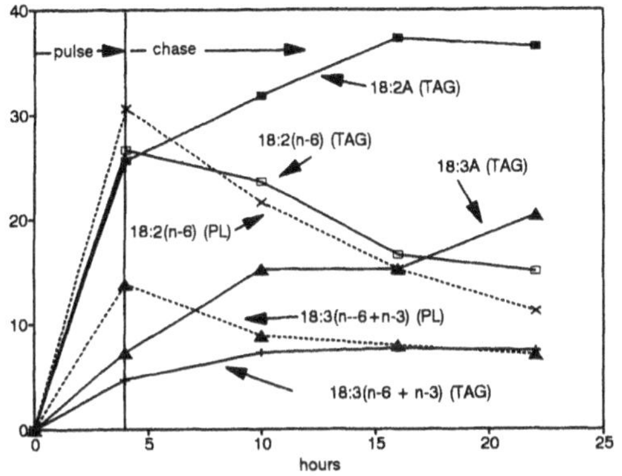

Fig. 1: [^{14}C]-Distribution from [1-^{14}C]-18:2(n-6) in the lipid fractions of the protonema cells. 50 mg (fr.wt.) of protonema was incubated with 20 nmoles of [1-^{14}C]-linoleic acid in 4 ml medium. The pulse time was 4 hrs. PL = phospholipid.

In [^{14}C]-acetate and [^{14}C]-18:2(n-6) feedings (fig.1), and at short incubation times the radioctivities recovered in 18:2A were several times higher than in 18:3A. This is amazing, since 18:3A is the dominating fatty acid of the cells and is accumulated to more than 60 mol% of the TAG, whereas 18:2A only reaches 15 to 20 mol%, and does not show a marked net synthesis in cells which are used for the incubation studies (fig.2).

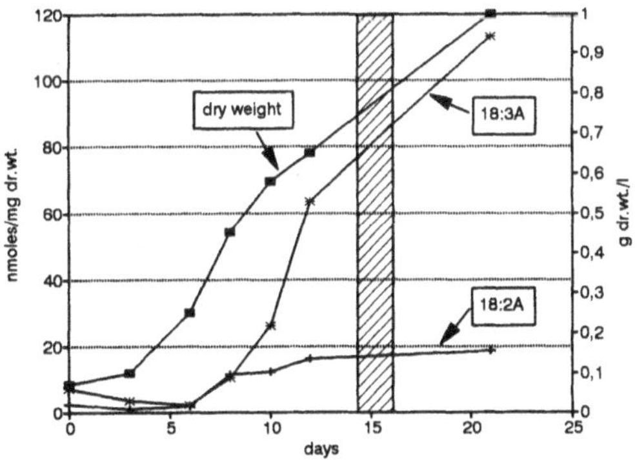

Fig 2: Growth and acetylenic fatty acid production of a protonema cell culture of *Ceratodon purpureus*. The shaded area represents the growth range of the cell material which was taken for tracer studies.

Upon prolonged incubation periods with [^{14}C]-acetate the label which initially accumulated in 18:2A decreased with a corresponding increase of radioactivity in 18:3A (fig.3), indicating a precursor/product relationship between 18:2A and 18:3A. More evidence for this relationship is provided by the observation that [^{14}C]-18:2A is metabolized to 18:3A by the protonema cells.

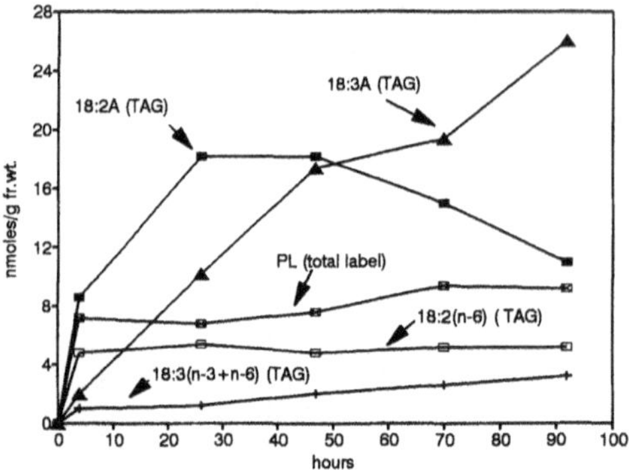

Fig. 3: Distribution of [^{14}C] from [1-^{14}C]-acetate in the lipid fractions of protonema cells in a long term study. 350 mg (fr.wt.) of protonema were incubated with 140 nmoles of [^{14}C]-acetate in 70 ml medium for 96 hrs. 10 ml samples of cell suspension which contained about 50 mg of protonema, were withdrawn every 24 hrs and analyzed.

18:3(n-3) and 18:4(n-3) are both converted to 18:3A by the moss cells, but 18:3(n-3) was not a better substrate for 18:3A formation than 18:3(n-6) or 18:2(n-6). In [^{14}-C]-acetate feeding experiments 18:2(n-6) and 18:3(n-6) incorporated much more label than 18:3(n-3) and 18:4(n-3), the latter being detected only in minute amounts.

Conclusion

We therefore conclude that 18:3A, the main fatty acid of the TAG fraction is primarily synthesized from 18:2(n-6) by two Δ6 desaturation steps, which are followed by a Δ15 desaturation. Since 18:2A was exclusively found in the TAG of the cell, the last desaturation step either takes place in the TAG molecule, or the 18:2A has to be detached, desaturated and reincorporated into TAG. These results which are confirmed by observations in sunflower seeds [4], present evidence that, in contrast to the general believe, plant TAGs are not metabolic inert molecules during the the cell stage of TAG deposition.

References

1. Bohlmann F, Burkhardt T, Zdero C. Naturally occurring acetylenes. New York: Academic Press, 1973;

2. Kohn G, Demmerle S, Vandekerkhove O, Hartmann E, Beutelmann P. Distribution and chemotaxonomic significance of acetylenic fatty acids in mosses of the Dicranales. Phytochmistry 1987;26:2271-2275.
3. Kohn G, Hartmann E, Stymne S, Beutelmann O. Biosynthesis of acetylenic fatty acids in the moss Ceratodon purpureus (Hedw.) Brid. J Plant Physiol 1994: in press)

4. Garces R, Sarmiento S, and Mancha M,: Oleate from triacylglycerols is desaturated in cold-induced developing sunflower (*Helianthus annuus* L.) seeds. Planta (in press)

GERMINATING RAPESEED AS BIOCATALYST FOR LIPOLYTIC AND SYNTHETIC REACTIONS*

Kumar D. Mukherjee and Iván Jachmanián
Institute for Biochemistry and Technology of Lipids, H. P. Kaufmann-Institute, BAGKF, Piusallee 68, D-48147 Münster, Germany

Introduction

The storage triacylglycerols of oil-rich seeds are hydrolyzed during germination of such seeds by the action of the endogenous lipase (triacylglycerol acylhydrolases, EC 3.1.1.3) and the lipolysis products, i. e. fatty acids and glycerol, are metabolized further in glyoxysomes and other organelles. It is well known that germinating seedlings of many oilseeds are rich source of lipases which have been isolated, purified and characterized from such tissues and some of these purified enzyme preparations have been used as catalyst in biotransformation of lipids [1,2]. Especially, the lipase from germinating seedlings of the very common oilseed - rape (*Brassica napus* L.) - has been isolated [3,4] partially purified [5,6] and used as biocatalyst after immobilization [7-10].

Here we report the use of germinating rapeseed homogenates - without any isolation or partial purification of the lipase - as biocatalyst for the hydrolysis of endogenous or exogenous triacylglycerols as a possible biotechnological alternative to conventional fat splitting at high temperatures and pressures for the production of fatty acids. Moreover, we show the possibility of using acetone powder from rape seedlings as biocatalyst for esterification. Substrate selectivity of rape seedling preparations towards fatty acids of less common structure is also reported with the view of enrichment of such fatty acids from mixtures via kinetic resolution.

Materials and Methods

Seeds of low-erucic rape (*Brassica napus*) cultivar Ceres and high-erucic oriental mustard (*Brassica juncea*) of Chinese origin were germinated in the dark on moistened filter paper for various periods at 20 to 25 °C. The whole seedlings were harvested and routinely 1 g seedlings were homogenized with 1 ml tris-HCl buffer (50 mM) at different pH. The homogenate was either used in hydrolysis reactions as such or after delipidation with pentane.

*Part of the doctoral thesis of Iván Jachmanián to be submitted to Facultad de Química, Universidad de la República, Montevideo, Uruguay.

J.-C. Kader and P. Mazliak (eds.), Plant Lipid Metabolism, 549–551.

Routinely, the homogenate of the rape seedlings of different age containing the endogenous triacylglycerols was incubated for various periods by magnetic stirring. Alternatively, 1 g of the various oils containing predominantly either triacylglycerols, e. g. low-erucic and high-erucic rapeseed oil, linseed oil, castor oil, borage oil, coriander oil, *Hydnocarpus wightiana* oil and hake (*Merluccius hubbsi*) liver oil, or wax esters, e. g. jojoba oil and orange roughy (*Hoplostethus atlanticus*) oil, was added to the delipidated homogenate prior to incubation.

Acetone powder from seedlings of low-erucic rape (*Brassica napus*) cultivar Ceres (Saaten-Union GmbH, Hannover, Germany) was prepared as described elsewhere [3].

Esterification reactions were carried out using 50 mg equimolar mixture of oleic acid, the reference standard, and the fatty acid examined with two times the stoichiometric amount of *n*-butanol or oleoyl alcohol in the presence of 50 mg acetone powder and 1 ml hexane by magnetic stirring at room temperature for various periods. Alternatively, 100 mg fatty acid mixtures derived from various oils was used in esterification.

Samples were analyzed by thin-layer chromatography and gas chromatography as described earlier [11]. Specificity constants of the esterification reactions ($1/\alpha$) were calculated from the composition of fatty acids and butyl esters [12].

Results and Discussion

The homogenates of rape seedlings - as such or after delipidation with pentane - were found to catalyze the hydrolysis of the triacylglycerols of various oils in the following order: low-erucic and high-erucic rapeseed oil > linseed oil > *Hydnocarpus wightiana* oil > borage oil > hake (*Merluccius hubbsi*) liver oil > castor oil > coriander oil, however, very little hydrolysis of the wax esters contained in jojoba oil and orange roughy (*Hoplostethus atlanticus*) oil occurred. In the hydrolysis of borage oil, coriander oil, *Hydnocarpus wightiana* oil and hake liver oil the homogenates of rape seedlings were found to discriminate against γ-linolenoyl, petroselinoyl, gorlioyl and n-3 docosahexaenoyl moieties, respectively, which are acyl moieties having Δ^4, Δ^6 or Δ^8 *cis*-double bonds. Consequently, such acyl moieties were enriched in the unhydrolyzed acylglycerols.

Acetone powders from germinating rape seedlings were found to catalyze the esterification of a wide variety of fatty acids with long chain and short chain alcohols. In esterification reactions with *n*-butanol, for example, fatty acids having Δ^4, Δ^6 and Δ^8 *cis*-double bonds (e.g. petroselinic, γ-linolenic, stearidonic, homo-γ-linolenic, docosahexaenoic and gorlic acid) were strongly discriminated against ($1/\alpha < 0.1$), whereas most of the common long chain fatty acids, such as oleic acid ($1/\alpha$ 1.0) were readily accepted as substrates. Several less common fatty acids, such as ricinoleic, 12-hydroxystearic, *trans*-epoxystearic, chaulmoogric and hydnocarpic acids were distinctly preferred as substrates ($1/\alpha$ 1.5 to 2.5) as compared to oleic acid ($1/\alpha$ 1.0).

The above findings show the potentials of germinating rapeseed as a cheap and easy to obtain biocatalyst for complete hydrolysis of oils for the preparation of fatty acids and for partial selective hydrolysis of oils or selective esterification of fatty acids for the enrichment of definite fatty acids from mixtures via kinetic resolution.

Conclusions

In hydrolysis and esterification reactions, catalyzed by homogenates and acetone powder, respectively, of germinating rapeseed fatty acids/acyl moieties having Δ^4, Δ^6 and Δ^8 *cis*-double bonds (e.g. petroselinic, γ-linolenic, stearidonic, homo-γ-linolenic, docosahexaenoic and gorlic) are strongly discriminated against, whereas ricinoleic, 12-hydroxystearic, *trans*-epoxystearic, chaulmoogric and hydnocarpic acids are distinctly preferred as substrates.

References

1. Mukherjee KD. Plant lipases and their application in lipid biotransformations. Prog Lipid Res 1994; 33:165-174.
2. Mukherjee KD, Hills MJ. Lipases from plants.In: Woolley P, Petersen SB, editors. Lipases - their structure, biochemistry and application. Cambridge: Cambridge University Press, 1994:49-75.
3. Hassanien FR, Mukherjee KD. Isolation of lipase from germinating oilseeds for biotechnological processes. J Am Oil Chem Soc 1986; 63:893-897.
4. Theimer RR, Rosnitschek I. Development and intracellular localization of lipase activity in rapeseed (*Brassica napus* L.) cotyledons. Planta 1978; 139:249-256.
5. Hills MJ, O'Sullivan JN. Partial purification of lipase from cotyledons of oilseed rape (*Brassica napus* L.). Biochem Soc Trans 1989; 17:480.
6. Weselake RJ, Thomson LW, Tenaschuk D, Mackenzie SL. Properties of a solubilized microsomal lipase from germinating *Brassica napus*. Plant Physiol 1989; 91:1303-1307.
7 Hills MJ, Mukherjee KD. Triacylglycerol lipase from rape (*Brassica napus* L.) suitable for biotechnological purposes. Appl Biochem Biotechnol 1990; 26:1-10.
8. Hills MJ, Kiewitt I, Mukherjee KD. Synthetic reactions catalyzed by immobilized lipase from oilseed rape (*Brassica napus* L.). Appl Biochem Biotechnol 1991; 27:123-129.
9 Hills MJ, Kiewitt I, Mukherjee KD. Lipase from *Brassica napus* L. discriminates against *cis*-4 and *cis*-6 unsaturated fatty acids and secondary and tertiary alcohols. Biochim Biophys Acta 1990; 1042:237-240.
10. Hills MJ, Kiewitt I, Mukherjee KD. Enzymatic fractionation of evening primrose oil by rape lipase: Enrichment of γ-linolenic acid. Biotechnol Lett 1989; 11:629-632.
11. Mukherjee KD, Kiewitt I, Hills MJ. Substrate specificity of lipases in view of kinetic resolution of unsaturated fatty acids. Appl Microbiol Biotechnol 1993; 40:489-493.
12. Rangheard M-S, Langrand G, Triantaphylides C, Baratti J. Multi-competitive enzymatic reactions in organic media: a simple test for the determination of lipase fatty acid specificity. Biochim Biophys Acta 1989; 1004:20-28.

LIPOLYTIC ACTIVITY ISOLATED FROM SUNFLOWER *(HELIANTHUS ANNUUS L.)* GERMINATING SEED SPHEROSOMES

Sellema BAHRI[1], Annette OURSEL[2], Jeannette BEN HAMIDA[1] and Jean-Claude KADER[2].
1 : Département des Sciences Biologiques, Faculté des Sciences de Tunis. Tunisie.
2 : Laboratoire de Physiologie Cellulaire et Moléculaire, Université Pierre et Marie Curie. France.

Introduction
Lipolytic activity was well studied in animals and microorganisms. More recently plant lipases were investigated; only two have been purified to homogeneity : the lipase from the scutella of corn associated to " spherosomes" has a molecular weight of 65 kDa (1) lipase from castor bean (*Ricinus communis*) associated to " glyoxysomes". This enzyme has a molecular weight of 62 kDa (2). What about sunflower ? As we expected, we have found a lipolytic activity in germinating sunflower seeds' spherosomes and the "true" lipase was partially purified(3). In this paper we report a purification procedure to homogeneity of the sunflower spherosomal lipase.

Materiels and methods
Sunflower seeds (Helianthus annuus L. var. Albena) were obtained from the Centre Technique Interprofessionnel des Oléagineux Metropolitains, Paris. The cotyledons of 3 day old sterilized seedlings were used. After spherosomes' extraction using the method of Qu et al(4). adapted to the sunflower, the washed lipid bodies were extracted with diethylether to remove the triacylglycerols. The suspension was centrifuged at 100.000 g in presence of 1 mM CHAPS for 90 min. The supernatant was containing the solubilized lipase.
Lipase activity was measured by a colorimetric method adapted from Duncomb (5). The substrate (sunflower oil) was purified , controlled by thin layer chromatography and emulsified in a 10 % gum arabic. The reaction was carried out at 30°C in a shaker water bath.
Sodium dodecyl sulfate/ polyacrylamide gel electrophoresis was carried out in slab gels The silver staining was used to reveal proteins in the separating gel (10 % polyacrylamide).

Results
After ether extraction isolated lipid bodies presented lipolytic activity. Lipase preparation was applied to a DEAE trisacryl M column exchange chromatography (2 X 20 cm) equilibrated with the elution buffer (Tris/HCl 20 mM,1 mM DTT, 1mM benzamidine, pH 7,5) and eluated with a gradient of 0 to 0,8 M NaCl in the same buffer. Lipase activity was found in 2 bound fractions by DEAE column at about 0,5 M NaCl. (fig.1). These fractions were pooled dialyzed filtred and applied on FPLC system (fig.2).

J.-C. Kader and P. Mazliak (eds.), Plant Lipid Metabolism, 552–554.

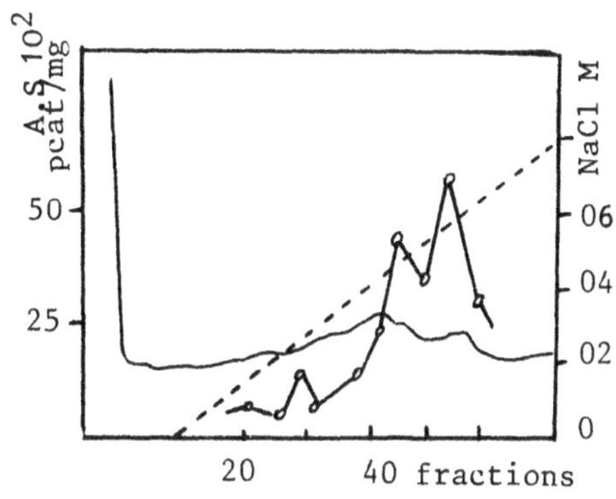

Figure 1 : Elution pattern by DEAE ion-exchange chromatography. 67 mg of solubilized proteins were deposited on the DEAE trisacryl M column (2 x 20 cm) equilibrated with the elution buffer (Tris/HCl 20 mM, 1 mM DTT, 1mM benzamidine, pH 7,5) and eluated with a gradient of 0 to 0,8 M NaCl in the same buffer.
(---) NaCl gradient, (—) Absorbance at 280 nm, (-∘-) Specific Activity, Flow rate 35 ml/h, 3 ml/Fraction.

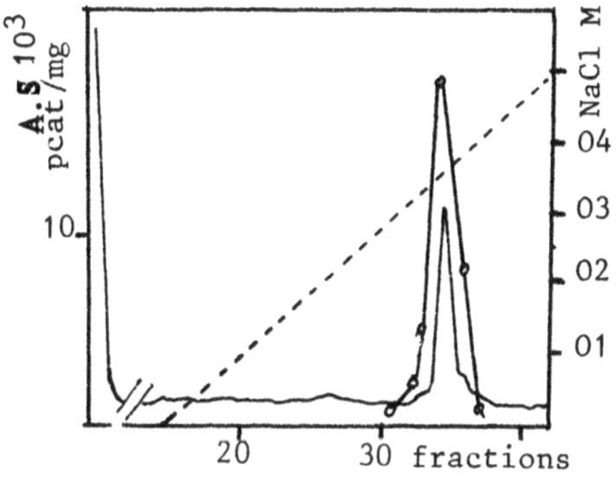

Figure 2 : Elution profil of trisacryl M active fraction from a Mono Q FPLC chromatography. 2 ml of dialyzed fractions were filtred than injected on the column equilibrated the washed 10 mn with the buffer A (Tris/HCl 20mM,3 mM NaN_3, pH 7,5). The Mono Q was eluated with a gradient of 0 to 0,8 M NaCl in the same buffer.
(---) NaCl gradient, (—) Absorbance at 280 nm, (-∘-) Specific Activity, flow rate 35 ml/h, 1ml/fraction.

Step	Protein (mg)	Specific Activity (pKat/mg)	Purification factor
Spherosomal membranes	175,2	173	1
Spherosomal extract (CHAPS)	67,5	380	2,2
DEAE trisacryl fractions	0,45	4610	26,6
FPLC fraction	0,09	12000	69,3

Table 1 : Purification tentative of spherosomal lipase from sunflower seedlings.

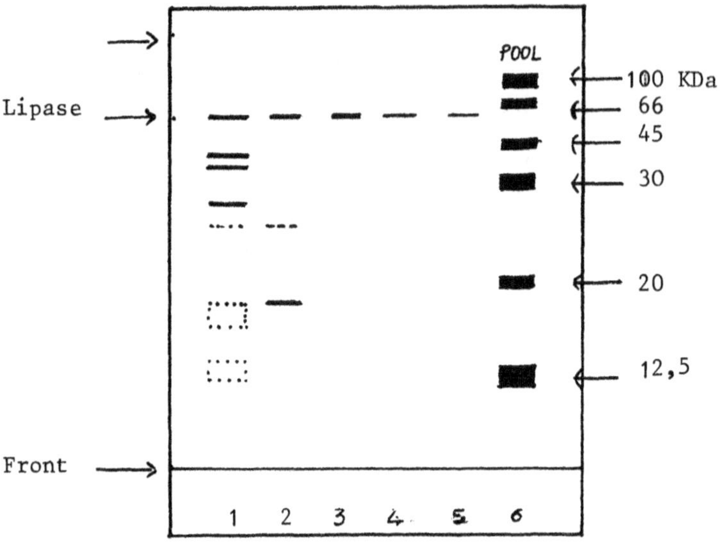

Figure 3: Sodium dodecyl sulfate/polyacrylamide gel electrophoregram of the final purification steps was carried out in slab gels . The silver staining was used to reveal proteins in the separating gel (10 % polyacrylamide). Lane 1 : Solubilized spherosomal membranes. Lane 2 : Dialyzed fractions. Lane 3, 4 and 5 : FPLC active fraction. lane 6 : molecular weight markers (phosphorylase A, M_r 100 kDa ; bovin serum albumin, M_r 66 kDa ; carbonic anhydrase , M_r 30 kDa).

Conclusion

Electrophoresis of the FPLC final step has shown the presence of a protein band of Mr 61 kDa in a moleculair range characteristic of lipases . The enzyme was purified 70 fold (table 1).
What's the interraction type between oil bodies and the lipase ?

References

1- Lin Y.H. & Huang A.H.C. Purification and initial characterisation of lipase from the scutella of corn seedlings. Plant physiol. 1984 ; 76 : 719-722.

2 - Maeshima M. & Beevers H. Purification and propartis of glyoxysomal lipase from castor bean. Plant physiol. 1985 ; 79 : 489-93.

3 - Bahri S., Oursel A, Ben Hamida J., Kader JC. in : The metabolism, structure and utilisation of plant lipid. Proceedings of the 10th plant lipid meeting 1992. Djerba Tunisia.

4 - Qu R., Wang S.H., Lin Y.H., Vance VB & Huang, A.H.C. Characteristics and biosynthesis of membrane proteins of lipid bodies in the scutella of maïze (Zea maïs L.). Biochem. J. 1986 ; 235 : 57-65.

5 - Duncombe WC. The colorimetric determination of long chain fatty acids in the 0,05 - 0,5 mole range. Biochem J. 1962 ; 83.

FACTORS AFFECTING LIPASE ACTIVITY IN THE OIL PALM (*ELAEIS GUINEENSIS*) MESOCARP

Ravigadevi Sambanthamurthi[1], Oo Khaik Cheang[2] and Siti Hasnah Parman[1]

[1] Palm Oil Research Institute of Malaysia (PORIM)
[2] Department of Biochemistry, University of Malaya

Introduction

The mesocarp of the ripe oil palm fruit accumulates large amounts of palm oil, an economically desirable product. The level of free fatty acids (FFA) is an important determinant of oil quality and little is known of the factors affecting this parameter in the oil palm fruit. Evidence is presented here for the presence of an active endogenous lipase in the oil palm mesocarp. Lipase activity was located in the oil body fraction. Maximum *in vitro* activity was observed at 18°C but using an *in vivo* assay maximum activity was obtained at 5°C. Low temperature activation was less pronounced in a water-saturated environment.

Methods

PREPARATION OF MESOCARP EXTRACT

Oil palm fruits were sliced and homogenised in 0.1M potassium phosphate buffer, pH 7.5. The homogenate was centrifuged at 18,000xg for 15 min at 4°C. The fat layer was removed and washed four times in an equal volume of the same buffer containing 0.1M NaCl.

LIPASE ASSAY

In vitro - Lipase activity was determined by measuring [^{14}C] oleic acid released from an emulsion of glycerol tri-[1-^{14}C]oleate (54.3 mCi/mmol).

In vivo - The oil was physically extracted from the fruit and FFA content estimated by titration.

J.-C. Kader and P. Mazliak (eds.), Plant Lipid Metabolism, 555–557.

Results and Discussion

LOCALIZATION OF LIPASE ACTIVITY

In vitro lipase activity was only detected in the floating fat layer. There was no activity in the supernatant fraction.

EFFECT OF LOW TEMPERATURE ON FFA LEVELS IN THE OIL PALM FRUIT (*IN VIVO* LIPASE ACTIVITY)

A time course of FFA release at $5^{o}C$ was carried out with two sets of fruits: one set in air and one set in distilled water. Figure 1 shows that both sets of fruits exhibited increasing FFA levels with continued cold exposure. Fruits immersed in cold water showed a quicker FFA release probably because of a faster equilibration with the cold ambient temperature. However, the final FFA levels obtained were much higher for fruits left in the cold air (71 %) compared to those left in the cold water (22 %). It is possible that water stress has an effect on lipolysis.

EFFECT OF TEMPERATURE ON *IN VITRO* LIPASE ACTIVITY

The lipase associated with the fat layer showed maximum activity at $18^{o}C$. However, when measured under *in vivo* conditions, maximum FFA release in whole fruits was obtained at $5^{o}C$ [1]. This observation could be attributed to structural changes in mesocarp tissue induced by the cold leading to enzyme-substrate interaction.

EFFECT OF FRUIT RIPENING ON LIPASE ACTIVITY

Lipase activity was first observed around 16 weeks after anthesis (WAA) and showed a progressive increase until 21 WAA. Since triacylglycerol synthesis in oil palm mesocarp also starts at 16 WAA and reaches a peak at 21 WAA [2], lipase activity appears to be synchronised with triacylglycerol biosynthesis in this tissue.

CYCLOHEXIMIDE TREATMENT

Cycloheximide treatment was used to determine if the lipase was synthesized in response to the cold treatment since it inhibits protein synthesis. Cycloheximide had no effect on the cold induced FFA release and it can be concluded that the lipase activity observed was due to preexisting enzyme in the tissue. This fact is also consistent with the quick response to cold treatment (Figure 1).

Figure 1 Time course of free fatty acid release at 5°C.

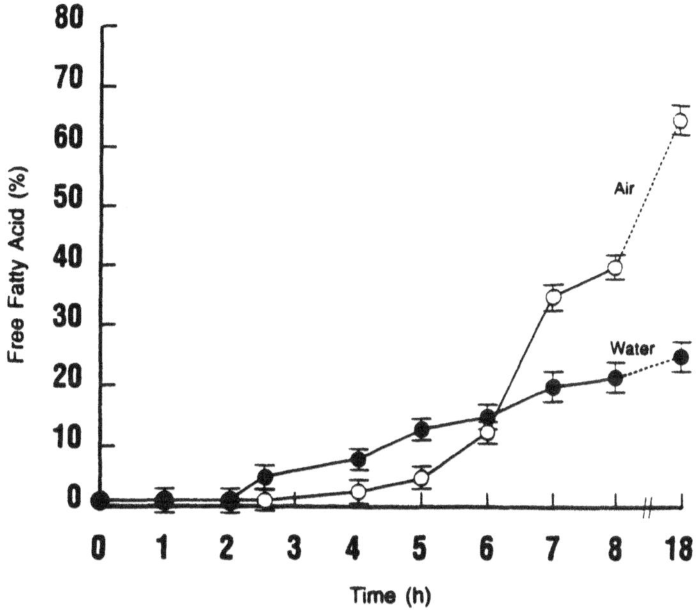

Conclusion

The oil palm mesocarp contains lipase that is activated at low temperature. The physiological significance of this lipase is not known.

Acknowledgement

The authors thank the Director-General of PORIM for permission to present this paper.

References

Oo, K.C., Teh, S.K., Khor, H.T. and Ong, A.S.H. Fatty acid synthesis in the oil palm (*Elaeis guineensis*): Incorporation of acetate by tissue slices of the developing fruit. Lipids 1985;20: 205-210.

Sambanthamurthi, R., Chong, C.L., Oo, K.C., Yeo, K.H. and Rajan, P. Chilling-induced lipid hydrolysis in the oil palm (*Elaeis guineensis*) mesocarp. J. Exp. Botany 1991;42: 1199-1205.

ARE OLEOSINS ONLY ASSOCIATED WITH OIL BODIES FROM DESICCATION TOLERANT PLANT TISSUES?

Denis J Murphy[1], Joanne H E Ross[1] and Hugh W Pritchard[2]
[1]Brassica & Oilseeds Research Department, John Innes Centre, Norwich, NR4 7UH, United Kingdom. [2]Jodrell Laboratory, Royal Botanic Gardens Kew, Wakehurst Place, Ardingly, RH17 6TN, United Kingdom.

INTRODUCTION

Oil bodies are associated with storage cells and tissues in many plant species, both in the gametophyte (eg pollen) and in the sporophyte (eg seed embryos and endosperm, fruit mesocarp). It is now clear that the oil bodies of many seeds do not consist of mere droplets of storage lipid, but are bounded by an annulus of specific interfacial proteins termed oleosins [1,2].

Very recently, a distinct but similar class of oleosin-like proteins also containing the characteristic central hydrophobic domain has been found in the pollen of rapeseed and corresponding genes have been isolated from both rapeseed and *Arabidopsis* [1]. Like seeds, pollen cells are propagules which carry their own storage compounds which are required to support their metabolism after germination. In the case of rapeseed and *Arabidopsis*, the seeds and pollen each contain oil bodies and are desiccation tolerant [1]. This established a correlation between desiccation tolerance and the presence of oil body proteins in both gametophytic and sporophytic plant tissues.

This correlation was investigated further by examining the ultrastructure and oil-body composition of seed and fruit tissues from olive and avocado. Olive seeds are desiccation tolerant and contain large numbers of oil bodies of similar diameter ($1\mu m$) to those found in rapeseed. In contrast, the mesocarp tissues of olive and avocado fruits are desiccation sensitive and contain very large oil bodies of $>20\mu m$. It was found that olive seeds contained the expected amounts of oil-body proteins that exhibited all the characteristics of oleosins, whereas oil bodies from the mesocarp of olive and avocado fruits contained no detectable proteins [3]. Therefore, oil-body proteins were only present in tissues which naturally underwent desiccation rather than being associated with all tissues that contain oil bodies.

Although the previous studies established a correlation between the presence of oil-body proteins and desiccation tolerance, the only desiccation-sensitive tissues studied were from fruits which are not true oil-storing organs; the function of oil-rich fruits is to entice herbivores to consume them, hence acting as vectors for propagation of the seeds which pass through the digestive system of the animals. It was therefore decided to compare directly the oil-body protein composition in desiccation tolerant (orthodox) and desiccation intolerant (recalcitrant) seeds from a range of sub-tropical and tropical oil-rich species.

J.-C. Kader and P. Mazliak (eds.), Plant Lipid Metabolism, 558–560.

MATERIALS AND METHODS

Desiccation tolerant seed was retrieved from long-term storage at Wakehurst Place; fresh samples of desiccation intolerant tissue were acquired immediately prior to use. Details of the species used are given in Table 1. For germination, 25 seed/embryos were incubated on 10 g dm^{-3} agar-water in plastic dishes at 26°C in the light (12 d^{-1} in 15μmol m^{-2} s^{-1} from Warm White fluorescent tubes). Germination was assessed as radicle emergence. Because of an insufficient number of seeds, *Mangifera indica* seed quality was assessed visually following dissection of the seed. Moisture content of the tissues used for the extraction of oil-bodies was determined gravimetrically on three to five replicates by drying at 103°C for 17 hours [4]. Oil bodies were isolated from homogenized seed and fruit tissues and all protein, lipid and ultrastructural analyses were performed as described previously [3].

Table 1 - Characteristics of the seeds used for oil body preparations

Species	Seed characteristics				Tissue extracted
	Storage[a] type	Oil content[b] (% dry weight)	Moisture content (% fresh weight)	Quality/ Germination (%)	
Agave angustifolia var. *marginata*	DT	27 S*	15	68	Whole seed
Mangifera indica	DI	6-12K (ref 6)	82	Fresh	Cotyledon
Swietenia humilis	DT	67 E (ref 5)	4	76	Embryo
S.macrophylla	DT	50 S (ref 6)	3	96	Embryo
Theobroma cacao	DI	50 S (ref 6)	37	100	Embryo

[a]DT, desiccation tolerant (orthodox); DI, desiccation intolerant (recalcitrant). [b]S, seed; E, embryo; K, kernel; *Pritchard, unpublished.

RESULTS AND DISCUSSION

The oil content, moisture content and germination of the seeds studied varied from 6-67%, 3-82% and 68-100% respectively (Table 1). The polypeptide profiles of oil-body preparations from these species are shown in Figure 1 with the preparation standards: rapeseed for desiccation tolerant material and oil palm (*Elaeis guineensis*) mesocarp for desiccation intolerant material. In each case, the amount of sample loaded per lane was normalized for the quantity of oil as determined by gravimetric and gas chromatographic methods. Therefore, the staining intensity of polypeptides reflects the ratio of protein:oil in the total oil-body fraction from each tissue. It is notable that the three desiccation tolerant species contain more or less equal amounts of total oil-body proteins, although the molecular masses of the various polypeptides does vary significantly. In contrast, the two desiccation sensitive seeds, mango and cocoa, contained very low levels of oil-body proteins, as did the desiccation sensitive fruit from oil palm.

The presence of low or undetectable amounts of oil-body proteins in some tissues, as assayed on SDS-PAGE gels, was also correlated with changes in the size of the oil bodies as determined by electron microscopy. Oil bodies from desiccation tolerant species tended to have diameters in

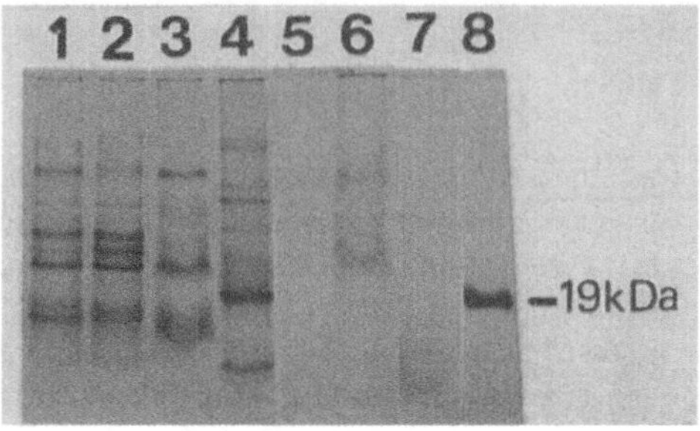

Figure 1. SDS-PAGE profile of oil body proteins from desiccation tolerant and desiccation sensitive* seeds and fruits. All lanes are loaded with delipidated oil bodies according to an equivalent amount of oil from total oil-body preparations. Lanes 1, *S.humilis*; 2, *S.macrophylla*; 3, *A.angustifolia*; 4, *B.napus*; 5, *M.indica**; 6, *T. cacao**; 7, *Elaeis guineensis* mesocarp*; 8, purified *B.napus* 19kDa oleosin.

the range of 0.5-1.5μm whereas those in desiccation sensitive species were in the range of 3 to 10-fold larger (data not shown). This implies that oleosins may have a role in maintaining oil-body size in seeds and that their presence is quantitatively correlated with desiccation tolerance, even in seed tissues. The function of oleosins in those lipid-storing plant tissues in which they are found may be to provide a continuous proteinaceous annulus around the storage oil bodies, in order to stabilise the small oil bodies in the relatively anhydrous environment found in many plant propagules. The large surface areas of these small oil bodies would then facilitate their mobilisation by lipases after germination and possibly after processing of the oleosin annulus to facilitate lipase penetration into the triacylglycerol matrix [1].

In conclusion, the data presented are consistent with the presence of oleosins in oil bodies of desiccation tolerant plant propagules in contrast to their relative dearth or complete absence from oil bodies in non-desiccation tolerant plant tissues, including the seeds of some plant species.

ACKNOWLEDGEMENTS

We thank Dr P Hadley (University of Reading, UK), Prof D J Osborne (The Open University, UK) and Dr P B Tompsett (Royal Botanic Gardens, Kew, UK) for the gift of the seed.

REFERENCES

1. Murphy DJ. Structure, function and biogenesis of storage lipid bodies and oleosins in plants. Prog Lipid Res 1993; 32:247-280.
2. Huang AHC. Oil bodies and oleosins in seeds. Ann Rev Plant Physiol 1992; 43:177-200.
3. Ross JHE, Sanchez J, Millan F, Murphy DJ. Differential presence of oleosins in oleogenic seed and mesocarp tissues in olive (*Olea europaea*) and avocado (*Persea americana*). Plant Sci 1993; 93:203-210.
4. International Seed Testing Association. International rules for seed testing. Seed Sci Technol 1985; 13:356-513.
5. Dickie JB, Ellis, RH, Kraak HL, Ryder K, Tompsett PB. Temperature and seed storage longevity. Ann Bot 1990; 65:197-200.
6. Eckey EW. Vegetable fats and oils. 1954; New York: Reinhold Publishing Corporation.

OIL BODY PROTEINS

Mark Millichip[1], Frances Jackson[1], Gareth Griffiths[1], Arthur Tatham[2], Alex Drake[3], Peter Shewry[2], Keith Stobart[1].
[1]Dept. Biological Sciences, Bristol University,Bristol, BS8 1UG.
[2]Dept. Agricultural Sciences, Long Ashton, Bristol, BS18 9AS.
[3]Dept. Chemistry, Birbeck College, 20 Gordon Street, London, WC1H 0AJ.

Introduction.

Oil body boundary proteins (so called Oleosins) have been extensively studied and many investigations have dealt with their molecular characterisation [1,2]. These proteins are highly hydrophobic. Little work, therefore, has been carried out on their purification for chemical and physical studies. Previous isolation procedures have utilised the differential density of the lipid in the oil-body to separate an oil-body fraction from the rest of the cell. This leads to contamination by many other cellular hydrophobic proteins. Here we present a protocol for the rapid purification of sunflower oil-body proteins, and assess some of their physical properties.

Purification.

8-10g of freshly dehusked developing sunflower seeds (17-21 D.A.F.) were homogenised in 50ml 100mM phosphate buffer, pH 7.2, containing 0.33M sucrose. The homogenate was passed through two layers of Miracloth. The filtrate was centrifuged at 20,000*xg* for 20 minutes. The oil-pad was resuspended in 8M urea, 20mM Tris buffer, pH 7.2. In a ratio of 1ml of oil-pad : 20ml of urea buffer. The suspension was centrifuged at 20,000*xg* for 20 minutes. The urea washing and centrifugation of the oil pad was repeated five times.

Results.

SDS-PAGE analysis of the crude homogenate, sucrose density washed, and urea washed oil-body preparations (Figure 1) clearly showed the removal of contaminating protein using the urea protocol. Incubation of the oil pad with 8M urea buffer removed large quantities of protein associated with the oil-body. Light microscopy indicated a uniform size of oil-body throughout the urea protocol. The proteins which remained associated with the oil-body fraction were of similar relative molecular weight, hydrophobicity, and amino acid composition as previously identified oleosins [2,3]. The urea protocol also produced clean oil-body preparations from other seed species (data not given).

Lipid analysis of the urea prepared oil-bodies showed that whilst TAG and DAG were the most abundant lipids (Table 1), little free fatty acids were present ($<$ 0.002% of the total mass). Polar lipid relative to neutral lipid was lower than previously reported [3,4]. Polar

J.-C. Kader and P. Mazliak (eds.), Plant Lipid Metabolism, 561–563.

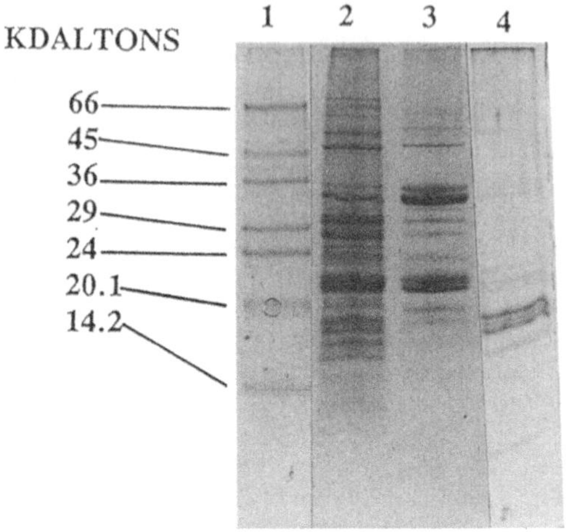

Figure 1) Tris-Tricine (12%) SDS-PAGE of Molecular Mass Markers (1), Sunflower Crude Homogenate (2), Sucrose Density Washed Oil-Body (3), Urea Washed Oil-Body (4).

lipid composition also differed from reported data in that the oil-bodies lacked PS (Table 1). Only a small quantity of negatively charged PI was present. It has been suggested previously [5], that negative phospholipids are required to stabilise the oil-body proteins. PS was also not observed in purified Rape and Maize oil-bodies (data not given).

Lipid	16:0	18:0	18:1	18:2	Mass (nmol)	%
TAG	5.0	7.4	61.4	25.6	10549	98.4
SE	11.1	N/A	50.6	38.4	8.4	0.1
DAG	5.0	4.8	64.4	25.9	91.0	0.8
Polar	9.2	8.7	54.4	27.8	72.0	0.67
PC	5.8	2.0	73.5	18.8	38.8	78.9
PE	18.6	5.5	42.9	32.6	6.6	13.4
PI	25.6	6.1	54.4	13.7	3.8	7.7

Table 1) Lipid Composition of Urea Washed Oil-Bodies. Neutral lipids separated in hexane:diethylether:acetic acid (70:30:1). Polar lipids separated by chloroform:methanol:acetic acid:water (85:15:10:3.5). Detection with I_2 and ninhydrin.

The urea purified proteins were analysed by Circular Dichroism. C.D.'s were carried out in a solvent with a dipole moment proported to represent the internal environment of a lipid bilayer. The spectra (Figure 2) showed a high percentage of α-helix (55-60%), the remainder being mainly random coil, with some ß-sheet configuration. As the dipole moment of the solution increased (with the addition of water) so did the ß-sheet content of the oil-body protein.

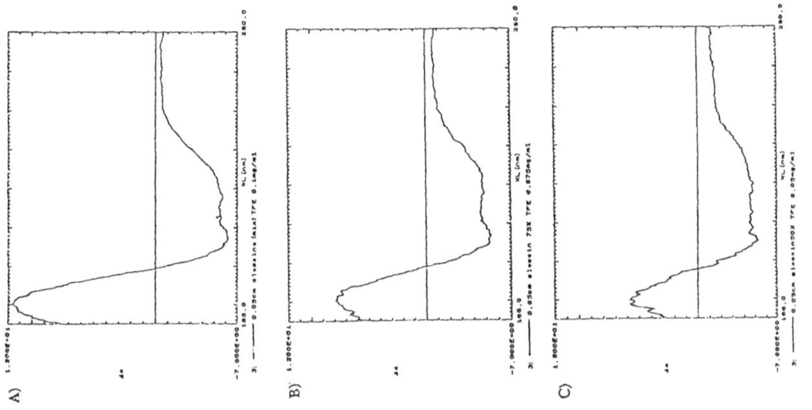

Figure 2) C.D. of purified Oil-Body Proteins in A) 100% Trifluoroethanol, B) 75% TFE, C) 50% TFE.

Conclusions.

1) A rapid and efficient procedure for the purification of oil-bodies has been developed which removes contaminating protein.
2) The polar lipid composition of the oil-bodies was different to that previously reported [5] and PS was almost absent.
3) Previous C.D. studies and amino acid predictions have indicated a high ß-sheet content of the oil-body protein [6]. However, here, C.D. studies using a "membrane mimic" solvent, showed a high α-helical component.

These results differ to previous data [1,2,5,6], and indicate that further modifications to the proposed models of oil-body structure may be necessary.

Acknowledgements.

The authors are grateful to A.F.R.C. (U.K.) for financial support.

References.

1. Huang AHC, Tzen JTC, Lee K, Bih FY, Ting JTL, Ratnayake C. Structure and ontogeny of seed oil-bodies in Maize and other species. Proceedings of the Phytochemical Society of Europe 1993;35:115-127.
2. Murphy DJ. Structure, function and biogenesis of storage lipid bodies and oleosins in plants. Prog Lipid Res. 1993;32:247-280.
3. Huang AHC. Oil bodies and oleosins in seeds. Ann Rev Plant Phys & Plant Mol Biol 1992;43:177-200.
4. Murphy DJ. Storage lipid bodies in plants and other organisms. Prog Lipid Res. 1991;29:299-324.
5. Tzen JTC, Lie GC, and Huang AHC. Characterisation of the charged components and their topology on the surface of plant seed oil bodies. J. Biol Chem 1992;267:15626-15634.
6. Li M, Smith LJ, Clark DC, Wilson R, Murphy DJ. Secondary structures of a new class of lipid body proteins from oilseeds. J. Biol Chem 1992;267:8245-8253.

OLIVE OIL BIOGENESIS. CONTRIBUTION OF FRUIT PHOTOSYNTHESIS

Juan Sánchez
Instituto de la Grasa, C.S.I.C.
Av. P. García Tejero, 4. 41012 Seville. Spain.

Previous studies have demonstrated that developing olive (*Olea europaea*) fruits can synthesize storage triacylglycerols (TAG) from a number of radiolabelled precursors, such as acetate, bicarbonate and CO_2, and that such a process is stimulated by light [1]. Those results showed the photosynthetic nature of developing olives, although they provided no insight on the relative contribution of fruit photosynthesis to the formation of storage TAG in this fruit. This question has been addressed by carrying out a field experiment in which olives were allowed to develop under either autotrophic or heterotrophic conditions, and their growth and lipid biosynthetic activities were compared with those of fruits developed under normal conditions. The results indicated that fruit photosynthesis might contribute significantly to TAG synthesis in olives.

Material and Methods

Olive trees, cv. Picual, 25 year old, were used; they were provided with drop irrigation since the time of full blossoming. Complete defoliation of whole branches was carried out 11 weeks after flowering (WAF), at the time when the lignification of the endocarp takes place. At the same time selected fruits, growing on non-defoliated branches, were enclosed in specially designed dark cages consisting of two pieces of tubing of 2 and 3 cm of diameter, made of black plastic, which were arranged one inside the other; small holes were made in the outer tube to allow ventilation of the enclosed fruit; the ends of each cage were covered with aluminium foil. Fruits growing on defoliated branches developed under autotrophic conditions, because they were deprived of the supply of photosynthates from adjacent leaves. On the other hand, fruits growing in the dark developed under heterotrophic conditions.

J.-C. Kader and P. Mazliak (eds.), Plant Lipid Metabolism, 564–566.

Results and Discussion

Olives were harvested at the middle of the developing period (21 WAF), when the rate of synthesis and accumulation of storage TAG was at its maximum, and at the beginning of the ripening period (28 WAF), when fruits were changing colour from green to purple, due to the loss of chlorophyll and synthesis of anthocyanins. As shown in Table 1 clear differences were observed between fruits developed under different trophic conditions. Olives growing in the dark (heterotrophic) reached the same size as the control, but they appeared etiolated, as it was to be expected for a green tissue growing in the dark, and the oil content was significantly reduced as compared with the control. On the other hand, fruits growing on defoliated branches (autotrophic) were significantly smaller than the control, but they were a bit greener and their oil content, expressed as percentage of fresh weight, reached a higher value than the control. These results show that both fruit photosynthesis and translocation of assimilates from the leaves contribute to the biogenesis of storage TAG in olives. Indeed the oil content per fruit was equally reduced in heterotrophic and autotrophic olives, suggesting that the contribution of those two processes to the biogenesis of olive oil is the same.

The biosynthetic activity of fruits growing under different trophic conditions was further investigated by carrying out labelling experiments with ^{14}C-acetate and ^{14}C-bicarbonate. Acetate is an immediate precursor of fatty acid synthesis which is incorporated into glycerolipids both in the light and in the dark, although the rate of incorporation is stimulated by light [2]. Lipid synthesis from bicarbonate, on the contrary, is essentially light dependent, because this precursor ought to be photosynthetically fixed through the Calvin cycle prior its incorporation into glycerolipids [3]. When tissue slices from the pulp of olives grown under different conditions were incubated with ^{14}C-acetate only minor differences were observed (Table 2): the rate of incorporation into lipids was reduced in heterotrophic olives to two thirds of that displayed by the

Table 1. **Characteristics of olives grown under different trophic conditions.**

Date	Growth Condition	Fresh Wt (g)	Chlorophyll (mg/g FW)	Oil Content % FW	Oil Content g/fruit
21 WAF	Control	3.01	0.11	18.3	0.42
	Heterotrophic	3.00	0.03	13.7	0.32
	Autotrophic	2.21	0.13	18.4	0.33
28 WAF	Control	5.52	nm*	19.7	0.95
	Heterotrophic	5.12	nm	15.4	0.68
	Autotrophic	3.26	nm	23.3	0.64

* nm: not measured.

Table 2. Biosynthetic capabilities of olives grown under different conditions. Olives used in these experiments were harvested 21 WAF; slices from the pulp were incubated with the labelled precusor in the light as described before [2].

Precursor	Growth Condition	Incorporation into lipids (nmol/h/g)	(%)
^{14}C-acetate	Control	18.2 ± 1.2	(100)
	Heterotrophic	12.1 ± 2.5	(66)
	Autotrophic	16.0 ± 0.1	(88)
^{14}C-bicarbonate	Control	7.2 ± 0.6	(100)
	Heterotrophic	0.2 ± 0.1	(3)
	Autotrophic	25.6 ± 0.3	(356)

control, whereas in autotrophic olives the reduction was still smaller. These results indicate that fatty acid and glycerolipid synthesis (i.e., the "dark metabolism") was not remarkably affected by the growth condition.

Lipid synthesis from ^{14}C-bicarbonate, on the other hand, was severely affected by the growth conditions. Heterotrophic (etiolated) olives displayed very low rates of incorporation into lipids as compared with the control, showing that in these fruits the contribution of fruit photosynthesis was insignificant: i.e., they indeed developed under heterotrophic conditions. In autotrophic olives, on the contrary, the rate of lipid synthesis from ^{14}C-bicarbonate was 3.5-fold higher than that of the control, indicating that these fruits, lacking of supply of nutrients from the leaves, are able to adapt themselves to the demands of photoassimilates by increasing their photosynthetic CO_2 fixing activity, thus allowing them to partially cope with the demands of growth and TAG synthesis.

Acknowledgment. Supported by the Comisión Interministerial Ciencia y Tecnología (Spain).

References.

1. Sánchez, J. (1994) Prog. Lipid Res. 33, 97-104.

2. Sánchez, J., de la Osa, C. & Harwood, J.L. (1990) Plant Lipid Biochemistry, Structure and Utilization. P.J. Quinn & J.L. Harwood, eds. Portland Press. London. pp. 390-392.

3. Sánchez, J., del Cuvillo, M.T. & Harwood, J.L. (1992) Metabolism, Structure and Utilization of Plant Lipids. A. Cherif et al., eds. C.N.P. Tunis. pp. 39-42.

COMPARATIVE STUDY OF OIL COMPOSITION IN FOUR OLIVE VARIETIES

Mondher KACHOURI ,Mongi M'SALLEM ,Mokhtar ZARROUK and Abdelkader CHERIF

Institut de l'Olivier, Ministère de l'Agriculture, 30 rue Alain Savary, 1003, Tunis, Tunisia

Introduction

Within physiological and biochemical evaluations of olive varieties,we have studied the biochemical characterization of lipogenesis and the fatty acids composition of olives.

Material and Methods

The four varieties were: "Chemlali","Chétoui" (local varieties) and "Arbequine", "Leccino" (introduced varieties). The olives were collected regularly from the fruit set to the full fruit maturity.
Lipid and fatty acids analysis and extraction have been described in a previous work (Marzouk and Chérif, 1981)

Results and Discussion

*Evolution of dry matter of olives during fruit maturation

The dry matter increased progressively in course of olive development stages and showed stabilisation tendancy at the end of maturity. Nevertheless, from beginning of July up to complete maturity,the evolution of dry matter (g/fruit) was more important in "Leccino" variety (1,4g/fruit at maturity) and lower in "Chemlali" (0,5g/fruit at the same stage). The two other varieties have shown intermediate level of progress ranging from 0,9 to 1,1g/fruit for the same stage.

*Evolution of lipids rate during fruit maturation

The evolution of lipids in olives of the four varieties were illustrated in figure 1. All varieties showed similar sigmoid shape evolution with the three conventionnal biosynthesis phases (Catalano and Sciancalepore,1976 ; Marzouk and Chérif,1981). Lipogenesis started really during the first fifteen days of July up to the begenning of November,then lowered. The studied varieties showed different lipid storage capacities. The lipid storage capacity was higher in "Arbequine" than "Leccino" variety. The local varieties had an intermediate synthesis capacity (Fig.1).

*Evolution of main fatty acid rate during fruit maturation

- Palmitic acid (C16:0)

The newly developed olive had the highest rate of palmitic acid; it varied from 27 to 30% according to the varieties.This percentage decreased

J.-C. Kader and P. Mazliak (eds.), Plant Lipid Metabolism, 567-569.

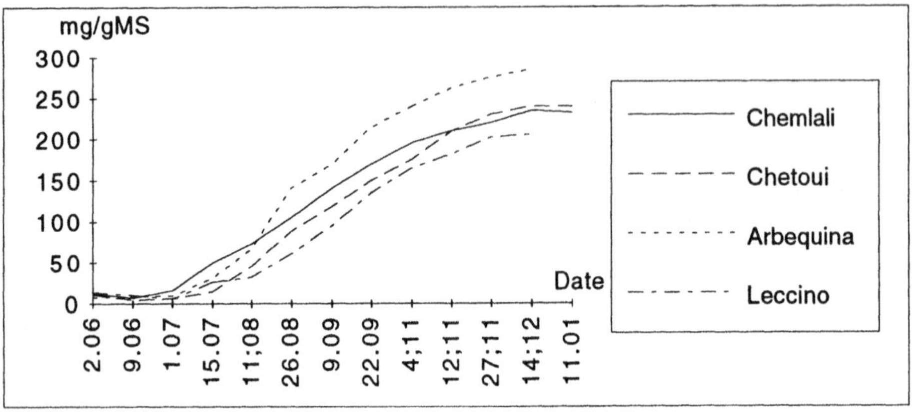

Fig.1 Evolution of lipids rate in olives during fruit maturation

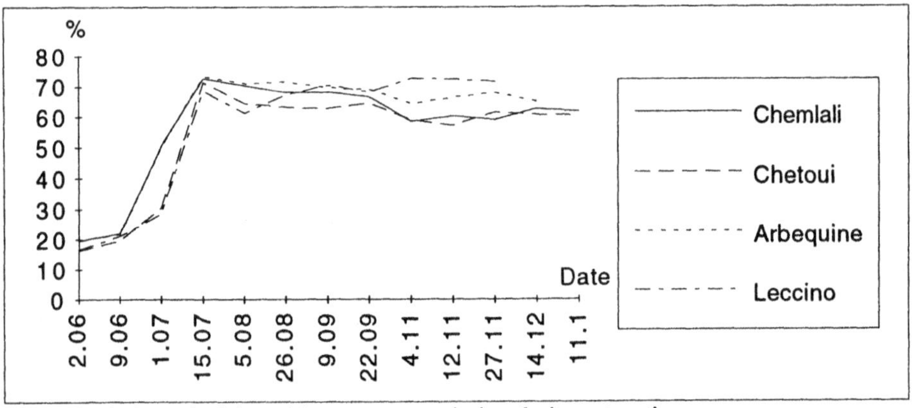

Fig.2 Evolution of oleic acid percentage during fruit maturation

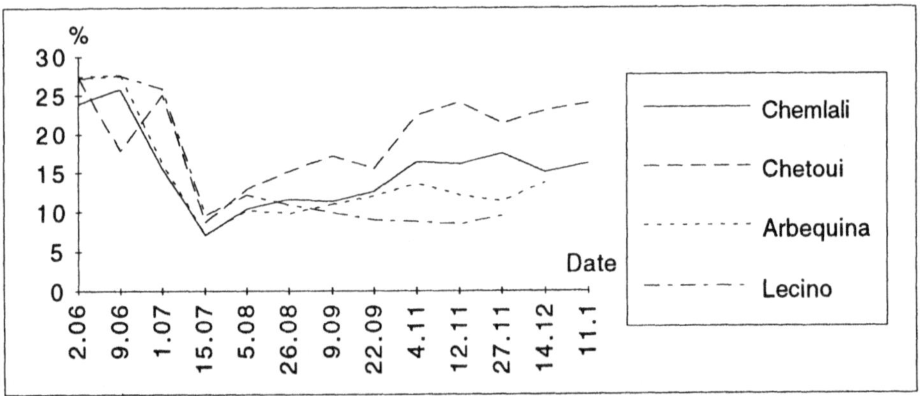

Fig.3 Evolution of linoleic acid percentage during fruit maturation

rapidly few weeks later to reach its minimum (14 to 16%) around mid-July. During maturation it increased slightly or remaind constant.

- Oleic acid (C18:1)

The figure 2 showed practically the same rate of oleic acid for the four varieties at the beginning of fruit set (16 to 19%). then,this percentage increased rapidly to reach,around mid-July, 68 to 73% according to the varieties. From that period until complete maturity, this percentage decreased slightly for the local varieties ("Chemlali" and "Chétoui") but it increased slightly or remaind constant for the imported ones ("Leccino" and "Arbequine"). At complete maturity,the percentage of this acid was higher for the introduced varieties (65 to 70%) and less than 60% for the local ones.

- Linoleic acid (C18:2)

From the first weeks of fruit formation to mid-July,the linoleic acid percentage decreased in the same way for the four varieties,then it increased with different speeds within varieties. At maturity, the percentage of this acid was respectively 16 to 23% for "Chemlali" and "Chétoui" varieties and only 9 to 13% for the introduced ones.

The results obtained demonstrated that local varieties had higher rate of linoleic acid and less oleic acid than european ones. Uzzan (1992) reported similar results.

Our study showed also that "Leccino" and "Arbequine" varieties cultivated under tunisian climate conditions had a fatty acid composition (mainly C18:1 and C18:2) similar to those cropped in their home site (Caselli and al, 1993 ; Tous and Romero,1992).

- Linolenic acid (C18:3)

At the beginning of fruit development cycle,the linoleic acid percentage varied from 22 to 26% according to the variety. This rate decreased rapidly and reached around 3% in mid-July; it kept on decreasing to reach less than 1% at complete maturity.

Conclusion

The present study has demonstrated that at maturity,the introduced varieties with regard to local ones, showed higher percentage of oleic acid and less linoleic acid. This quality can be used to improve some tunisian oil qualities through "blending" procedure. Furthermore, besides its adaptation to tunisian climate conditions, the "Arbequine" variety was found very performant (strengh, early production,high yield...)

References

Caselli S , Modi G , Nizzi Griffi F , Fiorino P. Variabilité de la composition en acides gras,en stérols et en alcools de l'huile d'olive de cultivars de la Toscane. Olivae 1993 ; 47: 46-51

Catalano M , Sciancalepore V. La composizione acidica dei lipidi neutri,glucolipidi e fosfolipidi setratti dalle olive. Nota III. Rivesta Italiana della Sostanze Grasses 1976 ; 53: 67-70

Marzouk B , Chérif A. La lipogenèse dans l'olive. I.Formation des lipides neutres. Oleagineux 1981 ; 36: 77-82

Toux J , Romero A. Fiche de la variété "Arbequine". Olivae 1992 ; 43: 28-29

Uzzan J. Olive et huile d'olive. In: Karleskind A,Wolff JP,editors. Manuel des corps gras.Paris: Lavoisier, 1992: 221-228

TOWARDS GENETIC ENGINEERING OF OIL PALM (Elaeis guineensis Jacq.)

Suan-Choo Cheah, Ravigadevi Sambanthamurthi, Siti Nor Akmar Abdullah, Abrizah Othman, Mohamad Arif Abdul Manaf, Umi Salamah Ramli and Ahmad Parveez Ghulam Kadir

Palm Oil Research Institute of Malaysia (PORIM)

Introduction

The high yielding oil palm has been identified as the most likely candidate for large-scale production of renewable plant oil-derived chemicals in the future [1]. Our efforts at genetic engineering this crop are directed towards the production of an oil with a high content of oleic acid. Such an oil will be industrially useful for producing chemical derivatives which can serve as an alternative to petrochemical feedstocks [2]. Our strategy for achieving this objective is to genetically modify the palm by *in vitro* means such that palmitate is diverted towards the formation of oleate during fatty acid synthesis in the mesocarp. We are approaching this in a concerted, multidisciplinary manner with studies being carried out in the areas of biochemistry, molecular biology and plant transformation.

Biochemical Studies

Our studies on fatty acid synthesis in the oil palm mesocarp have indicated that the enzymes acyl ACP thioesterase and beta-keto acyl ACP synthase II control the levels of palmitate and oleate. Thioesterase activity in the crude extract showed a marked preference for palmitoyl ACP [3]. Anion exchange chromatography (HR DEAE) resulted in the resolution of two peaks, one showing a marked preference for palmitoyl ACP and the other for oleoyl ACP. Both peaks were further purified using hydroxyapatite chromatography followed by ACP Sepharose chromatography.

The enzyme beta-keto acyl ACP synthase II was purified more than 10,000-fold. The activity of the enzyme in the mesocarp correlates positively with the level of unsaturation of the fatty acids present in the crude oil extract. Activity of synthase II increased with fruit ripening, reaching a maximum at 20 weeks after anthesis. The results of these investigations suggest that high levels of palmitate in palm oil is likely to be due to low synthase II activity and the specificity of the thioesterase for palmitoyl ACP.

J.-C. Kader and P. Mazliak (eds.), Plant Lipid Metabolism, 570-572.

Gene Isolation and Regulation

Regulatory sequences controlling tissue- and temporal-specific gene expression are required for *in vitro* manipulation to alter oil composition. For this purpose, we have also directed our efforts towards the identification and isolation of genes specifically expressed in the mesocarp during oil synthesis. As acyl carrier protein (ACP) activity has been shown to be induced just before the start of oil synthesis [4], we have chosen the ACP gene as a temporal marker. An attempt at isolating the ACP gene has picked up a cDNA clone from a mesocarp library which hybridises with the synthetic ACP gene probe, pPB269 [5]. DNA sequencing of the clone revealed a 50% homology between the cDNA insert and the spinach ACP-1 gene.

Our studies on differential gene regulation in the mesocarp employ the technique of comparing *in vitro* translated products of expressed genes during various stages of development. Stage-specific proteins have been identified [6]. Differential hybridisation of clones from a mesocarp library with first strand cDNA probes synthesized using mRNA of mesocarp and leaf as templates has also been carried out. Two cDNA clones were isolated as specific to mesocarp by this approach. However, Northern blot analysis showed that one of these was also expressed in seedlings. Plans are underway to use these clones as probes to screen a lambda genomic library for the isolation of their gene sequences.

Oil Palm Transformation

Much effort is being put into developing a reliable transformation protocol for the oil palm. We are employing both direct DNA uptake and biolistic bombardment to achieve this. Transformation of various tissues, including pollens, young leaves and embryogenic calli, has been attempted.

The ability to genetically transform oil palm via protoplasts is also desirable. Protoplasts have been successfully isolated from embryogenic cultures [7], inflorescences, seed embryos and vegetative apices of tissue culture plantlets. Embryogenic cultures were shown to be the best donor source in terms of protoplast viability and metabolic competence. Cell wall regeneration was routinely obtained for protoplasts isolated from embryogenic cultures. Limited cell division of the regenerated protoplasts to form microcalli has been observed, although at a low frequency.

Transient expression of the beta-glucuronidase (GUS) reporter gene has been observed following incubation of germinating pollens with DNA. In the effort to identify efficient promoters for expressing introduced genes in oil palm, young leaves and embryogenic calli were subjected to bombardment using the particle gun. Transient expression of the GUS reporter gene indicated the penetration of DNA into the tissues. Our next target is to regenerate the transformed tissues into whole plants and demonstrate stable transformation.

Acknowledgements

The authors thank the Director-General of PORIM for permission to present this paper. This work is supported by a grant from the Malaysian government's programme on the Intensification of Research in Priority Areas (IRPA).

References

1. Somerville C, Browse J. Plant lipids: metabolism, mutants, and membranes. Science 1991; 252:80-87.

2. Pryde EH. Utilization of commercial oilseed crops. Econ. Bot. 1983; 37:459-477.

3. Sambanthamurthi R, Oo KC. Thioesterase activity in the oil palm (*Elaeis guineensis*) mesocarp. In: Quinn PJ, Harwood JL, editors. Plant lipid biochemistry, structure and utilization. The proceedings of the ninth international symposium on plant lipids. London, U.K.: Portland Press Ltd., 1991: 166-168.

4. Slabas AR, Harding J, Hellyer A, Roberts P, Bambridge HE. Induction, purification and characterization of acyl carrier protein from developing seeds of oil seed rape (*Brassica napus*). Biochim. Biophys. Acta 1987; 921:50-59.

5. Beremand PD, Hannapel DJ, Guerra DJ, Kuhn DN, Ohlrogge JB. Synthesis, cloning, and expression in *Escherichia coli* of a spinach acyl carrier protein-1 gene. Arch. Biochem. Biophys. 1987; 256:90-100.

6. Abdullah SNA, Shah FH, Cheah SC. Detection of differentially expressed genes in the development of oil palm mesocarp. AsPac J. Mol. Biol. Biotechnol. 1994; 2:113-118.

7. Sambanthamurthi R; Oo KC, Ong ASH. Lipid metabolism in oil palm (*Elaeis guineensis* and *Elaeis oleifera*) protoplasts. Plant Science 1987; 51:97-103.

NEUTRAL LIPIDS OF MATURE AND DEVELOPING SEA BUCKTHORN (*HIPPOPHAË RHAMNOIDES* L.) FRUITS

Andrei G. Vereshchagin and Vladimir D. Tsydendambaev

Laboratory of Lipid Metabolism, Institute of Plant Physiology, Russian Academy of Sciences, Moscow, Russia

SYNOPSIS

Neutral lipid composition and metabolism of the fruit of sea buckthorn, a medicinal plant with an oil-bearing mesocarp, were studied. Genotypically distinct mesocarp and seeds of the fruit were shown to differ sharply from each other in their total fatty acid (FA) and triacylglycerol (TAG) content and composition, as well as in the TAG accumulation pattern, initial FA-forming enzyme systems being replaced during maturation by other ones. Various geographical climatypes of the plant (Siberian, Central Asian and Baltic) are similar in the composition of their mesocarp TAGs, which are formed according to 1,3-Random, 2-Random Distribution of FAs. In contrast, mesocarp TAGs of the Caucasian climatype differ significantly from the above ones in their composition pattern, which is best described by the theory of Restricted Random Distribution of FAs.

INTRODUCTION

Reserve oils of higher plants consisting for the most part of triacylglycerols (TAGs) are known to be concentrated almost exclusively in seeds. The biological role of seed oils is considered to be a source of organic matter and energy for the formation of seedling biomass during germination [1]. At the same time, there is a number of plant species (oil palm, avocado, olive, ivy, sea buckthorn etc.), which contain a considerable portion of their oil not only in seeds of a fruit, but also in its pulp. We designate this group as "plants with an oil-bearing mesocarp", and the investigation of their lipid metabolism is of considerable interest [2].

In the present study it is sea buckthorn that was used for such investigation, because it is widely distributed in our country. Moreover, its fruit oil is of great practical importance, being extensively used as a medicine. Anti-inflammatory effect of the oil is usually considered to be brought about by the presence of ascorbic acid, tocopherols, carotenoids, sterols etc. However, there is evidence that this effect can be enhanced by TAGs of the oil. Thus, investigation of neutral lipids of sea buckthorn is essential (3). Its fruits from Siberia are most important economically, and so they were mainly used in the investigation; nevertheless, fruits from other regions have also been studied.

J.-C. Kader and P. Mazliak (eds.), Plant Lipid Metabolism, 573-578.

Sea buckthorn fruit consists mainly of a juicy pulp (mesocarp) and a small seed. General characteristics of these organs are shown in table 1. It can be seen that mesocarp is different from seeds in its genotype, its share in the total fresh wt and oil content of the fruit, as well as its oil concentration and fatty acid (FA) composition of the oil. At the same time, oils of both organs consist of TAGs to the extent of 70 - 90% [3].

Table 1. General characteristics of mesocarp and seeds of mature fruits of Siberian sea buckthorn.

Parameter	Mesocarp	Seeds
Genotype	♀	♂ : ♀ = 1 : 1
Fresh wt distribution, % of total fruit wt	96-98	2-4
Oil distribution, % of total fruit oil	93	7
Oil concentration, % of fresh wt	2.5-4.0	13-22
TAG content in the oil, %	80-85	73-91
Major FAs of TAGs	$C_{16:1}$, $C_{16:0}$	$C_{18:1}$, $C_{18:2}$, $C_{18:3}$

FATTY ACID COMPOSITION OF SIBERIAN SEA BUCKTHORN MATURE FRUIT OIL

FA composition of mesocarp and seed TAGs is shown in figure 1. It is evident that mesocarp TAGs consist for the most part of $C_{16:1}$ and $C_{16:0}$ acids; there is also an appreciable amount of hexadecadienoic ($C_{16:2}$) acid, which was identified by us in the fruits for the first time [3]. The ratio between oleic and 11(*Z*)-octadecenoic (vaccenic) acid content in total $C_{18:1}$ acids varies from 0.3 to 3.0.

Seeds differ from mesocarp both in their genotype (table 1) and in the FA composition of their TAGs (figure 1), which are characterized by a high content of $C_{18:1}$, $C_{18:2}$ and $C_{18:3}$ acids. TAGs of such composition are usually considered to occur only in the seeds (*e.g.* flax, soybean etc.), which for the most part of the oil accumulation period include many chloroplasts actively forming polyenoic FAs. However, sea buckthorn cotyledons lack green plastids, are surrounded by an opaque seed coat and mature inside a fruit. Therefore, the

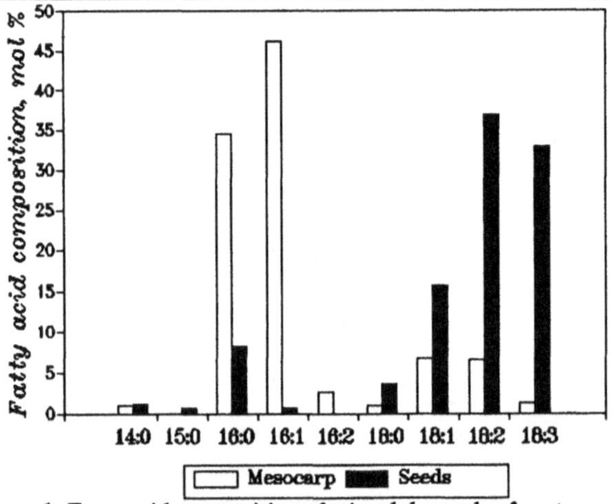

Figure 1. Fatty acid composition of triacylglycerols of mature Siberian fruit mesocarp and seed.

investigation of oil formation in these cotyledons is of considerable interest [2].

OIL ACCUMULATION IN SIBERIAN FRUITS

Mesocarp and seeds differ also from each other in the values of rate of their accumulation at successive stages of maturation. Mesocarp exceeds the seeds in this respect by more than one order of magnitude. Moreover, in developing mesocarp, the rate of oil formation constantly increases, whereas in the seeds, after reaching a maximum, it decreases sharply long before the technical maturity of fruits is achieved [2].

To determine the kinetics of accumulation of TAGs and their FAs during fruit ripening, the respective relative rate constants (k, d^{-1}) were calculated [4]. One can see (table 2) that in both parts of the fruit SE of k determination is rather modest. Thus, the time course of TAG and FA absolute content fits a sigmoid curve. It is known that k is directly proportional to the steepness of this curve, *i.e.* it is inversely proportional to the duration of oil and FA accumulation. When evaluating biosynthesis rate constants of individual FAs, as well as of saturated (S) and unsaturated (U) acids esterified in mesocarp and seeds, one can observe that in mesocarp TAGs they are 2-3 times lower than in TAGs of seeds. Thus, the latter differ from the mesocarp by a lower duration of the accumulation period of TAGs and their FAs. In addition, during TAG biosynthesis in mesocarp, S exceed U somewhat in their k value, and C_{16} FAs as a group are by far higher in this respect than C_{18} acids. It should be stressed that these differences embrace all components of a group regardless of an absolute and relative content of a given FA species in total TAGs of a mature mesocarp (figure 1). Therefore, the differences observed can be assumed to be brought about by the fact that C_{16} and C_{18} acid groups are formed with different rates and in various metabolic compartments of a cell. Finally, in the TAGs of seeds individual C_{18} acids are almost equal to each other in their k values.

Table 2. Relative rate constants (k, d^{-1}) of accumulation of TAGs and their FAs during fruit ripening. SE: standard error of k determination.

	Mesocarp		Seeds	
	$k \cdot 10^3$	$SE \cdot 10^3$	$k \cdot 10^3$	$SE \cdot 10^3$
Total TAGs	140	19	221	10
$C_{16:0}$	153	22	-	-
$C_{16:1}$	173	28	-	-
$C_{16:2}$	155	21	-	-
$C_{18:0}$	109	9	228	27
$C_{18:1}$	132	20	234	14
$C_{18:2}$	100	10	235	16
$C_{18:3}$	63	8	234	5
S	149	22	-	-
U	136	18	234	3

The genetically induced differences between mesocarp and seeds in their FA composition do not arise from the very onset of ripening. At 29 DAP these parts of the fruit differed little from each other in relative content of individual FAs in their lipids. Later on, there occured changes in the FA composition of mesocarp and seed TAGs.

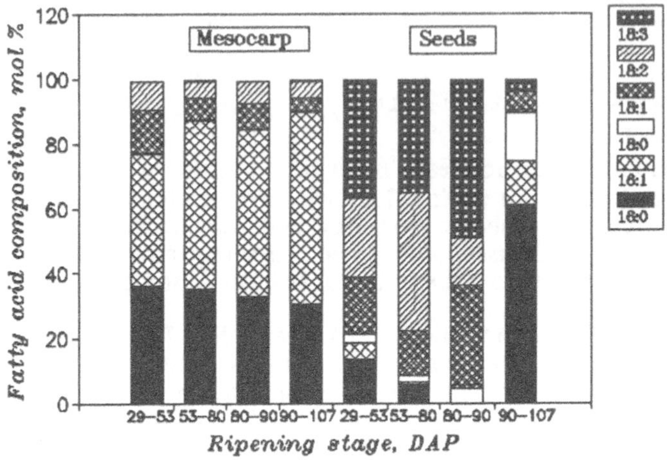

Figure 2. Fatty acid composition of triacylglycerol accretion formed *de novo* in mesocarp and seeds at each successive ripening stage. DAP - days after pollination.

As a result (figure 1), already at 80th DAP there was a sharp difference between the seeds and mesocarp in this respect [2].

To establish the composition specificity of TAGs formed at each successive ripening stage, we have performed calculation of this composition, and its results are presented in figure 2. One can see that during the development of mesocarp the biosynthesis of $C_{16:1}$ acid is becoming more intense, while the share of $C_{16:0}$ and **S** in successive TAG accretions decreases continuously. As regards the seeds, **S** share in the formation of total FAs of TAGs diminished during the period up to 90th DAP, and at the 3rd stage of ripening no C_{16} acids were formed. However, at the 4th stage, when a total TAG accretion is limited, the seeds, in their individual FA species biosynthesis pattern, became similar to mesocarp: $C_{16:0}$ and $C_{16:1}$ formation resumed dramatically, $C_{18:0}$ synthesis increased sharply and **S** comprised 76% of the total FAs formed *de novo*. To the best of our knowledge, the enhancement of **S** synthesis by the end of fruit ripening is observed here for the first time. All these data show that during the ontogenesis of seeds and mesocarp initial FA-forming enzyme systems are replaced by other ones [2].

TRIACLGLYCEROL COMPOSITION OF SIBERIAN FRUIT OIL

Distinctions between mesocarp and seeds of Siberian sea buckthorn in the FA composition of their TAGs could be assumed to affect the molecular species composition of these TAGs. To check this assumption, we determined the latter parameter (*mol* %) by using the standard technique of lipase hydrolysis [5], and the results of this determination are briefly presented below:

TAG species	$(C_{16})_3$	$(C_{16})_2C_{18}$	$C_{16}(C_{18})_2$	$(C_{18})_3$	Major TAG molecular species
Mesocarp	59-68	20-29	Traces	Traces	$C_{16:0}(C_{16:1})_2$, 27-31
Seeds	Traces	Traces	14-16	72-79	$(C_{18:2})_2C_{18:3}$, 11-14

To determine the extent to which TAG structure of Siberian fruits fits the 1,3-Random, 2-Random Distribution pattern predominant in most plant TAGs, we used empirical equations derived by Litchfield [6], which make it possible to calculate the content of a given acid A in 2-position of TAGs ($[A]_2$) without using lipase hydrolysis, but directly from a total FA composition of TAGs.

Found and calculated $[A]_2$ values of individual FAs of mesocarp and seed TAGs and the composition of the latter were shown to be in close agreement [5]. Thus, despite a sharp difference between mesocarp and seeds in their TAG composition, all of them are constructed after the same pattern.

TRIACYLGLYCEROL COMPOSITION OF FRUIT OIL IN VARIOUS SEA BUCKTHORN CLIMATYPES

Various geographical forms are now divided into four climatypes: Siberian (SI), Central Asian (CA), Baltic (BL) and Caucasian (CC). Total FA composition of TAGs of fruits of these climatypes is shown in table 3. One can see that all of them are similar to each other in TAG composition of seeds, but differ considerably as regards the mesocarp TAG composition. Thus, $C_{16:2}$ acid is present only in SI fruits, and CC differs significantly ($P > 99\%$) from other climatypes by an increased content of $C_{18:1}$ and $C_{16:0}$ acids and by a drastically reduced level of $C_{16:1}$ acid. To characterize the climatypes with respect to the TAG composition, coefficients of correlation (r) were established (table 4). One can see that there are two groups of climatypes as regards their TAG composition, *viz.* 1. SI, CA, BL, and 2. CC. Those of the 1st group are close to each other in this respect, but CC is quite different, and does not conform to the Litchfield theory [7]. This is caused by a high content of $C_{16:0}$ acid and considerable gaps between the found and calculated $[A]_2$ values for $C_{16:0}$ and $C_{16:1}$ acids. Therefore, we assumed that the composition of these TAGs can correspond to that calculated according to the Restricted Random Distribution (RRD) theory [6]. Indeed, there is a tight correlation ($r = 0.993$, $P > 99\%$) between the calculated TAG species composition and the found one.

Table 3. FA composition of TAGs, mol %, of fruits of various sea buckthorn climatypes

Fatty acids	Mesocarp				Seeds**			
	SI*	CA	BL	CC	SI	CA	BL	CC
$C_{16:0}$	32.3	34.2	32.6	47.0	9.0	8.6	7.1	7.9
$C_{16:1}$	53.8	54.8	42.4	16.4	1.2	2.5	1.1	0.9
$C_{18:1}$	4.7	11.0	20.8	35.0	17.3	24.0	21.1	16.3
$C_{18:2}$	5.7	-	3.5	1.6	39.6	36.2	34.9	45.2
$C_{18:3}$	0.4	-	0.7	-	36.9	25.9	33.2	25.7

* The climatype contains also $C_{16:2}$ acid (3.0).

** The climatypes contain also $C_{18:0}$ acid (4.5, 2.8, 2.6 and 3.7%, respectively).

Thus, mesocarp TAGs of SI, CA and BL climatypes are formed according to the 1,3-Random, 2-Random Distribution pattern, whereas the biosynthesis of those of CC climatype is best described by the theory of RRD. We are unaware of any other study, in which the presence, within the same organ of a same plant species, of TAG mixtures constructed according to different FA distribution patterns was demonstrated.

Table 4. Correlation pattern of mesocarp TAG composition in various sea buckthorn climatypes.

Climatypes	SI	CA	BL	CC*
	0.995	*0.993*	*0.947*	*0.981*, *0.537***
CA	**0.955**	-	-	-
BL	**0.798**	**0.914**	-	-
CC	**0.000****	**0.091****	**0.347****	-

Bold numbers: coefficients of correlation (r) between separate climatypes as regards their found TAG composition; italics: r between the found (by lipase hydrolysis) and calculated (according to Litchfield) TAG composition of a same climatype. * Left and right numbers: r coefficients for six TAG species devoid of C16:1 acid, and for eight TAG species containing C16:1 acid, respectively. ** No correlation ($P < 95\%$); all other r's are significanf with $P > 99\%$ confidence limit.

Acknowledgement

This work was partly supported by a grant from the "Frontiers in Genetics" Russian State Program.

REFERENCES

1. Hitchcock C, Nichols BW. Plant Lipid Biochemistry. New York: Academic Press, 1971: 241-42.
2. Berezhnaya GA, Yeliseev IP, Ozerinina OV, Tsydendambaev VD, Vereshchagin AG. Developmental changes in the absolute content and fatty acid composition of acyl lipids in sea buckthorn fruits. Plant Physiol Biochem 1993; 31: 323-32.
3. Berezhnaya GA, Yeliseev IP, Tsydendambaev VD, Vereshchagin AG. Determination of fatty acid composition and quantitative content of lipids in sea buckthorn fruits. Prikl Biokhim Mikrobiol 1988; 24: 568-72.
4. Vereshchagin AG. Comparative kinetic analysis of oil accumulation in maturing seeds. Plant Physiol Biochem 1991; 29: 385-93.
5. Ozerinina OV, Berezhnaya GA, Yeliseev IP, Vereshchagin AG. Composition and structure of triacylglycerols from sea buckthorn fruit mesocarp. Prikl Biokhim Mikrobiol 1988; 24: 422-29.
6. Litchfield CC. Analysis of Triglycerides. New York: Academic Press, 1972: 252-53.
7. Ozerinina OV, Berezhnaya GA, Yeliseev IP, Vereshchagin AG. Effect of climatype on the triacylglycerol composition of sea buckthorn mesocarp and seeds. In: Kalinina IP, editor. Proceedings of the International Symposium on Sea Buckthorn; 1993 Aug 23-25; Barnaul. Novosibirsk: Siberian Branch of RASKhN, 1993; 413-16.

AUTHOR INDEX

SUBJECT INDEX

GPSR Compliance

The European Union's (EU) General Product Safety Regulation (GPSR) is a set of rules that requires consumer products to be safe and our obligations to ensure this.

If you have any concerns about our products, you can contact us on ProductSafety@springernature.com

In case Publisher is established outside the EU, the EU authorized representative is:

Springer Nature Customer Service Center GmbH
Europaplatz 3
69115 Heidelberg, Germany

Batch number: 10809436

Printed by Printforce, the Netherlands